Environmental Change and Challenge

A Canadian Perspective

Environmental Change and Challenge

A Canadian Perspective

Philip Dearden and Bruce Mitchell

Toronto New York Oxford
OXFORD UNIVERSITY PRESS
1998

To Kate and Teresa Dearden, and
in memory of George and Jean Mitchell

Oxford University Press
70 Wynford Drive, Don Mills, Ontario M3C 1J9

Oxford   New York
Athens   Auckland   Bangkok   Calcutta   Cape Town   Chennai   Dar es Salaam   Delhi
Florence   Hong Kong   Istanbul   Karachi   Kuala Lumpur   Madrid   Melbourne
Mexico City   Mumbai   Nairobi   Paris   Singapore   Taipei   Tokyo   Toronto   Warsaw

and associated companies in
Berlin   Ibadan

Oxford is a trade mark of Oxford University Press

Pages 544–7 are a continuation of the copyright page

Canadian Cataloguing in Publication Data

Dearden, Philip
  Environmental change and challenge : a Canadian perspective

Includes bibliographical references and index.
ISBN 0-19-541014-9

1. Human ecology—Canada. 2. Man—Influence on nature—Canada.
3. Environmental protection—Canada. I. Mitchell, Bruce, 1944–   . II. Title.

GP511.D42 1997   333.7'2'0971   C97-931363-5

Design: Max Gabriel Izod
Formatting and Figures: Linda Mackey
Cartography: Paul Sneath, free&Creative
Illustrations: Julian Mulock
Figures: Ole Heggen

The publisher gratefully acknowledges the World Wildlife Fund, the International Fund for Animal
Welfare, and Earthroots for use of their photographs.

Main cover illustration: Earthroots
Small cover illustrations (l–r): David White/International Fund for Animal Welfare; Corel Photo CD–Sampler; © Hodson/Greenpeace; Corel Photo CD–Canada's East Coast; John & Karen Hollingsworth/US Fish & Wildlife Service; Corel Photo CD–Sampler; © Edwards/Greenpeace; Corel Photo CD–Canada's East Coast; P. Chivers/Earthroots; Corel Photo CD–Canada's East Coast.

1 2 3 4 — 01 00 99 98

This book is printed on permanent (acid-free) paper ∞ produced from 10 per cent postconsumer waste.

Printed in Canada

Oxford University Press Canada web site: http://www.oupcan.com

Contents Overview

Detailed Contents

Preface

The purposes of this book are to introduce concepts and methods to students in Canadian colleges and universities who are taking a first course in environmental science/studies, to impart an understanding of the biosphere's function, and to link basic environmental management principles to environmental and resource problems in a Canadian context.

The distinctive features of the book are an explicit link between environmental science and management and an emphasis on Canada. Detailed case-studies from various regions of Canada illustrate the concepts and ideas from science and management. After teaching this material for quite some time, we felt that Canadian students interested in environment deserved a text that was Canadian and that focused on their environmental interests.

Part A stimulates the students' attention and interest and provides context by giving some brief vignettes focused on different regions in Canada and beyond to highlight the opportunities and challenges in environmental science and environmental management, and to identify aspects, such as incomplete knowledge and understanding, conflicting interests and values, uncertainty, and trade-offs. Canada, like many other countries, has come to recognize the need for sustainable development that provides economic opportunities while safeguarding the natural environment. Part A provides some context for sustainable development at both international and national scales and also gives an overview of some of Canada's biophysical diversity, which is the focus of our management activities. The idea was not to provide a rigid biophysical description of each ecozone but to indicate the salient characteristics of each zone and highlight some of the main features of interest in accompanying text boxes.

Part B provides a basic primer on the environmental processes that constitute the earth's life-support systems. Primary emphasis is on energy flows (laws of thermodynamics, etc.), biogeochemical cycles, and biotic responses. Attention is given to the relevant processes and features of the Canadian landscape. These basic principles are linked throughout the book to actual problems, such as acid deposition, eutrophication, and global warming.

In Part C the concepts and methods discussed in Part B are combined with ideas about environmental management to illustrate how management is evolving from an expert-based, top-down, rational, comprehensive approach to one that is participatory, bottom-up, adaptive, and involves mutual learning. Various aspects of management—such as conflict resolution, impact assessment, ecosystem, and adaptive approaches—are considered within this context.

Part D provides detailed discussions of problem-solving situations in which environmental science and environmental management have to be interrelated. In parts B and C numerous brief examples are woven into the chapters. In Part D examples or case-studies are considered in greater detail to illustrate the concepts and methods outlined in parts B and C. To keep the presentation manageable, the chapters in this section focus on selected aspects of ecosystems. However, their connections with larger ecosystems are also considered.

In Part E a well-known case-study, Clayoquot Sound, illustrates how the biophysical processes and management approaches described in the preceding sections need to be brought together for effective environmental decision making. The book concludes with some ideas about what people can do to have some positive influence on the environmental challenges we now face.

Acknowledgements

We would like to thank the following people and organizations for arranging or providing information, maps, or photographs. Our sincere appreciation to: Paul Allen, Canadian Wildlife Service, Environment Canada; Lawrence Anukam, Department of Geography, Wilfrid Laurier University; Dave Cressman, Ecologistics Ltd; Arlin Hackman, World Wildlife Fund Canada; Garnet Gobert, Saskatchewan Water Corporation; Murray Haight, University of Waterloo; Harbourfront Village; Fred Heal, Meewasin Valley Authority; Sarah Kipp, Cal-Eco Consultants Ltd; Bill Lautenbach, Planning and Development, Regional Municipality of Sudbury; Stephen Naylor, University of Guelph; Steven Price, World Wildlife Fund Canada; Nigel Richardson, N.H. Richardson Consulting; and Brenda Sue, Department of Fisheries and Oceans.

We would also like to thank the following individuals who reviewed the chapters (in whole or in part) and provided suggestions to strengthen the book. We did our best to incorporate all comments, but the final content and structure of the book are our responsibility. Our deepest gratitude to: Dianne Draper, University of Calgary; Dave Duffus, University of Victoria; Murray Haight, University of Waterloo; Fred Heal, Meewasin Valley Authority; Sarah Kipp, Cal-Eco Consultants Ltd; Bill Lautenbach, Regional Municipality of Sudbury; Kevin McNamee, Canadian Nature Federation; George Mulamoottil, University of Waterloo; Stephen Naylor, University of Guelph; Maureen Reed, University of British Columbia; Nigel Richardson, N.H. Richardson Consulting; Dan Shrubsole, University of Western Ontario; John Sinclair, University of Manitoba; Graham Smith, University of Western Ontario; and Colin Wood, University of Victoria.

We also express our appreciation to the three reviewers who reviewed the entire manuscript. We believe the book is much improved as a result.

We would also like to express our deep appreciation to a number of people at Oxford University Press Canada for their work in transforming this manuscript into a book. In particular, we thank Valerie Ahwee, developmental editor, for her ongoing interest and assistance in so many aspects of the book, and especially for her careful copy-editing and success in obtaining photographs to illustrate the text. We also thank Sandra Brown for her word-processing skills, and Max Izod for his design work.

Environment, Resources, and Society

In this first part of the book we aim to do several things. First, we start with an example of a local resource management problem in Canada. It is probably not very well known outside its municipality. This is one reason why we chose it. Canadians across the country have become familiar with names such as Carmanah and Temagami. However, most environmental challenges are not so well known, but they occur on a daily basis in every province, territory, and municipality. These challenges are complex, full of uncertainty, and often generate conflict as they involve different stakeholders. By starting with such an example, we hope to illustrate the realities of the situation.

This local example is set within the larger context of national and global developments. Critical environmental problems are emerging at all scales, and our second goal in this part is to provide a brief overview of some of these challenges at the global and national scales. This leads into a review of the concept of sustainable development, a term made popular by the World Commission on Environment and Development in its landmark report, *Our Common Future* (1987). By providing some background on sustainable development on both the global and national stages, we hope to provide students with a useful framework for subsequent chapters that deal more with either biophysical or management aspects of environmental management. In reality, these two aspects are intimately related; sustainable development is one framework that reflects this link.

Finally, in this first section of the book we also wanted to convey some understanding of the biophysical background of Canada. Our goal is to make Canadians aware of and appreciate the immensity and beauty of the country and to provide a scientific description of it. For this reason we have provided many text boxes illustrating various aspects of Canada as a living land.

Reference

World Commission on Environment and Development. 1987. *Our Common Future*. Oxford: Oxford University Press.

◀ The beauty and intrinsic value of places such as Temagami are irreplaceable (*Earthroots/Terry Graves*).

Chapter 1

Environmental Change and Challenge

Change and *challenge*. Changes in environmental and resource systems occur due to the natural evolution of biophysical systems and/or to modifications in such systems because of human activity. Change can also occur in societies due to shifts in human values, expectations, perceptions, and attitudes, which may have implications for future interactions between those societies and natural systems.

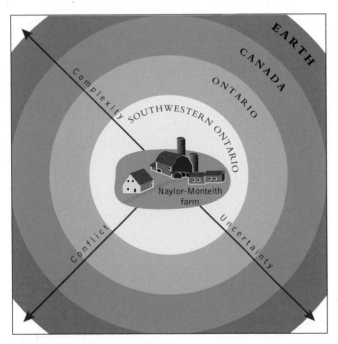

Figure 1.1 Complexity, uncertainty, and conflict occur on many levels. The case-study described here revolves around one family in southwestern Ontario and the challenge they face from past chemical dumping on their property. However, as the perspective increases to cover wider and wider areas, many more people become involved with a multiplicity of environmental problems and resulting conflicts. In general, the complexity, uncertainty, and conflict levels will also increase.

Changes in natural and human systems generate challenges. If we wish to protect the integrity of biophysical systems, yet also ensure that human needs are satisfied, there are questions about how to determine ecosystem integrity and basic human needs. Such questions force us to think about conditions *now* and in the *future*. Such questions also remind us that understanding environmental and resource systems requires both natural and social sciences. Neither, by itself, provides sufficient understanding and insight to guide decisions. Finally, such questions pose fundamental challenges as to whether we can realistically expect to manage or control natural systems, or whether we should focus on trying to manage human interactions with natural systems.

In this chapter, we start with a case-study to illustrate some of the opportunities and challenges for Canadians regarding resource and environmental systems. This example shows that management decisions often have to be made in the context of changing conditions, incomplete knowledge and understanding, conflicting interests and values, trade-offs, and uncertainty. These same conditions apply not only to the case-study or indeed to Canada but to the global picture. Most of the text is concerned with the national situation, but it is important to appreciate the broader context of the global situation (Figure 1.1). The next two sections of this chapter provide overviews of the global and national situations. The next chapter in this section continues the discussion begun in Chapter 1 on reconciling human demands and ecosystem protection with a discussion of sustainable development. Chapter 3, the final chapter in this section, provides an overview of the different ecoregions of Canada and their points of interest.

GROUNDWATER AND SOIL CONTAMINATION IN SOUTHWESTERN ONTARIO

In 1986 Stephen Naylor and Gabrielle Monteith bought a 40-ha farm from Donald and Ellen McDonald for $155,000 in Puslinch Township, 20 km southeast of Guelph, Ontario (Figure 1.2). They wanted to start a fish farm. In April 1992 they initiated a lawsuit against the McDonalds, and in late 1993 they sued Puslinch Township, the Ontario Ministry of Environment and Energy, a trucking firm, and a lawyer who handled the purchase of their property. The reason for the lawsuit? Underneath one of their fields is a 1-m deep pit, covering an area of about one-fifth of a hectare, full of refinery wastes that no one had mentioned when they purchased the property. Oil refinery wastes had been dumped on their farm between at least 1963 and 1966.

In 1993 Naylor found the old waste site when he dug into the ground and unearthed a black, oily sludge that smelled like oil wastes. Following that discovery, Naylor and Monteith became concerned about the health risks for themselves and their four children (then aged one to four years) of living in a house some 500 m from the abandoned dump. However, it soon became clear that no one would purchase the property from them because of the liability for cleaning up the waste site. Even if they abandoned the property, they could still be liable for future clean-up costs, which were estimated to be as much as $10 million.

People who had lived on an adjacent farm explained that tanker trucks used to arrive at the McDonalds' farm, always at night and usually with their headlights turned off, to dump their load in an old gravel pit. Whenever a dumping occurred, there was a strong and distinctive oily smell. The neighbours did not like this activity, and some had complained to the township council and to public health officials, who always claimed the dumping posed no health problems. The neighbours had also complained to the McDonalds about the dumping, but the owners claimed that they could not smell anything.

In 1964 Puslinch Township passed a by-law to stop the dumping of industrial wastes on rural properties. McDonald, who later sold the farm to Naylor and Monteith, was quoted in a report prepared in 1981 for the province as stating that wastes were dumped into his gravel pit until about 1972. No one had kept a record of the quantities or types of wastes dumped at the site. The 1981 report recommended that the sludge at the site should be sampled and, if necessary, removed to a licensed disposal area. Furthermore, the report recommended that the location of the site should be marked on the deed of the property to forewarn potential buyers. Such a warning was never placed on the deed.

The pit was covered about 1984, and when Naylor and Monteith purchased the farm in 1986, it looked like other fields on the farm. A study commissioned later by Naylor concluded that the site could pose a risk to human health. Contaminants had leached from the site into nearby surface water and a swamp. It was also believed that other nearby private wells could be contaminated.

As a result of the report and their experience, Naylor and Monteith wanted to leave the farm, but before doing so, they believed that they should

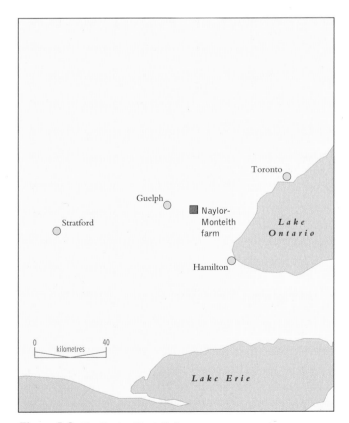

Figure 1.2 The Naylor-Monteith farm.

The Naylor-Monteith farm's barn and pasture (*Bruce Mitchell*).

The Naylor-Monteith farm. As a buyer, would you be able to tell if there was a waste pit in this field (*Bruce Mitchell*)?

be provided with a comparable farm in the same area and be reimbursed for their expenses (about $120,000) in fighting for a settlement. They also believed it was unfair for them to remain liable for any future clean-up costs. As Naylor is reported to have said, 'We didn't know about it, we didn't profit from it, we were innocent purchasers, but we're liable. When you buy a farm where you want to raise your kids and find you bought a toxic waste dump you're very angry' (*Kitchener-Waterloo Record*, 7 December 1993:A1). A former neighbour agreed. She commented that 'They didn't tell him the pit was there, and they should have. I think it's terrible that Naylor might be held responsible' (*Kitchener-Waterloo Record*, 7 December 1993:A2).

In mid-March 1995 Naylor and Monteith reached a $329,000 mediated settlement with four of the five other parties involved in the dispute: the Ontario government, Puslinch Township, a person who hauled the waste to their property, and their real estate lawyers. Dianne Saxe, their new lawyer, indicated that this settlement was probably the first

environmental case to be resolved in part through the use of a provincial pilot program, Alternative Dispute Resolution, and was reached after one-and-a-half days of mediation. In agreeing to the settlement, none of the four parties acknowledged responsibility for the oil waste problems on the property. Naylor subsequently explained that while their out-of-pocket expenses had totalled about $120,000, the settlement gave his family the opportunity to leave and look for another rural property.

However, an out-of-court settlement still had not been reached with the fifth party, Ellen McDonald, who, with her deceased husband had sold them the farm. However, the lawsuit initiated in 1992 against the McDonalds was finally settled in the spring of 1996. In addition, Naylor indicated that he wanted his name removed from the property deed of the farm with the contaminated site to ensure that he would not be responsible for future clean-up costs.

While Naylor and Monteith's experience was traumatic for them and their children, it highlights some of the issues this book addresses: *complexity*, *uncertainty*, *conflict*, and *rapid change*. Land or resource use considered acceptable at one time may not be tolerated later. Furthermore, our capability for detecting pollutants has steadily improved, thus changing conditions are constant. No one knows how many other similar waste dumps are buried under farms or urban areas in Ontario. Prior to Ontario's Environmental Protection Act in 1971, there was much unregulated dumping in the province. However, in 1991 the province completed an inventory of all known waste disposal sites in Ontario. The inventory included 1,358 disposal sites still in use and 2,334 that had been closed. However, the report also acknowledged that information was incomplete for many sites closed before 1971.

The problem of abandoned waste sites may or may not be widespread. No one can say for sure. As an area supervisor in the Ontario Ministry of Environment and Energy Office in nearby Cambridge commented, 'Do we know everything that's out there? No. We've tried our best to identify sites. There's always the possibility something has been missed' (*Kitchener-Waterloo Record*, 10 December 1993:A3). Furthermore, no one can explain conclusively what the impact of contamination from such sites may be on adjacent ecosystems, or on the health of people and other living things that live on or adjacent to the contaminated sites. Thus even if we understand the magnitude and extent of contamination from abandoned dump sites, significant questions remain regarding the implications and consequences. A particular challenge is determining who should pay to clean up such sites.

Improved understanding of affected ecosystems and social judgements will have to be combined to determine appropriate ways of dealing with old waste dump sites. But as the experience of Stephen Naylor, Gabrielle Monteith, and their four children illustrates, such environmental problems can have a direct and immediate impact upon people's lives and well-being. Such people do not want to wait for a long time for scientists to complete detailed research. They want action. There are similar situations throughout the world. Environmental change and challenge are not just abstract concerns for governments. They also have direct impacts upon billions of people. The next section provides an overview of some of the main global environmental trends.

THE GLOBAL PICTURE

The earth is different from all the other planets. As it hurtles through space at 107 200 km per hour, an apparently infinite supply of energy from the sun fuels life-support systems that should provide perpetual sustenance for the earth's passengers.

> ### Forecasting versus Backcasting
>
> *. . . Forecasting takes the trends of yesterday and today and projects mechanistically forward as if humankind were not an intelligent species with the capacity for individual and societal choice. Backcasting sets itself against such predestination and insists on free will, dreaming what tomorrow might be and determining how to get there from today. Forecasting is driving down the freeway and, from one's speed and direction, working out where one will be by nightfall. Backcasting is deciding first where one wants to sleep that night and then planning a day's drive that will get one there.*
>
> *The society that relies on forecasters worries that it does not like where the freeway is going, but it usually ignores all the exit signs because it can visualize little other than business-as-usual. The society that listens to backcasters has to decide first where it wants to go. It can dream dreams, and it can work to make its dreams come true. Neither society, of course, can guarantee that it will arrive where it expects, but the latter will travel more hopefully, at the price of having to think (Tinker 1996:xi).*

Our home, planet earth (© *1996 The Living Earth Inc.*).

Unfortunately, this does not seem to be the case. Organisms are now becoming extinct at rates unsurpassed for 65 million years. These extinctions cover all life forms and probably constitute the greatest number of extinctions experienced in the 4.5-billion-year history of the planet. Our seas are no longer the infinite sources of fish we thought they were. Our forests are dwindling at unprecedented rates. Even the atmosphere is changing in composition and making the spectre of significant climatic change a reality. Every raindrop bears the indelible stamp of the one species bringing about these changes—humans.

Why are these changes taking place? The answer is not simple, but it is clear that we are lost. We don't know where we are in terms of our relationship with the planet and our life-support systems. We don't know where we are going, or where we will end up if we continue as we have been doing. We haven't asked ourselves whether we want to go there, or whether we are on the right road. We haven't examined alternative destinations, and how we might be able to go elsewhere instead. We are lost. The situation is like the old joke about the pilot who tells his passengers that he has some good and bad news. The good news is that the winds are favourable and that we are ahead of schedule; the bad news is that we don't know where we're going.

To answer these questions, we must know where we are in the first place. One main variable that affects our impact on the planetary life-support systems is the number of passengers being supported. Although countless billions of passengers are on board, we are mainly concerned with humans, who seem to be having the most impact upon the system. This species, along with a few others, has experienced a staggering increase in population over the last century. In fact, the number of humans has doubled in the last forty years, and will probably double again in the next century as a result of exponential growth (Box 1.1). Ninety per cent of this future growth will occur in the lesser developed countries, which are the least able to afford it. The current world growth rate is about 1.7 per cent per year. This is lower than the peak rates of over 2 per cent in the late 1960s, but in absolute terms (the numbers added to the population total every year), growth has never been higher. Every fourteen weeks we add a population equal to that of Canada, an average increase of 10,700 people per hour. In one day, this amounts to the creation of a new city the size of Halifax or Victoria.

The steep curve of population increase shown in Figure 1.3 coincides with the time when humans learned how to exploit the vast energy supplies of past *photosynthetic* activity (coal and oil) in the earth's crust. Until then, energy supplies had been limited by daily inputs from the sun. The discovery of this new treasure trove of energy allowed humans to dramatically increase food supplies and improve our ability to process and transport materials. One result is that now some 5.8 billion humans are drawing upon the life-support systems for sustenance; before the Industrial Revolution, there were less than a billion. Another result is the pollution that now chokes the systems we depend upon.

It is not only the number of passengers that is important, however, for not all passengers have the same impact upon the life-support systems. First-class passengers get special meals three times a day, wine included; those in the economy section are lucky if they get one meal and must buy their own water, if it is available. Indeed, the 20 per cent of passengers in first class consume about 80 per cent of all resources used on the planet. Most Canadians, for example, are among the top per capita consumers of energy in the world (Figure 1.4), as discussed in more detail in Chapter 19. Energy is a good index of our planetary impact, reflecting our abilities to process materials and disrupt life-support systems through such pollution as *acid precipitation* and the production of *greenhouse gases*. Each Canadian consumes enough energy for sixty Cambodians. Does this mean, for example, that we in our relatively lightly populated first-class compart-

> *If current trends in resource use continue, and if the world population grows as projected, by 2010 per capita availability of rangeland will drop by 22% and the fish catch by 10%. . . . The per capita area of irrigated land, which now yields about one-third of the global food harvest, will drop by 12%. And cropland area and forestland per person will shrink by 21 and 30% respectively.*
>
> Sandra Postel,
> State of the World, *1994*

Box 1.1 Population and Exponential Growth

In 1798 a British clergyman, Thomas Malthus, pointed out that population growth was geometric in nature (i.e., 2, 4, 8, 16, 32, 64, etc.), whereas the growth in food supply was arithmetic (i.e., 1, 2, 3, 4, 5, etc.). This will inevitably, said Malthus, lead to famine, disease, and war. This was not a popular viewpoint in his day when population growth was considered very beneficial. For many years this Malthusian view was ignored. The opening up of new lands for cultivation in North America and the southern hemisphere and the later development of *green revolution* techniques allowed food supplies to increase rapidly.

Increasing numbers of experts, watching the decline in food supplies over the last few years (see Chapter 14) and the increase in population, particularly in less developed countries, now feel

that the Malthusian spectre is here once more. On the other hand, others, particularly economists such as Julian Simon, feel that more population simply furnishes more human resources upon which to build increases in wealth for the future. Political leaders from some of the lesser developed countries, which are experiencing the most rapid population growth rates, have also argued that population growth *per se* is not a problem, and that the main problem is over-consumption in the more developed countries. This distributive concern is echoed by women's groups who also are wary of coercive birth control programs, and think that most progress could be achieved by improving the status of women. Some religious groups oppose any attempts at birth control.

What do you think?

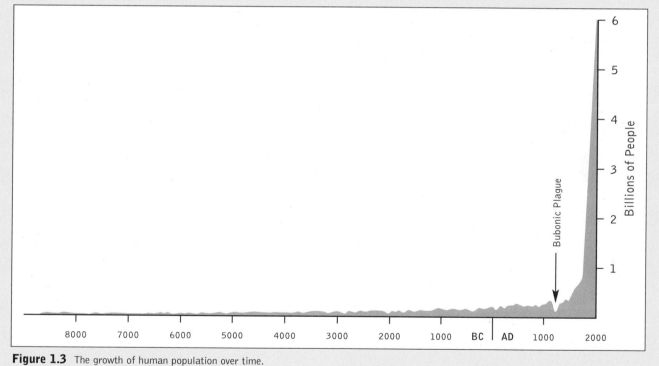

Figure 1.3 The growth of human population over time.

ment cannot or should not complain about over-population in Cambodia until their population reaches 1.6 billion and their energy consumption equals ours?

Obviously, there are very different kinds of passengers sharing our planet, and the differences among them have increased due to greater wealth

over the last twenty years. *Gross national product* (GNP) is an index that economists use to compare the market value of all goods and services produced for final consumption in an economy during one year. Over the last two decades the planetary GNP has risen by $20 trillion, but only 15 per cent of this has trickled down to the 80 per cent of

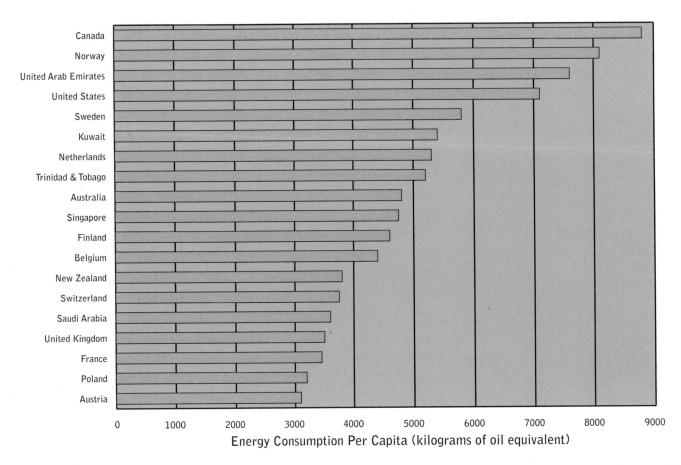

Figure 1.4 Energy consumption per capita in selected countries. The graph includes the countries with the highest energy consumption per capita. SOURCE: Resources Futures International, *Global Change and Canadians* (Ottawa: Canadian Global Change Program, 1993):12.

Children in the central highlands of Vietnam (*Philip Dearden*).

the passengers in the economy section of the spaceship. The rest has made the rich even richer. Each year the occupants of the economy section pay the first-class section interest payments that are more than four times the amount they receive in so-called aid programs. This has been likened to a hospital where the sick give blood transfusions to the healthy, rather than the other way round.

A global survey completed for UNICEF revealed that infant and child mortality rates in 1992 for some of the poorest countries were much higher than had been believed. For example, during 1992 nearly 1.1 million children under the age of five died in China. This number was 416,000 or 63 per cent more than what had been reported for 1991. Niger, a sub-Saharan country in West Africa with a population of 8.5 million, emerged as the world's most perilous place for a child. In that country, one in three children died before the age of one. In addition, almost one child in five died before reaching its fifth birthday. To put these figures in context, in the early 1960s a child in Niger was ten times more likely than a Canadian child to die before its first birthday. By 1992 the likelihood became forty times greater. Niger has suffered through drought and famine for years, and in such a situation it is difficult for people in that country to be concerned about long-term protection of the environment. For most of them, immediate survival is the overwhelming issue.

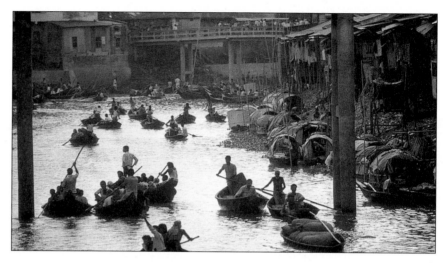

Flooding in Bangladesh (CIDA *Photo/Roger Lemoyne*).

In the summer of 1993 monsoon rains flooded nearly half of Bangladesh. An estimated 21 million people were left homeless and without food, safe drinking water, or medicine. In late July the government of Bangladesh sent nearly 20,000 doctors, nurses, and paramedics to the flooded areas to fight water-borne diseases. Early in 1993 a new strain of cholera appeared in Bangladesh and India and killed thousands of people. The flood waters prevented many people at risk from the new type of cholera from reaching hospitals and clinics.

The government could not estimate how many people were dying from the new strain of cholera, but by late July it was apparent that more people were dying from diarrhoea as a result of drinking unsafe water than from drowning. In addition, approximately 4,000 people drowned or disappeared during the four weeks of flooding during July in Bangladesh, northern India, and Nepal.

The deaths from the monsoon flooding in 1993 were far less than the 100,000 people killed in Bangladesh during a 1991 cyclone. However, the 1993 flooding covered a much broader area and affected more farm land. The flood waters destroyed millions of hectares of crops in the three countries, a major blow for nations in which the majority of people depend upon farming for their food and livelihood. Wildlife were also affected. In India the one-horned rhinoceros is an endangered species because poachers kill it for its horn, meat, blood, and bone. Due to the flooding, hundreds of the one-horned rhinoceros fled national parks, especially Kaziranga Park in the northeastern state of Assam, and abandoned their calves.

Annual monsoons often cause serious problems in the densely populated Indian subconti-

nent. As a result, they represent a chronic or ongoing problem, as well as a crisis when they occur. In 1993 the disruption to people and the environment was intensified due to unusually heavy rainfall at the start of the rainy season. Furthermore, other factors may be making people more vulnerable to monsoon flooding. These include overpopulation on the flood plains adjacent to major rivers, destruction of forests and other vegetation that protect soil from erosion during heavy rainfall, and massive dam systems designed to alter the paths of major waterways. As in Niger, in sub-Sahelian Africa many people in Bangladesh, India, and Nepal have difficulty wor-

---

Water and Health

Although we have a renewed appreciation for water's many functions, we are also more aware of our failure to manage the resource properly. During the 1980s major efforts were made to extend water and sanitation throughout the developing countries. However, the majority of poor people are still without safe drinking water and adequate sanitation services. . . . In urban areas, the number of people without access to sanitation actually increased by about 70 million in the 1980s. The health consequences of such service shortfalls make water a life-and-death issue for millions of people. Water-related diseases account for 8 percent of all illnesses in developing countries, affecting some 2 billion people annually. It is estimated that 2 million children die from such diseases each year, deaths that could be averted if water supply and sanitation were adequate (Serageldin 1995:3).

Big Numbers

There is no honest way around the reality that the big numbers having to do with population growth, disruption of the earth's biogeochemical cycles, species extinction, and the health of soils, forests, and water are running against us. No one of these is necessarily fatal to our prospects. Taken together, however, they point inescapably to the conclusion that we do not have much time to set things right if we are to avoid major traumas in the decades ahead. The momentum of big numbers is sweeping us toward a precipice, but the words, concepts, theories, and stories essential to comprehend our situation are not yet part of our political language or public mindset (Orr 1994:124).

---

rying about the long term when day-to-day survival is a major challenge.

The deaths caused by unclean water following the monsoon floods highlight the fact that while most Canadians take clean drinking water for granted, more than one-third of all people in the poorest nations do not have access to safe water, and 80 per cent do not have adequate sanitation facilities. Indeed, in developing countries, water-based diseases are the single greatest killer of infants and the main cause of illness in adults. For example, it has been estimated that diarrhoea kills some 4 million babies annually in the Third World.

Lack of safe water led the United Nations to designate the 1980s as the 'Clean Water Decade', with the goal of providing safe drinking water and adequate sanitation facilities for all people. Despite initiatives taken during the Clean Water Decade, by the early 1990s it was calculated that 1.3 billion people still did not have access to safe drinking water and 1.7 billion did not have access to adequate sanitation.

The stresses on the planetary life-support systems are a result of overpopulation and poverty, as well as overconsumption and pollution. Together, they result in *carrying capacity* pressures (a term discussed in more detail in chapters 2 and 5) at all different scales. Although in the past there have been many cultures that have violated the carrying capacities of their local environments with dire results, never before have we approached these limits at the global scale.

If we return to our original metaphor, although we may not be able to see every twist and turn of the path, we can at least get a fairly good idea of the direction in which we are heading. The prospects are not good. Fortunately, some of these trends have been given public attention at, for example the Earth Summit in Rio de Janeiro during 1992, which is discussed in more detail in chapters 2 and 20. The question remains: As we now know the direction we are heading in, will the international community come together for the common good to change direction and make the necessary sacrifices to do so? The next decade will determine the answer to this question.

The Canadian Picture

Where does Canada fit within this global picture? We, of course, are some of the privileged first-class passengers. In fact, we are some of the most privi-

---

Box 1.2 Globe Facts

- About 170 000 km^2 of tropical forests are destroyed every year. Almost half the original area has already been cleared.
- Topsoil is disappearing at a rate faster than accumulation on one-third of the world's crop lands.
- World economic activity has grown an average of 3 per cent per year since 1950. At this rate, by the year 2050, global output will be five times larger than it is today.
- Land degradation costs the world about $42 billion a year in lost crop and livestock output.
- 60 000 km^2 of new desert are formed every year, mainly as a result of overgrazing.
- Since preindustrial times, the concentration of carbon dioxide has increased by 27 per cent and methane by 144 per cent.
- Over 3 per cent of the stratospheric ozone layer has been depleted, leading to a 6 per cent increase in ultraviolet light on the earth's surface.

Box 1.3 Canada Facts

- We generate about 360 kg of solid waste per capita per year, ranking seventh in the world.
- We generate almost 6 tonnes of hazardous waste for each US $1 million of goods and services produced; Japan generates less than a quarter of a tonne.
- We have one of the highest per capita uses of water in the world (about 15 m³), roughly three times that of Sweden and Japan.

- We have one of the highest per capita energy uses in the world.
- We use our cars nearly 10 per cent more than residents of other industrialized countries.
- We rank eighth in terms of our production of greenhouse gases. The five top countries are all small oil-producing countries where gas leakage is high. Among comparable countries, only the US and Australia surpass Canada.

leged occupants. Our land is vast (about 13 million km³) and our population small (just over 30 million in 1996). Population density is 0.03 people per hectare, compared with eight people per hectare in Bangladesh. Canada would have to have a population of over 8 billion to equal this density. However, some people feel that Canada is already over-populated. Between 1971 and 1991 the population increased from 22 million to 28 million, an

increase of 27.7 per cent, roughly double that of other industrialized nations. As shown in Figure 1.5, most of this growth was in the western provinces, although the province that experienced the highest absolute level of growth was Ontario. Most population growth in Canada has occurred recently as a result of immigration rather than natural increases. Is this concurrent with the economists' view that the more people we have, the bet-

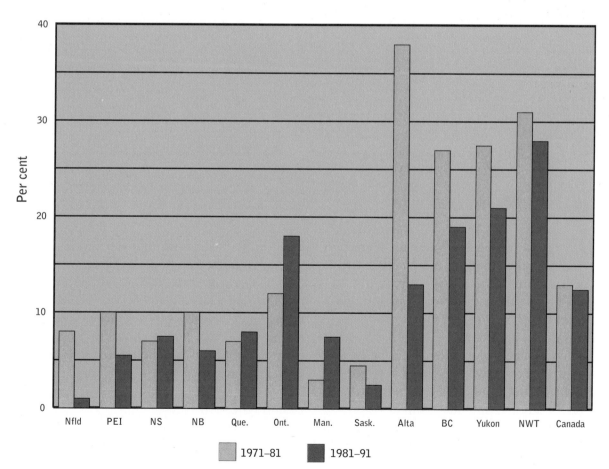

Figure 1.5 Population growth rate, 1971–91. SOURCE: Statistics Canada, Census of Population in *Human Activity and the Environment* (Ottawa: Statistics Canada, 1994):44.

Box 1.4 Your Ecological Footprint

In Chapter 3 we will discuss tracking and the pawprints left by animals. Have you ever thought about your ecological footprint? By this we mean the land base required to meet your needs. We derive all our energy and matter requirements from this land base, and also use it to dispose of our wastes. Can nature provide enough of these services on an ongoing basis to meet the needs of an individual, community, or nation?

These are the kinds of questions addressed by a group of researchers at the University of British Columbia. Human consumption was translated into the areas of productive land required to meet those needs. For example, the average Canadian requires a total of 4.8 ha to support current consumption (Table 1.1).

On a global scale, only 1.6 ha is available per person, an amount that is shrinking every year, largely as a result of population growth. To provide for everyone at Canadian standards would require three earths, not one. By the year 2030, estimates suggest that only 0.9 ha will be available per person. The demands differ greatly, however. In the Lower Fraser Valley, for example, the UBC team calculated that the local population needs an area twenty times greater than what is available to support it in terms of food, forest products, and energy. The ecological footprint of the area thus far exceeds the size of the area.

Table 1.1 The Ecological Footprint of the Average Canadian in Hectares Per Capita

	Energy	Built Environment	Agricultural Land	Forest	TOTAL
Food	0.4		0.9		1.3
Housing	0.5	0.1		0.4	1.0
Transport	1.0	0.1			1.1
Consumer goods	0.6		0.2	0.2	1.0
Resources in services	0.4				0.4
TOTAL	2.9	0.2	1.1	0.6	4.8

Source: M. Wackernagel et al., 'How Big Is Our Ecological Footprint? A Handbook for Estimating a Community's Appropriate Carrying Capacity' (Vancouver: Department of Family Practice, University of British Columbia, 1993).

ter it will be? Are more people good for Canada when we consider that there is more to Canada than just the economy? Does Canada have a moral obligation to accept migrants from overcrowded countries elsewhere? These are some important questions you might consider.

In terms of numbers alone, there can be little doubt that Canada is not overpopulated compared to virtually any other country. However, as discussed earlier, it is not simply the number of people but the impact they have that is critical. Canadians are among the top producers per capita in the world of industrial and household garbage, hazardous wastes, and greenhouse gases. Some think that the size of the country, the cold in win-

ter, and the heat in summer are the reasons behind our remarkable energy consumption, but it is clear that Canadians can contribute substantially in reducing impacts on planetary life-support systems. There are suggestions throughout the book on how you can help reduce the impacts on the environment. This is a theme that will be returned to in Chapter 22.

Canada's Green Plan

One way in which the Canadian government sought to address the environmental challenges facing the nation was to announce a five-year, $3 billion program in 1990 called the Green Plan. It had the following objectives:

- clean air, water, and land
- sustainable use of renewable resources
- protection of our special spaces and species
- preserving the integrity of our North
- global environmental security
- environmentally responsible decision making at all levels of society, and minimizing the impacts of environmental emergencies

Each objective was broken down into specific plans for action and given a budget (Environment Canada 1990). Targets—such as a 50 per cent reduction in Canada's generation of waste, the completion of the national parks system, and the stabilization of carbon dioxide and other greenhouse gas emissions at 1990 levels by the year 2000—were established. However, the plan was soon criticized as a mere political exercise, despite the environmental rhetoric. The *Globe and Mail* obtained documents through the Access to Information Act, which showed that the government had spent $12.6 million alone on the production and promotion of the plan. Nonetheless, a survey conducted for Environment Canada one year later found that 65 per cent of Canadians had little or no knowledge of the plan. The funds were also slashed by $500 million and frozen for a while, which also slowed implementation schedules. Four years into the plan, for example, there was still no agreement between the federal government and the provinces as to how greenhouse gas emissions would be stabilized.

Two political scientists (Hoberg and Harrison 1994) analysed the plan and were critical. They found that most of the forthcoming reduced funds were directed into education and research activities, with very little actually spent on direct environmental action, such as regulations or enforcement. They suggest that this allowed the government to seem environmentally conscious without actually doing anything to regulate environmentally unsound business practices, change government procedures, or provoke jurisdictional confrontations with the provinces. They conclude that 'by encouraging Canadians to think they can have environmental protection without economic sacrifice, the Green Plan denies the difficult choices that are called for today' (Hoberg and Harrison 1994:135. '[T]he Government now conducts environmental assessments of all proposed program and policy ini-

> *The . . . Green Plan is the most important environmental action plan ever produced in Canada. It is the source for more than 100 important and well-funded initiatives over the next five years. It is a comprehensive plan that deals with our environment as inter-related and whole. . . . Canada's Green Plan is an investment in our planet, our nation and ourselves.*
>
> Robert R. De Cotret,
> minister of environment

tiatives . . .' (Environment Canada 1990). Unfortunately, this did not extend to the single largest initiative undertaken by the government: the signing of the Free Trade Agreement with the US. No environmental impact assessments were completed.

However, Canada is making some progress. Some parts of the Green Plan have moved forward, as will be described in more detail in subsequent chapters. The government has also passed the Canadian Environmental Assessment Act and created the position of federal commissioner of the environment and sustainable development, who will report annually to Parliament on:

- federal policies, laws, regulations, and programs affecting sustainable development
- the extent to which federal legislation and policies comply with Canada's international commitments
- Environment Canada's capacity to lead the shift towards sustainability
- the government's progress towards sustainable development

It remains to be seen how effective these new mechanisms will be in helping to make the necessary changes in Canada to ensure that we help maintain the global life-support systems.

IMPLICATIONS

Most experts agree that where we are heading does not look very attractive. We appear to be exceeding the global carrying capacities of the earth's life-support systems. We have ceased to live off the interest and started to consume the capital at such a rate that it threatens the future viability of the system. Many species reach the limits of their environment's carrying capacity, overshoot them,

> *The earth's environmental assets are now insufficient to sustain both our present pattern of economic activity and the life-support system we depend on. . . . The roots of environmental damage run deep. Unless they are unearthed soon, we risk exceeding the planet's carrying capacity to such a degree that a future of economic and social decline will be impossible to avoid.*
>
> *Sandra Postel,*
> State of the World, *1994*

and have their numbers drastically reduced by environmental factors (see Chapter 5). Overall, humans have been able to avoid this process because we have used technology to increase our carrying capacities, but can we continue to increase indefinitely? Will even we have to accept some limits to our activities and population numbers?

If the answer is yes, then identifying the necessary changes is not that difficult. We need to balance birth and death rates, restore climatic stability, protect our atmosphere and waters from excessive pollution, curb deforestation and replant trees, protect the remaining natural habitats, and stabilize soils. The challenge, however, is charting the course necessary to fulfil these objectives. Before its demise, the Soviet Union had possibly the most stringent and comprehensive environmental pro-

tection regulations in existence (Peterson 1993), yet still ended up as one of the most polluted environments on earth. The regulations were simply not enforced. The secret is having a plan that not only addresses the goals mentioned earlier, but is actually able to achieve them through strategies that can be implemented.

In this book we try to provide some suggestions on how this can be achieved, with particular reference to the Canadian situation. In this chapter we started with an example illustrating the complexity of the environmental challenges at the local level. They are characterized by uncertainty, rapid change and conflict, and the need to appreciate the scientific/technical aspects of a problem and the social dimensions. This book provides an introduction to both. Part B outlines some of the main processes of the *ecosphere*, the basic functionings of the earth's life-support systems, and the ways in which we are disturbing them. Part C details some of the main Canadian management approaches used to address environmental challenges. Part D provides a thematic assessment of the challenges associated with urban development, forestry, agriculture, wildlife, water and energy management, and use of the global commons (such as fisheries and the atmosphere).

The relationship between these different aspects is illustrated in Figure 1.6. Natural systems

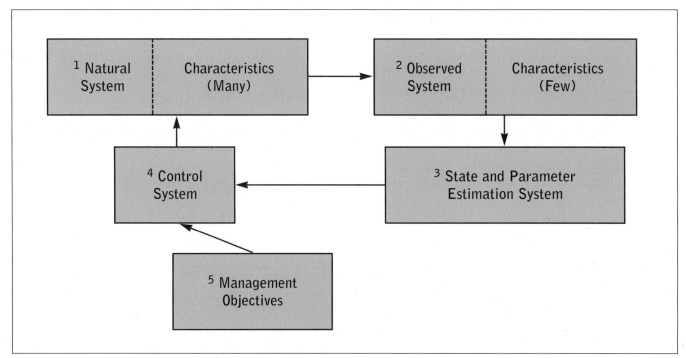

Figure 1.6   Simplified model of interaction of biophysical and social systems in resource management.

Box 1.5 Systems

You are in the educational system. You use the transportation system to go to your college or university, which is warmed by a heating system. What is a system?

Systems are composed of sets of things (for example, educational institutions, buses, heating components) that are all related and linked together in a functional way. Between these different components there is the flow of material (such as students, passengers, or heat) subject to some driving force (such as thirst for knowledge, the need to get somewhere, or the need to get warm). Systems are generalized ways of looking at these processes.

form the basis for all human activity (Box 1 in Figure 1.6). They range in scale from the giant atmospheric and oceanic circulation systems to the processes underway in a single living cell. The pollination of a flower by a bee, the melting of a glacier, and the biological fixation of nitrogen from the atmosphere are all parts of such systems, which are infinitely complex. There is a great deal of uncertainty as to how they function. One of the goals of natural science is to understand this complexity by constructing simplified models of how we think they work (Box 2 in Figure 1.6). The models presented in chapters 4 and 6 relating to energy flow through the biosphere and the nature of biogeochemical cycles are examples of these kinds of simplified representations of natural systems.

We do not know all the facts relating to these systems. We do not know all the components, let alone their functional relationships. Many species, especially insects, even in well-explored temperate countries such as Canada, still await discovery. Of those we know about, we have to select which ones we think are important and worth representing in our simplified models. Only recently, for example, have we become aware of the critical role played by various lichens in the circulation and retention of nitrogen in temperate rainforests (Chapter 6). Furthermore, not all characteristics are measurable, even if we are aware of their existence. Thus the simplified models that we use to understand natural systems are fraught with uncertainty.

On the basis of these models we try to estimate the status of a given system (Box 3 in Figure 1.6). How many fish spawn in a certain river? What proportion of the landscape supports commercial tree growth? What soil characteristics are suitable to support a given crop? If we understand the current status of the system, we can also ask what will be the result of changing certain parameters. What

would happen to the system, for example, if we took a certain number of fish from the river before they spawned, if we removed tree growth from a portion of the landscape, or if we grew a given crop for a particular time period? In other words, we try to assess the impacts of various changes to the system. Quite formal processes of impact assessment have taken place on social systems (as described in chapter 10) as well as on natural systems.

On the basis of this understanding, we try to replace natural systems with control systems in which the main decision regulators are not nature but humans (Box 4 in Figure 1.6). Instead of natural forces determining the number of fish that reach the spawning grounds, or the age of trees before they are replaced by other trees, or the type of species that grow in a particular location, humans make these decisions as we constantly modify the environment to our own advantage. These control systems are considered under topics such as forestry, water, energy and agriculture in Part D of the book.

Control systems are implemented on the basis of the social, economic, technological, and management constraints of society (Box 5 in Figure 1.6). These factors influence the demands for various outputs from the system and the speed of extraction. The environmental management strategies in Part C outline some of the main approaches to mediating between society's social and economic demands and the system's productive capacity. As with the natural systems, control systems are characterized not only by complexity and uncertainty but also by conflict among different societal groups regarding the rate of outputs and distribution of benefits from the control systems. Just deciding which social groups have a legitimate interest in a particular environmental issue is quite complex, as described in Chapter 11. There is a

Box 1.6 Yes, People Can Make a Difference!

In 1987, prompted by an eloquent *Equinox* article by tropical biologist Adrian Forsyth, student groups and naturalist groups across Canada started fundraising to help save tropical forests by selling the threatened lands surrounded Monteverde Cloud Forest Reserve in Costa Rica for $25 per acre. Within two years, World Wildlife Fund Canada had raised over $500,000 from these groups and thousands of individual Canadians. The funds were used by the Monteverde Conservation League to purchase 9000 ha of virgin cloud forest adjacent to the existing reserve. At the time, this tripled the size of the reserve from 4050 ha to 12 140 ha. By 1995, with the support of Costa Rica's government and funds raised worldwide, the reserve stood at 34 400 ha, forming the largest contiguous expanse of tropical forest along the forest divide of northern Costa Rica. Two more exciting developments followed. In the first, the Canadian International Development Agency is sponsoring reforestation, organic farming, skills training, and other community-based, sustainable-use projects in the wider region surrounding Monteverde. In the second, the Monteverde donors became 'Guardians of the Rainforest'. Their $25 donations, which sponsor protection of tropical forest, fund over 100 WWF conservation projects throughout Latin America. All this evolved because students in Canada chose to act on what they had learned!

wide range of dispute resolution mechanisms (as discussed in Chapter 12) that address the conflicts arising from resource allocation decisions.

The purpose of the framework shown in Figure 1.6 is to illustrate the complexity of the challenges in environmental management and the necessity of taking an integrated approach, such as the ecosystem approach described in Chapter 8, to understand this complexity. Both natural and social systems are fraught with uncertainty, making an adaptive approach (Chapter 9) necessary, with strong adherence to the *precautionary principle*, as discussed further in Chapter 2. Furthermore, our present predicament is largely the result of modifying natural systems to control systems before taking the time and making the effort to understand the consequences of our actions. The fisheries on both the Pacific and Atlantic coasts of Canada are in trouble (see Chapter 20) because our simplified models of the systems for decision making were inadequate. However, in some instances, even when the long-term implications of activities on the future viability of a resource are clear, the activities still continue for political and economic reasons. The overharvesting of timber (Chapter 15) is a good example.

Perhaps the most important message underlying Figure 1.6 and the environmental challenges we face is the need for a fundamental change in the way we view our relationship with nature, as discussed in more detail in chapters 2 and 21.

These changes must take place at all levels: international agencies (such as the World Bank), national and regional governments, and household and individual initiatives. Part of this book's goal is to motivate *you* to become more involved in making these changes happen. Suggestions for actions that you can take are given in many chapters, with a more extensive description in the last chapter.

SUMMARY

1. Ours is a rapidly changing world in terms of both environmental and human systems. Such changes generate considerable challenges for environmental management. The experience of the Ontario couple who bought a farm on which hazardous wastes had been buried illustrates the complexity, uncertainty, and conflict typical of many environmental management situations.

2. At the global scale there is undeniable evidence of unprecedented environmental degradation as a result of human activities. Human population numbers are a main challenge as are the consumer demands of people in the wealthier countries.

3. Conditions continue to deteriorate in many poorer countries. In Niger, for example, in the 1960s a child was ten times more likely to die than one in Canada; by 1992 the likelihood became forty times greater. An estimated 4 million babies a

year are killed by diarrhoea in lesser developed countries. Floods in Bangladesh in 1993 left 21 million people homeless and without food, safe drinking water, or medicine.

4. Canada is one of the most privileged countries, covering some 13 million km^2 and with a population of around 30 million. However, our environmental impacts are considerable. Our per capita consumption of water and energy is among the highest in the world. We also have some of the highest production per capita of waste products, including greenhouse gases.

5. Political response to these challenges has been varied with impressive initiatives in some areas, but virtually none in others, including regulations to meet global commitments to reduce emissions of greenhouse gases.

6. Many of these problems will reach crisis proportions over the next couple of decades and challenge our abilities to understand the earth's life-support systems and our management of the human activities causing these changes.

REVIEW QUESTIONS

1. What information is available in your municipality or province regarding environmental hazards, such as the kind that Stephen Naylor and Gabrielle Monteith encountered on their farm?

2. Who should be responsible for dealing with environmental hazards left over from earlier resource use and environmental standards that are no longer acceptable today?

3. If you were the judge hearing the Naylor-Monteith case, what information would you need to make a decision about the potential health risk for people living on their farm? How would you determine the monetary compensation for any potential risk?

4. Outline the main arguments for considering population growth either as a threat to global carrying capacities or as a building-block for future economic growth.

5. What moral obligations, if any, do Canadians have to assist people in the Third World whose standards of living do not meet basic human needs?

6. Is population growth or environmental degradation the major problem in the Third World? Which is cause and which is effect?

7. What are some of the Canadian government's main initiatives to address environmental problems?

8. What is a system? Outline the components of a system that you use on a regular basis.

REFERENCES AND SUGGESTED READING

Aptekar, L. 1994. *Environmental Disasters in Global Perspective.* New York: Macmillan.

Bird, P.M., and D.J. Rapport. 1986. *State of the Environment Report for Canada.* Ottawa: Minister of Supply and Services Canada.

Bongaarts, J. 1994. 'Can the Growing Human Population Feed Itself?' *Scientific American* (March):36–42.

Brown, L.R. 1993. 'The World Transformed: Envisioning an Environmentally Safe Planet'. *The Futurist* (May/June): 16–21.

Environment Canada. 1990. *Canada's Green Plan.* Ottawa: Minister of Supply and Services Canada.

Government of Canada. 1996. *The State of Canada's Environment.* Ottawa: Minister of Public Works and Government Services.

Hoberg, G., and K. Harrison. 1994. 'It's Not Easy Being Green: The Politics of Canada's Green Plan'. *Canadian Public Policy* 20:119–37.

IUCN/UNEP/WWF. 1991. *Caring for the Earth: A Strategy for Sustainable Living.* Gland, Switzerland: International Union for the Conservation of Nature.

Meyer, N. 1993. *Ultimate Security: The Environmental Basis for Political Stability.* New York: W.W. Norton.

Mitchell, B., ed. 1995. *Resource and Environmental Management in Canada: Addressing Conflict and Uncertainty*, 2nd ed. Toronto: Oxford University Press.

_____. 1997. *Resource and Environmental Management.* Harlow: Addison Wesley Longman.

Organisation for Economic Co-operation and Development. 1993. *OECD Environmental Data Compendium 1993*. Paris: Organisation for Economic Co-operation and Development.

Orr, D. 1994. *Earth in Mind*. Washington, DC: Island Press.

Peterson, D.J. 1993. *Troubled Lands: The Legacy of Soviet Environmental Destruction*. Boulder: Westview Press.

Postel, S. 1994. 'Carrying Capacity: Earth's Bottom Line'. In *State of the World*, edited by L.R. Brown. New York: W.W. Norton.

Resources Futures International 1993. *Global Change and Canadians*. Ottawa: Canadian Global Change Program.

Serageldin, I. 1995. *Toward Sustainable Management of Water Resources*. Directions in Development Series. Washington, DC: The World Bank.

Statistics Canada. 1986. *Human Activity and the Environment: A Statistical Compendium*. Ottawa: Minister of Supply and Services Canada.

_____. 1994. *Human Activity and the Environment*, cat. no. 11–509E. Ottawa: Statistics Canada.

Tinker, J. 1996. 'Introduction'. In *Life in 2030: Exploring a Sustainable Future for Canada*, edited by J.B. Robinson et al., ix–xv. Vancouver: University of BC Press.

Wackernagel, M., et al. 1994. 'How Big Is Our Ecological Footprint?' *Videa Journal* (December):2–3.

_____, et al. 1993. *How Big Is Our Ecological Footprint? A Handbook for Estimating a Community's Appropriate Carrying Capacity*. Vancouver: Department of Family Practice, University of British Columbia.

Worldwatch Institute. 1992. *Vital Signs: The Trends That Are Shaping Our Future*. New York: W.W. Norton.

_____. 1994. *State of the World, 1994*. New York: W.W. Norton.

Chapter 2

Sustainable Development

Canada is a fortunate country whose standard of living and style of governance are envied by people in many other nations. However, Canada is not without serious problems and challenges. As Pollard and McKechnie (1986:1) commented:

Our development, however, has had its own consequences, typical of industrialized, developed nations around the world. In our haste to attain personal, corporate and national wealth, we have not always questioned the meaning of wealth. We have been less than prudent in the way we dispose of unwanted materials. In using our resources, we have often ignored their relationships to the natural settings we call ecosystems. And we have not fully recognized that these ecosystems are but components of larger, inter-connected systems. . . .

As a consequence of such shortcomings, Pollard and McKechnie worried that Canadians were facing numerous important environmental and resource issues that 'threaten our social and economic well-being'. Assessing the situation in the mid-1980s, they concluded that the following were of particular concern at regional or national scales:

- loss of prime crop land, particularly around urban centres
- degradation of soils through loss of organic matter and nutrients; through contamination by toxic chemicals, airborne pollutants, and

pesticides; and through erosion and other factors
- water shortages, particularly in the Prairies, and deteriorating water quality in the Great Lakes basin and elsewhere
- environmental contamination, especially the transboundary movements of airborne pollutants

Added to that list today would be challenges associated with waste management, ozone depletion, global warming, and biodiversity in Canada, as well as Canada's role in dealing with poverty and protection of biodiversity in other world regions.

Such challenges are formidable, but cannot be ignored. The problems are characterized by complexity, uncertainty, conflict, and rapid change. One idea or concept that has been proposed to deal with these challenges is *sustainable development*. This chapter reviews the emergence of this concept, examines some of the debate associated with it, and presents a framework for the use of sustainable development.

EMERGENCE OF SUSTAINABLE DEVELOPMENT

This term was popularized by the World Commission on Environment and Development's report in 1987, which is frequently referred to as the Brundtland Report after its chairperson, Gro Harlem Brundtland, then the prime minister of Norway. However, the concept was not new, and various ideas and events preceded it.

Poverty and overpopulation are widespread in Third World countries (CIDA *Photo/Roger Lemoyne*).

The idea of carrying capacity, introduced in Chapter 1, has been used for decades, first for wildlife management, and subsequently to examine the relationship between a population and its environment. The challenge has been to estimate the numbers of animals or people that can be sustained in a given environment, how close the population is to a limit defined by nutritional and environmental indicators, and what actions should be taken to avoid exceeding specified limits (see Chapter 5).

Carrying capacity has not been just an academic exercise. In Canada, carrying capacity studies have been used to justify killing wolves, beaver, deer, and other animals when it is believed that their numbers have become too high relative to the amount of available food (Chapter 16). In Indonesia, when human carrying capacity appeared to be exceeded in Java, the government implemented a transmigration program in which people were relocated from overpopulated Java to other islands in the country. In all of these situations, populations were judged to be beyond the carrying capacity of the environment, or to be no longer sustainable. On that basis, policies and programs were developed to deal with what was considered an unacceptable problem. At a global scale, a report by the Club of Rome in 1972 drew attention to the possibility that worldwide resource shortages were imminent, and that countries needed to modify their policies and activities if they were to avoid exceeding the carrying capacity thresholds (Meadows et al. 1972).

The first major international meeting relevant to sustainable development was the United Nations Conference on the Human Environment, held in Stockholm in June 1972 (see Chapter 20). Attended by representatives of 114 governments, this was the first international meeting to focus upon environmental issues. However, as a result of the insistence of representatives from developing countries, development issues were also considered. This conference helped to raise awareness about environmental problems in the world. Outcomes included the Declaration on the Human Environment (included in Chapter 19) and the Action Plan for the Human Environment. More specifically, it led to the establishment of the United Nations Environment Programme, with head offices in Nairobi, to encourage international cooperation in reducing marine pollution and creating a global monitoring network.

Later in the same decade in Canada, a benchmark analysis was released by the Science Council of Canada (1977), which advocated a *conserver society*. The Science Council suggested that Canada and other developed nations were *consumer* societies, in which the emphasis was on material acquisition, undue consumption, and inappropriate waste disposal methods. In contrast, a conserver society would place less emphasis upon material

---

Box 2.1 Path to Sustainable Development

1972	UN Conference on the Human Environment, Stockholm
1977	Science Council of Canada Conserver Society Report
1980	IUCN World Conservation Strategy
1980	Brandt Report, *North-South: A Program for Survival*
1987	Brundtland Report, *Our Common Future*
1992	Earth Summit, Rio de Janeiro

aspects, and would seek to live more in harmony with its environment. In that regard, the Science Council was reflecting the ideas presented by Schumacher (1973) when he argued that 'small is beautiful'.

In 1980 the International Union for the Conservation of Nature (IUCN, but now called the World Conservation Union) published its World Conservation Strategy, in which it argued that conservation and development had to be integrated, an idea that would become one of the building-blocks of sustainable development. Emphasizing conservation, the IUCN maintained that conservation strategies should have three objectives: (1) maintenance of ecosystems, (2) preservation of genetic diversity, and (3) sustainable use of resources. Only in that manner, according to the IUCN, would it be possible to maintain essential ecological processes and life-support systems.

While the IUCN report stimulated discussion about sustainable development, it did not deal with the economic and political realities underlying environmental degradation: the effects of poverty, trade, colonialization, corruption, and war. It was another inquiry, the Independent Commission on International Development Issues (1980), chaired by Willy Brandt from Germany, that examined some of these realities in the context of North-South differences and called for the developed countries to accept some responsibility for problems in developing countries.

In the mid-1980s the United Nations created the World Commission on Environment and Development, chaired by Gro Harlem Brundtland of Norway. The commission had members from nations in both the North and the South, and visited many countries, including Canada, to hear suggestions on how development and environmental concerns could be balanced. Its report, *Our Common Future* (WCED 1987), drew worldwide attention to the concept of sustainable development. The most frequently cited definition of sustainable development from that report, along with several other perspectives, are shown in Box 2.2.

The Brundtland Report also emphasized that *basic human needs* had to be met, especially the essential needs of the world's poor. A major concern in the report was that increasing numbers of people were poor and vulnerable, and were degrading the environment in their ongoing struggle to obtain food and shelter because of their bleak situation. The challenge therefore was to ensure that growth could meet the needs of the next century's world population, which could be twice the present size, yet still be relying on the same environment. The Brundtland Commission concluded that it could not be 'business as usual' in the future.

The Brundtland Commission identified a number of general principles to guide development and environmental management in the future (Box 2.3), and to implement such general principles, the strategies shown in Box 2.4 were recommended.

The principles and strategies are very general. Since the Brundtland Report was prepared for the

---

Box 2.2 Perspectives on Sustainable Development from the Brundtland Report

Sustainable development meets the needs of the present without compromising the ability of future generations to meet their own needs. It contains within it two key concepts:

- the concept of 'needs', in particular the essential needs of the world's poor, to which overriding priority should be given; and
- the idea of limitations imposed by the state of technology and social organization on the environment's ability to meet present and future needs (WCED 1987:43)

'Scientists bring to our attention urgent but complex problems, bearing on our very survival: a warming globe, threats to the earth's ozone layer, deserts consuming agricultural land. We respond by demanding more details, and by assigning the problems to institutions ill equipped to cope with them' (WCED 1987:xi).

The rate of change is outstripping the ability of scientific disciplines and our current capabilities to assess and advise. It is frustrating the attempts of political and economic institutions, which evolved in a different, more fragmented world, to adapt and cope (WCED 1987:22). '. . . in the final analysis, sustainable development must rest on political will' (WCED 1987:9).

Box 2.3 Principles of Sustainable Development from the Brundtland Report

- Meet basic human needs.
- Reduce injustice and achieve equity.
- Increase self-determination.
- Maintain ecological integrity and diversity.
- Keep options open for future generations.
- Integrate conservation and development.

Box 2.4 Strategies for Sustainable Development from the Brundtland Report

- Revive economic growth.
- Change quality of growth.
- Meet essential needs for jobs, food, energy, water, and sanitation.
- Ensure a sustainable level of population.
- Conserve and enhance the resource base.
- Reorient technology and manage risk.
- Merge environment and economics in decision making.

United Nations, such generality is not surprising. Its authors hoped that their report would be accepted and used by many nations. As a result, it was never suggested that the Brundtland Report should provide a detailed blueprint regarding how to achieve sustainable development. Instead, it provided a more general vision or direction for the future, leaving individual countries to determine

how to achieve this vision. In contrast, the later IUCN (1991) follow-up report, *Caring for the Earth*, developed a much more detailed, recipe-type approach.

While there is inevitably ambiguity in parts of the Brundtland Report, certain aspects are clear: addressing poverty is a priority, basic human needs must be met, protection of environmental integrity is essential, economic and environmental aspects of development should be integrated, and a longer-term perspective to development planning is required. Regarding sustainable development, Canadians need to decide which of the initiatives proposed in the Brundtland Report should receive priority in Canada, and which should be our priorities as members of the global community. Regarding the latter, Canada needs to determine which actions it will take unilaterally, and which it will take multilaterally (see Chapter 20).

Sustainable development offers a major challenge to Canada and societies everywhere since it requires us to re-examine our current practices and procedures. Sustainable development has three strategic aspects. At one level, it represents a *philosophy*, in that it presents a *vision* or *direction* regarding the nature of future societies. Sustainable societies would be ones in which attention is given to meeting basic human needs, realizing equity and justice, achieving self-empowerment, protecting integrity of biophysical systems and other living things, integrating environmental and economic considerations, and keeping future options open.

Box 2.5 The Environment-Poverty Connection

In Canada we often think of economic growth as the antithesis of environmental protection. In wealthier nations, such as Canada, there is some truth in this, as the main threats to environmental integrity usually originate from activities undertaken to contribute to economic growth. In poor countries, however, this may not be the case, and the main threats to the environment come not from too much economic growth but too little. The resulting poverty drives people to exploit their environments in a non-sustainable fashion. It is difficult to think about food for tomorrow when you haven't eaten for several days already!

The data in Figure 2.1 clearly indicate, for example, that as incomes rise, so does the percentage of the population with safe drinking water supplies and sanitation. They also indicate, however, that other environmental threats, particularly from pollution, rise with income. A main challenge for sustainable development is to generate the environmental benefits of increasing wealth while implementing policies to ensure that environmental conditions don't deteriorate as a result of extra income.

Box 2.5 The Environment-Poverty Connection (continued)

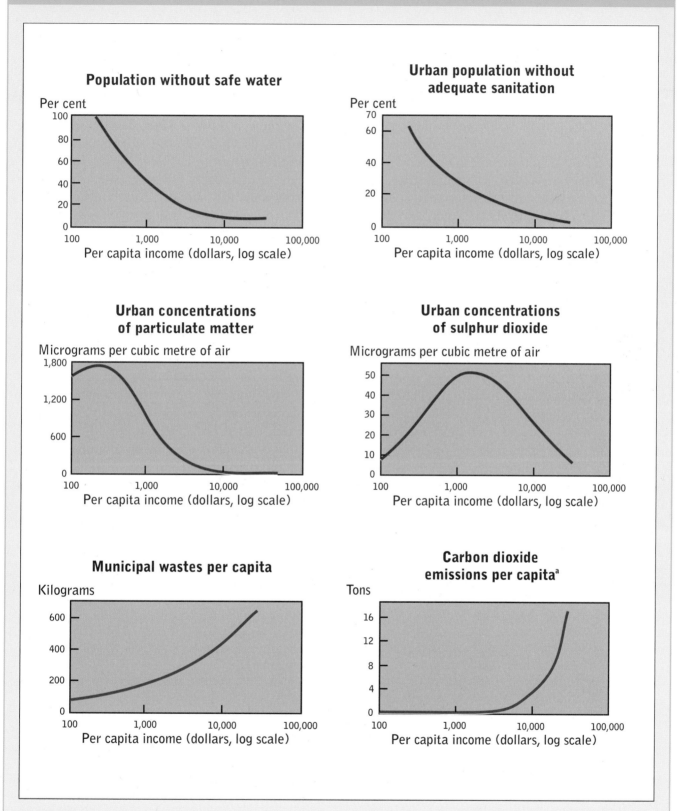

Figure 2.1 Environmental indicators at different country income levels. Estimates are based on cross-country regression analysis of data from the 1980s. [a]Emissions are from fossil fuels. SOURCE: World Bank, *World Development Report 1992* (Oxford: Oxford University Press, 1992):11. Copyright © 1992 by the World Bank. Used by permission of Oxford University Press, Inc.

As a *process*, sustainable development emphasizes a system of governance and management characterized by openness, decentralization, and accessibility (see Chapter 11). It accepts the legitimacy of local or indigenous knowledge, and seeks to incorporate such understanding along with science-based knowledge when developing strategies and plans. It also recognizes that conditions change, and that it is necessary to be flexible and adaptable, thereby being able to modify policies and practices in the light of experience (see Chapter 9). As a *product* related to specific places or resource sectors, sustainable development seeks to ensure that economic, environmental, and social aspects are considered together, and that trade-offs are evident to those who will be affected.

DEBATES ASSOCIATED WITH SUSTAINABLE DEVELOPMENT

The concept of sustainable development has generated both enthusiasm and frustration. The enthusiasm has come from those who believe that it provides a new direction or vision for the next century, one in which more attention will be given to longer-term implications of development and to balancing economic and environmental considerations. The phrase 'think globally and act locally' reminds us that while ultimately the globe is a single system in which actions in one part can often have implications for other parts, resolution of problems also requires significant action at the local level, thereby stimulating self-empowerment, partnerships, and cooperative approaches to management and development (see Chapter 11).

The frustration has come from those who believe that 'sustainable development' is so vague that it can be twisted and defined in ways to suit different interests, and that its general acceptance occurs only because different interests can use it to justify their positions. Thus developers are perceived as liking it because they can argue that growth must continue if basic human needs are to be met and standards of living raised. In contrast, environmentalists support the concept because they can use it to argue that environmental integrity must be given priority if there is to be long-term and sustained development.

However, some distinctions can be made. One is between *growth* and *development*. 'Growth' implies increase in size by addition of material. Thus it emphasizes quantitative change. In contrast, 'development' implies realization of potential or qualitative change. Although the Brundtland Commission did not make this distinction, and indeed it supported the concept of growth, sustainable development is interpreted by many to support an approach that is broader than growth. If this perspective is accepted, then many indicators of well-being and progress, such as gross national product, are no longer adequate, and it becomes necessary to incorporate other aspects into our measurements of progress. National accounting procedures

---

Box 2.6 'Too Much Green, Too Little Growth'

It is obvious from this article (*Globe and Mail* 3 July 1993:B2), reproduced from Canada's most widely circulated newspaper, that conflict between 'growth' and 'no-growth' visions of Canada still exist. Read the article carefully and identify the main assumptions, facts, and conclusions presented. Which do you agree with, and which do you disagree with?

Too Much Green, Too Little Growth
Terence Corcoran

While most of the population struggles to make ends meet and build a growing economy, governments appear to be gradually giving up control over economic policy making and turning it over to growth stoppers such as the Sierra Club. Look at the evidence.

In Washington, a federal district court judge says the Clinton administration must file an environmental impact study regarding North American free trade. The ruling, of mind-numbing intellectual density, plays into the hands of free-trade opponents who filed the suit, including Friends of the Earth and the Sierra Club.

In Vancouver, the government of British Columbia just announced it will join the U.S. government in setting up a

Box 2.6 'Too Much Green, Too Little Growth' (continued)

giant world wilderness area called Tatshenshini and ban all resource development within a chunk of territory larger than Ireland. The loss of jobs and economic growth fazes no one within the environmental movement, least of all the founding members of the Tatshenshini Wild lobby organization, a group that includes the Sierra Club.

South of Vancouver, in Portland, Ore., Sierra Club activists were instrumental in stirring up demonstrations and media fever over logging in the U.S. Pacific Northwest and over the fate of the spotted owl. In response, U.S. President Bill Clinton this week issued a major rollback in logging allowances and announced a $1-billion fund to retrain thousands of loggers.

Meanwhile, the Sierra Club figured in Canada Day demonstrations in Victoria and other cities around the world. They were organized to protest plans to allow limited logging in the Clayoquot Sound area of Vancouver Island. 'This issue is not going to die,' Sierra Club president Vicki Husband told the crowd in Victoria.

Obviously, the Sierra Club is not acting alone and unabetted in its attempt to freeze-dry economic development. Nor would it be accurate to single out the Sierra Club as if it were the all-powerful gang leader.

The point is that the Sierra Club and groups like it—Friends of the Earth, Greenpeace, and scores of others—represent a force that is gaining increasing control over a large and growing portion of North America's gross national product.

The same groups are buzzing around just about every sector of the economy, from the auto industry to garbage, tying up decision making, preventing economic development and forcing governments and the private sector—individuals and corporations—to incur billion of dollars in wasteful spending.

While roads are not being built, mines not developed, and industrial expansions are stalled, governments are being stampeded into ghastly recycling systems and bureaucrats are drafting regulations by the thousands:

The environmental movement, as manipulated by the professionals at groups such as the Sierra Club and Greenpeace, represents interest group politics at its extreme and at its worst. These groups routinely misrepresent facts and fabricate information, they abuse the media process and mislead their members and the public. But through it all, they seem to achieve a high degree of political credibility.

Politicians, who now spend their time gauging the relative strength of each interest group appearing before them, approach the green movement as if the environmentalists represented a valid constituency worthy of honest attention. The politicians appear to accept the notion that environmentalists are genuinely interested in working out reasonable agreements. But as we've seen in case after case, there are no grounds for agreement, because they will accept no compromise. And they will accept no compromise because they have nothing to gain by compromise. On the contrary, they thrive on continuing conflict and the gradual expansion of their spheres of influence.

The U.S. court ruling on NAFTA is being dismissed as a minor irritant that should not prevent passage of the agreement through Congress. Clearly, however, the ruling is bound to have an impact. Even if it is overturned in a higher court, the ruling represents another water mark in the rising tide of environmentalism's control over the economy.

In one of the more absurd sections in the ruling, the judge said NAFTA could produce lower farm prices that would create 'pressure to intensify domestic production methods which will have a detrimental effect on the environment.' Therefore, an environmental impact study is needed—to determine the effects of lower prices. Where will it end?

Eventually, the people whose lives are being affected by the environmentalists—a Growth Club whose standard of living is diminished and whose livelihood is being damage—are going to have to be heard. At the moment, however, the Sierra Club is in charge.

traditionally have not accounted for depletion of resources or pollution. Under a sustainable development approach, such aspects will have to be included in our measurements and judgements.

Sustainable refers to limits, and suggests that there are thresholds related to growth, which we exceed at our peril. There is tension here, however, as *sustainability* refers to the capability to persist and be robust and resilient. This is why some people prefer the term *sustainability* over *sustainable development* as they believe the former provides a more appropriate vision for the future.

Notwithstanding these distinctions, no single definition of sustainable development has become universally acceptable. The Brundtland Commission's statement that sustainable development requires meeting the needs of the present without compromising the ability of future generations to meet their own needs identifies an ultimate goal, but does not provide a clear definition or blueprint for achieving it. What has been accepted is that sustainable development does not imply one single pattern of development. Each country, region, and people will quite properly assign different weights to various aspects of sustainable development and to various ways of achieving it.

The different interpretations of sustainable development are highlighted by alternative perspectives in countries from the North and the South. In the North, or more economically developed countries, the focus for sustainable development has been upon environmental and economic dimensions. More specifically, the primary interest has been upon devising ways to integrate environmental costs into economic decision making while protecting jobs and industrial competitiveness. In contrast, in the South, or less economically developed countries, sustainable development has emphasized meeting basic human needs and achieving economic growth. Without satisfying basic needs and stimulating growth, it is believed that reversal of environmental degradation will not be possible. If people are too preoccupied with their survival, they are unlikely to be concerned about negative impacts on the environment.

It is important for Canadians to appreciate that their interpretation of sustainable development, with its emphasis upon integrating the environment and the economy, is likely not shared by over 80 per cent of the people in the world. Instead, such people are more concerned about realizing growth to satisfy needs for basic food and shelter.

Box 2.7 Indicators

Indicators are used to judge the status of a particular phenomenon. In Chapter 1 we introduced the idea of assessing the status of biophysical systems, and in chapters 5 and 16 we discuss *indicator species*. There are also socio-economic indicators. Gross national product (GNP), also introduced in Chapter 1, is one of the most commonly accepted. It measures the total value of all goods and services produced for final consumption in an economy, and is often taken as a symbol of the state of development in different countries. There are some fundamental problems, however, with using GNP as an indicator of societal well-being:

- The stock or capital of environmental assets, or increments to it, such as tree growth, is not taken into account.
- The more we deplete a resource, the more that enters into the flow of goods and services for the economy, and the higher the

GNP. The more we overcut our forests or overexploit our fisheries, the better this looks. Many lesser developed countries appeared to have significant economic growth in the past as a result of this illusion.
- Adoption of environmentally beneficial practices, such as cycling rather than taking the car to work, would result in a decline in GNP since you would be consuming less goods and services.
- Environmental disasters, such as the oil spill associated with the Exxon Valdez in Alaska, can be very beneficial for the GNP, given the goods and services required to clean up after the spill. Using GNP as an indicator of societal welfare might suggest that we need more oil spills!

These deficiencies have prompted the search for better indicators of societal welfare and sustainability.

Furthermore, since environmental management increasingly requires international cooperation to resolve issues related to global warming, ozone depletion, and biodiversity, it is essential that Canadians appreciate that their understanding of concepts such as sustainable development may not be shared universally. Such fundamental differences were highlighted at the Earth Summit in Rio de Janeiro during June 1992 when representatives of countries from the North and the South had difficulty in reaching agreements due to different perspectives about problems and solutions. Box 2.8 illustrates some of the pessimism regarding progress in achieving sustainable development since the Earth Summit. The events at the Earth Summit are considered in more detail in Chapter 20.

A FRAMEWORK FOR SUSTAINABILITY

Some basic ideas related to sustainable development are reviewed in this section. First, attention is given to a shift in 'world views' implied by sustainable development. Second, the implications of sustainable development at different spatial scales are examined. Third, some basic values, principles, and design criteria are explored.

A Shift in World Views

Kuhn (1992) has suggested that a different 'world view', or set of values and attitudes, will have to become dominant in Western societies such as Canada if sustainable development is to be achieved. He has contrasted this new environmental world view with what he calls the dominant social world view (Table 2.1).

Kuhn illustrated the difference in these two world views by drawing a parallel between *hard* and *soft* energy paths. The hard energy path emphasizes use of complex and sophisticated technology, as well as centralized decision making and control. It relies upon non-renewable energy resources, and is primarily oriented to the *supply*

Food distribution (*CIDA Photo/Roger Lemoyne*).

Children play among makeshift shelter in a slum (*CIDA Photo/David Barbour*).

side of energy policy and development. In contrast, the soft path emphasizes renewable energy resources and gives priority to improving energy efficiency and conservation in combination with decentralizing energy generation and control (Chapter 19).

Adoption of the soft energy path would require acceptance of the basic values of the new environmental world view. In other words, key ideas underlying the soft path are the need to reduce resource and material consumption, to recognize limits to the biosphere, and to become less complacent about accepting technological answers to issues that are basically social in origin and character. Solutions to energy shortages would thus not be confined to searching for new energy sources. Instead, equal or greater attention would be

**Box 2.8 Pessimism Regarding Progress Towards Achieving
Sustainable Development Since the 1992 Earth Summit**

In the foreword to the book *Compass and Gyroscope*, Philip Shabecoff (1993:vii) made the following comments:

> The 1992 Earth Summit in Rio de Janeiro could have been a turning point in the increasingly confrontational relationship between human beings and the natural world. . . .
>
> Little has changed, however, in the aftermath of Rio, and there are few signs of political will to reach the ambitious goals set

there. . . . there have been no major government actions, no major shifts of policy or economic priorities . . . to start implementing the mandate of the Earth Summit. Nor are there likely to be. The unpleasant truth is that the human community still does not know how to go about the business of creating a way of living that will sustain us and other life-forms with which we share the planet into future generations. We do not even really know yet what we mean by sustainable development.

Table 2.1 Major Characteristics of Competing Western World Views

	Dominant Social World View	New Environmental World View
Humankind and nature	Domination of nature Natural environment valued as a resource	Harmony with nature Natural environment intrinsically valued
Growth and technology	Continual economic growth Market forces Supply orientation Confidence in science and technology	Sustainable development Public interest Demand orientation Limits to science
Quality of life	Centralized Large scale Authoritative (experts influential) Increased material consumption	Decentralized Small scale Participative (citizen involvement) Decreased material consumption
Limits to the biosphere	Unlimited resources Non-renewable resources No limits to growth	Limits to resource extraction Renewable resources Limits to growth

Source: Adapted from R.C. Kuhn, 'Canadian Energy Futures: Policy Scenarios and Public Preferences', *Canadian Geographer* 36, no. 4 (1992):350–65.

directed to increasing the efficiency of, or achieving an absolute reduction in, energy use.

A shift to sustainable types of development will require some major changes. For example, Pierce (1992) has argued that there are five preconditions for a transition towards sustainable development. These include: (1) a *demographic transition* to stabilize population size; (2) an *energy transition* to high efficiency in production; (3) a *resource transition* to reliance on nature as income rather than capital (or, we should live off the interest rather than the

capital); (4) an *economic transition*, including a broader and more equitable sharing of benefits; and (5) a *political transition* requiring changes in traditional national and international divisions. Kuhn and Pierce's arguments imply that it cannot be 'business as usual' if sustainable development is to be achieved.

Implications of Different Spatial Scales

Reed and Slaymaker (1993) have argued persuasively that the interpretation of sustainable development should vary depending upon the spatial scale. They identified four scales for illustration (Table 2.2).

At a *planetary scale*, they argued that the stability of life-sustaining processes must be of prime concern as nothing less than planetary survival is at stake. Principal interest must therefore centre upon questions related to cycling, balance, and stability (see chapters 4, 5, and 6). Humans are only one among many species and life processes, and all are interdependent. As a result, questions of dominion, stewardship, and partnership are irrelevant. It is the health of the planet that matters at this scale, not the health of some individual species or organism. Thus humanity is *dependent* upon the existence of planetary processes and the challenge is to ensure their integrity.

At a *global scale*, the gap between rich and poor nations is the primary threat to sustainability. The appropriate strategy is therefore one of *partnership* among nations in the quest to address global environmental problems. At this scale, other questions related to equitable distribution of benefits and costs from resource use become important. At the global scale, environmental problems are explicitly related to problems of poverty, especially to the uneven distribution of population, resources, and wealth. The overriding criteria for sustainability at this scale include satisfaction of basic human needs, equity in access to and benefits from resource use, and protection of cultural self-determination. Whereas at the planetary scale priority should be given to protecting the integrity of biogeochemical processes to ensure planetary survival, at the world or global scale concern must be given to all human beings regardless of class, race, gender, culture, or geographic locality, as well as to other species that share the earth.

The *regional scale* focuses upon subnational levels. Depletion of the natural regenerative capacity of renewable resources, the use of the environment to assimilate waste products, and changes in patterns of urban, rural, and hinterland development are typical problems for sustainability at this scale. Issues linking environmental quality and social equity are often contentious at this level. At the regional scale, the need is for *partnerships* between culture and environment, as well as for *stewardship* of the environment. These concerns have led to community involvement in environmental management, and have been reflected in the use of public meetings, local round tables, and cooperative approaches for resource management (discussed in detail in Chapter 11). Finally, at the *local scale* the concept of *stewardship* best reflects an approach involving concern for care, respect, and responsibility. Stewardship requires maintaining the capacity of the environment to provide for non-consumptive values, such as aesthetic values, maintenance of genetic diversity, and spiritual needs. It also requires that 'resource development' does not deplete the ecosystem's capacity to reproduce over the long term. By incorporating stewardship and thereby maintaining the resource base and related ecological processes, resource development becomes one component of long-term social and economic development.

The four scales (planetary, global, regional, local) and associated interpretations of sustainability (dependence, partnership, partnership and stewardship, stewardship) are meant to be illustrative rather than exhaustive. However, the main point is that while sustainable development should not be interpreted in the same manner in each country, neither should it be interpreted the same way at different spatial scales. Furthermore, actions taken to achieve priorities at one scale might be diametrically opposed to the realization of priorities at

---

Table 2.2 Sustainable Development and Scale

Scale	Definition	Ethical Expression
Planetary	10^{10} km^2	Dependence
Global	10^8 km^2	Partnership
Regional	10^5 km^2	Partnership and Stewardship
Local	10^2 km^2	Stewardship

Source: Adapted from M.G. Reed and O. Slaymaker, 'Ethics and Sustainability', *Environment and Planning A* 25, no. 4 (1993):723–39.

another scale. For these reasons, it is appropriate for communities at the local level, as well as provincial or national governments, to develop sustainable development or conservation strategies that are interrelated.

Values, Principles, and Design Criteria

There are at least two moral or ethical reasons for trying to achieve sustainable development. First, humans and other life forms depend upon the biophysical environment for their survival. Without air, water, food, and other material resources, we could not survive. Furthermore, the environment often serves as a sink to absorb our wastes. As a result, the survival of human and other life forms depends upon an ongoing healthy biophysical environment. Second, we do not understand very much about the world around us. Risk, uncer-

tainty, ignorance, or indeterminacy often prevail (Table 2.3).

Such conditions suggest that humankind should be cautious when intervening in or interfering with biophysical processes. Indeed, this has led to a growing acceptance of the *precautionary principle* (see Box 2.9), which emphasizes the need to be cautious when outcomes are unknown and the potential consequences could be extremely adverse for future generations.

Against this background of uncertainty, Robinson et al. (1990:39) defined sustainability as 'the persistence over an apparently indefinite future of certain necessary and desired characteristics of the socio-political system and its natural environment'. Before examining what these necessary and desired characteristics are, it is appropriate to consider some other aspects of this definition.

Table 2.3 Different Types of Knowledge

Risk:	System behaviour is well known. The range of outcomes and probabilities associated with them can be predicted.
Uncertainty:	System characteristics are known, but we do not know the odds.
Ignorance:	Scientists will be surprised by the outcome. They do not know, but, with hindsight, can usually explain it.
Indeterminacy:	Scientific knowledge is inadequate. Causal chains and networks are open and not understood.

Source: Adapted from B. Wynne, 'Uncertainty and Environmental Learning: Reconceiving Science and Policy in the Preventive Paradigm', *Global Environmental Change: Human and Policy Dimensions 2*, no. 2 (1992):114, and M. Young, *For Our Children's Children: Some Practical Implications of Inter-Generational Equity and the Precautionary Principle*, Occasional Publication Number 6, Resource Assessment Commission (Canberra: Australian Government Publishing Service, 1993):3.

Box 2.9 The Precautionary Principle

The precautionary principle means trying to take into account in our decision making how little we might know about a certain phenomenon and what the consequences might be if we are wrong in our predictions. It has several implications:

- Regulatory action is often required before absolute cause-and-effect scientific proof is available when the potential for serious or irreversible change to the planetary life-support systems is possible. Climatic change is a good example.
- Allow ecological space to compensate for our lack of knowledge. In other words, do not employ maximum sustainable yield models, as is normally done in fisheries

management, unless scientific knowledge is impeccable.
- Link the burden of proof to the proposed development rather than to the status quo. In many situations, the onus of proof is on those trying to prevent a development or introduction of a new chemical or other environmentally damaging occurrence to establish that the alteration will damage the environment, rather than the other way round. This was the case with the spraying of the spruce budworm in eastern Canada (discussed in more detail in Chapter 15). Opponents of the spray had the difficult scientific task of proving damage by the spray, rather than the proponents proving that it wasn't damaging.

First, the phrase 'apparently indefinite future' reminds us that we cannot and likely would not want to try to guarantee 'persistence'. Such a guarantee is not practical, since we often do not understand what might happen in the short-term, let alone the long-term, future. In addition, such a guarantee is undesirable since we would not wish to preclude the opportunity for positive change, even regarding present attributes we now enjoy.

Second, attention to 'characteristics' of a system and its environment highlights that enhancing a system's sustainability is not the same as maintaining a system in its present state. Indeed, attempts to maintain existing systems may lead to unsustainability. In that regard, some ideas used in Chapter 9 can be introduced here. The goal in sustainability is not to increase the *reliability* (or resistance to breakdown) of systems but to increase their *resilience* (or the capacity to recover from disturbance). An analogy can be made by comparing *safe-fail systems*, which can fail gracefully without catastrophic implications, to *fail-safe systems*, which are less likely to fail initially but are likely to collapse totally when breakdown does occur. A challenge, of course, is successfully combining these two approaches, which can be mutually exclusive.

Third, the definition emphasizes both necessary and desirable characteristics, since it is difficult to separate necessary characteristics from desirable ones. Furthermore, it is easy to imagine types of sustainability that we would consider undesirable and therefore unnecessary. For example, slavery, poverty, or widespread crime might be sustainable in a society, but hopefully most people would not find them desirable or necessary. This point raises a more general question, however, regarding who decides what the desirable characteristics are and how such decisions are made. Indeed, this aspect has caused much disagreement among representatives of economically developed and developing countries, as well as business and environmental interests, about what the nature of sustainability should be.

Fourth, Robinson et al.'s (1990) definition of sustainability emphasizes that there is no one correct version of a sustainable society. This should not be viewed as a major problem. It simply recognizes that various means can be used to achieve an end, and that the end itself may take on different attributes depending upon conditions in a time and

place. Thus there is no absolute sustainability that can be identified. Indeed, it can be viewed as a moving target, requiring adjustment as conditions change. As a result, this definition is compatible with the concept of adaptive management (discussed later in Chapter 9). This situation also implies that it is sometimes easier to identify what is not sustainable than to specify what is sustainable.

On the basis of this definition, Robinson et al. (1990) presented their principles of sustainability (Box 2.10), which they separated into two categories: *environmental/ecological* and *socio-political*. These principles are wide-ranging, and touch upon topics well beyond environmental management. That is most appropriate, however, as it provides an excellent reminder that *sustainability* or *sustainable development* affects almost all aspects of natural and human systems.

Box 2.10 highlights the idea that political activity should involve allocating power that reflects 'environmentally meaningful jurisdictions'. The Brundtland Report also recognized this issue, and commented that too many international and national institutions were 'established on the basis of narrow preoccupations and compartmentalized concerns'. Furthermore, the Brundtland Report noted that most of the institutions responsible for dealing with the integrated problems of development and environmental management tended to be independent, fragmented, and worked along relatively narrow mandates.

The Canadian Constitution reflects such fragmentation and disregard for the need for environmentally meaningful jurisdictions. As shown in Box 2.11, the division of responsibilities between the federal and provincial governments in many instances creates as many problems as it solves. For example, responsibility for fish is assigned to the federal government, while water within provinces is assigned to provincial governments. Since fish can only live in water, their management requires collaboration between the two levels of government. Allocation of responsibility for environmental and resource management usually reflects the jurisdictions of the federal, provincial, and territorial governments rather than areas based on ecosystem criteria. This situation creates numerous challenges for managers who have to consider both aspects.

Box 2.10 Principles of Sustainability

Principles of Environmental/Ecological Sustainability

- Life-support systems must be protected. This requires decontamination of air, water, and soil, and reduction of waste flows.
- Biotic diversity must be protected and enhanced.
- We must maintain or enhance the integrity of ecosystems through careful management of soils and nutrient cycles, and we must develop and implement rehabilitative measures for badly degraded ecosystems.
- Preventive and adaptive strategies for responding to the threat of global ecological change are needed.

Principles of Socio-Political Sustainability

Derived from Environmental/Ecological Constraints

- The physical scale of human activity must be kept below the total carrying capacity of the planetary biosphere.
- We must recognize the environmental costs of human activities and develop methods to minimize energy and material use per unit of economic activity, reduce noxious emissions, and permit the decontamination and rehabilitation of degraded ecosystems.
- Socio-political and economic equity must be ensured in the transition to a more sustainable society.
- Environmental concerns need to be incorporated more directly and extensively into the political decision-making process through such mechanisms as improved environmental assessment and an environmental bill of rights.
- There is a need for increased public involvement in the development, interpretation, and implementation of concepts of sustainability.
- Political activity must be linked more directly to actual environmental experience by giving political power to more environmentally meaningful jurisdictions and promoting greater local and regional self-reliance.

Derived from Socio-political Criteria

- A sustainable society requires an open, accessible political process that puts effective decision-making power at the level of the government closest to the situation and the people affected by a decision.
- Everyone should have freedom from extreme want and from vulnerability to economic coercion as well as the positive ability to participate creatively and autonomously in the political and economic system.
- There should be at least a minimal level of equality and social justice, including equality of opportunity, to achieve one's full human potential, recourse to an open and just legal system, freedom from political repression, access to high-quality education, effective access to information, and freedom of religion, speech, and assembly (Adapted from Robinson et al. 1990:44).

In Part D we will look at how the principles outlined in Box 2.10 are being applied in environmental management in Canada. Where they are not being used, we will look at why that is so. The main question of interest will be whether managers believe the principles are flawed and therefore should not be used, or whether ways to translate them into practical guidelines are still being sought.

IMPLICATIONS

Achieving sustainable development will undoubtedly require a fundamental change in Canadians'

basic values and attitudes, not just regarding policies and practices in this country but also regarding our relationships with other countries, especially those in the South or developing regions. It is becoming increasingly clear that too many of our current activities and practices are *unsustainable* in the long term. As Rees and Roseland (1991) have concluded, sustainable development requires us to look ahead to future opportunities and constraints, and not just focus on those facing us today. Furthermore, we are only as secure as the planet's capacity to provide us, other life forms, and subse-

Box 2.11 Constitutional Responsibility for Resources in Canada

Federal Government

BNA Act, 1867, Section 91 (now the Constitution Act, 1982)

Section 91(12) Sea Coast and Inland Fisheries

Section 91(24) Indians and lands reserved for Indians

Section 91(29) Such classes of subjects as are expressly excepted in the enumeration of the classes of subjects by this Act assigned exclusively to the Legislatures of the Provinces

Provincial Governments

Section 92(5) The management and sale of Public Lands belonging to the Province, and of the timber and wood thereon

Section 92(13) Property and civil rights in the Province

Section 92(16) Generally all matters of merely local or private nature in the Province

Section 109 All lands, mines, minerals, and royalties belonging to the several provinces of Canada. . . .

Source: From the British North America Act, 1867, and repeated in the Constitution Act, 1982.

A deep concern for the environment and the resourcefulness of the Canadian people have brought about a strong commitment to achieve sustainable development in a uniquely Canadian way. . . . Even when faced with insurmountable obstacles, Canadians seem to find innovative ways to overcome the most difficult challenges. . . . Our hope is that we can share with the world our working model of sustainable development.

Alan Emery, president of the Canadian Committee of IUCN

the use of chemicals at home? These and many other similar questions have a direct impact upon maintaining environmental quality for future generations. In the final chapter, we will return to a discussion of some of the things that you can do to help.

SUMMARY

1. Sustainable development can be considered a philosophy, a process, and a product.

2. The International Union for the Conservation of Nature (now the World Conservation Union) argued in 1980 that 'conservation' and 'development' had to be integrated, and that conservation should (1) maintain ecosystems, (2) preserve genetic diversity, and (3) ensure sustainable use of resources.

3. Sustainable development was popularized in *Our Common Future*, the 1987 report by the World Commission on Environment and Development, or the Brundtland Commission.

4. The most frequently cited definition of sustainable development is 'development that meets the needs of the present without compromising the ability of future generations to meet their own needs.' Sustainable development also recognizes that there are limits and thresholds that we should not exceed.

5. The generally accepted principles for sustainable development include: (1) meeting basic human needs, (2) reducing injustice and achieving equity, (3) increasing self-determination, (4) maintaining ecological integrity and diversity, (5) keeping

quent generations with healthy air, water, food, energy, and other life-support essentials. Sustainable development pushes us to examine how we can ensure that these life-giving essentials can be provided on a continuous basis.

We should not be lulled into thinking, however, that sustainable development is the concern only of governments and large industries. Each of us can have a greater or lesser impact upon the environment, depending upon the numerous decisions we each make daily. How do you travel to university? How much water do you use every day? What do you buy new when you really don't need it? Do you recycle? Do you try to minimize options open for future generations, and (6) integrating conservation and development.

6. Sustainable development differentiates between 'growth' and 'development'.

7. There is no single model or framework for a sustainable development strategy. Each country, region, or people needs to custom design a strategy that reflects its conditions and needs. A sustainable development strategy may also need to be modified as time passes.

8. Hard and soft energy paths highlight key differences between sustainable and unsustainable development. Soft energy paths reflect what is known as the new environmental world view.

9. Spatial scale is important for preparing sustainable development strategies, and attention should be given to emphasizing independence at a planetary scale, partnerships at a global scale, partnerships and stewardship at a regional scale, and stewardship at a local scale.

10. Two moral reasons support the search for sustainable development: (1) all life forms depend upon the biophysical environment for their survival, and (2) many biophysical processes and interactions are not well understood, suggesting we need to respect limits in the natural environment.

11. The precautionary principle emphasizes the need to be cautious when outcomes are unknown and the potential for negative consequences are high. However, lack of knowledge is not an adequate justification for not taking action.

REVIEW QUESTIONS

1. What are the most important economic and environmental issues in Canada and in your province today? Why are they prevalent? What do you think will be the most important economic and environmental issues in the next ten to twenty years?

2. What are the similarities and differences between carrying capacity and sustainable development?

3. What is the difference between a conserver society and a consumer society?

4. What were the main principles and strategies presented in the Brundtland Report?

5. What is the difference between 'growth' and 'development'?

6. What are the characteristics of the new environmental world view? What are the implications for sustainable development?

7. What are the preconditions for achieving sustainable development?

8. What are the major obstacles to the achievement of sustainable development? For example, which of the following are the most significant: population growth, environmental degradation, or consumption patterns?

9. Why is awareness of spatial scale important in developing strategies for sustainable development?

10. What are the implications for achieving sustainable development of 'globalization' and movements towards free trade regions?

11. What moral reasons support the pursuit of sustainable development?

12. What is the precautionary principle, and what are its implications for sustainable development?

13. In your opinion, what is the best way to define sustainable development for Canada, for your province, and for your community?

14. What are the major disagreements or debates regarding sustainable development?

15. What are the key principles related to sustainable development?

16. Does your community or province have a sustainable development strategy? If so, what are its characteristics? If not, what might be the key elements of such a strategy?

REFERENCES AND SUGGESTED READING

Brandt, W. 1980. *North-South: A Programme for Survival.* Report of the Independent Commission on International Development Issues. London: Pan Books Ltd.

Dearden, P. 1989. 'Wilderness and Our Common Future'. *Natural Resources Journal* 29:149–57.

Emery, A. 1996. 'Building a Model of Sustainable Development'. *World Conservation* 3:4–5.

Government of Canada, 1997. *Building Momentum: Sustainable Development in Canada.* Canada's submission to the Fifth Session of the United Nations Commission on Sustainable Development, 7–25 April, Dept of Foreign Affairs and International Trade, Info Centre. Ottawa: Minister of Public Works and Government Services Canada.

International Institute for Sustainable Development. 1994. *Trade and Sustainable Development Principles.* Winnipeg: International Institute for Sustainable Development.

International Union for the Conservation of Nature and Natural Resources. 1980. *World Conservation Strategy.* Gland, Switzerland: IUCN.

———. 1991. *Caring for the Earth.* Gland, Switzerland: IUCN/World Conservation Union, United Nations Environment Programme, and World Wide Fund for Nature.

Keating, M., and the Canadian Global Change Program. 1997. *Canada and the State of the Planet.* Toronto: Oxford University Press.

Kuhn, R.C. 1992. 'Canadian Energy Futures: Policy Scenarios and Public Preferences'. *Canadian Geographer* 36, no. 4:350–65.

Meadows, D.H., D.L. Meadows, J. Randers, and W. Behrens. 1972. *The Limits to Growth.* London: Earth Island.

Pierce, J.T. 1992. 'Progress and the Biosphere: The Dialectics of Sustainable Development'. *Canadian Geographer* 36, no. 4:306–20.

Pollard, D.F.W., and M.R. McKechnie. 1986. *World Conservation Strategy: A Report on Achievements in Conservation.* Ottawa: Minister of Supply and Services Canada.

Reed, M.G., and O. Slaymaker. 1993. 'Ethics and Sustainability'. *Environment and Planning A* 25, no. 4:723–39.

Rees, W.E., and M. Roseland. 1991. 'Sustainable Communities: Planning for the 21st Century'. *Plan Canada* 31, no. 3:15–26.

Robinson, J.B., D. Biggs, G. Francis, R. Legge, S. Lerner, S.D. Slocombe, and C. Van Bers. 1996. *Life in 2030: Exploring a Sustainable Future for Canada.* Vancouver: University of BC Press.

———, G. Francis, R. Legge, and S. Lerner. 1990. 'Defining a Sustainable Society: Values, Principles and Definitions'. *Alternatives* 17, no. 2:36–46.

Schumacher, E.F. 1973. *Small Is Beautiful.* London: Blond and Briggs.

Science Council of Canada. 1977. *Canada as a Conserver Society: Resource Uncertainties and the Need for New Technologies.* Report No. 27. Ottawa: Science Council of Canada.

Shabecoff, P. 1993. 'Foreword'. In *Compass and Gyroscope: Integrating Science and Politics for the Environment,* by K.N. Lee, vii–ix. Washington, DC: Island Press.

Sitarz, D. 1993. *Agenda 21: The Earth Summit Strategy to Save Our Planet.* Boulder: Earth Press.

Walters, M. 1991. 'Ecological Unity and Political Fragmentation: The Implications of the Brundtland Report for the Canadian Constitutional Order'. *Alberta Law Review* 29, no. 2:420–49.

WCED (World Commission on Environment and Development). 1987. *Our Common Future.* Oxford: Oxford University Press.

World Bank. 1992. *World Development Report 1992.* Oxford: Oxford University Press.

Wynne, B. 1992. 'Uncertainty and Environmental Learning: Reconceiving Science and Policy in the Preventive Paradigm'. *Global Environmental Change: Human and Policy Dimensions* 2, no. 2:111–27.

Young, M. 1993. *For Our Children's Children: Some Practical Implications of Inter-Generational Equity and the Precautionary Principle.* Occasional Publication No. 6, Resource Assessment Commission. Canberra: Australian Government Publishing Service.

Chapter 3

Environmental Context for Canada

Canada is the second largest country in the world, with a land area of 9.2 million km² and a further 700 000 km² in fresh water. Within its borders lies 6.7 per cent of the world's land area, 9 per cent of the forests, 13 per cent of wilderness, and 3 per cent of crop land. It has the largest collection of freshwater lakes and 9 per cent of the global total of fresh water. Canada also shares the Great Lakes, which contain 22 800 km³ of water, or about 18 per cent of the fresh water in lakes on the planet's surface. The Canadian coastline, at an estimated 244 000 km, is the longest of any country in the world.

Within this vast land there is great diversity, ranging from some of the oldest rocks in the world in the Canadian Shield to the relatively youthful crests of the Rocky Mountains. Some of the flattest lands in the world are in the Prairie region, and there are major mountain chains in the west. There are deserts, temperate rainforests, massive ice-caps,

---

Canada

Canada is rocky seas of mountains and magnificent tables of plain, thousands of leagues of spruce woods and fertile miles of farm, frozen white oceans and cities dominating the earth as far as the eyes can see. Canada is foggy wet coasts and dry cold deserts, rolling golden grasslands and valleys ablaze with autumn leaves, lonely surf-girt islands and towns teeming with people. This land is many lands each worth knowing. To glimpse this diversity is to feel some of the meaning of being Canadian (Edwards 1979:i).

---

and vast areas of boreal forest. Yet overall, there is one dominant characteristic when compared with most other countries: the high latitude location. This creates our rigorous climate and resulting ecozones, and has a profound influence on human activities. It helps influence land-use activities and settlement patterns. Forests cover 45 per cent of the land area, about half of which are commercially exploited. Canada has less than 0.5 per cent of the world's population, and most of this lies within a narrow strip along the United States' border, leaving the rest of the landscape relatively sparsely populated.

We tend to think of Canada as immense, solid, and unchanging. Within the brief perspective of a human life span it is, but if we step back in time we gain a more dynamic perspective. The great continental plates have heaved and swayed over the millennia, and at one time the mountains of eastern Canada were higher than the Rockies are today. About half of Canada is composed of the rocks of the Canadian Shield, dating back as far as 3 billion years, although most of them are a mere 200 million years old. Some 180 million years ago an immense trough filled with a shallow sea stretched from the Arctic Ocean to the Gulf of Mexico. Rivers carrying sediment flowed into the trough, and since then two main periods of uplifting have occurred, which produced the Rocky Mountains.

These processes are reflected in the main geological regions of Canada (Figure 3.1), showing the Canadian Shield surrounded by four main platforms of sedimentary rocks: the Arctic, Interior, Hudson, and St Lawrence platforms. Although

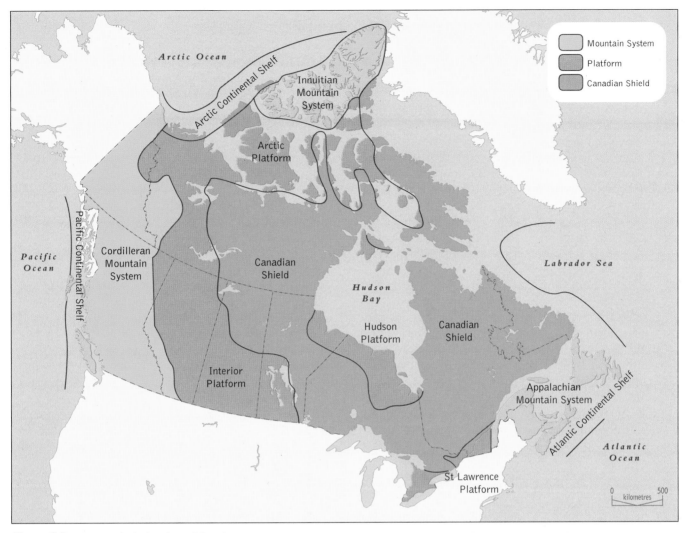

Figure 3.1 Major geological regions of Canada.

Only 10,000 years ago, most of Canada was covered in a thick layer of ice (*Philip Dearden*).

these regions appear as homogeneous units on the map, there are subtle internal differences. The Canadian Shield, for example, consists of seven structural components that have been subjected to different processes over various time periods. The three mountain systems shown in Figure 3.1 are of very different ages, with the Western Cordillera as the youngest. They have all been exposed to intense folding and faulting and contain great variability in rock types.

Not only has the land been rearranged by these earth-building processes, it has also been sculpted by ice ages caused by climate changes. The present Canadian

Box 3.1 Minerals

The diverse geological background described in the text has given Canada a wide variety of different types of rocks. Rocks are made up of minerals. If a large quantity of commercially important minerals are found in a rock, it is often called *ore*. These minerals may be exploited through mining. There are over 500 operating mines across Canada exploiting more than thirty different metals; over twenty non-metals; fuels such as coal, natural gas, and crude oil; and structural materials. Most mines, however, are extraction sites for low-value, high-volume building materials, such as sand, gravel, and crushed stone.

The geographical pattern of extraction depends upon the underlying geology. For most provinces and territories, metals such as copper, zinc, nickel, gold, uranium, and iron ore are the main products. Non-metals, such as potash and asbestos, are produced largely for export. The Canadian Shield is the greatest source of metals, whereas industrial minerals, such as potash and salt, are found largely in the sedimentary rocks of the Prairies.

The mines used to extract these minerals have major local and international impacts. They are important sources of local employment in many areas, especially in mining regions, such as the Abitibi area of Quebec and neighbouring areas in Ontario and the southern interior of BC. Mines also produce substantial environmental impacts, particularly water and air pollution. Some examples of these challenges, such as the pollution produced by the smelters at Trail in southern BC and at Sudbury in Ontario, are examined in later chapters.

In 1993 the total value of mineral production in Canada was $36 billion, 59 per cent of which was fuel production. Canada is a major global mineral producer and has a large proportion of the world's reserves of potash, gypsum, asbestos, and zinc. However, reserves of the main fuels and metals have been declining over the last decade, and are generally assessed with reserve lives of less than fifteen years. Reserve life is the stock of remaining reserves, for unlike most of the resources discussed in this book, minerals are non-renewable resources.

landscape is largely the handiwork of the last great ice age, which occurred some 10,000 years ago. Vast sheets of ice, over 3 km thick in some locations, covered the country, gouged out valleys and depressions, and left huge deposits when they finally retreated. An enormous inland sea, Lake Agassiz, which was larger than all the Great Lakes combined, once covered the Prairies, the result of glacial melt water trapped between the retreating glaciers to the north and blocks of ice to the south. It drained away as the ice melted. Only when these changes took place could the land be recolonized by the plants and animals that we know now. Indeed, some of these plants and animals are still recolonizing 10,000 years later!

It is important to appreciate the main biophysical characteristics of Canada in order to understand the environmental challenges we now face. This chapter provides an overview of these characteristics. Several approaches could have been used. There are national maps of physiographic regions, geology, soils, climate, and forests, for example. Recently, however, several government organizations have agreed upon an ecological land classification approach that combines these different attributes in ecozones. There are fifteen such ecozones in Canada (Figure 3.2), which will be used as the framework for this chapter. These zones represent the largest spatial units in a hierarchical classification system that becomes more and more detailed as the scale decreases. The fifteen ecozones can thus be divided into forty-seven ecoprovinces, which in turn are made up of 177 ecoregions composed of 5,395 ecodistricts. For the purposes of this overview, attention will be directed at the ecozone level. There are five marine ecozones in addition to these terrestrial ecozones. These will be described after the terrestrial zones.

TERRESTRIAL ECOZONES

1. Tundra Cordillera

Tundra is the treeless terrain, dominated by mosses, herbs, and shrubs, which grow between the forests of more southerly latitudes and the ice deserts farther north. It is distributed across the northern hemisphere, with small amounts found on the sub-Antarctic islands. In Canada, tundra stretches from the Western Cordillera, or mountainous area, all across the country in the Southern Arctic zone (zone 13). The Tundra Cordillera zone is at the northern extent of the Rocky Mountains and lies mainly in the Yukon, with steep, mountainous topography carved from the *sedimentary* bedrock.

Plant growth is limited by the harsh climate, low temperatures, and low precipitation, with a short growing season averaging seventy days per year. Most of the zone is outside the area of continuous permafrost, but the soils are heavily affected by freeze-thaw activity, or *cryoturbation*, giving rise to patterned ground features, such as polygons. Vegetation varies with altitude, but is mainly alpine tundra, with species such as mountain avens, saxifrages, dwarf shrubs, and lichens. Where trees are found, they are likely to be dwarf deciduous species, such as willows and birch that grow out along the ground rather than vertically and seldom reach over 1 m in height.

Wildlife is diverse and abundant in this zone and includes both barren ground and woodland caribou (Box 3.10), moose, mountain goat, black

and grizzly bears, wolf, marten, lynx, arctic ground squirrel, brown lemming, and the largest concentration of wolverines in the country. Oil and gas reserves are offshore in the Arctic Ocean. One of the main resource questions in this zone is whether these fuels can be produced and marketed without disturbing the fragile Arctic ecosystems. As with the rest of the tundra in Canada, many bird species migrate here in the summer for nesting and raising their young before returning south for the winter. These include waterfowl, such as canvas-backs, greater and lesser snow geese, and tundra swans. Few birds remain for the winter, though. Those that do include ptarmigan, snowy owls, and gyrfalcons.

2. Boreal Cordillera

The Boreal Cordillera, to the south and west of the Tundra Cordillera, is found in northern BC and southern Yukon. The mountains remain, but the wetter climate gives rise to more tree growth, hence the boreal designation. Physiographically, this zone is made up of the mountains in the west

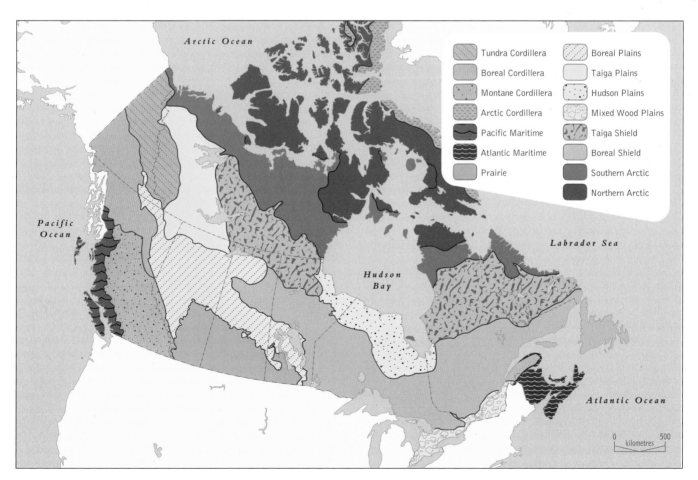

Figure 3.2 Terrestrial ecozones of Canada. SOURCE: E. Wiken, *Terrestrial Ecozones of Canada* (Ottawa: Environment Canada, 1986).

and east, separated by intermontane plains. Canada's highest mountains are in this zone, including Mount Logan at 5951 m, the highest in Canada. Much of the terrain in these high mountains, such as within Kluane National Park, is predominantly snow and ice. *Permafrost* is common at higher altitudes.

Predominant vegetation in the valleys is tree cover with species such as trembling aspen, balsam poplar, white birch, black and white spruce, alpine fir, and lodgepole pine. Alpine tundra dominates at higher elevations. Typical animal species include moose, woodland caribou, black and grizzly bears, lynx, Dall sheep, hoary marmot, and arctic ground squirrel. Birds include the rock, white-tailed, and willow ptarmigans. The Tatshenshini River drains out of these mountains and through the Alaska Panhandle into the Pacific Ocean. It was the site of one of Canada's most well-known environmental conflicts over the plans to develop a giant open-pit copper mine in the watershed. In 1994 the BC government decided against proceeding with the mine on environmental grounds and designated the watershed as a provincial park, which now forms part of the largest protected wilderness area in North America, linking Kluane National Park in Canada with Glacier Bay and St Elias national parks in the US.

Patterned ground, such as these polygons, is characteristic of cryoturbation, which is a common process in the Tundra Cordillera (*Philip Dearden*).

The black bear is primarily nocturnal, but will occasionally forage at midday (© *Mark Hobson*).

3. Pacific Maritime

Stretching along the entire British Columbia coast, the Pacific Maritime zone is influenced by the Pacific Ocean. Westerly winds sweeping across the ocean pick up considerable amounts of water. Most of this is precipitated on the windward side of the coastal mountains, giving the highest rainfall figures in Canada. For example, Prince Rupert, with 2552 mm of rainfall, has the highest total annual precipitation of any city in Canada, more so

Boreal

Boreal is a term that literally means 'of the North'. It is derived from Boreas, the name of the Greek god of the north wind. It is now applied to many northern phenomena, perhaps the most famous being the aurora borealis, *or northern lights. Many animal and plant species that live in the North have* borealis *as part of their Latin name, such as the delicate twinflower* Linnaea borealis, *which is found all across the country. It is also the name used to characterize the great northern forests that stretch not only across Canada but all across the northern hemisphere.*

Box 3.2 The Zen of Tracking

For most of us, tracking brings to mind old western movies showing Native scouts following the trail of desperados being brought to justice. At one time, however, the ability to track, to read the signs in the landscape left by others, was more common in society as communities depended upon their abilities to read such signs to secure food supplies. That skill is no longer so crucial for most of us, but tracking can still play a role in modern society, and in Canada we have more opportunity to benefit from this than in most other countries.

Tracking is not just looking for pawprints, trying to catch a quarry. Tracking involves reading all the signs left by others: the bent twig, crushed grass, a bush that shows signs of browsing, a scarred tree, smell, sound, and droppings. As we read and interpret these signs, we feel the same winds, hear the same sounds, and smell the same earth as the animal we are interested in. We begin to see the environment as the animal sees it, and the things that are important to it. We become that animal and start to transcend our human perspectives and limitations. Try it. Try it alone, and see if you can begin to perceive the environment and think like another animal.

Out west, one of the questions you may be interested in is whether the fresh bear print in front of you is that of a grizzly or a black bear. The chances of being seriously injured or killed are much greater with a grizzly

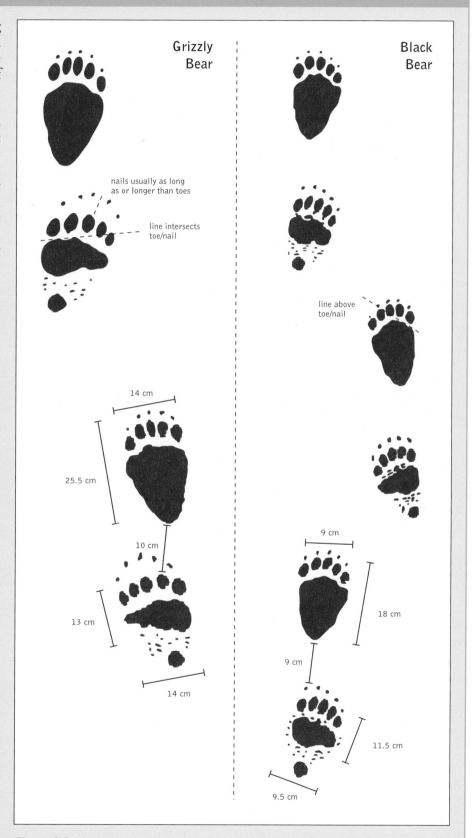

Figure 3.3 Grizzly bear and black bear tracks.

Box 3.2 The Zen of Tracking (continued)

than with a black bear! Although all bears can be unpredictable, the unpredictability of a grizzly is likely to have more serious consequences than that of a black bear! Tracking a bear, for example, also requires a close study of these signs in detail.

There are three main ways to distinguish between the tracks of the two species of bears, although none of them is infallible. First, grizzlies are usually bigger than black bears. If the foreprint is over 16 cm in length and the hindprint over 20 cm, then in all likelihood it is an adult grizzly. Black bears' prints are rarely over this size, although this does not rule out the possibility of an immature grizzly. A couple of other clues can also help, however. The nails of a grizzly are usually as long as, or longer than, its toes. Finally, in a soft substrate the toes of a grizzly look more pinched together and have a different curvature. If you draw a straight line from the base of the big toe (which is on the *outside* of the foot in bears) and across the leading edge of the palm pad, most of the little toe will be above this line in the pawprint of a griz-

zly; in a black bear most of the little toe will be below this line.

Tracks form the book of nature, a book that our ancestors knew by heart, but one that few of us can now read. But in Canada we are lucky because we still have the opportunity. The next time you have the chance, browse through a few pages and see what you can learn about life and about yourself.

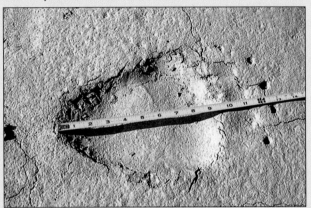

A grizzly bear track from the Tatshenshini Valley in northern British Columbia (*Philip Dearden*).

than the next highest, St John's, with 1482 mm. However, rain-shadow areas, such as Victoria, can receive a yearly average of as little as 619 mm. The maritime influence also results in the warmest average temperatures in the country, with mild winters averaging 4°–6° C, and relatively cool summers with July temperatures in the 12°–18° C range. Victoria is the city with the lowest annual average snowfall in Canada.

These climatic characteristics provide ideal growth conditions, with up to 220 frost-free days in the south. The temperate rainforests at lower elevations on deep alluvial soils have the most productive growth conditions in the country. Here grow the giant Sitka spruce and red cedar, with the Douglas fir in drier conditions. Western hemlock is the most abundant tree species, although it does not attain the age and size of these others. These trees form the backbone of British Columbia's lumber industry, but (as will be discussed in Chapter 15) they are no longer the infinite supply that they were once assumed to be.

The trees and the logging industry based on them have been the source of numerous environmental challenges, especially through the last

decade. Conflicts have arisen over whether particular areas should be logged or protected as parks. Various aspects of this issue are discussed at greater length in chapters 15 and 21. Wood processing has also been of concern, with many pulp and paper mills found to be in violation of pollution regulations. Both the federal and provincial governments have now started to enforce regulations, and rapid improvements have been obtained in water and air quality.

This productivity is also reflected in the aquatic environment as five species of Pacific salmon make their yearly migrations through rivers in this zone

Chum salmon, one of the five species of Pacific salmon, die after spawning, unlike the Atlantic salmon (*Philip Dearden*).

Box 3.3 Soils

It is not only the abundant moisture but also the deep, rich soils of the coastal valleys in British Columbia that have promoted such spectacular tree growth. Soils are not something we pay a lot of attention to, but without soil, life as we know it would not exist on this planet. It is from the soil that plants derive the water, nutrients, and structural support to grow. Organisms in the soil help break down these same plants and recycle their constituents, as described in Chapter 6. In fact, most of the earth's biodiversity is made up of soil organisms.

Just as we can define ecozones, so can soil scientists define soil zones to group soils that are relatively similar in terms of their measurable characteristics. A glance at the soil map of Canada (Figure 3.4) will reveal a close resemblance to the ecozone map, as at this scale both tend to reflect the gross climatic and geological conditions of the region. The Canadian System of Soil Classification includes nine orders, the largest category of classification:

Brunisols cover 8.6 per cent, and are brown soils found mainly under forests.

Chernozems cover 5.1 per cent, occur under grasslands, and are some of the most productive soils.

Cryosols are the dominant soils in Canada (covering some 40 per cent), and are associated with permafrost.

Gleysols only cover 1.3 per cent and are found in areas that are often waterlogged.

Luvisols cover 8.8 per cent and occur in a wide variety of wooded ecosystems. They have a higher clay content than brunisols.

Organics form in wetland ecosystems where decomposition rates are slow and cover 4.1 per cent of Canada.

Podzols are found beneath heathlands and coniferous forests, are relatively nutrient poor, and cover 15.6 per cent of Canada.

Regosols cover under 1 per cent of Canada and vary little from their parent material.

Solonets are saline soils, covering 0.7 per cent of Canada, and found mostly in grassland ecosystems.

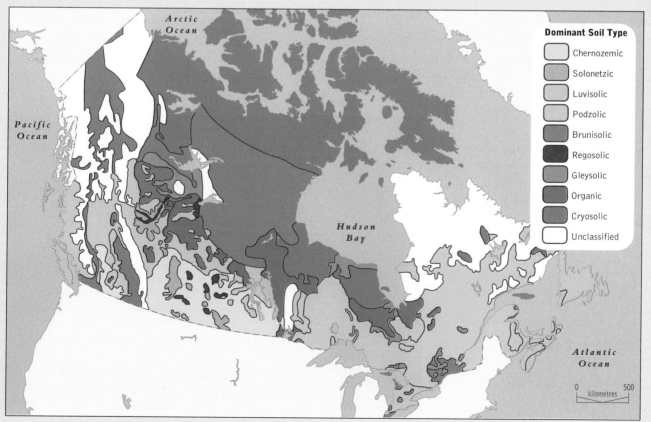

Figure 3.4 Soil zones of Canada.

Grizzly Country

But there is plenty of grizzly country in the mountains—country that is high and wild and rugged, a place where birds and streams and wind still blend in a song of the wilderness that lifts and falls in a cadence of freedom as sweet as life and as old as time among the proud gnarled trees and the rocky pinnacles. The same wind blows across the plains to the east; the same water flows down the valleys there, and the birds still nest and sing in the coverts. But there the song of freedom is muted now. It is not grizzly country any more. It is man country (Russell 1967:5).

to spawn. Some spawn in short coastal steams and estuaries in the zone, while others migrate great distances up into the Montane Cordillera zone. The numbers of fish returning to spawn have declined rapidly over the last few years. Habitat destruction on the spawning streams from activities such as logging is one reason (see Chapter 15), but overfishing and the failure of Canada and the US to agree upon adequate conservation measures through fishing allocation (see Chapter 20) is generally recognized as the main problem. Pacific salmon, unlike their Atlantic relatives, die after spawning, and their decaying bodies return nutrients to the sea and form a food bounty for scavengers, such as bears, raccoons, and bald eagles, which gather in great concentrations in the late fall to take advantage of the food supply. These species, along with cougars, whitetail deer, and wolves, are characteristic of this zone. The Steller's jay is the provincial bird of BC, and is common in the zone, as are the bald eagle and northwestern crow. Many waterfowl, such as loons, grebes, and many ducks, overwinter here before heading north in summer for nesting.

4. Montane Cordillera

Like the previous zone, the great altitudinal variation in this zone leads to considerable contrast between the summits of the snow-bound peaks through to high mon-

tane valleys, rolling plateaux, and deeply entrenched, desertlike conditions in the interior of British Columbia. The climate is similarly varied, but is generally characterized by long, cold winters and short, warm summers. Kamloops, the climate station with the highest average daytime summer temperature in Canada (27.2° C), is in this zone. Precipitation ranges from highs of over 1200 mm along the mountain summits to as little as 205 mm in the valleys in the rain shadow. Snowpack in this region gives rise to rivers flowing west to the Pacific (such as the Fraser and Columbia), east to Hudson Bay (such as the North Saskatchewan), and north to the Arctic Ocean (such as the Peace).

Vegetation varies according to these conditions, and can be thought of as a series of vertical zones increasing in altitude. At the summit the vegetation is alpine, characterized by lichens, herbs, and small shrubs. In the lower subalpine environment, trees such as alpine fir, Engelmann spruce, and lodgepole pine become more common. Below this zone there is considerable variation depending upon local conditions. Towards the north these tend to be characterized by ponderosa and lodgepole pines, Douglas fir, and trembling aspen. In the southeast's more moist conditions, the same species found on the Pacific coast—western hemlock, red cedar, and Douglas fir—may also be found.

Animals associated with this zone include bear, mule deer, elk, moose, woodland caribou, bighorn

Many Canadians are familiar with the spectacular mountain landscapes of the Montane Cordillera ecozone, which includes Banff, Canada's first and most visited national park (*Philip Dearden*).

sheep, coyote, and mountain goats. Many of these will be familiar to visitors of the famous national parks, such as Banff and Jasper, which are in this zone. It is also the zone where most people are likely to see a glacier. The Columbia Icefields in Jasper National Park draw thousands of visitors every year as the glacier is near the Icefields Parkway, a road between Banff and Jasper. Some of the challenges associated with national parks management at Banff are discussed in more detail in Chapter 17.

This zone also has some spectacular geology, which has led to the establishment of gold, copper, and coal mines. Many of the coal mines in the Rockies are now at high elevations. A main challenge for resource managers is the re-establishment of vegetation at such altitudes, which have limiting growth conditions and often highly toxic soils.

5. Boreal Plains

This northernmost extension of the Great Plains of central North America extends from the southern part of the Yukon in a wide sweeping band down into southern Manitoba. The underlying *glacial moraine* and *lacustrine* deposits give a generally flat to undulating surface similar to the Prairie zone to the south. Climatic differences that make it wetter and cooler than the Prairie zone produce a vegetation dominated by trees rather than grasses. However, the zone is still in the rain shadow of the Rockies and precipitation is modest, averaging 400 mm per annum over much of the zone. The growing season is also fairly short, with eighty to 130 frost-free days. Summers can be warm, with average July temperatures between 12.5°–17.5° C.

Coniferous trees include tamarack, jack pine, and black and white spruce, although deciduous trees such as trembling aspen, white birch, and balsam poplar are common, especially at the transition zone into the true prairie. This would have been prime habitat for wood bison (discussed in greater detail in Chapter 17) and still retains populations of woodland caribou, mule deer, moose, coyote, black bear, and lynx, among other species. This zone is now under increasing resource extraction pressure from forestry as technology is finding new ways to profitably harvest and process the deciduous trees there.

6. Taiga Plains

Taiga, a Russian word, is used to describe coniferous forests in that country. In North America it describes that portion of the boreal forest between the southern boundary of the tundra and the closed crown coniferous forest to the south. Topography is gently rolling with a high proportion of surface water storage, wetlands, and organic soils. The climate is cold and relatively dry, with as little as 200 mm of precipitation in the northern sections. Winters are long and cold, with mean daily January temperatures ranging from -22.5° C to -30° C. The annual number of frost-free days ranges between sixty and 100, while July temperatures are between 10°–15° C.

These conditions, plus the topography, give rise to large areas of wetlands dominated by species such as Labrador-tea, willows, dwarf birch, mosses, and sedges. On better-drained localities and uplands are mixed coniferous-deciduous forests containing white birch, trembling aspen, balsam poplar, lodgepole pine, tamarack, and black and white spruce. Animals in this zone include moose, woodland caribou, wood bison (in protected areas), black bear, wolf, lynx, and marten. Birds include the fox sparrow, northern shrike, sharp-tailed grouse, and red-throated loon.

7. Prairie

Prairie, or steppe as it is known in Europe, is a grassland found in relatively dry temperate climates. In North America these grasslands extend from central Alberta in the north to the Gulf of Mexico in the south and from the Rocky Mountains to the Mississippi River. Throughout this area the grassland is further subdivided into long-grass, short-grass, and mixed-grass prairie, depending upon growth potential, which generally increases towards the north and east as precipitation levels climb. The climate is continental, with cold winters, hot summers, and dry conditions, causing a pronounced summer water stress for plants. The ten cities with the most hours of sunshine per year in Canada are all in this zone.

The northern portion has slightly more precipitation and less heat in summer, giving more favourable growth conditions that may permit growth of trembling aspen and balsam poplar. The extent of these tree species and shrubs, such as wolf willow, has been able to increase as a result of fire

Box 3.4　Glaciation

At one time, glaciers covered all of Canada. Glaciation is one of the most important forces that has helped shape the landscape of the country. Glaciers form when there is an excess accumulation of melting snow. Over time, a glacier is created when this excess changes from snow into ice and starts to move under the force of gravity. There are several different types of glaciers, ranging from small ones on cliffs to continental ice sheets. The continental ice sheet that covered most of Canada has now melted. Most remaining alpine glaciers are in the Montane Cordillera.

Ice shapes the landscape through both the erosive action of its movement and through depositional activities as it melts and returns to water. Main features in glaciated areas include ice-fields, valley glaciers, ice-caps, and cirques. Valleys have steep sides with a broad U-shaped cross-section. They tend to be straighter than river valleys, as the ice erodes away ridges that may protrude into the valley. *Fiords* occur where these valleys have been flooded by the sea, as has happened on the coast of British Columbia and Labrador, for example. Peaks separating valleys are usually sharply eroded. *Cirques* form where small glaciers erode basins on the sides of moun-

tains. The erosive action of cirques and valley glaciers often produces sharp mountains known as *horns*, such as the Matterhorn in Europe or Mount Assiniboine in Canada. Continental glaciation produces few of these distinctive features, but tends to produce a smoother abraded surface, often pocketed by many small eroded depressions that later become lakes, like those on much of the Boreal Shield.

Depositional land-forms can be formed from material (*till*) deposited directly by the ice, by melt water flow (*glaciofluvial deposits*) and in lakes (*glaciolacustrine deposits*). Land-forms can be produced both under the glacier and at the glacial margins. Examples of the former include *drumlins*, which are large, streamlined hills up to 50 m high and 1000 m long, that appear to be the result of several different depositional processes. *Eskers* are another well-known depositional feature produced by glaciofluvial deposition in subglacial channels, and may extend hundreds of kilometres as long, sinuous ridges. At the glacial margins, moraines are a common depositional feature, and have a variety of forms depending upon where and how they were deposited. Where lakes form as a result of ice melt, sediments are laid down in strata known as *kames*.

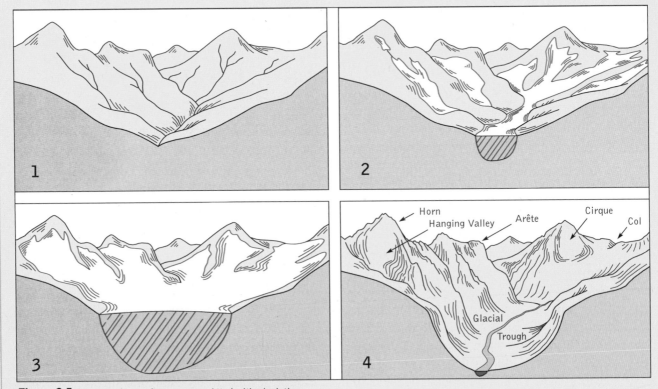

Figure 3.5 Some landscape features associated with glaciation.

suppression. Naturally occurring fires, sparked by lightning storms, would have been a frequent sight on the Prairies and would have helped maintain grass cover. To the south where conditions are drier and hotter, grass species dominate. Estimates suggest that prior to colonization, there were over 50 million ha of such prairie land in Canada. Much of this has now been converted into agricultural use, whether for cereal crops or ranching. Overgrazing has changed the composition of the remaining grasslands, and prickly pear cactus and sagebrush are now much more dominant.

Alkali lakes and sloughs, high in mineral salts, dot much of the landscape of the southern Prairie. Most plants cannot tolerate the extremely salty conditions around these lakes, and the margins become dominated by the salt-tolerant members of the glasswort family, which occupy similar positions on coastal marshes. The sloughs clearly stand out by late summer when all the water has evaporated, leaving only a salty white crust. A little farther from the slough, other salt-tolerant species, such as the dark green greasewood, a shrub common to the semidesert areas of Colorado and Utah, can be found. Although poisonous to domestic livestock, this species may be eaten with no ill effects by native species, such as pronghorn, ground squirrel, and many birds.

Typical mammals of the Prairie ecozone include mule and whitetail deer, coyote, prong-

> *. . . in the central portion of the continent there is a region, desert, or semi-desert in character, which can never be expected to become occupied by settlers . . . although there are fertile spots throughout its extent, it can never be of much advantage to us as a possession. . . . Knowledge of the country on the whole would never lead me to advocate a line of communication.*
>
> *Captain John Palliser,
> British North American
> Exploring Expedition,
> 1857–60*

horn, badger, northern pocket gopher, the white-tail jackrabbit, and one of its main predators, the recently reintroduced swift fox (discussed in more detail in Chapter 16). At one time these grasslands would have been home to gigantic herds of plains bison, pronghorn, and elk, preyed upon by thriving populations of plains grizzly bears and wolves. The vast herds have now disappeared, and many species in this ecozone are classified as Canadian species at risk. The wetlands in this zone constitute the main breeding area for many North American ducks, whose numbers have also declined precipitously over the years as wetlands have been converted into agricultural usage and dry conditions become more predominant. Costly rehabilitation efforts are now underway to try to restore the

Box 3.5 Canada's Olympic Champion, the Pronghorn

Pronghorn are reckoned to be the fastest middle-distance runners on earth, having been clocked at speeds of up to 98 km per hour for over 24 km. They are superbly adapted for speed with their small stomachs and big lungs, long strides, small feet, flexible spines (like the cheetah), and additional ligaments and joints, all designed to make this animal the fastest thing on the Prairies. Speed is the logical defence mechanism for a vulnerable animal easily seen by predators on the open landscape. Confidence in their ability to outrun predators makes them rather curious, and early settlers found them easy to attract and shoot. Initial herds estimated to contain 20–40 million animals in North America were soon reduced to under 30,000 animals as a result of this hunting pressure and

land-use change. There are now some half a million pronghorn remaining. They are the sole survivors of a family of North American antelopes, unrelated to the African antelopes, that at one time included over two dozen different species.

Like the bison (see Chapter 17), the pronghorn is also migratory, moving north in summer and south in winter when they may gather together in large numbers. Their main predators (other than humans, the plains wolf, and the grizzly) have been made extinct, leaving the coyote and, farther south, the bobcat. Unlike all other members of the bovid family (sheep, oxen, and goats), pronghorn shed their horns after the rut. Populations of pronghorn are now protected in Grasslands National Park in Saskatchewan.

Prairies for waterfowl species. Distinctive bird species of this zone include the endangered greater prairie-chicken, sage grouse, ferruginous hawk, American avocet, and the endangered burrowing owl, which is discussed in more detail in Chapter 14.

8. Taiga Shield

This zone has a taiga vegetation, but unlike the Taiga Plains, it occurs on the Precambrian Shield bedrock, which stretches across the country from Great Slave Lake in the west to Newfoundland in the east. These bedrock outcrops and glacial moraines are the most common surficial materials of the generally undulating plain. Low relative relief leads to large areas of waterlogged soils. Permafrost is also widespread.

The climate is very dry and cold. Annual precipitation is generally under 200 mm, with January temperatures ranging from -17.5° C to -27.5° C, and July temperatures from 7.5°–17.5° C. Frost-free days average between seventy and 100 per year. At the northern extremity of the zone lies the tree line as it merges gradually into the tundra. As conditions become less extreme towards the south, black spruce, tamarack, trembling aspen, balsam fir, white birch, and alder become more common.

Animals typical of this zone include moose, barren ground and woodland caribou, wolf, snowshoe hare, arctic fox, black bear, and lynx, while birds include northern phalarope, northern shrike, gray-cheeked thrush, and arctic and red-throated loons. Phalaropes are interesting bird species in that, unlike most other birds, the female is the more brightly coloured and initiates courtship. After laying her eggs, she flies south, leaving the male to incubate them and care for the chicks.

Environmental conditions in this zone are coming under greater scrutiny recently as a result of mining exploration and questions regarding the impacts of such activities. Impact assessment is discussed in more detail in Chapter 10.

9. Boreal Shield

The Boreal Shield is the largest ecozone in Canada, stretching along the Canadian Shield from Saskatchewan to Newfoundland. It is also part of one of the largest forest belts in the world, the boreal forest, extending all across North America and Eurasia, encompassing roughly a third of the earth's forest land and 14 per cent of the world forest *biomass*. This belt generally separates the treeless tundra regions to the north from the temperate deciduous forests or grasslands to the south. Winters are cold and summers are warm to hot, with moderate precipitation. In Canada the zone is influenced by cold Hudson Bay air masses, which give relatively high precipitation levels rising from 400 mm annually in the east to over 1000 mm in the west. Average January temperatures range from -10° C to -20° C with the July range of 15°–18°

The boreal forest in Newfoundland has bedrock outcrops, lakes, and muskeg (*Philip Dearden*).

Trembling aspen is characteristic of the boreal plains (*Philip Dearden*).

Box 3.6 Moose, the Largest Species of Deer

Several subspecies of moose are recognized in Canada. All migrated at one time or another across the Bering land bridge when sea levels fell. Early migrants were slightly smaller and darker and migrated farther south into the boreal forest proper, whereas later arrivals were larger, lighter in colour, and preferred the more open terrain of the taiga country. However, all subspecies are the largest members of the deer family, and a bull moose may grow up to 800 kg, second only to the bison in terms of weight.

The moose is predominantly a solitary creature of these open woodlands, well adapted to the snowy winters with their long legs and preference for browsing twigs. In fact, the name is said to derive from the Algonkian word *moos*, meaning 'eater of twigs'. They also have a big appetite for salt, and twigs are low in sodium content. The moose makes up for this by feeding heavily on aquatic plants, which are rich in sodium, in spring and summer. Moose may often be seen feeding in these preferred locations, sometimes diving as deep as 4 m to feed on particularly sodium-rich species. They also lose a tremendous amount of nutrients every year when they shed their massive antlers, which may stretch over 2 m and weigh over 30 kg.

Unlike other members of the deer family, the female moose plays the most vocal role in the rut. She emits a loud bawling to attract a male. Unfortunately, imitations of this bawling by humans are also quite successful in attracting prime males, which then become easy targets for hunters. Other than humans, adult moose have few enemies. They are just too big to be safely dealt with other than by the most skilled and desperate wolf pack or grizzly. As they get older and weaker, however, they become easier prey. Calves are vigorously but not always successfully defended by their mothers.

If you are in terrain where both moose and caribou (Box 3.10) occur and want to identify a track, caribou have proportionately broader feet than the moose, and their tracks appear as two widely separated crescents facing each other, usually under 12 cm in length (not including the dew claws) and 14 cm in width. Moose tracks are more heart-shaped, of greater length, and are longer than they are wide.

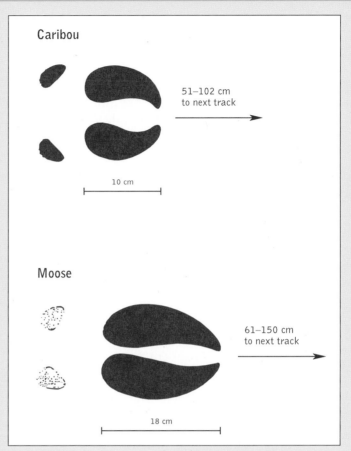

Figure 3.6 Caribou and moose tracks.

A bull moose (© *Wayne Lynch*).

Box 3.7 'The Rock'

Although Newfoundland, or 'the Rock' as it is sometimes affectionately known to the locals, is part of the Boreal ecozone, there are also some significant differences between Newfoundland and the mainland part of the zone, mainly due to the insular characteristics. Newfoundland has an interesting geological history. The theory of *continental drift*, for example, suggests that the easternmost part of the island was at one time part of North Africa and slowly drifted across the Atlantic some 200 million years ago where it came up against the North American *tectonic plate*. The resulting geological zones can be traced along the full length of the Appalachian Mountains in the United States, making Newfoundland the northernmost extension of the mountain chain. The climate is also harsh. Argentia, on the southern coast of the Avalon Peninsula, has the record for the highest average annual hours of fog in Canada at 1,890 hours.

Most islands have fewer animal species than the adjacent mainland, and Newfoundland is no exception. There are only fourteen native mammals, including woodland caribou, black bear, and other species typical of the Boreal zone. However, several authorities have recognized ten of these species as subspecies different from their mainland counterparts. Wolves were extirpated in the early part of the century. Another mammal that was introduced in 1878, the moose, has done quite well in the absence of one of its major predators. There are no deer on the island. However, there are spectacular sea bird colonies, particularly off the east and northeast coast as a result of the once-abundant fish populations offshore (see Chapter 20).

C. The average frost-free period is generally between sixty and 100 days, although some areas have as few as forty.

The terrain is characteristically rolling, with bedrock outcrops, glacial moraine, and many lakes dominated by coniferous forests of black and white spruce, balsam fir, tamarack, and jack pine, as well as significant cover of aspens and white birch. This forest cover is interspersed by large areas of wetlands, particularly moss-dominated bogs. Overall, the boreal forest is characterized by its lack of diversity of tree species; large areas are covered with just one or two species, particularly the spruces. Balsam fir becomes more dominant in areas of heavier precipitation towards the east. Wildfire is a natural component of the boreal forests. Species such as jack pine, an important species in well-drained areas of the Boreal zone, are particularly adapted to fire. The cones of the pine stay closed on the tree for many years and when they are opened by the heat of fire, the seeds propagate and the forest re-establishes itself. The cones stay viable in this manner even after the tree itself has died. Fires tended to occur every seventy to 100 years before fire suppression started with increasing logging of the boreal forest. Two influences—fire suppression and logging—have altered the species composition considerably to favour species such as balsam fir over black spruce, and have made the forests more vulnerable to attack by pests such as spruce budworm, which actually prefers the fir over the spruce as a host tree, as will be discussed in more detail in Chapter 15. This zone contains the highest timber volume of all the zones, largely due to its area. Individual trees are quite small, with black spruce generally between 16–26 cm in diameter and 9–15 m in height.

Animals in this zone include woodland caribou, whitetail deer, moose, black bear, wolf, lynx, raccoon, marten, red squirrel, and fisher, although the really dense forests are not used as permanent habitat by most large animals. Small mammals, such as squirrels, mice, voles, and shrews, may be abundant, however. Ticks are also common in this zone,

The Canadian Shield

The Canadian, or Precambrian, Shield is as central in Canadian history as it is to Canadian geography, and to all understanding of Canada. . . . So strong that not even the contraction of the globe itself has buckled its rigidity, it remains with its naked granite ridges, its multitudinous waters and sodden muskegs, an enduring contrast to the wide and fertile lands, the gentle slopes and hardwood forests of the Mississippi valley. The heartland of the United States is one of the earth's most fertile regions, that of Canada one of earth's most ancient wilderness and one of nature's grimmest challenges to man and all his works (Morton 1967:4).

surviving for years with no food until they are able to attach themselves to a host animal. They begin to suck blood immediately and may grow from 4 mm to almost 2 cm in size as a result of this bounty before they drop off the host to lay their eggs. Ticks, which also occur in other ecozones, spread various diseases to their hosts, such as tick paralysis. Representative birds include the spruce grouse; boreal chickadee; northern three-toed woodpecker; and great gray, horned, and boreal owls.

10. Hudson Plains

The most extensive wetland in Canada is in this ecozone, which extends from western Quebec through northern Ontario and up into northeast-

The extensive peatlands of the Hudson Plains (*Philip Dearden*).

Box 3.8 Canada's National Symbol, the Beaver

The beaver cannot really be considered the symbol of any one ecozone, since it is found all the way from Mexico to the Arctic, and from Vancouver Island to Newfoundland. It is, however, mostly associated with the northern woods and their waterways where it is well known for its water engineering. Many different species of beaver could once be found throughout the northern hemisphere. A Eurasian counterpart remains in small populations, but it is the North American beaver that has flourished and become one of the continent's most successful mammals. It also played a critical role, as did the sea otter on the West Coast, in attracting Europeans to the resources of North America.

The beaver is the second largest rodent in the world. All rodents are distinguished by their sharp incisor teeth, designed to gnaw through

The beaver, the national animal of Canada, can have a very significant impact on other species through its dam-building activities (*Philip Dearden*).

bark, crack nuts, or chew any other edible vegetable matter in a similarly efficient manner. Rodents comprise nearly 40 per cent of all mammal species. The speciality of the beaver, of course, is its ability to fell trees, some of which are as large as 1 m in diameter, which it can then use as a food supply and also to build its familiar dams and family lodges. Trees, particularly hardwoods such as poplars, are felled close to the water so that they can be dragged into the water, the beaver's preferred medium. With their broad flat tails, sleek coats, and powerful webbed hind feet, they are well equipped to undertake their aquatic construction activities. Their dams impede the flow of water, giving them greater access to trees and making them less vulnerable to terrestrial predators. They can use the pond so created as a low-energy way of transporting food to their lodge, which, surrounded by water, is virtually impregnable to predators.

Probably no other animal, except humans, has the ability to cause such radical and deliberate change to the environment. Beaver dams benefit not only beavers but also other water-oriented organisms, such as waterfowl, otter, muskrat, and frogs and other amphibians. With an estimated precolonization population of over 60 million animals, their ecological impact on the landscape would have been substantial. Although they were trapped out of large areas, they are now starting to recolonize as a result of conservation activities. Industrious, clever, effective, and resilient but not spectacular: is this the Canadian character?

ern Manitoba. Although the overall climatic characteristics of this zone are not very different from the other Boreal zones, the vegetation is different as a result of the poor drainage and low topography. The acidic, poorly drained and nutrient-deficient peat soils cannot be tolerated by most tree species, and the area is dominated by dense sedge, moss, and lichen covers. Better-drained areas may support black spruce and tamarack, the only conifer in Canada that is also deciduous. It loses its leaves every fall and regrows them again in the spring, bright green clusters of needles alongside the red cones from the previous season. The animals of this zone include woodland caribou, black and polar bears, fisher, marten, and arctic fox. There are also substantial populations of breeding waterfowl in summer, particularly Canada geese.

11. Mixed Wood Plains

This ecozone is the most urbanized in Canada, spreading up from the Lower Great Lakes north through the St Lawrence Valley. The topography is gentle, resulting mainly from the lacustrine, marine, and morainic deposits. The climate is continental, with warm, humid summers and cool winters. Mean daily July temperatures range from 18°–20° C, and January temperatures from -3° C to -12° C. There are between 120–80 frost-free days per year. Precipitation is from 720–1000 mm per year.

These conditions have produced the most diverse tree coverage in Canada (Figure 3.7) with

The mixed wood plains have the highest tree diversity in Canada (*Philip Dearden*).

over sixty-four species, according to Schueler and McAllister (1991). However, few intact areas of natural vegetation remain. In the northern part of the ecozone, the mixed coniferous-deciduous forest is dominated by red and white pine, oaks, maples, birches, and eastern hemlock. Further south, the warmer zones contain deciduous species, such as sugar maple, beech, white elm, basswood, and red and white oaks. The sap of all maples contains some sugar, and during pioneering days when sugar was scarce, most were tapped for their sugar content. However, in commercial production only the two species with the highest sugar contents, the sugar and black maples, are tapped. The white elm is also native to this zone. Unfortunately, many of them have been devastated by an imported fungus that causes Dutch elm disease. First noticed in 1944, the fungus is spread from tree to tree by beetles. It is a good example of the problems created by foreign organisms, to be discussed in more detail in Chapter 5.

Due to the amount of disturbance, large mammals, such as moose and bear, are no longer numerous in this zone. However, whitetail deer, raccoon, striped skunk, eastern chipmunk, eastern cottontail, and gray and black squirrels can still be found in abundance. Coyotes, not known in this area originally, are now becoming a problem as they interbreed with domestic dogs, lose their fear of humans, and increasingly prey upon livestock. Birds include the blue jay, whip-poor-will, Baltimore oriole, eastern bluebird, Carolina wren, bobwhite, cardinal, and green heron.

12. Atlantic Maritime

The Atlantic ocean influences this ecozone, which stretches from the mouth of the St Lawrence River across New Brunswick, Nova Scotia, and Prince Edward Island. However, conditions vary considerably between the upland masses of hard crystalline rocks, such as the Cape Breton and New Brunswick highlands, through to the coastal lowlands that support most of the population. Mean annual precipitation is as high as 1425 mm on the coast, but falls to less than 1000 mm further inland. Temperatures are

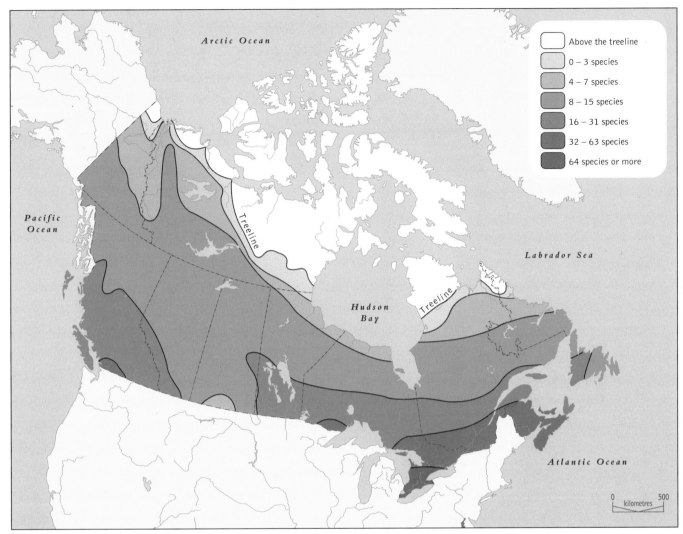

Figure 3.7 Number of tree species in Canada. SOURCE: F. W. Schueler and D.E. McAllister, map from 'Maps of the Number of Tree Species in Canada: A Pilot GIS Study of Tree Biodiversity', *Canadian Biodiversity* 1, no. 1 (1991):23.

also moderated by the ocean, with mean daily January temperatures of -2.5° C to -10° C and a mean July daily temperature of 18° C. In the highland areas, the number of frost-free days can be as few as eighty, whereas on the coast this increases to 180.

Forests are generally mixed stands of deciduous and coniferous species, such as balsam fir, red spruce, yellow birch, sugar maple, eastern hemlock, and red and white pine, mixed with boreal species, such as black and white spruce, white birch, jack pine, and balsam poplar. Mammals and birds are generally those found elsewhere in eastern Canada. There seem to be increasing signs, however, that the cougar, which was once found from the tip of South America all the way north to the Yukon, still survives in very small numbers in isolated populations in these eastern forests.

Secretive by nature, the cougar is often difficult to detect unless it kills livestock or its numbers become so great that immature animals are forced into contact with civilization as they attempt to establish territories, as often happens on Vancouver Island. Although the cougar is capable of killing a wide range of prey, including porcupines (another widely distributed species), abundant populations of deer make up a substantial part of its food source. Given this food supply and protection from hunting and further loss of habitat, hopefully the eastern cougar will continue to make a comeback.

13. Southern Arctic

Like the Taiga Shield ecozone, this zone is split into two by Hudson Bay. The terrain has generally rolling plains punctuated by bedrock outcrops of

Box 3.9 Carolinian Canada

Carolinian Canada is the wedge of land stretching from Toronto west to Windsor that contains 25 per cent of the country's population. It also contains the greatest number of tree species in the country (Figure 3.7) as the transition zone between the eastern deciduous forests to the south and mixed coniferous-deciduous forests to the north. This southernmost part of the country, with the southern Great Lakes' mediating effects on climate, enables semitropical tree species, such as the cucumber and sassafras, to spread into Canada. Windsor ranks as the third warmest city in Canada after Vancouver and Victoria, and is the most humid city in the country. It is little wonder that in summer the humidity and southern vegetation can give one the illusion of being much farther south.

Besides the distinctive vegetation, the Carolinian zone also supports a noteworthy bird population. Point Pelee is one of the top birding spots in North America. Birds, exhausted from crossing Lake Erie on their migrations north in spring, like to rest here. It is also part of the Car-olinian forest and is nesting habitat for many species that are rare in Canada, particularly warblers. Of the 360 bird species seen, about ninety stay to nest, and in spring there may be twenty-five to thirty different warblers seen on a good day. In the fall, the birds are joined by thousands of monarch butterflies as they, too, pause here before heading south on their 36 000-km journey to the Gulf of Mexico for the winter.

Point Pelee is a national park. Most of the rest of the Carolinian forest is not so well protected and is heavily fragmented by agriculture and urban developments. As a result, it is estimated that close to 40 per cent of Canada's rare, threatened, and endangered species are primarily Carolinian. The Carolinian Canada Programme, started in 1984, has coordinated the efforts of government agencies and private land-owners to protect the remaining forest. About half of the thirty-eight targeted sites now have some degree of protection, but biologists still worry that these fragments are too small and isolated to protect this most diverse area of Canada.

Precambrian Shield. Lakes are common. The climate is arctic with long, cold winters and short, cool summers. Mean daily temperatures are around 10° C in July and –30° C in January. Precipitation in the northern reaches is quite low, around 200 mm per annum, which doubles in the southern part of the ecozone. Similarly, the number of frost-free days doubles from forty to eighty in more southerly climes.

The vegetation is dominated by shrubs (such as dwarf birch and arctic willows) and various heath species (such as bearberry, bog rosemary, and Labrador-tea). Wetlands are common and are dominated by mosses and sedges. These support large herds of caribou and other species, such as moose, wolf, grizzly and polar bears, arctic fox, wolverine, musk ox, arctic hare, and arctic ground squirrel. Brown and collared lemmings are also locally abundant. Many birds—including large numbers of waterfowl, such as tundra swans, snow geese, and various species of ducks—also come here to nest in summer.

14. Northern Arctic

Due to its more northerly position, the Northern Arctic is somewhat less biologically diverse than the land further

In the short Arctic summer, richer areas are ablaze with brightly coloured flowers, as on Herschel Island, off the north coast of the Yukon (*Philip Dearden*).

Box 3.10 Caribou

Throughout vast areas of Canada the trails across the landscape are not those of humans but of the great herds of caribou as they migrate between their winter and summer feeding grounds. There are four subspecies of caribou in Canada. The northernmost and smallest subspecies is the Peary caribou, unique to Canada and found only on the Arctic islands. Grant's caribou is found mainly in Alaska, but one large herd, the porcupine herd, winters in the Yukon before migrating to the Alaskan coastal plain for calving and summer feeding. The most widely ranging subspecies are the woodland and barren ground caribou. The latter is the most numerous and the most important culturally and economically to the northern people. The barren ground caribou range over the tundra from Hudson Bay to the east end of Great Slave Lake and north from Great Bear Lake to the Arctic coast. The woodland caribou inhabit the woodlands south of this zone down to the Great lakes, and have been greatly depleted in numbers as a result of land-use changes, such as logging and hunting. There was once a fifth subspecies, a resident of the Queen Charlotte Islands, but it is now extinct.

Caribou are particularly well adapted to survive in harsh conditions. Their feet are the widest of all deer species, enabling them to move more easily over deep snow and springy muskeg. Their hair is hollow, which traps the air and acts as an insulator from the cold, and also helps them stay afloat as they cross the many rivers and lakes that they encounter on their migration routes.

Their well-furred, blunt muzzle allows them to search through the snow for the so-called reindeer mosses (really lichens) that they feed on in the winter. This preference for lichens also has some drawbacks; lichens collect pollutants, such as radioactive fallout, which can then concentrate in the bodies of the caribou. Native peoples who rely on caribou meat for their diets have much higher levels of radioactivity than people in urban environments. Given that over 80 per cent of resident Inuit eat caribou (Wong 1985), this is obviously cause for concern.

Caribou travel in great herds of 10,000–100,000 animals as they move between their winter and summer feeding grounds, although some herds, such as those on Peary Island and in the Shickshock Mountains of the Gaspé Peninsula, are non-migratory. These large concentrations of animals also attract predators, such as wolves and bears, which follow the migratory herds, picking off weaker individuals and calves. Traditionally the people of the North have also relied extensively on the caribou for their survival, using them for meat, clothing, tents, dog harnesses, and the antlers for tool-making. Unlike other deer, both females and males may grow antlers and, given the number of animals that discard their antlers every year, one might think that the tundra would be littered with antlers. This is not the case, however. The antlers are rich in calcium and are soon nibbled away by everything from the caribou themselves to mice, voles, and lemmings.

A herd of porcupine caribou at Vuntut National Park in the Yukon (*World Wildlife Fund Canada/Douglas Harvey*).

south. This zone extends over most of the non-mountainous terrain of the Arctic Islands and down into northern Quebec. The climate is dry and cold. Precipitation rarely reaches 200 mm per annum, and the long winters have mean January temperatures down to -35° C. Mean daily July temperatures are 5°–10° C , with an annual average of twenty frost-free days. Cape Warwick on Resolution Island has the highest recorded average annual wind speed in Canada; Arctic Bay has the lowest recorded annual rainfall.

In view of these conditions, it seems ironic that scientists have discovered the remains of rhinoceros, tapirs, alligators, and tortoises on Ellesmere Island in the far north of the Arctic. The latter two animals are especially surprising since they are reptiles that can only inhabit frost-free zones, and all four species seem more suited to the tropical forests than the frozen expanses of the Arctic. Other strange discoveries were to follow, such as the skull of an ancient flying lemur, now found only on Madagascar and in Southeast Asia, and the remains of early redwood trees. Obviously, conditions in the Arctic have not always been as cold as they are today.

However, the tropical connection still exists. Many birds from the Arctic migrate to tropical lands during the winter. Lesser golden plovers nest on the islands of the High Arctic and then fly south over 8000 km of open sea to Surinam and then on to South America. The northern wheatear travels across the Atlantic from southern Ellesmere Island and Baffin Island to central Africa. The long-distance record, however, is held by the Arctic tern, which eschews the tropical latitudes in favour of flying straight through to Antarctic, covering over 30 000 km in migration every year!

However today's vegetation and animal communities are very different from the tropical history. Ground cover, where vegetation exists, is dominated by a single layer of herbs and lichens. These arctic-alpine herb species, such as purple saxifrage, mountain avens, and arctic poppies, have evolved to take full advantage of the short arctic summer to bloom and seed before the long, cold winter starts again. Sedges, such as cotton-grass, may dominate where conditions are more moist. The Arctic is, however, rich in wildlife, with indigenous species, such as barren ground caribou, musk ox, arctic fox, polar bear, arctic hare, and brown and collared lemmings. Rock and willow ptarmigan; gyrfalcon; snowy owls; snow buntings; red phalarope; red-throated loons; and pomarine, parasitic, and long-tailed jaegers birds are typical of this zone.

Box 3.11 Musk Ox

Perhaps no animal epitomizes the Arctic more than the prehistoric-looking musk ox. With its great shaggy head and flowing mane pointing into an Arctic storm, it seems determined to survive in even the most inhospitable environments. They are, in fact, superbly adapted to the rigours of the North. They have a long, coarse outercoat covering a fine underlayer of soft wool. All their extremities, such as the tail and ears, are covered by this coat. Their keen senses allow them to smell food in the winter darkness, they have low metabolic rates, and they minimize energy loss by remaining relatively inactive.

Musk ox are not widely distributed throughout the Arctic, and they concentrate in areas where growth conditions are more hospitable to graze and browse on most Arctic herbs, grasses, and shrubs. The herds are comprised of females with a single bull, the leader, which finds and repels predators, mainly arctic wolves. Their well-known defensive strategy against such predators is to form a circle around the calves and point their formidable horned heads outward. Although effective against wolves, this strategy proved much less effective against humans with guns, who could stand at a distance and shoot this rather large target at will. Over 16,000 hides were shipped out of Canada between 1864 and 1916. In addition, Arctic explorers used them as a main source of meat supply. This hunting, combined with their low birth-rate (one calf every second year) and a few hard winters, drastically reduced their numbers. They are now protected, but may be hunted by licensed and indigenous hunters. There are now about 8,000 musk ox left in the Canadian Arctic, ranging from the Arctic islands south to the Thelon Game Sanctuary and the coast of Hudson Bay.

15. Arctic Cordillera

This is the arctic mountain zone covering eastern Baffin and Devon islands, and parts of Ellesmere and Bylot islands. The mountains, up to 2000 m in height, are draped in ice-caps. Long fiords penetrate the coastlines. The climate is very cold, with mean January daily temperatures of -25° C to -35° C, and July temperatures rising only up to a mean of 5° C. Precipitation is minimal, around 200 mm, with higher amounts at lower latitudes. Frost-free days vary between zero and twenty per year. Vegetation is very sparse, due to the large areas of permanent snow and ice. On more favourable sites at lower altitudes, there are plant communities similar to those in the Northern Arctic ecozone. Lichens are found throughout the extensive rock fields. Terrestrial mammals are largely absent. Polar bears frequent coastal areas. Birds include snow buntings, northern fulmars, hoary redpoll, and ringed plover in more hospitable areas.

MARINE ECOZONES

1. Pacific Marine

From the continental shelf, with depths of less than 200 m, this zone descends to the Abyssal Plain offshore, with depths in excess of 2500 m. Sea temperatures and salinities are higher than in the Arctic Ocean, making this zone a transition between cold polar waters and the temperate waters of mid-latitudes. The zone is particularly important for the feeding, staging, and wintering of waterfowl, sea birds, and shorebirds. Pollution is a constant concern. Ten pulp and paper mills discharge effluent into adjacent marine coastal waters, and large areas have been closed for harvesting shellfish as a result. One small oil spill off the Washington coast in 1988 resulted in the death of a least 56,000 birds off the BC coast (Burger 1993).

The Arctic Cordillera, Ellesmere Island (*Bruce Downie*).

The rich waters off the BC coast attract many pelagic (open ocean) bird species, such as this black-footed albatross wheeling off the west coast of the Queen Charlotte Islands (*Philip Dearden*).

The mammoth bones of the California Gray lie bleaching on the shores of those silvery waters, and are scattered along the broken coasts, from Siberia to the Gulf of California; and ere long it may be questioned whether this mammal will not be numbered among the extinct species of the Pacific.

Captain Charles Scammon,
The Marine Mammals of the
Northwestern Coast of North America, *1874*

This dovekie was killed by an oil spill. Oil penetrates through the feathers to the layer of down beneath and decreases the effectiveness of the feathers' insulating properties (*John W. Chardine*).

Killer whales, the highly intelligent toothed whales at the top of the marine food chain, can be seen in several locations along the West Coast (*Adrian Dorst*).

The coastal waters also support large populations of sea lions, seals, and a variety of whale species. However, many of these populations have yet to recover from hunting activities. The Pacific (or California) gray whale is the most abundant whale, and can be readily seen on its yearly migrations from its calving grounds in Baja California to the feeding areas in the Arctic Ocean as it passes, lingers, and feeds in BC's coastal waters. The Atlantic gray whale is now extinct, and the Pacific gray whale has recovered twice from excessive hunting pressure to become one of the most populous whales, with a current population of around 25,000. Humpback whales are also sometimes seen, but the most well-known are the killer whales that ply the waters round Vancouver Island, taking advantage of the salmon coming back to spawn. The sea otter was also once common on this coast, but was *extirpated* through hunting in the last century and is only now beginning to recover in numbers as a result of reintroductions.

2. Arctic Basin

This zone covers the Arctic Ocean Basin between Canada and Russia, Greenland, and Norway. Depths are mostly greater than 2000 m. It is covered largely in permanent pack ice. Biological productivity and diversity is low. Three species of whale—the bowhead, beluga, and narwhal—are found up to the edge of the pack ice. The ringed seal, the smallest of all the pinnipeds or fin-footed mammals (seals, sea lions, and walruses), is the most abundant marine mammal. The seals have claws on their front flippers to help scrape breathing holes through the ice. To insulate them against the cold, their bodies are composed of up to 40 per cent blubber in the winter. They are mainly krill eaters, but also dine on small fish. So well adapted are they to their ice environment that even mating occurs under the ice. In spring, the female finds a crevice through the pack ice and excavates a den for pupping in the snow above. Polar bears are their main predator.

3. Arctic Archipelago

This maze of channels and islands extends from Baffin Island and Hudson Bay in the east to the Beaufort Sea in the west. Ice cover is not permanent, and open water exists in most areas for two to three months every year. There are also areas of permanent open water, known as *polynyas*, that have significant concentrations of marine life. Resident birds, such as black guillemots, thick-billed murres, ivory gulls, and dovekies congregate in large numbers at these sites in winter to obtain food in the open leads. The polynyas vary greatly in size from 60–90 m in diameter to ones as large as the North Water polynya between Ellesmere Island and Greenland, which may cover as much as 130 000 km^2. These areas remain ice-free as a result of various combinations of tides, currents, ocean-bottom upwellings, and winds, which promote constant water movement. Their attraction for marine mammals also made polynyas very attractive sites for Inuit settlements and later for whalers, who battled to get through to large polynyas, such as the North Water, as early as possible every year to kill the whales. The bowhead whales' pre-exploitation population of about 11,000 in the area dropped to only a few hundred today. It is doubtful whether the population will ever recover.

Perhaps contrary to expectations, arctic marine mammals live in a less extreme environment than their terrestrial counterparts. Polar waters seldom drop below freezing. The temperature gradient between external and internal temperatures that animals living in this environment must cope with is therefore only about half that of land animals. Animals, such as whales, seals, and walruses, have virtually no hair but instead rely upon a thick layer of blubber to protect them from the cold. Their skin is directly exposed to the water, and they therefore need a very finely tuned thermal system to maintain a skin temperature warm enough to resist freezing but cool enough so as not to waste excessive energy.

Inuit people hunt and fish throughout the zone. There is potential for increased exploitation of hydrocarbons. This could cause serious resource conflicts between traditional ways of life, the populations of wildlife upon which they depend, and the desire for the development of oil and gas reserves.

A beater, like this harp seal pup, is one that has moulted its white fur (*International Fund for Animal Welfare*).

Icebergs, such as this one outside the harbour of St John's, are characteristic of the northwest Atlantic (*Philip Dearden*).

Whitecoat seal pups gain weight rapidly, increasing by approximately 2 kg per day (*International Fund for Animal Welfare/David White*).

4. Northwest Atlantic

Extending from Hudson Strait and Ungava Bay down to the St Lawrence, this zone includes mostly continental shelf where depths are less than 200 m, although in the Labrador Sea, depths may reach 1000 m. In winter, ice formations move along the coast as far as the south coast of Newfoundland. Sea temperatures in this zone may be 20° C higher than in the Arctic due to warm ocean currents. Marine life is generally abundant. A deep trench channels cold oceanic water up the St Lawrence as far as the entrance to the Saguenay Fiord, creating a very productive feeding zone for

beluga and various baleen whales in the upwelling currents. There are twenty-two species of whales and six species of seals in this zone.

This zone was also the site of the controversial seal hunt, which used to occur annually off these shores. Since the mid-eighteenth century, the harp seal had been the target of hunting, mainly for its pelt. Between 1820 and 1860, for example, about half a million were killed every year. In 1831 there were over 300 ships and 10,000 sealers involved in the hunt; 687,000 pelts were taken. Public outrage over the hunt in the early 1980s led to bans on the importation of sealskins into Europe, and the hunt collapsed.

5. Atlantic

This zone includes the highly productive waters of the Grand Banks, Scotian shelf, St Lawrence trough, and Bay of Fundy, as well as the Northwest Atlantic basin. Generally ice-free, this zone is a mixture of warmer waters moving north from the tropics and the cold water of the Labrador Current. Waters are generally shallow, with large areas under 150 m in depth. There is an abundant and diverse population of marine mammals, such as the grey, hooded, and harp seals, and the Atlantic pilot, killer, and northern bottlenosed whales. The fisheries along these coasts were once some

Small outport communities in Newfoundland have always relied heavily on harvesting marine products from seals to fish (*Tourism Newfoundland and Labrador*).

Box 3.12 Leviathan

Canada not only has the longest coastline in the world, it also has some of the richest waters. Cold oceanic waters from the north and deep upwelling currents mix with warmer waters from the tropics to create a rich abundance of life. Rich nutrient supplies, accompanied by shallow seas and long daylight hours in summer, have created some of the richest waters off the East Coast. Plankton flourish and provide the base for a diverse *food web* that supports three main groups of marine mammals: the Odontoceti (or toothed whales), the Mysticeti (or baleen whales), and the Pinnipedia (or seals and walruses). The baleen whales all feed on plankton, small fish, and marine algae using the plates of baleen in their mouths that filter these organisms from the water. Surviving on these small food items, the baleen whales have evolved as the largest creatures on this planet. The blue whale is the largest of the baleen whales. The largest recorded was a female over 30 m in length and weighing 140 tonnes. The calf is 7 m long at birth and may weigh almost 3 tonnes. Blue whales as well as other baleen whales, such as the fin and sei whales, were once abundant off the East Coast, as was the largest toothed whale, the sperm whale. All these whale populations were decimated by hunting, and are only now starting to recover.

In the late fifteenth century, Europe's need for oil to light its lanterns increased. Marine mammals, with their thick layer of blubber, were the solution. When exposed to high heat, the blubber can be rendered down to oil. The Basques, that great seafaring race from northern Spain, discovered the rich whaling grounds off the East Coast of Canada, and the slaughter began. The abundance of whales in these waters at this time is difficult to imagine. Early mariners complained of whales as a navigational hazard because they were so numerous; one missionary in the Gulf of St Lawrence reported that the whales were so numerous and loud, they kept him awake all night! Whales could be harpooned from shore. The limiting factor was not the number of whales but the ability to process them. Increasing numbers of shore stations were established, and thus began another toehold of colonization.

The rest of the story is fairly well known. As the human population increased, the whale populations declined. First to go was the one most hunted, the black right whale. One whale could be rendered into over 16 000 L of oil. The whale was slow and floated when killed. Yes, it was the *right* whale to kill. Besides the oil, the baleen was used for other indispensable purposes, such as corsets, brush bristles, sieves, and plumes for military helmets. A Basque shipowner could pay off his ship and all his expenses and still make a good profit in one year from such whales. Although the whales gained some respite when the Spanish Armada was destroyed by England, other nations finished off the whales. Other whales—the sperm, humpback, blue, fin, sei, and minke—soon joined the right whale as commercially, if not biologically, extinct. The destruction is a classic example of what often happens to unregulated exploitation of commons resources, a topic discussed in more detail in Chapter 20.

The two current main whaling nations are Japan and Norway. Although there has been a non-binding moratorium on commercial whaling since 1986, both these nations continue to hunt whales, using a clause that allows whales to be killed for scientific purposes, according to the regulations of the International Whaling Commission, which oversees the fishery. In 1995, for example, Japan killed 400 whales for so-called 'research' purposes. Meat from the kills can be sold. In Japan it is considered a delicacy and fetches a high price. In 1996 the Japanese planned to extend their hunt into the 20 million km² Antarctic whale sanctuary.

Removing such a large *biomass* from the top of the *food chain* obviously has ecological repercussions for other creatures. Unfortunately, we have also interfered so much with populations of other animals that the results are not always clear. Other krill eaters would benefit from removal of these large and efficient competitors. For example, in the Antarctic it has been suggested that increases in the populations of krill-eating seals, such as fur and crab-eating seals, came about as a result of this bounty. In turn, these higher numbers of seals require more breeding sites, and colonize areas previously used by birds, such as the albatross, for nesting. Did declining whale numbers also lead to a decline in the albatross? We cannot say for sure, but the example does illustrate the complexities of changes in food webs.

Box 3.12 Leviathan (continued)

Although whaling has now been controlled throughout most of the world, including Canada's coasts, resurgence of numbers is occurring very slowly for some species, such as the right whales. The largest whale, the blue, is also in trouble. Although numbers still exist, finding a mate is difficult for this wide-ranging species. Scientists have now found evidence that the blues are interbreeding with the more common fin whales, the second largest whale in the world, and fear that this hybridization will result in a loss of genetic identity for the blue, which could then disappear as a species. Unfortunately, the rapacious killing by our ancestors denied all future generations the privilege of seeing our seas full of these mighty creatures. Are we denying future generations other such opportunities by our selfish and irresponsible actions today?

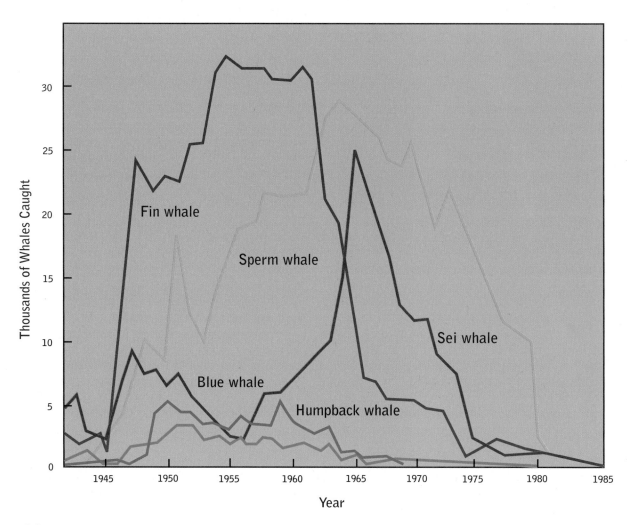

Figure 3.8 The catch figures for all whales show the same kind of pattern when graphed: a rapid growth followed by a sharp decline, a shape characteristic of unsustainable activities.

of the most abundant on earth (see Chapter 20), and a large population of over 300 different sea birds was also supported by the marine bounty. These include the northern gannet, which almost followed the great auk into extinction (see Chapter 16) as a result of hunting before protective laws were invoked in the early part of the century. Other species include the Atlantic puffin, murres, great and double-crested cormorants, guillemots, greater black-backed gulls, and many others.

Thick-billed murres nest along sea cliffs (*Newfoundland Parks and Natural Areas Division/Ned Pratt*).

Northern gannets range over the north Atlantic and are often seen well offshore (*Newfoundland Parks and Natural Areas Division/Ned Pratt*).

Major gas and oil discoveries have been made throughout the zone. The Hibernia field on the eastern edge of the Grand Banks, some 300 km east-southeast of St John's, is the only one currently likely to go into production. The field could have major economic implications for Newfoundland, as Storey (1995) outlines. Major delays have already occurred and the expected December 1997 production date may be even further postponed.

IMPLICATIONS

From this overview it is apparent that Canada is a very large and varied country in terms of the biophysical characteristics and the human use of different ecozones. A different perspective on land use is shown in Figure 3.9, summarizing the physical make-up of the country and some of the main land uses. From this it can be seen that Canada is

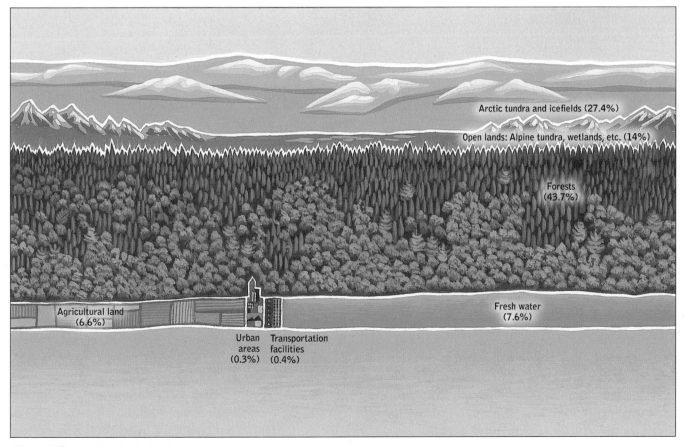

Arctic tundra and icefields (27.4%)

Open lands: Alpine tundra, wetlands, etc. (14%)

Forests (43.7%)

Fresh water (7.6%)

Agricultural land (6.6%)

Urban areas (0.3%)

Transportation facilities (0.4%)

Figure 3.9 Canada's physical make-up.

indeed a forest nation, with over 43 per cent of the land area classified as forests. The next largest category includes the Arctic tundra and ice-fields of the North. Combined with the third largest category of alpine tundra, wetlands, and other open lands, these amount to over 85 per cent of the total area of the country. Urban areas and infrastructure make up less than 1 per cent. These kinds of proportions help give Canada its distinctive character and underline the importance of careful environmental management.

SUMMARY

1. Canada is the second largest country in the world, encompassing 6.7 per cent of the world's land area, 9 per cent of the forests, 13 per cent of wilderness, 3 per cent of crop land, and 9 per cent of the global total of fresh water. The Canadian coastline is longer than that of any other country in the world. Our population is less than 0.5 per cent of the world total.

2. Canada has four main geological regions: the Canadian (Precambrian) Shield; the four sedimentary platforms that surround the shield; the Western Cordilleran, Innuitian, and Appalachian mountain systems; and the Atlantic and Pacific continental shelves. There are mineral deposits across the country, and Canada is a main exporter of them.

3. Glaciation has been a major process that formed the landscape into its present topography across most of Canada. It is still a dominant process in many areas at higher altitudes and latitudes.

4. The fifteen ecozones of Canada represent an ecological land classification system based on geology, physiographic regions, soils, climate, forests, and wildlife.

5. The Tundra Cordillera is a treeless zone that encompasses the mountains of Yukon and northern British Columbia between the forests of the south and the ice deserts farther north. Wildlife is diverse and abundant, including caribou, moose, bear, wolf, and goat. The fossil fuels in the offshore zone in the Arctic Ocean pose a main issue for environmental management as to whether these reserves can be exploited without disrupting the wildlife and human culture of the area.

6. The Boreal Cordillera lies to the south of the tundra, with wetter conditions permitting tree growth. This zone contains Canada's highest mountain, Mount Logan, at 5951 m. It also contains one of the largest copper deposits in the world, in the Tatshenshini watershed. In 1994 the BC government decided to designate the area a park rather than allow development of the deposit.

7. The Pacific Maritime zone stretches along the entire Pacific coast of Canada. Relatively warm onshore winds, which rise as they meet the coastal mountains, result in the greatest amount of rainfall in the country as well as the largest trees. These trees form the backbone of the Canadian lumber industry.

8. The Montane Cordillera encompasses great variation from the interior valleys of BC up to the heights of the Rocky Mountains. This zone includes Canada's oldest and most famous national parks, such as Banff, established in 1885. It also has one of the most accessible glaciers in Canada at the Columbia Icefields in Jasper National Park.

9. The Boreal Plains extend from the southern Yukon down into Manitoba. They are generally flat to undulating, with enough moisture to allow tree growth of species such as black and white spruce and jack pine. Deciduous trees, particularly trembling aspen and birch, are common at the transition zone into the true prairie to the south.

10. The Taiga Plains lie between the southern boundary of the tundra to the north and the closed-crown canopy of the boreal forest to the south. Wetlands are common.

11. The Prairie is the grasslands in the interior of the continent, which extend from Alberta to the Gulf of Mexico. It is too dry to support tree growth. Summers are hot; winters are very cold. Much of this area has been transformed into agricultural land. Many species from this zone are on Canada's list of rare and endangered species.

12. The Taiga Shield has a very dry and cold climate. It occurs on the Precambrian bedrock, which stretches from Great Slave Lake in the west to Newfoundland in the east. To the north, the zone merges into the tundra; farther south, tree growth is possible, with species such as black spruce and tamarack.

13. The Boreal Shield is the largest ecozone in Canada, stretching from Saskatchewan to Newfoundland, and separating the treeless tundra regions in the north from the temperate deciduous forests and grasslands in the south. It contains some of the oldest rocks in the world and is the site of most mineral mining in Canada.

14. The Hudson Plains are at the southern edge of Hudson Bay. A more moist climate and poor drainage combine to produce extensive areas of wetlands.

15. The Mixed Wood Plains, spreading from the Lower Great Lakes up through the St Lawrence Valley, have the greatest diversity of trees in Canada. Unlike the rest of the country, the warmer, more humid conditions in the southern part of the zone permit the growth of many deciduous species. This is also the most densely populated zone, and areas of natural vegetation are rare.

16. The Atlantic Maritime zone is profoundly influenced by the Atlantic Ocean, producing higher precipitation levels, warmer winters, and cooler summers than zones to the west. Forests are generally mixed deciduous and coniferous species.

17 The Southern Arctic is cool in summer and cold in winter. Precipitation levels are very low. Large herds of caribou and other species—such as musk ox, moose, wolf, and polar bear—may be locally abundant.

18. The Northern Arctic, including most of the Arctic islands, is less biologically rich than the land to the south. However, in summer many bird species migrate this far north for nesting.

19. The Arctic Cordillera is the mountainous part of the far north, covering eastern Baffin and Devon islands and parts of Ellesmere and Bylot islands. Vegetation is very sparse due to the large areas of permanent snow and ice. Lichens are found throughout the extensive rock fields. Large mammals are largely absent.

20. The Pacific Marine zone extends along the length of the Pacific coast. Sea ice is absent, and salinity and temperature levels are higher than in the Arctic. Marine mammals include Steller's sea lions, sea otters, and killer whales. There are ten pulp and paper mills that discharge effluent into adjacent marine waters. Municipal waste discharges are also significant.

21. The Arctic Basin encompasses the northern Arctic beyond the island archipelago. It is the polar sea separating Canada from Russia and Europe and is almost totally covered in ice. The bearded seal, beluga, and polar bear are found along the ice margins.

22. The Arctic Archipelago covers the maze of islands and channels that extend from Baffin Island and Hudson Bay in the east to the Beaufort Sea in the west. Ice cover is not permanent. Polynyas, areas of permanent open water, are very important biologically and may support regionally significant numbers of wildlife.

23. The Northwest Atlantic extends from Hudson Strait down to the St Lawrence. The waters may be 20° C warmer than the Arctic Ocean due to warm ocean currents. Marine life is abundant, with twenty-two different species of whales and six species of seals.

24. The Atlantic zone includes the highly productive waters of the Grand Banks, Scotian shelf, St Lawrence trough, and Bay of Fundy, as well as the Northwest Atlantic basin. The waters are less than 150 m deep over large areas. Tidal ranges are up to 16 m in the Bay of Fundy. At one time these waters supported one of the most productive fisheries on earth.

REVIEW QUESTIONS

1. What is an ecozone, and how many are there in Canada?

2. What ecozone are you in, and what are the defining characteristics of the zone? How does it differ from surrounding zones? How is it similar?

3. Discuss how one particular animal is adapted to the conditions of its environment.

4. Why are the trees in coastal BC bigger than those elsewhere in the country?

5. What is Carolinian Canada, and what are its defining characteristics?

6. What are the main differences among the marine ecozones?

REFERENCES AND SUGGESTED READING

Briggs, D., P. Smithson, T. Ball, P. Johnson, P. Kershaw, and A. Lewkowcz. 1993. *Fundamentals of Physical Geography*, 2nd Canadian ed. Toronto: Copp Clark Pitman Ltd.

Burger, A.E. 1993. *Interpreting the Mortality of Seabirds Following the Nestucca Oil Spill of 1988–1989: Factors Affecting the Seabirds Off Southwestern British Columbia and Northern Washington.* Technical Report Series No. 178. Vancouver: Canadian Wildlife Service, Pacific and Yukon Region.

Edwards, Y. 1979. 'Introduction'. In *The Land Speaks*. Toronto: National and Provincial Parks Association of Canada.

Forsyth, A. 1985. *Mammals of the Canadian Wild*. Camden: Camden House Publishing.

Gayton, D. 1990. *The Wheatgrass Mechanism: Science and Imagination in the Western Canadian Landscape*. Saskatoon: Fifth House Publishers.

Harker, P. 1995. 'Energy and Minerals in Canada'. In *Resource and Environmental Management in Canada: Addressing Conflict and Uncertainty*, edited by B. Mitchell, 286–309. Toronto: Oxford University Press.

Henry, D. 1986. *Red Fox: The Catlike Canine*. Washington, DC: Smithsonian Press.

Hirvonen, H.E., L. Harding, and J. Landucci. 1995. 'A National Marine Ecological Framework for Ecosystem Monitoring and State of the Environment Reporting'. In *Marine Protected Areas and Sustainable Fisheries*, edited by N.L. Shackell and J.H.M. Willison, 117–29. Wolfville, NS: Science and Management of Protected Areas Association.

Hoyt, E. 1984. *The Whale Watcher's Handbook*. Toronto: Penguin.

Lopez, B. 1986. *Arctic Dreams*. New York: Charles Scribner.

Morton, W.L. 1967. *The Canadian Identity*. Madison: University of Wisconsin Press.

Mowat, F. 1984. *Sea of Slaughter*. Toronto: McClelland and Stewart.

Murie, O. 1975. *Field Guide to Animal Tracks*, 2nd ed. Boston: Houghton Mifflin.

Rowe, S. 1972. *Forest Regions of Canada*. Ottawa: Environment Canada.

_____. 1990. *Home Place: Essays on Ecology*. Edmonton: NeWest Press.

Russell, A. 1967. *Grizzly Country*. New York: Ballantine Books.

Scammon, C.M. 1874. *The Marine Mammals of the Northwestern Coast of North America*. New York: G.P. Putnam's Sons.

Schueler, F.W., and D.E. McAllister. 1991. 'Maps of the Number of Tree Species in Canada: A Pilot Project GIS Study of Tree Biodiversity'. *Canadian Biodiversity* 1:22–9.

Scott, G.A.J. 1995. *Canada's Vegetation: A World Perspective*. Montreal and Kingston: McGill-Queen's University Press.

Storey, K. 1995. 'Managing the Impacts of Hibernia: A Mid-Term Report'. In *Resource and Environmental Management in Canada: Addressing Conflict and Uncertainty*, edited by B. Mitchell, 310–34. Toronto: Oxford University Press.

Wiken, E. 1986. *Terrestrial Ecozones of Canada*. Ecological Land Classification Series No. 19. Ottawa: Environment Canada.

Wong, M. 1985. *Chemical Residues of Fish and Wildlife Harvested in Northern Canada*. Environmental Studies Report 46. Ottawa: Indian and Northern Affairs Canada.

The Ecosphere

On Spaceship Earth,
there are no passengers;
we are all members of the crew.

—Marshall McLuhan

The first section of the book provided you with an overview of the global and Canadian situations regarding the relationship between humans and the earth, and a more detailed introduction to some of the regional variation within Canada. The above quote by Marshall McLuhan emphasizes that we are not passive bystanders in this interaction. We are members of the crew, helping to determine what happens to our planet. As crew members we need some idea of the workings of our spaceship, particularly the nature of the life-support systems. Imparting such an overview is the prime purpose of this next section. It describes the natural systems through simplified models, following the conceptual framework outlined in Chapter 1 (Figure 1.6). All the chapters deal to some extent with human disruptions of these natural systems, but the last chapter, Chapter 7, focuses exclusively on this topic as it

traces how interruptions of biogeochemical cycles have led to some of the most pressing environmental problems today, such as acid precipitation and global climatic change.

Most of what follows is derived from just one form of environmental knowledge: science. Science tries to understand and find order in nature and predict the outcome, given changes that may occur. Scientists collect data or facts about the environment and then try to make some order out of those facts. When we are not quite so sure about a relationship, we might construct a *scientific hypothesis* that we can then test using the data we collect. A *scientific theory* emerges when large numbers of scientists agree on how or why a certain phenomena occurs. A *scientific law* represents the most stringent form of understanding and lays down a universal truth that describes what happens in certain circumstances in all cases. In this next chapter you will be introduced to some such laws. They are very useful because they provide a firm building-block for our scientific understanding.

◄ The natural beauty of Temagami was first celebrated in a poem by Canadian poet Archibald Lampman (*Earthroots/B. Back*).

We should not, however, feel that any of the hypotheses, theories, or even laws advanced by science are unquestionable. The whole purpose of science is to ask questions, and science advances by continually changing and modifying previous knowledge. Being kind to scientists (and teachers!), but ruthless in your questions, is not a bad dictum to adhere to! Debate and disagreement in science are normal. In the following chapters, for example, there are some topics on which ideas are changing rapidly, and even some concepts (e.g., keystone species and climax vegetation) that are being questioned by some scientists. This active debate is often misunderstood by those outside the academic community who think it implies that scientists do not know anything. It can also be used for political purposes to support inaction on measures that might be unpopular, such as limits on industrial emissions to reduce acidic precipitation. It is, however, a sign of the vitality and strength of scientific thinking rather than the reverse.

Scientists try to follow certain rules and procedures that constitute *scientific methodology*. This process tries to link, with minimum error, the testing of hypotheses with the existing body of knowledge. It prescribes a series of steps that, over time, scientists have found are most likely to provide understanding about phenomena under study regardless of who is undertaking the observations or experiment. There is no one scientific method that all scientists have to follow. Many important scientific insights have come about through quite irregular approaches. It is, however, wise to learn from previous experience and to know the kinds of procedures that are usually followed in any given area of enquiry.

Science, especially environmental science, is also an increasingly collaborative undertaking. This involves not only workers from one discipline but from many disciplines, who together contribute their understanding of a particular phenomena. Understanding of acid precipitation, for example, involves chemists, biochemists, climatologists, geologists, hydrologists, geographers, biologists, health specialists, economists, and political and legal experts, to name a few. Each discipline has its own expertise and methods of approach. As you can see, understanding current environmental problems is very complex. The next four chapters will provide a very basic overview of the main processes that maintain the life-support systems on the planet.

A simple model of the planet would look like the layers of an onion (Figure 4.1). We are most concerned with the outer layers, the ecosphere, which consists of three main layers:

- The *lithosphere* is the outer layer of the earth's mantle and the crust, and contains the rocks, minerals, and soils that provide the nutrients necessary for life.
- The *hydrosphere* contains all the water on earth.
- The *atmosphere* contains the gases surrounding the lithosphere and hydrosphere. It can be further divided into three main sublayers. The innermost layer is the *troposphere*, which contains 99 per cent of the water vapour and up to 90 per cent of the air, and is responsible for our weather. Two gases, nitrogen (78 per cent) and oxygen (21 per cent), account for 99 per cent of the gaseous volume. This layer extends, on average, to about 17 km before it gives way to the second layer, the *stratosphere*, wherein lies the main body of ozone, which blocks out most of the ultraviolet radiation from the sun. About 50 km above the earth's surface is the *mesosphere*, and above that is the *thermosphere*. As distance from earth increases, the pressure and density of the atmosphere decreases as it melds into space.

These three layers combine to produce the conditions necessary for life in the *ecosphere*, stretching from the depths of the oceans up to the highest peaks, a layer some 20 km in depth, which contains some 30 million different organisms. We hope that the following chapters in Part B will give you some idea of the ecosphere's main environmental processes, and that you will also gain a greater understanding of how human activities interrupt these processes.

The arrival of the capelin at the beaches of New-foundland to spawn on the first full moon of June had long been a bounty not only for many animal and bird species but also for the settlers, who collected the fish for food and fertilizer. The capelin is a small fish of the north Atlantic that is the main food supply for many other species, including cod, salmon, halibut, mackerel, seals, various whale species, and many species of sea birds, such as puffins. Each year the capelin move to inshore waters to spawn. In the late 1970s and early 1980s the numbers of capelin arriving to spawn declined markedly.

The colourful bill is the most striking feature of the Atlantic puffin, which breeds among the rocks of sea islands (*Newfoundland Parks and Natural Areas Division/Homer Green*).

Spawning capelin on a beach in Newfoundland (*Philip Dearden*).

Normally 80–90 per cent of the Atlantic puffin's diet consists of capelin. In the early 1980s scientists noticed a large decline in this proportion, down to 13 per cent, resulting in severe malnutrition of puffin chicks and subsequent declines in the population. Although no direct action was taken to reduce the puffin population, the numbers fell as a result of the decline in their food base, the capelin. The energy flow between the species had been interrupted by the establishment of an offshore capelin fishery, which removed the capelin from the food chains that nourish so many marine species. The

Box 4.1 Traditional Ecological Knowledge

Most of the concepts presented in this book are the result of the 'scientific approach' to understanding different phenomena. There are other approaches, however, and one that is gaining increasing attention is traditional ecological knowledge (TEK). Scientists around the world have found that *indigenous* peoples often have an encyclopaedic knowledge of their local environments, which is expected for peoples gaining their sustenance directly from that environment. They have also found that indigenous peoples tend to undertake the same kinds of tasks as Western scientists, such as classifying and naming different organisms, and studying population dynamics, geographical distributions, and optimal management strategies. Unlike Western science, however, this knowledge is rarely recorded in written form, but is handed down orally from generation to generation.

Only now are we becoming more aware of this body of knowledge. In Canada, this has come about particularly as a result of increasing industrial interest in northern regions and the potential impacts upon Native communities. Inevitably, this has given rise to discussions about which form of ecological knowledge (Western or traditional) is the 'best'. Both have their advantages and disadvantages. Modern science is informed by developments around the world, but is limited in terms of its knowledge of changes over time in a particular place, an aspect that is especially strong in traditional knowledge. Scientists also tend to concentrate on information that can be tested by replication and ignore idiosyncratic and individual behaviour, which is given substantial weight by indigenous hunters, as Stephenson (1982) notes in his study of the relationship between the Nunamiut of northern Alaska and wolves, for example.

Management systems also differ. The traditional system is self-regulating, based on communal property arrangements. Conservation practices, such as rotation of hunting areas, were commonly practised, as described by Feit (1988) for moose hunting and beaver trapping by the James Bay Cree. However, the system is not infallible, especially under the onslaught of outside influences and commercialization. Similarly, the modern system of private property rights and state allocation of harvesting rights does not always work. Scientists and indigenous peoples are now benefiting from both systems of knowledge and management approaches, and are trying to use both through comanagement arrangements.

puffins were one noticeable victim of this appropriation, but undoubtedly other species suffered the same consequences. The example illustrates the importance of understanding how energy links species and flows through ecosystems. Changing the energy available at one part of the food chain will have repercussions throughout the ecosystem.

Reading this book, taking notes in class, even snoozing at home all require energy. That energy comes ultimately from the radiant energy of the sun, which is transformed into chemical energy in the form of food supplies before being turned into mechanical energy that you use to perform your activities. In this chapter you will gain some appreciation of energy in relationship to such transformations, how energy flows through ecosystems, and the ecosystem structures that result.

ENERGY

Energy is the capacity to do work and is measured in *calories*. A calorie is the amount of heat necessary to raise 1 g or 1 mL of water by $1°$ C, starting at $15°$ C. Energy comes in many forms: radiant energy (from the sun), chemical energy (stored in chemical bonds of molecules), as well as heat, mechanical, and electrical energy. Energy differs from matter in that it has no mass and does not occupy space. It affects matter by making it *do* things—work. Energy derived from an object's motion and mass is known as *kinetic energy*, whereas *potential energy* is stored energy that is available for later use. The water stored behind a dam is potential energy, which becomes kinetic energy when it pours over the dam. The gas in a car is potential energy before it is poured into the engine to create mechanical energy for propulsion.

Much of the energy available for use is *low-quality energy* that is diffuse, dispersed, at low temperatures, and difficult to gather. The oceans, for example, contain an enormous amount of heat, but it is very costly to harness this energy for use. *High-quality energy*, such as a hot fire or coal or gasoline,

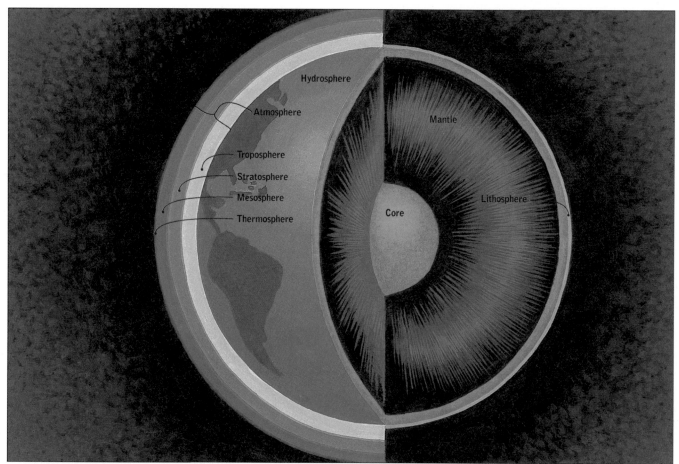

Figure 4.1 A simplified model of the earth.

is very easy to use, but the energy disperses quickly. Much of our economy and technology are now built around the transformation of low-quality energy into high-quality energy for human use.

It is important that we match the quality of the energy supply to the task at hand. In other words, we should not use high-quality energy for tasks that can be fuelled by low-quality supplies.

Organisms require energy for growth, tissue replacement, movement, reproduction, and indeed all aspects of life. To gain a comprehensive perspective on life we must understand energy.

Laws of Thermodynamics

Two laws summarize the way in which trillions of energy transformations per second take place all over the globe. These are known as the *laws of thermodynamics*. According to the first, the *law of conservation of energy*, energy, in any non-nuclear transformation, can neither be created nor destroyed; it is merely changed from one form into another. Organisms do not create energy; rather, they obtain

Traditional societies in many lesser developed countries, such as the Lisu people in the highlands of northern Thailand, still rely on wood as the main source of energy. Although this can lead to deforestation if demands are greater than growth, it also limits the amount of energy that can be transformed by these societies (*Philip Dearden*).

it from the surrounding environment. When an organism dies, the energy of that organism is not 'lost'. It flows back into the environment and is transformed into different types of energy, the total sum of which equals the original amount. Similarly, cars obtain their energy from gasoline. The fuel gauge's drop from full to empty does not indicate that energy has been consumed; it has merely been transformed from chemical energy into other forms of energy, including the mechanical energy to move the car.

According to the second law, when energy is transformed from one form into another, there is always a decrease in the quality of usable energy. During any transformation, some energy is lost as lower-quality, dispersed energy is dissipated to the surrounding environment, often as heat. The amount of energy lost varies depending upon the nature of the transformation. In a car, for example, only about 10 per cent of the chemical energy of the gasoline is actually converted into mechanical energy to turn the wheels. The remainder is dispersed into the environment. Put your hand onto the hood of a car that has just stopped running. The heat you feel is a result of this *second law of energy* or *law of entropy*. Entropy is a measure of the disorder or randomness of a system. High-quality, useful energy has low entropy. As this is dispersed through transformation, the entropy increases.

The second law is important for organisms because they must continuously expend energy to maintain themselves. Whenever energy is used, some of that energy is lost to the organism, creating a need for an ongoing supply that must exceed these losses if the organism is to survive. If losses exceed gains for an extended period, the organism dies.

There are also many other important ramifications of this law. For example, energy cannot be recycled. It flows through systems in a constantly degrading manner. We think of 'advanced' societies as energy consumers (see chapters 19 and 21). Large dams and nuclear power stations are very visible signs of a modern economy much sought after by lesser developed nations. As we become more economically developed, we find increasing ways to transform energy. Cars, telephones, electric can-openers, blenders, microwaves, hot tubs, computers, and compact-disc players are all energy transformers. Yet the more energy is transformed, the more it

is dispersed into the atmosphere as entropy increases. This dispersion can be likened to a bar of soap in a bowl of water. As the soap is used over time, it dissolves into the water, making it less and less useful for washing. Energy may be thought of in a similar fashion as use gradually disperses useful and concentrated forms into the atmosphere.

Some of the key transformations required for achieving a sustainable society (as discussed in Chapter 2) are to regard high-energy consumption as undesirable, to reduce energy waste, and to switch from the non-renewable sources of energy that now dominate (particularly coal and oil) to renewable sources, such as those discussed in Chapter 19. Until the Industrial Revolution, the speed of processing raw materials was limited by the energy available, supplied largely by human and animal labour, combined with wood, wind, and water power. These sources were in turn limited by the input of solar energy over a relatively short time. The use of coal and oil to fuel steam engines removed these limitations. Acid rain, climatic change, and many of our current environmental problems are a direct result of the transformation of society's energy base from a renewable to a non-renewable one. We have released the energy input of millions of years in the last 250 years, which is the blink of an eye in geological terms. Many problems are a result of this increase in entropy.

ENERGY FLOWS IN ECOLOGICAL SYSTEMS

Energy is the basis for all life. The source of virtually all this energy is the sun. Over 150 million km away, the sun, a giant fireball of hydrogen and helium, constantly bombards the earth with radiant energy. This energy, although it is only about 1/50 millionth of the sun's output, fuels our life-support systems, creates our climate, and powers the cycles of matter (to be discussed in Chapter 6). The atmosphere reflects about a third of the received energy back into space (Figure 4.2). Of the remainder, about 42 per cent provides heat to the earth's surface, 23 per cent causes evaporation of water, and less than 1 per cent forms the basis of our ecological systems.

Figure 4.2 illustrates the law of conservation of energy. The total amount of energy received by the earth is equal to the total amount that is lost.

Box 4.2 What Is Life?

We have said that energy is essential for all life, but what is life? Living organisms have a number of common characteristics:

1. They use energy to maintain internal order.
2. They increase in size and complexity over time.
3. They can reproduce.
4. They react to their environment.
5. They regulate and maintain a constant internal environment.
6. They fit the biotic and abiotic requirements of a specific habitat.

We think we have a fairly good idea of what constitutes life, but there is still much debate as to how life developed on earth. Over eighty years ago two scientists proposed the Big Bang theory, which explains the origin of the universe as the result of a massive explosion some 15 billion years ago. The solar system came about from the resulting matter. As the chunks of matter grew in size, they heated up. As the earth cooled, warm seas formed and precipitation helped create a nutrient-rich environment. Over time, the constant bombardment of this nutrient-rich soup by high-energy levels from the sun created

chemical reactions producing small organic compounds, such as amino acids and simple organic compounds. Scientists have managed to recreate several organic compounds necessary for life from inorganic molecules by bombarding them with energy.

Over billions of years larger organic molecules came to be synthesized until the first living cells, probably bacteria, developed 3.6–3.8 billion years ago. These cells passed through several stages over billions of years, with increasingly complex development. This activity was in the ocean environment, protected from ultraviolet radiation. Between 2.3–2.5 billion years ago a major change occurred when photosynthetic bacteria produced oxygen into the atmosphere as they manufactured carbohydrates from carbon dioxide. Over time, the oxygen reacted with the abundant and poisonous methane in the atmosphere, reducing levels of this gas and resulting in the atmosphere we know today. Some oxygen was also converted to ozone in the lower stratosphere, which would protect evolving life from ultraviolet radiation and allow the emergence of life from deeper to shallower waters, and eventually onto land itself.

Human activity delays the amount of heat lost back to space by trapping it in the atmosphere through increased levels of heat-trapping gases, such as carbon dioxide and methane. The implications of this are discussed more fully in chapters 7 and 20.

Producers and Consumers

Plants transform the sun's energy into matter through *photosynthesis* (photo = light, synthesis = to put together). Using energy from the sun, plants combine carbon dioxide and water into high-energy carbohydrates, such as starches, cellulose, and sugars. The light energy from the sun is absorbed by the plants' green pigments (called chlorophylls). Photosynthesis also produces oxygen. Some of this is used by plants in various metabolic processes. The rest goes into the

Plants that grow on the forest floor have differing strategies to obtain enough light to survive. Most, such as many ferns, can survive on relatively low light levels. Some, such as the devil's club shown here, grow very large leaves (over 40 cm wide for the devil's club) in order to expose as much photosynthetic surface as possible to the low light levels. The devil's club is a member of the ginseng family and well known among Native peoples in western North America for its medicinal properties (*Philip Dearden*).

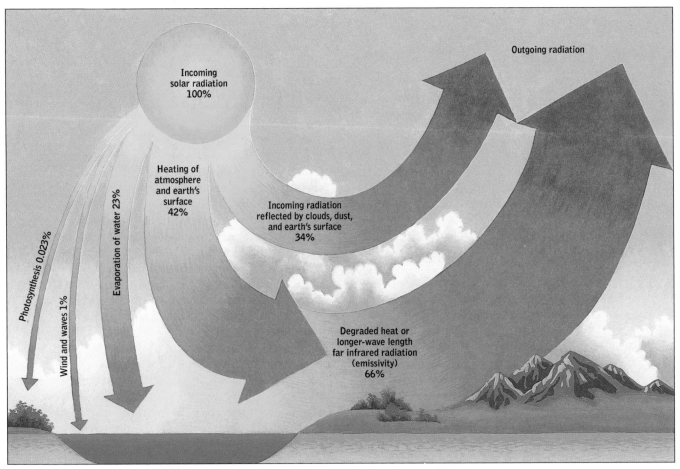

Figure 4.2 The earth's energy input and output, a good example of the first law of thermodynamics.

atmosphere. Hundreds of millions of years of evolution have produced the oxygen in the atmosphere that we depend on for life.

Organisms with the ability to capture energy and manufacture matter are known as *autotrophs* (auto = self, trophos = feeding) or *producers*. All other organisms obtain their energy supply by eating other organisms, and are known as *heterotrophs* (heter = different) or *consumers*. There are two kinds of autotrophs: *phototrophs* and *chemoautotrophs*. Phototrophs obtain their energy from light, while chemoautotrophs gain their energy from chemicals available in the environment. Although most of us are aware of the critical role of plants (the phototrophs), the chemotrophs play an equally critical yet not so visible role in our life-support systems. Most of them are bacteria and play a fundamental role in the *biogeochemical* (nutrient) cycles, to be discussed in more detail in Chapter 6.

Phototrophs convert the sun's light energy into chemical energy, using carbon dioxide and water to produce carbohydrates. Following the second law of thermodynamics, some energy will be lost in this transformation: the efficiency rate is only between 1–3 per cent. In other words, up to 99 per cent of the energy may be lost. Nonetheless, this conversion is sufficient to produce billions of tonnes of living matter, or *biomass*, throughout the globe.

Besides photosynthesis, *cellular respiration* is another essential energy pathway in organisms. In both plants and animals, this involves a reversal of the photosynthetic process in which energy is released rather than captured. High-energy, organic carbohydrate is broken down through a series of steps to release the stored chemical bond energy. This produces carbon dioxide (the inorganic molecule), water, heat (because of the law of entropy), and energy, which the organism can use for various purposes, such as transmitting messages, changing in cellular shape or structure, moving ions across membranes, or even moving the organism itself where this is possible. Unable to obtain energy from photosynthesis, this is how we and all other organisms get energy supplies.

Box 4.3 Chemoautotrophs in the Deep

We think of the deep sea floor as a biological desert. In the 1970s, however, scientists discovered that rich biological communities were supported by hydrothermal vents on the sea floor, mainly bacteria, which derive their energy from sulphide emissions. It has now been discovered that there are some similar chemoautotrophic-based communities on the remains of whale skeletons found at depth, nourished by sulphides produced as the carcasses decay. Discoveries of fossils suggest that falling dead whales may have provided dispersal stepping-stones for these communities for over 30 million years. The question is: What was the impact on these communities when whalers virtually eliminated whales from the oceans? Scientists do not yet have the answer to this question.

Box 4.4 Carnivorous Plants

Not all plants are autotrophs. Carnivorous plants, such as the pitcher-plant (the floral emblem of Newfoundland), for example, gain their energy by ingesting the insects that become trapped in their funnel-shaped leaves. The plant, which grows in boggy areas across Canada, has no photosynthetic surfaces. Instead the leaves act as 'pitchers' to hold a soapy liquid from which the hapless insect cannot escape. It is now thought that the plant breaks down the dead insects with help from other insects that have developed immunities to the decomposing enzymes produced by the plant. In fact, the plant plays host to several insects, which seem to thrive on the environment provided by the pitcher-plant. These are examples of *mutualism*, in which both species benefit from a relationship.

The carnivorous pitcher plant, the provincial flower of Newfoundland, grows in abundance (*Philip Dearden*).

For cellular respiration to occur, most organisms must have access to oxygen or they will die. Such organisms are known as *aerobic organisms*. *Anaerobic organisms*, such as some bacteria, can survive even without oxygen. This makes them very useful in the breakdown of organic wastes, such as sewage.

Food Chains

The energy captured by autotrophs is subsequently passed on to other organisms, the consumers, by means of a *food chain* (Figure 4.3). Products are eaten by the *herbivores*, which are, in turn, the source of energy for higher-level consumers or *carnivores*. All these organisms will be fed upon by *decomposers* after they die. Each level of the food chain is known as a *trophic level*. A giant Douglas fir tree on the Pacific coast and a minute arctic flower on Baffin Island would both be on the same trophic level as autotrophs. Herbivores, on the second level, range in size from elephants to locusts. It is the role in energy transformation, rather than the size of the organism, that is the important factor in determining trophic level.

Some organisms—such as humans, raccoons, and cockroaches—are *omnivores* and can obtain their energy from different trophic levels. When we eat vegetables, we are acting as *primary consumers*. When we eat beef, we are at the third trophic level, acting as *secondary consumers*. When we eat fish that have derived their energy from eating smaller organisms, we may be *tertiary* or even higher-level consumers on the food chain. The level at which

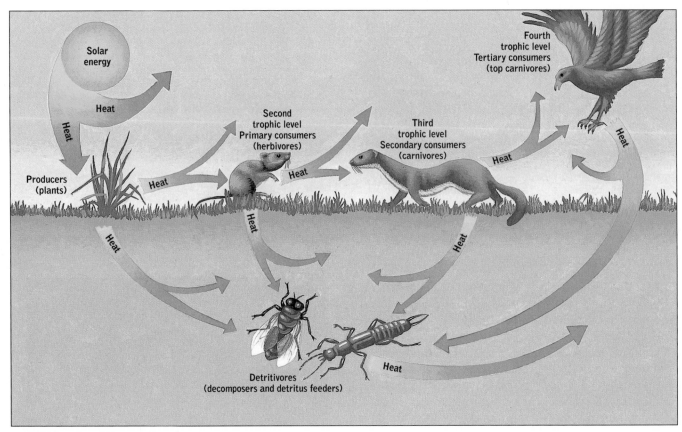

Figure 4.3 A food chain.

food energy is obtained has some important implications (to be discussed later).

Also important are the decomposer food chains (Figure 4.4). These are based on dead organic material or *detritus*, which are high potential energy but difficult to digest for the consumer organisms. However, various species of microorganisms (bacteria and fungi) are able to digest this material as their source of energy. Indeed, many large grazing animals, such as cows and moose, have this bacteria in their stomachs to help break down the cellulose in plant material. These *decomposers* (or *saprotrophs*) derive their energy from dead matter (sapro = putrid). They are joined by consumers, such as earthworms and marsh crabs, known as *detritivores*, which may consume both plant and animal remains.

These decomposer food chains play an integral role in breaking down plant and animal material into products, such as carbon dioxide, water, and inorganic forms of phosphorus, nitrogen, and other elements. For example, dead wood is first broken down by fungi, which consume simple carbohydrates, such as glucose. After that, other fungi, bacteria, and

This rather stunned looking red fox has just escaped from the clutches of a large female golden eagle that managed to lift it some 3 m from the ground before one of the authors happened on the scene and unwittingly rescued one carnivore from becoming the prey of another one at the next trophic level (*Philip Dearden*).

organisms, such as termites, break down the cellulose, which is the main constituent of the wood. Were it not for these organisms, wood and other dead organisms would accumulate indefinitely on the forest floor.

The relative importance of grazing and detritus food chains varies. The latter often dominate in forest ecosystems where less than 10 per cent of the tree leaves may be eaten by herbivores. The remainder die and become the basis for the detritus food chain. In the coastal forests of British Columbia, for example, there are about 140 species of birds, mammals, and reptiles through which energy can flow. By way of contrast, there are over 8,000 species involved in breaking down the soil litter. The same is often true in freshwater aquatic systems where there may be relatively little plant growth but abundant detritus from overhanging leaves and dead insects. However, the converse is true in marine ecosystems (see Box 4.5) in which 90 per cent of the photosynthetic *phytoplankton* (phyto = plant, plankton = floating) may be grazed by the primary consumers, the *zooplankton*.

Bracket fungi such as these are an important component of decomposer food chains (*Philip Dearden*).

In general, theory suggests that the more components there are to the ecosystem, the more alternative pathways there are available for energy flow and the more able the ecosystem will be to withstand stress. In the Arctic, for example, a simple food chain may be phytoplankton to zooplankton to cod to ringed seal to polar bear. All these species are very heavily dependent upon the species at the preceding trophic level. If one of these species were drastically reduced in number or made

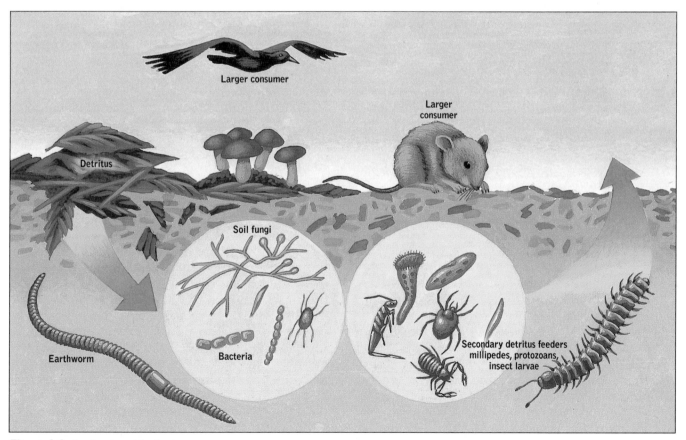

Figure 4.4 Detritus-based food web.

The diversity of marine ecosystems is astonishing. Here are some different species of sea stars at Burnaby Narrows in Gwaii Hanaas National Park Reserve, BC (*Philip Dearden*).

extinct, the chances of another species fulfilling the first species's role is quite low, and the whole food chain may well collapse. This situation can be compared to that at the other extreme, say in a tropical forest where there are many more species and the chance of other species combining to fulfil the ecological role of a depleted one is therefore much higher. In practice, however, many other factors are involved, such as the relative degree of specialization of the various organisms, and examples of relatively unstable complex systems abound, so care must be exercised before applying the theory as a universal truth (e.g., see Schulze and Mooney 1993).

Box 4.5 Oceanic Ecosystems

Looking at earth from space we see a blue, not a green, planet, reflecting the fact that 71 per cent of the earth's surface is covered by oceans. Life originated in the sea perhaps 3.5 billion years ago, and only came onto land some 450 million years ago, hence much of our biological ancestry lies within these waters. Although the total number of known species is higher on land than in the oceans, the number of *phyla*, distinguished by differences in fundamental body characteristics, is higher in the oceans. Of the thirty-three different animal phyla, for example, fifteen exist exclusively in the ocean, and only one exclusively on land. We share the same phylum as the fishes (the *chordata*), characterized by a flexible spinal cord and a complex nervous system.

Through their photosynthetic activity, the early marine bacteria helped create the conditions under which the rest of life evolved. Current photosynthetic activity is no less important to our survival. Scientists estimate that the phytoplankton in the *euphotic* zone of the oceans produce between a third and one half of the global oxygen supply. In doing so they also extract carbon dioxide from the atmosphere. Some 90 per cent of this carbon dioxide is recycled through marine food webs, but some also falls into the deep ocean as the detritus of

decaying organisms and is stored as dissolved carbon dioxide in deep ocean currents, which may take over 1,000 years to reappear on the surface. The oceans contain at least fifty times as much of the gas as the atmosphere and are playing a critical role in helping delay the so-called greenhouse effect, which is discussed in more detail in chapters 7 and 19.

These phytoplankton, so important to atmospheric regulation, are also the main autotrophic base for the marine food web. From the tiny zooplankton to the great whales, almost every marine animal has phytoplankton to thank for its existence. Phytoplankton flourish best in areas where ocean currents return nutrients from the deep ocean back to the euphotic zone. This occurs in shallow areas, such as the Grand Banks, where deep ocean currents meet the coast, or where two deep ocean currents meet head on. Such areas are the most productive in what is generally an unproductive ocean, and they are the best sites for fisheries. Ninety per cent of the marine fish catch comes from these fertile nearshore waters. Unfortunately, these are also the sites of the greatest pollution. The blueness of most of the rest of the ocean is a visible symbol of the low density of phytoplankton, which is the result of low nutrients. That is why the sea is blue, not green.

Box 4.5 Oceanic Ecosystems (continued)

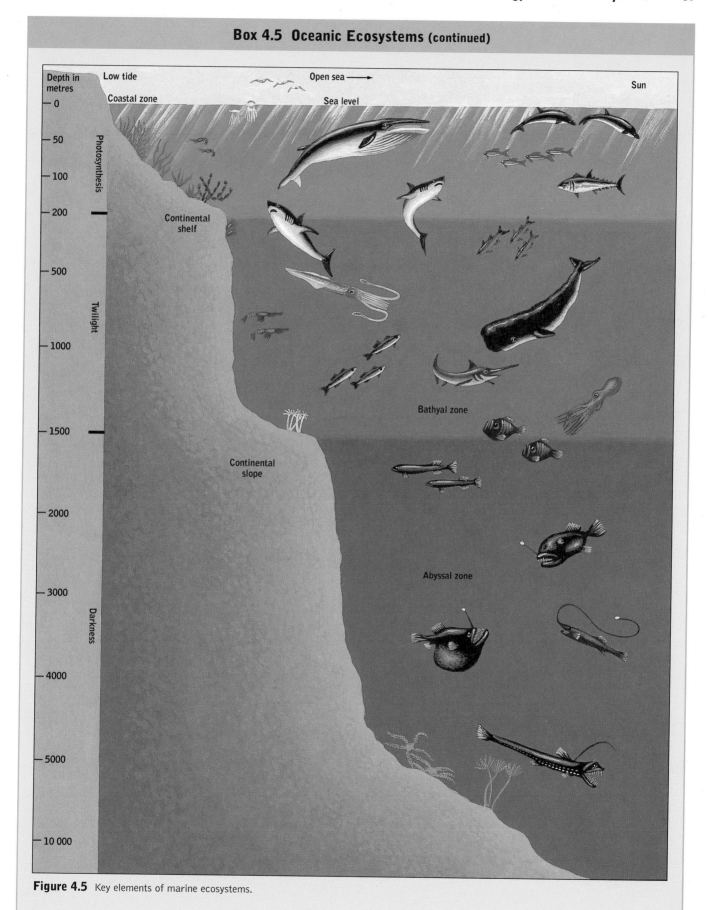

Figure 4.5 Key elements of marine ecosystems.

Rarely are food chains as simple as the one shown in Figure 4.3. Usually there are many competing organisms and energy paths. Thus scientists tend to think of food chains more as complex *food webs* in which there are numerous alternative routes for energy to flow through the ecosystem (Figure 4.6). Ecologist Bristol Foster has described one of the most unusual examples of this complexity when, while on the BC coast, he witnessed a garter snake foraging in a tidal pool and swallowing a small fish, while at the same time being held by a green sea anemone.

The number of species increases as conditions become more amenable for life from the poles to the tropics (Figure 4.7). In the Arctic, for example, there are relatively few species, and therefore relatively few alternative pathways for energy flow. If a prey species, such as the arctic hare, decreases in number, then so too will the organisms dependent upon it higher in the food chain, such as the lynx, because there are few other species upon which these organisms can feed. This gives rise to the familiar population cycles in the North as predator numbers closely reflect the availability of dominant prey species (Figure 4.8).

Biotic Pyramids

The second law of thermodynamics describes how energy flows from trophic level to trophic level, with a loss of usable energy at each succeeding transformation. In natural food chains, the *energy efficiency* or amount of total energy input

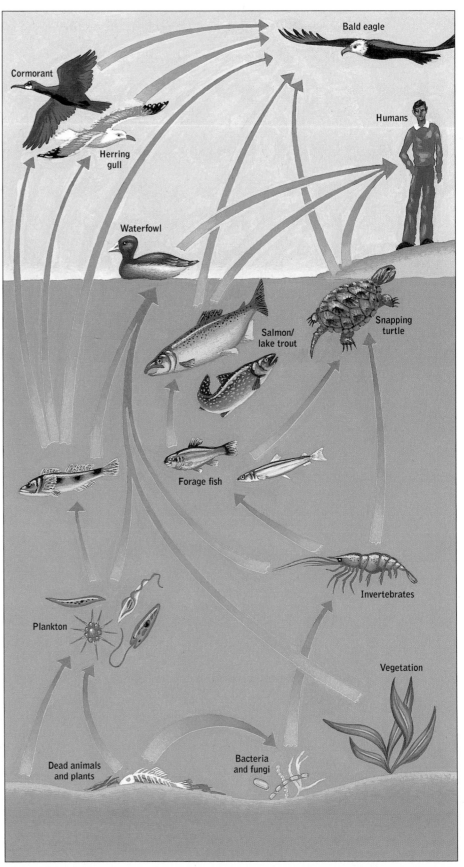

Figure 4.6 A simplified Great Lakes food web. SOURCE: Environment Canada, *Toxic Chemicals in the Great Lakes and Associated Effects* (Toronto: Department of Fisheries and Oceans; Ottawa: Health and Welfare Canada, 1991).

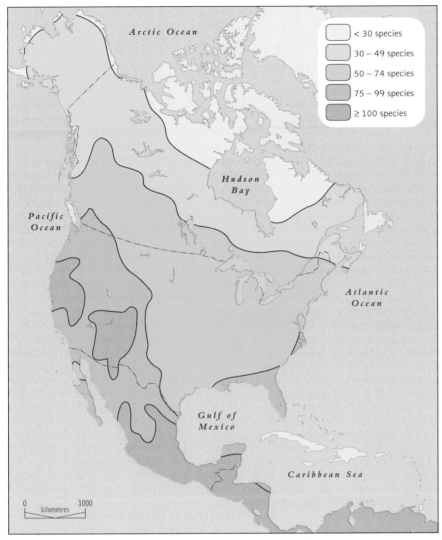

Figure 4.7 The number of species per latitude. SOURCE: After C.G. Simpson, 'Species Density of North American Recent Mammals', *Systematic Zoology* 13 (1964):15–73.

Legend:
- < 30 species
- 30 – 49 species
- 50 – 74 species
- 75 – 99 species
- ≥ 100 species

The arctic fox follows the polar bear in winter and feeds on scraps, but also eats lemmings, hares, berries, and birds (*Ontario Ministry of Natural Resources. Copyright 1997 Queen's Printer Ontario*).

of a system that is transformed into work or some other usable form of energy may be as low as 1 per cent. In general, we expect about 90 per cent of the energy to be lost at each level (Figure 4.9). Similar losses are experienced in biomass and numbers of organisms at each trophic level. This explains why there are fewer secondary than primary consumers and fewer tertiary than secondary. Carnivores must always have the lowest numbers in an ecosystem in order to be supported by the energy base below. This is what happened in the case of the Atlantic puffins, as described in the introduction to this chapter. The biomass of carnivores (puffins) was no longer able to be supported by the energy from the preceding trophic level, the capelin.

There are several reasons for the low energy efficiencies of natural food chains. First, not all the biomass at each trophic level is converted into food for the next trophic level. Many organisms have developed characteristics to avoid being eaten by something else! Generally, only 10–20 per cent of the biomass of one trophic level is harvested by the next level. Furthermore, of that which is consumed, not all of it is ingested. Humans, for example, are not well equipped to break down and consume the bones or fur of animals. Of the biomass that is ingested, not all of it is digested and assimilated into the organism. The proportion of ingested energy actually absorbed by an organism is the *assimilated food energy*. Finally, as cellular respiration liberates energy for the growth, maintenance, and reproduction of the organism, then energy is further released as heat.

The longer the food chain, the more inefficient it is in transforming energy, thus reflecting the second law of thermodynamics. An arctic marine food chain—which goes from the producers (phytoplankton) to primary consumers (zooplankton) and subsequently to the whales, the largest animals ever to exist on earth—is a very efficient food chain because it is so short, and there are only three energy transformations wherein energy is lost. Longer food chains involve a proportionately larger loss of energy due to the greater number of energy transformations.

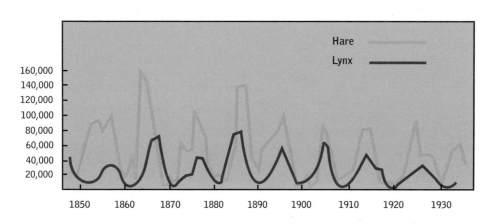

Figure 4.8 Lynx and hare cycles. Predator-prey relationship creates balance between predator and prey populations. Data based on pelts received by Hudson's Bay Company. SOURCE: D.A. MacLulich, University of Toronto Studies, Biological Series No. 43, 1937.

The *energy pyramid* also has important implications for humans. For example, it takes 8–16 kg of grain to produce 1 kg of beef. This means that more land must be cultivated to provide people with a high-meat diet as opposed to one based on grains. As humans are one of the species that can access food energy at several different trophic levels, in terms of energy efficiency, it would be better to operate as low on the food chain as possible; that is, as primary consumers or as vegetarians. This topic is discussed in more detail in Chapter 14.

Productivity

Productivity in ecosystems is measured by the rate at which energy is transformed into biomass, or living matter, and is usually expressed in terms of kilocalories per square

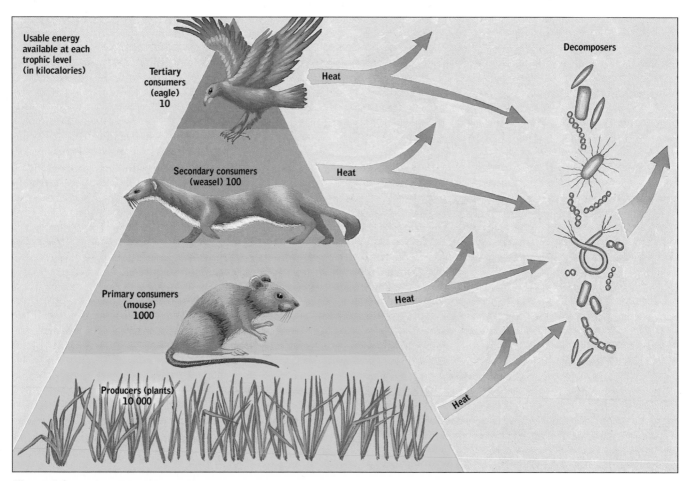

Figure 4.9 Generalized pyramid of energy flow.

Box 4.6 Arctic Population Fluctuations

Populations of many animals in the North show distinctive fluctuations in number over regular time periods. A three-to four-year cycle is usual for smaller animals, such as lemmings and meadow voles, and a longer one of nine to ten years for larger animals, such as snowshoe hares and several species of grouse and ptarmigans. The predators of these herbivores show similar fluctuations in numbers as their energy source becomes critically depleted.

One of the points this book illustrates is the uncertainty regarding many aspects of environmental management. In this case, there is a fair amount of certainty about the dates of fluctuations in these populations, but there is considerable uncertainty as to why these fluctuations occur. Four ideas have been advanced:

1. The seasonality of the Arctic environment means that plants grow for only a few months every year, yet herbivores must eat throughout the year. Thus when population levels rise, overexploitation of the food supply can occur rapidly and for a relatively long period, but the plants will take a long time to recover. Thus the relationship between the herbivores and their food supply is the determining factor, and the predator numbers simply reflect those of the prey.
2. . A second hypothesis builds upon this idea, but suggests that as populations of small animals, such as lemmings, increase, then more nutrients vital to plant productivity become tied up in this higher trophic level. This lack of nutrients causes reductions in plant productivity and quality, leading to starvation for the herbivores.
3. Other ideas postulate more of an interaction between the predators and prey. Keith et al. (1984), for example, studied the ecology of snowshoe hares in northern Alberta to test a food supply-predation hypothesis. This suggests that food supply shortages halt populations of herbivores, which are subsequently caught by predators until numbers fall enough to permit plant recovery. Keith et al. reported that malnutrition of hares was evident, but that 80–90 per cent of deaths were caused by predators, thereby supporting the hypothesis.
4. The final idea suggests that food supplies play a negligible role in herbivore population cycles, and that such cycles would not occur in the absence of predators (Trostel et al. 1987). Some support for this hypothesis has been forthcoming as scientists have found that hare population crashes occur despite supplementary feeding programs (Smith et al. 1988).

As you can see, ecological systems are not easy to understand. In many cases, several factors may contribute to changes such as population cycles.

metre per year. *Gross primary productivity* (GPP) is the overall rate of biomass production, but there is an energy cost to capturing this energy. This cost, *cellular respiration*, (R) must be subtracted from the GPP to reveal the *net primary productivity* (NPP). This is the amount of energy that is available to heterotrophs.

All ecosystems are not the same in terms of their abilities to fix biomass as rates of photosynthesis are regulated by light levels, nutrient availability, temperature, and moisture. The most productive per unit area are estuaries, swamps, marshes, and tropical rainforests (Figure 4.10). Recent data indicate that the temperate rainforests, such as those that grow in the Pacific Maritime ecozone described in chapters 2 and 15, are as equally productive as the tropical forests. Other ecosystems are more limited due to deficiencies in one or more of the characteristics noted earlier. A desert, for example, lacks water, the Arctic lacks heat, and the ocean lacks nutrients.

Unfortunately, in Canada some of our most highly industrialized and polluted lands are adjacent to estuaries. There is not one sizeable estuary on the east coast of Vancouver Island, for example, that is not used by the logging industry either as a mill site or a log-storage area. The estuary of the Fraser River has been extensively dyked, industrialized, and polluted. There are few data on the effects of these intrusions on our most productive ecosystems.

Humans already take about 40 per cent of terrestrial NPP for our own use. The remainder supports all the other organisms on earth, which in turn maintain the environmental conditions that keep us alive. The human population is projected

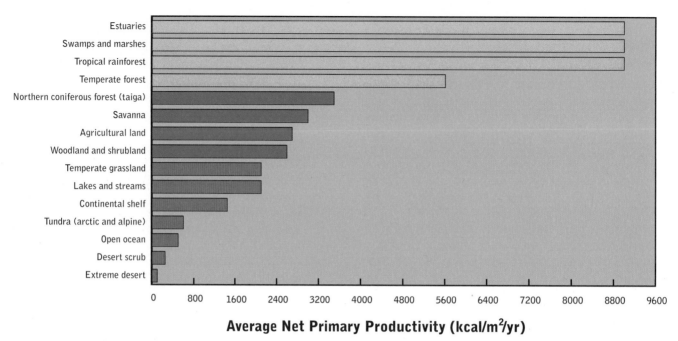

Average Net Primary Productivity (kcal/m²/yr)

Figure 4.10 Estimated annual average net productivity of producers per unit of area in principal types of life zones and ecosystems. Values are given in kilocalories of energy produced per square metre per year.

to double over the next forty years. It is highly doubtful that the earth's systems could withstand a concomitant increase in the amount of NPP appropriated for human use. Once again, this is indicative of the carrying capacity challenge that we face and which was discussed in Chapter 1.

In addition to primary productivity, measurements can also be made of *net community productivity (NCP)*, including heterotrophic as well as autotrophic respiration. Measurements indicate that as communities mature, although GPP and NPP rise, an increasing proportion of the energy of the community is devoted to heterotrophic respiration (Table 4.1). In mature communities, the amount of respiration may be sufficient to account for all the

Estuaries are among the most productive ecosystems. Unfortunately, they are also very convenient sites for industrial activity, such as the log boom storage shown here, which inhibits productivity (*Philip Dearden*).

energy fixed by photosynthesis. There is thus no net gain, leading to the characterization of such communities as 'decadent' by resource managers, such as foresters, who are mainly interested in the productivity of the autotrophs.

Over time, natural systems mature towards maximization of NCP. On the other hand, humans are often concerned with maximizing NPP. This is an example of the decision regulators discussed in Chapter 1. Natural forest system decision regulators may allow trees to grow for over 1,000 years before they die. The control system exerted by forest management determines that the trees will maximize NPP before considerable amounts of energy are devoted to heterotrophic respiration (Figure 4.11). The age of the trees in systems managed for forestry will hence be much younger, often about 100 years old, rather than several centuries old.

Auxiliary energy flows allow some ecosystems and sites to be especially productive. For example, in natural systems the tidal energy in an estuary is an auxiliary energy flow that helps to bring in nutrients and dissipate wastes, so that organisms do not have to expend energy on these tasks and can devote more energy to growth. Agriculture relies extensively on the inputs of auxiliary energy in the form of pesticides, fertilizer, tractor fuel, and the like to supplement the natural energy from the sun

Table 4.1 Production and Respiration as (kcal/m²/yr) in Growing and Climax Ecosystems						
	Alfalfa Field (US)	Young Pine Plantation (England)	Medium-Aged Oak-Pine Forest (NY)	Large Flowering Spring (Silver Springs, Fla)	Mature Rain-forest (Puerto Rico)	Coastal Sound (Long Island, NY)
Gross primary production	24 400	12 200	11 500	20 800	45 000	5700
Autotrophic respiration	9200	4700	6400	12 000	32 000	3200
Net primary production	15 200	7500	5000	8800	2500	2500
Heterotrophic respiration	800	4600	3000	6800	13 000	2500
Net community production	14 400	2900	2000	2000	Little or none	Little or none

Source: Adapted from E.P. Odum, *Fundamentals of Ecology*, 3rd ed. (Toronto: Holt, Rinehart and Winston, cbs College Publishing, 1971:46). Copyright © 1971 by Saunders College Publishing, reproduced by permission of the publisher.

to augment crop growth, as discussed in Chapter 14. In many cases this subsidy (mostly derived from fossil fuels) is more than the sun's energy input. Without this subsidy, productivity would be much reduced. There is a cost to this subsidy, however, in terms of the high energy costs of providing it, and the environmental effects it creates as it is dispersed into the environment in the form of pollution.

ECOSYSTEM STRUCTURE

The energy flows described earlier are all part of the ecosphere, which can be broken down to smaller units. At the smallest level is the individual *organism*. A group of such individuals is a *population*. Together with other organisms in a particular environment, they form a *community*. The *ecosystem* is a collection of communities interacting with the environment. However, ecosystems represent a somewhat abstract conceptualization of the environment, which can range greatly in scale. Due to the highly interactive nature of the relationship between organisms and their environment, it is often difficult to define precisely the boundary of an ecosystem (as discussed in Chapter 8). Furthermore, ecosystems are *open systems* in that they exchange material and organisms with other ecosystems. Ecosystems and communities thus provide useful abstractions for the study of the environment, but should not be regarded as precise categories that will be agreed upon by all scientists.

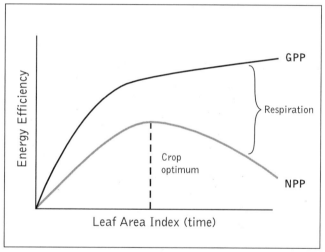

Figure 4.11　Diagram to illustrate general relationship between productivity and time as a forest matures. Foresters might consider the optimal stage of the forest to be at maximum NPP, even though GPP continues to increase over time.

Many ecosystems taken together and classified according to their dominant vegetation and animal communities form a *biome*. Canada, due to its size, has as many biomes as any country in the world. These were discussed in more detail in Chapter 3. Water availability and temperature are the main factors that control biome distribution. How these influence biomes at the global scale is summarized in Figure 4.12.

Abiotic Components

The food chains described earlier constitute the living or biotic component of ecosystems. The abi-

Figure 4.12 Influence of temperature and rainfall on biome.

otic components have an important role in determining how these biotic components are distributed. Important abiotic factors include light, temperature, wind, water, nutrient availability, and many other soil characteristics (see Box 4.7). All these factors influence different organisms in various ways. It is the interaction among these characteristics, the organisms, and among the organisms themselves that determines where and how well each organism can grow.

Just as the laws of thermodynamics explain energy flows, some useful principles help us understand how organisms react to different situations. The first of these is known as the *limiting factor* principle. This tells us that all factors necessary for growth must be available in certain quantities if an organism is to survive. Thus a surplus of water will not compensate for an absence of an essential nutrient or adequate warmth. In other words, a chain is only as strong as its weakest link. The

weakest link is known as the *dominant limiting factor*. A major goal of agriculture is to remove the effect of the various limiting factors. Thus auxiliary energy flows are employed to ensure that a crop has no competition from other plants (weeding), or that water supply is adequate (irrigation), or that the plant has optimal nutrient supply (fertilizer).

The corollary of the above is that all organisms have a range of conditions that they can tolerate and still survive. This is known as the *range of tolerance* for a particular species. This range is bounded on each side by a zone of intolerance for which limiting factors are too severe to permit growth (Figure 4.14). For example, there may be too much or too little water. As conditions improve for a particular factor, then certain individuals within the population can tolerate those conditions, but they are not optimal and relatively few individuals can exist. This is known as the *zone of physiological stress*. Additional improvement results in a range where

Box 4.7 Soils

Soils are critical in determining the vegetation growth of an area. Soil is a mixture of inorganic materials (such as sand, clay, and pebbles), decaying organic matter (such as leaves), and water and air. This mixture is home to billions of micro-organisms that constantly modify and develop the soil. In the absence of these organisms, the earth would be a sterile rock pile rather than the rich, life-supporting environment it is now. Most of these organisms are in the surface layer of the soil, and one teaspoon may contain hundreds of millions of bacteria, algae, and fungi. In addition, there are many larger species—such as roundworms, mites, millipedes, and insects—that play vital roles in this very complex ecology.

Most soils form from the parent material where they are found. This may originate from the weathered remains of bedrock or where sediments have been deposited from elsewhere by water, ice, landslides, or the wind. Over centuries this mixture is modified by ongoing physical and chemical weathering and organic activities, so that quite different soils may result depending upon the location. These different processes result in the formation of different layers in the soil called *soil horizons*. A view across these horizons is called a *soil profile*. Figure 4.13 shows a generalized profile. However, not all soils have all these different horizons.

Soils also differ in their *texture* or size of different materials. Clay is the finest, followed by silt, sand, and then gravel, the coarsest. Soils that contain a mixture of all these with decomposed organic material (or *humus*) are called *loams* and often make the best soils for vegetation growth. Texture is a main determinant of *soil permeability*, or the rate at which water can move through the soil. Water moves very slowly through soils composed mainly of the smallest particles (clay) and the soil becomes easily waterlogged. On the other hand, the large spaces between particles of sand or gravel lead to rapid drainage, and the

soils may be too dry to support good vegetation growth. Plants obtain their nutrient supply necessary for growth from ions dissolved in the soil water, hence permeability is critical.

Soil has many different chemical characteristics. One of the most important is the pH value (see Chapter 7), measuring the acidity of the soil, which helps determine what minerals are available and in what form. Different plants have different mineral requirements. Farmers often try to change the acidity of their soils by adding lime, for example, if the soil is too acidic, or sulphur if the soil is too alkaline.

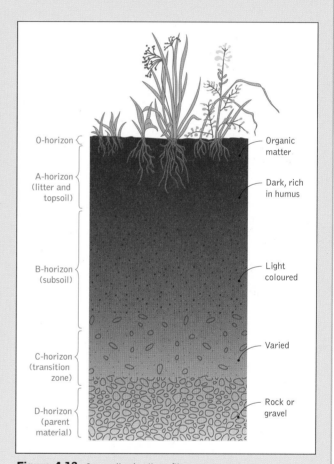

Figure 4.13 Generalized soil profile.

conditions are ideal for that species, which is the *optimum range*. Here, in theory, there will be the highest population of a particular organism.

The concepts of limiting factors and range of tolerance can be illustrated by an example from Saskatchewan where smallmouth bass are introduced every year into lakes for sport fishing. These hatchery-raised fish survived from year to year (if

they weren't caught), proving that they were within their range of tolerance for survival as individuals. They did not, however, breed successfully because the hatchlings needed a slightly warmer temperature to develop than the mature fish did to survive. Thus there are different levels of tolerance for species at different life stages. The dominant limiting factor for the smallmouth bass in this

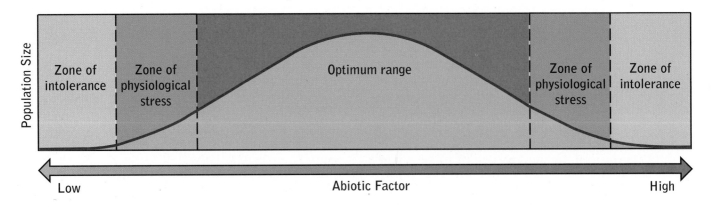

Figure 4.14 Range of tolerance.

environment was the low temperatures during hatching.

Water availability is often the critical factor that determines differences between communities. Where precipitation is in excess of about 1000 mm per year, for example, trees will usually dominate the landscape if other factors are suitable. Below 750 mm, the range of tolerance for trees is exceeded and grasses will dominate because they have a much lower range of tolerance for water stress (about 100 mm per annum). If the water availability is below that level, even grasses run into their zone of intolerance and then cacti, sagebrush, and other very drought-resistant species dominate.

Organisms are not, however, reacting to just one abiotic factor, such as water availability, but to all the factors necessary for growth. Sometimes the optimum range for one factor will not overlap with the optimum range for other factors, which would place the organism in the zone of physiological stress for that factor, which becomes the dominant limiting factor. Organisms may also be out-competed for a particular factor in their optimum range by another organism and again be forced into a zone of physiological stress. In other words, the simple single-factor model represented in Figure 4.14 is more complicated because of the numerous abiotic and biotic influences that must also be taken into account. However, that model does provide a useful conceptual tool to help understand the spatial distribution of organisms.

Biotic Components

Other species also have an important role in influencing species distribution and abundance. Species interact in several ways, including competing for scarce environmental resources. Each species needs a specific combination of physical, chemical, and biological conditions for its growth. This is known as the *niche* of that species. Where the species lives is known as the *habitat*.

The principle of *competitive exclusion* tells us that no two species can occupy the same niche in the same area. Most species have a *fundamental niche* representing the potential range of conditions that it can occupy, and also a narrower *realized niche* representing the range actually occupied. The physical conditions for growth exist throughout the fundamental niche, but the species may be out-competed in parts of this area through the overlapping requirements of other species. *Specialist* species have relatively narrow niches and are more susceptible to population fluctuations as a result of environmental change. A classic example of such a specialist is the panda, which eats only one plant, bamboo. Whenever the bamboo supply falls, as it does after it flowers, then this specialist species has few alternative sources of food. When bamboo was abundant, the pandas would simply move to a new area. However, as the animals have become increasingly restricted to smaller and more isolated reserves, the availability of their food supply has become a major problem.

In Canada specialist species include the burrowing owl, woodland caribou, and the whooping crane (discussed later in chapters 14, 15, 16, and 17). *Generalist* species, on the other hand, like the black bear and coyote, may have a very broad niche with a wide variety of food sources. Such generalist species have adapted most successfully to the new environments created by humans.

Intraspecific competition occurs among members of the same species, whereas *interspecific competition* occurs between different species. Both forms of competition are a result of demands for scarce resources. Intraspecific competition occurs particularly when individual species densities are very high. Interspecific competition occurs when species's niches are similar. Competition may be reduced through *resource partitioning* when the resources are used at different times or in different ways by species with overlapping fundamental niches. For example, hawks and owls both hunt for similar types of prey, but at different times as owls are mainly nocturnal.

Intraspecific competition may lead certain individuals to dominate a specific area. The area is known as a *territory* and may be aggressively defended against intruders. Dominant male grizzly bears establish such territories, which may be as large as 1000 km^2, although it is unlikely that grizzly bears can always effectively defend such a large territory from intruders. During the breeding season male robins also establish and defend nesting territories, the boundaries of which are proudly advertised in song. Establishment of a territory requires sufficient resources in order for breeding pairs to be successful. Ultimately it regulates population size in areas where favourable

Some species, such as the black bear, are very adaptable and have a relatively broad range of tolerance. On the Pacific coast, black bears are frequent scavengers of the intertidal zone where many items are considered potential food. The year-round availability of an abundant and varied food source results in very large individuals (*Philip Dearden*).

The burrowing owl—whose habitat includes open grassland, prairies, farm land, and airfields—nests in burrows in the ground and feeds on various rodents and arthropods (*Manitoba Wildlife Branch*).

Box 4.8 Species Relationships

There are other kinds of relationships between species besides competition. In *predation*, for example, a predator species benefits at the expense of a prey species. The lynx eating the hare and the osprey eating the fish are familiar examples of this kind of relationship, although in a broader sense we should also consider the herbivore eating the plant. Predation is a major factor in population control, and usually results in the immediate death of the prey species.

A special kind of *predator-prey relationship* is *parasitism*, where the predator lives on or in its prey (or host). In this case the predator is often smaller than the prey and gains its nourishment from the prey over a more extended time period, which may lead to the eventual death of the host. This may cause the death of the parasite, too, although some parasites, such as dog fleas and mosquitoes, can readily switch hosts.

Tapeworms, ticks, lamprey, and mistletoe are all examples of parasites.

Not all relationships between species are necessarily detrimental to one of the species. *Mutualism* is when the relationship benefits both species. These benefits may relate to enhanced food supplies, protection, or transportation to other locations. The relationship between the nitrogen-fixing bacteria and their host plants (described in Chapter 6) is an example of such a relationship, which results in enhanced nutrition for both partners. Other examples include the relationships between flowering plants and their pollinators, which result in the distribution of pollen to other plants and the protection that ants give to aphids in return for the food the aphids extract from plants. Interactions that appear to benefit only one partner but not harm the other are examples of *commensalism*.

habitat is limited because those individuals unable to defend territories must move to less favourable areas where their likelihood of breeding success is limited.

Species with a strong influence on the entire community are known as *keystone* species. In Canada, our national symbol, the beaver, is a good example of such a species. Beavers can have a profound impact upon their environments through their construction activities, which raise and lower water levels. This in turn affects the limits of tolerance of other species in the community, which may suddenly find themselves submerged under a beaver dam or facing lower water levels downstream. Different species will have different reactions to this change, depending on their range of tolerance to water.

It is, of course, especially significant when a keystone species is removed from an area, or *extirpated*, by human activity. Such changes may take some time before they become noticeable. For example, changes to soil characteristics as a result of the extermination of major herbivores, such as bison, may take centuries before they become noticeable. The same is true for the other large grazers, such as the great whales (discussed in Chapter 3), which we have greatly decimated over the last couple of centuries.

IMPLICATIONS

From this discussion it should be apparent that ecosystems are very complicated. There are complex interrelationships among organisms, and between organisms and their environment. A change in part of this matrix will result in corresponding changes throughout. Humans are now such a dominant influence on global environmental conditions at all scales that there are significant changes as a result of human activities. There is considerable uncertainty as to how ecosystems and the life-support systems of this planet will react to these changes. However, even under natural conditions, ecosystems are not static. The next chapter will focus on how ecosystems change over time.

SUMMARY

1. Understanding energy flows is critical to understanding the ecosphere and environmental problems. The laws of thermodynamics explain how energy flows through systems. The first law is that energy can neither be created nor destroyed, but merely changed from one form to another. The second law is that at each such energy transformation, some energy is converted to a lower-quality, less useful form.

2. Energy is the basis for all life. Through the process of photosynthesis, certain organisms use sunlight to transform carbon dioxide and nutrients into organic matter. This matter forms the basis of the food chains through which energy is passed from one trophic level to another. At each trans-

---

Box 4.9 Species Distributions

There are important implications for species distributions:

- A species may have a wide range of tolerance for some factors, but a very narrow range for others.
- Species with the largest ranges of tolerance for all factors tend to be the most widely distributed. For example, species such as cockroaches and rats enjoy a virtually global distribution.
- Many weed and pest species are successful because of their wide range of tolerance. Eurasian water milfoil, which is a significant nuisance in many waterways in Canada, is a non-native plant that can grow in conditions from Canada to Bangladesh.
- Response to growth factors is not independent. Grass, for example, is much more susceptible to drought when nitrogen intake is low.
- Tolerance for different factors may vary through the life cycle. Critical phases often occur when organisms are juveniles and during the time of reproduction.
- Some species can adapt to gradually changing conditions for some factors, up to a point. However, after this *threshold* of change is reached, the population will collapse.

ference, the second law of thermodynamics dictates that some energy is lost, often as much as 90 per cent.

3. Productivity is a measure of the abilities of different communities to transform energy into biomass. The most productive communities are found in estuaries, wetlands, and rainforests.

4. The ecosphere is the thin, life-supporting layer of earth characterized by interactions between the biotic and abiotic components. It can be further subdivided into communities, ecosystems, and biomes.

5. The concepts of limiting factors and range of tolerance help us understand the interaction between the biotic and abiotic components of the ecosphere.

6. Each species needs a specific combination of the physical, chemical, and biological conditions for its growth. This is the niche of that species.

7. According to the principle of competitive exclusion, no two species can occupy the same niche in the same area at the same time.

8. Species compete for scarce resources in any given habitat. However, there are many other forms of relationships between species, such as predation, parasitism, mutualism, and commensalism.

9. Species with a strong influence on the entire community are known as keystone species.

REVIEW QUESTIONS

1. What are the main biotic and abiotic components of ecosystems?

2. How do the laws of thermodynamics apply to living organisms?

3. How do the laws of thermodynamics apply to environmental management?

4. What are chemoautotrophs, and what role do they play in ecosystem dynamics?

5. On what trophic level is a pitcher-plant? Why?

Are there plants on the same trophic level in your area? What are they and where do they grow?

6. In what kinds of ecosystems do detritus food chains dominate?

7. What roles do phytoplankton play in maintaining ecospheric processes?

8. What are the management implications of recognizing concepts, such as specialist, generalist, and keystone species? Can you think of any examples in your area?

9. What are the dominant limiting factors for plant communities in your area?

10. Draw a line from east to west and another from north to south on a map of your province or territory and show the main environmental gradients and the vegetational response in each quadrant.

REFERENCES AND SUGGESTED READING

Barbour, M.G., J.H. Burk, and W.D. Pitts. 1987. *Terrestrial Plant Ecology*. Toronto: Benjamin/Cummings.

Brewer, R. 1994. *The Science of Ecology*. Philadelphia: Saunders College Publishing.

Colinvaux, P. 1980. *Why Big Fierce Animals Are Rare*. London: Penguin.

Ehrlich. P.R. 1986. *The Machinery of Nature*. New York: Simon and Schuster.

Environment Canada. 1991. *Toxic Chemicals in the Great Lakes and Associated Effects*. Toronto: Department of Fisheries and Oceans; Ottawa: Health and Welfare Canada.

Feit, H.A. 1988. 'Self-Management and State-Management: Forms of Knowing and Managing Northern Wildlife'. In *Traditional Knowledge and Renewable Resource Management in Northern Regions*, edited by M.M.R. Freeman and L.N. Carbyn, 72–91. Edmonton: Boreal Institute for Northern Studies.

Gayton, D. 1990. *The Wheatgrass Mechanism: Science and Imagination in the Western Canadian Landscape*. Saskatoon: Fifth House.

Keith, L.B., J.R. Cary, O.J. Rongstad, and M.C. Brittingham. 1984. 'Demography and Ecology of a Declining Snowshoe Hare Population'. *Wildlife Monographs* 90:1–43.

Krebs, C. 1994. *Ecology*, 4th ed. New York: Harper-Collins.

Odum, E.P. 1971. *Fundamentals of Ecology*, 3rd ed. Toronto: Holt, Rinehart and Winston, CBS College Publishing.

_____. 1989. *Ecology and Our Endangered Life-Support Systems*. Sunderland, Mass.: Sinauer Associates.

Okasen, L. 1990. 'Exploitation Ecosystems in Seasonal Environments'. *Oikos* 57:14–24.

Rowe, S. 1990. *Home Place*. Edmonton: NeWest Press.

Schulze, E.D., and H.A. Mooney, eds. 1993. *Biodiversity and Ecosystem Function*. New York: Springer-Verlag.

Simpson, C.G. 1964. 'Species Density of North American Recent Mammals'. *Systematic Zoology* 13:15–73.

Smith, C.R. 1992. 'Whale Falls: Chemosynthesis on the Deep Sea Floor'. *Oceanus* 35:74–8.

Smith, J.N.M., C.J. Krebs, A.R.E. Sinclair, and R. Boonstra. 1988. 'Population Biology of Snowshoe Hares: II. Interactions with Winter Food Plants'. *Journal of Animal Ecology* 57:269–86.

Stephenson, R. 1982. 'Nunamiut Eskimos, Wildlife Biologists and Wolves'. In *Wolves of the World: Perspective of Behaviour Ecology and Conservation*, edited by F. Harrington and R.C. Paquet, 434–9. Parkridge, NJ: Noyes Publications.

Trostel, K., A.R.E. Sinclair, C.J. Walters, and C.J. Krebs. 1987. 'Can Predation Cause the 10 Year Hare Cycle?' *Oecologia* 74:185–92.

Chapter 5

Ecosystem Change

The interaction between the ecosystem components (the abiotic and biotic factors discussed in the preceding chapter) is a dynamic one. Communities and ecosystems change over time. The speed of change depends upon the factors creating the change and the organisms' range of tolerance to the new environment. Some changes are very rapid, such as a landslide that totally destroys the existing ecosystem. Others, such as climate change, occur over long periods and allow communities to adjust slowly to the new conditions. As vegetation communities change, so also do the heterotrophic components that depend upon plants for food. Similarly, if the components of the food web change, this may well cause a change in vegetation.

In this chapter several aspects of change in ecosystems will be examined, starting with the process of ecological succession. This will be followed by a discussion of ecosystem homeostasis and its contributing factors, as well as different aspects of population growth. Finally, longer-term changes—such as evolution, speciation, extinction, and biodiversity—will be considered.

ECOLOGICAL SUCCESSION

Ecological succession is an example of slow adaptive processes. It involves the gradual replacement of one assemblage of species by another as conditions change over time. *Primary succession* is the colonization and subsequent occupance of a previously unvegetated surface, such as when a glacier retreats or a landslide destroys the previous ecosystem (Figure 5.1). There is little soil, and the first species to occupy the area, known as *primary col-*

onizers, must be able to withstand high variability in temperatures and water availability, and a difficult-to-access nutrient supply from the rocks. Few species can tolerate such conditions. Lichens are often the first.

Lichens can exist on bare rock surfaces (Box 5.1). Over time, in combination with other physical and chemical processes, lichens break down these rocks. They trap water and nutrients. Biomass increases. Over generations, these changes make it possible for other species, particularly mosses, to colonize. Mosses grow faster than lichens, resulting in greater accumulation of biomass and soil over time. Eventually the lichens are out-competed by the faster-growing mosses.

The next stage is invasion by herbaceous plants, such as grasses and weeds. Such species are able to colonize a wide range of habitats and have efficient reproductive strategies to allow them to do so. Fireweed is a good example, and the numerous wind-borne seeds of this species are a common sight in many parts of Canada in the autumn.

Over time the growth and decay of these early herbaceous species allow certain hardy shrubs to grow, which in turn further ameliorate conditions until more demanding tree species become established. In areas where precipitation and temperature are adequate, tree growth will often represent the final stage of this successional process. Each stage along the way is known as a *seral* stage, with the final stage known as the *climax* community.

Ecological succession should not be thought of as an inevitable linear progression. It is more a guideline to help us understand the changes that

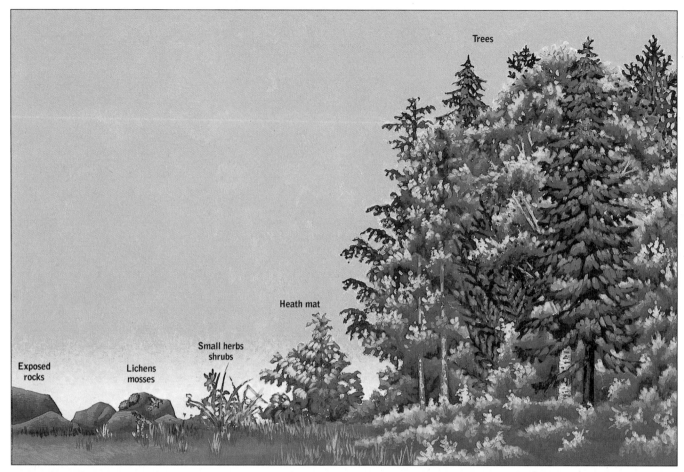

Figure 5.1 Succession.

Fireweed, seen here growing alongside the Tatshenshini River in northern British Columbia, is a common herb in early successional sites throughout Canada (*Philip Dearden*).

The term tree line is used to describe areas where vegetation communities dominated by trees give way to those dominated by other types of vegetation, such as herbs and grasses. Rarely, however, is there a sharp line but rather an ecotone where patches of both tree- and grass-dominated communities exist together (*Philip Dearden*).

Box 5.1 Lichens

Some environments have such harsh growing conditions that virtually nothing can survive. However, one of the few organisms that can be found in such places are lichens. Lichens are composed of fungi and *photosynthetic* algae in a *mutualistic* relationship. The fungi cling to rocks or trees with their filaments and soak up water. The algae, in turn, produce food for both partners through photosynthesis. This may include the fixation of nitrogen from the atmosphere by *cyanobacteria*. Lichens can survive intense cold and drought and have been evolving for over 1 billion years, making them one of the most primitive living things. Individual lichens may be over 4,000 years old. Over centuries the growth of lichens may produce the thinnest of soils, enabling other species that can tolerate harsh conditions to colonize. Lichens are thus very important as *primary colonizers*.

Over 18,000 species of lichens throughout the world have been identified, and undoubtedly many others have yet to be discovered. They have different life forms: dust, crust, scale, leaf, club, shrub, and hair lichens. The most well known in Canada are the leafy kind found growing on trees, an encrusting kind that grows on rocks and sometimes trees, and the so-called (and misnamed) reindeer mosses found throughout northern Canada. Besides providing an essential food supply for caribou and other animals, lichens have also been used by humans as a flour when dried and ground up, and also as a dye for wool and other fabrics. Because lichens receive their mineral requirements directly from the air, they are very efficient accumulators of pollution. They can absorb concentrations of pollutants that exceed their own *tolerance levels*, and hence are excellent *indicator species* for air pollution, as lichens will be absent from heavily polluted areas. Wong and Brodo (1992) analysed current versus past collections of lichens in Ontario, and found that of the 465 species collected prior to 1930, forty-two had not been collected since that year, and another ten not since 1960.

Lichen (*Philip Dearden*).

may take place in communities. In some instances—for example, in recently glaciated terrain—very hardy species of trees, such as willows and alder, may become established in favoured sites that had little previous colonization. Cyclic succession may also occur where a community progresses through several seral stages but is then returned to earlier seral stages by natural phenomena, such as fire (Box 5.2) or intense insect attack. The different seral stages are also not discrete but may blend from one zone into another. These blending zones, known as *ecotones*, tend to be the areas with the highest species diversity since they contain species from both communities and are relatively richer zones between communities.

Sand-dune succession is also a common form of primary succession in which the primary colonizers are grasses that can withstand high variability in temperature and water, as well as the constantly shifting sand. The grasses help to stabilize the sand until mat-forming shrubs invade. Later conditions may become suitable for hardy trees, such as pines, which may in turn be replaced by species such as oaks.

Climax should be thought of as a relative rather than absolute stage. Communities do not

Box 5.2 Fire, Management, and Ecosystem Change

In many areas, fire is a natural occurrence that has a profound impact upon plant and animal communities. In some communities it may be the dominant influence, and if suppressed by human interference, those communities will change in composition. Humans have used fire since earliest times to manipulate ecosystems to produce desired effects, such as removing forests to facilitate agriculture, burning grasslands to generate new grass growth, and as a tool in hunting to scare animals into running in certain directions. Weber and Taylor (1992) outline five current uses of fire in forest management: hazard reduction, silviculture, insect and disease control, wildlife habitat enhancement, and range burning.

Fire has several important ecological implications:

- It favours the growth of certain species over others. Some species are quite fire-resistant (such as the Douglas fir), while others may be aided in their germination by the heat from fire (as with many pine species). Fire may result in the death of other species. Thus where fire is common, it may be the dominant influence on the composition and structure of some communities.
- It tends to increase the diversity of species in a community over the short term.
- It releases nutrient from the biomass into the soil and atmosphere; some may be lost from the site, but the remainder helps stimulate growth of some species.
- It stimulates the growth of various grasses and herbs that provide fodder for herbivores, which may in turn increase carnivore populations.
- Soil temperatures are increased not only during the fire but also afterwards as the site is more open to the sun and has a lower albedo. This also influences chemical and biological changes in the soil, stimulating microbial activities and enhancing decomposition.
- Fires that are too intense or frequent may cause such nutrient impoverishment of a

site that further growth of trees is prevented, and the vegetation may become dominated by grasses and low shrubs. This is known as a *plagioclimax*. Many of the heathlands of northern Europe were created in this manner, and clear-cutting and fire in nutrient-poor black spruce forest sites in Canada can have the same effect.

Fire is also a highly emotive topic for management. Early concepts of forestry and conservation encouraged policies of total fire suppression, with little consideration given to the role of fire in various ecosystems. This led to unanticipated changes in some ecosystems and a build-up of organic debris, which led to the large fires in Yellowstone National Park in 1988 that burned despite the efforts of 10,000 firefighters and the US Army (Bath 1991). As often happens, however, there is a strong reaction against inadequate policies, sometimes leading to calls for no fire suppression at all. Some fires may be ecologically appropriate. Others may be caused by human carelessness or a lack of ecological understanding. Furthermore, it is not possible to ignore the potentially destructive effects of fires.

In Canada, as discussed in more detail in Chapter 7, the warmest years (since records began to be kept in 1861) have been 1981, 1983, 1987, 1988, 1989, 1990, 1991, 1994, and probably 1995. Many scientists think this may be due to global warming caused by the emission of greenhouse gases. Before this warming trend, Canada normally lost 1 million ha of forest per year to fire. Since the mid-1980s this has now averaged 2.2 million ha (Figure 5.2). In so doing, of course, millions of tonnes of carbon that were biologically fixed in the biomass of the trees are released as carbon dioxide, thereby further exacerbating the build-up of greenhouse gases in the atmosphere. This is another example of a positive feedback loop. The hotter it gets, the drier it gets, the more fires we have, the more carbon dioxide is released, and the hotter it gets.

change up to the climax and then cease to change. Rather, the nature of the species assemblage is more constant over time once a climax stage is reached. The climax vegetation for most areas is

controlled by the prevailing climate, and thus is known as a *climatic climax*. In some areas, other factors, such as soil conditions, may be more important than climate in determining community com-

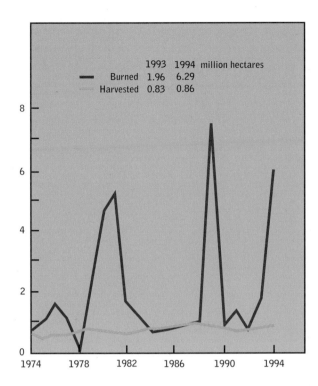

Alder is a common pioneer of secondary succession on logged-over sites. The smaller species in the foreground is broom, a non-native plant from Europe. Both these species biologically fix nitrogen into the soil (discussed in Chapter 6), and help fertility levels to recover (*Philip Dearden*).

Sand dunes are a good place to observe the successional changes over time, as shown here at Shallow Bay on the west coast of Newfoundland. With increasing distance from the sea, the communities change to those in later seral stages representing the build-up and colonization of the sand (*Philip Dearden*).

Figure 5.2 Forest area burned in Canada 1974–94. SOURCE: Canadian Forest Service, *The State of Canada's Forests 1996* (Ottawa: Natural Resources Canada, 1996):78.

position and structure. These are known as *edaphic climaxes* (see Box 5.3).

In addition to the primary succession described earlier, successional processes also occur on previously vegetated surfaces—such as abandoned fields, avalanche tracks, or burned-out areas where soil is already present. This process is known as *secondary* succession. The earlier soil-forming stages of primary succession do not have to be repeated, so the process is much shorter with the dispersal characteristics of invading species as a main factor in determining community composition. Annual weeds play a main role until perennial weeds, such as goldenrod, become established. Eventually the community will be invaded by shrubs and ultimately tree species. A major challenge for agriculture and forest managers is to prevent this natural recolonization from taking place by species that may not yield the required products. As a result, chemical herbicides (as discussed in greater detail in chapters 14 and 15) are often used to arrest secondary successional processes.

Similar kinds of processes also occur in aquatic environments. Here the natural ageing is called *eutrophication* (eu = well, trophos = feeding) as

nutrient supplies increase over time with inflow and the growth and decay of communities. The process can be relatively rapid in shallow lakes as the nutrients (one of the auxiliary energy flows discussed in Chapter 4) promote increased plant growth, which leads to more biomass and nutrient accumulation. The lake becomes shallower over time with less surface area of water, and the aquatic communities may eventually be out-competed by marsh and ultimately terrestrial plants. This process is a good example of a *positive feedback loop* (the shallower the lake gets, the stronger the forces to make it shallower), which will be discussed in more detail in the next section. Eutrophication may also constitute a significant management problem as the species being replaced often have greater values to humans than the species replacing them. This problem and some Canadian examples will be discussed in more detail in Chapter 7.

Indicators of Immature and Mature Ecosystems

As successional changes take place in communities, several trends emerge. For example, productivity declines as the slower-growing species move in. Diversity increases as more specialized species come to dominate the community and more finely subdivide the resources of a particular habitat. Thus despite the flexibility of the successional concept, certain differences between immature and mature systems are generic (Table 5.1).

In general, mature ecosystems tend to have a high level of community organization among a greater number of larger plants and have a well-developed trophic structure. Food chains are dominated by decomposers, with a high efficiency in nutrient cycling and energy use. Net productivity is low. Immature ecosystems tend to have the opposite of these characteristics.

Effects of Human Activities

Humans have a profound influence on ecological succession. Many activities keep certain communities in early seral stages. In other words, humans seek to

maintain the characteristics of the immature ecosystems shown in Table 5.1, as opposed to those of the mature ecosystems that would result if natural processes were allowed to proceed. Agriculture, for example, usually involves large inputs of auxiliary energy flows (such as pesticides) to ensure that succession does not take place as weeds try to colonize the same areas

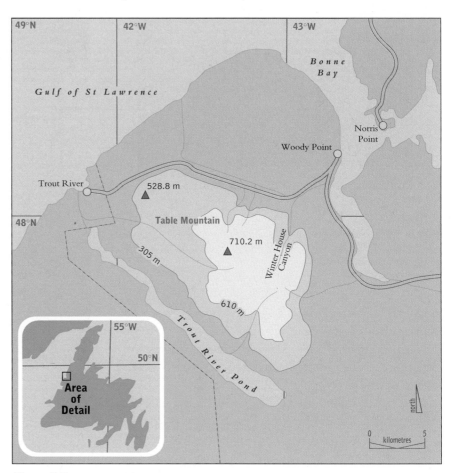

Figure 5.3 Location of Table Mountain.

Table 5.1 Characteristics of Immature and Mature Ecosystems

Characteristic	Immature Ecosystem	Mature Ecosystem
Food chains	Linear, predominantly grazer	Weblike, predominantly detritus
Net productivity	High	Low
Species diversity	Low	High
Niche specialization	Broad	Narrow
Nutrient cycles	Open	Closed
Nutrient conservation	Poor	Good
Stability	Low	Higher

Source: Modified from E. Odum, 'The Strategy of Ecosystem Development', *Science* 164 (1969):262–70. Copyright 1969 by the American Association for the Advancement of Science.

Box 5.3 Edaphic Climax: Table Mountain, Newfoundland

The west coast of Newfoundland, like most of the rest of the island, is dominated by the boreal forest, as described in Chapter 2. In Gros Morne National Park (Figure 5.3), however, and at other locations on the west coast, this greenery (white spruce, paper birch, balsam fir) is interspersed with practically treeless, orange-coloured outcrops that bear little if any similarity to the surrounding vegetation. These outcrops are the result of the distinctive chemical composition of the bedrock known as serpentine. Along with three other serpentine outcrops in western Newfoundland, the Table Mountain massif in Gros Morne was formed on the floor of the Atlantic Ocean millions of years ago and raised to its present position through the process of continental drift, as described in Box 3.7.

This geological history has given the serpentines a distinctive chemical composition characterized by high levels of nickel, chromium, and magnesium and low levels of calcium. Most of the climax species of the surrounding forests cannot tolerate these conditions; if they grow at all, they are stunted. Instead the serpentines are host to communities of tough arctic-alpine species that have survived since the retreat of the glaciers. They were not displaced through succession like many species on the surrounding bedrock (Dearden 1979). These serpentine communities are edaphic climaxes in which the underlying geology has been more important than climate in determining plant cover.

The difference between the dominant vegetation of the edaphic climax of Table Mountain and the surrounding boreal forest can be clearly seen along the geological boundary (*Philip Dearden*).

The inhospitable soil chemistry has allowed rare species, such as this *Lychnis alpina*, to continue to grow in the area as relicts from the ice age (*Philip Dearden*).

being used to grow crops. The same can increasingly be said for commercial forestry.

Maintaining ecosystems in early successional stages has several implications:

- As discussed earlier, the productivity of early successional phases is often greater than that of later phases.
- Nutrient cycling (to be discussed in more detail in the next chapter) is often more rapid in early stages. Trees, for example, not only hold nutrients in their mass for a longer time

than herbaceous plants but also protect the soil from high temperatures. High temperatures result in more rapid breakdown of organic material and release of nutrients into the environment. Plants' absorption and storage of nutrients are also much reduced. Consequently, disturbance of an ecosystem may result in a significant loss of nutrient capital from a site. There is usually a higher nutrient concentration in the soil water and more water leaving the disturbed ecosystem in streamflow than in the previously undisturbed ecosystem.

- Biodiversity overall tends to be reduced.
- The species most adversely affected are often highly specialized ones at higher trophic levels.
- The species that benefit the most are usually pioneer species (weeds and pests) that have broad ranges of tolerance and efficient reproductive strategies for wide dispersal.

ECOSYSTEM HOMEOSTASIS

In the early 1970s, residents of the Okanagan Valley in British Columbia began to complain to the government about excessive weed growth in some

Signs warning of the spread of Eurasian water milfoil were placed at boat loading ramps throughout British Columbia, but did little to stem the colonization (*Philip Dearden*).

In the tourist economy of the Okanagan Valley where resorts rely on water-based activities to attract clientele, there was considerable conflict among different stakeholders regarding the most appropriate means of controlling the spread of milfoil (*Philip Dearden*).

of the lakes in the valley. Several popular beaches were becoming virtually unusable due to the weeds, which spread rapidly. This was of considerable concern to the residents not only because their recreational activities were affected but also because of the impact on the economy of this tourist area in which water-based recreation was the main attraction.

Eurasian water milfoil had arrived. Over the next couple of decades it would spread not only to all the lakes in the Okanagan but also to many other lakes in southern BC and other provinces. The government spent large amounts of money trying to control the spread of the species, but to no avail. Originating in Eurasia, the milfoil had reached the eastern shores of North America probably a century ago and has since spread across the continent, replacing native aquatic plants in many water bodies.

This biophysical event, the spread of a Eurasian plant into North America, also illustrates the dynamic relationship among the biophysical, socio-economic, and management systems that is the main focus of this book. In BC, for example, the dependence of local economies, such as that of the Okanagan Valley on water-based tourism, resulted in a strong response to milfoil that involved the use of the chemical 2,4-D. This created considerable conflict (see Chapter 12) between different stakeholders (Chapter 11) regarding the relative impacts (Chapter 10) of the plant versus the control mechanism. Critics claimed that management had failed to consider the broader perspectives that would have been included with an ecosystem approach to the problem (Chapter 8), and that management failed to adapt to the changing parameters of the situation (Chapter 9).

Situations such as this are not uncommon. We tend to think of ecosystems as having relatively constant characteristics, of being in a balance where internal processes adjust for changes in external conditions. This is known as *ecosystem homeostasis* and implies not a static state but one of *dynamic equilibrium*. James Lovelock (1979, 1988), in the Gaia hypothesis, postulated that the ecosphere itself is a self-regulating homeostatic system in which the biotic and abiotic components interact to produce a balanced, constant state (Box 5.4). This is an example of a highly *integrated* system in which there is a strong interaction between the different parts of the system. Other systems may not be so highly dependent upon one another. Cells in a colony of single-celled organisms, for example, may be removed and have little effect upon the remainder due to the low integration of the system.

Not all ecosystems are equal in their abilities to withstand perturbations. *Inertia* is an ecosystem's ability to withstand change, whereas *resilience* is the ability to recover to the original state following disturbance. Some ecosystems can have low inertia and high resilience or any other possible combination. In terms of human usage, it is best to work with systems that have both high inertia and high resilience. This means that they are relatively difficult to disturb, and even when disturbed they will recover quickly. Such systems are relatively stable. The best growth sites for forestry (alluvial sites in nutrient-rich areas at low elevations) would fit into

this category. Many tropical and Arctic sites would fit into the opposite combination where sites are easily disturbed and recover only very slowly, if at all.

Ecosystems have various attributes—such as feedback, species interactions and population dynamics—which maintain this equilibrium. Occasionally, as in the case with the milfoil described earlier, these mechanisms cannot cope with a perturbation and radical change can occur. In the case of the milfoil, this non-native plant replaced a variety of native aquatic species. Similarly, other non-native invaders—such as the purple loosestrife, sea lamprey, and zebra mussels—are not affected by the normal control factors in their habitats and violate ecosystem homeostasis.

Alien or Invader Species

Canada has felt the impact of thousands of non-native plants introduced into the country. It is estimated that over 500 species of alien plants in Canada have developed into agricultural weeds that cost farmers millions of dollars every year to control, such as the various species of knapweed introduced to Canada and the US from the Balkan states, probably in shipments of alfalfa. The diffuse knapweed causes the most problems; it has a wide range of tolerance and a very effective seed dispersal system, which it has used to colonize vast areas of range land in western Canada. It is also thought

---

Box 5.4 The Gaia Hypothesis

As humans probed deeper and deeper into space and looked back at earth from this unique perspective, it became increasingly clear that our planet was significantly different from all the millions of others. It seems like a happy coincidence that of the vast range of temperatures that could be experienced, those on earth are just right for life, between the freezing and boiling points for water, even though the energy output of the sun has increased by over 30 per cent during the last 3.6 billion years. The gaseous composition of the atmosphere is also just what we need to breathe.

The Gaia hypothesis, named after the ancient Greek goddess of the earth by James Lovelock, the originator of the idea, postulates that the earth is one giant self-regulating superorganism

that maintains these conditions necessary for life. Organisms act like cells in a body, with integrated functions to promote the health of that body, or, in this case, optimum conditions for life. Active, automatic feedback processes among the atmosphere—lithosphere, hydrosphere, and ecosphere—maintain this homeostasis.

In reading this section of the book, you should gain some appreciation of these complex interactions. There is no doubt that since the beginning of life on earth, organisms have not only adapted to existing conditions but also modified them in ways that are beneficial to life. However, most scientists do not believe that the earth will adapt and compensate in ways that are beneficial to humans for all changes that are now being caused by human activity.

A roadside sign describes the knapweed problem (*Philip Dearden*).

that this species may be *allelopathic*; that is, it can directly inhibit the growth of surrounding species by producing chemicals in the soil. The species displaces native species and considerably reduces the rangelands' carrying capacity. Cattle will eat it only as a last resort, and the nutritive content is less than 10 per cent of that of the displaced native species. Initial control efforts relied on chemical sprays. A more integrated approach now involves using biological control and limiting its spread through stricter controls on vehicular access to rangelands, which is one of the main means of seed distribution.

Besides plants, many other species have also proved troublesome. Two fungi, chestnut blight and Dutch elm disease, for example, have had significant impacts upon the landscape of eastern Canada. Both attack native trees that were once significant parts of the eastern deciduous forests. The American chestnut was attacked by an Asian pathogenic fungus introduced on stocks of Japanese chestnuts in the last century, and the elm by a European fungus transmitted among trees by beetles. Researchers at the University of Manitoba and the University of Toronto are now exploring the potential of using a natural toxin called man-

sonone, which occurs in some elms, to breed seedlings that are more resistant to Dutch elm disease.

One notable non-native invasion is that of the zebra mussel, which has spread into the Great Lakes. The mussel, named for its striped shell, joins numerous other alien species in the Great Lakes, including the sea lamprey, alewife, and rainbow smelt. It is thought that the mussel, a native of the Black and Caspian seas, was introduced from the ballast of freighters in the mid-1980s. It was first found in 1988 in a sample of aquatic worms collected from the bottom of Lake St Clair. Evidence from Europe indicated that the species was an aggressive colonizer able to displace most native species.

The mussels start their three- to six-year life span as free-swimming larvae before attaching themselves to a hard surface, usually in the top 3–4 m of water, although they can live as deep as 30 m. By the end of 1988 the mussels had colonized half of Lake St Clair and two-thirds of Lake Erie at densities as high as 30,000 per square metre. The mussels have now spread extensively through the lakes where they appear capable of colonizing on any hard surface. They have encrusted water intake and discharge pipes, severely reducing their efficiency and necessitating significant expense to remove the mussels. Water intake may be reduced by as much as 50 per cent. Many different approaches are being tried to screen out the mussels, but they seem to pass through most physical barriers. At the moment chlorination is the most common measure, but this also creates the potential problem of toxic organochlorine formation. The mussels also colonize fish spawning sites for other fish, with as yet undetermined impacts on their populations.

Population levels of other species are likely to be more indirectly affected through the food chain. The mussels are filter feeders that remove phytoplankton from the water, thereby affecting all the species higher in the food chain, such as walleye, bass, trout, and perch. In some European locations, the mussels' invasion has led to clearer water as a result of their feeding on the phytoplankton. These changes may also benefit some species, even fish species. Bottom feeders (such as carp and whitefish) and invertebrates (such as crayfish) may benefit as more nutrients are returned to the lake bottoms, either in the form of dead mussels or mussel faeces. It remains to be seen whether species higher in the food chain, such as scaup and other waterfowl, can help control the spread of the mussel. Already the numbers of some of these species stopping to feed during their migration appear to have risen considerably. Although the ducks may have some impact on the mussel population, as they have in Europe, the infestation may be just too large and the number of ducks too small to control the problem.

If controlling the zebra mussels seems impossible, what about trying to ensure that the same kind of problem doesn't arise again? Scientists have known for a long time that ballast water was a major threat to aquatic communities. In 1980 a sampling of fifty freighters entering the Great Lakes uncovered fifty-six different species of exotic aquatic invertebrates, as well as over 100 phytoplankton. A more recent study of Japanese freighters in Oregon found 367 taxa of organisms. Given the magnitude of these introductions, it is inevitable that a few of these species will have a population explosion. The coastguard now asks freighters to exchange their ballast water for seawater before entering the St Lawrence. Unfortunately, as is so often the case, the program is voluntary, there is no monitoring to encourage compliance, and it is geographically restricted. Scientists are now considering ways to kill unwelcome organisms by chemicals or heat before they are released. A satisfactory solution has yet to be found.

Species Removal

It is not only the introduction of species to new habitats that can disturb ecosystem homeostasis but also the removal of species from food webs. The reduction of some species, such as the keystone species discussed in the last chapter, may be particularly disruptive. One well-known example relates to the extirpation of the sea otter from the Pacific coast.

When James Cook anchored at Nootka Sound on the west coast of Vancouver Island in 1778, he reported that the fur of the sea otter 'is softer and finer than that of any others we know of; and, therefore, the discovery of this part of the continent of North America, where so valuable an article of commerce may be met with, cannot be a matter of indifference'. Indeed, it was not. The British, seeking trading goods to barter with the Chinese in exchange for tea, discovered that sea otter pelts were in great demand in China, and therefore made every effort to ensure that the West Coast became British (rather than Spanish or Russian) Columbia!

The sea otter is a large aquatic weasel-like mammal of the outer coasts that thrives in the giant kelp beds. They lack a protective layer of blubber, but have very fine fur that effectively traps air and insulates them from the cold Pacific waters. They also need lots of food (up to 9 kg per day) to fuel the fast metabolism that counteracts energy loss to the environment. Favourite prey are sea urchins, crabs, shellfish, and slow-moving fish.

The otters were easy to catch and Russian and Spanish hunters, aided by local Native populations, finished off what the British had begun. Within forty years, populations were reduced from over half a million to 1,000–2,000. It is likely that they were completely extirpated off the coast of British Columbia. However, there were relict populations to the south around Monterey, and to the north in the Aleutian Islands. Otters from this latter population have now been reintroduced to the coast of British Columbia, where small but vibrant populations now thrive again.

A sea otter off Vancouver Island (*World Wildlife Fund Canada*/© *Mark Hobson*).

Scientists discovered the otters' key role in maintaining ecosystem homeostasis after studying two groups of islands off Alaska. They noticed that although the islands were very similar in terms of location and physical conditions, one group of islands had much more life—bald eagles, seals, kelp beds, and otters—than the other. Otters play a critical role in controlling sea urchin populations (Estes et al. 1989). Sea urchins are voracious eaters of kelp (large, brown seaweed), which may be the world's fastest growing plant, with increments of up to 60 cm per day. Given the support of the ocean, kelp does not need to invest much energy in developing heavy support structures, leaving more energy for growth.

Kelp plays a major role in coastal communities. It provides food and habitat for many other species. Diatoms, algae, and microbes grow on the fronds of the kelp, along with colonies of filter-feeding bryozoans and hydroids. Predators abound. Fish come to feed off the colonists or seek protection from open-water predators, such as seals, sea lions, and killer whales. When kelp is overgrazed by sea urchins, this productive habitat disappears. The urchins eat through the holdfasts that anchor the kelp to the ocean floor and the kelp is soon washed away into the open ocean or onto land. As the kelp disappears, so do all the species dependent upon it. The otter population on one of the two islands in Alaska had managed to escape the fur hunters, who eliminated otters elsewhere. This island had the kind of rich coastal community that should extend all along the outer coast of the north Pacific. Otter populations are hence critical in maintaining the productivity of the entire community, including bald eagle populations, through their control of the urchin population. The fact that the fashion tastes of Chinese mandarins 200 years ago, met by traders from the other side of the world who wanted to enjoy afternoon tea, is still reflected in bald eagle populations 5000 km away on the BC coast indicates the complex interactions between biophysical and human systems that are the focus of this book.

Feedback

Feedback is an important aspect of maintaining stability in ecosystems whereby information is fed back into a system as a result of a change. Feedback initiates responses that may exacerbate (*positive feedback*) or moderate (*negative feedback*) the change. There is, for example, considerable debate regarding the role of feedback loops in global change, as discussed in more detail in Chapter 7. One positive feedback loop that may have a strong influence in Canada is the effect of increased temperatures in the North. This would increase the area of land that is free of snow in summer. Snow has a high *albedo*; in other words, it reflects rather than absorbs much of the incoming radiation. As temperatures rise, the area covered by snow will be replaced by areas covered in rocks and vegetation, which have lower albedo values. This will cause more heat to be absorbed, which in turn will contribute to global warming. A similar contribution of forest fires was noted in Box 5.2.

Negative feedback loops may counteract such positive feedback loops. One example is the role of phytoplankton in global warming. Phytoplankton produce a gas called dimethyl sulphide. When seawater interacts with this gas, sulphur particles are formed in the atmosphere, which serve as condensation nucleii for cloud droplets. As the planet heats up, the productivity of the phytoplankton should increase, leading to an increase in the production of gas and cloud droplets. This will increase cloud cover and reflect solar radiation, leading to a cooling of the earth and the maintenance of a dynamic equilibrium.

Almost all the examples in this chapter can be used to illustrate some aspect of feedback mechanisms. The allelopathic qualities of the diffuse knapweed, for example, show a positive feedback loop that helps spread the species. The more the species spreads, the more conditions are created into which only it can spread. The sea otters provided a negative feedback loop on the sea urchin-kelp relationship. If the urchins become too numerous and overgraze the kelp beds, then increases in otter populations would help reverse this change. When this negative feedback loop was removed from the system, there was nothing to maintain its dynamic homeostasis.

Similar examples of feedback loops occur at all scales, even down to the regulation of temperatures in individual organisms. Sometimes these feedback messages can be rapid, as in the case of organism thermoregulation. In other cases, considerable delays can occur between the stimulus for change and the resulting feedback response. Unfortunately, as the example of the positive feedback loop and snow melt described earlier indicates, sometimes the delay between the stimulus and the response may be so long that we are not aware of it. By the time we are aware, it may be too late to moderate the stimulus and a powerful positive feedback loop may already, albeit slowly, have been set in motion. This is one reason why many scientists support immediate actions to reduce the emission of greenhouse gases (see Chapter 20), even though we do not yet have a clear understanding of all the relationships involved.

We are also becoming more aware of the chaotic nature of many systems whereby a slight perturbation becomes greatly enhanced by positive feedback. The so-called butterfly effect, for example, traced how the turbulence of a butterfly flapping its wings in South America might, through cascading effects on air flows, influence the weather in North America. Further research has revealed the existence of similar phenomena in many different systems where very small changes in systems can have great influence on outcomes. *Chaos theory* tries to discern pattern and regularity in such systems and allow for greater predictability.

Synergism

Synergism is also an important characteristic that may influence change in ecosystems. A *synergistic* relationship occurs when the effect of two or more separate entities together is greater than the sum of the individual entities. For example, in Chapter 7 some attention is devoted to the problem of acid deposition. The effects of acid deposition are often exacerbated by the presence of other pollutants, such as ground-level ozone. Individually, both these forms of pollution may cause a certain amount of damage to an ecosystem. However, when combined, their effects are magnified.

At an individual scale, we can see the same effects for smokers. The potential for getting lung cancer is increased by both smoking and the inhalation of radon gas. However, if someone who smokes also inhales radon gas, the chances of getting lung cancer are greater than the sum of the individual exposures to each hazard. This is because the particles of tobacco ash that accumulate in the lungs as a result of smoking also act as anchors for radon particles. Together, the two exposures are far more dangerous than the sum of the individual exposures.

POPULATION GROWTH

The number of individuals of a species is known as the *population*. When calculated on the basis of a certain area, such as the number of sea otters per hectare, it reflects the *population density*. The number of organisms in a population is a very important characteristic. Low numbers will make a species more vulnerable to extinction. Changes in population characteristics are known as *population dynamics*.

Populations change as a result of the balance among the factors promoting population growth and those promoting reduction. The most common response is through adjustments in the birth and/or death rates to the factors shown in Figure 5.4, although emigration and immigration can be important factors in some species. Population change is calculated by the formula $I = (b - d)\,N$, where I is the rate of change in the number of individuals in the population, b is the average birth

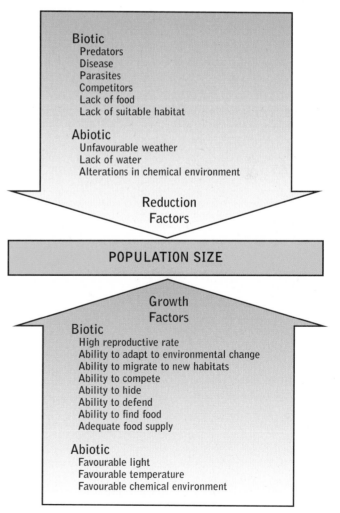

Figure 5.4 Factors affecting population growth.

rate, *d* is the average death rate, and *N* is the number of individuals in the population at the present time. As long as births are greater than deaths, the population will increase exponentially over time (Figure 5.5) until the environmental resistance of the factors shown in Figure 5.5 start to have an inhibiting effect that will flatten out the curve.

The *carrying capacity* of an environment is the number of individuals of a species that can be sustained in an area indefinitely, given a steady resource supply and demand. Most species will grow rapidly in numbers up to this point and then fluctuate around the carrying capacity in a dynamic equilibrium (Figure 5.6). The carrying capacity is not one fixed figure, however, but will vary along with changes in the other abiotic and

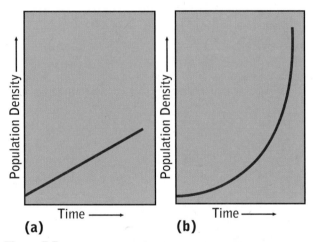

Figure 5.5 Geometric (a) and arithmetic (b) growth patterns.

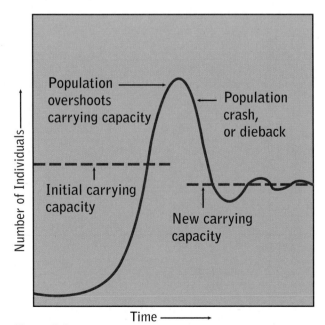

Figure 5.6 Carrying capacity and population growth rates.

biotic parts of the ecosystem. In the puffin–capelin example in the introduction to Chapter 4, the carrying capacity of the north Atlantic waters to support the puffin population was severely reduced due to a reduction in their food supply, which was caused by competition from humans. Management inputs, such as providing supplementary feeding or other habitat requirements, are often used to change an area's capacity to meet human demands.

Organisms that demonstrate the kind of S-shaped growth curve of Figure 5.6 are *density dependent*, and as the population density increases, the rate of population growth decreases. Not all organisms share this characteristic. Some are *density independent*, and the population operates with a positive feedback loop: the more individuals there are in the population, then the more that are born and the population grows at an increasing rate in a J-shaped curve. At some point, this population meets environmental resistance, which causes the population to crash back to, or below, the carrying capacity. The algae that bloom on ponds in the late spring or early summer are a result of this kind of growth. In reaction to the increased nutrient availability after winter, the algae grow at a great rate until their food supply is exhausted and the algae population crashes.

In some locations in Europe, this is what has happened with the zebra mussels discussed earlier. Mussel populations have peaked and stayed there for a few years before exhausting the food supply and crashing to 10–40 per cent of the original numbers. However, there are other locations, such as Sweden, where the expected crash has yet to occur. Given the enormous food supplies of the Great Lakes and the low numbers of predators, such as waterfowl, it may be a very long time before any natural mussel population crash happens.

The capacity of species to increase in number is known as their *biotic potential*, the maximum rate at which a species may increase if there is no environmental resistance. Different species, however, have different reproductive strategies. Some species, known as *r-strategists* (such as the zebra mussel), produce large numbers of young starting early in life and over a short period, and invest little parental energy in their upbringing. Most of their energy is spent on reproduction and they have few resources left to devote to maintaining a longer life span. Such species are usually small,

These jellyfish and Pacific white-sided dolphins are good examples of marine r and K species respectively (*Philip Dearden*).

short-lived, and respond to favourable conditions through rapid reproduction. They are opportunists and their reproductive strategy is essentially based upon *quantity*. Such species tend to dominate the early seral stages of the successional process.

K-strategists, on the other hand, produce few offspring but devote considerable effort into ensuring that these offspring reach maturity. Their strategy is based upon quality. Individuals live longer and are usually larger. Often populations of K-strategists reach the carrying capacity of an environment and are relatively stable compared to r-strategists, which may experience large variation in population size. The characteristics of these different strategists are summarized in Table 5.2. Typical examples of r-strategists include insects, rodents, algae, annual plants, and fish. A mature codfish, for example, may produce over 9 million eggs in one season. However, fewer than 5 per cent of these may mature and survive their first year. Most K-strategists are larger organisms, such as the larger mammals (including humans). Their lower biotic potential and ability to disperse often means that they are more restricted to the later seral stages of succession. Many endangered species (see Chapter 16) are K-strategists. The great whales (Box 3.12), which have perhaps only one offspring every three years, are good examples. When the conditions to which they have become accustomed and under which they evolved their reproductive strategy change dramatically—for example,

if new predators, such as humans, appear—they have little capacity to increase their reproductive rate.

EVOLUTION, SPECIATION, AND EXTINCTION

Over the long term, populations adapt to changing conditions through *evolution*, a change in the genetic make-up of the population with time. Within any population there will be some variation in the genetic composition that may predispose a certain segment of the population to adapt to certain conditions. If conditions change to favour those variations, then the success of those that are genetically better adapted to the new conditions will be improved. In this way, over time natural selection can lead to changes in the characteristics of a population. *Phyletic evolution* is the process in which a population has undergone so much change that it is no longer able to interbreed

Table 5.2 Characteristics of K-Strategists and r-Strategists

K-Strategists	r-Strategists
Few, larger young	Early reproductive age
More care of young	Many small young
Slower development	Little care of young
Later reproductive age	Rapid development
Greater competitive ability	Limited competitive ability
Longer life	Short life
Larger adults	Small adults
Live in generally stable environments	Live in variable or unpredictable environments
Emphasis on efficiency	Emphasis on productivity
Stable populations usually close to its carrying capacity	Large population fluctuations usually far below carrying capacity

with the original population, and a new species is formed.

Genetic diversity helps protect species from extinction. The resilience of a species depends partly on the magnitude of the environmental change, how rapidly it takes place, and the capacity of the species's gene pool to respond to these changes. In general, the broader the gene pool, the greater the capacity to adapt to change.

It is not only changes in the abiotic environment that promote evolutionary change. Species may also change through *co-evolution*, in which changes in one species leads to changes in another. Each species may become an evolutionary force affecting the other. For example, a prey species evolves to be more effective in avoiding a predator. In turn, the predator may evolve more efficient hunting techniques to detect the prey. Many such relationships have evolved in the tropical forests, especially between specific plants and animals, due to the long period of evolutionary change in such environments. However, Canada also has many examples, particularly relating to pollination, where various insects, birds, and bats have evolved to pollinate flowering plants, resulting in a great diversity of plant shapes, sizes, and colours.

Evolution results in the formation of new species as a result of divergent natural selection responding to environmental changes. This is the process of *speciation*. It occurs most often when members of the same species become geographic-

ally isolated so that they can no longer interbreed. If conditions differ in the respective environments of the different breeding groups, then over time natural selection will favour those individuals that adapt the best to those conditions. Thus it is thought that the polar bear evolved as a separate species from the grizzly bear some 10,000 years ago. Bears with characteristics that aided them in hunting seals on ice flows, such as a lighter-colour fur and greater strength, would be relatively more successful in the far North as opposed to elsewhere in the range where a brown pelt and greater mobility might be an advantage. In this way, a single bear species became two bear species through adaptation to different environments and the process of natural selection. This process of local adaptation and speciation is known as *adaptive radiation*.

Extinction, the opposite of this process, is the elimination of a species that could no longer survive under new conditions. The fossil record leads scientists to estimate that perhaps almost 99 per cent of the species that have lived on earth are extinct. The fact that we still have up to 90 million species, more than ever before, indicates that speciation has exceeded the extinction level. However, speciation takes time. Even with r-strategists it may take hundreds and thousands of years; with K-strategists this may be extended to tens of thousands of years. Evidence suggests that human activities have tipped the scale recently in favour of extinction over speciation, as will be discussed in more detail in Chapter 16. Table 5.3 gives some examples of species that once existed in Canada but that are now extinct.

Extinction, like speciation, should not be considered just a smooth, constant process; it is interrupted by relatively sudden and catastrophic changes. Multicellular life, for example, has experienced five major and many minor mass extinctions. Scientists think that the age of the dinosaurs, a remarkably successful dynasty that effectively relegated mammals to minor ecological roles for over 140 million years, was brought to an end 65 million years

Polar bears feed mostly on seals, but will also eat birds and their eggs, as well as vegetation (*National Film Board of Canada*).

Box 5.5 The Burgess Shales

Stephen Jay Gould, the famous Harvard paleontologist, calls the Burgess Shales in Yoho National Park in British Columbia the single most important scientific site in the world because of the extensive bed of fossils on the flanks of Mount Wapta. They are fossils from the Cambrian era, some 530 million years ago, when diverse life forms flourished. The site's fossils from this era are preserved in great detail, even down to the soft body parts, such as stomach contents.

The story revealed by this detail is one of great diversity during which all but one phylum of animal life made a first appearance in the geological record. The site also contains many body patterns for which there are no current counterparts. Thus it seems as if life could be characterized as 3 billion years of unicellularity, followed by this enormously diverse Cambrian flowering in a brief 5 million-year period, and a further 500 million years of variations on the basic anatomical patterns set in the Cambrian era. Why or how this flowering took place is uncertain. There seems to be a combination of explanations. First, there was literally an open field available for colonization, an environment ripe to support life, but with little life in it. Therefore species did not have to be particularly good competitors to survive. Virtually anything could survive. Since this time, even after mass extinctions, sufficient species have remained to make it pretty tough competition for any newcomers. Second, it seems as if the early multicellular animals must have maintained a flexibility for genetic change and adaptability that declined as greater specialization arose and organisms concentrated on refining the successful designs that had already evolved. Furthermore, we have little idea why most of these early experiments in life died out and yet others remained. There seems to be no common traits shared by the survivors to indicate that they were the victors of Darwinian strife. Perhaps just the lucky ones survived.

In his fascinating book, *Wonderful Life*, Gould suggests that this challenges many of our established views of evolution as an inevitable progression from the primitive and few to the sophisticated and many over time. It also radically challenges our view of ourselves as the logical end-point of evolutionary change, the rightful inheritors of the world. In Gould's own words: 'If humanity arose just yesterday as a small twig on one branch of a flourishing tree [of evolution], then life may not, in any genuine sense, exist for us or because of us. Perhaps we are only an afterthought, a kind of cosmic accident, just one bauble on the Christmas tree of evolution' (1989:44). In other words, we should be humble!

The site of the Burgess Shales World Heritage Site on the slopes of Mount Wapta in Yoho National Park (*Philip Dearden*).

ago by the impact of a large extraterrestrial object, and then the mammals took over. This chance occurrence not only led to the demise of the cold-blooded dinosaurs but favoured the survival of the rodentlike mammals with their smaller body size, less specialization, and greater numbers. Small body size was probably a sign of the mammals' inability to challenge the dinosaurs during the normal evolutionary process, but became a positive feature favouring survival under the new conditions.

There are other examples of this non-random impact of mass extinctions on life. The features that

Box 5.6 Humans and Extinction

Extinction is a natural process. Scientists can try to estimate the average rate of natural species extinction by examining the fossil record. This suggests that extinctions among mammals might be expected to occur at the rate of about one every 400 years, and among birds, one every 200 years. Current extinction rates are difficult to estimate because we do not have a full inventory of all the species that might exist, so we don't know what we are losing. Based upon current rates of habitat destruction for tropical forests, figures of over 100,000 extinct species per year are often quoted. Many of these species would be unidentified arthropods, as these comprise the majority of species in tropical forests. However, even more conservative estimates suggest that about 2–8 per cent of the planet's species will become extinct over the next twenty-five years, many times what might have been expected in prehuman times. Although there may be disagreements over the rate of extinctions, there is widespread consensus that humans have vastly increased the rate over natural extinction rates. In 1996 the most comprehensive evaluation of species status ever attempted was completed by the World Conservation Union (IUCN), incorporating the work of over 7,000 scientists. The report concluded that one-quarter of the world's species of mammals are threatened with extinction, and that half of these may be gone within the next decade.

Human impacts on biodiversity will be discussed in more detail in Chapter 16 on endangered species. However, evidence suggests that humans have been having a major impact upon biodiversity for quite some time. Paul Martin (1967), for example, was one of the first to suggest that humans may have been a major factor in causing the extirpation of several species of large mammals from North America at the end of the last ice age some 10,000 years ago. At this time at least twenty-seven genera comprising fifty-six species of large mammals, two genera and twenty-one species of smaller mammals, and several large birds became extinct. The extinctions included ten species of horses, four species of camels, two species of bison, a native cow, four elephant species, the sabre-toothed tiger, and the American lion. Although this was also a time of global climatic change, no such extirpations were associated with the same time period in Eurasia. This period also saw a substantial in-migration of humans from the Asian continent, who began to prey upon animals unfamiliar with, and therefore not adapted to, human hunting. This hunting, combined with environmental stresses and repercussions through the food chain, were sufficient to extirpate the species.

Charles Kay has also conducted research into the subsequent impact of Native Americans on ungulate populations before the arrival of European influences. He concludes (Kay 1994) that even then humans were the main limiting factor on ungulate populations in the inter-mountain west, and that elk, in particular, were overexploited. The people had no effective conservation strategies and hunted to maximize their own welfare, regardless of environmental impacts. Thus the image of North America as a vast wilderness unaffected by human activities before the coming of the Europeans appears to be a myth. Even that mightiest symbol of the wild, the grizzly bear, was apparently under pressure from Aboriginal hunters in Alaska (Birkedal 1993).

It seems difficult to believe that what we now consider primitive weapons, such as this fishing spear from the Warao people of the Orinoco delta, may have enabled humans to hunt many other species to extinction (*Philip Dearden*).

Table 5.3 Some Canadian Vertebrate Species That Are Now Extinct

Species	Distribution	Last Recorded	Probable Causes
Great auk (*Alca impennis*)	Canada, Iceland, UK, Greenland, Russia	1844	Hunting
Labrador duck (*Camptorhynchus labradorius*)	Canada, US	1878	Hunting, habitat alteration
Passenger pigeon (*Ectopistes migratorius*)	Canada, US	1914	Hunting, habitat alteration
Deepwater cisco (*Coregonusjohannae*)	Canada, US	1955	Commercial fishing, introduced predators
Longjaw cisco (*Coregonus alpenae*) (*Great Lakes*)	Canada, US	1978	Commercial fishing, introduced predators

make some species successful during normal times may not enable them to adapt successfully to the new conditions, making life a somewhat chaotic and unpredictable process rather than the smooth path that evolutionary theory might suggest.

Finally, we should not (as emphasized in Box 5.5 on the Burgess Shales) think that evolution is fundamentally a story that demonstrates the benefits of complexity over simplicity. Humans, the most complex organisms, are *not* the pinnacle of life's achievement, nor the most successful. That distinction is accorded to the other end of the complexity spectrum, the bacteria. From the beginning of the fossil record until present times, bacteria have provided the most stable presence. According to Stephen Jay Gould (1994:87):

> Bacteria represent the great success story of life's pathway. They occupy a wider domain of environments and span a greater range of biochemistries than any other group. They are adaptable, indestructible and astoundingly diverse. We cannot even imagine how anthropogenic intervention might threaten their extinction, although we worry about our impact on nearly every other form of life. The number of *Escherichia coli* cells in the gut of each human being exceeds the number of humans that has ever lived on this planet.

Dinosaur Provincial Park in Alberta is a World Heritage Site that holds the world record for the number of different species of dinosaurs found there (*Philip Dearden*).

BIODIVERSITY

Over billions of years, interaction between the abiotic and biotic factors on life through the process of evolution has produced many different life forms. The main classification is the *species*, life forms that resemble one another and can interbreed successfully. Two species are created from one as a response of natural selection to changes in environmental conditions, as explained in the previous section. Biodiversity is the sum of all these interactions, and is usually recognized at three different levels:

- *Genetic Diversity*: The variability in genetic make-up among individuals of the same species.

- *Species Diversity*: The total number of species in an area.
- *Ecosystem Diversity*: The variety of ecosystems in an area.

Scientific knowledge of biodiversity is very primitive. There may be over 100 million species, although most scientific estimates suggest between 3–30 million, of which we have identified some 1.8 million (Figure 5.7). Some 56 per cent of these are insects, and 14 per cent are plants, with vertebrates such as mammals, birds, and fish comprising just 3 per cent. Even new mammals are still being discovered, such as the giant muntjac and saola discovered recently on the borders of Vietnam and Laos. However, most species awaiting discovery are probably invertebrates, bacteria, and fungi from the tropics. We also know relatively little about the ocean. Only about 15 per cent of described species are from the oceans and most biologists agree that there are fewer species to be found there than on land. On the other hand, there are thirty-two phyla in the oceans, compared with only twelve on land. Species identification is only the first building-block in biodiversity. We also need to understand the differences in genetic diversity within species, and how species interact in ecosystems to understand how the life-support systems of the planet work.

Biodiversity in Canada

Biodiversity is not evenly distributed around the world. Some biomes, mainly tropical forests, are extremely diverse (Box 5.8), while temperate latitudes are much less so. Overall, as discussed in Chapter 4, species numbers decline in a gradient from the tropics to the poles. Latin America, for example, is home to over 85,000 plant species. North America has 17,000, of which only 4,000 are in Canada. A similar gradient for birds is shown in Table 5.4. Several reasons have been advanced to account for the latitudinal gradient in species richness. Following a review of these different reasons, Rohde (1992) concludes that many factors contribute at different scales, but the primary cause appears to be the effect of solar radiation (i.e., temperature), which increases evolutionary speed at lower latitudes.

It is estimated that Canada has over 71,000 different species, including flora, fauna, and micro-

---

Box 5.7 Counting Critters

Since actually discovering and describing new species is a very slow process, scientists have devised several means to help estimate just how many species we might have on earth.

1. Species–area curves are one of the most popular approaches. As an area increases in size, the number of species there also increases, but the number of new species gradually levels off as the same species are encountered repeatedly. This relationship has enabled scientists to construct species–area curves that show the number of species likely to be found in areas of different size and hence predict how many species might be found in larger unsampled areas.
2. Most of the species we have yet to describe are probably rainforest insects. Studies indicate that there are up to 1,200 beetle species in the canopy of a single tree in Peru. Since we know that 40 per cent of insects are beetles, this suggests over 3,000 species in the canopy alone. About half this number will be found lower down and on the trunk, leading to estimates of 4,500 species on one tree alone. Given that there are over 50,000 species of tropical trees, the numbers that may be found in total will be vast. Some researchers predict that there are over 30 million insect species.
3. Ecological ratios can be used to predict the populations of little-known groups from their relationships with better-known groups. For example, the ratio of fungus to plant species in Europe is 6:1. If this holds worldwide, then there should be over .6 million species of fungus. At the moment, under 70,000 have been described.

None of these approaches (and these are just three of the more popular ones) can deliver an accurate estimate of the numbers of species on the planet. Each one, however, is consistent in indicating that we have a long way to go before we can claim that we really know the nature of life on this planet.

Number of Living Species Known and Estimated: World

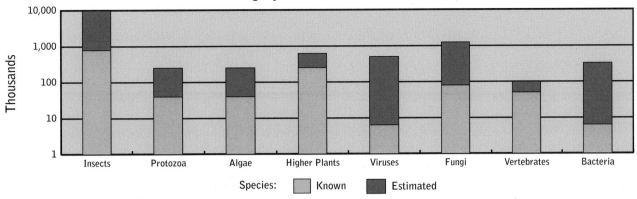

Number of Living Species Known and Estimated: Canada

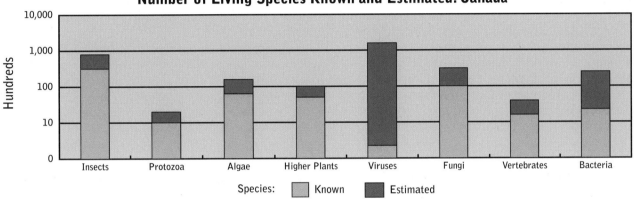

Figure 5.7 Numbers of known and estimated living species in the world and in Canada. SOURCE: B. Groombridge, *Global Biodiversity: Status of the World's Living Resources* (London: Chapman and Hall, 1992):17).

Box 5.8 The Tropical Forests

Charles Darwin, who is attributed with developing our ideas on the mechanisms for evolution in his famous book, *The Origin of Species*, published in 1859, originated most of his ideas in the tropics. He was astounded by what he saw: 'nothing but the reality can give any idea how wonderful, how magnificent the scene is.' It was here, a place of high energy inputs and abundant moisture, that evolution could be most readily appreciated, where adaptation is at its most complex and intricate, and the struggle for survival most dramatic.

The diversity of the tropical forests is astounding and estimates suggest that at least half of the world's species are within the 7 per cent of the globe's surface that the tropical forests cover. In 100 m² in Costa Rica, researchers found more than 233 tree species; one tree in Venezuela was home to over forty-

seven different species of orchids; there are 978 different species of beetles that live on sloths, and over 1,750 different species of fish in the Amazon basin. In general, the rainforests of South America are the most species rich, followed by Southeast Asia and then Africa. There are several reasons that account for this abundance:

1. The tropical rainforests have been around for over 200 million years, since the time of the dinosaurs and before the evolution of the flowering plants. At this time, it is thought that there was just one gigantic land mass before continental drift started to form the continents as we now know them. The vegetation of many areas was subsequently wiped out by succeeding glacial periods, which had minimal impact upon the rainforests. Hence evolutionary forces and speci-

Box 5.8 The Tropical Forests (continued)

ation have had a long time to occur in the tropics.

2. The tropics receive a higher input of energy from the sun than other areas of the globe. Not only are they closer to the sun, they also have little or no winter. The flux in solar input at the equator between the seasons is 13 per cent, compared with 400 per cent at latitude 50.

3. The tropical rainforests are so called not only because they are in the tropics but because they also receive a minimum of 2000 mm of precipitation that is evenly distributed throughout the year. Moisture is hence not a limiting factor, allowing constant growth. There is a strong correlation between diversity and rainfall.

4. Over the long period of evolution there is a positive feedback loop. As more species developed and adapted, there were further adaptations as more species sought to protect themselves against being eaten and to improve their harvesting of available food supplies. It is thought that plant diversity, in particular, has been partly the result of the need to adapt defences against the myriad of insects that feed upon them. As the plants develop their defences, so do insects constantly adapt to the new challenge. Given the relative speed of evolutionary processes in these groups, this positive feedback loop has given them very high biodiversity. In a system where most plants are immune to most insects, but highly susceptible to a few, it pays to be a long way from your nearest neighbour. Should an individual plant be attacked, it would be more difficult for the insects to find the next suitable target if the plants are far apart. Successful trees are hence widely distributed, which allows more opportunity for speciation to occur.

The tropical rainforests are the most diverse ecosystems on earth. They are also characterized by very specific examples of co-evolution and mutualism, where two species are absolutely codependent upon one another. Over 900 species of wasp, for example, have evolved to polinate the same number of fig trees. Each wasp has adapted to just one species of fig. Should anything destroy the food supply in such finely tuned systems, then obviously the other species will meet its demise.

There are also drawbacks to living in this lush environment. While evolution has benefited from most of the characteristics cited earlier, the soils have suffered. They have been exposed to weathering processes for a very long time, with no renewal and remixing from glaciation. The warm temperatures and abundant moisture are perfect for chemical weathering to great depths, and most tropical soils in this zone have long since had their nutrients washed out. A fundamental difference between tropical and temperate ecosystems is that in the tropics, unlike more temperate climes, most of the nutrients are stored in the biomass and not in the soils. When tropical vegetation is removed by logging, for example, then most of the nutrients are eliminated. Although the tropical rainforests are our most diverse expression of life on this planet, they are also one of the most fragile.

Species that evolved among the complexity of tropical forests have developed many adaptations to protect themselves. Rubber, for example, is produced by rubber trees to protect themselves from being eaten by insects, whereas the camouflage of the stick insect in this photo gives it some protection from predators (*Philip Dearden*).

organisms (but not viruses). Specialists estimate that there are slightly fewer than 68,000 organisms still awaiting discovery. The taxonomic groups containing the most numbers of species are shown in Figure 5.8. Mosquin, Whiting, and McAllister (1995) point out that these groups, although not as well known as groups such as birds and mammals, undertake key functions in ecosystems, often functions that we are only just becoming aware of and that support the more familiar and larger organisms. Beneficial insects, for example, fertilize flowers and control pests; crustaceans provide food for fish; bacteria recycle nutrients; and fungi produce bread, beer, and penicillin.

Another important element of biodiversity is the concept of *endemism*. Endemic species are found nowhere else on earth. In Canada we have relatively few endemic species compared, for example, to southern Africa where some 80 per cent of the plants are endemic, or southwest Australia where 68 per cent are endemic. In Canada it is estimated that 1–5 per cent of our species may be endemic.

Table 5.4 Changes in the Numbers of Breeding Birds in Areas of Comparable Size with Latitude

Location	Approx. Median Latitude	No. of Species of Breeding Birds
Greenland	70° N	56
Labrador	55° N	81
Newfoundland	49° N	118
New York State	43° N	195
Guatemala	15° N	469
Colombia	5° N	1,525

Source: E.O. Wilson, *The Diversity of Life* (Cambridge: Harvard University Press, 1992):196. Copyright © 1992 by E.O. Wilson. Reprinted by permission of Harvard University Press.

Examples include the Vancouver Island marmot (Canada's only endangered endemic mammal species), the Acadian whitefish, and twenty-eight species of plants from the Yukon. Reasons for our low endemism include the recent glaciation over most of the country, which effectively wiped out localized species, and the wide-ranging nature of many of our existing species. In terms of protecting biodiversity, it is especially important that endemic species are given consideration.

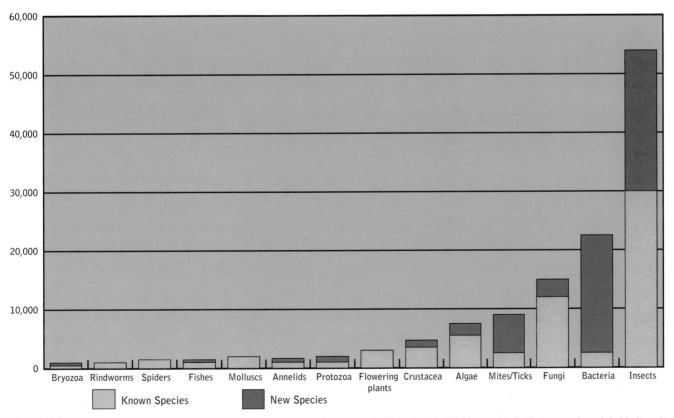

Figure 5.8 Groups with the most species in Canada (excluding viruses). SOURCE: T. Mosquin, P.G. Whiting, and D.E. McAllister, *Canada's Biodiversity: The Variety of Life, Its Status, Economic Benefits, Conservation Costs and Unmet Needs* (Ottawa: Canadian Museum of Nature, 1995):58.

Box 5.9 Biological Uncertainty

Attention has been drawn to the high degree of biological uncertainty regarding the numbers of species in Canada and elsewhere, let alone the ecological functions of each species. Few things in the natural world are absolutes. Even sex differences blur. We have known this for some time for more primitive species, such as slugs and earthworms, which are *hermaphrodites*; that is, one individual has both male and female sexual functions. It was somewhat of a surprise, however, to Charles Francis, a biologist with the Canadian Wildlife Service, when undertaking field research in Malaysia, to find the first free-ranging wild male mammals that lactate. He was collecting Dayak fruit bats when he found that several males also had breasts and milk. He speculates that this may have evolved among males that are monogamous. Another theory suggests that something in the bats' diet contains steroids that mimic female hormones.

There is also uncertainty regarding many of the basic ecological principles described in these last two chapters. Concepts such as succession, ecosystems, and communities have been found unsatisfactory for addressing many real-life ecological problems. Ecosystems often exhibit many different states that can be reached by various paths and that may produce no identifiable local climax (Pickett et al. 1992). Furthermore, species distributions along environmental gradients may overlap broadly (Levin 1989) and be subject to many natural and human-induced disturbances. Heterogeneity is the norm rather than the exception. These observations have changed the ways ecologists look at ecological systems to recognize the openness of systems, the importance of episodic events, and the numerous possibilities for intervention in ecological processes. Nonetheless, the classical concepts provide a useful background against which to organize this new, more flexible, 'non-equilibrium' approach.

IMPLICATIONS

This chapter has emphasized that ecosystems are dynamic entities that change over time. Without such change we would not have evolved and the dinosaurs would not have become extinct. The main implication of this is to accept and try to understand the nature of these changes and be able to distinguish between those that are essentially the results of natural processes and those that are the result of human activities. We cannot impose static management regimes on dynamic ecosystem processes without causing ecological disruptions. A visible reminder of this was the fire-suppression policy characteristic of many national park services, which often ignored the natural role of fire in these ecosystems. When fires did start in such ecosystems, the build-up of fuel was often so great as to cause a major and very damaging fire, as happened in Yellowstone. Most park services have now abandoned such practices for a more dynamic approach that tries to mimic the role of natural fires through prescribed burning programs.

Unfortunately, the temporal and spatial scales of ecosystem change are often so great that they are very difficult to observe in the human life span. Scientists are only now beginning to unravel the mysteries of some of these dynamic interactions between the different

The Vancouver Island marmot is Canada's only endangered endemic mammal species (*Philip Dearden*).

components of the ecosphere. There are complicated feedback loops and synergistic relations. In some cases, positive feedback loops are strengthened to accelerate undesirable changes that underlie some of the most serious environmental challenges facing humanity, such as global warming, a topic that will be discussed in more detail in chapters 7 and 20. When faced with such dynamic ecosystem changes, we must use equally dynamic thinking to face the challenges of the future.

SUMMARY

1. Ecosystems change over time. The speed of change varies from very slow (over evolutionary time scales) to rapid (caused by events, such as landslides and volcanic eruptions).

2. Ecological succession occurs as a slow adaptive process involving the gradual replacement of one assemblage of species by another as conditions change over time. Primary succession occurs on surfaces that have not been previously vegetated, such as surfaces exposed by glacial retreat; secondary succession occurs on previously vegetated surfaces, such as abandoned fields.

3. Fire is an important element in ecosystem change. Some ecosystems, such as much of the boreal forest, have evolved in conjunction with periodic fires. Fire suppression in such ecosystems can be detrimental to these natural processes.

4. Ecosystem homeostasis is a state of dynamic equilibrium in which the internal processes of an ecosystem adjust for changes in external conditions. Not all ecosystems are equal in their abilities to withstand perturbations. Inertia is an ecosystem's ability to withstand change; resilience is the ability to recover to the original state following disturbance. Both contribute to the stability of the system.

5. Important causes of loss of ecosystem homeostasis include the introduction of alien species or the removal of native keystone species.

6. Feedback mechanisms in ecosystems may either exacerbate (positive feedback loops) or mitigate (negative feedback loops) change.

7. Population change occurs as a result of the balance between factors promoting growth (e.g., increase in birth rates or reduction in death rates) and those promoting reduction (e.g., declines in birth or survivorship rates or increase in death rates).

8. Different species have different reproductive strategies. K-strategists produce few offspring but devote considerable energies to ensure that these offspring reach maturity. In comparison, r-strategists produce large numbers of young starting early in life and over a short period, and devote little or no energy to parental care.

9. Populations adapt to changing conditions over the long term through evolution. Evolution results in the formation of new species as a result of divergent natural selection responding to environmental change. This is speciation. Extinction results in the elimination of species that can no longer survive under new conditions.

10. Biodiversity involves the variety of life at three different scales: genetic, species, and landscape. Estimates suggest that Canada has a total of 71,000 known non-viral species, with 68,000 yet awaiting discovery.

REVIEW QUESTIONS

1. What are the different kinds of succession? Can you identify different seral stages in your area?

2. What is an edaphic climax? Can you find some local examples and identify the dominant limiting factor?

3. How does the concept of succession relate to environmental management?

3. How important was fire in the development of vegetation patterns in your region? Is there a fire management plan in your region? If so, what are its management goals?

4. Identify the main non-native plant and animal species in your region. What effect are they having on the local ecosystems? What are the implications for management?

5. Can you think of any other examples of negative and positive feedback loops in the ecosphere besides those mentioned in the text?

6. Are K-strategists or r-strategists most vulnerable to environmental change?

7. How does genetic diversity help protect a species from extinction?

8. What area in Canada has been called the most important scientific site in the world and why?

REFERENCES AND SUGGESTED READING

Bath, A.J. 1991. 'Yellowstone National Park Visitor Attitudes toward Fire Management Issues in the Park'. In *The Dauphin Papers: Research by Prairie Geographers*, edited by J. Welsted and J. Everitt, 151–63. Brandon: Brandon Geographical Studies I.

Biodiversity Science Assessment Team. 1994. *Biodiversity in Canada: A Science Assessment for Environment Canada*. Ottawa: Environment Canada.

Birkedal, T. 1993. 'Ancient Hunters in the Alaskan Wilderness: Human Predators and Their Role and Effect on Wildlife Populations and the Implications for Resource Management'. In *Partners in Stewardship: Proceedings of the 7th Conference on Research and Resource Management in Parks and on Public Lands*, edited by W.E. Brown and S.D. Veirs, Jr, 228–34. Hancock, Michigan: The George Wright Society.

Canadian Forest Service. 1996. *The State of Canada's Forests 1996*. Ottawa: National Resources Canada.

Carlton, J.T., and J.B. Geller. 1993. 'Ecological Roulette: The Global Transport of Nonindigenous Marine Organisms'. *Science* 198:394–6.

Colinvaux, P. 1980. *Why Big Fierce Animals Are Rare*. London: Penguin.

Darwin, C.R. 1859. *On the Origin of Species*. London: John Murray.

Dearden, P. 1979. 'Some Factors Influencing the Composition and Location of Plant Communities on a Serpentine Bedrock in Western Newfoundland'. *Journal of Biogeography* 6:93–104.

_____. 1983. 'Anatomy of a Biological Hazard: *Myriophyllum spicatum* L. in the Okanagan Valley, British Columbia'. *Journal of Environmental Management* 17:47–61.

_____. 1984. 'Public Perception of a Technological Hazard: A Case Study of the Use of 2,4-D to Control Eurasian Water Milfoil in the Okanagan Valley, British Columbia'. *The Canadian Geographer* 28:324–40.

_____. 1985. 'Technological Hazards and "Upstream" Hazard Management Strategies: The Use of the Herbicide 2,4-D to Control Eurasian Water Milfoil in the Okanagan Valley, British Columbia, Canada'. *Applied Geography* 5:229–42.

_____, and C. Hall. 1983. 'Non-consumptive Recreation Pressures and the Case of the Vancouver Island Marmot (*Marmoto vancouverensis*)'. *Environmental Conservation* 10:63–6.

Ehrlich, P.R. 1986. *The Machinery of Nature*. New York: Simon and Schuster.

Estes, J.A., D.O. Duggins, and G.B. Rathbun. 1989. 'The Ecology of Extinctions in Kelp Forest Communities'. *Conservation Biology* 3:252–64.

Gillis, P.L., and G.L. Mackie. 1994. 'Impact of the Zebra Mussel, *Dreissena polymorpha*, on Populations of Unionidae (Bivalvia) in Lake St Clair'. *Canadian Journal of Zoology* 72:1260–71.

Gould, S.J. 1989. *Wonderful Life: The Burgess Shale and the Nature of History*. New York: W.W. Norton.

_____. 1994. 'The Evolution of Life on Earth'. *Scientific American* 271:84–91.

Groombridge, B. 1992. *Global Biodiversity: Status of the World's Living Resources*. London: Chapman and Hall.

Kay, C.E. 1994. 'Aboriginal Overkill: The Role of Native Americans in Structuring Western Ecosystems'. *Human Nature* 5:359–98.

Levin, S.A. 1989. 'Challenges in the Development of a Theory of Community and Ecosystem Structure and Function'. In *Perspectives in Ecological Theory*, edited by J. Roughgarden, R.M. May, and S.A. Lewis, 242–55. Princeton: Princeton University Press.

Lovelock, J.E. 1979. *Gaia: A New Look at Life on Earth*. Oxford: Oxford University Press.

_____. 1988. *The Ages of Gaia*. New York: W.W. Norton.

Martin, P.S. 1967. 'Prehistoric Overkill'. In *Pleistocene Extinctions: The Search for a Cause*, edited by P.S. Martin and H.E. Wright, Jr, 75-120. New Haven: Yale University Press.

Mosquin, T. 1994. 'A Conceptual Framework for the Ecological Functions of Biodiversity'. *Global Biodiversity* 4:2–16.

_____, P.G. Whiting, and D.E. McAllister. 1995. *Canada's Biodiversity: The Variety of Life, Its Status, Economic Benefits, Conservation Costs and Unmet Needs.* Ottawa: Canadian Museum of Nature.

Odum, E.P. 1969. 'The Strategy of Ecosystem Development'. *Science* 164:262–70.

_____. 1989. *Ecology and Our Endangered Life-Support Systems.* Sunderland, Mass.: Sinauer Associates.

Pickett, S.T.A., V.T. Parker, and P.L. Fiedler. 1992. 'The New Paradigm in Ecology: Implications for Conservation Biology above the Species Level'. In *Conservation Biology*, edited by P.L. Fiedler and S.K. Jain, 66-88. New York: Chapman and Hall.

Primack, R.B. 1993. *Essentials of Conservation Biology.* Sunderland, Mass.,: Sinauer Associates.

Rohde, K. 1992. 'Latitudinal Gradients in Species Diversity: The Search for the Primary Cause'. *Oikos* 65: 514–27.

Schulze, K.A., and H.A. Mooney. 1993. *Biodiversity and Ecosystem Function.* New York: Springer-Verlag.

Weber, M.G., and S.W. Taylor. 1992. 'The Use of Prescribed Fire in the Management of Canada's Forested Lands'. *The Forestry Chronicle* 68:324–33.

Wilson, E.O. 1992. *The Diversity of Life.* Cambridge: Harvard University Press.

Wong, P.Y., and I. Brodo. 1992. 'The Lichens of Southern Ontario'. *Syllogeus*, vol. 69. Ottawa: Canadian Museum of Nature.

Chapter 6

Ecosystems and Matter Cycling

The collapse in the Atlantic puffin population, described in Chapter 4, was a result of interference with energy flow through the ecosystem. There are implications, however, for other aspects of ecosystem functioning. Puffins and most other sea birds play an important role in recycling *nutrients*, particularly phosphorus, from marine to terrestrial ecosystems. If these systems are disturbed, then the efficiency of the recycling mechanisms can be greatly reduced. As the phosphorus cycle has very limited recycling capabilities from aquatic to terrestrial systems, the impact of interference could be substantial. This chapter explains how *matter*, such as phosphorus, cycles in the ecosphere, and some of the implications of disturbing these cycles. So upsetting are these disturbances that Chapter 7 will be devoted to some of the management implications of biogeochemical cycling.

MATTER

Everything is either matter or energy. However, in contrast to the supply of energy, which is virtually infinite, the supply of matter on earth is limited to that which we now have. Matter, unlike energy, has mass and takes up space. It is what things are made of and is composed of the ninety-two natural and seventeen synthesized chemical elements, such as carbon, oxygen, hydrogen, and calcium. *Atoms* are the smallest particles that still exhibit the characteristics of the element. Subatomic particles include protons, neutrons, and electrons, which have different electrical charges. At a larger scale, the same kinds of atoms can join to form *molecules*. When two different atoms come together, they are known as a *compound*. Water (H_2O), for example, is a compound made up of two hydrogen atoms (H) and one oxygen atom (O). Four major kinds of organic compounds—carbohydrates (sugars and starches), fats (lipids, hormones, etc.), proteins (enzymes, etc.), and nucleic acids (DNA, RNA, etc.)—make up living organisms.

Matter also exists in three different states; solid, liquid, and gas, and can be transformed from one

Water is the only substance that occurs in all three phases of matter at the ambient temperatures of the earth's surface (*Philip Dearden*).

to the other by changes in heat and pressure. At the existing temperatures on the earth's surface, we have only one representative of the liquid state of matter, water. We can also readily see water in its other two states as ice (solid) or clouds (vapour).

Just as energy flow is explained by the laws of thermodynamics, the *law of conservation of matter* helps us understand how matter is transformed. According to this law, matter can be neither created nor destroyed, but merely transformed from one form into another. Thus matter cannot be consumed so that it no longer exists; it will always exist, but in a changed form. When we throw something away, it still exists on this planet as matter somewhere. All pollution results from this law. The huge superstacks on large smelters, such as at Inco in Sudbury, do not dispose of wastes; they just disperse those wastes over a much larger area (Chapter 13). The matter being dispersed is the same, and ultimately falls as acid deposition somewhere. The same is true for all the wastes that we wash down our sinks. They do not disappear, but collect in larger water bodies and create pollution problems.

BIOGEOCHEMICAL CYCLES

For millions of years matter has been moving among different components of the ecosphere. These cycles are as essential to life as the energy flow described in the previous chapter. About thirty of the naturally occurring elements are a necessary part of living things. These are known as *nutrients*, which may be further classified into *macronutrients* (needed in relatively large amounts by all organisms) and *micronutrients* (required in lesser amounts by most species) (Table 6.1). About 97 per cent of organic mass is composed of six nutrients: carbon, oxygen, hydrogen, nitrogen, phosphorus, and sulphur. These nutrients are continually being cycled among different components of the ecosphere in characteristic paths known as *biogeochemical* cycles.

Figure 6.1 shows a generalized model of such a cycle. Like all the other diagrams of cycles in this chapter, it is a good example of the kind of simplifying models that scientists construct to represent the vast complexity of earth processes, as described in Chapter 1. Nutrients can be stored in the different compartments shown in Figure 6.1 for varying amounts of time. In general, there is a

According to the law of matter, emissions from stacks such as these do not simply disappear but end up somewhere else, often with undesirable consequences, such as acid deposition or global warming (*Philip Dearden*).

> We know from studies of chemistry that our bodies are re-organized star-dust, recycled again and again, so that, truly, our bones are of corals made.
>
> Stan Rowe,
> 'In Search of the Holy Grass', 1993

large, relatively slow-moving abiotic pool that may be in the atmosphere or lithosphere, and is chemically unusable by the biotic part of the ecosystem or is physically remote. There is a more rapidly interacting exchange pool between the biotic and abiotic components. Nutrients move at various speeds from the biotic to abiotic pools. For example, very rapid exchange takes place through respiration as carbon and oxygen move rapidly between the biotic and atmospheric components. The elements that now make up your body have undergone millions of years of recycling through these various compartments. You are a product of recycling!

Table 6.1 Relative Amounts of Chemical Elements That Make Up Living Things

Major Macronutrients (> 1% dry organic weight)		Relatively Minor Macronutrients (0.2–1% dry organic weight)		Micronutrients (<0.2% dry organic weight)	
Name of Element	Symbol	Name of Element	Symbol	Name of Element	Symbol
Carbon	C	Calcium	Ca	Aluminium	Al
Hydrogen	H	Chlorine	Cl	Boron	B
Nitrogen	N	Copper	Cu	Bromine	Br
Oxygen	O	Iron	Fe	Chromium	Cr
Phosphorus	P	Magnesium	Mg	Cobalt	Co
		Potassium	K	Fluorine	F
		Sodium	Na	Gallium	Ga
		Sulphur	S	Iodine	I
				Manganese	Mn
				Molybdenum	Mo
				Selenium	Se
				Silicon	Si
				Strontium	Sr
				Tin	Sn
				Titanium	Ti
				Vanadium	V
				Zinc	Zn

Source: C.E. Kupchella and M.C. Hyland, *Environmental Science: Living within the System of Nature* (Toronto: Allyn and Bacon, 1989):45.

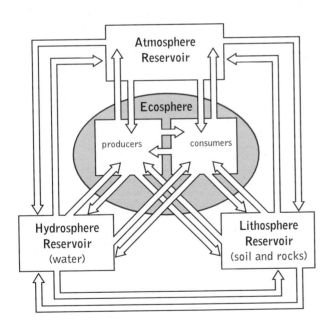

Figure 6.1 Generalized model of the biogeochemical cycle.

Ecosystems also vary substantially in the speeds of cycling and the relative proportions of nutrients in each compartment. Some systems have nutrient-poor soils, for example, and have developed different mechanisms to store nutrients in other compartments. Tropical forest ecosystems are classic examples. Most of the nutrients are stored in the biomass as opposed to the soil system. When leaves fall to the ground, they are rapidly mined for nutrients by plant roots before those nutrients have a chance to be leached out of the system. In contrast, many temperate forests have soils of high fertility. Removal of the nutrients in the biomass through logging, for example, does not remove as high a proportion of the site nutrient capital as removal in tropical ecosystems. (This is discussed in more detail in Chapter 15.) Lack of appreciation for these differences (see Table 6.2) is one reason why forestry methods developed in Europe and North America have not proven successful in the tropics.

Speed of cycling may also change within a cycle depending upon the season. In the carbon cycle, for example, there is greater uptake of CO_2 in spring and summer as deciduous trees grow leaves. In fall there is a correspondingly greater release as the leaves fall off and decompose. On average, a carbon dioxide molecule stays in the atmospheric component of the cycle for five to seven years. This is known as the *residence time*. It takes an average of 300 years for a carbon molecule

Box 6.1 The Decomposers

In Chapter 4 attention was drawn to the importance of decomposer organisms and detritus food chains. These are the main means by which nutrients in the biotic component of the ecosphere are returned to the abiotic, so that plants can use them once again. Photosynthesis has been described as the process of making a complicated product out of simple components; decomposition is the reverse process of making simple components out of that complicated product.

Even while still on the plant, leaves may be attacked by decomposer organisms, such as fungi, which release products, such as sugars, which are then washed to the ground by rainfall. Once leaves fall to the ground, they are progressively broken down by various groups of organisms. Larger organisms—such as earthworms, slugs, snails, beetles, ants, and termites—help break up the leaf material initially. Many gardeners are fully aware of how slugs, for example, can devour green leaves in great quantities.

Fungi and heterotrophic bacteria further break down the organic matter, releasing more resistant carbohydrates and then cellulose and lignin. The humus, the organic layer in the soil, is composed mainly of products that can resist rapid breakdown. A chemical process, oxidation, is mainly responsible for the decay of this material.

As anyone who has seen leaf decay in autumn knows, the process can occur quite rapidly. The speed varies depending upon the environment. Warm environments tend to promote more rapid microbial activity. Leaf decay in the tropics takes place in a matter of weeks. In the boreal forest, however, where conditions are cold and the leaves (such as spruce and pine needles) are quite resistant, recycling of the nutrients in the leaves may take decades.

to pass through the lithosphere, atmosphere, hydrosphere, and biotic components of the carbon cycle. By way of contrast, it may take a water molecule 2 million years to make a complete cycle. The speed of cycling is influenced by such factors as the chemical reactivity of the substance. Carbon, for example, participates in many chemical reactions. It also exists as a gas. In general a gaseous phase speeds up a cycle because gas molecules move more quickly than molecules in the other states of matter.

The rapid global circulation of gases is well illustrated by the inert gas, argon, which makes up about 1 per cent of the atmosphere. Millions of argon atoms were just exhaled in that breath you took. Some of them may feel familiar, as seventeen of them were breathed in by you on this very day last year, no matter where you were. And you will encounter another seventeen tomorrow. Scientists calculate that

Table 6.2 Approximate Distributions of Carbon and Nitrogen in Temperate and Tropical Rainforests

	Tropical Rainforest	Temperate
Carbon in vegetation	75%	50%
Carbon in litter and soil	25%	50%
Nitrogen in biomass	50%	6%
Nitrogen in biomass above ground	44%	3%

Slash-and-burn agriculture is a common way to transfer nutrients from the biomass to the soil to increase agricultural productivity. This photograph shows several swiddens (fields cut in the forest) by the hill tribe people of northern Thailand. The soils in the swiddens rapidly lose fertility caused by the burning of the biomass, and they are then abandoned for secondary succession to occur (*Philip Dearden*).

by the time you are sixty, you will have inhaled one argon atom breathed in by every creature that is living or has ever lived on this planet.

Cycles can be classified according to the main source of their matter. *Gaseous* cycles, as the name would suggest, have most of their matter in the atmosphere. The nitrogen cycle is a good example. *Sedimentary* cycles, such as the phosphorus and sulphur cycles, hold most of their matter in the lithosphere. In general, elements in sedimentary cycles tend to cycle more slowly than those in gaseous cycles and the elements may be locked into geological formations for millions of years.

Recycling rates between components under natural conditions achieve a balance over time if inputs and outputs are equal. Human activity changes the speed of transference between the different components of the cycles. Many of our pollution problems result from a human-induced build-up in one or more components of a cycle that cannot be effectively dissipated by natural processes. The cycles become acyclic.

In addition to the biogeochemical cycles, some attention will also be given in this chapter to the *hydrological* cycle. This cycle is critical in all other cycles as water plays a main role in the mobilization and transportation of materials. The energy for this, as with all other aspects of the cycles, ultimately comes from the sun. Photosynthesis powers the biotic aspects of the cycles, and atmospheric circulation, fuelled by the sun's energy, controls the water power that is so important for weathering and erosional processes.

Sedimentary Cycles

Sedimentary cycles involve the mobilization of materials from the lithosphere to the hydrosphere and back to the lithosphere. Some, such as sulphur, may involve a gaseous phase, while others, such as phosphorus, do not. These cycles rely essentially on geological uplift over long time periods to complete the cycle. Lack of a gaseous phase means that the cycle is missing one potential route for more rapid recycling, which can lead to problems when mobilization rates are increased through human activity. Phosphorus and sulphur will be discussed here, but other elements—such as calcium, magnesium, and potassium—follow similar pathways.

Phosphorus (P)

Phosphorus, a macronutrient incorporated into many organic molecules, is essential for metabolic energy use. It is relatively rare on the earth's surface compared with biological demand, and hence it is essential that phosphorus cycles efficiently between components. Many organisms have devised means to store this element in their tissues. Phosphorus moves very readily within plants from older tissues to more active growth sites. Deciduous trees may recirculate up to 30 per cent of their phosphorus back to their more permanent components before the leaves fall in an effort to preserve this nutrient.

Phosphorus is a prime example of a nutrient held in a *tight circulation* pattern under natural circumstances between the biotic and abiotic components. Replenishment rates through weathering and soil availability are limited, so the amount retained by the biomass is quite critical. Phosphorus can stay in terrestrial systems for up to 100 years before it is leached into the hydrosphere. Phosphorus is often the dominant limiting factor for plant growth (Chapter 4) in terrestrial soils, and agricultural productivity relies heavily upon augmenting this supply (auxiliary energy flow) through fertilizer application.

The availability of phosphorus in the soil is influenced by soil acidity. Acidity is measured on the pH scale, which is discussed in more detail in the next chapter. Below pH 5.5, for example, phosphorus reacts with aluminium and iron to form insoluble compounds. Above pH 7, the same thing happens in combination with calcium. Obviously, things that change soil pH, such as acid precipitation (discussed in more detail in the next chapter), can have a critical effect upon phosphorus availability. This is an example of the kind of synergistic reaction discussed in Chapter 4 in which the combination of either high or low pH values with low phosphorus availability can have a stronger effect than the sum of the two individually because of the chemical reactions.

The main reservoir of phosphorus is in rocks in the earth's crust (Figure 6.2). Geological uplift and subsequent weathering make phosphorus available in the soil where it is taken up by plant roots. Many higher plants have a mutualistic relationship with soil fungi, *mycorrhizae*, which helps

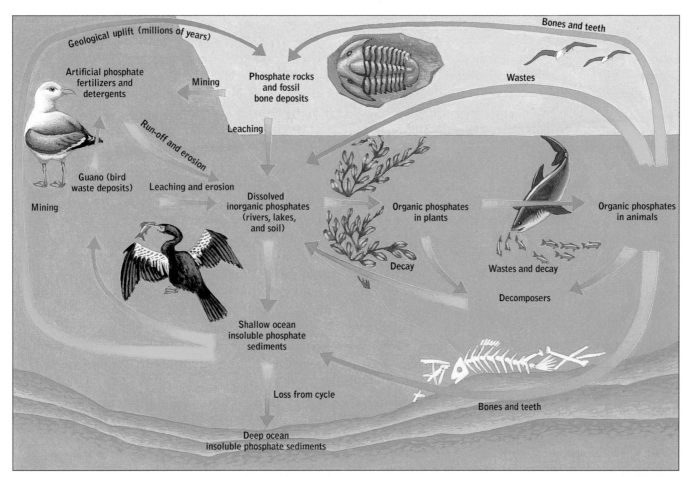

Figure 6.2 The phosphorus cycle.

them gain improved access to phosphorus in the soil. Once incorporated into plant material, the phosphorus may be passed on to other organisms at higher trophic levels. Animal wastes are a significant source of phosphorus return to the soil. All these organisms eventually die and the organic material is broken down by the decomposer food chains. This may take some time as a considerable amount of the phosphorus is within animal bones. In the past, concentrated sources of animal bones, such as those at the bison jumps used by Native peoples on the Prairies, have been used by farmers as a source of phosphate fertilizer. One of these, Head-Smashed-In Buffalo Jump in south-

The Head-Smashed-In World Heritage Site in southern Alberta is rich in phosphorus as Native peoples used the cliff to kill stampeding bison (*Philip Dearden*).

ern Alberta, has been recognized as a World Heritage Site (see Chapter 17), although this is partly because the site was not exploited by farmers for phosphorus and is relatively undisturbed.

Following breakdown in the soil, the phosphorus is then either taken up again by plants or removed by water transport. Bacteria mineralize the returned organic phosphorus into inorganic forms so it can once more be taken up by plants. Most of the water transport occurs in particulate form by streams, which is one reason to be concerned about excessive sedimentation through land-use activities, such as agriculture and logging, described in chapters 14 and 15.

Stream transport ultimately ends up in the ocean. One of the reasons why estuaries have such high productivities, as discussed in Chapter 4, is due to this nutrient input from upstream. The circulation patterns within estuaries tend to trap nutrients, but some phosphorus finds its way into the shallow ocean areas of the coastal zone. It may be fixed in biomass by phytoplankton or other aquatic plants in these areas in the *euphotic* (eu = well, photos = light) zone and once again be incorporated into the food chain. The coastal zones, with this plentiful supply of nutrients and photosynthetic energy from the sun, cover less than 10 per cent of the ocean's surface, but account for over 90 per cent of all ocean species.

Beyond the coastal zone and the continental shelves, water depth increases into the open ocean. Phosphorus and other nutrients that have not been incorporated into food chains, plus elements from dead oceanic organisms, filter through to the *bathyal* and ultimately the *abyssal zones* (Figure 4.5). Here uptake by organisms is extremely limited, and the nutrients must either be moved back to the euphotic zone by upwelling currents or wait to be geologically uplifted over millions of years to move into another component of the cycle. Where

---

Box 6.2 Weathering, the Rock Cycle, and Plant Uptake

The weathering of the rocks in the earth's crust plays an important role in supplying long-term inputs to biogeochemical cycles. Weathering is part of the rock cycle whereby rocks that have been uplifted are eroded into different constituents. The rock cycle involves the transformation of rocks from one type to another, such as when volcanic rocks are eroded and washed into the ocean. Over millions of years the resulting sediments are turned into sedimentary rocks. In turn, these sedimentary rocks may be compressed within the earth's crust and altered by heat and pressure before being uplifted once more through the process of continental drift.

Weathering involves numerous different processes. In Canada mechanical weathering involves the physical breakup of rocks as a result of changing temperatures. The action of water is important. Chemical processes, such as hydration and carbonation, further the process by removing elements in solution. Secondary clay minerals are produced from primary rock minerals by hydrolysis and oxidation. These clays are very important in keeping the nutrients in the soil. The soil can be thought of as a giant filter bed in which each particle is chemically active. As water containing many different nutrients in solution percolates through, the clays hold some of these nutrients, which become available for plant uptake.

Plants constantly lose moisture from their leaves. This creates a moisture gradient within the plant that draws up water to replace what was lost. Water then moves from the roots in replacement, and more nutrient-laden water is taken in by the roots. It is the roots' job to keep the plant supplied with water. As nutrients are removed from the soil water around the plants, new nutrients move within the soil water to replace them.

These sedimentary rocks along the Alsek River in the Yukon have been compressed and folded as part of the rock cycle (*Philip Dearden*).

Box 6.3 Human Impacts on the Phosphorus Cycle

Humans intervene in the phosphorus cycle in several ways to accelerate the mobilization rate by:

- mining phosphate-rich rocks for fertilizer and detergent production, creating excessive run-off into aquatic environments
- removing biomass, leading to accelerated erosion of sediment and solutes into streams
- concentrating large numbers of organisms—such as humans, cattle, and pigs—which create heavy burdens of phosphate-rich waste materials

- removing phosphorus from oceanic eco-systems through fishing, and returning phosphorus again (through the dissolution of wastes) to fresh water and ultimately to the marine system

All these interventions result in excessive phosphorus accumulation in fresh water systems. Human activity now accounts for approximately two-thirds of the phosphorus reaching the oceans. The environmental impacts of this nutrient enrichment will be discussed in more detail in the next chapter.

such upwellings occur, such as off the west coasts of Africa and South America, there are very rich fisheries due to the combination of high nutrient and energy levels.

Two other recycling mechanisms also occur. One is the biotic one described earlier as marine birds—such as puffins, cormorants, and other fish-eating birds—return phosphorus to land in the form of their droppings, which contain all the phosphorus from the marine food chain. This phosphorus, known as *guano*, constitutes the largest source of phosphorus for human use and is heavily mined for fertilizer production. A small amount of phosphorus is also returned to land through the atmosphere as sea spray.

Sulphur (S)

Like phosphorus, sulphur is a sedimentary cycle, but differs from phosphorus in two important ways. First, it has an atmospheric component, and hence better recycling potential. Sulphur is not often a limiting factor for growth in aquatic or terrestrial ecosystems. Second, like most of the other cycles (but unlike phosphorus), sulphur is very dependent on microbial activity. Sulphur is a necessary component of all life and is a building component of proteins.

Sulphur is like phosphorus in another aspect. It is not available in the lithosphere and must be transformed into sulphates to be absorbed by plants. Bacteria are critical here, changing sulphur into various forms in the soil (Figure 6.3). The exact form depends upon factors, such as the pres-

ence (*aerobic*) or absence (*anaerobic*) of oxygen (which usually reflects the relationship of the transformation site to the water table) and the presence of other elements, such as iron. These microbial transformations by chemoautotrophs release gases, such as hydrogen sulphide (H_2S), to the atmosphere, giving the familiar rotten egg smell we associate with marshlands, or produce sulphate salts (SO_4). The sulphates can then be absorbed by plants through their roots, sulphur enters the food chain, and the same processes occur as in the biotic components of the other cycles.

The complexity of these cycles is illustrated further by some of the interactions that occur between cycles. For example, the phosphorus cycle

Box 6.4 Human Impacts on the Sulphur Cycle

Humans intervene in the sulphur cycle mainly by burning sulphur-containing coal, largely to produce electricity, and smelting metal ores that also contain sulphates. Almost 99 per cent of the sulphur dioxide and about one-third of the sulphur compounds reaching the atmosphere are a result of these activities. These sulphur compounds react with oxygen and water vapour to produce sulphuric acid (H_2SO_4), a main component of acid deposition. This will be discussed in more depth in the next chapter.

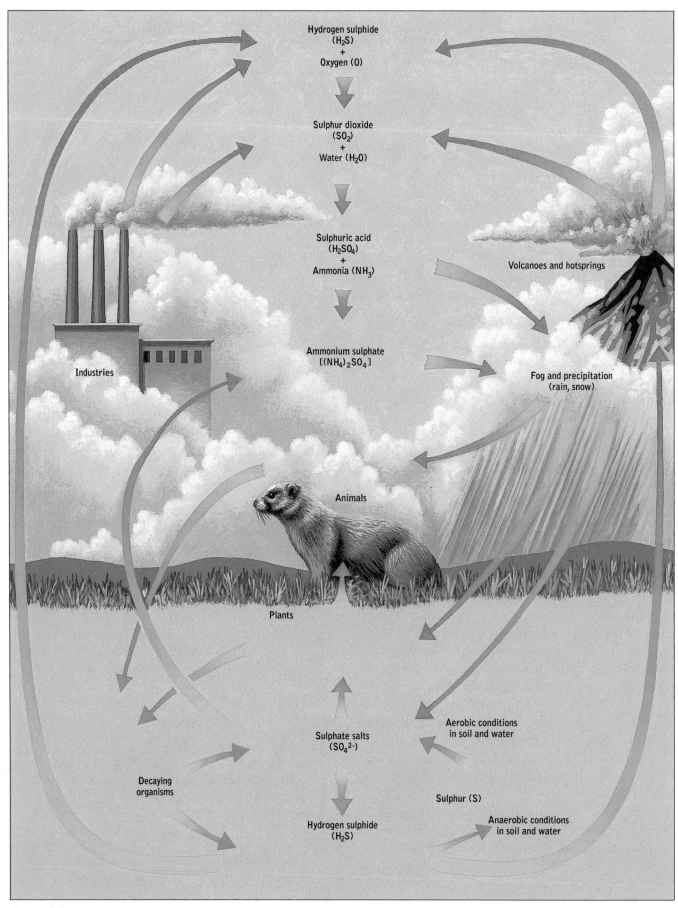

Figure 6.3 The sulphur cycle.

benefits when iron sulphides are formed in sediments and phosphorus is converted from insoluble to soluble forms where it becomes available for uptake.

The upward movement of the sulphur cycle's gaseous phase is also important as significant quantities of sulphur are returned to the atmosphere, thereby shortening the long sediment uplift time that characterizes the phosphorus cycle. This is fortunate since average ocean residence times are quite long and sulphur is continually lost to the ocean floor. From the upper reaches of the oceans, sulphur can be returned to the atmosphere by phytoplankton or photochemical reactions. However, unlike phosphorus, a relatively small proportion of sulphur is fixed in organic matter and availability is not usually a problem.

Gaseous Cycles

Nitrogen

Nitrogen is a colourless, tasteless, odourless gas required by all organisms for life. It is an essential component of chlorophyll, proteins, and amino acids. The atmosphere is over 78 per cent nitrogen gas, yet most organisms cannot gain access to the necessary nitrogen from the atmosphere. Instead, the nitrogen is obtained from the soil as nitrates. The atmospheric reservoir is linked to the biotic components of the food chain mainly through *nitrogen fixation* and *denitrification*, both mediated through microbial activity (Figure 6.4).

Biological nitrogen fixation occurs as bacteria transform atmospheric nitrogen (N_2) into ammonia (NH_3). The most important fixers are bacteria of the *Rhizobium* family, which grow on the root nodules of certain plants, such as members of the pea or legume family, like peas, beans, clover, and alfalfa. The bacteria and roots of the plant communicate through chemical stimuli, which results in the bacteria infecting root cells. Once infected, the cells swell into the nodules that you can see on the roots of the peas or beans in your garden. In a remarkable example of co-evolution, the plant supplies the products of photosynthesis to the relationship, and the bacteria transform the atmospheric nitrogen into nitrates. It is one of the few examples known where two organisms cooperate to make one molecule.

This resulting enrichment is why farmers grow crops such as alfalfa and clover as part of a crop rotation to help build up nitrates in the soil. About half of the nitrogen circulating in agricultural ecosystems comes from this source. The increasing costs of fertilizer worldwide have focused more attention on biological nitrogen fixation as part of meeting the global food challenges of the future (see Chapter 14). Through genetic engineering, for example, it may be possible to inject other crops, such as cereals, with similar symbiotic unions between plants and nitrogen-fixing bacteria. It may not be that simple, however. For example, research indicates that species involved in dinitrogen fixation may also be particularly susceptible to phosphorus deficiencies, given their high phosphorus and energy requirements (Chapin et al. 1991).

Some wild species—such as alder, lupins, and vetch—also have similar bacteria and hence play a valuable role when they act as primary colonizers in the successional process or help recolonize sites that have been logged (Chapter 15) or otherwise disturbed. These relationships are *mutualistic* in that both organisms benefit. The plant gets enhanced nutrient supply and the organisms find a home in which the plant supplies them with various sugars. One of the most celebrated examples of such a partnership is that between the fungi and algae in lichens, discussed in more detail in Chapter 5.

Other bacteria and algae that fix nitrogen are not attached to specific plants. In the Arctic, where nitrogen is a limiting factor on plant growth, an important source of nitrogen fixation is by *cyanobacteria*, or blue-green algae, such as *Nostoc commune* (Lennihan et al. 1994). Sometimes combinations of various cyanobacteria and other organisms form crusts over the soil surface, known as *cryptogamic* crusts, which are important in facilitating successional processes. Other important free-living nitrogen-fixing micro-organisms are also found in the ocean. It is estimated that in terrestrial ecosystems about double the amount of nitrogen is fixed by the mutualistic relationships described earlier as these free-floating relationships. The major supply of nitrogen, however, comes from the breakdown of existing biomass by decomposer food chains. Nitrogen is tightly circulated in most ecosystems between the dead and living biomass.

In addition to these biological mechanisms, some nitrogen is also fixed through *atmospheric fixation*, which occurs largely during thunderstorms. Lightning causes extremely high temperatures that

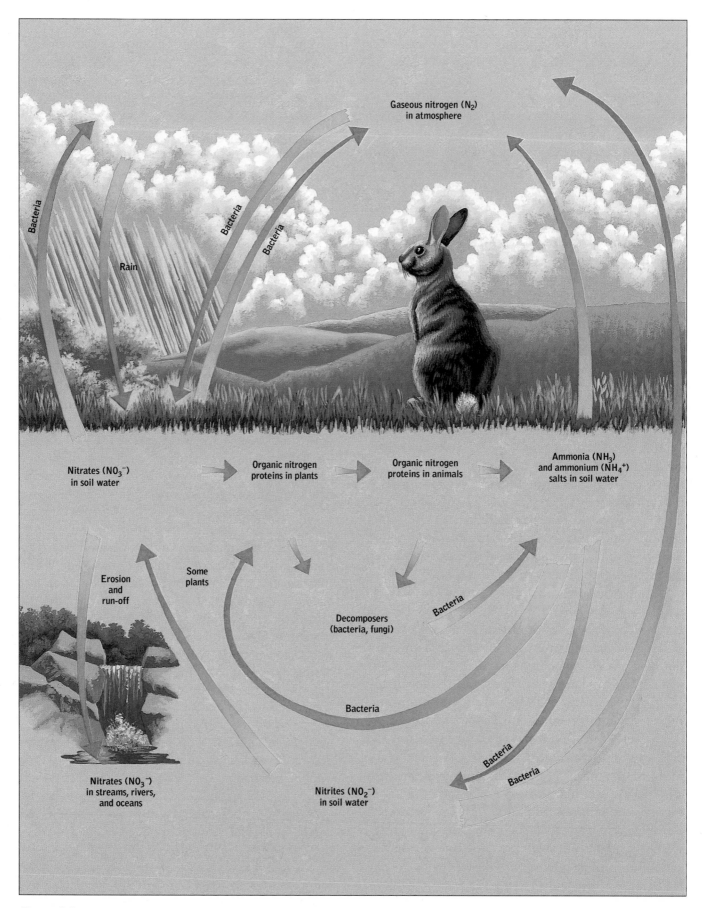

Figure 6.4 The nitrogen cycle.

unite oxygen and nitrogen, which eventually combine to form nitric acid (HNO_3). Nitric acid is subsequently carried to earth as precipitation and converted into the nitrates (NO_3) that can be taken up by plant roots. Estimates on the importance of atmospheric fixation vary, but 10 per cent of total fixation would be a maximum figure and most estimates are about half of this.

Once fixed in the soil, ammonia and nitrates are incorporated into plant matter and then through the food chain. On death, the biomass is converted back to ammonia gas (NH_3) and ammonium salts (NH_4) by bacterial action and returned to the soil. Some ammonia and ammonium is changed to nitrites and then to nitrates by chemotrophic bacteria, such as *Nitrosomonas* and *Nitrobacter*, in a process known as *nitrification*. There are also other bacteria, anaerobic bacteria, which through the *denitrification* process convert nitrates into nitrogen gas and back into the atmosphere (Figure 6.4)

Nitrates are highly soluble in water and if not held tightly by the soil, they may be lost to the ecosystem by surface run-off. Ammonia is also susceptible to loss by soil erosion as it tends to adhere to soil particles. Like phosphorus, nitrogen is often a limiting factor for growth. When excessive concentrations occur in water, it is a major contributor to the process of eutrophication (to be discussed in the next chapter). Unlike phosphorus, however, nitrogen is not immobilized in deep ocean sediments, but has an effective feedback mechanism to the atmosphere from the ocean through microbial denitrification.

Carbon

Although carbon dioxide gas (CO_2) constitutes only 0.03 per cent of the atmosphere, it is the main reservoir for the carbon that is the building-block for all necessary fats, proteins, and carbohydrates that constitute life. Plants take up carbon dioxide directly from the atmosphere through the process of photosynthesis and release oxygen at the same time. The carbon becomes incorporated into the biomass and is passed along the food chain. Residence times can vary greatly, but older forests constitute a significant repository for centuries' worth of carbon. Organisms' respiration transforms some of this carbon back into carbon dioxide (Figure 6.5), and the cellular respiration of decomposers helps to return the carbon from dead organisms into the atmosphere. Thus the cycling of carbon and flow of energy through the food chain are intimately related.

Besides this relatively rapid exchange, some carbon can also be stored in the lithosphere for extended periods as organisms are buried before they decompose. Through geological time, millions

---

Box 6.5 Human Impacts on the Nitrogen Cycle

Humans disrupt the nitrogen cycle in many ways. The cycle is complex and dependent upon numerous bacterial reactions. We know little of the effects of human activity on these various bacterial groups. Significant interventions occur through the following:

• Chemical fixation supplies nitrates and ammonia as fertilizer. The amount fixed is estimated to be about one-third of that produced by natural processes. The main impacts are through run-off of excess fertilizer (contributing to eutrophication) and denitrification (contributing to climatic change), both discussed in more detail in the next chapter. There are also health concerns related to excessive nitrate levels in water supplies. These nitrates may be transformed to nitrites in the digestive systems of babies, which in turn may lead to an oxygen deficiency in the blood known as *methaemoglobinaemia*, or blue-baby syndrome. There are numerous instances in Canada of well closures in agricultural areas for this reason.

• Nitrate and ammonium ions are removed from agricultural soils through the harvesting of nitrogen-rich crops.

• High temperature combustion produces nitric oxides, which combine with oxygen to produce nitrogen dioxide. Nitrogen dioxide reacts with water vapour to form nitric acid, a main component of acid deposition, a topic also discussed in more detail in the next chapter.

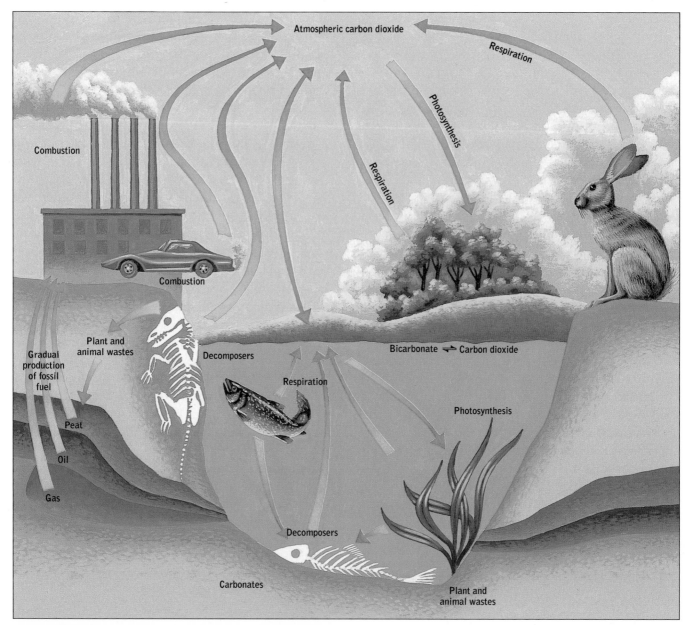

Figure 6.5 The carbon cycle.

Large amounts of carbon are stored in the lithosphere, such as in these coal-beds, the product of millions of years of photosynthetic activity (*Philip Dearden*).

of years of photosynthetic energy have been transformed into fossil fuels by this process as a result of heat and compression. The highly productive forests and marine environments of the distant past have become the coal, oil, and natural gas that fuel the world's economy today.

Some of the carbon dioxide is dissolved into the shallower ocean before re-entering the atmosphere. Residence time is about six years in these shallow waters, but much longer (up to 350 years) when mixed with deeper waters. These residence times are now of considerable scientific interest due to the rising levels of carbon dioxide in the atmosphere (as will be discussed in the next

Box 6.6 Human Impacts on the Carbon Cycle

As human populations increase, there have been two major changes to the carbon cycle:

- Natural vegetation, usually dominated by tree growth, has been replaced by land uses, such as urban and agricultural systems, which have reduced the capacity to take up and store carbon.
- For the last 200 years or so, human activities, particularly industrial activity, have mobilized large amounts of fossil fuels from the lithospheric component of the cycle to the atmospheric component.

Both these impacts have significant implications for global climatic change, which will be discussed in the next chapter.

chapter) and the potential for the oceans to absorb these increases.

Large amounts of carbon are also stored for much longer periods in the ocean. When marine organisms die, their shells of calcium carbonate ($CaCO_3$) are cemented together to form rocks, such as limestone. Over millions of years, the limestone may be uplifted to become land and is then slowly weathered to release the carbon back into the carbon cycle.

THE HYDROLOGICAL CYCLE

Water, like the nutrients discussed earlier, is necessary for all life. You are made up of 90 per cent water. Although other planets, such as Venus and Mars, have water, only on earth does it occur in the liquid state. Water also occurs in a fixed supply that cycles between various reservoirs driven by energy from the sun. The ocean is by far the largest reservoir, containing over 97 per cent of the water on earth. Most of the rest exists in the polar ice-caps, with only a small amount readily available as the fresh water that sustains terrestrial life (Table 6.3). Water travels continuously

between these various reservoirs through the main processes of evaporation and precipitation known as the *hydrological cycle* (Figure 6.6).

Average residence times in each reservoir vary greatly (Table 6.3). In the deep ocean, it may take 37,000 years before water is recycled through evaporation into the atmosphere, whereas once in the atmosphere, average residence time is about nine to twelve days. These figures have special relevance regarding the effects of pollution. Although many major rivers have been seriously polluted, the flushing action of the river, combined with the short residence time of the water, means that a relatively rapid recovery is often possible. This is not the case, however, with groundwater pollution, especially deep groundwater pollution.

As water demands grow, we have become increasingly dependent upon such groundwater sources. Once they become polluted with agricultural biocides or industrial wastes, however, they may be unsuitable for use by any life form for centuries. The importance of this is underscored by the fact that Canada's water in underground sources is estimated to be thirty-seven times that in surface sources. One-quarter of the Canadian population relies on groundwater for domestic use, while some communities, such as Fredericton, are almost totally dependent on groundwater.

The hydrological cycle transports water from the oceans to the atmosphere, through terrestrial and subterranean systems and back to the ocean, all fuelled by energy from the sun. Eighty-four per cent of the water in the atmosphere is evaporated directly from the ocean surface. The remainder comes from evaporation from smaller water bodies or from the leaves of plants (*transpiration*) or the soil

Table 6.3 Global Water Storage

Reservoir	Average Renewal Rate	Per cent of Global Total
World oceans	3,100 years	97.2
Ice sheets and glaciers	16,000 years	2.15
Groundwater	300–4,600 years	0.62
Lakes (fresh water)	10–100 years	0.009
Inland seas, saline lakes	10–100 years	0.008
Soil moisture	280 days	0.005
Atmosphere	9–12 days	0.001
Rivers and streams	12–20 days	0.0001

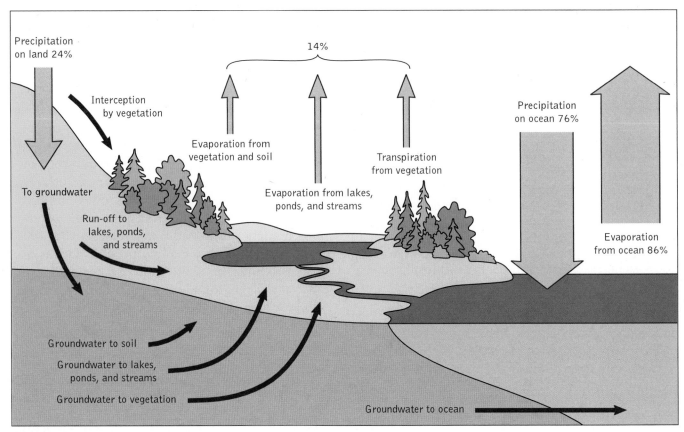

Figure 6.6　The hydrological cycle.

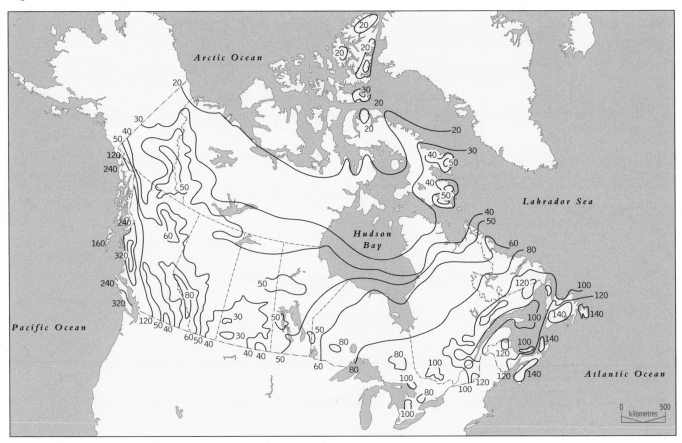

Figure 6.7　Average annual rain and snow for Canada (cm). SOURCE: After D. Phillips, *The Climate of Canada* (Ottawa: Minister of Supply and Services Canada, 1990):210.

Box 6.7 Precipitation

Precipitation occurs in several forms: rain, snow, hail, dew, fog, and rime ice. It occurs when the accumulated particles of condensed water or ice in clouds become large enough that they overcome the uplifting air currents and fall to earth as a result of gravity. Some of this precipitation may never reach the ground. Lower air layers may be warmer and drier and re-evaporation may occur as the precipitates pass through these layers, giving an excellent example of the speed of some of these mini cycles that occur as part of the larger earth cycles. Distribution of precipitation is one of the main factors influencing the nature and location of global biomes and the ecozones of Canada, as described in Chapter 3. In Canada precipitation varies from almost nothing in the Arctic to over 3000 mm on the West Coast (Figure 6.7).

Differences in precipitation occur for various reasons. At the global scale, heating of equatorial regions causes air to rise. As the air rises it cools, condenses, clouds form, and precipitation occurs. As a result, equatorial regions tend to have consistently high rainfall. When this air falls as it cools over subequatorial regions, it tends to be dry, such as in the Sahara Desert.

In Canada much of the precipitation comes from low-pressure systems, large cells of rising air that form along the boundary between warm and cold air masses. Moisture-laden winds cross oceans and are forced to rise as a result of mountain barriers, thus influencing precipitation levels. Westerlies coming across the Pacific meet the Western Cordillera and create the highest precipitation levels in the country. As the air warms up in its descent from the mountains, it can hold more moisture, and precipitation levels fall considerably to produce a *rain-shadow effect*, which accounts for the small amounts of precipitation across the Prairies, as little as 300–400 mm. Precipitation levels rise again in central Canada due to disturbances bringing moisture from the Gulf of Mexico and Atlantic Ocean. In southern Ontario, precipitation increases to 800 mm and over 1000 mm in the lee of the Great Lakes, a major source of moisture for downwind local-

ities. In Atlantic Canada, exposure to maritime influences increases once more, giving up to 1500 mm on the south coast of Newfoundland. The Arctic is very dry because the prevailing winds from the north are very cold and hence have little moisture-carrying capacity. They pass over terrain that has relatively few sources of water evaporation (sources often remain frozen for a good proportion of the year) and an absence of low-pressure systems in winter. Topography is also an important determinant of whether the precipitation falls as snow or rain.

More localized precipitation may be produced by convection as warmer and lighter air rises and cooler, heavier air sinks. This mechanism is important for localized storms in summer when the air is heated by the warm ground sending columns of moist, warm air to great elevations, resulting in thunderstorm activity.

Desert-like conditions occur in some areas of Canada largely due to rain-shadow effects. Although we associate the Fraser River with the high rainfalls of the West Coast, the interior of British Columbia is quite dry (*Philip Dearden*).

and plants (*evapotranspiration*). As water evaporates, it leaves behind accumulated impurities. The most common dissolved substance in the ocean is sodium chloride, or table salt, which also contains many other elements in trace amounts. Evapora-

tion acts as a giant purification plant until further pollutants are encountered in the atmosphere.

Once in the atmosphere, the water vapour cools, condenses around tiny particles called *condensation nuclei*, forms clouds, and falls to earth as

Relative humidity is high most of the year in Atlantic Canada and produces some beautiful atmospheric effects (*Philip Dearden*).

rain, snow, or hail. The warmer the air, the more water it can hold. Moisture content can be expressed in terms of *relative humidity*, the amount of moisture held compared with how much could be held if fully saturated at a particular temperature. At a relative humidity of 100 per cent, the air is *saturated* and cloud, fog, and mist form. Clouds are moved around by winds and continue to grow until precipitation occurs and the water is returned to earth.

About 77 per cent of precipitation falls into the ocean. The remainder joins the terrestrial part of the cycle in ice-caps, lakes, rivers, groundwater, and moves between these compartments. Gravity moves water down through the soil until it reaches

the *water table*, where all the spaces between the soil particles are full of water. This is the groundwater. Lakes, streams, and other evidence of surface water occur where the land surface is below the water table. Surface water is a major factor in sculpting the surface of the earth. At greater depth, the groundwater may penetrate to occupy various geological formations known as *aquifers*.

As mentioned in Box 6.9, water is unique in that it is the only substance that exists in all three phases of matter (solid, liquid, and vapour) at the ambient temperatures and pressure of the earth's surface. Water is stored in all three forms within the hydrological cycle and moves among these forms by the processes shown in Figure 6.9. *Sublimation* is the process for direct transfer between the solid and vapour phases of matter, regardless of direction. This explains why on bright, sunny winter days when the air is dry, snowbanks may decrease in size without any visible melting. About 75 per cent of the world's fresh water is stored in the solid phase, and it may stay in this phase for a long time. Analysis of ice in Antarctic, for example, indicate that some of it is over 100,000 years old.

Although there are relatively constant amounts of water in the different storage compartments over the short term, over the long term these can

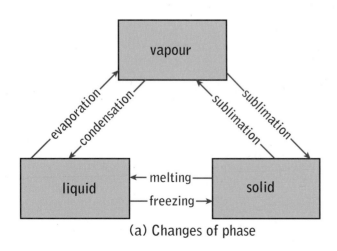

(a) Changes of phase

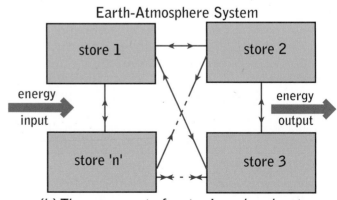

(b) The movement of water in a closed system

Figure 6.9 Changes of phase in the hydrological cycle.

Box 6.8 Groundwater

Groundwater is found within spaces between soil and rock particles and in crevices and cracks in the rocks below the surface of the earth. Above the water table is the *unsaturated* zone where the spaces contain both water and air. In this zone water is called *soil moisture*. Like surface water, groundwater moves downhill, but rarely as quickly, and not at all through *impermeable* materials, such as clay. *Permeable* materials allow the passage of water, usually through cracks and spaces between particles. An *aquifer* is a formation of permeable rocks or loose materials that contains usable sources of groundwater. They vary greatly in size and composition. *Porous media* aquifers consist of materials such as sand and gravel through which the water moves. *Fractured* aquifers occur where the water moves through joints and cracks in solid rock. If an aquifer lies between layers of impermeable material, it is called a *confined* aquifer, which may be punctured by an *artesian well*, releasing the pressurized water to the surface. If the pressure is sufficient to bring water to the surface, the well is known as a *flowing artesian well*.

Areas where water enters aquifers are known as *recharge* areas. *Discharge* areas are ones in which water appears once more above ground. Such discharge areas can contribute significantly to surface water flow, especially in periods of low precipitation. As pointed out in the text, groundwater is also a very significant part of the Canadian water supply. Dependencies range from 100 per cent in Prince Edward Island down to 17 per cent in Quebec, with Ontario having the largest total consumption of groundwater.

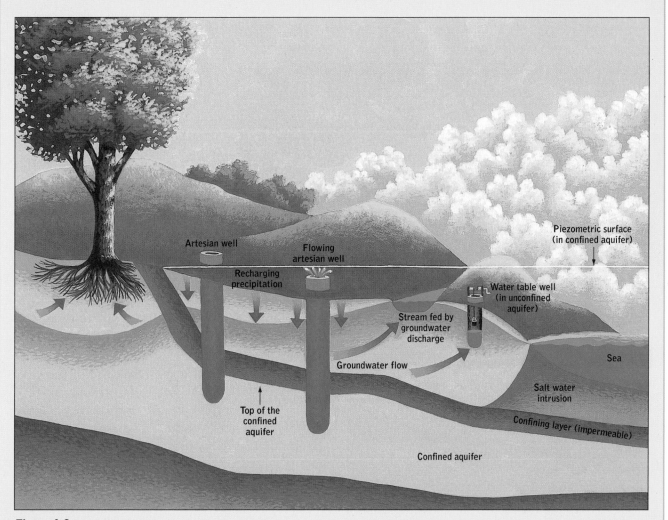

Figure 6.8 Groundwater flow.

Box 6.9 Some Important Properties of Water

Water has several properties that make it unique:

- Water is a molecule (H₂O) that can exist in a solid, gaseous, or solid state.
- These molecules have a strong mutual attraction, promoting high surface tension and high capacity to adhere to other surfaces. These properties allow water to move upwards through plants.
- Water has a high heat capacity, enabling it to store a great deal of heat without an equivalent rise in temperature. This is why the oceans have such a moderating influence on climate.
- It takes a lot of heat to change water from its liquid to gaseous form. This is why evaporation results in a cooling effect.
- There are few solids that do not undergo some dissolution in water. This allows water to carry dissolved nutrients to plants, but it also means that water is easily polluted.
- Unlike other substances, when water passes from the liquid to solid phase of matter it becomes less rather than more dense. This is why ice floats on top of water, which permits aquatic life to exist in cold climates.

Most of these properties spring from the fact that although the water molecule is electrically neutral, the charges are distributed in a bipolar manner. In other words, there is a positive charge at one end of the molecule and a negative charge at the other. This means that water molecules have a very strong attraction to each other, and also explains why water is such a good solvent as the charges increase the chemical reactivity of other substances.

Water in the solid phase of ice may be in these large glaciers of the St Elias range in western Yukon for thousands of years before melting and flowing to the ocean (*Philip Dearden*).

A river is water in its loveliest form; rivers have life and sound and movement and infinity of variation, rivers are veins of the earth through which the life blood returns to the heart.

Roderick Haig-Brown,
A River Never Sleeps, *1946*

change markedly. Large amounts of water are evaporated from the oceans and precipitated on land as snow during glacial periods, for example. Over time the snow accumulates and builds ice-fields that may be over 1 km thick. This effectively removes water from the oceanic component, causing the sea level to fall. During glacial times, the area of land will increase relative to the ocean, and the area covered by ice may increase up to 300 per cent.

In Canada the solid phase of water is particularly important. Canada is estimated to have perhaps up to one-third of the world's fresh water, but most of this is held in the solid phase. Canada's 100,000 glaciers' volume of fresh water is over 1.5 times that of the surface supply. Furthermore, over 95 per cent of the country is covered with snow for part of the winter. Spring melt is hence a critical part of the hydrological cycle in Canada as water moves from the solid to liquid phase. This creates a run-off regime for many Canadian rivers, characterized by very low late winter flows and a

Box 6.10 Human Impacts on the Hydrological Cycle

Human impacts on the hydrological cycle have been large scale and long-standing. Humans depend on water for survival and to provide food and other products. Impacts on the hydrological cycle include:

- the storage and redistribution of run-off to augment water supplies for domestic, agricultural, and industrial uses
- the building of storage structures to control floods
- the drainage of wetlands
- the pumping of groundwater
- cloud seeding
- land-use changes, such as deforestation, urbanization, and agriculture, which affect run-off and evapotranspiration patterns
- climatic change caused by interference with biogeochemical cycles

Water is also universally appreciated for its aesthetic qualities and gives us some of the most beautiful and well-known natural attractions, such as Niagara Falls (*Philip Dearden*).

ter) on this already critical situation could be of major proportions. Already severely water-deficient areas could become more stressed, and the irrigation water needed to feed the world's population may no longer be available. Competition for scarce water resources will undoubtedly increase between different users and nations in the future.

IMPLICATIONS

This chapter deals with some of the most fundamental aspects of how the planetary ecosystems work: the cycling of matter between the different components. Without such cycling, life would not be possible, yet we rearrange the paths of these cycles with impunity and drastically reallocate quantities between components. As you read this book, you are taking part in global biogeochem-

very high spring melt flow that slowly diminishes over the summer into the winter lows as water is stored in the solid phase once more. This marked seasonality is one of the reasons why Canada has developed considerable expertise in the construction of water-storage facilities.

Canada also has abundant storage of fresh water in lake systems, covering almost 8 per cent of the country's area. These lakes are replenished by river flow, which contains approximately 9 per cent of the world supply (Table 6.4). About 60 per cent of the discharge in Canada drains north to the Arctic Ocean (Figure 6.10), whereas 90 per cent of the Canadian population lives within 300 km of the US border, creating the potential for water deficits in this water-rich country (see Chapter 18).

Nevertheless, on a global scale, Canada is exceedingly fortunate regarding water resources. Many countries already face critical shortages, as discussed in Chapter 1. The implications of climatic change (discussed in the next chap-

Table 6.4 Mean Annual Stream Discharge to the Oceans for Selected Canadian Rivers

River	Area (km²)	Discharge (m³ s⁻¹)
Saguenay	90 100	1820
St Lawrence	1 026 000	9860
Churchill	281 300	1200
Nelson	722 600	2370
Albany	133 900	1400
Koksoak	133 400	2550
Yukon (at Alaska border)	297 300	2320
Fraser	219 600	3540
Columbia (at Washington border)	154 600	2800
Mackenzie	984 195	10 800

Source: D. Briggs et al., *Fundamentals of Physical Geography*, 2nd Canadian ed. (Toronto: Copp Clark Pitman, 1993):206.

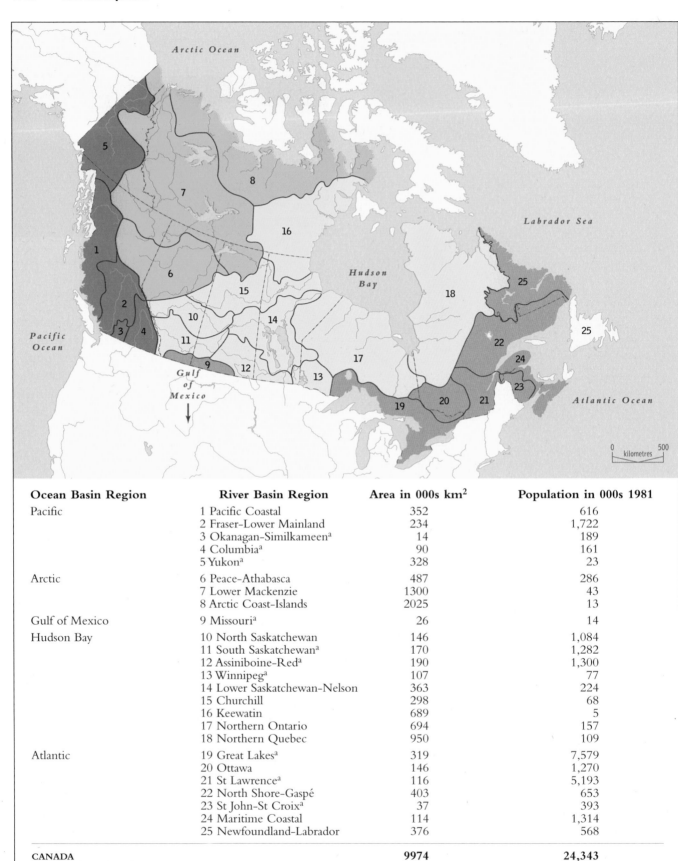

Ocean Basin Region	River Basin Region	Area in 000s km²	Population in 000s 1981
Pacific	1 Pacific Coastal	352	616
	2 Fraser-Lower Mainland	234	1,722
	3 Okanagan-Similkameen[a]	14	189
	4 Columbia[a]	90	161
	5 Yukon[a]	328	23
Arctic	6 Peace-Athabasca	487	286
	7 Lower Mackenzie	1300	43
	8 Arctic Coast-Islands	2025	13
Gulf of Mexico	9 Missouri[a]	26	14
Hudson Bay	10 North Saskatchewan	146	1,084
	11 South Saskatchewan[a]	170	1,282
	12 Assiniboine-Red[a]	190	1,300
	13 Winnipeg[a]	107	77
	14 Lower Saskatchewan-Nelson	363	224
	15 Churchill	298	68
	16 Keewatin	689	5
	17 Northern Ontario	694	157
	18 Northern Quebec	950	109
Atlantic	19 Great Lakes[a]	319	7,579
	20 Ottawa	146	1,270
	21 St Lawrence[a]	116	5,193
	22 North Shore-Gaspé	403	653
	23 St John-St Croix[a]	37	393
	24 Maritime Coastal	114	1,314
	25 Newfoundland-Labrador	376	568
CANADA		9974	24,343

Figure 6.10 Drainage regions of Canada. [a]Canadian portion only; area and population on US side of international basin regions are excluded from totals. SOURCE: Environment Canada, *Currents of Change: Final Report, Inquiry on Federal Water Policy* (Ottawa: Environment Canada, 1985):35. Reproduced with the permission of the Minister of Public Works and Government Services Canada, 1997.

ical cycling, as is every other organism, but most humans think little of their impacts on these cycles. We will deal with this topic in more detail in the next chapter.

SUMMARY

1. Matter has mass and takes up space. It is composed of ninety-two natural and seventeen synthesized chemical elements. According to the law of conservation of matter, matter can neither be created nor destroyed, but merely transformed from one form into another. Matter cannot be consumed.

2. Elements necessary for life are known as nutrients. They cycle between the different components of the ecosphere in characteristic paths known as biogeochemical cycles.

3. Humans disturb these cycles through various activities resulting in environmental problems, such as acid rain and global warming.

4. Cycles can be classified into gaseous or sedimentary depending upon the location of their major reserves.

5. Phosphorus is an example of a sedimentary cycle. The main reservoir of phosphorus is the earth's crust. Phosphates are made available in the soil water through erosional processes and are taken up by plant roots and passed along the food chain. There is no atmospheric component to the cycle, making it especially vulnerable to disruption. The main human use for phosphorus is as fertilizer. It is a main cause of eutrophication (discussed in more detail in the next chapter).

6. Sulphur is also a sedimentary cycle, but differs from phosphorus in that it has an atmospheric component. Like phosphorus, it is an essential component of all life. Bacteria enable plants to gain access to elemental sulphur by transforming it to sulphates in the soil. Sulphur is a main component of acid deposition.

7. Nitrogen is a gaseous cycle. Almost 80 per cent of the atmosphere is composed of nitrogen gas, yet most organisms cannot use it as a source of nitrates. Instead, various bacteria help transform nitrogen into a form that can be used by plants. As with the other cycles, these nitrates are then passed along the food chain. Nitrates are used as fertilizers and contribute to eutrophication. Various nitrous oxides also contribute to acid deposition and the catalytic destruction of ozone.

8. Carbon dioxide constitutes only 0.03 per cent of the atmosphere, but is the main source of carbon, which is the basis for life through the process of photosynthesis. Carbon is incorporated into the biomass and passed along the food chain. Organisms' respiration transforms some of this carbon back into carbon dioxide, and the cellular respiration of decomposers helps to return the carbon from dead organisms into the atmosphere. Carbon dioxide emissions from burning fossil fuels are a main contributor to global climatic change.

9. Water travels between the different components of the ecosphere through the hydrological cycle, fuelled by energy from the sun. Ninety-seven per cent of water is in the oceans. Less than 1 per cent is readily available for human use.

10. Canada has up to one-third of the world's fresh water. Most of this is in the solid form of ice. Canada also has high storage capacities for liquid water, with lakes covering an estimated 8 per cent of the country. These are replenished by a river flow containing approximately 9 per cent of the world's total river discharge.

REVIEW QUESTIONS

1. Summarize some of the key differences and similarities between energy and matter.

2. Why is life dependent upon biogeochemical cycles?

3. Explain why decomposer organisms are important in biogeochemical cycling.

4. What are some of the important implications of biogeochemical cycling for forestry and agricultural activities?

5. Outline the main characteristics of the hydrological cycle in Canada.

REFERENCES AND SUGGESTED READING

Bowen, R. 1986. *Groundwater*. London: Elsevier Applied Science Publishers.

Briggs, D., P. Smithson, T. Ball, P. Johnson, P. Kershaw, and A. Lewkowicz. 1993. *Fundamentals of Physical Geography*, 2nd Canadian ed. Toronto: Copp Clark Pitman.

Chapin, D.M., L.C. Bliss, and L.J. Bledsoe. 1991. 'Environmental Regulation of Nitrogen Fixation in a High Arctic Lowland Ecosystem'. *Canadian Journal of Botany* 69:2744–55.

Environment Canada. 1985. *Currents of Change: Final Report, Inquiry on Federal Water Policy*. Ottawa: Environment Canada.

Haig-Brown, R. 1946. *A River Never Sleeps*. New York: W. Morrow.

Hare, F.K., and M.K. Thomas. 1974. *Climate Canada*. Toronto: Wiley.

Howarth, R.W., J.W.B. Stewart, and M.V. Ivanoy. 1992. *Sulphur Cycling on the Continents: Wetlands, Terrestrial Ecosystems and Associated Water Bodies*. Toronto: Wiley.

Lennihan, R., D.M. Chapin, and L.G. Dickson. 1994. 'Nitrogen Fixation and Photosynthesis in High Arctic Forms of *Nostoc commune*'. *Canadian Journal of Botany* 72:940–5.

Odum, E.P. 1989. *Ecology and Our Endangered Life-Support Systems*. Sunderland, Mass.: Sinauer Associates.

Rowe, S. 1993. 'In Search of the Holy Grass: How to Bond with the Wilderness in Nature and Ourselves'. *Environment Views* (Winter):7–11.

Chapter 7

Human Activity and Biogeochemical Cycling

Despite the apparent sophistication of human society, the humble fact remains that society could not exist without biogeochemical cycles and those unpretentious bacteria that make them work. All the cycles are also susceptible to perturbations by human activity. Such is the scale of human actions that the major transfers between some of the reservoirs in the cycles are human induced. Some of society's most well-known environmental challenges result from these transfers. This chapter covers three of these in more detail: eutrophication, acid deposition, and global climatic change.

EUTROPHICATION

What Is Eutrophication?

Eutrophication is a natural nutrient enrichment of water bodies that makes them more productive. It is an appropriate place to start the discussion on disruption of biogeochemical cycles because we have relatively more experience with this perturbation, it is an important aspect of water pollution in Canada, and it also demonstrates that (at least with some problems) it is possible to have a positive impact when sufficiently strong action is taken.

Phosphorus and nitrogen are often the two main limiting factors for plant growth in aquatic ecosystems. Systems with relatively low nutrient levels, *oligotrophic* ecosystems, have quite different characteristics than those with high nutrient levels (*eutrophic*), as summarized

in Table 7.1. Natural terrestrial ecosystems are relatively efficient in holding nutrient capital. The progression from an oligotrophic to eutrophic condition (through the process of succession discussed in Chapter 5) may take place over thousands of years. This rate is influenced by the geological make-up of the catchment area and the depth of the receiving waters. Catchments with fertile soils will progress more quickly than those with soils lacking in nutrients. Depth is important because shallower lakes tend to recycle nutrients more efficiently.

How Is Eutrophication Caused?

Cultural eutrophication (eutrophication caused by human activity) speeds up this process considerably by a few decades, mainly through the addition of excess phosphates and nitrates to the water body. As the lake becomes shallower because of this input, nutrients are used more efficiently, productivity increases, and eutrophication progresses. This

Table 7.1 Characteristics of Oligotrophic and Eutrophic Water Bodies

Characteristic	Oligotrophic	Eutrophic
Nutrient cycling	low	high
Productivity (total biomass)	low	high
Species diversity	high*	low
Relative numbers of undesirable species	low	high
Water quality	high	low

*Lakes that are extremely non-productive (e.g., high mountain lakes) will have low species diversity.

Animal feedlots are a major source of nutrients, such as phosphates and nitrates, which speed up eutrophication (*Philip Dearden*).

is a classic example of a *positive feedback loop* as change in the system promotes even more change in the same direction. Additional phosphates and nitrates come from many different sources (Table 7.2), and in accord with the law of conservation of matter (Chapter 6), they do not simply disappear but accumulate in aquatic ecosystems.

What Are the Effects?

This enrichment promotes increased growth of aquatic plants, particularly favouring the growth of floating phytoplankton over the *benthic* plants rooted in the substrate. As the phytoplankton get more light than the benthic plants, the plants produce less oxygen at lower depths. Oxygen is critical for the maintenance of more diverse, oxygen-demanding fish species, such as trout and other members of the salmonid family, which also start to decline in number. The oxygen produced by the phytoplankton through photosynthesis tends to stay in the shallower water and returns to the atmosphere rather than replenish supplies at greater depths.

Oxygen depletion is further exacerbated by the decay of the large mass of phytoplankton produced. Dead matter filters to the bottom of the lake where it is consumed by oxygen-demanding decomposers. Once broken down, nutrients may be returned to the surface through convection currents and provide more food for more phytoplankton and algae. Blue-green algae replace green algae in eutrophic lakes, which further exacerbates the problem as most blue-green algae are not consumed by the next trophic level, the zooplankton.

These effects of oxygen depletion in a water body also occur whenever excess organic matter is added. Under natural conditions, water is able to absorb and break down small amounts of organic matter; the amount broken down depends upon the size, flow, and temperature of the receiving water body. The greater the size and flow and the lower the temperature of the water, the greater the ability to absorb organic materials and retain oxygen levels.

When organic wastes are added to a body of water, the oxygen levels fall as the number of bacteria rises to help break down the waste. This is known as the *oxygen sag curve* (Figure 7.1), which is measured by the *biological oxygen demand (BOD)*, the amount of dissolved oxygen needed by aerobic decomposers to break down the organic material in a given volume of water at a certain temperature over a given period. At the discharge source, the oxygen sag curve starts to fall and there is a corresponding rise in the BOD. As distance from the input source increases and the bacteria digest the wastes, then the oxygen content returns to normal and the BOD falls.

Table 7.2 Main Nutrient Sources Contributing to Cultural Eutrophication

Run-off from	fertilizers (N and P)
	feedlots (N and P)
	land-use change, such as cultivation, construction, mining natural sources
Discharge of	detergents (P)
	untreated sewage (N and P)
	primary and secondary treated sewage (N and P)
Dissolved nitrogen oxides	(from internal combustion engines)

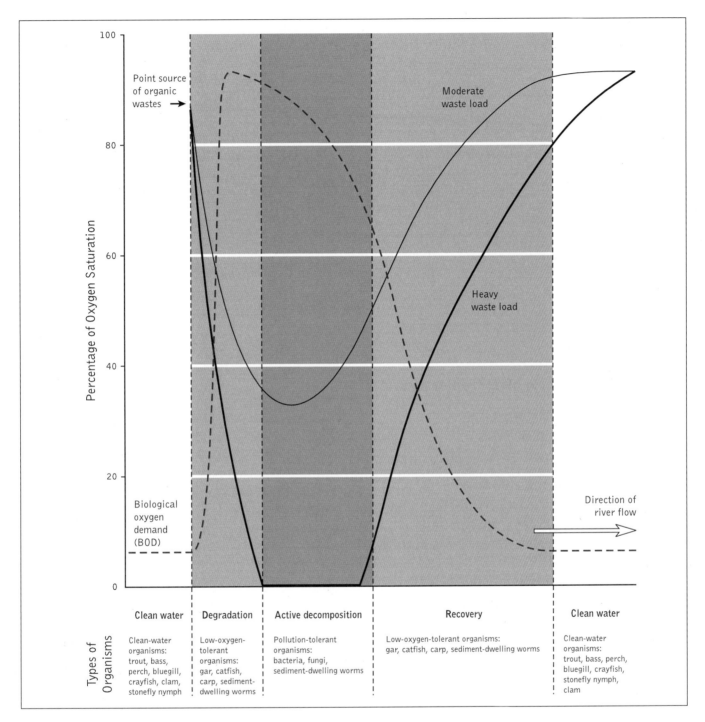

Figure 7.1 Oxygen sag curve and biological oxygen demand (BOD).

Major sources of nitrates and phosphates, such as run-off from feedlots and sewage discharge, also contain large amounts of oxygen-demanding wastes. Heat is another source of oxygen stress. The overall result is a progression to a less useful and less healthy water body. The composition of the fish species changes to those less dependent on high oxygen levels, species that are generally less desirable for human purposes. Populations of waterfowl may fall as aquatic plants die off, as was reported for Lake Erie (Prince 1985). The water becomes infested with algae, aquatic weeds, and phytoplankton, making swimming and boating unpleasant and giving off unpleasant odours. Water treatment for domestic or industrial purposes becomes more expensive.

What Can We Do About It?

The main way to control eutrophication is by limiting the input of nutrients into the water body (Table 7.2). Domestic and animal wastes must be treated to remove phosphates. Advanced treatment can remove up to 90 per cent. More difficult problems occur with diffuse, *non-point* sources, such as run-off from urban areas and agricultural land. Such flows must be controlled at the source since they enter the water body, by definition, in so many different locations. In the past, measures to control water pollution have been directed largely towards *point* sources of pollution, or single discharge points, such as effluent discharges from sewage plants or industrial processes. By and large, these sources are easy to identify and monitor because of their high visibility, and pollution from such sources has fallen as a result. Increasing attention is now being directed towards the non-point sources.

Lake Erie: An Example of Eutrophication Control

In all parts of the country, there are many examples of eutrophic lakes. Lake Erie (Figure 7.2) is a particularly well-known case because of its size and importance. It is the second smallest and also the shallowest of the Great Lakes, which together contain almost 20 per cent of the world's fresh water. Lake Erie has experienced considerable changes in fish species composition since the early explorers

described a highly diverse community. Gone are the lake sturgeon, cisco, blue pike, and lake whitefish as the human population in the basin has increased and water quality has declined. There are some 13 million people in the Erie drainage basin, and 39 per cent of the Canadian and 44 per cent of the American shore is taken up by urban development. There is also considerable industrial use around the shore and intensive agricultural use throughout the basin.

In the past, up to 90 per cent of the bottom layer of the lake's central zone became oxygen deficient in summer. Huge algae mats over 20 m in length and 1 m deep were common. Beaches were closed. The natural eutrophication that might have taken thousands of years was superceded by cultural eutrophication in the space of fifty years.

In 1972 the Great Lakes Water Quality Agreement was signed by Canada and the US to try to come to terms with this problem. International efforts were further strengthened by the signing of the 1985 Great Lakes Charter in which it was agreed to take a cooperative and ecosystem approach to the lakes. Since the 1970s, phosphorus controls, implemented under the Canada Water Act, have led to significant reductions in the phosphorus concentration of the water (Figure 7.3). Phosphate-based detergents were banned and municipal waste treatment plants upgraded.

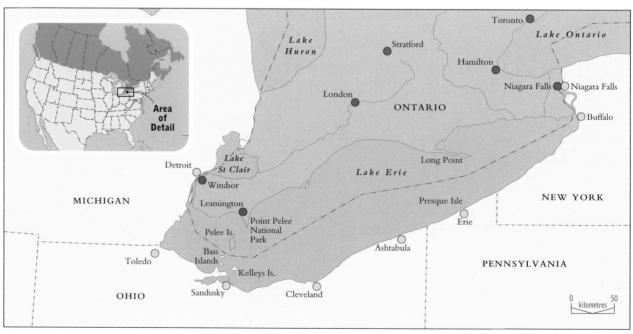

Figure 7.2 Lake Erie.

These measures have led to improvement in water quality, but significant problems still remain. The controls are largely on point-source pollution: discharges that have a readily identifiable source, such as waste treatment plants and industrial complexes. However, much of the remaining nutrient load comes from non-point sources, such as run-off from agricultural fields, lawn fertilizer, and construction sites, which are much more difficult to regulate. Furthermore, the levels of nitrite and nitrate concentrations have been rising, leading to fears that although phosphorus controls might have been effective, there is still potential for increased eutrophication because of other nutrients. Another issue not addressed by the controls on phosphorus relates to toxic wastes; separate measures are needed to deal with them.

ACID DEPOSITION

In 1966 fisheries researcher Harold Harvey was puzzled to find that the 4,000 pink salmon he introduced to Lumsden Lake in the La Cloche Mountains southwest of Sudbury, Ontario, the previous year had all disappeared. Their passage upstream and downstream of the lake had been blocked. What could have happened to them? To unravel the mystery, he began to take more measurements of the lake and look into its past history. The results were startling. It was not only the salmon that had disappeared but also many other species of fish indigenous to the lake (Table 7.3). The reason soon became apparent. Between 1961 and 1971, Lumsden Lake had experienced a 100

per cent increase in the acidity of its waters, along with many other lakes in the same region (Table 7.4). The changes had shifted the lakes beyond the species' limits of tolerance, as discussed in Chapter 4. They and many of the species upon which they depended for food simply could not tolerate the new conditions and had perished. They were victims of the effects of acid deposition.

Table 7.3 Disappearance of Fish from Lumsden Lake

1950s	Eight species present
1960	Last report of yellow perch
1960	Last report of burbot
1960–5	Sport fishery fails
1967	Last capture of lake trout
1967	Last capture of slimy sculpin
1968	White sucker suddenly rare
1969	Last capture of trout-perch
1969	Last capture of lake herring
1969	Last capture of white sucker
1970	One fish species present
1971	Lake chub very rare

Source: H. Harvey, 'The Acid Deposition Problem and Emerging Research Needs in the Toxicology of Fishes', *Proceedings of the Fifth Annual Aquatic Toxicology Workshop*, Hamilton, Ontario, 7–9 November 1978. Reprinted with author's permission.

Table 7.4 Lake Acidification in the La Cloche Mountains, 1961–1971

Lake	pH 1961	pH 1971
Broker	6.8	4.7
David	5.2	4.3
George	6.5	4.7
Johnnie	6.8	4.8
Lumsden	6.8	4.4
Mahzenazing	6.8	5.3
O.A.S.	5.5	4.3
Spoon	6.8	5.6
Sunfish	6.8	5.6
Grey (1959)	5.6	4.1
Tyson (1955)	7.4	4.9

Source: H. Harvey, unpublished speech, based on R. Beamish and H. Harvey, 'Acidification of the La Cloche Mountain Lakes, Ontario, and Resulting Fish Mortalities', *Journal of the Fisheries Research Board of Canada* 29, No. 8 (1972):1135. Reprinted with the author's permission.

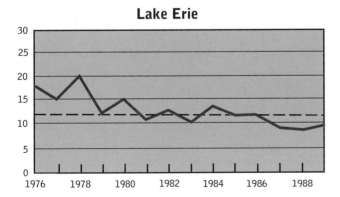

Lake Erie

Figure 7.3 Estimated total phosphorus loadings in thousands of tonnes in Lake Erie. SOURCE: Environment Canada, *The State of Canada's Environment* (Ottawa: Minister of Supply and Services Canada, 1991):12.

What Is Acid Deposition?

Acids are chemicals that release hydrogen ions (H+) when dissolved in water, whereas a *base* is one that releases hydroxyl ions (OH-). These two neutralize each other when they come together to form water (H_2O). Acidity is a measure of the concentration of hydrogen ions in a solution, and is measured on the *pH scale*, which goes from 0 to 14 (Figure 7.4). The midpoint of the scale, pH 7, represents a neutral balance between the presence of acidic hydrogen ions and basic hydroxyl ions. The pH scale is logarithmic. A decrease in value from pH 6 to pH 5 means that the solution has become ten times more acidic. If the number drops to pH 4 from pH 6, then the solution is 100 times more acidic.

Precipitation, either as snow or rain, tends to be slightly acidic, even without human interference, due to the chemical reaction in the atmosphere as carbon dioxide in the atmosphere combines with water to form carbonic acid. Generally, a pH value of 5.6 is accorded to 'clean' rain. Acid rain is defined as deposition that is more acidic than this. In Canada rainfall has been recorded with pH levels much lower than this. Acidic deposition is a more generic term that includes not only rainfall but also snow, fog, and dry deposition from dust.

How Is It Caused?

The increases in acidity shown in Figure 7.5 and reflected in the pH levels of the lakes in Table 7.4 are due to human interference in the sulphur and nitrogen cycles. The largest sources of interference are through the smelting of sulphur-rich metal ores and the burning of fossil fuels for energy. These processes change the distribution of the elements between the various pools that were shown in figures 6.3 and 6.4, such that natural processes are inadequate to deal with the build-up of matter. Increased amounts of sulphur and various forms of nitrogen are released into the atmosphere where they may travel thousands of kilometres before being returned to the lithosphere as a result of depositional processes. Human activities account for more than 90 per cent of the sulphur dioxide and nitrogen oxide emissions in North America.

Excessive sulphur is produced when ore bodies, such as copper and nickel, are smelted at high temperatures to release the metal. Unfortunately, such ores often contain more sulphur than metal, and the sulphur is released into the atmosphere as a waste product of the process. A similar effect is created when sulphur-containing coal is burned as the energy source in powerplants. In Canada the smelting of metal ores accounts for half of the sulphur dioxide emissions east of Saskatchewan, with power generation and other sources accounting for 20 per cent and 30 per cent respectively. In the US electrical utilities are the largest source of sulphur dioxide, accounting for 72 per cent of emissions in 1993. The effects of these emissions became obvious around smelting plants such as those at Inco in Sudbury and Trail in BC (Box 7.1). Trees were destroyed over large areas. Now there are very encouraging signs of rehabilitation (as discussed in more detail in Chapter 13).

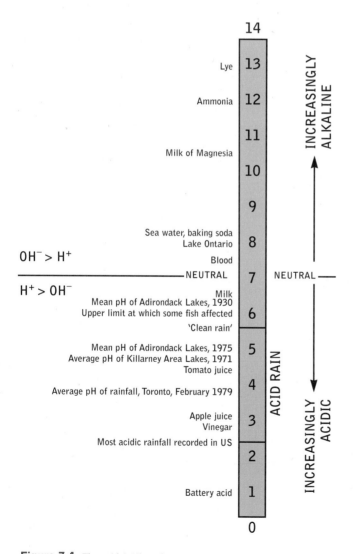

Figure 7.4 The acid (pH) scale.

Box 7.1 Trail, British Columbia

The Trail lead-zinc smelter opened in 1896 and released as much as 9450 tonnes of sulphur dioxide annually by 1930. By 1936 all conifers within a 16-km radius had been destroyed by acid deposition, with signs of damage extending over 60 km away. Some recovery is now underway as emissions have fallen.

This is not the only concern regarding the smelter, however. Tests in the early 1990s revealed that lead deposition in the community was so high that three-quarters of the preschoolers have levels high enough to be of medical concern. At this level, the US Department of Health and Human Services concluded that such children may have lower intelligence and slower neurobehavioural development. Expected declines of between three to ten IQ points would be expected, which would put three times more children into the retarded end of the IQ spectrum. Regular monitoring programs of

lead levels in children have now been started. However, the smelter still turns out over 300 kg of lead per day. Residential areas have soil levels that are six times the Canadian suggested guideline. Cominco Metals is now developing a new smelting process by which it hopes to reduce these levels.

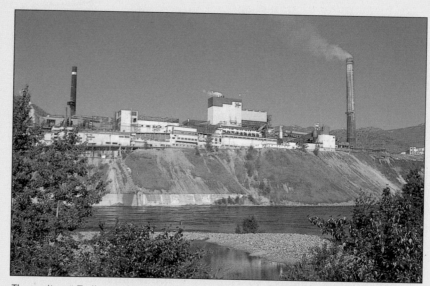

The smelter at Trail caused extensive damage to the surrounding vegetation (*Philip Dearden*).

These obvious signs of ecological damage were ignored for many years as they were considered an unavoidable cost of industrial development. However, as the ecological implications became more well known, governments responded by encouraging industry to build higher stacks to eject the waste farther into the atmosphere. Inco, for example, built a 381-m superstack in 1972. This improved matters somewhat at the local scale, but merely created problems elsewhere, especially in Quebec, when entire air masses were acidified and dropped their acid burdens over a larger area. Because weather patterns are not random, these acidified air masses tend to travel in the same kinds of patterns. In central and eastern Canada, as air masses travel from southwest to northeast, they bring heavy pollution from the heavily industrialized Ohio Valley in the US, which falls mostly in Canada. It is estimated that approximately half of the sulphate falling in Canada originates in the US.

These point sources of pollution are, however, easier to monitor and control than the other main

source of acids, nitrogen emissions, 35 per cent of which comes from various means of transport as a result of high-temperature combustion. The remainder is split between emissions from thermoelectric generating stations and other industrial, commercial, and residential combustion processes.

The emission stack of the Inco smelter in Sudbury was once the largest anthropogenic source of sulphur emissions in the world (*Canapress Photo Service*).

What Are the Effects of Acid Deposition?

Aquatic Effects

The effects of acid deposition on the fish of Lumsden Lake and other aquatic ecosystems are one visible sign of some of the impacts of acid deposition. Other species are also affected as the pH of the water body declines. Indeed, as can be seen in Figure 7.5, fish are often not the most sensitive species and are really more of an indicator of the damage that has already occurred. When insects, such as mayflies, are eliminated, species higher in the food chain that feed on them are also affected through food depletion. The same is true of fish-eating birds, such as loons, whose young have been shown to have a lower chance of survival on acidified lakes due to starvation.

Unfortunately, some of these impacts are permanent. Examination of historical angling records in Nova Scotia, for example, indicate that of sixty main salmon rivers, thirteen runs of salmon are extinct and a further eighteen are virtually extinct.

Acid rain kills many fish (*Ontario Ministry of Environment and Energy*).

It is estimated that over half of the total salmon production has been lost as a result of declining pH levels. It has also been suggested by Edge (1987) that the endangered Acadian whitefish is threatened by declining pH levels.

Even if fish manage to survive, they may be grossly disfigured with twisted backbones and flattened heads because their bones have been deprived of the nutrients necessary for strength. Reproductive capacities may be sufficiently impaired to lead to eventual population declines. Generally, the time of reproduction is the most sensitive part of the life cycle. The critical factor is often the lower pH level of the water as the snow melts. At this time the build-up of acids over the winter can result in even higher acidity than that experienced during the rest of the year. This pulse of acidity is called the *acid shock* and it may also be one of the causes of stress on amphibious creatures, such as frogs, which often use small temporary pools of water for breeding in the spring following run-off. At Lumsden Lake, for example, spring run-off produced a pH as low as 3.3, over 100 times more acidic than the 1961 levels.

In addition to more acidic water and food chain effects, other chemical changes are also cause for concern. The increased acidity, for example, releases large amounts of

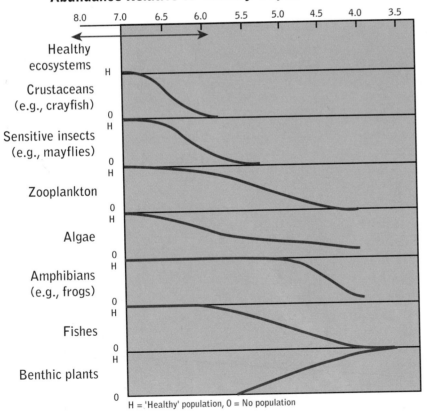

Abundance Relative to 'Healthy' Populations

Figure 7.5 Sensitivity of various aquatic organisms to pH level. SOURCE: Environment Canada, *The State of Canada's Environment* (Ottawa: Minister of Supply and Services Canada, 1991).

aluminium from the terrestrial ecosystem into the rivers and lakes. Here it forms a toxic scum, lethal to many forms of aquatic life.

The extent of acidification in Canadian aquatic ecosystems is not easy to determine, partly due to the huge area involved. One technique is to assess the relative proportions of sulphates and bicarbonates. Where the ratio of bicarbonates to sulphates is less than 1:1, then serious acidification is occurring. Tests on over 8,500 lakes in central and eastern Canada indicate the areas in which the problem is most severe (Figure 7.6).

Terrestrial Effects

Terrestrial effects of acid deposition first became visible around emission sources, such as at Trail and Sudbury (described earlier), as trees began to die. The acids eat away at the sensitive photosynthetic surfaces of the leaves before joining the soil water. Similar effects have been recorded elsewhere. In Germany, for example, foresters estimate that over half the forests are dying through the effects of various pollutants. In the eastern US, severe impacts have been noted, especially at higher elevations where growth conditions may already be stressful. Broad-leaved trees, such as sugar maples (the source of Quebec's $40 million annual maple syrup industry), are thought to be particularly susceptible because of the large surface area of their leaves.

Once in the soil, the acids leach away the nutrients required for plant growth, leading to nutrient deficiencies. The high levels of aluminium that are released by the acids also inhibit the uptake of nutrients. The bacteria that are so critical to the biogeochemical

cycles, as described in the previous chapter, are also adversely affected and cause changes in natural soil processes. Decomposition and humus formation are retarded. The actual physical contact of the acids with the plant roots can also inhibit growth and lower resistance to disease.

These impacts are now visible over much wider areas than those surrounding high-emission sources. Extensive areas of damage have been recorded in Europe and in the eastern United States. At these larger scales, however, it is often difficult to single out one cause, such as acid deposition. It is likely that other factors, such as climate change and high levels of ozone brought about by excessive nitrogen oxides, are also important in placing stress on these communities in synergistic reactions.

In central and eastern Canada, 89 per cent of the high capability forest land receives more than 20 kg per hectare of acid deposition, which is considered an acceptable level of deposition. Severe impacts have been noticed. A survey of white birch around the Bay of Fundy, for example, found 10

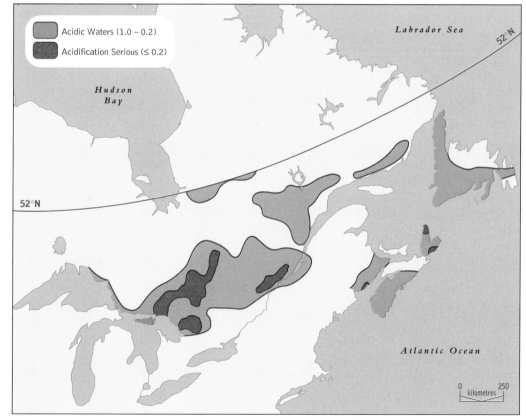

Figure 7.6 Acidification of surface waters in eastern Canada. SOURCE: D.S. Jeffries, 'Southwestern Canada: An Overview of the Effects of Acid Deposition on Aquatic Resources', in *Acid Deposition and Aquatic Ecosystems: Regional Case Studies*, edited by D.F. Charles and S. Christie (New York: Springer-Verlag, 1991):273–89. Reprinted by permission of Springer-Verlag New York Inc.

Box 7.2 Soil Organisms and Acid Deposition

The soil contains many organisms, including mites, insects, worms, bacteria, fungi, and many others involved in the critical ecological functions of biogeochemical cycling and energy flow. There may be over 100 million bacteria and several kilometres of fungal hyphae, for example, in a single gram of healthy soil. What happens to these organisms when they are subject to increased acidity is one of the key questions about the impacts of acid deposition.

Mycorrhizal fungi, for example, are very important to the growth of many plants as they help transport nutrients from the soil water into the roots. Browning and Hutchinson (1991) undertook research on jack pine to find out the potential impacts of increased acidity on these mutualistic relationships. They found that changes in calcium to aluminium ratios caused by increased acidity influenced the succession of mycorrhizal fungi on tree-root systems. The potential effects of these kinds of changes on tree growth, the ecological health of the community, and forest yields are still uncertain.

per cent of the trees already dead and virtually all others showing signs of damage (RMCC 1990).

The impacts on plant life are not restricted to natural ecosystems. Significant changes can also occur on agricultural lands as direct damage to crops or as more indirect changes through alterations to soil chemistry. The growth of crops—such as beets, radishes, tomatoes, beans, and lettuce—is inhibited, and biological nitrogen fixation is diminished at a pH of 4. In central and eastern Canada, 84 per cent of the most productive agricultural lands receive more than 20 kg per hectare. In some areas the application of lime to the soil has become a routine agricultural procedure to neutralize the acids by adding more basic ions.

Heterotrophs can also be affected. As the forest cover diminishes, so does the habitat for many species. Toxic metals—such as cadmium, zinc, and mercury—released from the soil by the acids may be concentrated in certain species of plants and lichens and accumulate in the livers of the species eating them, such as moose and caribou.

Sensitivity

Not all ecosystems are equally sensitive to the effects of acid deposition. Some areas have a high capacity to neutralize the excess acids due to the high base capacity of the bedrock and soils. The Prairies are such a region, where underlying carbonate-rich rocks, deep soils, and other factors combine to provide high *buffering* capacity. On the other hand, areas with difficult-to-weather rocks with low-nutrient content, such as granite, and with thin soils following glaciation, often have very low buffering capacities. Much of central and east-

ern Canada and coastal British Columbia are in this latter category (Figure 7.7). The provinces most vulnerable to acid deposition in terms of the amount of area ranked as highly sensitive are Quebec (82 per cent), Newfoundland (56 per cent), and Nova Scotia (45 per cent). The spatial coincidence of low buffering capacities and high deposition rates explains why most attention in Canada has centred on the central and eastern parts of the country.

Socio-economic Effects

The environmental effects described earlier have socio-economic implications, and acid deposition has direct effects on human health. In aquatic ecosystems, for example, declines in fish populations obviously have implications for those involved in fishing, whether commercially or for sport. One study suggests that the increased angler activity made possible as a result of emission control actions in central and eastern Canada would be worth $380 million between 1986 and 2015 (DPA Group 1987). The impacts of acid deposition on tree growth and the forest industry are also of considerable concern. Some studies indicate that tree-growth reductions of up to 20 per cent might be experienced. An economic study suggests a reduction in the forest industry of $197 million per year (Minister of the Environment 1991). The full effects, however, are likely to be much greater than this. European research indicates that a time lag of some twenty to thirty years is likely before the effects of acid rain are reflected in reduced tree growth. As outlined in Box 7.2, relatively little is known about many of the complex relationships in

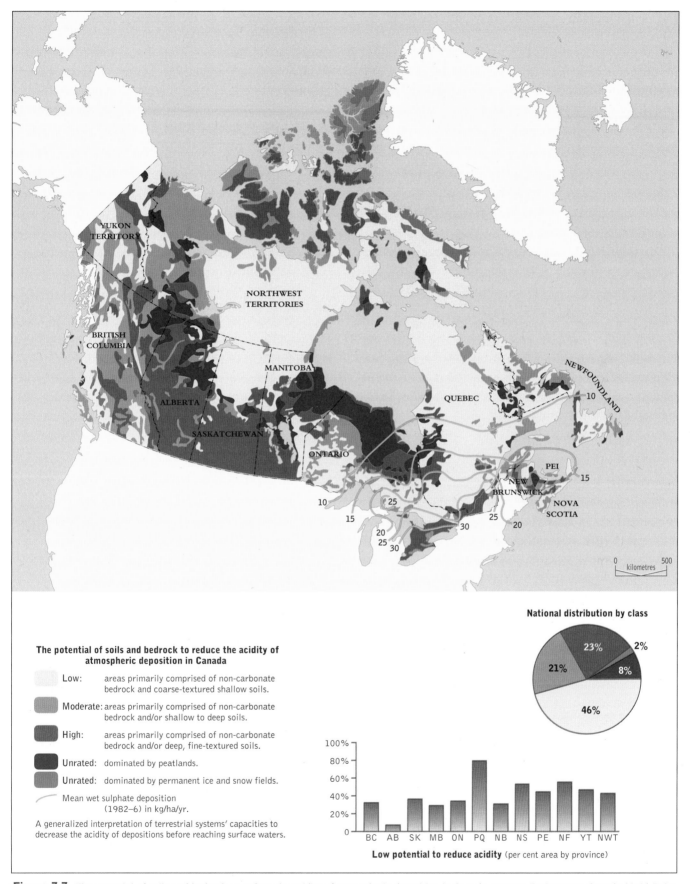

Figure 7.7 The potential of soils and bedrock to reduce the acidity of atmospheric deposition in Canada. SOURCE: Environment Canada, 'Acid Rain: National Sensitivity Assessment', Environment Fact Sheet 88–1 (Ottawa: Environment Canada, 1988).

the soil community that may be affected by increased acidity.

Many values, however, are difficult to express in monetary terms. Thousands of Canadians, for example, maintain a lakeside cottage. It is challenging, and some would say impossible, to ascribe a value to the changes that might occur as the lakes become devoid of life. For example, common loons, which are for many the quintessential symbol of the Canadian wilderness, are quite sensitive to lake acidification. Below pH levels of 4.5, loons do not seem able to find enough food to feed their young (Alvo et al. 1988).

It is not only the natural environment that is damaged as a result of acid deposition but also the human-built environment. The acids eat away at certain building materials. The effects have been most damaging in Europe where there are many old monuments, but can also be seen on the Houses of Parliament and nearby statues in Ottawa. Additional annual maintenance costs in Canada as a result of acid deposition have been estimated at $830 million.

The most direct impacts on human health are thought to result from breathing airborne acidified particles, which can impair respiratory processes and lead to lung damage. There appears to be significant relationships between air pollution levels in southwestern Ontario, for example, and hospital admissions for respiratory illnesses. Comparative studies between heavily polluted areas in Ontario and less polluted areas in Manitoba found diminished lung capacity of about 2 per cent in children in the more polluted area. Further studies have confirmed the relationship, although whether high sulphate or ozone levels are responsible has yet to be conclusively determined.

Humans can also be affected by ingesting some of the products of acid rain. For example, in some areas where older delivery systems for drinking water are in place, the increased acidity can corrode pipes and fittings and result in elevated levels of lead in the water supply. This is why in certain areas, such as Victoria, it is recommended that schools flush their water fountains early in the morning to spill the water that has been held overnight and that might have elevated lead levels.

Acidified water may also have other substances that are deleterious to human health. Excessive levels of aluminium have already been noted.

When metals such as mercury, chromium, and nickel are leached from the substrate into water, they may be taken up and concentrated along the food chain and eventually cause a human health problem.

What Can We Do About It?

Had emission levels continued at 1980 levels, it was predicted that 600,000 lakes in Canada would eventually be virtually lifeless due to excess acidity. Fortunately, this has not happened due to various measures to combat the problem (Box 7.3). Critical early measures were to undertake more scientific research to ascertain the scale and dimensions of the problem, and then to raise public and political awareness to implement the necessary changes. Non-governmental organizations—such as the Sierra Club of Ontario, Pollution Probe, and the Coalition on Acid Rain—played a fundamental role, particularly in this last task.

One of the main challenges associated with acid deposition is that the areas generating the emissions causing the problem are not the sole recipients of the deposition. This was brought to light in Europe where many of the most damaging effects of acid deposition were being felt by the

Box 7.3 Killarney Lakes Recovering

The lakes in the area of Killarney Provincial Park, southwest of Sudbury, have begun to recover. Two decades ago, most life in the lakes had been killed due to acid rain from the Sudbury mining smelters.

An Ontario Ministry of Natural Resources biologist explained in the summer of 1995 that the recovery can be linked directly to a 90 per cent reduction in harmful sulphur emissions from the Sudbury mining operations of Inco and Falconbridge since 1985 (see Chapter 13).

In Killarney Provincial Park, scientists had documented the disappearance of sixteen species of lake trout, sixteen populations of smallmouth bass, and numerous populations of other fish.

Insects and small fish have reappeared in significant numbers in lakes that had been devoid of many aquatic species. Lake trout and great northern pike are also present.

lightly industrialized Scandinavian nations, which were downwind from the more heavily industrialized areas of Britain and Germany. The same is true in Canada where estimates suggest that over half of the acid deposition originates in the US. It has also been found recently that the acidic fog in the Canadian Arctic in the spring is largely a result of emissions from Europe and Russia. To address such problems, international efforts are required.

Canada has addressed the problem through efforts at the national, bilateral, and multilateral scales. In 1983 the Canadian Council of Resource and Environment Ministers agreed upon an annual target deposition of 20 kg per hectare as an acceptable goal, taking political and economic costs into account. Two years later an agreement was reached among all the provinces east of Saskatchewan, the area defined as 'eastern' Canada within the context of acid rain, to set specific emission reductions to reach this target (Table 7.5). Figure 7.8 shows the progress that has been made towards this goal. The area of eastern Canada receiving 20 kg per hectare or more of wet sulphate per year has declined by nearly 59 per cent from 0.71 million km² in 1980 to about 0.29 million km² in 1993.

Table 7.5 Sulphur Dioxide Emissions

	1980 Base Case (tonnes)	1994 Objectives (tonnes)
Manitoba	738 000	550 000
Ontario	2 194 000	885 000
Quebec	1 085 000	600 000
New Brunswick	215 000	185 000
Prince Edward Island	6 000	5 000
Nova Scotia	219 000	204 000
Newfoundland	59 000	45 000
	4 516 000	2 474 000

Source: Government of Canada and Government of the United States of America, *Canada/United States Air Quality Agreement Progress Report* (March 1992):16.

The federal government has committed itself to ensuring that a cap of 2.3 million tonnes of SO_2 emissions annually is maintained in eastern Canada to the year 1999, and that a permanent national cap of 3.2 million tonnes annually will be in place by 2000. This target was reached in 1993 mainly as a result of industrial process changes, installation of

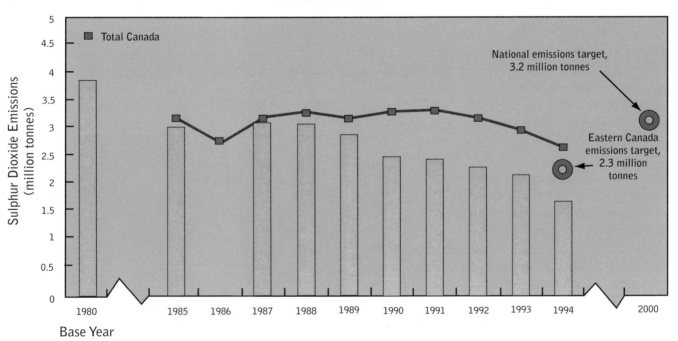

Figure 7.8 Sulphur dioxide emissions in eastern Canada. source: Environment Canada, 'Acid Rain', soe Bulletin no. 96–2 (Ottawa: Environment Canada, 1996).

Automobile emissions reduce the air quality in many urban centres (*Courtesy of the Toronto Transit Commission*).

scrubbers, and fuel-switching. By 1994 SO_2 emissions were down to 2.7 million tonnes, a 41 per cent reduction from the 1980 level of 4.6 million tonnes.

Bilaterally, early efforts to convince decision makers in the US to control emissions met with little success. However, in 1990 intense lobbying efforts came to fruition. The US Clean Air Act was revised to reduce the 1980 sulphur emission levels by half by the turn of the century. It is estimated that this will cost some $4 billion per year, or an increase of roughly 1.5 per cent for each family utility bill. Whether the US will actually meet its target of a 9.1 million tonne reduction by the year 2010 is open to question. However, overall, the amount of acid rain falling on eastern Canada originating in both countries is down by an estimated 33 per cent since controls were implemented.

Efforts to control the emissions of nitrogen oxides have not been quite so successful. Between 1980 and 1990 both Canadian and US emissions remained fairly constant. Canada is committed, under the terms of the Canada–US Air Quality Agreement, to a 10 per cent reduction from stationary sources by the year 2000. Annual emissions in the US must be

reduced by 1.8 million tonnes below that of the 1980 level by the year 2000. Mobile emission reductions are being sought by introducing more stringent performance standards on exhaust emissions from new vehicles. Programs announced in 1996 aim to cut hydrocarbon emissions by about 30 per cent and nitrogen oxide by 60 per cent on 1998 model cars.

Canada has also been active multilaterally, signing the 1979 Convention on Long-Range Trans-Boundary Air Pollution and then taking international leadership in signing the 1985 Helsinki protocol, which committed member countries to a 30 per cent reduction in emissions by 1994, a target that was met. Canada has also been active in conferences and organizing international monitoring networks on acidification.

These measures appear to be having some benefits. A recent study in the Ontario, Quebec, and Atlantic region shows that of 202 lakes monitored since 1983, 33 per cent have improved, 56 per cent have not changed, and 11 per cent experienced increased acidity. However, it was only in the Sudbury region where the majority of monitored lakes showed some improvement, an improvement attributed to the reduced emissions from the Inco smelter. Elsewhere there was little change, due to the ongoing transboundary flows from the US (Figure 7.9). While these studies are reassuring in

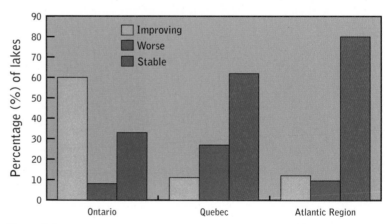

Figure 7.9 Trends in lake acidity by region, 1981–94. SOURCE: Environment Canada, State of the Environment Bulletin No. 96-2 (Ottawa: Environment Canada, Spring 1996).

confirming the relationship between declining emissions and lowering acidity levels, they do not indicate an end to the acid deposition problem for several reasons.

- Scientists fear that emission declines are a result not only of controls but also of the economic recession, and that if the economy picks up, so too will emissions.
- It is still uncertain whether clean-air regulations will be met in the US. Utilities that surpass their emission targets are allowed to trade their excess polluting capacities to heavier polluting facilities. How effective this will be in overall controls remains to be seen.
- Unlike the US, Canada does not have a Clean Air Act that can be enforced by law, creating the possibility of further non-compliance with targets.
- Emission controls have focused largely on point-source control of sulphate emissions. It is more difficult to address the more diffuse nitrogen derivatives coming mainly from the transportation sector.
- Concern has focused on eastern Canada, but there are pockets of acidity all across the country.
- Although some lake pH levels are showing signs of recovery, it is still too early to assess the biotic implications in terms of community health.
- Even if the target of under 20 kg per hectare is reached, this will still exceed the tolerance levels of many species in more vulnerable habitats where half this amount may constitute a critical loading.
- The pH levels are still low enough in many areas to have damaging effects on sensitive species.

CLIMATIC CHANGE

The two environmental problems discussed earlier both show some signs of amelioration. Such is not the case with the final issue to be discussed, global environmental change, of which we may have only just seen the beginning. Like acid deposition, climatic change is of particular concern for Canada. Many of the most drastic effects of global change will be felt at high latitudes. Given the high proportion of the country north of the Arctic Circle,

this could have a major impact on Canada, and is a main reason behind the considerable investment in finding out more about the problem. Furthermore, Canadians are, per capita, among those most responsible for the changes that are taking place.

What Is Climatic Change?

Although we know from examining historical records, such as tree rings and glacial ice, that the earth has experienced considerable climatic variability, things seem remarkably constant in terms of human life and memory. Now, however, there are definite signs of change and, unlike before, these changes are not natural fluctuations over long periods but human-induced changes that will occur very rapidly on the geological time scale.

Climate is the long-term weather pattern. It is characterized in terms of averages and variability around those averages. Climatic change may involve change in averages, variability, or both. Records indicate that over the last 3 million years, average global temperature swings have been less than 7° C. Ten thousand years ago, when most of Canada was covered by ice sheets, the average temperature was only 5° C less than it is now. Small changes in average temperatures can obviously cause quite drastic changes in environmental conditions.

The earth's surface and atmosphere are heated differentially by short-wave radiation from the sun. The differences in heat and pressure between the poles and the tropics fuel the global circulation system as heat and moisture are redistributed around the world. The temperature balance of the earth is maintained through the return of the continually absorbed solar radiation back to space as infrared

This coral was found near the Arctic Ocean and illustrates how climates have changed in the past (*Philip Dearden*).

Box 7.4 The Canadian Climate

The perception of Canada is of a cold, northern country, and the average annual temperature of -3.7° C compared to the global average of 15° C would seem to bear this out. However, as usual, averages mask a lot of variation, especially in a country as big as Canada. Vancouver, for example, has an average temperature of 10° C , Toronto 8° C, and Halifax 6° C, compared with Alert in the Arctic at -18° C.

Precipitation also varies widely (as discussed in Chapter 6), with over 3200 mm along parts of the West Coast and less than 200 mm in the Arctic. This variation, spanning over forty degrees of latitude between the northern and southern extremities, is reflected in the ecozones that are described in more detail in Chapter 3.

Climate is the *average* weather condition of a particular place. The fact that the Arctic climate is cold does not mean that there are not warm or even hot days there. It just means that, over the long term, there are far more cold days than hot days.

radiation, consistent with the first law of thermo-dynamics. Long-term temperature changes are a result of shifts in the amount of energy received or absorbed. These may be caused over long cycles (100,000 years) by factors such as the shape of the earth's orbit around the sun, wobbles of the earth's axis, and the angle of tilt. Such a 100,000-year cycle of glaciation can be traced over 600,000 years, using evidence from sources such as glacier ice and the chemical characteristics of marine sediments. It is more difficult, however, to explain some of the shorter-term fluctuations that occur on a shorter time scale.

Natural events, such as the eruption of large volcanoes and changes in ocean currents like El Nino, are known to have an influence. Volcanoes, such as the eruption of Mount Pinatubo in 1991, eject large quantities of dust and sulphur particles into the atmosphere, which reduce the amount of solar radiation reaching the earth's surface. In 1815 the eruption of a giant volcano, Mount Tambora in Indonesia, had climatic repercussions throughout the world, leading to recurrent summer frosts in many parts of the northern hemisphere. This resulted in crop failures and increased starvation in areas such as Ireland, which, in turn, contributed to increasing immigration to North America. Scientists calculate that these effects were caused by a reduction of as little as 0.7° C in the average surface temperature in the northern hemisphere. Even this was dwarfed in comparison by the eruption of Mount Toba in Sumatra some 75,000 years ago, which may have lowered surface temperatures in the northern hemisphere by as much as 3–5° C and helped trigger another ice age.

Changes in ocean currents can also be influential. El Nino represents a marked warming of the waters in the eastern and central portions of the tropical Pacific as westerly winds weaken or stop blowing, usually two to three times every decade. In normal years, the trade winds amass warm water in the western Pacific. As the winds slacken, this water spreads back eastward and towards the pole into the rest of the Pacific. This triggers weather changes in at least two-thirds of the globe, causing droughts and extreme rainfall in countries along the Pacific and Indian oceans, including Africa, eastern Asia, and North America.

It now appears, however, that climatic change may occur more quickly than ever before due to human activities. Temperatures over the next century may be higher than at any time over the last million years. Most scientists think that these changes may have already begun (Figure 7.10). Globally, eight of the ten warmest years on record were during the 1980s. The hottest year to date was 1995, edging out the second hottest, 1990, and beating the 1950–80 average by 0.4° C, according to NASA and researchers at the University of East Anglia in England. This has been interpreted as a return to global warming following the cooling effects of the eruption of Mount Pinatubo in 1991 in the Philippines, which cast tonnes of dust into the atmosphere and screened out the sun. In Canada there has been a statistically significant rise in average temperature of 1.1° C between 1895–1991. The warmest year on record in this time period was 1981, followed by 1987. The warming has been most pronounced in central Canada (Figure 7.11).

Box 7.5 Measuring Climate Change

An essential step in assessing climatic change is to compare present variations in climate with those of the past. Current data are largely instrument-based weather observations or the instrumental record. Even here there are difficulties. More modern and accurate data, of the upper atmosphere, for example, or satellite data are available only for the last two decades or so. This period also coincides with the greatest human impacts on climate, and does not provide any type of control for climate change in the absence of industrialization.

Scientists reconstruct former climates using proxy information, which comes from many different sources. For example, examining historical records of climate-influenced factors—such as the price of wheat in Europe over the past 800 years, the blooming dates of cherry trees in Kyoto, Japan, since AD 812, the height of the Nile River at Cairo since AD 622, the number of severe winters in China since the sixth century, the information in sailors' and explorers' logs and other such sources—all contribute to constructing a picture of past climates.

There are also climate-sensitive natural indicators, such as tree rings and glacial ice. Cores obtained from ancient ice in Greenland and Antarctica have been analysed using the ratio of two oxygen isotopes, which indicate the air temperature when the original snow accumulated on the glacier surface. Tree rings are also very useful. Outside the tropics where there are noted differences in seasons, the width and density of tree rings reflect growth conditions, including climate. Some species (red cedar in coastal BC, for example) may live for well over 1,000 years and can provide valuable indicators as to past climates. The same kinds of rings also characterize the growth of many long-living corals in the tropics, which may be over 800 years old and provide valuable evidence for previous El Nino events.

Although most scientists think that this warming trend will continue, there is a wide divergence of opinion as to how quickly warming will occur due to the complexities involved in the atmospheric system and the number of positive and negative feedback loops involved. At a conference convened by the World Meteorological Organisation and the United Nations Environment Programme, leading scientists suggested that average temperatures will rise 1° C by 2025 and another 3° C by the end of the next century. Changes will be greatest in higher latitudes and more pronounced in winter than summer.

How Is Climatic Change Caused?

One of the main factors in maintaining the relative thermal balance over the long periods discussed earlier is the mixture of gases in the

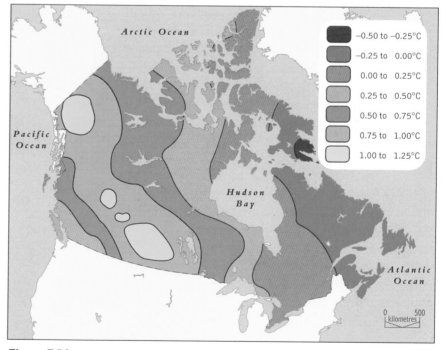

Figure 7.10 Average annual temperature departures from 1951–80 average for 1980–9. The 1980s experienced tremendous warming on almost a national scale, with a broad area of well above normal temperatures covering the entire central and western portions of the country. The central core of this warming, with values greater than 0.8° C above normal, extended from the Yukon in the northwest to southern Manitoba in the southeast. Only the extreme eastern area, consisting of the Atlantic Region, northern Quebec, and southern Baffin Island, showed below normal values. Similar annual temperature patterns have persisted into 1990 and 1991. SOURCE: D.W. Gullett and W.B. Skinner, *The State of Canada's Climate: Temperature Change in Canada 1895–1991*, State of the Environment Report 92-2 (Ottawa: Environment Canada, 1992):33.

Indicator: Global and Canadian Average Temperatures

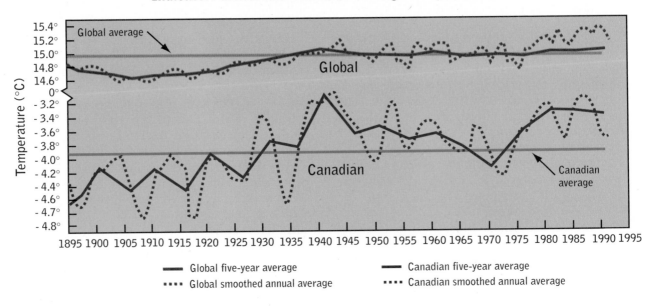

Figure 7.11 Global and Canadian average temperatures, 1895–1992. SOURCE: *Climate Change* (May 1994):4.

atmosphere. Certain gases, known as the *greenhouse gases*, absorb the infrared radiation emitted from the earth. Their impact on incoming radiation is minimal due to the different wavelengths on which energy enters and leaves the atmosphere. However, the gases trap outgoing heat and thereby help to heat the lower atmosphere. The earth's surface radiates at an average temperature of 15° C, whereas the planet radiates to space at –18° C. Due to this difference the earth's surface is about 33° C warmer than it would be otherwise. Were it not for these gases, it is doubtful whether life could have evolved on earth.

The effect is thus similar to that of a greenhouse into which energy can enter but the rate of return is reduced due to the trapping effects of the glass (Figure 7.12). The main greenhouse gases responsible for this effect are carbon dioxide, methane, water vapour, ozone, and nitrous oxide. Past records show a clear relationship between temperature fluctuations and the concentrations of carbon dioxide and methane (Figure 7.13). The concentration of these two gases has been rising steadily over the last century, particularly over the last few decades (Figure 7.14). The concentration of atmospheric carbon dioxide has increased from roughly 280 ppm (parts per million) before the Industrial Revolution to today's level of 360 ppm. At this rate of change, concentrations of CO_2 will double from preindustrial levels by the year 2075

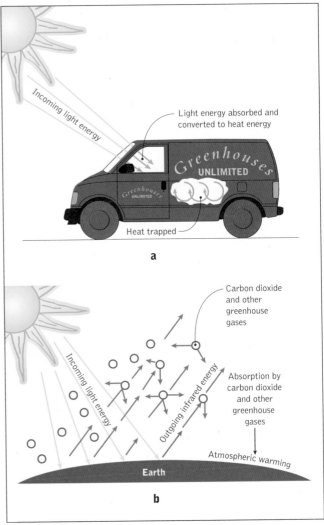

Figure 7.12 The greenhouse effect.

The forest industry is a major contributor to the rising levels of carbon dioxide in the atmosphere not only through deforestation but also through emissions from processing plants (*Philip Dearden*).

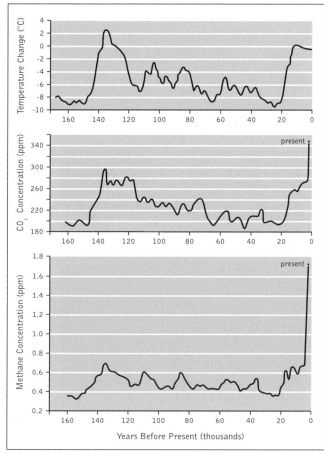

Figure 7.13 Variation of temperature over Antarctica and of global atmospheric carbon dioxide and methane concentrations during the last 160,000 years, as inferred from the Vostok ice core from Antarctica. SOURCE: J. Chappellaz, J.M. Barnola, D. Raynaud, Y.S. Korotkevich, and C. Lorius, 'Ice Core Record of Atmospheric Methane over the Past 160,000 Years', *Nature* 345, no. 6271 (1990):127–31.

and reach a level not experienced for over 160,000 years. We interfere with the carbon cycle in two main ways:

- We release large amounts of carbon from the lithospheric to the atmospheric reservoir by burning fossil fuels. These products, coal and oil, had been laid down in the earth's crust as a result of millions of years of accumulation of the sun's energy in the form of plant and animal remains. Before the Industrial Revolution, virtually none of this was used. Human activities were regulated by the rate of incoming energy fixation through wood production and the work of animals. The discovery of ways to use this huge bonanza of stored energy paved the way for the Industrial Revolution and all the subsequent developments that have enabled society to process matter so quickly. Now some 2 billion tonnes of fossil fuel are burned every year, adding 5.5 billion tonnes of CO_2 to the atmosphere. Canada, in per capita terms, is a leading contributor in this regard (Table 7.6). Figure 7.15 shows the main contributors to these emissions in Canada.

Box 7.6 Canada Facts

- In 1992 Canadian emissions of CO_2 from fossil fuel use were more than twice as high as that in 1958.
- Fossil fuel use accounted for 98 per cent of Canada's human-induced emissions of CO_2 in 1992.
- Net annual carbon dioxide emissions in Canada are about 2 per cent of the world total, and Canada's population is less than 0.5 per cent of the world total.
- In 1994 Canada emitted 2.5 per cent more CO_2 from fossil fuel burning than in 1993, up from the average annual 1.3 per cent increase over that of the previous decade. Including cement and lime production raised this total to 4.5 per cent higher than the Canadian target for the year 2000 (State of the Environment Directorate 1994:2).

Atmospheric Concentration of Carbon Dioxide (1958–92)

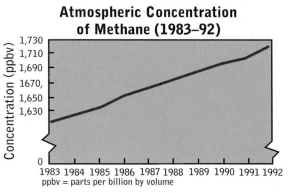

ppmv = parts per million by volume

Concentrations of Other Greenhouse Gases

Methane increased by 6.2 per cent between 1983 and 1992, while nitrous oxide rose by 2.3 per cent. Chlorofluorocarbons (CFCs) and tropospheric (ground-level) ozone are also known to influence the global climate system, although at present their net effect is unclear.

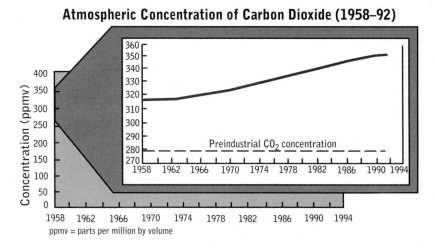

Atmospheric Concentration of Methane (1983–92)

ppbv = parts per billion by volume

Atmospheric Concentration of Nitrous Oxide (1977–93)

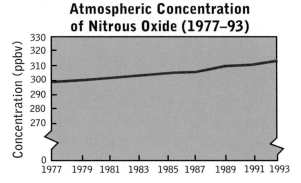

Figure 7.14 Global atmospheric concentration of carbon dioxide, methane, and nitrous oxide. SOURCE: Scripps Institute of Oceanography, Mauna Loa, Hawaii station; National Oceanic and Atmospheric Administration, Climate Monitoring Laboratory.

Carbon Dioxide Emissions from Fossil Fuel Use in Canada (1958–92)

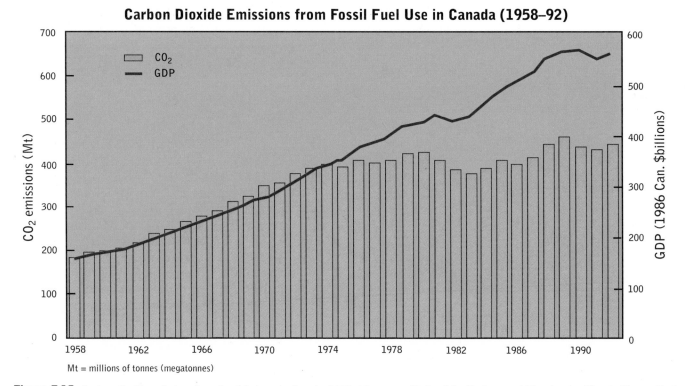

Mt = millions of tonnes (megatonnes)

Figure 7.15 Carbon dioxide emissions from fossil fuel use in Canada, 1958–92. SOURCE: State of the Environment Directorate, *Climate Change Environmental Indicator Bulletin,* State of the Environment Bulletin 94-4 (Ottawa: Environmental Canada, 1994).

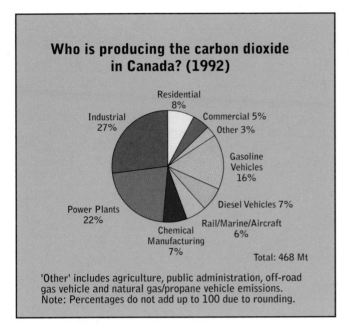

Who is producing the carbon dioxide in Canada? (1992)

Residential 8%
Commercial 5%
Industrial 27%
Other 3%
Gasoline Vehicles 16%
Power Plants 22%
Diesel Vehicles 7%
Chemical Manufacturing 7%
Rail/Marine/Aircraft 6%
Total: 468 Mt

'Other' includes agriculture, public administration, off-road gas vehicle and natural gas/propane vehicle emissions.
Note: Percentages do not add up to 100 due to rounding.

Figure 7.15 (continued). SOURCE: State of the Environment Reporting, *Climate Change*, SOE Bulletin 94–4 (Ottawa: Environment Canada, 1994).

• Deforestation is also a main cause of rising atmospheric CO_2 levels. Trees are a main repository for carbon. When they are cut down, removed, and processed, the carbon in the trees is also removed from that site. If the site is allowed to regrow trees, then over time the amount of carbon on the site will increase again. However, the rapid deforestation in this century, with many sites being used for other purposes, means that there is far less carbon

now held by this biological component of the cycle than a century ago. Much of the carbon has been returned to the atmosphere as carbon dioxide. Forest fires (Box 5.2) have the same effect. Deforestation thus leads to increased amounts of CO_2 in the atmosphere and reduced uptake. Deforestation may account for up to one-third of the rise in CO_2 in the atmosphere.

There is still much that we do not know, however, about increasing CO_2 levels in the atmosphere. Scientists calculate, for example, that over 6.5 billion tonnes are being pumped into the atmosphere by human activities every year. Unfortunately, we can only measure an extra 2.7 billion in the atmosphere, and there is considerable scientific debate as to the whereabouts of the 3.8 billion tonnes that are unaccounted for. Some scientists think it is being absorbed by the oceans, where it could be absorbed for another thousand years or more without undue side-effects. Others think it is being absorbed by increased uptake by forests, particularly the vast areas of temperate and boreal forests of the northern hemisphere, where its storage potential would be considerably shorter than in the oceans. Experiments are now taking place to see whether both or just one of these 'sinks' are repositories for the missing carbon. The results have obvious implications for how serious the global warming problem becomes.

Although CO_2 is the single main influence on rising temperature, the combined effects of the other greenhouse gases may be greater in the future. Methane, for example, occurs in much smaller quantities, but traps twenty-five to thirty times the heat per molecule of CO_2. Its major sources are organic, such as from organic decay in swamps and other anaerobic environments (see Box 7.7), and from the waste products of ruminants, such as sheep and cattle. Humans have encouraged increases in methane emissions mainly through land-use changes, and by vastly increasing the number of domesticated ruminants and the

Table 7.6 CO^2 Emissions for Selected Countries from Industrial Processes, 1991

World	Carbon Dioxide Emission (000 metric tonnes) Total	Per Capita Carbon Dioxide Emissions (metric tonnes)
United States	4 931 630	19.53
Canada	410 628	15.21
Australia	261 818	15.1
Germany	969 630	12.13
United Kingdom	577 157	10
Japan	1 091 147	8.79
Sweden	53 498	6.23
China	2 543 380	2.22
India	703 550	0.81

Source: World Resources Institute, *World Resources 1994–95* (New York: Oxford University Press, 1994).

Box 7.7 The Experimental Lakes Research Area

Generating enough electricity to fuel society's demands while minimizing environmental impacts is one of the main challenges we face today. Hydroelectric power has generally been regarded as preferable to the acid deposition-linked coal- and oil-generating stations or nuclear stations with their associated difficulties in disposing of the wastes. It was somewhat of a surprise, therefore, when researchers at the Experimental Lakes Research Area (ELA) in northwestern Ontario discovered that reservoirs created for hydroelectricity generation were responsible for releasing large amounts of carbon dioxide and particularly methane to the atmosphere. The emissions occur as a result of bacterial decomposition of flooded peat and forest biomass.

This was not the ELA's first finding of global significance. Since it was established in 1968, researchers have also made significant contribu-tions to the understanding of eutrophication and acid deposition. Fifty-seven small lakes and their watersheds are used as experimental sites to help determine the impacts of various en-vironmental perturbations.

Research on eutrophication at the ELA was instrumental in developing the phosphorus-control strategies in the Great Lakes Water Quality Agreement. In 1987 research on acidi-fication was initiated and contributed to new estimates regarding damage to aquatic ecosys-tems. Again the results were used as the basis for international accords to limit emissions. The lakes have also provided a valuable function as monitoring sites to track changes in environ-mental conditions over time, such as global warming. In view of the national and global importance of the research at the ELA, it is unfortunate that its future government funding looks very uncertain.

amount of decay in waste disposal sites. Changes in methane concentrations over time are shown in Figure 7.14. Methane concentrations are rising at a rate of between 0.75–1.0 per cent every year.

Another important greenhouse gas is nitrous oxide, which increased in atmospheric concentra-tion by 3.7 per cent between 1984–93. It is pro-duced naturally from soils and water. Major human influences are the application of nitrogen fertilizers, increases in agricultural land, and the burning of fossil fuels and biomass. When nitrogen fertilizers are added to agricultural land, some of the nitrogen is denitrified by bacteria and becomes nitrous oxide. Thus global warming also involves disruption of the nitrogen as well as car-bon cycle. This kind of interaction also illustrates some of the complexities of the modern environ-mental challenge. Fertilizers are necessary to feed a burgeoning world population, but, although essential for food supplies over the short term, their use also contributes to global cli-matic change that may ultimately lead to a much greater reduction in food supplies.

Chloro-fluorocarbons (CFCs) are also powerful greenhouse gases that have many times the warming potential of carbon dioxide, but due to their relatively low concen-trations, CFCs do not have as much impact. One CFC molecule may have from 10,000–20,000 times the impact of a CO_2 molecule. Due to CFCs' interaction with stratospheric ozone (Box 7.9), efforts have been made to eliminate their use by the

Although this rural scene may look bucolic, the rising numbers of cattle on the earth contribute significantly to two of the problems discussed in this chapter: eutrophication and increased methane levels leading to global climatic change (*Philip Dearden*).

Box 7.8 Canada Facts

- Canadian domestic supply of ozone-depleting substances decreased by 53 per cent between 1987 and 1992.
- Manufacture of CFCs in Canada ceased in 1993; new supplies could be imported until 1996.
- Canada accounted for just under 2 per cent of the world's supply of CFCs and halons in 1986 (State of the Environment Directorate 1993).

end of this decade as discussed in more detail in Chapter 20.

What Are the Effects of Climatic Change?

There is much uncertainty about the effects of global change due to the complexity of the biophysical and socio-economic systems involved. Biological systems are often characterized by positive feedback loops that produce extreme and unanticipated effects. The relationships between water vapour, sea ice, clouds, and oceans are also poorly understood, as are their roles as sources and sinks for carbon dioxide. It is estimated, for example, that the oceans have already absorbed more than a third and perhaps as much as a half of the excess CO_2 emitted. However, we are not sure of their future absorption capacities. As temperatures rise, for example, more CO_2 will be released from the oceans than is absorbed. On the other hand, the same rise in temperature may increase photosynthesis by phytoplankton, which would tend to have the reverse effect. Increasing amounts of water vapour in the atmosphere also need to be taken into account. An increase of 2–3° C in air temperature will allow it to hold up to 33 per cent more water vapour. Since water vapour is also a greenhouse gas, this will also lead to more global warming.

Complex mathematical models (Global Climate Models) have been developed to predict some of the effects of these changes. Models for Canada give changes in summer and winter temperatures (as indicated in figures 7.17 and 7.18) if there is a doubling in CO_2 concentrations. Such large changes would obviously have major effects that would differ among seasons and regions.

Although global climatic change is often called global warming, the effects on weather will be

The role of the oceans in helping to mitigate the impacts of global warming through absorption of carbon dioxide is still uncertain, as is the oceanic response to warmer temperatures. Scientists are already detecting larger wave swells in many parts of the world that may be linked to these changes (*Philip Dearden*).

much more variable than just warming. Indeed, in some areas cloud cover may increase to produce more cooling effects. The East Coast of Canada, for example, has experienced overall cooling since about 1950, despite the evidence for increases elsewhere. Evidence also points to more extreme climatic events. Higher frequencies of extreme rainfall have already been documented in Japan, the United States, Russia, and China. Between 1963–92, the number of floods causing major damage increased from two every five years to twenty-four, and the number of floods causing at least 100 deaths more than doubled. There were also significant increases in serious tropical storms and droughts. Of course, these cannot be directly attributed to climatic change, but they fit the predictions of the models assessing the impacts of global change, and have persuaded the usually fiscally conservative insurance industry to seek greater government commitment to curb emissions. The following are some of the anticipated impacts.

Ocean and Coastal Zone Impacts

Sea temperatures will rise. Since 1945 sea temperatures in the Antarctic have risen at a rate equiva-

Box 7.9 Ozone Depletion

Ultraviolet radiation from the sun causes some oxygen molecules to split apart into free oxygen atoms. These may recombine with other oxygen molecules to form ozone (O_3) in the outer layer of the atmosphere, known as the *stratosphere*. This layer of ozone helps filter out ultraviolet (UV) radiation from penetrating to the earth's surface where it destroys protein and DNA molecules. Without this protective layer, it is doubtful whether life could have evolved at all on earth.

Although there are natural causes of variation in ozone levels, recent observations have indicated that this layer is being broken down by the emission of various chemicals from earth. Since 1979 the amount of stratospheric ozone over the entire globe has fallen per decade by about 4–6 per cent in mid-latitudes and by 10–12 per cent in higher latitudes. Levels sank to all-time lows in 1992 and 1993 due to the after-effects of the eruption of Mount Pinatubo in June 1991. These decreases have led to average increases in exposure to UV-B of 6.8 per cent per decade at 55° N and 9.9 per cent at the same latitude in the southern hemisphere. In general, penetration of UV-B radiation increases by 2 per cent for every 1 per cent decrease in the ozone layer.

Gradual thinning has occurred over Canada in the last decade. In the first part of 1993, for example, ozone levels averaged 14 per cent below normal, and 7 per cent below normal during the summer of that year. Levels in 1995 again achieved these record lows (Figure 7.16). The additional penetration of ultraviolet rays as a result of this depletion will lead to increased cases of skin cancer in humans. Scientists suggest that a sustained 1 per cent decrease in stratospheric ozone will result in a 2 per cent increase of non-melanoma skin cancer. It will also inhibit growth throughout the food chain. Researchers have found that increases in UV radiation severely inhibit phytoplankton growth, for example, and the growth of many insect larvae that form the base of aquatic food chains. Terrestrial systems are also affected, including agricultural productivity.

One of the major catalysts for ozone destruction is chlorine atoms. The *chloro-fluorocarbons* (CFCs) mentioned as a contributor to global warming are a main source of these atoms. These chemicals are used for foam insulation, refrigeration, and air conditioning. International protocols, as described in Chapter 20, have been formulated to phase out their use. In Canada new supplies of ozone-depleting substances fell from a high point of 27.8 kilotonnes in 1987 to 5.7 kilotonnes in 1994. Other substances, such as hydrofluorocarbons (HCFCs), are now used in the air-conditioning systems of most new cars manufactured in Canada. Although these HCFCs still destroy ozone, they are only about 5 per cent as destructive as CFCs. Of course, if consumers could get by without having air conditioning in their cars, this would reduce the need for HCFCs as well!

Global CFC production in 1994 was also down by 77 per cent from its peak in 1987. Unfortunately, although CFC use in developing countries is still half of that in the US, consumption rates are increasing by 3–7 per cent every year. Canada has also taken action to limit the impacts of other ozone-destroying substances, such as the biocide methyl bromide, for which levels have been frozen at the 1991 levels, with a further commitment to reduce consumption by 25 per cent by January 1998.

Even with the successful implementation of these treaties, lag effects will cause the amount of ozone-depleting chemicals to rise in the atmosphere until at least the turn of the century. After that it may take more than forty years for the chemical equilibrium to be restored and ozone levels to recover. In the meantime, you should try to stay out of the sun as much as possible, wear a good sunscreen, and use sun-glasses.

lent to 1.3° C per century. The Japanese archipelago has had a rise of 0.7° C between 1984–91. It is estimated that global sea levels could rise between 30 cm and 1.5 m before the end of the next century. This would lead to the following:

• There would be devastating consequences for island states, such as the Maldives, and for densely populated coastal countries, such as Bangladesh. A 50-cm sea level rise would increase the worldwide number of people subject to coastal inundation from 46 million to 92 million. In Canada coastal cities (such as Vancouver, Charlottetown, and Saint John) and coastal infrastructure (such as sewage plants and industrial facilities) in many other areas may be flooded.

Stratospheric Ozone Levels Over Canada

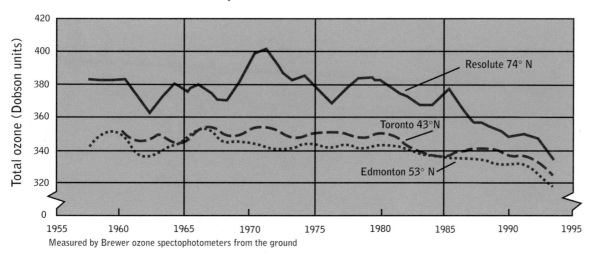

Resolute 74° N

Toronto 43°N

Edmonton 53° N

Measured by Brewer ozone spectophotometers from the ground

Figure 7.16 Stratospheric ozone levels for three Canadian cities. SOURCE: Atmospheric Environment Service, Environment Canada.

• There would be reductions in the area and thickness of sea ice, changes in freeze-up and thaw dates, an increase in iceberg numbers, and changes in oceanic currents and wave patterns. These changes will have a profound effect on marine transportation and safety, ice-breaking requirements, and weather patterns. Major impacts could be felt by the offshore energy sector due to these changes. Wave heights in the north Atlantic have risen by 25 per cent over the last quarter of a century, although average wind speeds have remained the same. Many scientists suspect that global warming is the cause, although this has not been proven.

• There would be an increase in the frequency and intensity of tropical storms. Such storms require vast amounts of energy, derived primarily from warm ocean waters. As temperatures rise, so will the amount of energy available. The cost of hurricane Andrew, which struck the US Gulf Coast in 1992, was in excess of $20 billion, and the total economic losses from wind storm damage in the 1980s was three times greater than those of the 1960s. It is also likely that such storms will also move outside traditional areas as energy levels continue to rise.

• There will be changes in the abundance, distribution, and migration routes of fish species. Little is known about the factors affecting fish distributions and abundance, as the failure to manage the fish resource on Canada's coasts so

Canada's north would experience major changes as a result of global warming as water levels change, as the amount and duration of snow cover change, and as permafrost melts. This could have drastic effects on areas such as the Mackenzie Delta (*Philip Dearden*).

clearly demonstrates (see Chapter 20). It is hence especially difficult to predict the potential impacts of these gross physical changes. One suggestion is that salmon production will fall due to increased coastal temperatures and lowering salinities as a result of increased precipitation. On the other hand, expanded ranges of some species, such as squid and mackerel on the Atlantic coast and hake and tuna on the Pacific coast, might open up new fisheries.

Freshwater Impacts
Freshwater impacts will vary considerably from region to region. In some areas, such as the West Coast, it is anticipated that increased cloud cover and precipitation will be one result of global

Figure 7.17 Projections for climate warming in summer, June–August (Celsius). SOURCE: Canadian Climate Program Board, *Climate Change and Canadian Impacts: The Scientific Perspective* (Ottawa: Environment Canada, 1991):12.

warming. On the other hand, it is predicted that the already-dry areas of the Prairies will become even drier. Anticipated impacts include:

- a net basin run-off decrease of 25–50 per cent in the Great Lakes–St Lawrence system, resulting in lower lake levels, navigational problems, and decreased hydro power production
- over thirty new species could enter the Great Lakes from warmer waters, while many existing species would decline or disappear

- a longer ice-free shipping season in the Great Lakes and other major lakes
- water shortages in many of the drier regions of the country
- changes in the abundance and location of wetlands; existing wetlands may dry up completely while others may form as permafrost melts

Terrestrial Impacts

As discussed in chapters 4 and 5, climate is a major determinant of growth conditions. Many species

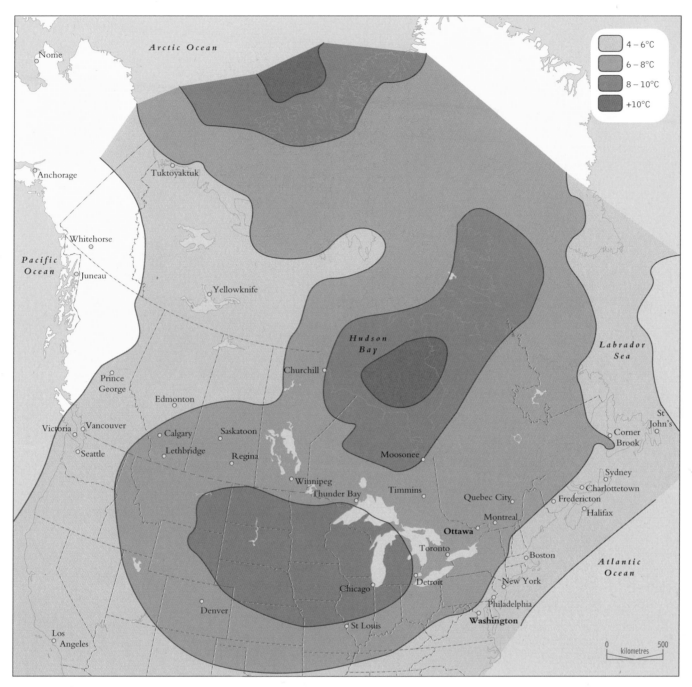

Figure 7.18 Projections for climate warming in winter, December–February (Celsius). SOURCE: Canadian Climate Program Board, *Climate Change and Canadian Impacts: The Scientific Perspective* (Ottawa: Environment Canada, 1991):13.

exist within fairly narrow ranges of tolerance. As temperatures and precipitation levels change, this will result in substantial changes in growing conditions for both natural and agricultural systems. Changes could include the following:

• There may be a major reorientation of the ecozones of Canada (Figure 7.19) as the warmest ecozones (cool temperate, moderate temperate, and grassland) increase in size while

the cooler ones (boreal, subarctic, and arctic) contract. This scenario is for a doubling of CO_2, which could occur as early as 2025, with the climatic impacts felt during the subsequent decade. However, most species, especially trees, could not migrate quickly to the areas with suitable growth conditions; a rate of 100–200 km per decade would be required, far in excess of most natural rates of 20–50 km per century. Furthermore, soil conditions might not be

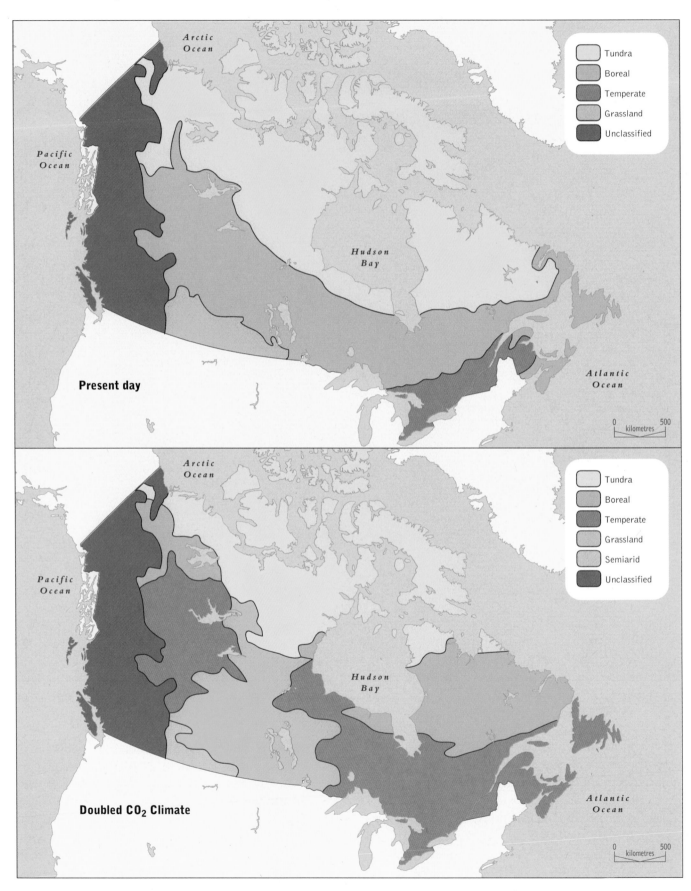

Figure 7.19 Changes in forest and grassland boundaries resulting from a typical doubled CO_2 climate. SOURCE: H. Hengeveld, *Understanding Atmospheric Change*, State of the Environment Report 91–2 (Ottawa: Environment Canada, 1991):44

suitable where climatic conditions are. Massive die-off is likely to be the result over the long term.

- There may be a profound impact on the forest industry as the die-off occurs. The boreal forest will be drastically reduced in size. Replanted forest areas may be planted with species unable to grow there in the future. Fires will increase in severity and frequency (Flannigan and Van Wagner 1991), as will attacks by insect pests.

- There may be longer and warmer growing seasons for agriculture, with southern Ontario becoming like current southern Illinois, Saskatchewan like Nebraska, and Alberta like Colorado. A corresponding northern extension of agriculture is unlikely due to soil and economic conditions. There would be a threat of very serious drought in drier areas, lower soil moisture overall, and a greater need for irrigation.

- There may be a major reduction in the area of permafrost. Currently about half of the country is underlain by permafrost. Melting could cause terrain instability for northern communities, transportation, and pipelines. Studies in the 1.8 million km^2 Mackenzie basin (which spans parts of the Northwest Territories, Yukon, British Columbia, and the Prairies) indicate that this melting is already underway, with the permafrost retreating about 100 km over the last century. The depth of the active layer of permafrost (that which freezes and melts every year) has also increased considerably. Records indicate that temperatures have increased about 1° C over the past century. As more land is released from permafrost, organic decay will increase, resulting in heightened levels of methane emissions, another good example of a positive feedback loop as discussed in Chapter 5.

- There may be a major challenge to our park and protected area system. Current parks have been selected specifically to represent certain ecological conditions. Climatic change makes it likely that these conditions will change substantially, meaning that important ecological areas may be unrepresented by the park system, and some current parks may be redundant.

- There may be wildlife displacement as changes in vegetation occur. Local extinctions could occur if suitable migration routes are not available. One of the world's largest caribou herds, the porcupine caribou, in northern Yukon and Alaska, would have to contend with increased snowfall, which could lead to a significant decline in successful pregnancies due to difficulties in foraging for food during the winter. Mosquito harassment would also increase in summer in the interior, forcing more animals onto the coastal plain.

- There may be many other socio-economic impacts as society tries to adjust to the situation. In some rural situations, especially for First Nations peoples, villages may have to be abandoned as animal populations change. In urban areas, the elderly may be especially vulnerable to summer heat waves. The Canadian economy's dependence on forestry and agriculture will mean that strong effects will be felt throughout the economy, depending upon what happens to these two very climatically sensitive industries.

International Impacts

Besides the very serious implications mentioned earlier, the international ramifications should not be overlooked. There are serious health implications, for example. Malaria, already undergoing a strong resurgence in the tropics as the mosquito vectors develop immunity to the biocides used to control them, will claim about 50–80 million additional cases annually, according to one model, against an assumed global background total of 500 million by 2100. Areas of potential disease contraction would spread from the tropics into less well-protected temperate area populations. Countries much poorer than Canada will be faced with the collapse of their agricultural systems as water distributions change. What will happen to the 50 million people who depend upon the Nile as a water source if the rains shift from the Ethiopian highlands in the headwaters, as seems likely? What will happen if Israel's surface water supplies decline to the point of non-existence? What will happen when high sea levels and tropical storms combine to flood Bangladesh? What kinds of responsibilities does Canada have to the international community in situations such as these? Will Canada become a

Canada will face major challenges as a result of global warming, but other countries will also be affected. In many areas of the world, such as here in Israel, water supplies are already minimal. Even small changes in the frequency and amount of rainfall could have severe impacts (*Philip Dearden*).

main destination for environmental refugees? What do you think should be Canada's policy regarding issues such as these?

What Can We Do About It?

Clearly, there is considerable uncertainty regarding the precise effects of the anticipated climatic changes. However, anticipated impacts are sufficiently serious to warrant a serious response—the application of the *precautionary principle*, whereby if potentially disastrous but uncertain consequences

can be avoided at a reasonable cost, then every effort should be made to do so (see Chapter 2). Unfortunately, as with acid deposition, this requires both a national response and a coordinated international response. The increase in greenhouse gases results from human demands generated by burgeoning population levels and greater consumer demands. As with so many environmental problems, the real challenge is to find effective ways to address these demands.

Over 100 national governments have endorsed the 1996 report of the Intergovernmental Panel on Climate Change, which represents the efforts of some 2,500 top scientists throughout the world. According to the report, human activities are having a discernible effect on global climate. International mitigation efforts have focused on eliminating the production of human-made greenhouse gases, such as CFCs, and trying to limit the emissions of others, such as CO_2. These efforts are reviewed in more depth in Chapter 20. Canada, for example, has agreed to stabilize net carbon dioxide emissions at 1990 levels by the year 2000. Other countries have made more stringent cuts than this; others are more lenient. The United Nations'

Box 7.10 What You Can Do

As with so many of the challenges discussed in this book, global change is a truly international problem and requires the coordinated efforts of governments throughout the world. However, there are still ways in which individuals can try to have some positive influence:

1. Both acid rain and global change are profoundly influenced by our love affair with the automobile. Find something else to love, like a bicycle or a bus. Reduce your consumption of fossil fuels.
2. Use energy more efficiently in your home and all other aspects of life. Make sure your

appliances, lighting, and heating are energy efficient.
3. Encourage power utilities to invest more effort in developing renewable energy sources (Chapter 19).
4. Try to get by without air conditioning in your car and at home. Plant shade trees to reduce the summer heat levels in your house.
5. Plant trees to help absorb atmospheric carbon dioxide.
6. Let your political representatives know that you are in favour of mandatory measures to curb emissions, even if this costs you more money.

Intergovernmental Panel on Climatic Change has estimated that an absolute reduction in CO_2 emissions of 60–80 per cent is needed to stabilize atmospheric concentrations at current levels.

Despite the Canadian commitment to reducing CO_2 emissions, neither the federal nor provincial governments have adequate policies in place that will yield the required 20 per cent reduction in emissions. In fact, by the year 2000 emissions could increase over 1990 levels by as much as 13 per cent. Since 1990 Canada has added twelve power stations that use fossil fuels, which will add 19.4 million tonnes of CO_2 into the atmosphere every year. Natural gas production has also increased by 40 per cent, yielding another 16 million tonnes of CO_2.

However, several options are being considered for a national action program on climate change, including:

- residential and commercial building energy consumption initiatives, such as the adoption of an energy code and preferential mortgages for new structures meeting such a code
- measures to improve the energy efficiency of industry
- improved standards of energy consumption for appliances
- transportation sector energy conservation, such as a national inspection program for vehicles, an accelerated retirement program for inefficient vehicles, and improvements in mass-transit services
- energy generation changes to reduce emissions, a search for further supply-side efficiencies, and an emphasis on demand-side management, a reduction in methane and carbon dioxide emissions from the oil and gas supply system, and support for the development of alternative energy supplies
- mitigation measures aimed at non-energy sources, such as reducing emissions from industrial processes and planting trees

Little was achieved in early 1995 when these initiatives were considered at a provincial-federal meeting. The Voluntary Challenge and Registry Program was announced with no legislative teeth, which encourages corporations and communities to set their own goals and choose their own initia-

The threat of climate change is real and present, and the cost in human discomfort and suffering is incalculable. . . . Our efforts to limit greenhouse gas emissions are not working as well as we had expected. It is clear that we are not moving fast enough or far enough in our collective effort. . . . There is clearly a clarion call for governments and politicians to take the next step. . . . If we can't do that, we will have failed. Not the scientist, but we the politicians will have failed.

Sergio Marchi,
minister of environment
for Canada, 17 July 1996

tives. Unfortunately, a similarly weak and ineffective stance has also dominated at the global level, as described further in Chapter 20.

Among the provinces, British Columbia appears to have taken the most aggressive position. Vehicles account for almost half of the CO_2 emitted in urban areas in the province. The government introduced legislation in 1994 to set regulations to:

- set acceptable emission standards for air pollution from new vehicles
- develop standards for vehicle fuels
- specify the regions of the province to which the regulations apply

Over 130 municipalities across the country have also taken action to try to limit the impacts of global change. Almost two-thirds of Canadian households have at least one member who commutes to work. Almost 79 per cent of these drive to work, and the number is increasing every year. Vancouver has made new requirements for existing and future building by-laws to favour low-pollution modes of travel, such as walking or using car pools or bicycles. Toronto retrieves gas from its landfill sites to turn into electricity, estimating that the volume of greenhouse gases diverted is equivalent to removing half a million cars from the roads every year. Regina estimates that the 21,000 trees planted since 1989 have reduced CO_2 emissions by 2400 tonnes per year.

Even if all the proposed measures are successfully implemented, experts warn that some warming is inevitable. Therefore, in addition to these actions, we will need to adapt to the changing conditions, keeping in mind the implications discussed earlier. These might include reforestation programs favouring species that can tolerate warmer conditions, or considering the way in which coastal infrastructure is planned. Scientists are also continuing to explore technical solutions to the problem. One such approach involves collecting CO_2 from concentrated sources, such as emissions from powerplants, and liquifying or solidifying it and injecting it into the deep ocean where it could stay buried for hundreds of years. These approaches are already being tried in Japan and Indonesia, but many questions remain as to their feasibility on a large scale and the potential impacts on ocean circulation and life. Dumping industrial products into the ocean also contravenes international treaties that would have to be adjusted before larger-scale efforts are mounted. However, scientists point out that since the 1992 Rio conference during which the pledges were made to limit CO_2 emissions to 1990 levels by the turn of the century, levels have risen annually by 2 ppm, and by the end of the century the level will increase by 7 per cent over present levels rather than decrease. In light of the seeming inability to curb emission levels, scientists feel that technical solutions, such as ocean injection, might be a critical part of our efforts to limit the impacts of global change.

IMPLICATIONS

The last four chapters should have given you some appreciation for the major processes that support life on earth. The environmental challenges described in the last chapter are among the most serious ones facing humanity. If predictions relating to the impacts of global climatic change are borne out, for example, then the speed of ecological change will create considerable problems for our highly interrelated global economic system. Although climate change is a natural and constant process, never before will humanity have faced such a rapidly changing climate with such large numbers to feed. The advanced technological capabilities that allow us to maintain such high standards of living and to feed most of the world's population will also be part of the solution to global change, but, more fundamentally, we need to increase our appreciation of humanity's relationship to the planet.

As should be clear by now, the main challenge overall is not 'resource' management but people management, and the ways in which we can try to modify our activities to reduce impacts. The next section of the book introduces different ways of approaching environmental management, with a main emphasis on the 'human' side of the management equation.

SUMMARY

1. Major pollution problems—such as eutrophication, acid deposition, and global warming—are the result of human disruption of biogeochemical cycles.

2. Eutrophication is the nutrient enrichment of water bodies over time. Although it is a natural process, human disruption of the phosphorus and nitrogen cycles has caused a marked acceleration in the rate of eutrophication. This promotes excessive plant growth, which leads to oxygen depletion when the plants die and start to decay. Over time this leads to changes in the composition of fish species and makes water treatment more expensive. The rate of eutrophication can be slowed down by limiting the inputs of nutrients into water bodies.

3. Precipitation tends to be naturally acidic. However, as a result of disturbances in the sulphur and nitrogen cycles, the acidity has increased dramatically over much of Canada during the last few decades. The greatest impacts are caused by burning sulphur-rich fossil fuels or by smelting sulphur-rich metal ores. The sulphur dioxide that is produced mixes with water in the atmosphere to produce sulphuric acid. Emissions of various nitrogen oxides as by-products of high-temperature combustion account for most of the remainder. The resulting increase in acidity has a damaging effect on both aquatic and terrestrial ecosystems. Emission controls have now been agreed upon to try to limit these impacts, and some improvements have been noted.

4. Although climatic fluctuations over time are normal, many scientists feel that the current global warming trend is too widespread and occurring too quickly to be a natural occurrence. The main cause of this warming, they suggest, is the emis-

sion of various greenhouse gases such as carbon dioxide, methane, CFCs and nitrous oxide as a result of industrial activity and land-use change. These gases act in a similar way to a greenhouse in that they help prevent the return of solar radiation to the atmosphere, thereby causing temperatures to rise. Predictions suggest that temperatures may rise between 1–3° C over the next century, with the greatest changes in high-latitude countries such as Canada. This would result in widespread changes in ecozones, agriculture, forestry, fisheries, water distribution, permafrost, and many other characteristics that have a profound influence on the socio-economic well-being of Canadians. International protocols have been established to limit the emissions of greenhouse gases, but to date little progress has been made.

REVIEW QUESTIONS

1. Which biogeochemical cycles are most responsible for eutrophication, and in what ways are they changed by human activities?

2. Are there any eutrophic lakes in your region? If so, what are the main inputs that are causing eutrophication and where do they come from?

3. How can eutrophication be controlled? Discuss one example.

4. What is the pH scale and what is it used for?

5. Which biogeochemical cycle is most responsible for acid deposition and how do disruptions occur as a result of human activities?

6. What are the effects of eutrophication and acid deposition on aquatic ecosystems? What do you think might be their combined impact?

7. Are all areas equally sensitive to the impacts of acid deposition? If not, what influences the relative vulnerability of different areas?

8. What are the main socio-economic impacts of acid deposition likely to be?

9. What is the greenhouse effect?

10. What is the contribution of the carbon cycle to this effect, and how has it been influenced by human activities?

11. Why is Canada more vulnerable than many other countries to the impacts of global warming? Explain some of the likely impacts.

12. Why do you think that progress on curbing the emission of greenhouse gases has been more difficult to achieve than progress on eutrophication or acid deposition?

REFERENCES AND SUGGESTED READING

Alvo, R.Z., D.J.T. Russell, and M. Berrill. 1988. 'The Breeding Success of Common Loons (*Gavia immer*) in Relation to Alkalinity and Other Lake Characteristics in Ontario'. *Canadian Journal of Zoology* 66:746–52.

Briggs, D., P. Smithson, and T. Ball. 1993. *Fundamentals of Physical Geography*, 2nd Canadian ed. Toronto: Copp Clark Pitman.

Brown, L., N. Lensen, and H. Kane. 1995. *Vital Signs 1995*. New York: Worldwatch Institute, W.W. Norton and Co.

Browning, M.H.R., and T.C. Hutchinson. 1991. 'The Effects of Aluminium and Calcium on the Growth and Nutrition of Selected Ectomycorrizal Fungi of Jack Pine'. *Canadian Journal of Botany* 69:1691–9.

Canadian Climate Program Board. 1991. *Climate Change and Canadian Impacts: The Scientific Perspective*. Ottawa: Environment Canada.

Canadian Global Change Program. 1993. *Global Change and Canadians*. Ottawa: Royal Society of Canada.

Chappellaz, J., J.M. Barnola, D. Raynaud, Y.S. Korotkevich, and C. Lorius. 1990. 'Ice Core Record of Atmospheric Methane over the Past 160,000 Years'. *Nature* 345, no. 6271:127–31.

Clair, T.A., et al. 1995. 'Regional Precipitation and Surface Water Chemistry Trends in Southeastern Canada (1983–1991)'. *Canadian Journal of Fisheries and Aquatic Sciences* 52:197–212.

Cohen, S.J. 1991. 'Possible Impacts of Climatic Warming Scenarios on Water Resources in the Saskatchewan River Subbasin, Canada'. *Climatic Change* 19:291–317.

DPA Group. 1987. *Assessing Historical and Future Economic Impacts and Net Economic Effects Related to Acidic Deposition on the Sports Fishery of Eastern Canada*. Unpublished contract report. Halifax: Department of Fisheries and Oceans.

Edge, T.A. 1987. 'The Systematics, Distribution, Ecology and Zoogeography of the Endangered Acadian Whitefish (*Coregonus canadensis*, Scott, 1967) in Nova Scotia, Canada'. M.Sc. thesis, Department of Biology, University of Ottawa.

Environment Canada. 1988. 'Acid Rain: National Sensitivity Assessment'. Environment Fact Sheet 88-1. Ottawa: Environment Canada.

_____. 1991. *The State of Canada's Environment*. Ottawa: Minister of Supply and Services Canada.

_____. 1995. 'Stratospheric Ozone Depletion'. Ottawa: State of the Environment Bulletin No. 95-5.

_____. 1996. 'Acid Rain'. State of the Environment Bulletin No. 96-2. Ottawa: Environment Canada.

Flannigan, M.D., and C.E. Van Wagner. 1991. 'Climate Change and Wildfire in Canada'. *Canadian Journal of Forestry Research* 21:66–72.

Gullett, D.W., and W.R. Skinner. 1992. *The State of Canada's Climate: Temperature Change in Canada 1895–1991*. State of the Environment Report 92-2. Ottawa: Environment Canada.

Hare, K. 1995. 'Contemporary Climatic Change'. In *Resource and Environmental Management in Canada: Addressing Conflict and Uncertainty*, edited by B. Mitchell, 10–28. Toronto: Oxford University Press.

Harvey, H. 1978. 'The Acid Deposition Problem and Emerging Research Needs in the Toxicology of Fishes'. In *Proceedings of the Fifth Annual Aquatic Toxicology Workshop*, Hamilton, Ontario, 7–9 November.

Hengeveld, H. 1991. *Understanding Atmospheric Change*. State of the Environment Report 91-2. Ottawa: Environment Canada.

Jeffries, D.S. 1991. 'Southwestern Canada: An Overview of the Effects of Acid Deposition on Aquatic Resources'. In *Acid Deposition and Aquatic Ecosystems: Regional Case Studies*, edited by D.F. Charles and S. Christie, 273–89. New York: Springer-Verlag.

Kupchella, C.E., and M.C. Hyland. 1989. *Environmental Science: Living within the System of Nature*. Toronto: Allyn and Bacon.

Longcore, J.R., H. Boyd, R.T. Brooks, G.M. Haramis, D.K. McNicol, J.R. Newman, K.A. Smith, and F. Stearns. 1993. *Acidic Deposition: Effects on Wildlife and Habitats*. Technical Review 93–1. Bethesda: The Wildlife Society.

Malley, D.F., and K.H. Mills. 1992. 'Whole-Lake Experimentation as a Tool to Assess Ecosystem Health, Response to Stress and Recovery: The Experimental Lakes Area Experience'. *Journal of Aquatic Ecosystem Health* 1:159–74.

Minister of the Environment. 1991. *The State of Canada's Environment*. Ottawa: Minister of the Environment.

Mungall, C., and D.J. McLaren, eds. 1990. *Planet under Stress*. Toronto: Oxford University Press.

Phillips, D. 1990. *The Climate of Canada*. Ottawa: Minister of Supply and Services Canada.

Prince, H.H. 1985. 'Avian Communities in Controlled and Uncontrolled Great Lakes Wetlands'. In *Coastal Wetlands*, edited by H.H. Prince and S.M. D'Itri, 99–119. Chelsea, Michigan: Lewis Publishers.

RMCC (Federal/Provincial Research and Monitoring Coordinating Committee). 1990. *The 1990 Canadian Long-Range Transport of Air Pollutants and Acid Deposition Study*. Downsview, Ontario: Environment Canada, Atmospheric Environment Service.

Robinson, J.B., et al. 1993. *Canadian Options for Greenhouse Gas Emissions*. Ottawa: Royal Society of Canada.

Rudd, J.W.M., et al. 1993. 'Are Hydroelectric Reservoirs Significant Sources of Greenhouse Gases?' *Ambio* 22:246–8.

Schindler, D.W. 1987. 'Detecting Ecosystem Responses to Anthropogenic Stresses'. *Canadian Journal of Fisheries and Aquatic Science* 44 (Supplement 1):6–25.

_____. 1988. 'Effects of Acid Rain on Freshwater Ecosystems'. *Science* 239:149–57.

Smit, B., ed. 1993. *Adaptation to Climatic Variability and Change*. Guelph: Canadian Climate Program, University of Guelph.

State of the Environment Directorate. 1993. 'Stratospheric Ozone Depletion'. State of the Environment Bulletin 93–2. Ottawa: Environment Canada.

_____. 1994. 'Climate Change Environmental Indicator Bulletin'. State of the Environment Bulletin 94–4. Ottawa: Environment Canada.

World Resources Institute. 1994. *World Resources 1994–95*. New York: Oxford University Press.

Environmental Management Strategies

The chapters in the previous section focused on the science pertinent to resources and environment in Canada. In this section, attention is focused upon concepts, methods, and approaches that may be applied in resource and environmental management. However, a key qualification should be made. Often it is inappropriate to think that we actually 'manage' the environment or resources. Instead, we often attempt to manage the *interaction* between humans and the environment. This is why resource and environmental management involves more than the application of technical expertise. It also requires sensitivity to various interests and values.

Part C contains five chapters, each of which provides a perspective on how we can make decisions to move towards sustainable development or sustainability. In the preceding section, Chapter 5 addressed issues related to ecosystem change, and Chapter 6 considered matters regarding ecosystems and material cycling. Building on chapters 5 and 6, the first chapter in this section focuses on the ecosystem approach as it has been interpreted and applied by resource and environmental managers. Particular attention is given to how systems may be defined, recognizing that there is a hierarchy of systems. The advantage of the ecosystem approach is that it focuses attention upon entire systems, their component parts, and the linkages among those parts, rather than directing attention to individual sectors (forests, water, wildlife) or components (water quantity and quality) in isolation from one another. At the same time, the ecosystem approach creates challenges for managers regarding how to define ecosystem integrity or health. It also requires managers to think carefully about how comprehensively they will define a management system if they are to develop timely and effective strategies and solutions.

We know that ecosystems are dynamic and evolve with or without human intervention. As a result, there is always some uncertainty about what the future will be like, and, indeed, unanticipated surprises often occur. The concept of adaptive management, (discussed in Chapter 9) is one way to try and deal more systematically with the reality that our understanding of biophysical and human systems is incomplete and imperfect, and that changes and surprises are ongoing. Adaptive management is based on the belief that we can learn from trial-and-error approaches, yet not cause irreversible degradation when we make errors. An adaptive approach suggests the need for humility as we try to understand the processes and patterns in ecosystems, and reminds us that we can still expect to be surprised and to find that our solutions need to be modified.

One way to determine what adjustments might be made to strategies and developments is to identify, predict, monitor, and evaluate the impacts associated with resource and environmental management decisions. As a result, in Chapter 10 the focus is upon impact assessment, with consideration given to both biophysical and social impacts. The concept of environmental impact assessment was formally introduced by the United States National Environmental Policy Act of 1969, and now many countries throughout the world require impact assessment as part of the development process. While the major focus has been on determining the impacts from projects, increasing attention is turning to identifying the impacts related to programs and policies. Some of the most difficult issues in impact assessment relate to dealing with cumulative effects (the sum of many impacts resulting from numerous small initiatives), and designing compensation or mitigation strategies for unavoidable negative impacts.

For several decades after the Second World War, the approach to resource and environmental management was 'expert driven' and 'top down'. In other words, technical experts from government agencies or private firms made decisions about development, with little consultation with local people who had to live with the consequences of these decisions. Beginning in the late 1960s and early 1970s, there were efforts to incorporate public participation into resource and environmental management. This normally involved providing information to the public about issues and possible solutions, and in some instances consulting with the public when defining issues and solutions. By the 1990s, the participatory process had evolved to include approaches that facilitate empowerment of local people and communities, and that give significant power to local groups. This shift led to the concept of stakeholders, meaning the importance of including in planning and management those who have pertinent responsibilities, whose interests could be affected, or who could intervene to block or impede decisions. The concepts of partnerships and stakeholders are examined in Chapter 11.

The inclusion of stakeholders and a more participatory approach can be seen in such initiatives as cooperative processes to locate noxious facilities (such as ones to treat toxic industrial wastes), comanagement (especially for forestry and fisheries and involving First Nations peoples), the inclusion of non-governmental organizations as partners in planning and management, the creation of round tables as a forum for a wide range of interests, and the use of intergovernmental partnerships, such as the Canadian Council of Environment Ministers.

Conflict is part of life and living, and indeed can be desirable when it highlights something that is not working well. Given the many different but legitimate interests in natural resources and the environment, it is hardly surprising that from time to time there are disagreements about the significance of some aspect of the environment, or about the best way to allocate scarce resources among competing uses. In Chapter 12, consideration is given to alternative ways to resolve disputes. We have traditionally relied upon political, administrative, or judicial processes to resolve disputes, which often serve us very well. However, there is growing interest in alternative dispute resolution (ADR) approaches, particularly alternatives to judicial or court-based approaches. The legal approach is based upon processes created and tested over centuries, but is often adversarial and costly. Alternative dispute resolution methods avoid the courts and the adversarial legal process, and achieve resolution based upon collaborative problem solving and the use of consensus. Alternative dispute resolution is not intended to replace the judicial approach, but is an alternative. ADR also highlights that resource and environmental management often becomes conflict management among people with different values and interests.

The chapters in Part C provide an insightful overview of the ecosystem approach, adaptive management, impact assessment, partnerships and stakeholders, and conflict resolution. Each is an important element in resource and environmental management. Each also represents one aspect that deserves our attention in designing and implementing policies, strategies, and plans.

Chapter 8

The Ecosystem Approach

In its *Conservation Strategy* published in 1992, the Saskatchewan Round Table on Environment and Economy stated that 'ecosystems consist of communities of plants, animals and micro-organisms, interacting with each other and the non-living elements of their environment'. In the same year, the final report of the Royal Commission on the Future of the Toronto Waterfront argued that an ecosystem approach emphasizes that human activities are interrelated, and that decisions made in one area affect all others. The royal commission concluded that 'dealing effectively with . . . environmental problems . . . requires a holistic or ecosystem approach to managing human activities'.

Such interpretations of ecosystems and an ecosystem approach reflect the ideas presented in chapters 4, 5, 6, and 7, in which the emphasis was on systems, interrelationships or linkages, energy flows, and ongoing change. In this chapter, attention turns to the characteristics, opportunities, and challenges provided by the ecosystem approach for environmental management.

CHARACTERISTICS OF THE ECOSYSTEM APPROACH

The Royal Commission on the Future of the Toronto Waterfront identified the following characteristics of an ecosystem approach, noting that it:

- includes the whole system, not just parts of it
- focuses on the interrelationships among the elements

- recognizes the dynamic nature of the ecosystem, presenting a moving picture rather than a still photograph
- incorporates the concepts of carrying capacity, resilience, and sustainability, suggesting that there are limits to human activity
- uses a broad definition of environments: natural, physical, economic, social, and cultural
- encompasses both urban and rural activities
- is based on natural geographic units, such as watersheds, rather than on political boundaries
- embraces all levels of activity: local, regional, national, and international
- understands that humans are part of nature, not separate from it
- emphasizes the importance of species other than humans, and of generations other than the present
- is based on an ethic in which progress is measured by the quality, well-being, integrity, and dignity it accords natural, social, and economic systems (Royal Commission on the Future of the Toronto Waterfront 1992:xxi)

OPPORTUNITIES THROUGH THE ECOSYSTEM APPROACH

If these attributes are accepted as defining an ecosystem approach, they present some basic challenges to contemporary environmental and resource management. First, in a Western industrialized society such as Canada, most humans believe that they have a dominant role relative to nature, and that the environment and natural resources exist to satisfy human needs and wants. By emphasizing that humans are part of nature rather than separate from it, and by recognizing the inherent value of non-human species and things, the ecosystem approach questions such a belief (see Table 2.1 in Chapter 2 to review some of these key ideas).

Second, by taking a holistic perspective focusing on interrelationships, the ecosystem approach reminds us of the need to consider management problems and solutions in the context of linked 'systems'. It forces us to appreciate that decisions made about one system, such as land, can have consequences for other systems, such as water or wildlife, and vice versa. Thus many flooding problems are often associated with land-use practices, such as removing vegetation, which accelerates run-off, or allowing development in areas subject to regular inundation. Concentrating only on the aquatic system to reduce flood damages is unlikely to be effective, since many land-based activities exert some influence on this system.

In contrast, the conventional approach to environmental or resource management has often focused upon systems in isolation from one another, as reflected by one government agency being responsible for forestry, another for wildlife, another for water, another for agriculture, and another for urban development. All of these agencies have a shared interest in resources such as wetlands, but if they each focus only upon their own goals or interests, they can undermine the activities or values of other agencies.

> *At the most basic level, problems derive from the definition of the units for which planning and management are undertaken. Management units often bear no relationship to the realities of ecological problems (even the home-range of the species for which protection is sought), their connections to economic and social processes, or local peoples' cultural and political identity. Instead, they are arbitrary units defined by lines drawn on the map—lines often drawn by someone who has never been to the region and who, for example, decides a river would make a good boundary.*
>
> S.D. Slocombe,
> *'Implementing Ecosystem-Based Management', 1993*

Third, the ecosystem approach demands that the links between natural and economic or social systems be considered. Such a focus is also one of the basic thrusts behind sustainable development, as discussed in Chapter 2. When such linkages are recognized, it becomes apparent that certain thresholds normally exist in natural systems, which, if exceeded, lead to deterioration and degradation. For example, agricultural production can be increased by adding chemical inputs (fertilizers, pesticides, and herbicides). However, the cumulative effects of agrochemicals may eventually cause concern about the safety of the product being grown (e.g., fruits or vegetables) for human consumption. Other concerns may also be created by introducing chemicals into adjacent environments, resulting in eutrophication and pollution, as discussed in chapters 7 and 14. This aspect reinforces the idea that while sustainable development accepts the need for development to meet basic human needs, it also reminds us that some kinds of growth are not sustainable if they lead to degradation of natural, economic, or social systems.

Fourth, the holistic perspective reminds us that in environmental management it is important to recognize that decisions made at one place or scale can have implications for other places or scales. If an industry allows toxic substances to enter a river, people and communities downstream will bear some consequences related to that action, as illustrated in Chapter 1. Or, if a community, state, or nation is unwilling to reduce emissions from factories, some of the costs will be borne by the people, states, or nations downwind since air pollutants are often carried well beyond the borders of the community or state in which the pollutants are generated, as is the case with acid deposition (discussed in chapters 7 and 13).

Fifth, given the impacts of decisions in one place for people and activities in other places, the ecosystem approach raises questions about the most appropriate areal or spatial unit for planning and management. The conventional management unit has been based on administrative boundaries, such as municipal, regional, provincial, or national boundaries. In contrast, the ecosystem approach suggests that areas identified on the basis of other units, such as watersheds or airsheds, may have more functional value. For example, in managing for migratory birds that travel from the Gulf of

Mexico to the Canadian Arctic, national boundaries have little relevance (see also Chapter 16). The management area in this situation transcends at least three nations (Canada, the United States, and Mexico). If managers in each nation develop plans and strategies without regard to what is being done in other jurisdictions, their separate initiatives are not likely to be very effective.

Sixth, an ecosystem approach emphasizes that systems are dynamic or continuously changing. An ecosystem, whether a local wetland, prairie grassland, boreal forest, or urbanizing area, is not static. In addition to daily, seasonal, and annual variations, ongoing longer-term changes occur, as illustrated by the transition of natural grasslands to cultivated crop land, or of farm land to urban land use. In the next section we will consider the implications of this aspect of changing conditions for managers who must decide what the indicators of ecosystem health are, and what type of ecosystem should be sought. This aspect of change also emphasizes why managers and management strategies must be capable of adapting or adjusting to evolving situations. This latter aspect has led to interest in *adaptive management*, which is examined in Chapter 9.

In summary, the ecosystem approach incorporates the key ideas that humans are part of nature rather than separate from it, that a perspective emphasizing interrelationships is needed, and that critical thresholds exist. When these aspects are combined, it can be appreciated why the Royal Commission on the Future of the Toronto Waterfront concluded that:

> . . . the ecosystem approach is both a way of doing things and a way of thinking, a renewal of values and philosophy. It is not really a new concept: since time immemorial, aboriginal

Grasslands National Park, Saskatchewan (*Parks Canada, Grasslands National Park*).

Temagami in Ontario (*Earthroots*).

peoples around the world have understood their connectedness to the rest of the ecosystem—to the land, water, air, and other life forms. But, under many influences, and over many centuries, our society has lost its awareness of our place in ecosystems and, with it, our understanding of how they function (Royal Commission on the Future of the Toronto Waterfront 1992:31–2).

The importance of the ecosystem perspective had also been noted in Canada's Green Plan in which it was concluded that Canadians must learn to plan and act in terms of ecosystems, a point also

Like all creatures, the mouse, raccoon, and osprey are part of an intricate, interdependent ecosystem (*Earthroots*).

highlighted in Chapter 2. Furthermore, the Green Plan noted that:

> We live in a complex and integrated environment. All creatures, including humans, interact with and depend on each other. They all draw on the materials and energy of the physical environment to obtain food and recycle wastes. They all affect each other's behaviour (Minister of Supply and Services 1990:18).

Past responses to environmental problems paid very little attention to these important interrelationships. Today the increasing number and complexity of environmental issues demand that we adopt a more integrated approach (Minister of Supply and Services 1990:18).

The need and rationale for adopting an ecosystem approach has been spelled out dramatically by the Conservation Authorities of Ontario, which concluded that in Ontario:

> The fundamental problem that exists in resource management today is . . . constraint that the current body of legislation, agency structures and mandates do not recognize the concept of ecosystem based management. The overlapping of mandates between the Ministries of Natural Resources, Environment and Energy, Agriculture and Food and Municipal Affairs and Conservation Authorities and Municipalities is evident to everyone.
>
> The situation has evolved over time as the provincial government reacted to specific problems with specific solutions. This issue by issue approach results in a situation that, when viewed from an ecosystem perspective, borders on the ludicrous (Conservation Authorities of Ontario 1993:2).

Thus there are many reasons as to why an ecosystem approach should be used more frequently in Canada for environmental management. However, as noted in Chapter 1, perfect approaches rarely exist. While concepts or methods all have advantages, they also usually have disadvantages. In the next section, attention turns to some of the problems in translating the ecosystem approach from theory into practice.

CHALLENGES IN IMPLEMENTING AN ECOSYSTEM APPROACH

Various aspects hinder the implementation of an ecosystem approach. In this section, the following are considered: imperfect science, establishing boundaries, determining limits, deciding upon scope, and compatibility with conventional values and traditions.

Imperfect Science

In chapters 4, 5, 6, and 7 discussion focused upon energy flows and material cycling, key aspects for the management of ecosystems. It became clear in these chapters that while there are impressive advances in our knowledge and understanding, there are also aspects over which scientists disagree or for which they recognize that our understanding is incomplete or inadequate. Thus it is important to appreciate that ecology as a science cannot always provide guidance for all of the questions managers have about ecosystems. In other words, the simple models we have constructed to represent reality, as outlined in Chapter 1, are as yet far from adequate. Having reached this conclusion, it is also appropriate to state that imperfect understanding is common to all fields of science and social science. The implication is that environmental managers usually have to make decisions with incomplete knowledge, conflicting advice, and considerable uncertainty about future conditions. Nevertheless, it is helpful to recognize some of the grey areas in ecological science so that we do not have unrealistic expectations.

Walker and Norton (1982) identified 'principles' that have been developed regarding ecosystems. Here are some key ones.

Stochastic Processes

The structure and behaviour of all ecosystems are, to a greater or lesser extent, the result of stochastic (or random) processes. These random effects preclude deterministic management or planning policies that assume the possibility of perfect prediction. This principle emphasizes the need for ecosystem management and planning to be flexible and to make allowances for unexpected events (discussed further in Chapter 9).

The Complexity of Ecosystems

. . . these demands on ecology are predicated on a vision of science which assumes that it can provide firm knowledge, and that the only way of obtaining this knowledge is the scientific method. The standard scientific method works well with billiard balls and pendulums, and other very simple systems. However, systems theory suggests that ecosystems are inherently complex, that there may be no simple answers, and that our traditional managerial approaches, which presume a world of simple rules, are wrongheaded and likely to be dangerous. In order for the scientific method to work, an artificial situation of consistent reproducibility must be created. This requires simplification of the situation to the point where it is controllable and predictable. But the very nature of this act removes the complexity that leads to emergence of the new phenomena which makes complex systems interesting. If we are going to deal successfully with our biosphere, we are going to have to change how we do science and management. We will have to learn that we don't manage ecosystems, we manage our interaction with them. Furthermore, the search for simple rules of ecosystem behaviour is futile (Kay and Schneider 1994:33).

Environmental Stability and Ecological Resilience

Stable environments permit relatively complex and finely balanced ecosystems to exist, whereas unstable environments result in more simple and resilient ecosystems. The implications are threefold. First, the more stable the environment is over time, the greater the ecological risk associated with changing that environment. Second, the attainment of a diverse biota with complex interactions requires a long–term stable environment, or one in which fluctuations are regular. Third, stabilizing a naturally fluctuating environment is likely to result in a decrease in the resilience of the ecosystem.

Environmental Fluctuation and the Ecological Niche

Directly competing species cannot coexist unless critical life phases are separated in time, such as flowering in spring versus autumn. However, a fluctuating environment makes possible the coexistence of complete or very close competitors. Stabilizing the habitat, such as by reducing water-level changes in an impoundment or river, can therefore lead to the loss of species.

Ecosystem Complexity and Ecological Resilience

Stable ecosystems are generally complex in terms of species and trophic structure. However, increasing the complexity of a plant community or of an ecosystem as a whole does not necessarily lead to increased stability. There is no clear evidence regarding the consequences of reducing complexity in a natural ecosystem.

Community Diversity and Stability

Community diversity or spatial variety of communities, as opposed to species, is a function of the number of different community or habitat types and the scale of this community pattern. A reduction in community diversity may lead to a reduction in the stability of the system as a whole.

Regarding ecological 'principles', Norton and Walker were surprised that so few could be identified. They also noted that the principles were neither complete nor universally applicable. Indeed, they stated that 'the relevance of ecological theory and practices to resource planning and development remains largely obscure', and that 'substituting appropriate ecological analysis for rhetoric has proved more difficult' (1982:309–10).

Several explanations have been offered as to why so few unambiguous and relevant principles had been developed. One explanation is that the 'principles' are of two kinds. One kind is general statements about basic concepts (diversity, stability, resilience) of ecology that are informative but not directly applicable in specific problem-solving situations. The other kind is developed for specific resource management situations, such as range management (carrying capacity) or forestry (sustainable yield). These ideas are helpful after a decision has been made about the use for a specific place, resource, or environment, but they are not helpful in deciding what the most appropriate use is or what the consequences of a particular use might be.

Another perhaps more significant explanation is the view that 'tight laws applicable under all conditions are unlikely to exist in ecology' (Norton and Walker 1982:311). The principles usually have too many ifs, buts, and maybes to allow for unequivocal guidelines. The reality is that ecological principles cannot be modified or adapted with regard to the specific situation under consideration. As a result, when dealing with complex ecosystems we should not concentrate upon trying to find laws or principles common for all situations, but instead look for information that will help us deal with the conditions in a particular place and time.

The main conclusion from this brief review is that the science of ecology is not yet in a position to offer scientific 'truths' or well-established principles upon which managers can always base decisions. As in most fields of science, there are disagreements and controversies about appropriate concepts and their implications for guiding human management initiatives. At the same time, however, many basic concepts of ecology do provide a basis from which to examine, question, and challenge many environmental policies and practices in Canada.

Defining Boundaries

One of the challenges for an ecosystem approach is defining the boundaries of an ecosystem. If a holistic perspective and emphasis on interrelationships are key attributes, then there is a danger that the analyst will conclude that 'everything is connected to everything else' and soon end up overwhelmed by the possible linkages requiring attention.

Figure 8.1 illustrates this dilemma. An analysis of the impacts of developing a new industrial estate in the upper part of the Don River catchment in Metropolitan Toronto would quickly make it clear that the downstream consequences regarding water quality and other aspects should be considered. The Don River, as a hydrological unit, would be one level of ecosystem to consider. However, as Figure 8.1 highlights, the Don River is a subsystem of other systems (which could be defined as either the Greater Toronto bioregion delimited by the Niagara Escarpment on the west, Oak Ridges Moraine on the north, and Lake Ontario on the south), or the Great Lakes system (which includes two Canadian provinces and eight US states), or the biosphere (which includes the entire globe). At the global biosphere level, concern could relate to the release of pollutants from the industrial estate into the atmosphere and the implications for global warming or the depletion of ozone.

The dilemma is that there is no obviously appropriate ecosystem to select. Rather, there are

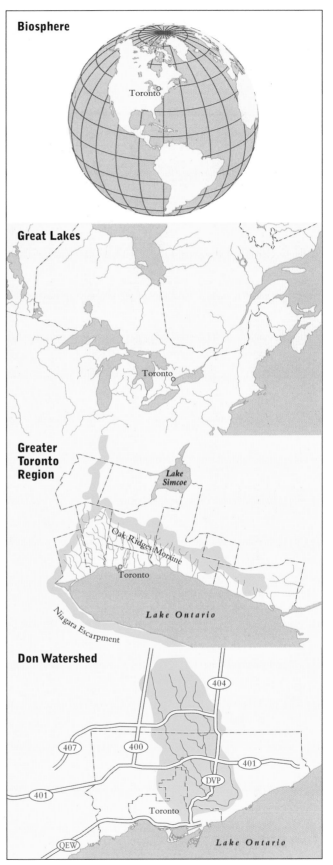

Figure 8.1 Ecosystem units. SOURCE: Royal Commission on the Future of the Toronto Waterfront, *Regeneration: Toronto's Waterfront and the Sustainable City, Final Report* (Ottawa: Minister of Supply and Services Canada; Toronto: Queen's Printer of Canada, 1992):41.

The Queen Charlotte Islands have lush, moss-covered trees (© Mark Hobson).

choices depending upon the problem and interests. To illustrate, management of forests might most logically be based on ecological systems, such as the coastal rainforest in British Columbia or the boreal forests in Alberta, Saskatchewan, and Manitoba. Such forest systems often extend across many river basins, yet water managers often believe that river basins are the most appropriate ecosystem for management since they allow interactions in the hydrological cycle to be captured. However, wildlife managers may conclude that neither forestry-based nor river basin ecosystems are appropriate for their needs, especially when dealing with migratory animals and birds that move across or over a number of such systems in their annual cycle of activity. The lesson is that different managers can provide excellent rationales to justify the use of different boundaries to define ecosystems. The boundary issue therefore becomes yet another example of the ambiguity and lack of clarity in making the transition from the concept to the practice of ecosystem-based management.

The problem of selecting appropriate boundaries of ecosystems has contributed, at least in part, to the recognition of *bioregions*. Bioregions are not defined by surveyors' lines but instead reflect a

place that makes sense to its inhabitants due to its natural and cultural characteristics. Bioregions are usually delineated to reflect the natural and cultural diversity and richness of a place. The associated *bioregional perspective* believes it is important to recognize that a place comprises interconnected parts, and that the role of those parts and their relationships must be understood and respected (Plant and Plant 1990, 1992; Sale 1985). This perspective does not provide a clear-cut way to determine the boundary of an ecosystem, but it does consider both natural and cultural aspects. For example, the Dunnville bioregion has been established in the most southerly portion of the Grand River in southern Ontario, near the point where the Grand River enters Lake Erie. The Dunnville bioregion includes both a geographic area and the activities within that area. It arbitrarily extends from immediately below a dam located on the Grand River in the community of Dunnville downstream to Lake Erie, as well as west and east to the limits of the catchment drained by the Grand River (Figure 8.2). As a result, the farm land, forests, and settled areas are all components of the bioregion, in addition to the river itself and adjacent wetlands and marshes.

The Dunnville Bioregion Association was established in the early 1990s as a not-for-profit, community-based association of people who live in, or have a special interest in, the bioregion. The primary stated objective of the Dunnville Bioregion Association is to promote and guide the integration of a wide range of activities and interests in the region. The association wants to focus on opportunities for future enhancement of the natural resources (including the soil, water, vegetation, and terrestrial and aquatic species) in the Dunnville bioregion into an overall plan that stimulates both economic prosperity and stability as well as protection of the integrity of natural and social systems in the area.

A particular focus of the Dunnville Bioregion Association has been to encourage tourism to create economic development and provide jobs. Despite the association's position that this focus is designed to ensure development compatible with protection of the area's natural integrity, the proposal has triggered sharp criticism from some residents who perceive it as a smokescreen for a few land-owners to promote development from which

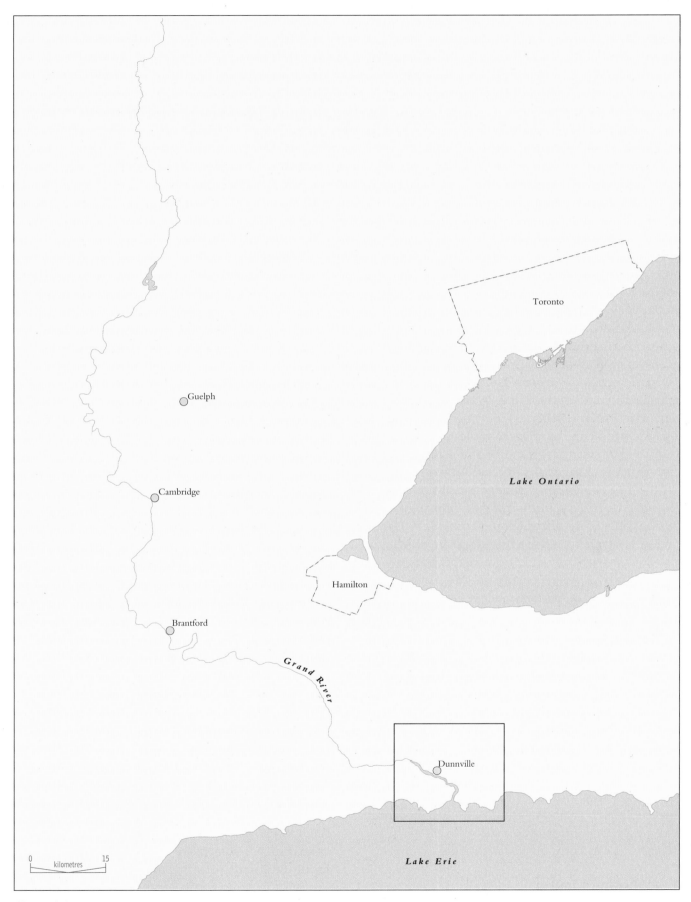

Figure 8.2 Dunnville bioregion.

they will be the main beneficiaries. The association members have argued that this alleged motive is not the case. Debate over this point has been vigorous and often acrimonious over several years. Such debate illustrates the discussion in Chapter 2 about the possibility of sustainable development being used by both proponents of development and proponents of environmental integrity.

Even if agreement can be reached about the most appropriate or functional ecosystem, there may be problems in actually using that spatial unit for management if it straddles more than one administrative jurisdiction (municipal, regional, provincial, national). For instance, in addressing water-quality problems in the Great Lakes, it seems logical to use the ecosystem in Figure 8.1, which includes all of the Great Lakes. Such an approach is a formidable challenge as it requires interaction among two national, two provincial, and eight state governments, as well as hundreds of municipalities. Even if that were achieved and the Great Lakes ecosystem were used for management, it would be apparent that the sources for some of the water-quality problems in that ecosystem come from outside. For example, some of the acid precipitation that has changed the pH level of water in lakes within the ecosystem is generated from coal-fired thermal electricity plants in the Ohio and Tennessee valleys, as discussed in Chapter 7. Unless action is taken to reduce the emission of sulphur dioxide and nitrous oxide from such plants, action to resolve the water quality in the Great Lakes basin will only be partially successful.

Thus analysts and managers usually have no clear-cut guidelines as to which spatial unit to use, even if there is agreement that an ecosystem approach will be used. Alternative ecosystems with different spatial extents are available. The science of ecology does not provide answers regarding which is most appropriate, and, even if consensus is achieved, practical problems still have to be handled regarding the relationship between the boundaries of ecosystem units and the boundaries of administrative units. A further complication is that in many respects, 'boundaries' do not even exist except as human constructs to simplify the analysis of our complex environment.

Establishing Limits

One of the characteristics of an ecosystem approach identified at the beginning of the chap-

ter was the incorporation of the concepts of carrying capacity, reliance, and sustainability. In other words, the ecosystem approach reminds us that there are limits or thresholds. Thus clear-cutting during logging creates the potential for erosion and other types of environmental degradation after the trees have been removed. Such degradation may inhibit the regeneration of trees in the logged area and may also create significant damage to fish habitat in adjacent streams, as discussed in Chapter 15. In draining wetlands to achieve increased crop production, farmers may change the habitat that supports migratory birds to such an extent that the numbers of birds may be threatened (discussed further in Chapter 14). As a result of overfishing, commercial fish stocks may be depleted to such a point that fish processing plants do not receive sufficient fish and have to operate only part time or close down, thus jeopardizing the livelihood of communities dependent on the fishery (discussed further in Chapter 20). These examples—logging, wildlife habitat, and fishing—all indicate that once limits or thresholds are crossed, there can be negative consequences for both human and non-human life.

If ecosystems are to be managed properly, then indicators of ecosystem health are required. In other words, just as baseline data are required in impact assessment (see Chapter 10) to estimate possible changes as a result of development, so ecosystem management requires baseline data and indicators to monitor the health of ecosystems.

However, it is often difficult to establish what baseline conditions for an ecosystem are appropriate, given that ecosystems are dynamic and therefore may change continuously *under natural conditions*. For example, there has been considerable debate regarding the appropriate management strategy if a prairie grasslands park were to be established in southwestern Saskatchewan. Dave Gauthier, a geographer at the University of Regina, investigated the feasibility of determining the nature of the prairie grasslands before the arrival of Europeans. He discovered that information from early explorers and surveyors was not detailed enough to reconstruct the ecosystem that existed in pre-European times. Their observations about 'long and plentiful grass' were not detailed enough for an ecologist to develop a plan to have the park returned to the types of vegetation that existed before the arrival of Europeans.

Logic suggests, however, that the presence of huge herds of bison influenced the prairie landscape (see Chapter 16). Many large grazing animals with sharp hooves ensured that the southern Prairies were covered in grasslands, and that woodland or forests did not appear, as might have been the case without the presence of the bison. Furthermore, natural fires also contributed to the maintenance of grassland areas. If fires were suppressed through fire-fighting initiatives, another natural element of the ecosystem would have been removed. Farmers with land adjacent to any park could reasonably be expected to support fire-suppression measures in the park to ensure that their crops were not damaged by fire.

Gauthier concluded that it was not possible to develop a management plan to maintain a prairie grasslands park in pre-European conditions because the necessary baseline data could not be established. Even if the baseline data were available, we would not be able to create an exact replica since conditions have changed. For example, without the vast herds of bison, it would be difficult to maintain the grassland conditions. The question left to be answered is this: What kind of ecosystem should a prairie grasslands park (or any other area) be based upon? The answer is often ambiguous.

The hot, dry summers and cold winters in Grasslands National Park support several species of rare herbs and grasses, as well as Canada's only black-tailed prairie dogs (*Parks Canada, Grasslands National Park/André Cornellier*).

Wood bison (*World Wildlife Fund Canada/Nic Larter*).

If managers can agree upon baseline data for an ecosystem, they also have to agree on what indicators should be used to track or monitor the health of that system, just as a doctor monitors the health of a patient by using blood tests and other diagnostic procedures. Development of environmental indicators is in its early stages, and as yet there is little consensus regarding indicators and associated thresholds (Box 8.1).

Canada has been one of the global leaders in such work, having produced two state of the environment reports at the national level (Bird and Rapport 1986; Minister of Supply and Services 1991), although it was announced in 1995 that federal budget cuts had resulted in the termination of any additional reports. In addition, the federal government has created the State of the Environment Office, whose task is to develop practical indicators regarding matters as diverse as trends of stratospheric ozone depletion (Environment Canada 1992) and toxic contaminants (Environment Canada 1993). There have also been many

Box 8.1 Indicators and Thresholds

Indicator: A measure of something of interest, such as the amount of heat, pressure, speed, safety, biodiversity, or ecological integrity.

Threshold: A point or limit beyond which something is unsatisfactory relative to a consideration such as health, welfare, or ecological integrity.

Example: Focus of concern: human welfare relative to quality of water.

Example of one water-quality indicator: Amount of nitrates.

Rationale: A link between infantile methaemoglobinaemia (blue-baby syndrome) and high levels of nitrate consumption is well documented. This condition can occur when a large amount of nitrate is consumed and subsequently reduced to nitrite. Nitrite in the bloodstream can convert the iron in haemoglobin into a ferric state, producing methaemoglobin. Conversion of more than 50 per cent of the haemoglobin to methaemoglobin, which is not capable of carrying oxygen, is often fatal.

Other evidence indicates a possible link between nitrate consumption and gastric cancer, liver disease, and respiratory defects.

Threshold: In 1970 formal standards for drinking-water quality were introduced by the World Health Organization. For nitrate, an upper limit of 100 mg per litre was established, and nitrate concentrations between 50 mg per litre and 100 mg per litre were classified as conditionally acceptable.

In 1984 the World Health Organization modified its recommendation for nitrate in drinking water to the equivalent of 10 mg per litre (44.3 mg of NO_3 per litre).

Thus thresholds may change with new scientific understanding. At the same time, different thresholds may be used in different places, as is highlighted in one of the case-studies in Chapter 16 (Based on Watson 1995:290–2).

---

regional state of the environment reports completed in Canada.

The establishment of baseline data and appropriate indicators is designed so managers can ensure that the carrying capacity of ecosystems is not exceeded. However, while carrying capacity is a concept we can intuitively understand, managers face major obstacles in identifying those thresholds beyond which it is appropriate to pass. Although science can establish thresholds relative to range

Carrying Capacity

. . . carrying capacity is not a simple, single, absolute value. There is no fixed figure we can point to for a particular . . . area and say 'This is the carrying capacity.' The . . . manager is faced with a complex set of conditions. He [or she] must consider a wide range of activities, . . . users . . . and values, many of which are incompatible with one another (Lime and Stankey 1971:175).

[carrying capacity] . . . a phrase delightful in its simplicity, complex in its meaning, and difficult to define, as in different situations and to different people it is understood in different ways (Barkham 1973:218).

carrying capacity for either cattle or deer, science by itself cannot establish whether it is appropriate to exceed such limits. Such a decision must reflect other subjective decisions about management objectives, assumptions, and values.

To illustrate, the carrying capacity of Toronto's SkyDome for a Blue Jays' baseball game is about 50,000 people, and most people attending a game have a more enjoyable experience if the baseball stadium is full. However, if that same number of people were placed in an equivalent space in the centre of a provincial or territorial wilderness park, most people would probably conclude that the park's carrying capacity had been exceeded. What has changed from SkyDome to the wilderness park is not the number of people but the management objectives and the expectations of the participants. This example illustrates why science alone cannot always provide the answer regarding 'the' carrying capacity for an area or place.

The implication is that carrying capacity figures are relative rather than absolute, and that they can also change markedly if management objectives or expectations are altered. In other words, different carrying capacities are possible. If the carrying capacity for one ecosystem is exceeded, that system will often collapse, resulting in a different kind of ecosystem. Thus when some commentators said in the late 1960s that Lake Erie 'was dying', they did not understand that the ecosystem was simply changing to one that was highly eutrophic due to the increased loadings of nitrates and phosphates, as discussed in Chapter 7. Such a eutrophic ecosystem is complex and rich, although probably less so than the more oligotrophic system it replaced; it is not dying nor dead. Whether or not a eutrophic system is desirable is a different question.

Scope

The issue of defining spatial boundaries discussed previously is one aspect of a more general challenge related to an ecosystem approach. That is, what is the most appropriate way to interpret and apply the holistic perspective and the concern for interrelationships? There are at least two alternatives.

The most conventional way is to define the scope in a *comprehensive* manner. In other words, once the boundaries of the ecosystem have been defined, attention is concentrated upon the totality of the system, its component parts, and their relationships. A comprehensive approach seeks to understand the entirety and complexity of a system, based on a belief that such an understanding is needed for effective management.

A comprehensive interpretation of an ecosystem approach is certainly consistent with the idea of addressing a system and the linkages among its components. However, in practical terms, the use of the comprehensive perspective has revealed some major weaknesses. With a focus upon complete understanding of a system, the comprehensive approach normally requires a significant period of time for data collection, analysis, and completion of a plan—three to four years has not been unusual. The problem has been that the completed plan often becomes more of a historical document than an action-oriented guide because conditions change over such an extended period of time. This has often resulted in disenchantment with a comprehensive approach among environmental managers, who conclude that it does not produce a practical product.

Another problem has been that the comprehensive approach often creates unrealistic expectations. Decision makers and the public alike believe that a comprehensive approach will result in a plan that will address all problems and provide recommendations to deal with them all. The product of many comprehensive analyses in Canada has been a mix of action-oriented recommendations and calls for more analysis to address aspects not yet understood. After a three- to four-year study, it is hardly surprising that policy makers and the public are not impressed when told that yet more research needs to be done because the complexity of the system and all of its parts and relationships could not be understood within the available time and resources.

A third barrier has been the lack of institutional arrangements to implement the recommendations even if a comprehensive analysis is done in a reasonable time period and most of the recommendations are action oriented. As noted earlier in this chapter (and in Chapter 2), many agencies are organized to deal with specific components of the environment or with specific resources. When recommendations focus upon the ecosystem as a whole and require an approach that can deal with multiple components and relationships, the administrative system is often inadequate.

As a response to the problems of a comprehensive approach, people have been turning increasingly to the *integrated* approach. This use of 'integrated' rather than 'comprehensive' has not just been an academic distinction of terms. An integrated approach maintains the concern with systems, their component parts, and their linkages, but is more *focused* and *selective*. In other words, rather than trying to understand the totality of the system, the environmental manager focuses only upon key variables believed to account for most of the variation in the system.

An analogy can be taken from the idea that 20 per cent of the people in a group will do 80 per cent of the work. In a similar manner, if 20 per cent of the variables in a system cause 80 per cent of the variation in a system (and this often seems to be the case), then all the extra effort required to identify and understand the other 80 per cent of variables causing the remaining 20 per cent of variation does not seem very cost effective. If we accept that we will never understand the complexity of an entire system, it seems reasonable to be satisfied with dealing with 80 per cent of the problem and reallocating the resources that might have been used to pursue the remaining 20 per cent of the problem to other societal needs.

The integrated approach can usually be completed in a shorter period than the comprehensive approach. Furthermore, by focusing upon selected key variables, it should result in more action-oriented recommendations rather than a long list of recommendations calling for more investigation. In addition, if effort is made throughout the analysis to determine which of the key variables are best handled by existing organizations, the chances of the recommendations being implemented will be greater. This final aspect will be considered in greater detail in Chapter 11. An example of an integrated approach is shown in Box 8.2.

The integrated perspective is not a perfect way to interpret the ecosystem approach, but it does offer a pragmatic way to ensure that the output of an analysis based on an ecosystem has practical value. Such a consideration is of critical importance. If an ecosystem approach does not produce a practical product for environmental managers,

then the concept will lose legitimacy or credibility. Once that happens, it may become extremely difficult to persuade decision makers to consider such an approach. It would be tragic if we discredited a concept (that in itself was not flawed) by using a flawed method of translating it from concept to practice. At the same time, it must also be appreciated that considerable time is often required to understand ecosystems. If the focus is always short term, rather than long term as advocated by the Royal Commission on the Future of the Toronto Waterfront, there will always be a challenge in applying the ecosystem approach.

Values and Traditions

A key characteristic of an ecosystem approach is that humans are part of nature rather than separate from it. If that concept is accepted, the ecosystem perspective challenges one of the dominant beliefs of many Canadians—that humans have the right and opportunity to dominate all other living and non-living things on the earth and to use them for the convenience of humankind. The ecosystem approach forces us to recognize the limits or

---

Box 8.2 Example of an Integrated Approach: Subwatershed Planning in Ontario

A subwatershed plan should present the following information:

Subwatershed boundaries, including the rationale for their establishment.

Relationship of the subwatershed plan to watershed plans (if available) and to other urban drainage, environmental land use, and planning studies and programs.

Identification of form and function of natural systems, including land uses, natural features, linkages, and surface and groundwater systems. Identification of existing systems should include aquatic and terrestrial features/habitats; the quantity and quality of surface and groundwater resources, relationships, and water-related dependencies; and factors influencing the viability of the resources.

Subwatershed objectives for public health, public safety, aquatic life, resource management, flood plain management, and urban, agricultural, and other land uses.

Plan recommendations: (1) Specify areas for protection, rehabilitation and/or enhancement.

It should be clearly noted where changes within the subwatershed should not occur, along with appropriate setbacks from natural areas, and recommended management strategies for these areas; (2) establish areas that can be developed in a manner compatible with subwatershed objectives; identify how this can be achieved through use of best management practices and drainage system design that will protect, enhance and/or rehabilitate natural areas and systems.

Implementation plan, outlining policy/guidelines to direct development planning and design; design, function, siting, and timing of facilities; funding of works, interagency review/approvals, and regulation requirements; recommendations and responsibilities for future studies; operation and maintenance responsibilities; monitoring program and responsibilities; approaches and responsibilities for information updating and corrective actions; and time frame for review/update of plan (Based on Table 1 'Example of an Integrated Approach: Subwatershed Planning', Ontario Ministry of Natural Resources. Copyright 1997 Queen's Printer Ontario).

thresholds that, if exceeded, lead to degradation. Even if there are major problems in establishing such limits or thresholds, it is intellectually difficult to deny them. However, the question posed about humans' relationship with other living and non-living things threatens many people's interests and values, and can therefore be expected to encounter resistance and opposition.

Other values and traditions are in question as well. As Caldwell (1970) explained, the reason why the ecosystem approach has not been widely adopted is not only because of the complexity and ambiguity, although these are not trivial factors. In his view, an ecosystem approach encounters resistance because it is inconsistent with the values, assumptions, institutions, and practices shaping the social arrangements in many countries that affect the custody and care of the environment.

Values and practices in banking, taxation, insurance, and property law, when combined with non-ecologically oriented public policies about the environment and resources, present very resistant barriers to an ecosystem approach. For example, property rights often do not provide incentives for land or resource owners to consider the implications of their activities for adjacent owners or users. As Caldwell concluded:

> To conceive an ecosystems approach to public land policy, one must have first arrived at an ecological viewpoint toward the world of man [*sic*] and nature. But this is not the viewpoint from which pioneers, land speculators, farmers, stockmen, lawyers, bankers, or local government officials have commonly seen the land. To institute an ecosystems approach to public land policy, a great many things besides land must be considered. An ecosystems approach is basically a total systems approach. It therefore includes in its purview many things omitted in less comprehensive systems. It would impose constraints upon single purpose approaches to the environment and would arouse hostility among individuals whose single purpose pursuits would thereby be constrained (Caldwell 1970:205).

Although Caldwell was writing in 1970 and was referring to the United States, much of what he referred to also exists today in Canadian society. It would be naive and unrealistic not to understand that some powerful individuals and many groups will regard an ecosystem approach as a threat to their interests. If an ecosystem approach is to be implemented, awareness of these other values and interests will be necessary.

IMPLICATIONS

The ecosystem approach is advocated as an alternative to conventional approaches to resource and environmental management. The ecosystem approach: (1) challenges the prevailing view, even if it is more implicit than explicit, that humans have a dominant position among other living and non-living things on the earth; (2) indicates that attention should be given to systems and their interrelationships; (3) suggests that management areas need not always be based on administrative or political boundaries; and (4) highlights that change is ongoing and continuous, requiring managers and their strategies to be flexible enough to accommodate change and adjustment.

Just as the conventional approach to environmental management is flawed, the ecosystem approach also has weaknesses. First, the science of ecology still has many concepts and principles over which scientists disagree. Such ambiguity means that ecologists cannot always provide the type of advice that managers would like to have to address problems. Second, while the concept of an ecosystem is relatively easy to understand, in transforming the idea into practice the manager encounters many choices as to which kind of ecosystem is most desirable. Once again, the science of ecology cannot always provide clear-cut answers as to which ecosystem is best. Third, the concept of ecosystems introduces the idea of limits or thresholds that should not be exceeded. However, decisions about such limits must ultimately be made primarily with regard to management objectives and assumptions, as well as public expectations. As a result, science by itself cannot define appropriate limits, although it should provide a key input into decisions about them. Fourth, the scope or interpretation of an ecosystem approach can be either 'comprehensive' or 'integrated'. Increasingly, the more focused *integrated* approach is proving to be more practical. Fifth, the ecosystem approach challenges many basic values, beliefs, and interests in

Canadian society. As a result, resistance and opposition to the introduction or implementation of an ecosystem approach can be expected.

At the same time, the ecosystem approach provides an exciting opportunity to move environmental management policies and practices in a direction that better reflects the balances and tensions among natural, economic, and social systems. While the ecosystem approach is not perfect, it does appear to address directly some of the problems that the conventional approach to environmental management has not seemed able to resolve. In Part D, attention will be given to considering how well the ecosystem approach has been applied in Canada, and where there may be new opportunities to implement it.

SUMMARY

1. Ecosystems consist of communities of people, plants, animals, and micro-organisms interacting with each other and the non-living elements in their environment.

2. An ecosystem approach involves defining and managing an entire system rather than only some parts of it, considering interrelationships among elements, accepting the dynamic or changing nature of a system, and emphasizing that people are a part of rather than separate from or dominant over nature.

3. An ecosystem approach challenges much of contemporary resource and environmental management practice since the latter has focused on resource sectors (such as forestry, agriculture, or minerals), considers people as separate from the environment, encourages specialization by resource and environmental managers, relies on unilateral rather than collaborative initiatives, and is based on administrative rather than natural spatial units.

4. Various barriers impede implementation of an ecosystem approach, including imperfect science; disagreement about appropriate scope, boundaries, and limits; and incompatibility with existing traditions and values. The latter are represented by such interests as banking, taxation, insurance, and property rights.

5. Principles important for an ecosystem approach are: (1) stochastic or random processes, (2) stability and resilience, (3) complexity and resilience, (4) diversity and stability, and (5) ecological fluctuations and niches.

6. There are relatively few unambiguous and relevant ecological principles because of disagreement about appropriate assumptions and the inability to generalize from specific experiences.

7. Bioregions are one means of defining an ecosystem for planning and management purposes.

8. Defining and measuring ecosystem health or integrity is a major challenge, and development of indices and related thresholds is still at a very preliminary stage.

9. Carrying capacity is a relative rather than an absolute concept, and is influenced relative to management objectives and values.

10. An ecosystem or holistic approach can be interpreted in either a *comprehensive* or *integrated* manner, with the latter seeming more practical for problem solving.

REVIEW QUESTIONS

1. What are the main characteristics of an ecosystem approach?

2. Why are random or stochastic processes important to consider in ecosystem management?

3. What are the relationships among complexity, diversity, stability and resilience, and what are their significance for ecosystem management?

4. How would you decide upon appropriate boundaries to define an ecosystem for the purposes of environmental management? Have any management boundaries in your region been created on the basis of an ecosystem? If so, what have been the strengths and weaknesses of the management experience?

5. What is a bioregion? Provide an example of a bioregion in *your* region of Canada and describe its key characteristics.

6. What is the significance of limits or thresholds for the ecosystem approach?

7. Are there similarities between the concepts of carrying capacity and sustainability? What are they? What are the implications?

8. What are the differences between the comprehensive and integrated interpretations of a holistic approach?

9. In your opinion, should humans be considered as part of or separate from nature?

REFERENCES AND SUGGESTED READING

Barkham, J.P. 1973. 'Recreational Carrying Capacity: A Problem of Perception'. *Area* 5:218–22.

Bird, P.M., and D.J. Rapport. 1986. *State of the Environment Report for Canada*. Ottawa: Minister of Supply and Services Canada.

Caldwell, L.K. 1970. 'The Ecosystem as a Criterion for Public Land Policy'. *Natural Resources Journal* 10, no. 2:203–21.

_____. 1994. 'Disharmony in the Great Lakes Basin: Institutional Jurisdictions Frustrate the Ecosystem Approach'. *Alternatives* 20, no. 3:26–31.

Conservation Authorities of Ontario. 1993. *Restructuring Resource Management in Ontario: 'A Blueprint for Success'*. Mississauga, ON: Credit Valley Conservation Authority.

Dorney, R.S. 1989. *The Professional Practice of Environmental Management*. New York: Springer-Verlag.

Environment Canada. 1992. 'Stratospheric Ozone Depletion'. Environmental Indicator Bulletin, no. 92–1, State of the Environment Reporting. Ottawa: Environment Canada.

_____. 1993. 'Toxic Contaminants in the Environment'. *Environmental Indicator Bulletin*, no. 93–1, State of the Environment Reporting. Ottawa: Environment Canada.

Holling, C.S., and M.S. Goldberg. 1971. 'Ecology and Planning'. *Journal of the American Institute of Planners* 37, no. 4:221–30.

Kay, J.J., and E. Schneider. 1994. 'Embracing Complexity: The Challenge of the Ecosystem Concept'. *Alternatives* 20, no. 3:32–9.

Kettle Creek Conservation Authority. 1994. *Conservation Strategy*. St Thomas, ON: Kettle Creek Conservation Authority.

Lime, D.W., and G.H. Stankey. 1971. 'Carrying Capacity: Maintaining Outdoor Recreation Quality'. In *Recreation Symposium Proceedings*, 174–84. New Darby, PA: US Department of Agriculture, Forest Service, Northeastern Forest Experiment Station.

Lopez, B. 1989. 'American Geographies'. *Orion* (September).

Minister of Supply and Services. 1990. *Canada's Green Plan*. Ottawa: Minister of Supply and Services Canada.

_____. 1991. *The State of Canada's Environment*. Ottawa: Minister of Supply and Services Canada.

Norton, G.A., and B.H. Walker. 1982. 'Applied Ecology: Towards a Positive Approach. I. The Context of Applied Ecology'. *Journal of Environmental Management* 14, no. 4:309–24.

Ontario Ministry of Environment and Energy/Ministry of Natural Resources. 1993. *Subwatershed Planning*. Toronto: Queen's Printer for Ontario.

Plant, C., and J. Plant. 1990. *Turtle Talk: Voices for a Sustainable Future*. Philadelphia: New Society Publishers.

_____, and J. Plant. 1992. *Putting Power in Its Place: Create Community Control*. Philadelphia: New Society Publishers.

Royal Commission on the Future of the Toronto Waterfront. 1992. *Regeneration: Toronto's Waterfront and the Sustainable City, Final Report*. Ottawa: Minister of Supply and Services Canada; Toronto: Queen's Printer of Ontario.

Sale, K. 1985. *Dwellers in the Land: The Bioregional Vision*. San Francisco: Sierra Club.

Saskatchewan Round Table on Environment and Economy. 1992. *Conservation Strategy for Sustainable Development in Saskatchewan*. Regina: Saskatchewan Round Table on Environment and Economy.

Slocombe, S.D. 1993. 'Implementing Ecosystem-Based Management: Development of Theory, Practice, and Research for Planning and Managing a Region'. *BioScience* 43, no. 9:612–22.

Walker, B.H., and G.A. Norton. 1982. 'Applied Ecology: Towards a Positive Approach. II. Applied Ecological Analysis'. *Journal of Environmental Management* 14, no. 4:325–42.

Watson, N. 1995. 'Nitrate Pollution in England: A Hazards Management Perspective'. *Managing the Water Environment: Hard Decisions from Soft Data*, vol. 1, 287–301. Proceedings of the 48th Annual Conference. Cambridge, ON: Canadian Water Resources Association.

Chapter 9

Adaptive Management

In previous chapters, it was emphasized that ecosystems are dynamic or subject to change, and therefore the ecosystem approach must be able to deal with change. Change in itself is not always daunting if it happens in a gradual or incremental manner. However, change can occur as a result of major surprises or unanticipated events. Such situations reflect the considerable uncertainty about the future, as well as the turbulent conditions under which environmental managers work. This chapter focuses upon the nature of surprise and turbulence, and then considers the idea of adaptive environmental management as a process of decision making that is a better response to uncertainty.

SURPRISE, TURBULENCE, AND CHANGE

Trist (1980) has provided some useful insights regarding surprise, turbulence, and change. In his view, there is no such thing as *the future*, but instead there are *alternative possible futures*. Which future actually occurs depends very much on the choices made and on the actions taken to implement those choices. He argued that 'the paradox is that under conditions of uncertainty one has to make choices, and then endeavour actively to make these choices happen rather than leave things alone in the hope that they will arrange themselves for the best' (Trist 1980:114).

A challenge in making choices and taking initiatives is what Trist refers to as *turbulent conditions*, which have become increasingly prevalent. For example, energy plans in Canada became obsolete in the early 1970s when OPEC countries rapidly quadrupled oil prices. Such an increase in prices had not been included in the forecasts and assumptions on which energy plans had been based. During the 1980s when decisions were being made about the East Coast fishery, no one expected that in the early 1990s the cod fishery would be effectively closed, and that thousands of people in Atlantic Canada would lose their jobs (discussed in detail in Chapter 20). These and other events came as surprises. They created bewilderment and anxiety, and raised doubts regarding the capability of science, planning, and planners since decision makers and their decision-making process apparently had not been able to anticipate or adapt to rapid change.

A wind farm in southern Alberta (*Philip Dearden*).

Trist suggested that in industrialized countries, four planning or management situations have followed a sequence from the simplest to the most complex and uncertain (Table 9.1). In the first two planning situations, many small organizations were involved, so the possibility for any one individual or group to affect outcomes significantly was small. As a result, individuals mostly functioned on their own and did not need or seek to establish linkages or partnerships with other groups or companies.

The third type, called the *disturbed reactive* stage, began to emerge in the late 1800s and early 1900s, and was characterized by large organizations emerging to take advantage of economies of scale and also by more direct and intense competition. One of the best examples was the development of Henry Ford's production process for the mass manufacture of automobiles. The outcome was the establishment of very large manufacturing plants and the domination of the automotive industry by a relatively few major firms. The most effective strategy in this type of situation was to assemble more and more human and financial resources in a centrally controlled system. The bigger one became, the better were the chances for survival. This led to the emergence of big government, big corporations, and big unions.

The fourth phase, which is the one that Trist believed began to characterize Western societies in the 1980s, is called *turbulent*. In this situation, large competing organizations, acting independently and oriented in diverse directions, generate unanticipated and often discordant consequences. The outcome is loss of stability. Trist concluded that many Western industrial societies, including Canada, were still approaching environmental management during the 1980s as if they were in the third planning situation (*disturbed reactive*) when in fact they had already moved into the fourth type (*turbulent*). In his view, it was not surprising that

many of our environmental management strategies were ineffective. He concluded that too often we organized ourselves to manage for conditions that no longer existed, and then we wondered why carefully developed structures, mechanisms, and processes did not work.

Trist then raised the question as to which approach was the most suitable for *turbulence*. While there was no perfect answer, he believed that preferred responses to all four decision-making situations could be identified (Table 9.2). He believed that for a *turbulent* situation, it was important to move towards what he called a 'negotiated order', which recognized that no organization was so big that it could conduct its affairs in isolation from what others were doing. The new guiding value is *collaboration*. However, as Trist emphasized, this new value is incompatible with the idea of competi-

---

Table 9.1 Types of Planning Situations

Type	Name	Characteristics
I	Placid random	Many small-size participants; no one participant can unduly influence or control outcomes; individuals and groups work alone.
II	Placid clustered	Some large-size groups form; specialization begins to occur; some participants can influence outcomes; individuals or groups still operate mostly alone.
III	Disturbed reactive	Large organizations begin to achieve economies of scale; competition becomes more direct and intense; centralization occurs.
IV	Turbulent	Competition leads to unanticipated and often unwanted outcomes; instability and commotion are common.

Table 9.2 Responses to Alternative Decision-Making Situations

Planning Situation	Mode	Preferred State	Posture
Placid random	Inactive	Present	Wait and see
Placid clustered	Reactive	Past	Put it back
Disturbed reactive	Preactive	Future	Predict and prepare
Turbulent	Interactive	—	Make it happen

Source: Based on E. Trist, 'The Environment and System-Response Capability', *Futures* 12, no. 2 (1980):117, 118.

tion, the dominant value in the third kind of planning situation (*disturbed reactive*).

Table 9.2 illustrates the modes of response associated with the four planning situations. The *inactive mode* assumes that the present is the ideal and preferable compared to either the past or the future. The prevailing wisdom would include sayings such as 'It is better to wait and see', or 'Let sleeping dogs lie.' In the *reactive mode*, the past is viewed as better than the present or the future. The goal thus becomes to restore the good old days. For the *preactive mode*, the future is regarded as better than the past or the present. The challenge is to determine where and when the best opportunities will appear. Attention focuses upon predicting and preparing to seize future opportunities. Technical forecasts, simulation, and modelling become the tools to pursue the art of the calculable.

However, in the *interactive mode*, the past, present, or projected future do not seem particularly good. The only worthwhile future is one that reflects our choices, and therefore the desirable future depends on our making it happen. This usually cannot be achieved by just one person, group, or organization but requires a collaborative and cooperative effort. Thus Trist concluded that the interactive mode seems to be the most appropriate response for a turbulent situation. Such a mode or style of response was referred to as *adaptive planning*.

ADAPTIVE PLANNING

Adaptive planning has the characteristics shown in Box 9.1.

Box 9.1 Characteristics of Adaptive Planning

- Collaboration of interest groups
- Identification of shared values
- Continuous learning
- Continuous evaluation and modification

In addition to the characteristics in Box 9.1, adaptive planning also reinforces the distinction between a comprehensive and integrative approach as outlined in Chapter 8. As Trist explained, comprehensive planning of the master plan or blueprint

type, which flourished during the third situation (*disturbed reactive*) for planning, is no longer appropriate in the fourth type (*turbulent*). In his view, the level of uncertainty created by the faster rate of change is too great and plans become outdated before they can be implemented. As a result, the new planning or management style becomes an *interactive* one. In other words, it is *continuous* because emerging changes require frequent modifications; it is *participative* since many stakeholders should be involved; it is *integrated* because all levels must make their inputs from their own perspectives; and it is *coordinated* due to the need to consider the interdependence of decisions.

Such an approach highlights the need for *process* and the *product* of management to be given attention. Some of these ideas will be explored in more detail in Chapter 11, which considers partnerships and stakeholders, and in Chapter 12, which addresses approaches for dispute resolution. Adaptive planning has also been recommended as a framework for forest management, as discussed in Chapter 15.

ADAPTIVE MANAGEMENT

The concept of adaptive management is not new. People in many parts of the world who depend upon subsistence agriculture for survival have practised adaptive management for thousands of years. For example, subsistence farmers often have a number of small plots scattered as much as a half day's walk from each other. While from a Western perspective this practice may seem inefficient, it has the benefit of providing the farmer with different soil and microclimatic conditions. For example, a plot at the bottom of a valley will likely have adequate water due to the proximity of a river, but may also be prone to damaging floods. A hillside plot may be susceptible to more slope erosion, but it will have better air drainage and therefore be less susceptible to frost damage than a lowland plot. By having numerous plots in different locations, the subsistence farmer has adapted to uncertain but likely future conditions, and has reduced the chances of losing an entire crop in a season or a year.

Adaptive management can also be observed over time in a given place, as subsequent groups or generations use an environment or harvest resources. Bennett (1969) analysed the activities of four groups of people (Plains Cree Indians, ranchers, farmers, and Hutterite Brethern) in southern Saskatchewan with respect to the natural and social resources available to them. Bennett showed how population mobility had been 'normal' for the nomadic Indians as they hunted bison and other hooved animals of the Plains and foraged for different vegetables and small game in forested uplands. In contrast, population mobility was 'abnormal' for the Euro-Canadian settlers due to their customs and sedentary occupations. They intro-

South Saskatchewan from the Cypress Hills (*Philip Dearden*).

duced different types of agricultural production, settlement patterns, and transportation networks, as well as social customs. Thus given different values, cultures, and technologies, various groups adapted in different ways to both the opportunities and the constraints in southern Saskatchewan.

Adaptation

Bennett developed some important ideas about adaptation. In his view, if coping or adapting is successful, people achieve their goals. In a market economy, success for a society is defined and measured relative to the quantity of *production*, *income*, and *consumption*. However, he emphasized that such an interpretation of success only reflects one dimension of adaptation. Another and equally important dimension is *conservation of resources*. Human activity that achieves economic gain but does so by exhausting or abusing natural resources may be adapting in one dimension but maladapting in another. In contemporary terms, such human activity is unlikely to be *sustainable* (Chapter 2).

From this perspective about adaptive behaviour, Bennett concluded that it implies movement or change of some kind since it is concerned with the process of achieving goals, but it usually creates new goals or problems in doing so. As a result, in Bennett's view, the concept of adaptation actually involves two components. The first is *adaptive strategies* or the patterns from the many individual adjustments people make to obtain and use resources and to solve the immediate problems confronting them. The second is *adaptive processes*

or the changes introduced over relatively long periods by the ongoing and repeated use of such strategies.

Adaptive Environmental Management

Although adaptive management is not a new concept, it was revitalized by work stimulated and led by Holling (1978, 1986). The intent of this work was to develop an adaptive approach for environmental impact assessment and management (see Chapter 10 for a discussion of impact assessment).

The main conclusion was that it is necessary to prepare policies and approaches capable of coping with the uncertain, the unexpected, and the unknown. Holling and his coworkers observed that humankind has always lived in uncertain conditions, and yet has generally prospered. The customary way of handling the unknown has been through trial and error. Existing experience and information are drawn upon. Errors or mistakes provide new information to allow subsequent activity to be modified. As a result, failures generate new information and insight, which lead to new knowledge. In this manner, accumulated experience and knowledge provide the departure point for new ideas and initiatives.

However, the effectiveness of the trial-and-error method requires some conditions. The experiment should not destroy the experimenter. Or, at a minimum, someone must remain to learn and benefit from the experiment. The experiment should also not cause irreversible changes in the environment. Furthermore, the experimenter should be able to start over, having perhaps been

A Perspective on Adaptive Management

Adaptive management is an approach to natural resource policy that embodies a simple imperative: policies are experiments; learn from them. In order to live we use the resources of the world, but we do not understand nature well enough to know how to live harmoniously within environmental limits. Adaptive management takes that uncertainty seriously, treating human interventions in natural systems as experimental probes. Its practitioners take special care with information. First, they are explicit about what they expect, so that they can design methods and apparatus to make measurements. Second, they collect and analyze information so that expectations can be compared with actuality. Finally, they transform comparison into learning—they correct errors, improve their imperfect understanding, and change action and plans. Linking science and human purpose, adaptive management serves as a compass for us to use in searching for a sustainable future (Lee 1993:9).

humbled and enlightened by a previous 'failure'. Finally, the experimenter must have the will and capability to begin again.

In Holling's view, it was becoming increasingly difficult to satisfy the minimum conditions required for the trial-and-error method using the traditional approaches to environmental management and development. Our trials in environmental management and development have been producing errors that are greater and more costly than society can afford. The concern about climate change, as discussed in chapters 7 and 20, reflects worry that we may not be able to reverse such change before serious problems have occurred. Moreover, even when errors were not, in theory, irreversible, the magnitude of the original capital investment and the associated prestige often made reversibility unlikely. A fundamental reason could be found in a basic value or attitude of industrialized people: they do not like to admit to or pay for past mistakes but prefer to try and correct them. The outcome of trying to correct an inappropriate initiative is often additional investment, additional costs related to control and maintenance, and progressive loss of future options.

Given the above, Holling argued that it is inappropriate to try and develop an alternative approach that tries to eliminate the uncertain and the unknown. Such an approach would be based on the illusion that there was sufficient knowledge to support greater monitoring, regulation, and control. Instead, he and his coworkers believed that the appropriate direction for improved management would be to allow the trial-and-error approach to work again.

While reducing rather than eliminating uncertainty is desirable, Holling concluded that equal attention had to be given to developing the ability to prepare for uncertainty and to obtain benefits from the unexpected. This perspective is at the heart of adaptive environmental management. By reducing uncertainty but also benefiting from it, the goal is to develop more resilient policies.

Resilience and Stability

Holling argued that the concept of resilience can be related to the structure and behaviour of ecological systems. Thus how a system responds to a planned or unexpected disturbance depends upon its *stability*. One perspective, implicit in many of humankind's approaches to environmental management, assumes that global stability exists. In other words, no matter what the magnitude of a disturbance, it is believed that the system will recover to its original, stable condition after the disturbance has been removed. Holling termed this view *Benign Nature*, which can accommodate trial and error on any scale. From this perspective, 'big' or 'large-scale' initiatives are always allowable and even desirable, since they can achieve economies of scale.

The Role of Adaptive Management

. . . active adaptive management can play a central role, because its premise is that knowledge of the system we deal with is always incomplete. Not only is the science incomplete, the system itself is a moving target, evolving because of the impacts of management and the progressive expansion of the scale of human influences on the planet. Hence, the actions needed by management must be ones that achieve ever-changing understanding as well as the social goals desired. That is the heart of active experimentation at the scales appropriate to this question. Otherwise the pathologies of management are inevitable—increasingly fragile systems, myopic management, and social dependencies leading to crises (Walters and Holling 1990:2067).

A different viewpoint sees a high degree of instability in ecological systems. From this interpretation, ecosystems are fragile and have a natural rhythm of small-scale extinctions. Ecosystems persist only because of diversity in their structure and spatial distribution. External sources generate recovery. This view, termed *Ephemeral Nature*, is associated with a 'small is beautiful' perspective, and emphasizes the importance of spatial variety, diversity of opportunity, and local autonomy.

In the real world, there is a combination of these two extremes. Natural systems can and do have more than one stable mode of behaviour. As long as variables, such as population density or amount of nutrients, stay within a certain range, small disturbances can be absorbed or accommodated. While quantities may change, qualitative behaviour does not. Consequently, small disturbances can be introduced gradually without threatening the integrity of the system. However, eventually one extra increment of change or disturbance can 'flip' or push the system across a boundary into some totally different type of behaviour. For example, a river becomes an open sewer, or a lake becomes eutrophic. In these circumstances, Holling suggested that managers should view nature not as benignly forgiving but more as *Practical Joker*.

With nature viewed as a *Practical Joker*, more caution is needed, especially with the 'big is necessary' approach. If boundaries differentiate between desirable and undesirable conditions or states, then the manager's challenge is to avoid a situation in which the system gets too close to a dangerous boundary or threshold.

One other view about stability is also important. All three conditions described so far (Benign Nature, Ephemeral Nature, and Nature as Practical Joker) have implicitly accepted that the rules of the game are structured and fixed. However, as noted in chapters 5 and 8, systems are not static or always deterministic. Variability and change are the rule, and chance events are important and even dominate some ecosystems. To illustrate, fire is often not a disaster but the source of

maintenance for many grassland ecosystems. However, when and where fire will occur under natural conditions in a grassland ecosystem is usually difficult to predict or anticipate.

Furthermore, the components in an ecosystem can shift, through internal processes and mechanisms, from one kind or phase of stability to another. For example, periodic insect outbreaks, such as the spruce budworm in New Brunswick, can be set off by chance patterns of weather, by dispersal of moths from other areas, or even by the natural growth of a forest (see Chapter 15). Populations can appear to explode from very low but stable numbers to high and apparently stable numbers, at least for a period of time. While such high numbers are stable for the insect, the forest cannot accommodate the defoliation on a sustainable basis. As the forest dies back, the insect numbers drop, and the forest begins the process of regeneration. Over time, this pattern can be viewed as stability, and the swings and movements of the pattern as essential for the renewal of the forest and for the maintenance of biodiversity.

As a result of such patterns, it is clear that the components or variables of ecosystems are not stable. Indeed, at a local scale or in a short time frame some species may even become extinct, to be reintroduced through contributions from other places. As Holling (1978:10) concluded, 'the variables are moving continually and the stability boundaries are being tested.'

The spruce budworm defoliates and kills spruce and other evergreens (*Ontario Ministry of Natural Resources. Copyright 1997 Queen's Printer Ontario*).

The conditions outlined earlier create an important situation that environmental analysts and managers must recognize: it is not only the variables that shift and move but also the boundaries between regions of stability. Understanding the changing pattern of variables and the evolving structure and boundaries of ecosystems are central to dealing with the unknown in adaptive environmental management. Holling argued that environmental policies too often try to reduce variability within partially understood systems, either as an objective in itself or to meet safety, health, or other standards. Using human intervention to constrain variability often results in shifting the balance in an ecosystem so that the regions of stability change. In Holling's view:

Paradoxically, success in maximizing the distance from a dangerous stability boundary may cause collapse, because the boundary may implode to meet the variables. If surprise, change and the unexpected are reduced, systems of organisms, of people, and of institutions can 'forget' the existence of limits until it is too late.

This final view is of Resilient Nature, where resilience is a property that allows a system to absorb and *utilize* (or even benefit from) change (Holling 1978:11).

The concept of adaptive environmental management is built on these ideas. Adaptive environmental management accepts that: (1) surprise, uncertainty, and the unexpected are normal; (2) it is not possible to eliminate them through management initiatives; and (3) management should provide allowance for them. The goal is to develop more resilient environmental policies and practices. To achieve this goal, Holling and his coworkers suggested a process that focused on those features shown in Box 9.2. In addition, Table 9.3

Box 9.2 Key Aspects of an Adaptive Environmental Management Process

- generation of a range of alternative objectives
- design of effective policies to achieve alternative objectives
- generation of indicators (social, economic, resource, environmental) relevant for decisions
- evaluation of each policy regarding the behaviour of the indicators over space and time
- partial compression of indicator information to allow screening of the most appropriate policies
- communication and interaction between and among those who design, choose, and endure the policies (staff, decision makers, citizens)

Table 9.3 Comparison of Conventional and Adaptive Approaches

	Conventional Management	Adaptive Management
Process	Answer oriented	Question oriented
Design strategy	Optimal solution to problem at hand	Multiple solutions (resilient mix)
Burden of proof	Bias towards study (e.g., acid rain)	Bias towards action plus monitoring (e.g., water budget)
Purpose of monitoring	Compliance and crediting	Learning and adjusting
	Problem curable	Continuing management
Range of utility	Project not repeatable	Project repeatable
	Experiments too risky (e.g., to individuals)	Experiments acceptable (e.g., populations more important than individuals)
	Project failure is a management failure	Failure can be productive

Source: Adapted from K. Lee and J. Lawrence, 'Adaptive Management: Learning from the Columbia River Basin Fish and Wildlife Program', *Environmental Law* 16, no. 3 (1986):431–60.

shows the differences between the adaptive and conventional approaches to environmental management.

In the case-studies presented in Part D, attention will be given to how ideas compatible with adaptive environmental management have been incorporated into policies and practices. Boxes 9.3 and 9.4 outline two experiences that used adaptive management. For the balance of this chapter, however, attention is given to a situation in Quebec in which the ideas from adaptive environmental management have been considered.

THE KATIVIK ENVIRONMENTAL QUALITY COMMISSION

Mulvihill and Keith (1989) have argued that relatively little attention has been given to the *organi-* *zational* or *institutional* aspects of adaptive environmental management. They believed that adaptive management should also be applicable to organizational or institutional arrangements, given Trist's argument that the context for organizations is 'seldom stable or predictable'. They sought to develop a set of principles that reflected the basic ideas of adaptive environmental management and that could be used to design effective organizations and processes.

Mulvihill and Keith noted that the James Bay and Northern Quebec Agreement of 1975 was the first Aboriginal land claim to be settled in northern Canada. It established different management arrangements for areas in Quebec north and south of the 55th parallel. The region north of the 55th parallel is known as northern Quebec or Kativik

---

Box 9.3 Fish and Wildlife Program in the Columbia River Basin

The Northwest Power Planning Council in the United States has initiated the Fish and Wildlife Program in the Columbia River basin. A centrepiece of this program is an ambitious effort to restore runs of salmon and steelhead to the river and its tributaries.

Lee and Lawrence (1986) have noted that the use of adaptive management for this program has been based on five principles:

1. Protecting and restoring fish and wildlife is a common objective for hunters, fishers, scientists, and naturalists. However, the short-term human interests are often poorly correlated with the needs of the natural system. As a result, the focus is on the shared, long-term interest in protecting and rebuilding stocks.
2. Projects are inevitably experiments. The choice becomes one of making them good or poor ones. Because of imperfect knowledge, no measure can be guaranteed to perform as anticipated or intended. Some will fail. Others will do better than expected.
3. Action is overdue. Action should not be deferred until 'enough' is known. Acting with *the expectation of surprise* is an important way to produce new

knowledge. Knowledge can be increased by incorporating learning into projects as experiments.
4. Information has value, not only as a basis for action but also as a product of action. Generating that information requires explicit consideration of the questions to be answered and careful experimental design to answer those questions.
5. Enhancement measures may be limited in time, but management is ongoing. Enhancement activities must include deliberately designed means for learning and remembering how to improve management of the resource.

The Hugh Keenleyside hydroelectric dam on the Columbia River (*Philip Dearden*).

Box 9.4 Fisheries Management in the Saint John River, New Brunswick

During 1981, it was proposed and approved to apply adaptive management ideas to fisheries management for waters behind the Mactaquac Dam near Fredericton. The intent was to initiate a ten-year program that would include manipulating stock harvest and escapement of fish to establish a broad range of data measurements about the fishery. The purpose was to streamline a process to determine appropriate levels of sustainable harvests.

However, opposition from commercial fishers, conservative thinking by fishery managers, and relatively high costs were encountered. During 1987 the federal Department of Fisheries and Oceans announced that it was abandoning the experiment due to budget cuts (Jessop 1990).

This experience reflects the observation by Collie and Walters that:

Adaptive management is difficult to implement because it often requires short-term pain for long-term gain; current yields must be sacrificed in the quest for larger long-term yields. Applied to single stocks, adaptive policies are poor experiments because there are no replicate or control stocks. Even if the population responds as expected, one cannot be sure that the stock responded to harvesting or to some other, perhaps environmental trend (Collie and Walters 1991:1273).

Box 9.5 Design Features and Capabilities of Adaptive Organizations and Processes

1. Minimum critical specification★
2. Referent organization★
3. Organizational redundancy★
4. Lean number of members
5. Strategic location★
6. Semiautonomy/multiple accountability
7. Appropriate legislation★
8. Flexible process
9. Open to public scrutiny★
10. Able to link diverse interests
11. Continuous learning/self-evaluation
12. Rhythm of change
13. Exploring new approaches★
14. Anticipatory scanning
15. Awareness of organizational life cycles★
16. Environmental knowledge within organization

★Features discussed in this chapter

Source: P.R. Mulvihill and R.F. Keith, 'Institutional Requirement for Adaptive EIA: The Kativik Environmental Quality Commission', *Environmental Impact Assessment Review* 9, no. 4 (1989):402.

(Figure 9.1). It has some 6,500 inhabitants, 5,000 of whom are Inuit living mostly in coastal villages. The nine-member Kativik Environmental Quality Commission was created to review and assess environmental and social impacts of projects proposed within northern Quebec. Four of its members are appointed by the Kativik regional government and the other four by the government of Quebec. The chairperson is an individual acceptable to both the Kativik and Quebec governments.

Aboriginal initiatives and a growing international circumpolar movement—along with a lack of a solid economic base, an increasingly mixed social structure, an absence of formal land-use

planning, and uncertainty related to development projects—collectively create at least what Trist described as a *disturbed reactive*, if not a *turbulent*, situation in that region. Against this background, Mulvihill and Keith were interested in examining whether the Kativik Environmental Quality Commission had been designed to use an adaptive approach.

In order to answer that question, they first had to determine the key design features of adaptive organizations and processes. They concluded that sixteen features should be incorporated into any organization and processes if they were to be adaptive (Box 9.5). In the following discussion, some of these design features and capabilities are considered.

Minimum Critical Specification

It is stipulated at the outset that there should be no more rules or guidelines than what are absolutely necessary to begin operating. The intent is to ensure that the organization can become involved in determining the most appropriate design for its own operations. This feature is most likely to keep

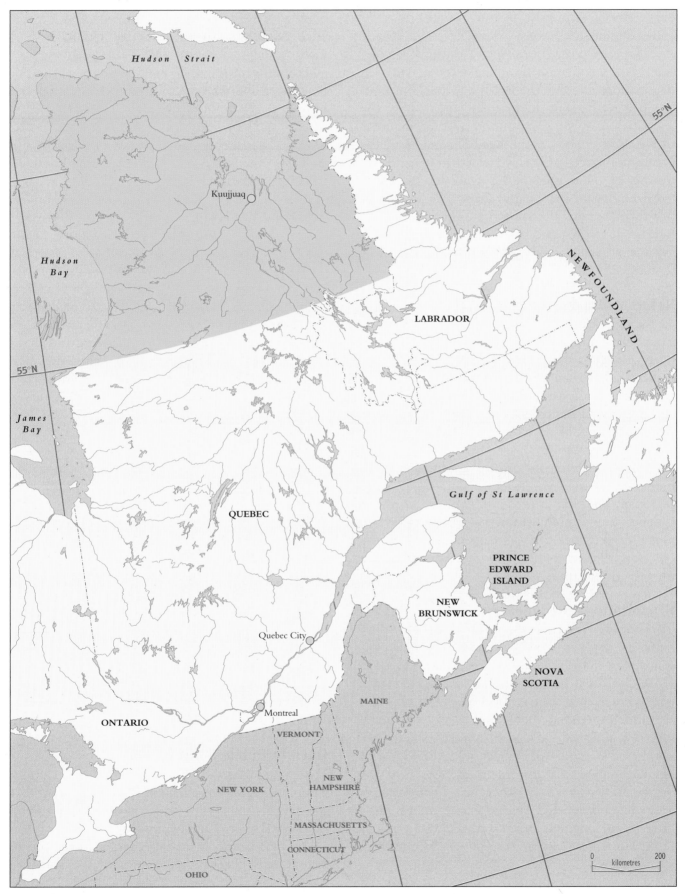

Figure 9.1 Katavik region in Quebec (Quebec north of 55° latitude).

organizations responsive to unexpected situations and changing conditions.

The Kativik Environmental Quality Commission (KEQC) emerged from the intense two-year negotiations to reach the James Bay and Northern Quebec Agreement. In that agreement, only the legislative base, the composition, and the mandate of the KEQC were established. Beyond that, the organization was primarily treated in a hands-off manner by the drafters of the treaty and the Quebec Ministry of the Environment. The result was that the KEQC was allowed to develop its own style, internal strategies, and rules, all of which made it capable of dealing with evolving conditions in its region.

Referent Organization

At the beginning of the chapter, it was noted that Trist believed turbulent conditions are most effectively addressed through interorganizational cooperation rather than by organizations operating in isolation from one another. He argued that locally centred and stakeholder-controlled organizations would be most likely to stay sensitive to existing and future conditions.

The KEQC has many of these qualities. It is not hierarchical either in structure or attitude. It is regionally focused since it functions in northern Quebec. Inuit representation was provided for, and indeed the James Bay agreement required the commission to recognize and give priority to the interests of the Inuit.

Organizational Redundancy

This concept is based on a belief that those organizations that systematically build into their *individual parts* those functions necessary for the *entire organization* will be more responsive and creative. The rationale is that individuals or groups who can see and understand the complete organizational system will operate in a more broadly based manner. In other words, individual tasks would always be determined with regard to the needs of the organization as a whole rather than the interests of a single component. This approach builds in a capacity for ongoing learning, as well as a capacity to monitor and question basic values and operating assumptions.

The KEQC satisfies most of these features. It usually has both generalists and specialists as members, with some overlap of interests and experience. Moreover, some deliberate redundancy of interests helps to ensure that in their deliberations all commission members will be aware of the full range of issues relevant to a problem or proposal. Furthermore, a mix of specialists, generalists, Aboriginal, and non-Aboriginal members has led to the use of an ecosystem or holistic perspective as well as to the acceptance of the value of continuous mutual learning.

Strategic Location

The physical location of an organization's headquarters can have both symbolic and practical implications. The location of the office for the KEQC in Kuujjuaq and its usual practice of holding its meetings in northern Quebec have enhanced the commission's adaptiveness and effectiveness. Its distance from the provincial government and proximity to local governments have helped to keep the commission sensitive to local conditions and needs, yet still maintain a direct link to the Quebec Ministry of Environment.

Appropriate Legislation

It has been argued that organizations such as the KEQC must be empowered to make significant decisions rather than act only as an advisory group. The counterargument is that legislated organizations and processes may lose flexibility because mandates and other aspects are prescribed by statute. However, on balance, it appears a legislative basis that provides real power is desirable since organizations have no way to influence policy or practice in the absence of such power.

Mulvihill and Keith concluded that rather than detracting from the KEQC, the legislation gave it credibility and power because those interacting with the KEQC understood that it could not be easily ignored or bypassed.

Open to Public Scrutiny

If an organization is to be sensitive to local and changing conditions, it should be designed so that the public can review its activities. While public consultation was not required by its mandate, the KEQC decided early on to use extensive public consultation. Perhaps even more important, the KEQC decided to open its files for public scrutiny and always provide reasons for its decisions. None of the characteristics that help to make the commission open to public scrutiny is revolutionary, but collectively they help to ensure that it remains aware of the needs and conditions in its region.

Exploring New Approaches

Holling has cautioned that non-adaptive organizations move towards stability, prefer to avoid risks, and are usually reactive. In contrast, an adaptive organization must continually consider new approaches. The KEQC has been open to adopting new approaches, as illustrated by the initiatives taken to ensure public openness and access, and by choosing to interpret its role on a case-by-case basis. The commissioners have been willing to experiment with untested approaches.

Awareness of Organizational Life Cycles

Adaptive organizations do not seek survival for its own sake. Instead, they accept that a time might come when their usefulness will end. The members of the KEQC appear to be aware that changing conditions could make the commission more or less relevant. They accept that the commission's role in the future could be substantially less than it has been, and do not seem to believe that the commission will exist indefinitely.

This review of some of the sixteen features identified by Mulvihill and Keith (1989) for adaptive organizations is meant to give a sense of what is required if an organization is to be adaptive. After their analysis, they reduced the sixteen features to four general principles (Table 9.4).

The key principles are that: (1) the mandates and processes must be well defined but not rigid; (2) organizations must be able to operate in complex, changing, and multiparticipant systems; (3) the appointed members or environmental organizations must be open to new ideas, willing to explore new approaches, and prepared to seize opportunities; and (4) related processes, such

as regional planning and environmental impact assessment, are integrated and coordinated. In examining case-studies in Part D, attention will be given to the extent to which these basic principles are being used by organizations responsible for environmental management in Canada.

IMPLICATIONS

As highlighted in Chapter 8, ecosystems evolve continuously, so therefore the ecosystem approach to environmental management must also be modifiable. Such change is not always gradual and incremental but sometimes occurs as major shifts or 'flips', which makes it a formidable challenge. This aspect creates the uncertainty, indeterminacy, and surprise, which Wynne and Young described and highlighted earlier in Table 2.3. Environmental managers must make decisions in what has been increasingly characterized as a turbulent decision-making environment, which has led to growing interest in adaptive environmental management.

Table 9.4 Principles and Criteria for Designing Adaptive Environmental Organizations and Processes

Principles	Criteria
Mandate and processes are well defined but not rigid; breadth and discretion	Minimum critical specification Appropriate legislation Rhythm of change Flexible process
Operate in complex, dynamic, multiparticipant systems	Anticipatory scanning Able to link diverse interests and functions Public scrutiny Semiautonomy/multiple accountability Strategic location
Innovative membership	Continuous learning/self-evaluation Environmental knowledge Exploring new approaches Awareness of organizational cycles Lean number of members with appropriate mix of perspectives
Integration and coordination of related processes	Reference organizations Organizational redundancy Able to link diverse interests and functions Flexible process

Source: P.R. Mulvihill and R.F. Keith, 'Institutional Requirements for Adaptive EIA: The Kativik Environmental Quality Commission', *Environmental Impact Assessment Review* 9, no. 4 (1989):409, Table 1.

Holling has outlined general guidelines for adaptive environmental management, arguing that managers must accept the inevitability of change and uncertainty. He believes that developing approaches that try to eliminate uncertainty is unlikely to be productive. Instead, managers need to develop approaches that reflect the high probability that they will always be surprised. In that regard, Holling argues that we should strive to develop what he calls 'resilient' approaches.

Adaptive environmental management can be used by individuals, local governments, provincial or federal organizations, First Nations, or international institutions. Adaptive environmental management can also be applied to both biophysical and institutional systems. Mulvihill and Keith believed that most attention has been directed to biophysical systems and therefore examined the Kativik Environmental Quality Commission in Quebec to show how the concepts of adaptive environmental management can also be applied to organizations and management processes. In later chapters we will determine the extent to which the concepts or principles of adaptive environmental management are being incorporated into policies and practices in Canada.

SUMMARY

1. Often resource and environmental managers must deal with surprise and turbulence because ecosystems are dynamic and changing.

2. Four different types of planning situation have been identified: (1) placid random, (2) placid clustered, (3) disturbed reactive, and (4) turbulent.

3. Organizations often approach planning as if they were in disturbed reactive conditions, whereas turbulent conditions are increasingly more dominant.

4. Planning for turbulent conditions requires partnerships, interaction, collaboration, coordination, and the ability to be adaptive.

5. Adaptive planning requires (1) collaboration of interest groups, (2) identification of shared values, (3) continuous learning, and (4) continuous evaluation and modification.

6. Adaptive environmental management strategies are not new. Subsistence resource users have practised adaptive management for centuries as one way to reduce their vulnerability to variability and uncertainty in nature.

7. Adaptive environmental management strategies are designed to ensure continuity in production, income, consumption, and the conservation of resources.

8. Adaptive environmental management is intended to help people deal with the uncertain, the unexpected, and the unknown.

9. Adaptive environmental management involves deliberate experimentation and systematic monitoring so that it is possible to learn from experiences. 'Failures' generate new information, which leads to new knowledge and understanding, which in turn become the departure point for new initiatives.

10. A central goal of adaptive environmental management is to develop policies and practices that result in greater resilience (capacity to absorb and use change in a positive way).

11. Ecosystems have stable and unstable characteristics. Adaptive environmental management focuses on dealing with conditions of instability.

12. One of the most systematic initiatives in applying adaptive environmental management has been in the US Fish and Wildlife Program in the Columbia River basin.

13. In Canada, the Kativik Environmental Quality Commission in northern Quebec has been incorporating many of the concepts of adaptive environmental management.

14. When applying adaptive environmental management to organizations, the key principles are that: (1) the mandates and processes must be well defined but not rigid; (2) organizations must be able to operate in complex, changing, and multi-participant systems; (3) the appointed members or environmental organizations must be open to new ideas, willing to explore new approaches, and prepared to seize opportunities; and (4) related processes, such as regional planning and environmental impact assessment, are integrated and coordinated.

REVIEW QUESTIONS

1. Why are ecosystems dynamic or continuously changing?

2. Why are turbulence, surprise, and uncertainty 'normal' regarding ecosystems?

3. What are the differences among the following decision-making situations in environmental management: placid random, placid clustered, disturbed reactive, and turbulent?

4. What are the differences among reactive, preactive, and interactive responses in environmental management?

5. What are the characteristics of adaptive environmental management?

6. What are some of the problems or disadvantages of adaptive environmental management?

7. What is the distinction between stability and resilience relative to ecosystems? What are the implications for adaptive environmental management?

8. What are the features of the Kativik Environmental Quality Commission in Quebec that make it a good model for an organization with capability for adaptive environmental management?

9. Can you identify an example in your region that illustrates effective use of adaptive environmental management? If not, can you identify a situation in which the outcome might have been improved if adaptive environmental management had been used? Are there situations in your region for which an adaptive approach would not be recommended?

10. It is often suggested that people living in the Third World have developed better adaptive environmental management strategies than people in the First World, such as Canadians. Why would such a conclusion be reached? Do you agree or disagree? Can you identify an example of effective adaptive environmental management in the Third World from which Canadians could learn?

REFERENCES AND SUGGESTED READING

Bennett, J.W. 1969. *Northern Plainsmen: Adaptive Strategies and Agrarian Life*. Chicago: Aldine.

Collie, J.S., and C.J. Walters. 1991. 'Adaptive Management of Spatially Replicated Groundfish Populations'. *Canadian Journal of Fisheries and Aquatic Science* 48:1273–84.

Holbert, C. 1993. 'How Adaptive Is Adaptive Management? Implementing Adaptive Management in Washington State and British Columbia'. *Reviews in Fishery Science* 1, no. 3:261–83.

Holling, C.S., ed. 1978. *Adaptive Environmental Assessment and Management*. Chicester: John Wiley.

_____. 1986. 'The Resilience of Terrestrial Ecosystems: Local Surprise and Global Change'. In *Sustainable Development in the Biosphere*, edited by W.C. Clark and R.E. Munn, 292–317. Cambridge: Cambridge University Press.

Jessop, B.M. 1990. 'Passage and Harvest of River Herring at the Mactaquac Dam, Saint John River: An Attempt at Active Fishery Management'. *North American Journal of Fisheries Management* 10, no. 1:33–8.

Lee, K.N. 1993. *Compass and Gyroscope: Integrating Science and Politics for the Environment*. Washington, DC: Island Press.

Lee, K., and J. Lawrence. 1986. 'Adaptive Management: Learning from the Columbia River Basin Fish and Wildlife Program'. *Environmental Law* 16, no. 3:431–60.

McLain, R.J., and R.G. Lee. 1996. 'Adaptive Management: Promises and Pitfalls'. *Environmental Management* 20, no. 4:437–48.

Miller, A. 1993. 'The Role of Citizen Scientist in Nature Resource Decision-making: Lessons from the Spruce Budworm Problem in Canada'. *The Environmentalist* 13, no. 1:47–59.

Mulvihill, P.R., and R.F. Keith. 1989. 'Institutional Requirements for Adaptive EIA: The Kativik Environmental Quality Commission'. *Environmental Impact Assessment Review* 9, no. 4:399–412.

Trist, E. 1980. 'The Environment and System-Response Capability'. *Futures* 12, no. 2:113–27.

Walters, C.J. 1986. *Adaptive Management of Renewable Resources*. New York: Macmillan.

_____, and C.S. Holling. 1990. 'Large-Scale Management Experiments and Learning by Doing'. *Ecology* 71, no. 6:60–8.

Chapter 10

Impact Assessment

There is much uncertainty surrounding the prediction of impacts. The many sources of uncertainty include climatic variability, a scarcity of descriptive information or data on aspects of the study area, and a base of scientific knowledge that is both incomplete and unreliable.

—Environmental Assessment Panel,
Rafferty-Alameda Project

Hydro power development at James Bay in Quebec; a dam and reservoir on the Oldman River, Alberta; runway expansion at the Vancouver airport; the Rafferty and Alameda dams and reservoirs in southern Saskatchewan; low-level military training flights over Labrador and Quebec; oil exploration in Lancaster Sound; disposal of low-level radioactive wastes at Port Hope, Ontario. All of these initiatives have had environmental impacts, and the preparation of environmental impact statements. The use of environmental impact assessments has been a response to the increasing size and complexity of projects, greater uncertainty in predicting impacts, and the general public and special interest groups' desire to become more involved in planning and decision-making processes.

What is environmental impact assessment? Many definitions have

been presented. For example, in their well-known report, Beanlands and Duinker (1983:18) defined it as 'a process or set of activities designed to contribute pertinent environmental information to project or programme decision making'. In a slightly different way, the Canadian Environmental Assessment Research Council stated that it is a

. . . process which attempts to identify and predict the impacts of legislative proposals, policies, programs, projects and operational procedures on the biophysical environment and on human health and well-being. It also interprets and communicates information about those impacts and investigates and proposes means for their management (Canadian

The top of the spillway of LG2, one of the three hydroelectric plants in the James Bay project (*Earthroots*).

Environmental Assessment Research Council 1988:1).

Initially, environmental impact assessment placed primary emphasis upon the physical and biological resources that might be affected by a project, with attention to reducing negative consequences. However, partially in response to public pressure, the focus of environmental impact assessment has gradually broadened to incorporate social and community concerns, leading to the concept of social impact assessment. In addition, more attention has been given to basic policy questions, such as establishing the appropriateness of the objectives a project is designed to satisfy, considering alternative projects that could also satisfy the same objectives, and examining how compensation could be provided for impacts or losses that are not mitigable.

Smith (1993) suggests that three types of impact assessment can be identified:

- *Technology Assessment*: The most general type of assessment, which concentrates upon the broad consequences of changing technology upon a society, people, or region.
- *Environmental Impact Assessment*: Such assessments initially focused upon the effects of development on natural ecosystems. Over time, the biophysical emphasis has been broadened to include social consequences of development.
- *Social Impact Assessment*: The main focus has been on the impacts of development on human communities and their welfare. Such assessments were initially done separately from environmental impact assessments. Today they are usually included as an integral part of environmental impact assessments.

Since each of these elements—technology, environment, and social—is important, Smith suggests that it is more appropriate to use the term impact assessment rather than environmental impact assessment. His approach is sensible since such assessments also incorporate economic implications of proposals, hence the title of this chapter and the use of the term impact assessment in this book.

EVOLUTION OF IMPACT ASSESSMENT

Environmental impact assessment was formally introduced by the United States government through its National Environmental Policy Act of 1969. This legislation required federal departments in the United States to consider the environmental consequences of development. More specifically, one section of the statute stipulated that all federal agencies had to:

. . . include in every recommendation or report on proposals for legislation and other major Federal actions significantly affecting the quality of the human environment, a detailed statement by the responsible official on
    (i) the environmental impact of the proposed action,
    (ii) any adverse environmental effects which cannot be avoided should the proposal be implemented,
    (iii) alternatives to the proposed action,
    (iv) the relationship between short-term uses of man's [*sic*] environment and the maintenance and enhancement of long-term productivity, and
    (v) any irreversible and irretrievable commitments of resources which would be involved if the proposed action should be implemented.

This approach, based on legislation, led to the challenge of many development proposals in the American courts. In Canada, at both the federal and provincial levels, the American experience was watched closely. Reacting against what they viewed as a growing and unfortunate backlog of cases in the US courts, Canadian officials considered alternative ways to include environmental assessment into development planning.

The Canadian government subsequently decided to use an administrative rather than a legal approach. In other words, rather than passing a law that required environmental impact assessments, the federal government stipulated that environmental assessments would be conducted at the discretion of the minister of the environment for projects on federal lands or involving federal departments as the proponents. (However, in Chapter 18 you will find that this position has been challenged successfully.)

During 1973 the Canadian government created the Environmental Assessment and Review Process (EARP) to oversee this policy. In explaining the rationale for this approach, the minister of the environment stated in 1974 that:

I hope, in the process, that we can avoid the delays and other pitfalls which a strictly legalistic approach would cause in this country. . . . We will not hold up important developments which are clean from an environmental point of view and, in contrast to the situation which has developed in the United States, we will not bring the environmental assessment process into disrepute. We will not be charged with blocking everything (House of Commons 1974:499).

Subsequently, each of the provinces developed its own approach to impact assessment. Most opted for the policy or administrative approach. However, some, such as Ontario with its Environmental Assessment Act of 1975, chose a legislatively based approach. However, even where legislation was used, considerable discretion was provided for the responsible minister to exempt development activities from the assessment process.

In a positive sense, discretionary powers could be viewed as a conscious attempt to use an adaptive management approach (as discussed in Chapter 9), with discretion allowing decisions to take into account the specific conditions in each case. In a negative sense, the discretionary provisions too

---

Environmental Assessment in Canada

In practice, however, environmental assessment laws and processes have not been automatically effective. Their purpose is difficult and delicate. They are intended to force open and careful consideration of a new and generally ill-understood set of concerns, and they are directed at decisionmakers who are generally hostile to greater openness and to additional, imposed duties. Failure is easy. Moreover, laws and processes that are weak, unclear or simply difficult to administer with consistency and efficiency do not just fail to foster greater environmental sensitivity in planning and decisionmaking; they tend also to undermine the general credibility of government efforts to encourage environmental responsibility (Gibson 1993b:12).

---

often led to disregard for impact assessment provisions and/or to the exclusion of projects from review for political reasons. Moreover, the discretionary power and differing requirements at the federal and provincial levels created the potential for inconsistency and conflicts. Not surprisingly, there were difficulties, as will be discussed later in this chapter.

COMPONENTS OF IMPACT ASSESSMENT

This section addresses two aspects: the essential elements of impact assessment, and the criteria that should be used to determine the effectiveness of impact assessment.

Essential Elements of Impact Assessment

Many would agree that impact assessment should include at least five elements. First, it is necessary to *identify potential impacts* as a result of development or other activity. This basic step is not always straightforward. For example, the potential impacts will be influenced by the size of an area and the time period under consideration. For this reason, increasing attention is being given to the question of *scoping* in impact assessment, which is the need to identify the spatial and temporal scales involved, the individual and cumulative impacts, as well as the aspects of the environment most at risk to damage.

Second, having identified potential impacts, it is then necessary to *predict their likely effects*. In addressing the challenge of prediction, analysts have to deal with all the problems of uncertainty and complexity highlighted in Chapter 1, as well as imperfect scientific understanding discussed in chapters 4, 5, 6, and 7. Predictions are usually influenced by assumptions made to simplify a situation, and it is important for the analyst to recognize such assumptions and to make them explicit.

A third element is to *interpret and assess the consequences or implications* of effects. Even if analysts can identify impacts and predict their effects, there can be disagreement over varying interpretations about their significance. For example, if it is predicted that the habitat of a threatened animal will be damaged, what conclusions should be drawn? What is the significance of such a negative impact compared to jobs that might be created by the proposed development? People with different values and interests might have quite different interpretations about the implications of damage to wildlife habitat.

If agreement can be reached about the implications of effects, then it is important to *devise mitigation and compensation* measures. In most situations, it is unrealistic to assume that development will not trigger impacts and effects. In order to gain some advantages, some negative consequences often occur. The issue, however, is to try to minimize or mitigate negative consequences, often by monitoring effects and taking a long-term perspective for management.

When mitigation cannot be devised, then society needs to consider how to compensate those who carry the burden of the development. As evident in the example at the beginning of Chapter 11, in Canada often a small number of people have carried most of the costs of resource development, and only recently has their compensation been given serious consideration. More problematic is how humans should compensate other forms of life affected by our modification of the environment.

A fifth and too often overlooked element of impact assessment is *communication of findings and conclusions*. This is a major challenge, as often impact assessments involve considerable technical analysis completed by scientific 'experts'. Often such people are not particularly skilled in conveying their information, interpretations, and conclusions clearly to non-technical people. Yet, as is discussed in Chapter 11, if there are to be more partnerships and involvement of stakeholders in environmental management, good communication is essential. That will range from ensuring the involvement of stakeholders in the planning and impact process, often by providing funding to support participation and by providing information and results in a way that is understandable. The key elements of impact assessment are summarized in boxes 10.1 and 10.2.

Determining Effectiveness of Impact Assessment

If impact assessment is to be effective, it is important that we continually evaluate what is being done so we can learn from experience and make modifications to improve future activity. Such learning and modification are key components of *adaptive environmental management*, which was dis-

---

Box 10.1 Key Elements of Impact Assessment

- Identifying potential impacts
- Predicting likely effects
- Interpreting and assessing consequences
- Devising mitigation and compensation measures
- Communicating findings and conclusions

Box 10.2 Main Steps in Impact Assessment, after the Ontario Process

1. Description of the purpose of the undertaking.
2. Description of and statement of the rationale for: (i) the undertaking; (ii) the alternative methods of carrying out the undertaking; (iii) the alternatives to the undertaking.
3. Description of: (i) the environment that will be affected or that might reasonably be expected to be affected, directly or indirectly; (ii) the effects that will be caused or that might reasonably be expected to be caused to the environment, and; (iii) the actions necessary or that may reasonably be expected to be necessary to prevent, change, mitigate, or remedy the effects upon or the effects that might reasonably be expected upon the environment, by the undertaking, the alternative methods of carrying out the undertaking, and the alternatives to the undertaking.
4. Evaluation of the advantages and disadvantages to the environment of the undertaking, the alternative methods of carrying out the undertaking, and the alternatives to the undertaking.

---

cussed in Chapter 9. In this subsection, two sets of criteria are presented to stimulate thinking about which questions should be posed when assessing experience with impact assessment.

The Canadian Environmental Assessment Research Council (1988a:1–2) suggested that the key criteria for evaluating impact assessment should be *effectiveness*, *efficiency*, and *fairness*. They also recognized that some of these criteria con-

> ## Box 10.3 Criteria for Evaluating Impact Assessment
>
> *Effectiveness*
>
> - Information generated in impact assessment contributes to decision making.
> - Predictions of the effectiveness of impact management measures are accurate.
> - Proposed mitigatory and compensatory measures achieve management objectives.
>
> *Efficiency*
>
> - Impact assessment decisions are timely relative to economic and other factors that determine project decisions.
> - Costs of conducting impact assessments and implementing their recommendations can be determined and are reasonable.
>
> *Fairness*
>
> - All interests or stakeholders have equal opportunity to influence the decision before it is made.
> - People directly affected by projects have equal access to compensation.

flict with one another. Therefore, it is always essential 'to determine the degree to which trade-offs and compromises must be made to satisfy the larger public interest'. Furthermore, it was accepted that such trade-offs and compromises involve value judgements rather than only technical judgements.

Another set of criteria or principles for guiding the design of environmental assessment has been developed by Gibson (1993b), based on a review of twenty years of experience at the federal level in Canada as well as in Ontario. His eight principles are:

- An effective impact assessment process must encourage an *integrated approach* to the broad range of environmental considerations and be dedicated to achieving and maintaining local, national, and global sustainability.
- Assessment requirements *must apply clearly and automatically* to planning and decision making in all undertakings that may have

significant environmental effects and implications for sustainability within or outside the jurisdiction.

- Assessment decision making must *identify the best options* rather than merely acceptable proposals. Thus it must require critical examination of purposes and comparative evaluation of alternatives.
- Assessment requirements *must be established in law* and must be specific, mandatory, and enforceable. Assessment requirements must be well understood, all central components of the process must be enshrined in law, and compliance with the requirements and products of the process must be legally enforceable.
- Assessment work and decision making must be *open*, *participative*, and *fair*.
- The assessment process must be designed to *facilitate efficient implementation*.
- Terms and conditions of approvals must be *enforceable*, and approvals must be *followed by monitoring of effects* and *enforcement* of compliance.
- The process must include provisions for *linking assessment work into a larger regime*, including the setting of overall biophysical and socio-economic objectives and the management and regulation of existing as well as proposed new activities.

The eighth principle deserves particular attention in Canada. It reflects a view that impact assessment is not an end in itself but a means to an end. In other words, impact assessment is an approach to planning and should be a tool to incorporate environmental considerations into decision making along with the more conventional technical, financial, and political considerations.

In Part D it will be useful to consider the extent to which impact assessments have been designed to satisfy the criteria set out by Gibson and the Canadian Environmental Assessment Research Council. Furthermore, if they were designed with regard to such criteria, to what extent has impact assessment processes been able to meet them? Particularly relevant case-studies are found in Chapter 18 (the Rafferty-Alameda project in Saskatchewan) and Chapter 19 (James Bay development in Quebec).

CHALLENGES IN IMPACT ASSESSMENT

People conducting impact assessment have to juggle technical matters and value judgements. Often there are no right or wrong answers but rather different answers depending upon the starting-point for the assessment and the assumptions made. Some of the challenges are discussed in this section.

What Types of Initiatives Should Be Assessed?

In Canada and most other countries, impact assessments have been conducted primarily for *projects*, especially major capital projects, such as dams and reservoirs, nuclear or other types of powerplants, oil or natural gas drilling or pipelines, waste disposal facilities, and runway expansions at major airports. The rationale has been that such development usually has the potential for significant environmental and social impacts, and there is a readily identifiable proponent or a group of people or communities that would be affected.

However, it has been argued for some time that impact assessments also could and should be completed for both *policies* and *programs*. The argument in favour of this approach (called *strategic assessment*) is that often projects are only the means of implementing policies and programs. Therefore, waiting to conduct impact assessments on projects means that they are not useful because they are too late. At the policy or program level, decisions may already have been taken to preclude or eliminate possible alternatives. Thus, for example, a policy that endorses the use of nuclear powerplants may restrict the focus to different types of nuclear options rather than considering nuclear and non-nuclear alternatives. At the time of assessing potential impacts of a proposed nuclear station, participants might suggest that a non-nuclear plant would have fewer impacts. However, the response could be that the energy policy endorsed the use of nuclear plants, and therefore that issue had already been decided.

While it is easy to accept the logic of having impact assessments completed for policies and programs as well as for projects, it is also easy to appreciate that such assessments can be difficult to do. As Bregha et al. (1990:3) have explained, policies can be general or specific, stated or implicit, incremen-

> *This process of predicting and minimizing the consequences of a single action has not adequately considered the accumulative nature of some effects, the non-linear response of some natural systems, nor the linkages between a single action and other related activities. . . . cumulative impact analysis evaluates the consequences of multiple activities and sources of impact on a larger set of environmental components.*
>
> C.K. Constant and L.L. Wiggins,
> 'Defining and Analyzing
> Cumulative Impacts', 1991

tal or radical, independent or linked to other policies. Furthermore, Canada's system of governance offers few incentives to politicians to define specific objectives. All of these characteristics of policies and programs can make it difficult to assess their impact.

Other challenges arise regarding which activities should have impact assessments completed. One issue is particularly difficult. On the one hand, society expects government regulations to be reasonable and efficient. In that regard, it is normal to have some lower limit or threshold below which assessments are not required. For example, it is unlikely that air quality in the community would be adversely affected if one home-owner were to build an outdoor barbeque in her backyard. However, if every home-owner in the community decided to build an outdoor barbeque, it is possible that air quality could be negatively affected. The issue then is balancing reasonableness and efficiency against the possibility that the impact of many small developments *in aggregate* may have significant implications. This problem has been described as one of *cumulative effects* and will be considered in more detail in a later subsection.

In order to handle the possibility of cumulative effects without creating unduly onerous impact requirements, Ontario has used *class assessments*. In other words, governments have established a blanket review and approval procedure for development such as municipal roads and conventional sewage treatment plants, which are frequent, recurring, and for which there is already considerable experience. However, there is always a grey area when one must decide whether an initiative should fall under a class assessment or a specific assessment.

In Ontario, for example, a hearing to examine the feasibility of a class assessment for forestry in the province lasted over four years. The initial idea was that many aspects of logging and other forestry operations are frequent, repeatable, and routine, and so could be readily handled under a class assessment. However, during the hearings stakeholders raised questions about the overall policy (or lack of it) for forestry in the province. The result was that the examination of class assessments for forestry became very wide-ranging, and touched on projects, programs, and policies.

When Should Impact Assessments Be Done?

The final report by the federal Environmental Assessment Panel (1991:2), which reviewed the Rafferty-Alameda project in southeastern Saskatchewan (discussed in detail in Chapter 18), stated that 'environmental impact assessment should be applied early in project planning. That is the intent of both provincial and federal processes.' The panel concluded that the Rafferty-Alameda project was 'well advanced, however, when both the first and this Panel became involved. This put some limits on the usefulness of the review.' Unfortunately, too often in Canada it appears as if developments have been well advanced before environmental impact assessments are conducted.

The creators of environmental impact assessment had intended for EIA to be used jointly with benefit–cost analysis and other technical analysis to determine the appropriateness of development

The excavation and gravel fill for the foundation of Rafferty Dam's spillway (*SaskPower*).

proposals and to design mitigatory measures. However, as with the Rafferty-Alameda project, environmental assessments have often been conducted *after* the decision has already been made regarding whether or not the project will be carried out. As a result, the impact assessment has rarely been used to determine whether or not a project should be approved, but rather to establish which mitigatory measures could be used to reduce or soften negative impacts.

If impact assessment is to weigh environmental considerations equally with economic, financial, and technical aspects, then the impact process must be designed so that assessments are completed as an integral part of the feasibility analysis prior to approval. Clearly, the mind-set for development must also change. Impact assessment should not be a regulatory hurdle that developers have to circumvent, but an incentive to encourage decisions that are consistent with the principles of sustainable development (which were discussed in Chapter 2).

How to Determine the Significance of Impacts and Effects

As already discussed in the section on components of impact assessment, a difficult challenge is determining the significance or implications of impacts and effects. This is not a scientific or technical issue. Significance is influenced by the prevailing values of a society, a place, and a time. What might be

Figure 10.1 The clash between Regina and Ottawa prompted the usual spate of editorial cartoons, such as this one by Brian Gable (*26 October 1990, courtesy* The Globe and Mail, *Toronto*).

considered significant in one place or time will not necessarily be viewed in the same way in another place or time. Furthermore, people from different cultural backgrounds in the same place and at the same time may have quite different perspectives regarding what is important enough to protect or preserve.

Accepting that judgements about significance take in account not only technical or scientific matters strengthens the rationale for extending partnerships in environmental management and for ensuring that key stakeholders participate in planning and assessments. This is why Chapter 11 is devoted to different aspects of public participation.

One of the major challenges in determining significance is that often the issues in a dispute have intangible features that are not readily described in financial terms. This same problem is encountered by analysts conducting benefit-cost analyses of different alternatives.

What is the value of an undisturbed stretch of white water in a river that might be inundated by a dam built to generate hydroelectricity? What is the value of a wetland that will be drained to allow

This blighted landscape was once a stand of trees (*Earthroots*).

It is difficult to assess the values associated with scenery or the worth of an outdoor recreation experience (*Philip Dearden*).

> *It was a very difficult process because one must compare the relative value or loss generated by one [highway] corridor against that generated by another. Examples of the difficulties involved include the following: does an ecologically unique bog system rate higher or lower than an 1840 mansion; or how does a bald eagle's nest compare with an Indian archeological site? This process is further complicated by a lack of knowledge about the status of many of our natural resources.*
>
> C.M. Kitchen,
> *'Ecology and Urban Development', 1976*

a subdivision to be built or to enable a farmer to increase food production? What is the cost to wildlife if their habitat is disturbed by a mine or an oil pipeline? What is the value of a stand of old white pine or of Douglas fir? Questions such as these pose major dilemmas for those involved in impact assessment. There is usually no obvious answer, and different interests may come to quite different conclusions about what is significant and what action is appropriate.

What Should Be Done Regarding Inadequate Understanding of Ecosystems?

Often the scientific understanding of ecosystems is incomplete. Furthermore, information about a specific ecosystem is usually inadequate to predict the potential effects of human intervention.

Even the most basic ecological concepts are not without problems. For example, some time ago McIntosh (1980) commented that ideas such as community, stability, succession, and climax were causing (and still cause) major disagreements among ecologists. In some instances, scientists disagree over terminology and definitions of these concepts, even after more than three-quarters of a century of ecological research. Some of these difficulties have already been considered in chapters 4, 5, and 6, and you may wish to review them. With uncertainty and disagreement over basic ecological concepts, it is understandable why predictions are difficult, or why there may be different scientific interpretations of the same data, especially when such data are incomplete. This situation partially explains why it is not unusual for proponents and opponents at environmental hearings to each have scientific experts reach opposite conclusions.

However, as indicated earlier, inadequate or incomplete information can also frustrate impact assessments, even for aspects for which knowledge is reasonably good. To illustrate, Nakashima examined the potential effects of oil drilling on eiders (a large sea duck) in Hudson Bay. He concluded that:

> The application of environmental impact assessment in the North is severely limited by a lack of fundamental baseline information on the natural environment. Where funds and commitment have been readily available, millions of dollars and countless hours of effort have been invested in hasty attempts to address the more substantial gaps (Nakashima 1990:23).

Nakashima continued, remarking that 'shortcomings in our scientific knowledge of northern ecosystems are staggering and financial and human resources which can be dedicated to their eradication are severely limited'. This view has been reiterated by Bregha et al., who remarked that:

> The successful anticipation of future problems is based on the ability to forecast the possible consequences of a given action. This ability, in turn, depends on our knowledge of environmental processes. Such knowledge is still often remarkably superficial . . . (Bregha et al. 1990:25).

To address partially the problem of incomplete information and understanding, Nakashima argued that more use should be made of *indigenous knowledge* (also referred to as traditional ecological knowledge [TEK], as discussed previously in Box 4.1). He illustrated his argument by discussing the Inuit's ecological understanding of sea birds in the coastal communities of Inukjuak and Kuujjuarapik, Quebec, and Sanikiluaq, NWT, along southeastern

> *It should be the experience that leads to a modification of knowledge, rather than abstract knowledge forcing people to perceive their experience as being unreal or wrong.*
>
> U. Franklin,
> The Real World of Technology, *1990*

Hudson Bay. He noted that Inuit hunters are well positioned to contribute to the protection of the environment since they have accumulated excellent information about where and when animals live. Such knowledge can be invaluable in estimating and assessing the potential impacts of an oil spill on wildlife. Nakashima concluded that:

> Traditional environmental knowledge presents us with a rare opportunity which should not be ignored. A cooperative approach to the management and protection of the northern environment, one which allows Native and scientific knowledge systems to interact in complimentary [sic] rather antagonistic fashion, provides a constructive alternative to the current state of affairs (Nakashima 1990:23).

Incomplete understanding and inadequate data will continue to be the reality for those involved in impact assessments in Canada. For this reason, use of *adaptive environmental management* approaches is attractive, given its emphasis upon learning by trial and error and its acceptance of uncertainty. In view of these problems, it is clear that indigenous knowledge can make a valuable contribution to environmental management.

How Should Cumulative Effects Be Handled?

Many serious global environmental problems—diminished biodiversity, build-up of carbon dioxide, depletion of the ozone layer in the stratosphere, and acid rain—are at least partially consequences of cumulative effects from many individual activities. Within Canada, other problems also reflect the consequences of cumulative effects, such as the loss of high-quality agricultural land adjacent to urban areas, soil erosion, increasing air pollution from cars, stress on waste disposal facilities, contamination of groundwater supplies, and traffic congestion.

Native peoples had to have detailed knowledge of their environment to survive. On the West Coast the cedar provided many essentials of life ranging from canoe blanks to clothing material. This culturally modified tree in Gwaii Hanaas has had part of its bark stripped off for use (*Philip Dearden*).

Constant and Wiggins (1991) have explained that environmental impact assessments have focused mainly upon the direct impacts of single actions on a particular set of critical environmental components. In their view, this approach has not adequately accounted for the cumulative nature of some effects. The Canadian Environmental Assessment Research Council has defined cumulative effects as those impacts on the natural and social environments that: (1) occur so frequently in time or so densely in space that they cannot be assimilated, or (2) combine with effects of other activities in a synergistic manner. Various types of cumulative effects are shown in Table 10.1.

A key aspect of cumulative effects is *synergy*. In other words, as mentioned in Chapter 5, the total cumulative effect of a set of individual activities may be considerably greater than the simple sum of their effects. Thus as the Environmental Assessment Panel (1991:53) for the Rafferty-Alameda project concluded, the environmental impacts of a project 'have to be considered within the context of the entire ecosystem in which the project is placed'. Such a suggestion is easy to make. To translate it into practice has been and will continue to be a major challenge.

Table 10.1 Types of Cumulative Effects

Issue Types	Main Characteristics	Examples
Time crowding	Frequent and repetitive impacts on a single environmental medium	Wastes sequentially discharged into lakes, rivers, or watersheds
Space crowding	High density of impacts on a single environmental medium	Habitat fragmentation in forests, estuaries
Compounding effects	Synergistic effects due to multiple sources on a single environmental medium	Gaseous emissions into the atmosphere
Time lags	Long delays in experiencing impacts	Carcinogenic effects
Space lags	Impacts resulting in some distance from source	Major dams; gaseous emissions into the atmosphere
Triggers and thresholds	Impacts to biological systems that fundamentally change system behaviour	Effects on changes in forest age on forest fauna
Indirect	Secondary impacts resulting from a primary activity	New road developments opening frontier

Source: N.C. Sonntag et al., *Cumulative Effects Assessment: A Context for Further Research and Development* (Ottawa: Minister of Supply and Services Canada, 1987):6.

What Should Be the Nature of Public Involvement?

As will be discussed in Chapter 11, several reasons are used to justify the role of public involvement. Garipey (1991) indicated that public involvement has at least three functions in an impact assessment process:

- First, it opens up the decision-making process to the public, and in so doing helps make the process a fair one. Furthermore, by involving the public in decision making, it enhances the credibility of the process.
- Second, public involvement helps to broaden the range of issues and alternative solutions considered, and allows the public to share in devising mitigation measures. In this manner, citizens share in establishing the conditions under which a proposal will be approved.
- Third, public participation can contribute to social change. That is, by participating in the decision-making process, people become more aware about conditions in their own environment. Such enhanced awareness can lead to new initiatives within the community to iden-

tify and address other problems and to a greater likelihood that the government will appreciate the concerns of a community.

Varying kinds of public involvement have occurred in environmental impact assessments in Canada. In Quebec, for example, Garipey (1991) reviewed five projects initiated by Hydro-Québec to determine what public participation had achieved. He found that aspects of the three rationales identified earlier were present in the five projects. In particular, he noted that one of the roles of public participation had been to move the projects out of the realm of 'mystery' (in which only technical experts had a role) into the realm of 'mastery' (in which laypeople could influence the outcome). In his view, the experience in Quebec demonstrated that public involvement can transform technological issues into social choices.

Due to public participation, Hydro-Québec was pushed to consider more aspects than it likely would have had it been able to operate without public scrutiny. For instance, in one of the projects public criticism led the provincial government to force Hydro-Québec to change a power line from an aerial to an underground crossing of the St Lawrence River, even though costs were increased by over $150 million. As well, public participation in this project undoubtedly led to a government decree that Hydro-Québec had to continue negotiating with Native communities about the power corridor and implement intensive monitoring and follow-up programs.

However, Garipey also concluded that the public participation process focused primarily upon questions relating to the actual siting of the facility rather than its feasibility. In addition, he found that public comments came mainly from specific, local, and land-owner interests. That is, people became involved generally because of the immediate territorial impacts rather than because of concerns about broader regional implications. Garipey's assessment of the experience with public participation in impact assessment in Quebec is positive and optimistic. A much more critical perspective from experience in Quebec is presented in Chapter 19, with reference to hydro power development in catchments draining into James Bay.

Some of the most significant changes have occurred in northern Canada. Environmental issues there have often centred on development of non-renewable resources, especially minerals and hydrocarbons. For many years, government policy regarding economic development in the North explicitly excluded Native peoples in that region from participating in decisions that had major effects on their lives. However, that situation started to change during the 1980s, even though there are still incidents in which Native peoples legitimately believe that they have not been included in development planning (see especially Chapter 19).

The Inuvialuit Final Agreement (IFA), passed by the federal Parliament in July 1984, was the first modern Native claims agreement in the territories. This agreement provided the 2,500 Inuvialuit of the western Arctic with rights and land ownership for over 91 000 km^2 of land. From the perspective of impact assessment, it also established new procedures for environmental impact screening and review of projects that affected these Inuvialuit lands.

As Reed (1990) explained, the IFA created a *legal* obligation for both the federal and territorial

governments to ensure that development proposals had environmental impact screening and review prior to the beginning of any development activity. Furthermore, development would have to be compatible with the basic goals of the agreement, which are: (1) to preserve Inuvialuit cultural identity and values within a changing North, (2) to enable the Inuvialuit to be equal and meaningful participants in the northern and national economy, and (3) to protect and preserve the Arctic wildlife environment and biological productivity.

The procedures of screening and review regarding environmental impacts specify joint Inuvialuit-government representation on the committees responsible for impact screening and review. Both the Screening Committee and the Review Board have an equal number of participants from government and Inuvialuit organizations. For example, the Screening Committee has seven members: three Inuvialuit, one government member from each territory, one federal government member, and a chairperson appointed by the federal government, with Inuvialuit agreement.

While under the agreement no additional provisions were made for public participation at the screening stage, the Screening Committee has organized its meetings to allow public input. In addition, Reed found that while there is no requirement for final recommendations to be made public, information about proposals and the

---

Public Participation and Impact Assessment

Three usual arguments have been put forth for emphasizing a public role in environmental assessments in Canada. First, environmental assessment is inevitably a value-laden business, and broad participation and scrutiny is the best means of combating narrow biases and encouraging careful attention to matters of public concern. Second, participative openness has become a political imperative because Canadians are now generally unwilling to rely on the assurances of authorities and increasingly expect to be allowed to participate effectively in the making of decisions that will affect their lives. Finally, broad involvement is appropriate because environmental assessment is for everyone a learning process (Gibson 1993b:19).

outcomes of meetings were discussed openly by all committee members.

As will be confirmed in Chapter 11, there has been an increasing movement to incorporate public participation into environmental management in Canada. These two examples from Quebec and the North illustrate that such participation has indeed been occurring. In Part D we will note the various ways in which public involvement is being used across the country.

What Should Be Done About Mitigation and Compensation?

The Canadian Environmental Assessment Research Council (1988b) has reported that the assessment process needs to establish ways to prevent or minimize adverse impacts as a result of development activity. It also concluded that the traditional impact assessment decision-making process has been adversarial because a proponent of a project is pitted against affected parties and the general public. Generally, mitigative measures have been presented by the proponent, supported by voluminous scientific data. The public then has the task of questioning the proposed measures by pointing out their shortcomings. As a result, the burden of proof has been on the intervenor to demonstrate inadequacy in the impact assessment. The final decision has usually been the responsibility of appointed or elected government officials. The Canadian Environmental Assessment Research Council concluded that alternative ways to develop mitigatory measures are needed, and that negotiation and mediation (discussed in Chapter 12) offer a real alternative.

A major challenge in determining mitigation and compensation is putting a value on the adverse impacts of a project because direct cause-and-effect relationships between development activity and impacts are often difficult to establish. This problem becomes even more difficult when a project is located adjacent or close to another project or projects; this, of course, is the dilemma of cumulative effects. The key point is that when direct relationships cannot be established between a project and its effects, the proponent will usually be reluctant to make provision for mitigation. Thus placing a value on adverse impacts is a serious problem for which there is no ready solution.

There has been a movement to monitor the impacts of developments and then to mitigate when appropriate. Such monitoring and mitigation have been most successful when they have been coordinated between the developer and the affected community. However, a significant challenge is to avoid having the impact assessment become a driving force for compensation agreements rather than focusing upon mitigation and other measures to avoid or reduce negative impacts.

How Should Monitoring Be Developed?

Chapters 4, 5, 6, and 7 made it clear that many ecological processes unfold over *decades*, and that therefore effects of some ecosystem changes may become apparent only over a lengthy time span. It has also been established that due to incomplete ecological science, it is often difficult to predict which changes may occur in ecosystem structures or processes as a result of development. If we are to improve our knowledge of the resilience, sensitivity, and recuperative powers of ecosystems, monitoring is needed. In addition, monitoring can help to ensure that mitigatory measures recommended in an impact assessment have actually been implemented. Monitoring can also track public concerns or fears regarding a project and thereby help to ensure that they are recognized and addressed.

Often such monitoring is not conducted in environmental and resource management in Canada. It is usually time consuming and expensive, and the results may not be useful until after many years of monitoring. When financial and human resources are scarce, it is tempting for managers to reduce or eliminate monitoring and redirect scarce resources to new development activity, as has happened with the Experimental Lakes Research Area in northwestern Ontario, as discussed in Box 7.7. However, by following that route we reduce the opportunity to learn from previous experience, and thus make it difficult to begin using the adaptive management approach discussed in Chapter 9. The role of monitoring has generally been neglected in Canada and most other countries. On the other hand, there have been some instances where the local community has been used to conduct monitoring activities. Such an approach seems very promising.

IMPLICATIONS

Experience with environmental impact assessment in Canada started in the early 1970s with the creation of the federal Environmental Assessment and Review Process. Each province also has procedures for assessment. As Gibson (1993a, 1993b) has argued, impact assessment is not only a vehicle for incorporating environmental considerations into planning, it is also a broader approach to determine which actions should be taken to make the best of opportunities. Impact assessment is also a key tool for achieving the sustainability goals outlined in Chapter 2. In practice, however, Gibson concluded that environmental impact assessment has not always been effective.

Some of the challenges of impact assessment have been presented in this chapter. It has been argued that impact assessment is not only a technical exercise. It also involves many value decisions, particularly regarding the interpretation of the significance or consequences of effects. At the same time, many scientific challenges remain. Ecologists continue debating about terminology, concepts, and measurement procedures. We have difficulty in establishing meaningful baseline and longitudinal data in order to understand the structures and processes of ecosystems. We have great difficulty in isolating the effects of development activities from other factors causing change in environmental systems, and are particularly challenged when dealing with cumulative effects. We are also still learning about the most appropriate ways to achieve public involvement. Thus the challenges in impact assessment are numerous and major at both scientific and policy levels.

SUMMARY

1. Environmental impact assessment (EIA) was introduced to try to ensure that environmental considerations were given equal attention relative to technological and economic ones.

2. EIA is a process to identify, predict, and evaluate the positive and negative impacts of resource development initiatives on the biophysical environment and on human health and well-being.

3. EIA has focused upon the consequences from *projects*, but it is argued that similar attention needs to be given to the consequences of *policies* and *programs*.

4. When significant negative impacts are unavoidable, EIAs should strive to identify what compensation should be provided to those bearing the costs. Determining appropriate compensation is a challenge because of the difficulty in putting a monetary value on negative, intangible impacts.

5. There are three types of impact assessment: technological assessment, environmental impact assessment, and social impact assessment. The latter two are often combined, and hence the more suitable generic term is impact assessment or IA.

6. EIA was formally introduced in the United States in 1969 through the National Environmental Policy Act, which required all federal departments to consider environmental impacts resulting from development. The Canadian federal government introduced a non-statutory EIA process in 1973 (Environmental Assessment and Review Process or EARP). Provinces followed with either legislatively or administratively based processes.

7. Essential components of IA are: (1) identifying potential impacts, (2) predicting likely effects, (3) interpretating and assessing implications, (4) devising mitigative and compensatory measures, and (5) communicating findings and conclusions.

8. Principles to guide IA include: (1) an integrated approach; (2) automatic and clear application; (3) identification of the best options; (4) a basis in legislation; (5) an open, participative, and fair approach; (6) facilitation of implementation; (7) enforceability and capacity to be monitored; and (8) ability to relate to regional planning and development.

9. Cumulative effects, or the aggregate results of many small and unrelated development activities, are a major challenge for IA. Cumulative effects are impacts that occur so frequently in time or so densely in space that they cannot be assimilated, or

that combine synergistically with effects from other activities.

10. IA should be initiated early in the planning or management process. However, in Canada IA often seems to start after the initial approval decision has already been made.

11. Determining the significance of impacts is a value-based rather than a technical exercise. Experts are not necessarily the only people who should be involved in determining significance of impacts.

12. Scientific knowledge and data are often incomplete and inadequate. That is why it is sensible to combine traditional ecological knowledge with scientific knowledge when attempting to identify possible impacts and determining their implications.

13. Public involvement is included in IA to open up the decision-making process to ensure it is (and perceived to be) fair, to broaden the range of issues and solutions considered, and to empower local people.

14. The Inuvialuit Final Agreement has provided for direct participation of the Inuvialuit people of the western Arctic in IA.

15. Monitoring programs are required to track anticipated impacts and recommended mitigatory and compensatory measures, and to generate new knowledge to facilitate adaptive environmental management.

REVIEW QUESTIONS

1. What are the characteristics of impact assessment?

2. What is the difference between a legal and an administrative approach to impact assessment?

3. What are the five key elements or stages in impact assessment?

4. What criteria should be used to determine whether an impact procedure is working well?

5. Is it possible to complete impact assessments of policies and programs as well as for projects?

6. How early in the development process should impact assessments be conducted?

7. What are some of the problems in determining the significance of impacts?

8. Why is our information about and understanding of ecosystems often inadequate?

9. What are cumulative effects? How can they be estimated?

10. What should be the role of public participation in impact assessment?

11. What is the difference between mitigation and compensation in impact assessment?

12. Why is monitoring of impacts and effects too often overlooked or not done?

13. Investigate the details of impact assessment in your province or territory. Can you find examples of successful and unsuccessful applications, and determine the important influences on the outcomes?

14. In your opinion, what are the most important changes that could be made to improve our ability to conduct effective impact assessments?

REFERENCES AND SUGGESTED READING

Beanlands, G.E., and P.N. Duinker. 1983. *An Ecological Framework for Environmental Impact Assessment in Canada*. Halifax: Institute for Resource and Environmental Studies, Dalhousie University.

Berger, T.R. 1977. *Northern Frontier, Northern Homeland: The Report of the Mackenzie Valley Pipeline Inquiry*, 2 vols. Ottawa: Minister of Supply and Services Canada.

Bregha, F., J. Benidickson, D. Gamble, T. Shillington, and E. Weick. 1990. *The Integration of Environmental Considerations into Government Policy*. Prepared for the Canadian Environmental Assessment Research Council. Ottawa: Minister of Supply and Services Canada.

Canadian Environmental Assessment Research Council. 1988a. *Evaluating Environmental Impact Assessment: An Action Prospectus*. Ottawa: Minister of Supply and Services Canada.

_____. 1988b. *Mitigation and Compensation Issues in the Environmental Assessment Process: A Research Prospectus*. Ottawa: Minister of Supply and Services Canada.

Constant, C.K., and L.L. Wiggins. 1991. 'Defining and Analyzing Cumulative Impacts'. *Environmental Impact Assessment Review* 11:297–309.

Delicaet, A. 1995. 'The New Canadian Environmental Assessment Act: A Comparison with the Environmental Assessment Review Process'. *Environmental Impact Assessment Review* 15, no. 6:497–505.

Dirschl, H.J., N.S. Novakowski, and M.H. Sadar. 1993. 'Evolution of Environmental Impact Assessment as Applied to Watershed Modification Projects in Canada'. *Environmental Management* 17, no. 4:545–55.

Environmental Assessment Panel. 1991. *Rafferty-Alameda Project: Report of the Environmental Assessment Panel*. Ottawa: Federal Environmental Assessment Review Office.

Franklin, U. 1990. *The Real World of Technology*. Toronto: CBC Enterprises.

Garipey, M. 1991. 'Toward a Dual-Influence System: Assessing the Effects of Public Participation in Environmental Impact Assessment for Hydro-Québec Projects'. *Environmental Impact Assessment Review* 11, no. 4:353–74.

Gibson, R.B. 1992. 'The New Canadian Environmental Assessment Act: Possible Responses to Its Main Deficiencies'. *Journal of Environmental Law and Practice* 2, no. 3:223–55.

_____. 1993a. 'Challenges for Environmental Assessment and Planning in Southern Ontario: Lessons from Case Experience'. In *Environmental Assessment and Planning in Ontario—Challenges and Perspectives*, edited by R.E. Stenson and J.G. Nelson, 7–14. Heritage Resources Centre Occasional Paper no. 24. Waterloo, ON: University of Waterloo.

_____. 1993b. 'Environmental Assessment Design: Lessons from the Canadian Experience'. *Environmental Professional* 15, no. 1:12–24.

House of Commons. 1974. *Debates*, vol. 1. Second session, twenty-ninth Parliament. Ottawa: Queen's Printer, J. Davis.

Kitchen, C.M. 1976. 'Ecology and Urban Development: The Theory and Practice of Ecoplanning in Canada'. In *Canada's Natural Environment: Essays in Applied Geography*, edited by G.R. McBoyle and E. Sommerville, 217–40. Toronto: Methuen.

McIntosh, R.P. 1980. 'The Relationship between Succession and the Recovery Process in Ecosystems'. In *The Recovery Process in Damaged Ecosystems*, edited by J. Cairns, 11–62. Ann Arbor: Ann Arbor Science Publishers.

Nakashima, D.J. 1990. *Application of Native Knowledge in EIA: Inuit, Eiders and Hudson Bay Oil*. Prepared for the Canadian Environmental Assessment Research Council. Ottawa: Minister of Supply and Services Canada.

Nikiforuk, A. 1997. *'The Nasty Game': The Failure of Environmental Assessment in Canada*. Report prepared for the Walter and Duncan Gordon Foundation of Toronto.

Partidário, M.R. 1996. 'Strategic Environmental Assessment: Key Issues Emerging from Recent Practice'. *Environmental Impact Assessment Review* 16, no. 1:31–55.

Peterson, E.B., Y.-H. Chan, N.M. Peterson, G.A. Constable, R.B. Caton, C.S. Davis, R.R. Wallace, and G.A. Yarranton. 1987. *Cumulative Effects Assessment in Canada: An Agenda for Action and Research*. Prepared for the Canadian Environmental Assessment Research Council. Ottawa: Minister of Supply and Services Canada.

Reed, M.G. 1990. *Environmental Assessment and Aboriginal Claims: Implementation of the Inuvialuit Final Agreement*. Prepared for the Canadian Environmental Assessment Research Council. Ottawa: Minister of Supply and Services Canada.

Shapcott, C. 1989. 'Environmental Impact Assessment and Resource Management, a Haida Case Study: Implications for Native People of the North'. *Canadian Journal of Native Studies* 9, no. 1:55–83.

Shoemaker, D.J. 1994. *Cumulative Environmental Assessment*. Department of Geography Publication Series no. 42. Waterloo, ON: University of Waterloo.

Smith, L.G. 1993. *Impact Assessment and Sustainable Resource Management*. Harlow: Longman Scientific and Technical; New York: John Wiley and Sons.

Sonntag, N.C., R.R. Everitt, L.P. Rattie, D.L. Colnett, C.P. Wolf, J.C. Truett, A.H.J. Dorsey, and C.S. Holling. 1987. *Cumulative Effects Assessment: A Context for Further Research and Development*. Prepared for the Canadian Environmental Assessment Research Council. Ottawa: Minister of Supply and Services Canada.

Chapter 11

Stakeholders and Partnerships

Attitudes and policies regarding public participation and consultation have changed dramatically since the Second World War. For example, in 1948 the Aluminum Company of Canada (Alcan) was invited by the government of British Columbia to consider building an aluminium smelter based on low-cost electricity from hydro power. Alcan and the provincial government reached a decision that led to the creation of the community of Kitimat in northwestern British Columbia and the building of an aluminium smelter there (Figure 11.1). The decision-making process used to reach this agreement was typical of the approach and style of much development during that time.

A key part of the aluminium smelter package was an agreement that Alcan could build a dam on the Nechako River and divert water at Kemano to generate hydroelectricity. About 334 km^2 of land were to be flooded by the reservoir, including ten reserves in the area belonging to the Cheslatta band. A survey indicated that the best reserve land, such as that for natural hay meadows, would be flooded. However, there was virtually no consultation with the Aboriginal people occupying the land to be inundated (Day and Quinn 1992:84–95).

In 1952 Alcan accelerated its development project so that aluminium could be provided for the Korean War. Subsequently, there was significant pressure on the Cheslatta band members to sell their land and leave the area. Day and Quinn found that Alcan applied for a licence to flood two lakes on 2 April 1952, and only six days later the Cheslatta Dam was closed and water storage started. The

first meeting with band members was on 3 April, a few days before the water started to rise. In mid-October, the federal government reported that the minister of northern affairs had approved the sale of the Cheslatta band land to Alcan seven months after the licensing approval from the government of British Columbia.

Day and Quinn reported that thirty years after these events, the Carrier Sekani Tribal Council stated that:

> The Indians of Cheslatta were never given an opportunity to discuss the merits of the . . . Dam. They were told about the dam after it had been built and after the flooding had already begun. . . . Most of the people were forced to live in tents between April and November, 1952, before DIA (Department of Indian Affairs) finally found property for them.
>
> They were forced to build a new life in a farming community with which they had little in common. Many were forced to abandon their traditional occupations of hunting, trapping and fishing (Day and Quinn 1992:95).

In contrast, by the 1980s policies and practices in Canada had shifted dramatically, at least in theory. Frequently used words were participation, collaboration, cooperation, sharing, consensus building, local empowerment, stakeholders, and partnerships. For example, writing in the mid-1980s, two people employed by Environment Canada stated that:

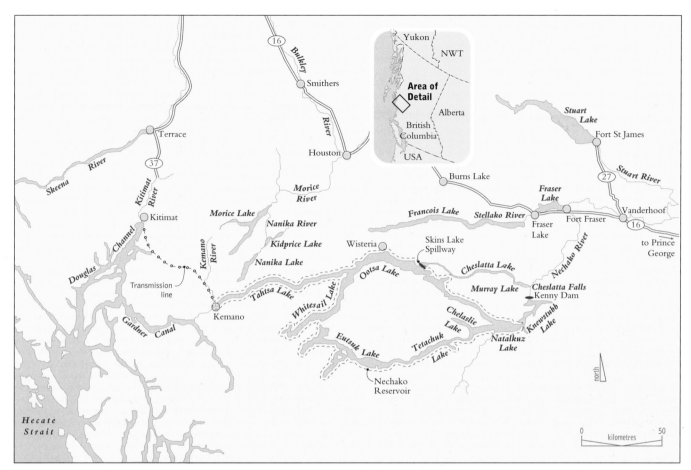

Figure 11.1 Kitimat, Kemano, and the Nechako River. SOURCE: J.C. Day and F. Quinn, *Water Diversion and Export: Learning from Canadian Experience* (Waterloo: Department of Geography, University of Waterloo, 1992):86.

Public participation has emerged as an indispensible element in effective resource planning and management. Originally conceived as a mechanism for obtaining local views on complex issues, public participation provides a safeguard against ill-informed decisions, educates the public and decision makers alike of each other's concerns, builds public confidence and understanding, and provides new and valuable information to planners and policy makers (Pollard and McKechnie 1986:32).

By the early 1990s, at least in theory the importance of public participation had been recognized by federal and provincial governments. In the federal government's Green Plan, it was stated that 'also needed is effective public participation' and 'decision-makers recognize that we can no longer rely solely on experts for the solutions to environmental problems. Instead, we need input from a wider cross-section of the population' (Government of Canada 1990:18). In a similar way,

the Ontario Ministry of Natural Resources stated that:

Partnership arrangements in natural resource decision-making and management must, and will, be significantly increased so that the public shares more fully and directly in the benefits and responsibilities of resource stewardship.

If public input is to be meaningful, the people involved must have access to information that is clearly understandable. If people who are to be affected by resource management decisions are to have a say, they must be presented with *all* the options. These must be explained in a coherent, non-technical manner so as to enable people to be active citizen-participants rather than passive consumers (Ontario Ministry of Natural Resources 1992:11).

The Manitoba Round Table on Environment and Economy (1992:5) argued that one of ten basic principles for sustainable development in that province should be *shared responsibility*, and that one of six guidelines to achieve sustainable development should be *public participation*. It concluded that it was essential to 'encourage and provide opportunity for consultation and meaningful participation in decision making processes by all Manitobans' as well as to ensure 'due process, prior notification and appropriate and timely redress for those affected by policies, programs, decisions and developments' (1992:5).

In the remainder of this chapter, attention will turn to considering the nature of public participation as a means to reallocate *power* among participants, as well as alternative ways to facilitate *empowerment* of people relative to the environmental management process.

DEGREES OF SHARING IN DECISION MAKING

The basic characteristics of sustainable development were outlined in Chapter 2. Several of them—reducing injustice and increasing self-determination—provide the philosophical rationale for sharing power in environmental management. Indeed, it is unlikely that sustainable development can be achieved unless a strong commitment is given to a management process that helps to reduce injustice and increase self-determi-

nation. However, redistribution of power for environmental management can challenge vested interests. As a result, it is not always easily or readily accepted.

Arnstein (1969) has provided a helpful perspective on power redistribution by identifying what she called 'rungs' on the ladder of citizen participation (Table 11.1). With reference to the agreement between the government of British Columbia and the Aluminum Company of Canada discussed at the beginning of this chapter, the Cheslatta band was not even on the first rung of Arnstein's ladder. In that and many other development situations during the period immediately after the Second World War, citizens were not involved in planning or decision making, even in the manipulative or therapeutic manner described by Arnstein. Citizens were basically excluded from decisions, while arrangements were reached between elected and technical officials from the public sector and representatives of the private sector.

By the late 1960s and early 1970s significant changes had occurred in Canada. Public involvement programs had moved up to Arnstein's rungs of informing, consultation, and placation. In other words, by that time, environmental managers often incorporated public participation into environmental management initiatives. However, because

Table 11.1 Rungs on the Ladder of Citizen Participation

Rungs	Nature of Involvement	Degrees of Power Sharing
1. Manipulation	Rubberstamp committees	
2. Therapy	Powerholders educate or cure citizens	Non-participation
3. Informing	Citizens' rights and options are identified	
4. Consultation	Citizens are heard but not necessarily heeded	Degrees of tokenism
5. Placation	Advice is received from citizens but not acted upon	
6. Partnership	Trade-offs are negotiated	
7. Delegated	Citizens are given management power for all or parts of programs	Degrees of citizen power
8. Citizen control		

Source: S. Arnstein, 'A Ladder of Citizen Participation', *Journal of the American Institute of Planners* 35, no. 4 (1969):216–24.

of concern that public agencies were ultimately accountable for resource allocation decisions and expenditure of public funds, public agencies usually took the position that information and advice received through public participation was only one of several sources to be considered.

Resource and environmental managers believed that they had the mandate, responsibility, and power to decide which trade-offs best reflected societal needs and interests and to make final decisions. This viewpoint was usually reinforced from conviction that there was no one public interest but many, and frequent conflicts among them were common. Giving citizens responsibility for decision making was often viewed as dangerous as it could too easily evolve into a form of anarchy in which no one was responsible or accountable for decisions.

During the 1980s there was increasing dissatisfaction with both the process and product associated with many resource and environmental management decisions. Growing numbers of Canadians rejected the idea that 'technically correct' answers could always be found for development decisions. Instead, there was a prevailing view that such decisions could ultimately only be based on weighing conflicting goals, aspirations, and values. In these situations, technical expertise was a legitimate but only one input to decision making. Furthermore, as decisions seemed to become more complex with more considerations and trade-offs, it became less likely that any one expert or organization could have all of the necessary insight to reach a 'wise' decision.

From these considerations arose the idea that stakeholders had a right to participate in decisions. In this context, *stakeholders* are those who should be included because of their direct interest, and that includes: (1) any public agency with pertinent management responsibilities, (2) all interests significantly affected by a decision, and (3) all parties who might intervene in the decision-making process or block or delay the process. Because more and more increasingly complex decisions had to be made, the traditional forum for public participation—the political process with elected representatives consulting and reflecting constituents' views—no longer seemed adequate. As a result, various individuals and non-governmental organizations began pressing for public involve-

ment, which moved higher up on Arnstein's ladder. While few believed that total 'citizen control' was feasible or desirable, they expected real 'partnerships' and 'delegated power' to be desirable and achievable.

This change from the earlier approach exemplified by the government of British Columbia and Alcan agreement for hydro-power development on the Nechako River is illustrated by more recent comments from both the BC government and the Saskatchewan Round Table on Environment and Economy (1992:54). Thus by the early 1990s the British Columbia Ministry of Environment, Lands and Parks (1992:4) was committed to 'shared responsibility and partnerships in reaching the goals of environmental protection and sustainability; consultation and discussion; . . . strong representation and advocacy for environmental protection and enhancement.' For its part, the Saskatchewan Round Table on Environment and Economy commented that:

Across Canada and in Saskatchewan, conflicts escalate when more and more people want to use limited resources for a greater number of purposes. Conflicts challenge traditional decision making processes and many decisions are protested, appealed or ignored. Misgivings are growing about the ability of governments to adequately mediate between competing interests. Appeals, court cases and civil disobedience create costs, a psychological drain and often additional uncertainty.

Conflicts have grown not only because of increasing demands for resources, but also because of narrow approaches to decision making. We have hesitated to broaden involvement in the planning, construction or operation of major projects.

Projects are often undertaken with the participation of only those with similar resource interests and conflicts are faced only when many options are already closed (Saskatchewan Round Table on Environment and Economy 1992:54).

As a result, the Saskatchewan Round Table concluded that there was 'value [in] a full and open declaration of diverse interests, early in the consultative process', which could be used to develop

Box 11.1 Who Are 'Legitimate' Stakeholders?

Some of the highest-profile environmental conflicts in the country over the last decade have been over alternative uses of forest land, as discussed in Chapter 15. Local communities that depend on the logging industry usually argue for cutting timber in these controversial areas, and claim that groups from farther away do not have a legitimate right to disrupt local economies. Environmental groups, often based farther away and with an urban-based clientele, usually argue for preservation for the benefit of

future generations. They claim that local communities care only about their own immediate economic interests and not about the overall good of society.

A classic example was the debate over the future of South Moresby Island or *Gwaii Hanaas*, as it is known to the Haida. The only community on South Moresby at the time was Sandspit whose economy was heavily dependent upon logging. The community, not surprisingly, argued against the creation of the national park. In the end, the local community's view did not prevail, a national park was created, and the community's economy declined. In contrast, during the autumn of 1996 the Ontario provincial government authorized the cutting of timber in the Temagami region of northeastern Ontario. One of the main motivations was to provide employment to the local community, which had a very high unemployment rate. Balancing the interests of stakeholders in these kinds of situations when there is a clear winner and a clear loser is one of the most difficult tasks for environmental decision making, as discussed in more detail in the next chapter.

A stand of Haida totem poles in Ninstints in South Moresby, the wilderness archipelago of the Queen Charlotte Islands (*World Wildlife Fund Canada/Mark Hobson*).

trust and build consensus (Saskatchewan Round Table on Environment and Economy 1992:54). The significance of decisions built on consensus is that they normally generate support and commitment. The Saskatchewan Round Table also recognized that such an approach can be 'time consuming, costly and sometimes frustrating' (1992:54). Nevertheless, it concluded that unilateral decisions that may seem efficient often end up in 'costly lengthy appeals or court action' (1992:54).

As a result of such thinking, the idea of *partnerships* became popular. The partnership concept has been implemented through *comanagement* initiatives and other approaches that reflect a genuine reallocation of power to citizens and away from

elected officials or technical experts. The siting of a waste treatment and disposal facility in Alberta is one of the earliest and most imaginative initiatives to incorporate stakeholders into a partnership approach.

Swan Hills, Alberta: Waste Treatment and Disposal Facility

Canadian society generates various demands that need to be satisfied, but individuals are usually not enthusiastic about having the necessary facility located nearby. For example, sand and gravel (or aggregate material) are essential for constructing roads and buildings. Because of the ratio of high weight to low value, sand and gravel pits are developed close to the source of demand (urban areas) and are therefore usually adjacent to other human activity. Because of the noise, dust, and increased truck traffic, most people are not pleased to have a sand and gravel operation near their property. For similar reasons, people often object to power corridors, sewage treatment plants, major airports, and similar facilities established nearby. Such facilities have become known as LULUs (locally unwanted land uses), and are often associated with the NIMBY (not in my backyard) reactions from people nearby.

Siting a facility for the treatment and disposal of wastes, especially hazardous wastes, qualifies as a LULU for most people. Often there is concern about leakage from a waste disposal site, which might contaminate ground or surface water, as highlighted in the example at the beginning of Chapter 1. Decomposition of organic and other material can generate methane and other dangerous gases. Yet Canadians generate significant amounts of waste, and urban communities and industries are facing increasing pressures to handle their wastes in a responsible manner. While it is usually believed that responsible waste management involves reducing, recycling, and reusing, it is also accepted that society will have some waste that must be treated or disposed.

Smith (1993:172–6) has provided a good overview of the innovative way in which Alberta incorporated stakeholders in a process that led to the creation of a hazardous waste treatment and disposal facility near the community of Swan Hills, northwest of Edmonton. The decision to build the waste treatment and disposal facility was made in 1984 after a three-year search process. It is generally acknowledged that this process was the *first* time in North America that 'social acceptability' was a primary criterion in siting such a facility.

The catalyst for the process was the recognition within Alberta during the late 1970s that a provincial facility was needed to handle the estimated 100 000 tonnes of industrial and hazardous wastes generated each year. The first stage in the search process was the organization of public hearings at sixteen centres across the province. These meetings provided information to the public regarding the types and amounts of wastes being produced in the province, the alternative methods for treating and disposing of such wastes, and the considerations deserving attention when siting a

waste treatment and disposal facility. The meetings also provided a forum for people in the province to indicate which considerations they thought needed attention.

After these public hearings, the province created a hazardous waste team that included some technical experts and the general public. The team developed the following approach. First, a constraint map was created at a scale of 1:1,000,000, which screened the entire province to eliminate areas that did not seem appropriate relative to specified physical, biological, land-use, and social criteria. Similar mapping was completed at a regional scale of 1:250,000. To this point, the approach was conventional.

A second component, however, made the search process innovative. During 1981 *every* municipal and local government in Alberta was briefed about the siting process. The briefing included an explanation of the implications of the provincial and regional screening maps, and an explanation of the criteria used to define the attributes of a desirable waste facility. Over 120 community meetings were held during this part of the process. Once these meetings were completed, the municipal and local governments were asked to choose from two alternatives. They could withdraw from further participation in the search process, or they could be included in more detailed regional studies. Fifty-two out of a potential seventy local-level governments indicated that they wished to continue into the next phase.

In the detailed regional analyses, all findings regarding the merits of and possible problems from a waste facility were presented at public meetings. If any public opposition was expressed in a community, that locality was eliminated from further consideration. Environmental constraints caused other areas to be eliminated. From this process, twelve communities indicated interest in being considered further, and each identified three possible sites. For the communities, the advantages of having such a facility included employment opportunities, enhanced tax base, economic diversification, as well as improved services and utilities.

Further screening eliminated seven of the twelve communities. Detailed drilling was then conducted in the remaining five communities to verify the appropriateness of the local geology for a waste facility. Seminars were held in each of the five communities. Attention was given to the rationale for the facility, the technology to be used, and the nature of the proposed treatment plant. *Public support* was a precondition for selecting a site, and public plebiscites were held in each community. Voter participation was high, and in the two leading candidate communities the support was very high (77 per cent in Swan Hills, 79 per cent in Ryley). The final decision was made by the provincial Cabinet in 1984, and Swan Hills was chosen. Construction began shortly afterward, and in September 1987 operations began at the Alberta Hazardous Waste Management Facility.

The Alberta process has been considered innovative for several reasons. Perhaps most impressive was the relatively short period of time (three years) for determining the location of a hazardous waste treatment facility. In many other provinces it has taken as long as ten years to find and approve a site. The short period of time in the Alberta process reflected public understanding and acceptance of the need for such a facility as well as a process that involved the stakeholders from the outset. The process in Alberta permitted communities to decide for themselves whether or not they wished to host such a facility. In this way, the local communities were empowered and so could legitimately view themselves as real partners. The voluntary participation in the process also meant that many of the characteristics of a LULU or NIMBY event were diffused, since any opposition to the facility in a community meant that it would no longer be considered a possible host.

As with all approaches, criticisms can be made (Box 11.2). Regarding the public participation aspects, perhaps the major reservation about a voluntary approach for siting a waste treatment and disposal facility is that it could take advantage of relatively remote and economically vulnerable communities or regions. The jobs, increased tax base, and economic diversification represented by a waste facility are likely to be more appealing to a community or area that is struggling economically than to one that is prosperous. A second concern could emerge depending upon the process used within a community to determine whether it might volunteer to host the facility, and how local agreement was reached.

A third concern, of course, would emerge if no community or area volunteers to be a host. This problem will be considered more directly in Chapter 12 when examining alternative mechanisms for

---

Box 11.2 Swan Hills: The Aftermath

Although the siting process for Swan Hills exemplifies the depth of stakeholder input that is desirable for these kinds of facilities, things have not gone quite so well for the facility since its opening. The facility was designed to accommodate Alberta's special wastes, particularly the backlog of PCBs and other toxic chemicals that had been stockpiled. When opened in 1987, it could treat 18 000 tonnes of waste a year. However, it lost a lot of money. This was of some concern to the provincial government since it owned 49 per cent of the plant and had guaranteed profits of 3 per cent over the prime interest rate on the capital (also borrowed) of the operator, Bovar, Inc. The company was also guaranteed a profit by the government whether the plant was profitable or not. These financial details are important, since they suggest that there was some advantage to the company to increase its capital investment.

This is exactly what happened. In 1990 the company argued successfully for an $80 million expansion, which increased capacity to 55 000 tonnes, on the basis that this would save money. The extra waste never materialized. In fact, the facility had to import wastes from other provinces just to test its new equipment. Estimates suggest that due to the high costs of disposal at the facility, some 80 per cent of Alberta's wastes are still stockpiled, which is allowed by law, or shipped to the US for cheaper disposal. Operating at less than 50 per cent capacity, the facility has now obtained government approval to import wastes. However, some local groups, including the Lesser Slave Lake Regional Native band, are not pleased about becoming a toxic waste dump for the continent, something not planned for in the original agreement.

This aftermath illustrates how complicated these issues can be, and that even processes designed and implemented to the highest standards are sometimes overcome by the force of circumstance and become less than ideal.

dispute resolution. Despite such potential problems, the approach developed for the waste facility near Swan Hills strove to avoid injustice and to ensure self-determination by the host community. In that regard, the approach was consistent with the tenets of sustainable development and was considerably ahead of its time.

EMPOWERMENT

The recognition of stakeholders and partnerships in environmental management represents an acceptance that those who will live with the long-term consequences of development decisions should have a voice in the management process.

A Native demonstration at Temagami (*Earthroots*).

If people are so empowered, they are more likely to be supportive of decisions and be willing to accept some negative consequences.

This section focuses on how stakeholders are incorporated into environmental management. Particular attention is given to the recognition of the importance of indigenous peoples, non-governmental organizations, comanagement initiatives, round tables, and intergovernmental partnerships.

Indigenous Peoples

Key stakeholders in environmental and resource management are the First Nations peoples. In Canada's Green Plan, the federal government indicated that it intended to strengthen partnerships with First Nations peoples. The plan states that:

Canada's aboriginal communities have long understood the importance of resource management and environmental stewardship. Native peoples depend upon nature for traditional and commercial activities and cultural well-being. In recent times, however, they have experienced rapid changes that increasingly threaten their natural environment.

Aboriginal peoples are more and more affected by clear cut logging, new roads, mines, pipelines, hydro-electric plants and other development pressures. The land, water and wildlife on some reserves contain high levels of toxic substances. Aboriginal peoples are not always opposed to development but would like the impact of development on their lives to be considered. . . .

For environmental matters affecting aboriginal Canadians to be resolved effectively and constructively, natives themselves must be active participants in decision-making processes as well as implementation activities that affect their communities (Government of Canada 1990).

This perspective represents a significant shift from the approach illustrated by the development in the early 1950s of the Nechako River for hydro power development. Part of the reason for this change has been the Constitution Act of 1982, which significantly altered the status of Aboriginal peoples' traditional rights.

A hierarchy of various rights and laws was established under the British North America Act of 1867. The traditional rights of Aboriginal peoples had the least recognition and were overridden by provincial statutes, treaty agreements, and federal statutes, in that order. However, in 1982 the Constitution Act became the dominant law of the country, and traditional rights were enshrined in that statute. The implication is that traditional rights of Aboriginal peoples now take precedence over federal and provincial laws or treaty agreements. This hierarchy of rights was confirmed by

This road sign in Temagami asserts the authority of the Anishnabai (*Earthroots*).

the *Sparrow* case in which a West Coast Native fisher's rights were upheld by the Supreme Court of Canada, which stated that the Crown can exercise its power regarding a natural resource used by Aboriginal peoples only to the extent necessary for conservation (Usher 1991).

In other words, if it can be demonstrated that Aboriginal peoples have been using, managing, and conserving a natural resource since time immemorial, it can be argued that such a resource falls beyond the jurisdiction of the Crown. It is unlikely that the courts would give Aboriginal peoples total control of the resource, but at least they would recognize a *dual* or *shared* interest. This legal leverage, combined with the growing recognition of the value of stakeholder involvement in environmental management, has increased the opportunity for First Nations peoples to manage natural resources or to become partners in management. Particularly important in this regard has been the increasing awareness that traditional knowledge about environmental systems should

be given serious consideration when developing management plans.

In later chapters, more attention will be given to the emerging role of Aboriginal or First Nations peoples in environmental management. At this stage, however, it is useful to identify the attributes of indigenous resource management systems. According to Cizek, there are six key features:

1. *Closed Access*: The access to resource harvesting is restricted to a specific group of people.
2. *Community of Harvesters*: The harvesters have a common kinship and geographic proximity, as well as common norms and values regarding resource use.
3. *Maintenance of Norms and Values*: Positive reinforcement for use patterns is provided through oral history, spirituality, and social celebrations. Violation of accepted use patterns is countered initially by counselling, and then by punishment involving shunning, ridicule, and gossip.
4. *Stewards of the Resource*: Recognized people act as stewards of the resource because they possess traditional ecological knowledge about the resource system and harvesting systems.
5. *Sustainability of Harvests*: Harvesting patterns are designed never to exceed sustainable yields.
6. *Historical Continuity*: For the legal basis of aboriginal rights to be applicable, the indigenous resource management system must have existed continuously for 'time immemorial' (Cizek 1993:30–1).

Consistent with the ideas of self-empowerment and social justice linked to sustainable development, the role of First Nations peoples in environmental management can occur through self-government or management, or through partnerships. A later section on comanagement considers one form of partnership.

Non-governmental Organizations

Non-governmental organizations (NGOs) have become increasingly involved in environmental management in Canada. As noted earlier, their emergence has reflected a growing concern about the capability of governments and corporations to make decisions that effectively balance economic development and protection of natural systems. Consequently, non-governmental organizations have identified several roles for themselves. Some groups have evolved because their members want to pool their skills and resources to provide hands-on environmental stewardship. Others have taken on the role of environmental watchdogs and pressure governments and corporations to make environmentally sensitive decisions. Of course, some groups take on both roles (Lerner 1993). Regardless, each year NGOs contribute thousands of unpaid hours in activities related to preservation, rehabilitation and restoration, monitoring, research, education, preparation of submissions, and fundraising.

Lerner (1993) has provided an excellent overview of the role and activities of what she has labelled environmental stewardship groups in Canada. The focus of such groups ranges from the protection of the fruit lands in the Niagara Peninsula in Ontario, protection and establishment of parkland in Montreal, the stewardship of the Tusket River and the St Mary's River in Nova Scotia, to the defence of old-growth forests in the Carmanah Valley in British Columbia.

In Part D, attention will be given to the role of such NGOs or environmental stewardship groups in environmental management. Particular interest will centre on how they have formed networks or partnerships with other similar groups, as well as partnerships with government agencies and the private sector.

Comanagement

Pinkerton (1993:37) has defined comanagement arrangements as ones that 'involve genuine power sharing between community-based managers and government agencies, so that each can check the potential excesses of the other'. Such arrangements have most often been developed with regard to fisheries and forestry since each is vulnerable to overharvesting by individuals, corporations, or even state-based organizations. When communities become involved in comanagement agreements, they can often play an effective role because they have already developed means to prevent overharvesting through both informal and formal methods of control.

Pinkerton (1993) has described a model forest practices act in British Columbia as a good example of an attempt to create a comanagement agreement among the provincial government, First

Nations communities, labour unions, small businesses, and environmental groups. A key element would be for the provincial government to allocate considerable power to bioregional boards for forest management. In Pinkerton's words, the new forestry act would be

> [the] culmination of a series of efforts by BC communities over two decades to reform forest management toward sustainable rates of logging on a regional basis, holistic management for all forest values, optimal wood utilization standards and the recognition of First Nations' rights and decision-making processes (Pinkerton 1993:35).

As mentioned earlier, a central feature would be decentralizing important forest land-use planning activities to local boards that would be partly elected and partly appointed by the province to ensure representation of local stakeholders. Furthermore, the proposed structure would 'curtail the influence of political parties and civil servants on the management process, although governments and bureaucracies would play an important oversight role if basic standards and procedures set forth in the act were violated.' Developments in Clayoquot Sound have highlighted local people's desire for such an arrangement (discussed in more detail in Chapter 21).

Round Tables

During May 1986 members of the World Commission on Environment and Development (the Brundtland Commission) visited Canada and met with members of the Canadian Council of Resource and Environment Ministers (CCREM) in Edmonton. As a direct consequence of that meeting, CCREM established the National Task Force on Environment and Economy in October 1986. As the task force commented in its first report the next year, it was created 'to initiate dialogue on environment-economy integration among Canada's environment ministers, senior executive officers from Canadian industry, and representatives from environmental organizations and the academic community' (1987:1). Its mandate was to 'foster and promote environmentally sound economic development' (1987:1). Having representatives of government regulatory agencies, the private sector, and environmental organizations in such a group was consistent with the Brundtland Commission's position that environmental and economic issues had to be considered together and not in isolation.

The task force endorsed various key ideas in its first report. With regard to stakeholders and partnerships, for example, it argued that basic principles for sustainable development had to include 'shared responsibility' and 'integrated decision making' (1987:2). They also recommended that it was desirable to 'increase public participation', to replace remedial and reactive approaches with an 'anticipate and prevent' approach, and to develop a 'multisectoral approach to defining and implementing sustainable economic development' (1987:2, 3). More specific recommendations are shown in Box 11.3.

The task force commented that 'of all our recommendations, we consider Round Tables to be

among the most important' (1987:1). The round tables were viewed to represent a 'new process of consultation' that would involve senior decision makers from diverse groups (1987:10). The task force believed that individuals invited to serve on the round tables should be senior decision makers who exercised influence over policy and planning in their own organizations. The deliberations of the round tables were to be oriented to achieving consensus and exerting 'direct influence on policy and decision makers at the highest levels of government, industry, and non-government organizations' (1987:10). The task force commented that 'Round Tables are not proposed to challenge the authority of any existing office or institution. Instead, they would exert influence, founded on their credibility, their independence and their access to the views of important sectors and levels of society' (1987:11).

The National Round Table on the Environment and the Economy was announced by the prime minister during October 1988 and met for the first time in June 1989. In a report during 1991, the National Round Table commented that it represented:

> . . . a new process that brings together some of the *competing and converging interests* with direct stakes in environmental, social and economic objectives. It then becomes a forum for those traditionally competing interests to find common ground on which they can take actions towards sustainable development (National Round Table on the Environment and the Economy 1991:4).

The National Round Table has taken various initiatives to promote integration of environmental and economic considerations. In addition, round tables have been established for the provinces and the territories. A key activity for all of them has been to promote or help develop a conservation strategy for their respective jurisdictions.

The round table concept is innovative because it brought together people of diverse backgrounds and interests and explicitly acknowledged that government politicians and technical experts had to work in partnership with people from the private sector, non-governmental organizations, and other interests. Round tables were also designed so

---

Box 11.3   National Task Force Recommendations Regarding Stakeholders and Partnerships

- Ensure that all levels and departments of government establish consultation processes to encourage and facilitate public involvement and influence in policy making and planning.
- Establish formal mechanisms to hold ministers and their departments accountable for promoting environmentally sound economic development.
- Each province and territory should form a multisectoral round table on environment and economy to bring together existing organizations to cooperate on environmental-economy integration at the provincial and territorial levels.
- Concurrent with the formation of provincial and territorial round tables, a national round table should be formed with representatives from these round tables and members from the federal cabinet, national non-governmental organizations, labour, academic, and business associations.

that they could influence policy and practice, particularly by having access to senior decision makers in various sectors.

On the other hand, the round tables can be considered élitist since they rely primarily upon very senior people who normally would already be 'well connected' to sources of power. However, it was never suggested that the round tables alone would be able to do everything required to facilitate more openness in decision making or to create desirable partnerships. As much as anything, the round tables represent an intent to consider environmental and economic issues together, an acceptance of the importance of a holistic approach, and an acceptance of the need to incorporate a wide range of perspectives in developing policies and practices.

Intergovernmental Partnerships

As noted in Chapter 2, responsibility for the environment has not been allocated to any one government in Canada. As a result, the history of environmental management has been characterized by joint or intergovernmental approaches. To illustrate, the federal, provincial, and territorial governments often share responsibility for interjurisdictional resources, such as rivers that extend from one jurisdiction to another, or migratory animals. The Canadian Council of Ministers of the Environment (CCEM), originally called the Canadian Council of Resource and Environment Ministers, was created to provide a national forum to consider matters of shared interest and responsibility.

Canada must also deal with other national governments to address international problems. For example, global warming and ozone depletion require collaborative efforts because unilateral initiatives are unlikely to be successful. For example, Canada is an energy-intensive country and ranks high in per capita production of greenhouse gases and in emissions per unit of gross domestic product (Table 11.2). However, due to its relatively small population, Canada is not a major *overall* contributor to the greenhouse gases, which are considered one of the primary triggers for global warming. Canada is the source of not more than 2 per cent of the world's carbon dioxide, 2 per cent of nitrous oxide, 1 per cent of methane, and 2 per cent of chloro-fluorocarbons (CFCs). At the same time, Canadians could be affected by global warm-

ing if it led to warmer temperatures in higher latitudes, rising sea levels, and coastal flooding. For instance, a 1-m rise in sea level would flood more than 250 buildings in Charlottetown, PEI.

The concern about greenhouse gases and ozone depletion has resulted in Canada joining representatives from other countries to develop multilateral approaches. An example is the 1987 Montreal Protocol on Substances That Deplete the Ozone Layer, which was developed through intensive negotiations among many nations. Under that protocol, Canada committed itself to reducing the use of ozone-depleting substances by at least 50 per cent of 1986 levels by 1999. At the subsequent London Conference in 1990, Canada ratified a stronger protocol, which stipulated the elimination of CFCs by the year 2000, and of other important ozone-depleting substances by 2005. In fact, Canada declared that it would phase out CFCs completely by 1997.

One partnership that Canada has developed over the years has been with the United States, particularly in dealing with transboundary environmental problems, such as air and water pollution or delineation of fishing areas. A key mechanism in this respect is the International Joint Commission (IJC), an organization created to address problems directed to it by the governments of Canada and the United States.

The various partnerships discussed in this section emphasize that it is often not possible for only one government or jurisdiction to deal with environmental issues. Not only must federal and

Table 11.2 Comparison of Energy Intensity in G-7 Countries, 1988

Country	Energy Requirements (Tonnes of Oil Equivalent Per Capita)
Canada	9.6
United States	7.8
West Germany	4.5
United Kingdom	3.7
France	3.7
Japan	3.3
Italy	2.6

Source: After Environment Canada, 'The State of Canada's Environment' (Ottawa: Ministry of Supply and Services Canada, 1991):12–30.

provincial governments in Canada work collaboratively with First Nations peoples, the private sector, non-governmental organizations, and the general public, they must also work with each other regarding environmental management. In addition, as a member of the global community, Canada must work in partnership with other nations in dealing with issues such as global warming and ozone depletion, or protection of whales and other marine species.

IMPLICATIONS

A significant shift in approach has occurred since the Second World War regarding public involvement in environmental management. Following the Second World War, it was common for development decisions to be based only on discussions between governments and the private sector (Figure 11.2). By the 1980s, at least in theory or on paper, provision had been made to involve the public in the planning and decision-making process. Initially, provisions for public involvement gave public agencies the real decision-making power on the basis that such an arrangement was the only way to ensure accountability for decisions.

However, with regard to Arnstein's ladder of public participation, it appears as if the legitimacy of delegated power and real partnerships has become recognized and accepted. The Swan Hills hazardous waste treatment and disposal facility in Alberta was one of the first to involve stakeholders in a decision-making process that explicitly addressed environmental, economic, and social issues. Since then, other means have been developed. Two of the more notable have been comanagement arrangements and round tables. Perhaps most significant has been the recognition in law of

Figure 11.2 Canadian interjurisdictional disagreement. SOURCE: Courtesy *The Interior News,* Smithers, BC.

the rights of First Nations peoples and the provision of opportunity for both self-management and partnerships involving them.

Experience has shown that public participation does not happen effortlessly. It can be time consuming, expensive, frustrating, and demanding. However, it seems to be increasingly accepted that the initial costs of public participation help to avoid costly delays, appeals, and court cases related to management and development decisions. Public participation is also accepted because of its consistency with some of the basic ideas related to sustainable development, especially the importance of self-empowerment and achievement of social justice.

Canada has developed domestic and international partnerships. Unilateral efforts will not be sufficient to deal with issues such as greenhouse gases or ozone depletion, and as a member of the global community, Canada must be willing to participate actively and effectively in global initiatives. Because of its lengthy common border, Canada has a strong incentive to work with the United States on transboundary problems, such as air and water pollution.

Public participation, stakeholders, and partnerships are not a panacea for environmental management. However, it is difficult to imagine future environmental management strategies not including them as key elements. In Part D we will consider the extent to which they are included in present policies and practices for environmental management in Canada, which conditions are required to make them effective, and which conditions create problems.

SUMMARY

1. Many federal and provincial government resource and environmental agencies have made a public commitment to involve the public in their management activities.

2. Public participation can range from token involvement of local people, to information and education programs, to consultation, to reallocation of real power to local people.

3. Ideally, public participation involves reallocating power to local people and empowering them.

4. Stakeholders are those who should be included in planning and management because of their direct interest. They include: (1) any public agency with pertinent management responsibilities, (2) all parties significantly affected by a decision, and (3) all parties who might intervene in the decision-making process to block or delay it.

5. Resource and environmental agencies have been creating partnerships and developing approaches for comanagement to involve local people in planning and development decisions.

6. The Alberta Waste Treatment and Disposal Facility at Swan Hills is an early example of a conscious attempt to use a collaborative and participatory approach to find a site for a noxious facility. The only communities considered were those that volunteered to host the waste treatment facility.

7. First Nations peoples are often among the most affected by resource development projects. Under the Constitution Act of 1982, however, Aboriginal traditional rights have been given precedence over federal and provincial laws and treaty agreements, a situation that was confirmed by the *Sparrow* case in British Columbia.

8. Indigenous resource management systems are characterized by: (1) closed access, (2) a community of harvesters, (3) maintenance of norms and values, (4) stewardship, (5) sustainability of harvests, and (6) historic continuity.

9. Comanagement arrangements involve genuine power sharing between community-based managers and government agencies, so that each can check the potential excesses of the other.

10. Round tables are multisectoral organizations that provide a forum for people from government, the private sector, labour, non-governmental organizations, and universities to facilitate cooperation in integrating environmental and economic issues.

11. Intergovernmental partnerships are important domestically and internationally. Domestically, Canada has established the Canadian Council of Ministers of the Environment to harmonize federal and provincial initiatives. The International

Joint Commission enables Canada and the United States to address transboundary environmental issues, especially ones related to air and water pollution and fishing agreements. Canada is a signatory to international agreements, such as the Montreal Protocol on Substances That Deplete the Ozone Layer (1987), under which the country is committed to reducing use of ozone-depleting substances by at least 50 per cent of 1986 levels by 1999.

REVIEW QUESTIONS

1. What are stakeholders?

2. Why is there increasing interest in creating partnerships for environmental management?

3. What is meant by empowerment?

4. What are LULUs and NIMBYs?

5. What are the main features of indigenous resource management systems?

6. What are the characteristics of comanagement? Are there any examples of comanagement in your province or territory? If so, have they been effective?

7. Why were round tables created, and what has been their main activities?

8. When was the round table created in your province or territory? In your opinion, what has it accomplished to date?

9. Is there a local round table in your community? If so, what has it accomplished? If not, what would be required to create such a local round table?

10. Why was the International Joint Commission established?

REFERENCES AND SUGGESTED READING

Aberley, D. 1991. 'The Hazelton Experience, a Community Takes Action to Protect Its Forest Legacy'. *Forest Planning Canada* 7, no. 3:18–20.

Arnstein, S. 1969. 'A Ladder of Citizen Participation'. *Journal of the American Institute of Planners* 35, no. 4:216–24.

Asch, M., and P. Macklem. 1991. 'Aboriginal Rights and Canadian Sovereignty: An Essay on *R. v. Sparrow*'. *Alberta Law Review* 29, no. 2:498–517.

Berg, L., T. Fenge, and P. Dearden. 1993. 'The Role of the Aboriginal Peoples on National Park Designation, Planning, and Management in Canada'. In *Parks and Protected Areas in Canada: Planning and Management*, edited by P. Dearden and R. Rollins, 225–55. Toronto: Oxford University Press.

Berger, T.R. 1977. *Northern Frontier, Northern Homeland: The Report of the Mackenzie Valley Pipeline Inquiry*, 2 vols. Ottawa: Minister of Supply and Services Canada.

Berkes, F., P. George, and R.J. Preston. 1991. 'Co-management: The Evolution in Theory and Practice of the Joint Administration of Living Resources'. *Alternatives* 18, no. 2:12–18.

British Columbia Ministry of Environment, Lands and Parks. 1992. 'New Approaches to Environmental Protection in British Columbia: A Legislation Discussion Paper'. Victoria: Ministry of Environment, Lands and Parks.

Brunton, J., and M. Howlett. 1992. 'Differences of Opinion: Round Tables, Policy Networks and the Failure of Canadian Environmental Strategy'. *Alternatives* 19, no. 1:25–8, 31–3.

Bush, M. 1990. *Public Participation in Resource Development after Project Approval*. Prepared for the Canadian Environmental Assessment Research Council. Ottawa: Minister of Supply and Services Canada.

Cassidy, R., and N. Dale. 1988. *After Native Claims? The Implications of Comprehensive Claims Settlements for Natural Resources in British Columbia*. Lantzville, BC: Oolichan Books.

Christensen, B. 1995. *Too Good to Be True: Alcan's Kemano Completion Project*. Vancouver: Talon Books.

Cizek, P. 1990. *The Beverly-Kaminuriak Caribou Management Board: A Case-Study of Aboriginal Participation in Resource Management*. Ottawa: Canadian Arctic Resources Committee.

_____. 1993. 'Guardians of Manomin: Aboriginal Self-Management of Wild Rice Harvesting'. *Alternatives* 19, no. 3:29–32.

Connor, D.M. 1986. 'A New Ladder of Citizen Participation'. In *Constructive Citizen Participation: A Resource Book* by D.M. Connor. Victoria: Development Press.

Day, J.C., and F. Quinn. 1992. *Water Diversion and Export: Learning from Canadian Experience.* Department of Geography Publication Series no. 36 and Canadian Association of Geographers Public Issues Committee no. 1. Waterloo: Department of Geography, University of Waterloo.

Dearden, P. 1981. 'Public Participation and Scenic Quality Analysis'. *Landscape Planning* 8:3–19.

Doering, R.L. 1993a. 'Canadian Round Tables on the Environment and the Economy: Their History, Form and Function'. Working Paper no. 14. Ottawa: National Round Table on the Environment and the Economy.

_____. 1993b. 'Canadian Round Tables on the Environment and the Economy'. *International Environmental Affairs* 5, no. 4:355–70.

Durning, A.T. 1992. 'Guardians of the Land: Indigenous Peoples and the Health of the Earth'. Worldwatch Paper no. 112. Washington, DC: Worldwatch Institute.

Elder, P.S., ed. 1975. *Environmental Management and Public Participation.* Toronto: Canadian Environmental Law Research Foundation.

Environment Canada. 1990. *A Framework for Discussion on the Environment.* Ottawa: Minister of Supply and Services Canada.

Filyk, G., and R. Cote. 1992. 'Pressures from Inside: Advisory Groups and the Environmental Community'. In *Canadian Environmental Policy: Ecosystems, Politics and Process*, edited by R. Boardman, 60–82. Toronto: Oxford University Press.

Finley, K. 1988. *Cross-cultural Exchange of Ecological Knowledge: Toward a Community-Based Conservation Strategy for the Bowhead Whale.* Ottawa: Minister of Supply and Services Canada.

Freeman, M. 1992. 'The Nature and Utility of Traditional Ecological Knowledge'. *Northern Perspectives* 20, no. 1:9–12.

Freeman, M.M.R., and L. Carbyn, eds. 1988. *Traditional Knowledge and Renewable Resource Management in Northern Regions.* Edmonton: Boreal Institute, University of Alberta.

Gadgil, M., and F. Berkes. 1991. 'Traditional Resource Management Systems'. *Resource Management and Optimization* 8, no. 3–4:127–41.

Gill, A.M. 1990. 'Women in Isolated Resource Towns: An Examination of Gender Differences in Cognitive Structures'. *Geoforum* 21, no. 3:347–58.

Government of Canada. 1990. *Canada's Green Plan.* Ottawa: Minister of Supply and Services Canada.

Hartig, J.H., and P.D. Hartig. 1990. 'Remedial Action Plans: An Opportunity to Implement Sustainable Development at the Grassroots Level in the Great Lakes Basin'. *Alternatives* 17, no. 3:26–38.

_____, and M.A. Zarrell, eds. 1992. *Under RAPS: Toward Grassroots Ecological Democracy in the Great Lakes Basin.* Ann Arbor: University of Michigan Press.

Hodgins, B.W., and J. Benidickson. 1989. *The Temagami Experience: Recreation, Resources, and Aboriginal Rights in the Northern Ontario Wilderness.* Toronto: University of Toronto Press.

Howlett, M. 1990. 'The Round Table Experience: Representation and Legitimacy in Canadian Environmental Policy-Making'. *Queen's Quarterly* 97, no. 5:580–601.

Jackson, T., and G. McKay. 1982. 'Sanitation and Water Supply in Big Trout Lake: Participatory Research for Democratic Technical Solutions'. *Canadian Journal of Native Studies* 2, no. 1:129–45.

Jacobson, J.L. 1992. 'Gender Bias: Roadblock to Sustainable Development'. Worldwatch Paper no. 110. Washington, DC: Worldwatch Institute.

Law, N., and J.H. Hartig. 1993. 'Public Participation in Great Lakes Remedial Action Plans'. *Plan Canada* (March): 31–5.

Lerner, S., ed. 1993. *Environmental Stewardship: Studies in Active Earthkeeping.* Department of Geography Publication Series no. 39. Waterloo: Department of Geography, University of Waterloo.

Loney, M. 1987. 'The Construction of Dependency: The Case of the Grand Rapids Hydro Project'. *Canadian Journal of Native Studies* 7, no. 1:57–78.

Manitoba Round Table on Environment and Economy. 1992. *Sustainable Development: Towards Institutional Change in the Manitoba Public Sector.* Winnipeg: Manitoba Round Table on Environment and Economy.

M'Gonigle, R.M. 1988. 'Native Rights and Environmental Sustainability: Lessons from the British Columbia Wilderness'. *Canadian Journal of Native Studies* 8, no. 1:107–30.

Nakashima, D.J. 1990. *Application of Native Knowledge in EIA: Inuit, Eiders and Hudson Bay Oil.* Prepared for the Canadian Environmental Assessment Research Council. Ottawa: Minister of Supply and Services Canada.

National Round Table on the Environment and the Economy. 1991. *A Report to Canadians.* Ottawa: National Round Table on the Environment and the Economy.

_____. 1992. *Annual Review, 1991–1992.* Ottawa: National Round Table on the Environment and the Economy.

_____. 1993. *Annual Review, 1992–1993.* Ottawa: National Round Table on the Environment and the Economy.

National Task Force on Environment and Economy. 1987. *Report.* Ottawa: National Task Force on Environment and Economy.

_____. 1988. *Progress Report.* Ottawa: National Task Force on Environment and Economy.

Neads, D. 1991. 'West Chilcotin Community Resources Board'. *Forest Planning Canada* 7, no. 6:10–14.

Nixon, B. 1993. 'Public Participation: Changing the Way We Make Forest Decisions'. In *Touch Wood: B.C. Forests at the Crossroads*, edited by K. Drushka, B. Nixon, and R. Travers, 23–66. Madeira Park, BC: Harbour Publishing Co.

Ontario Ministry of Natural Resources. 1992. *Direction '90s.* Toronto: Queen's Printer.

Ontario Round Table on Environment and Economy. 1990. 'Challenge Paper'. Toronto: Queen's Printer.

Parenteau, R. 1988. *Public Participation in Environmental Decision-Making.* Ottawa: Minister of Supply and Services Canada.

Pinkerton, E. 1989. 'Attaining Better Fisheries Management through Co-management Prospects, Problems, and Propositions'. In *Co-operative Management of Local Fisheries: New Directions in Improved Management and Community Development*, edited by E. Pinkerton, 3–33. Vancouver: University of BC Press.

_____. 1990. *The Future of Traditional Ecological Knowledge and Resource Management in Native Communities.* Prepared for the Canadian Environmental Assessment Research Council. Ottawa: Minister of Supply and Services Canada.

_____. 1992. 'Over-coming Barriers to the Exercise of Co-management Rights'. In *Growing Demands on a Shrinking Heritage: Managing Resource Use Conflicts*, edited by M. Ross and J.O. Saunders, 276–303. Calgary: Canadian Institute for Resources Law.

_____. 1993. 'Co-management Efforts as Social Movements: The Tin Wis Coalition and the Drive for Forest Practices Legislation in British Columbia'. *Alternatives* 19, no. 3:33–8.

Pollard, D.F.W., and M.R. McKechnie. 1986. *World Conservation Strategy—Canada: A Report on Achievements in Conservation.* Ottawa: Minister of Supply and Services Canada.

Quinn, F. 1991. 'As Long as the Rivers Run: The Impacts of Corporate Water Development on Native Communities in Canada'. *Canadian Journal of Native Studies* 11, no. 1:137–54.

Reed, M.G. 1993. 'Governance of Resources in the Hinterland: The Struggle for Local Autonomy and Control'. *Geoforum* 24, no. 3:243–62.

_____. 1994a. 'Local Politics in the Provincial Norths: Struggles in Resource Management and Economic Development'. In *Geographical Perspectives on the Provincial Norths*, edited by M.E. Johnston, 224-55. Thunder Bay, ON: Lakehead University Centre for Northern Studies and Copp Clark Longman.

_____. 1994b. 'Locally Responsive Environmental Planning in the Canadian Hinterland: A Case Study in Northern Ontario'. *Environmental Impact Assessment Review* 14, no. 4:45–69.

Saskatchewan Round Table on Environment and Economy. 1992. *Conservation Strategy for Sustainable Development in Saskatchewan.* Regina: Saskatchewan Round Table on Environment and Economy.

Sewell, W.R.D., P. Dearden, and J. Dumbrell. 1989. 'Wilderness Decision-Making and the Role of Environmental Interest Groups: A Comparison of the Franklin Dam, Tasmania, and South Moresby, British Columbia Cases'. *Natural Resources Journal* 29:85–96.

Smith, L.G. 1993. *Impact Assessment and Sustainable Resource Management.* Harlow: Longman.

Tester, F.J. 1992. 'Reflections on Tin Wis: Environmentalism and the Evolution of Citizen Participation in Canada'. *Alternatives* 19, no. 1:36–41.

Usher, P.J. 1987. 'Indigenous Management Systems and the Conservation of Wildlife in the Canadian North'. *Alternatives* 14, no. 1:3–9.

_____. 1989. 'Some Implications of the *Sparrow* Judgement for Resource Conservation and Management'. *Alternatives* 18, no. 2:20–1.

_____. 1991. 'Some Implications of the Sparrow Judgement for Resource Conservation and Management'. *Alternatives* 18, no. 1:20–1.

Wolfe, J. 1994. 'First Nations' Strategies for Reintegrating People, Land Resources and Government'. In *Public Issues: A Geographic Perspective*, edited by J. Andrey and J.G. Nelson, 239–70. Department of Geography Publication Series no. 41. Waterloo, ON: Department of Geography, University of Waterloo.

Chapter 12

Conflict Resolution

For far too long we have convinced ourselves and the public that we are the practitioners of order and certainty. Implicit in our education, our methods, even our regulatory authority is the promise that our purpose is to make plans which will give direction toward an ordered environment, made certain by the surety of our forecasts of future behaviour. Let's change the paradigm from order to disorder, from certainty to uncertainty. Let's accept conflict as legitimate, let's define planning as a process of resolving conflict. . . .

—L. Sherman and L. Livey,
'The Positive Power of Conflict'

Conflicts and disputes occur for many reasons. They may result from clashing or incompatible values, interests, needs, or actions. In an environmental context, conflicts may arise due to substantive or procedural issues. At a substantive level, concerns may arise about the effects of resource use or project development; about multiple use of resources and areas; about policies, legislation, and regulations; or over jurisdiction and ownership of resources. At a procedural level, there may be concerns about who should be involved, when, and how.

Conflict: *A disagreement, dispute or quarrel. The clashing of opposed principles, statements, etc.: To be directly opposed, disagree, clash.*

World Book Dictionary, *1975*

However, conflict is not necessarily bad or undesirable. It can help highlight aspects in a process or system that hinder effective performance. Conflict can also be positive if it clarifies differences due to poor information or misunderstandings. Approached in a constructive manner, conflict can result in creative and practical solutions to problems. On the other hand, conflict can be negative if it breeds lack of trust or misunderstanding, or reinforces biases. It can also be negative if it is ignored or set aside, thus escalating a problem or creating stronger obstacles to be overcome later.

Whether conflict is positive or negative, it is usually always present. As Johnson and Duinker (1993) explain, conflict persists because people see things differently, want different things, have different beliefs, and live their lives in different ways. Basic differences among people and their values, interests, needs, and activities cause conflicts. Such differences can be exacerbated by other factors, which include lack of understanding of other people or groups; people using different kinds and sources of information; differences in culture, experience, or education; as well as different values, traditions, principles, experiences, perceptions, and biases.

Conflicts are a normal part of life, and we need to devise ways to deal with them. Rather than ignore conflict, we need to follow the advice in the quote by Sherman and Livey at the beginning of this chapter. We should accept that conflict is legitimate, and that environmental planning and management is often a process for resolving conflicts.

THE DYNAMICS OF DISPUTES

Johnson and Duinker (1993) have suggested that conflicts have a *life cycle* with five stages. These stages are:

Absence: At this initial stage, conflict is not apparent. The conditions for a conflict may be emerging, but actual conflict does not occur. For example, a city council decides to increase water rates significantly to encourage more careful water use in the community. The rates will take effect in eighteen months. Because it will be some time before the new prices are implemented, some industrialists who are worried about the impact of these new charges on their operating costs set this matter aside for the moment because they are preoccupied with other more immediate concerns.

Emergence: At this point, actual conflict becomes explicit. It is often characterized as the period when normal activity and progress related to resource and environmental management are impeded or disrupted. An example might be a public meeting that resource managers use to provide information about proposed resource use or development. During this meeting, participants become upset and leave with a vow to change what is scheduled to happen.

Growth: A conflict may become more serious. The competing participants strengthen their positions and aggressively criticize the values, beliefs, and views of the other side in the dispute. Letters of complaint may be submitted, people may stop meeting with or talking to the other side, and decisions are delayed. If non-governmental organizations are involved, letter-writing campaigns and demonstrations may start. Media interest begins, and some initial

> ### Addressing Conflict
>
> *We shouldn't expect that all disputes would go away if everybody would just agree to get along better and be reasonable in the demands they place on resources. That will not happen. . . . for many reasons there will be an increasing number of situations which can lead to conflicts, even bitter conflicts. What we can hope for, though, is that people and organizations in disagreement with each other can find productive and insightful ways to resolve their disputes, to the mutual satisfaction of all parties (Johnson and Duinker 1993:1–2).*

stories appear in the newspapers, radio, and television. Thus it can be seen that the life cycle evolves, with the incompatibility of interests becoming more pronounced, interests continuing to diverge, and more groups becoming involved.

Maturity: By this stage, a full-blown conflict exists. It is highly visible and attracts considerable media attention. Likely politicians have become involved. Strong statements are made, and mistrust and bitterness develop on all sides. The resource or environmental management process is almost paralysed as managers focus on damage control or crisis management and have little time to devote to management of the resource or environment under dispute.

A Temagami Wilderness Society demonstration (*Earthroots*).

Agreement and Implementation: In a positive scenario, eventually an agreement about the dispute is reached and implemented. As a positive outcome, the parties in the dispute have gained a greater understanding and respect about the interests and concerns of those with whom they were in conflict. It is also possible that new information has become available to all participants as a result of the need to analyse the issue before reaching an agreement. However, if the agreement does not have reasonable support, it is very likely that the conflict will revert to one of the earlier stages in the life cycle, to flare up again sometime later (Box 12.1).

This concept of a life cycle for conflict can be useful in understanding what point a dispute has reached and anticipating how it might evolve. Indeed, as Figure 12.1 indicates, Downs (1972) has argued that many public policy issues or conflicts go through an issue-attention cycle. Furthermore, resource and environmental managers rarely deal with only one dispute or conflict at a time. Usually they handle several disputes simultaneously. By identifying the conflict's life cycle stage, a manager may better anticipate what issues may arise next, what the level of emotions or feelings may be, and what resources may need to be allocated to the different conflicts competing for attention.

APPROACHES TO HANDLING DISPUTES

Disputes usually centre upon three main aspects: *rights*, *interests*, and *power*. The traditional means used to deal with disputes are *political*, *administrative*, and *judicial*. The third is the most well known and involves the courts. In litigation, the main issues of concern are *fact*, *precedent*, and *procedure*. Attention focuses upon establishing a winner or punishing an offender.

The judicial or litigation approach uses a process that has evolved over centuries. Standards for procedure and evidence are well established. Accountability is ensured through appeal mechanisms and the professional certification

---

Box 12.1 Life Cycle of Disputes

It has been suggested that disputes have a life cycle involving five stages:

1. absence
2. emergence
3. growth
4. maturity
5. agreement and implementation

With regard to a conflict in resource or environmental management in your area, does this concept of a life cycle help you understand how the conflict evolved? Does it allow you to anticipate or predict what might happen next in the dispute?

---

of lawyers. On the other hand, often the judicial process is viewed as unduly adversarial, time consuming, and expensive. Emond (1987) has also noted that an adversarial and adjudicative process encourages participants to exaggerate their private interests, conceal their bottom line, withhold information, and try to discredit their opponents. An alternative to the judicial approach is *alternative dispute* or *conflict resolution*. In contrast to the judi-

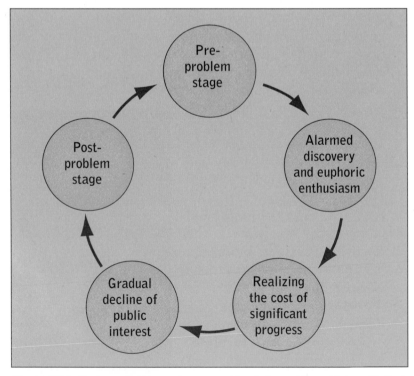

Figure 12.1 The issue-attention cycle. SOURCE: After A. Downs, 'Up and Down with Ecology—the "Issue-Attention" Cycle', *Public Interest* 28 (1972):38–50.

cial approach's emphasis on fact, precedent, and procedure, alternative dispute resolution emphasizes the *interests* and *needs* of the parties involved. While the judicial approach concludes with either a winner or a loser, or identifies a party to be punished, alternative dispute resolution focuses on reparation for harm done and improving future conduct. Another key distinction between the two is that the judicial approach emphasizes *argument*, while the alternative dispute resolution approach stresses *persuasion*.

It has never been suggested that the alternative dispute resolution approach should replace the judicial approach, but that Canadian society has relied too much on the latter, although perhaps less so than in the United States. The argument is that we should identify situations for which an alternative dispute resolution approach might be appropriate. In the following section, some of the general attributes of the alternative dispute resolution approach are considered.

ATTRIBUTES OF THE ALTERNATIVE DISPUTE RESOLUTION APPROACH

Characteristics of Alternative Dispute Resolution

Alternative dispute resolution involves *collaborative* processes through which stakeholders explore and try to resolve their differences. As noted by Sadler and Armour (1987) and summarized in Box 12.2, alternative approaches have several main features. In subsequent subsections, attention is given to the strengths associated with such an approach, the preconditions for its success, and the factors that have led to effective agreements.

Box 12.2 Main Features of Alternative Dispute Resolution

- The disputants themselves are in the best position to identify the issues and how to settle them.
- There is a voluntary commitment to joint problem solving.
- Direct fact-to-face negotiation among the stakeholders will be effective.
- There is a genuine desire to build consensus and reach a mutually acceptable agreement.

Strengths of Alternative Dispute Resolution

At least six strengths or advantages can be identified (Shaftoe 1993). These include:

An emphasis upon the issues and interests rather than the procedures: In court cases or impact assessment hearings dealing with environmental issues, too often the focus is upon procedures or process instead of on the fundamental issues of the dispute. We are all familiar with court cases in which charges are dropped or withdrawn due to a technicality, which may range from a form not being dated properly to participants not being informed of their legal right to participate in development of a plan. In alternative dispute resolution, the emphasis is on developing a mutually agreed-upon process that recognizes and satisfies the interests of all participants.

An outcome that usually results in a greater commitment to the agreement: Involving the stakeholders in the issue from the outset ensures that they end up satisfied that they had a genuine role in creating the agreement or decision. The stakeholders are empowered, an aspect of sustainable development discussed in both chapters 2 and 11. If stakeholders believe they have some ownership in the agreement or decision, they are more likely to be committed to and supportive of its implementation.

The attainment of a long-lasting settlement: A court decision may leave people dissatisfied. Such a situation may lead to appeals and counterappeals, as well as to various strategies to thwart the implementation of the decision or the future activities of opponents. If alternative dispute resolution is successful in addressing the actual interests of the stakeholders, the outcome is more likely to be accepted and supported. The stakeholders would not need to prolong the conflict if they believe that their real needs and interests had been addressed.

Constructive communication and improved understanding: In the adversarial process associated with judicial and quasi-judicial approaches, communication is usually deliberately limited and information is not shared. The intent is not so much to understand and appreciate the viewpoints of the other stakeholders but to discredit their perspective, credibility, evidence, and arguments. Alternative dispute resolution seeks to achieve constructive interaction among stakeholders so they can better appreciate the interests and needs of other participants. It is not assumed that greater understanding will result in acceptance or approval of those other interests and needs. However, it is hoped that such understanding will lead to an acceptance of their legitimacy and a willingness to develop creative solutions that can incorporate the basic interests of the competing stakeholders.

Effective use of information and experts: Since environmental disputes often arise from disagreements over facts or their interpretation, different stakeholders usually produce technical information and arguments to support their positions. Furthermore, the various stakeholder groups usually find experts who will argue that their position is the 'correct' one. In a judicial situation, a judge or arbitrator has to decide whose evidence and arguments are the most credible. In this situation, the opponents have no incentive to share information or the views of the experts. In alternative dispute resolution, the stakeholders are encouraged to work together to assemble information and to accept only the evidence they can all agree on. The outcome is that everyone works from the same set of data, and less time is wasted by arguing over assumptions and methods related to the data.

Increased flexibility: In the judicial approach, the process and standards have evolved over centuries. Such processes and standards are designed to ensure equity and accountability. At the same time, they are also designed to deal with a vast range of situations and therefore the 'fit' may not be good in a specific situation. In contrast, in alternative dispute resolution, participants develop procedures and ground rules to reflect the circumstances of their situation and what they believe is essential to lead to an agreement acceptable to all.

These six strengths of the alternative dispute resolution approach also highlight some of the limitations of the judicial or court-based approach. However, this does not mean that judicial approaches are inappropriate for environmental issues. As with most things, no approach is perfect, and alternative dispute resolution is no exception. Thus while in some circumstances an alternative dispute resolution approach may be more effective than a court-based one, the key is to recognize the strengths and weaknesses of each and to determine when one or the other will be the most effective. In the next section, some of the preconditions for successful application of alternative dispute resolution are considered. These preconditions become weaknesses when they are not satisfied but an alternative dispute resolution is used nevertheless.

Preconditions for Effective Alternative Dispute Resolution

Jeffery (1987) and Shaftoe (1993) have identified the preconditions that must be met if the alternative dispute resolution approach is to be effective. They are the following:

Recognition of a dispute: Parties must recognize that there is disagreement about one or more issues. They must acknowledge that they are involved in a dispute. Without such acknowledgement, use of an alternative dispute resolution mechanism is not likely to be effective.

Motivation or incentive to search for a solution: Participants must be willing to find a solution. If there is no such motivation, a negotiated solution is unlikely.

Definition of issues: The issues under dispute must be sufficiently well defined so that the parties are informed about them and prepared to begin negotiating. If the issues are poorly defined or ambiguous, the likelihood of a successfully negotiated agreement is low.

Identification of interests: It is essential that all parties relevant to the dispute are identified and included in discussions. This includes the stakeholders: those with responsibility or authority regarding the issue, those who will be affected by the decision (which can be a major challenge if future generations will be most affected), and those who oppose implementation of any agreement.

Representation of interests: While it is not usually possible to have every interest represented at the negotiating table, it is essential that the range of interests involved in the dispute be represented. It is also essential that the representatives of various interests be able to communicate with their group members regarding progress on both substantive and procedural issues.

Scope for compromise: Negotiation usually requires give and take or compromise by all stakeholders. When disputes are centred upon basic values or principles, compromise may be unacceptable to participants and negotiation will likely fail. In such a situation, the court-based approach may be the only way to find a solution.

Capability for implementation: There must be the capability to implement any agreement. If all stakeholders are involved and accept the agreement, the likelihood of implementation should be high.

These preconditions highlight the fact that the alternative dispute resolution approach is not perfect. There will be situations in which the preconditions cannot be satisfied, and therefore the prospects for a successfully negotiated settlement are low. In such situations, it may be necessary for a decision to be reached through one of the other political, administrative, or judicial processes.

Factors Contributing to Effective Negotiated Agreements

According to Shaftoe (1993), alternative dispute resolution is most effective when the following factors are present.

Equal opportunity: All participants become involved on an equal basis. Imbalances in power or resources (technical expertise, information, funding) need to be recognized and

addressed. This factor has led to limited recognition of the concept of *intervenor funding* or providing money to groups so that they can acquire expert guidance and information (Box 12.3). However, some believe that this recognition has emerged only in the face of ongoing resistance from government agencies and private firms, and is still highly variable in its application.

Participant design: The participants take a lead role in designing the process used in the negotiations. This factor ensures that the participants have a central role in establishing the agenda, influencing the ground rules (such as how agreement is to be reached), and arranging logistical aspects. Having the participants involved in the design ensures that they have some ownership and control over what will happen.

Common interpretation of consensus: This is one of the key elements of a negotiated approach, and must be determined before discussions about substantive and procedural issues begin. There is no single most appropriate definition for consensus, especially in crosscultural situations, such as when First Nations and non-Native

---

Box 12.3 Intervenor Funding

In the mid-1970s, Justice Thomas Berger arranged for funding so that members of the public could participate in the inquiry regarding a Mackenzie Valley oil pipeline. The money was provided by the private sector and the government.

Many hailed the Berger inquiry as an innovative benchmark for its efforts to enable the public to participate in the inquiry. Others viewed that inquiry as unduly expensive and time consuming.

Since the Berger inquiry, intervenor funding has usually been made available for major projects.

What is your view about intervenor funding? If members of the public receive such support, to whom should they be accountable?

Natives build a blockade to protest logging (*Earthroots*).

Consensus: *Agreement in opinion; the opinion of all or most of the people consulted; general agreement.*

World Book Dictionary, *1975*

groups interact. It can range from 100 per cent support, to lack of dissension (with silence meaning acceptance), or to agreement by a strong majority (for example, 70 per cent). If a situation arises in which consensus on a package cannot be reached, provision may be made to identify those aspects for which agreement is possible, and to set aside those aspects for which disagreement persists. Alternatively, the parties may agree to have a third party become an arbitrator for the outstanding issues.

Principled Approach: A distinction is often made between principled and positional approaches, based on the ideas of Fisher and Ury (1981). In a positional approach, participants identify and stick to a preferred solution, and attempt to coerce their opponents into accepting their choice. When negotiations are concluded, each participant has compromised part of his or her preferred solution, and the final agreement is less than satisfactory to all.

In contrast, the principled approach emphasizes interests and needs rather than pos-

itions, and separates the people from the problem. The focus is on developing an agreement that meets the interests of the opponents. A basic belief is that interests are often similar, although positions may be quite different.

The difference between interests and positions is important, and an example may help to clarify the distinction. Dorcey (1986) illustrated this difference in discussing alternative ways of viewing a water-quality management issue. In his example, a regulating agency takes a position that a company would not be allowed to discharge effluent from an expanded manufacturing plant unless it constructed extra treatment capacity for waste water. The company, on the other hand, argues that the small amount of extra waste water created by the modest expansion does not justify the expense of adding to the treatment capacity.

Given these conflicting positions, a third party is invited to explore how the problem might be resolved. The third party discovers that the regulating agency has become increasingly worried that the discharge from the plant's current production is already degrading the quality of the receiving waters beyond permissible standards. The agency does not have hard data to confirm this belief, but its officials have come to this conclusion based on their experience with another manufacturing plant nearby. In contrast, company officials are convinced, based on limited sampling from their waste water, that the discharge from the plant is well within standards. Indeed, the company is pleasantly surprised at how good the conditions are, and is pleased because it genuinely wants to maintain good relations with nearby communities and people in a commercial fishing industry.

Having identified some of the reasons for the two conflicting *positions*, the third party investigator discovers that in fact the *interests* of both parties are remarkably similar: they both want to maintain good water-quality conditions. Their disagreement arises from different conclusions drawn from the information available to them. Based on this conclusion, the investigator suggests some studies to determine more conclusively what the present water-quality conditions are, and then to determine whether further waste treatment is needed if the production capacity is expanded.

It might turn out that the existing water conditions were good due to unusually vigorous tidal flushing in the receiving waters, indicating that a modest additional amount of waste water was unlikely to be a problem. On the other hand, if the studies showed that the company's preliminary data about the condition of the waste water discharge underestimated the pollution, then the company could examine alternative ways to improve the quality of its discharge both for the current and for the desired increase in production capacity. In this manner, the interests of both parties would be satisfied. Table 12.1 summarizes the characteristics of the positional and principled approaches.

Continued incentive to participate: Stakeholders must be motivated to seek a mutually acceptable solution for the alternative dispute resolution approach to work (Box 12.4). In addition, the motivation or commitment must continue for the duration of the dispute. This normally happens when the participants believe that a negotiated solution provides a significantly better option than any other alternative. If negotiations reach a stage where one or more participants conclude that another approach will give greater gains, then they are likely to withdraw from a negotiated approach.

If one or more of the factors—equal opportunity, design by participants, common interpretation of consensus, principled discussions, or continued incentive for participation—are absent, the probability of alternative dispute resolution leading to a satisfactory conclusion is slim.

Table 12.1 Characteristics of Positional and Principled Negotiating

Positional Approaches		Principled Approach
Soft	**Hard**	
Participants are friends.	Participants are adversaries.	Participants are problem solvers.
The goal is agreement.	The goal is victory.	The goal is a wise outcome reached efficiently and amicably.
Make concessions to cultivate the relationship.	Demand concessions as a condition of the relationship.	Separate the people from the problem.
Be soft on the people and the problem.	Be hard on the problem and the people.	Be soft on the people, hard on the problem.
Trust others.	Distrust others.	Proceed independently of trust.
Change your position easily.	Dig into your position.	Focus on interests, not positions.
Make offers.	Make threats.	Explore interests.
Disclose your bottom line.	Mislead as to your bottom line.	Avoid having a bottom line.
Accept one-sided losses to reach agreement.	Demand one-sided gains as the price of agreement.	Invent options for mutual gain.
Search for the single answer: the one *they* will accept.	Search for the single answer: the one *you* will accept.	Develop multiple options to choose from; decide later.
Insist on agreement.	Insist on your position.	Insist on objective criteria.
Try to avoid a contest of will.	Try to win a contest of will.	Try to reach a result based on standards independent of will.
Yield to pressure.	Apply pressure.	Reason and be open to reason; yield to principle, not pressure.

Source: After R. Fisher and W. Ury, *Getting to Yes: Negotiating Agreement without Giving In* (Harmondsworth: Penguin, 1981):13.

Box 12.4 Negotiating Involves People

Fisher and Ury remind us that negotiating requires sensitivity to the other people at the table. In their words:

- . . . you are dealing not with abstract representatives of the 'other side', but with human beings. They have emotions, deeply held values, and different backgrounds and viewpoints; and they are unpredictable.
- Ultimately, however, conflict lies not in objective reality, but in people's heads. . . . Facts, even if established, may do nothing to solve the problem.
- As useful as looking for objective reality can be, it is ultimately the reality as each side sees it that constitutes the problem in a negotiation and opens the way to a solution.
- The ability to see the situation as the other side sees it, as difficult as it may be, is one of the most important skills a negotiator can possess. It is not enough to know that they see things differently. If you want to influence them, you also need to understand empathetically the power of their point of view and to feel the emotional force with which they believe in it.
- Understanding their point of view is not the same as agreeing with it (Fisher and Ury 1981:19–25).

TYPES OF ALTERNATIVE DISPUTE RESOLUTION

Earlier in this chapter, the general differences between traditional dispute resolution approaches (political, administrative, judicial) and alternative approaches were outlined. The features of specific types of alternative dispute resolution approaches are presented in this section. These include public consultation, negotiation, mediation, and arbitration.

Public Consultation

Various aspects of public consultation were reviewed in Chapter 11 in the context of partnerships and stakeholders. Public participation or citizen involvement has been used since the late 1960s for resource and environmental management in Canada. Such consultation initially focused on having the public help in identifying key issues and reviewing possible solutions. However, this public input was simply one of many inputs considered by the managers, who determined which trade-offs were appropriate and then made the final decisions. The public had no real power or authority in the management process.

By the mid-1980s this approach was modified as public participation moved towards the concepts of partnership and delegated power. Some of the comanagement initiatives outlined in Chapter 11 illustrated the shift to give real power to the public. However, public consultation is not considered one of the emerging types of alternative dispute resolution, as in the latter *all* decisions concerning the dispute are the exclusive domain of stakeholders.

Negotiation

Negotiation is one of the two main types of alternative dispute resolution. Its distinctive characteristics include two or more parties involved in a dispute joining in a voluntary, joint exploration of issues with the goal of reaching a mutually acceptable agreement. Because participation is *voluntary*, participants can withdraw at any time. Through *joint exploration*, the parties strive to identify and define issues of mutual concern and develop solutions that are *mutually acceptable*. The normal procedure is to reach agreements by *consensus*.

Mediation

Mediation is the second main type of alternative dispute resolution technique. The distinguishing feature is the inclusion of a *neutral third party* (called a mediator) whose task is to help the disputants overcome their differences and reach a settlement. The third party has no power to impose any outcome. The responsibility to accept or reject any solution remains exclusively with the stakeholders in the dispute. In addition, the third party has to be acceptable to all of the parties in the conflict.

Mediators have a variety of roles. They assist the parties in the dispute to come together. In this role, mediators act as *facilitators*. They can also help the parties with *fact finding*. The mediators do not necessarily have the expertise to provide the needed information, but they can help identify what information is needed and then assist in seeking out the necessary data. In the actual mediation role, often the third person first meets separately with the parties in the dispute, and subsequently holds joint meetings of all parties. Mediators try to assist each stakeholder in understanding the interests and objectives of the other stakeholders, help to find points in common, and assist the stakeholders in settling differences through negotiation and compromise. Key roles for the mediator are to maintain momentum in the negotiations, keep the parties communicating with each other, and ensure that proposals are realistic.

Arbitration

Arbitration differs significantly from negotiation and mediation because the stakeholders normally accept a third party responsible for making a decision regarding the issue or issues in conflict. In mediation, the third party has no power to impose a settlement. In arbitration, normally the arbitrator's decision is *binding* upon the parties. However, there are instances of non-binding arbitration in which the arbitrator makes a decision regarding the conflict, but the stakeholders may accept or reject it. In some ways, the judicial or court-based approach has many of the elements of binding arbitration since the judge reaches a decision, which is then imposed upon those involved in the dispute. The main difference is that in arbitration the stakeholders usually have a voice in selecting the arbitrator. In the judicial procedure, the participants do not have a role in deciding which judge will hear their case.

As already noted, public consultation or public participation has been used in Canada for environmental management for almost three decades, so there is considerable experience with it. Judicial or court-based approaches have also been used ever since the country was settled by people from Europe. The newly emerging approaches are negotiation and mediation, although aspects of negotiation have certainly been included as part of public consultation and judicial approaches. In Part D

some of the case-studies provide further details regarding how negotiation and mediation are used for addressing environmentally based conflicts.

CASE-STUDY: GRASSY NARROWS AND WHITE DOG OJIBWAY BANDS

After nine years of controversy and negotiations, in June 1986 a benchmark settlement was approved by the Supreme Court of Canada. The settlement was between Reed Paper Company of Dryden and the Ojibway people from the Grassy Narrows and White Dog reserves in northwestern Ontario regarding mercury pollution in the English-Wabigoon River system.

Inorganic mercury had been discharged into the English-Wabigoon River from the Dryden Paper Company Ltd and Dryden Chemicals Ltd, both subsidiaries of Reed Paper Ltd of England. Scientists estimated that from the early 1960s until 1970, some 9–13.5 kg of mercury had been released daily into the river system. Their judgement was that it would take the river system up to seventy years to clean itself by natural processes.

Inorganic mercury deposited into rivers and lakes is a serious pollutant because bacteria can change it into toxic methyl mercury, which then enters the food chain and bioaccumulates in fish (see also Chapter 14 regarding bioaccumulation). By 1970, the Fisheries Research Board of Canada had documented mercury levels in English-Wabigoon River fish comparable to those found in fish from Minamata Bay in Japan, the site of a well-known methyl-mercury poisoning in 1956.

This pollution was catastrophic for the 1,200 Ojibway people on two reserves (Grassy Narrows and White Dog) located on the lower part of the English River because fish from the river was the main source of protein and a key part of their diet. In addition, they worked as fishing guides, and once the mercury problem became known, their income from guiding was lost.

Blood tests became controversial. They indicated high levels of mercury in the bodies of the Ojibway. Medical experts from Japan who examined the Ojibway people concluded that they were victims of mercury poisoning. However, Canadian medical authorities were unwilling to diagnose mercury poisoning, partly because the symptoms were similar to those of other diseases (especially

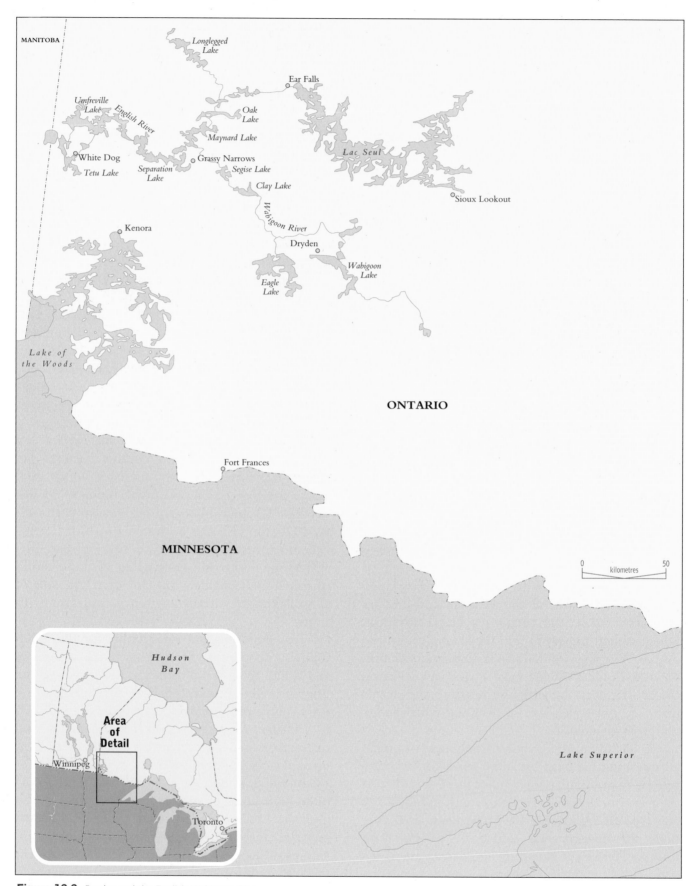

Figure 12.2 Dryden and the English-Wabigoon River.

alcoholism and other non-environmental illnesses), and because toxic blood levels from chronic exposure to mercury had not been officially established. While the scientific and medical evidence was thus open to debate, there was little doubt regarding the major negative impact of the alleged poisoning of the people on the two reserves.

The Reed Paper Company was charged in 1976 by the Ontario government under its Environmental Protection Act. Reed was convicted and fined $5,000. In 1976 the Ojibway began civil action against Reed. For nine years after the lawsuit was filed against Reed, many obstacles were encountered in trying to bring the case to trial. Lack of a practical procedure for a class-action suit required identification of all injured people, which necessitated a long and costly process. However, in 1985 the federal government appointed Mr Justice Emmett Hall, a respected former Supreme Court of Canada judge, to examine the situation. Justice Hall concluded that any court case would be extremely complicated, time consuming, expensive, and have an uncertain outcome. Determining liability and damages would be difficult, even though the Reed Paper Company agreed that compensation was necessary. As a result, he recommended that the Ojibway people seek a negotiated, out-of-court settlement. The courts were simply not a good forum to address an environmental issue in which large numbers of people had been affected.

Under the settlement, the two Indian bands would receive just under $16.7 million. Both the federal and provincial governments would pass related legislation. Once the bands ratified the settlement, they would forfeit rights to further claims.

The purpose of the legislation was to create the Mercury Disability Fund, intended to compensate both present and future band members who experienced ill health that could be traced to mercury poisoning. The money for the fund would come from the two bands, each of which would allocate $1 million from their compensation package. If the fund were inadequate to compensate all of the victims, the Ontario government would contribute. If any money were left after all injured persons had been compensated, the bands could reallocate the fund to other needs in their communities.

What were some of the factors that allowed a negotiated settlement to be reached? West (1987) suggested the following:

Choice of mediator: A person with the stature and competence of Mr Justice Hall was crucial.

Political pressure: Both the federal and provincial governments exerted considerable pressure for a settlement. The governments were also prepared to pass legislation to facilitate implementation of an agreement.

Willingness to settle: The two Indian bands were frustrated and tired after nine years and wanted a settlement. Reed Paper had acknowledged the pollution and expected to pay some compensation, but did not accept liability for all of the damages associated with the mercury contamination. As a result, both parties were prepared to participate voluntarily in mediation, which was probably one of the key factors for success.

Given these three aspects specific to the English-Wabigoon situation, West cautioned that the settlement might not represent 'a model for future pollution dispute resolution' (1987:146). He noted that 'it is important to recognize that a unique set of circumstances existed and provided the impetus for this settlement' (West 1987:146). Nevertheless, he concluded that this early experience indicated that alternative dispute resolution approaches could be effective.

> *Environmental conflicts require determining issues which take into account not only individual interests, but also the interests of the wider community. Where large groups suffer environmental damage, there must be some mechanism available to allow plaintiffs to present their claims quickly and with a minimum of legal hassle. Archaic technical rules and cumbersome procedures are both inappropriate and detrimental to achieving justice in environmental dispute resolution.*
>
> *L. West, 'Mediated Settlements of Environmental Disputes', 1987*

PREPARING FOR DISPUTE RESOLUTION

In this concluding section, attention is given to some issues that must be considered in preparing for negotiation or mediation-based approaches. These considerations will then be reviewed in the later case-studies. Much of this discussion is based on a report from the British Columbia Round Table on the Environment and the Economy (1991a).

Ripeness and Consensus

Key considerations are to ensure that a dispute is mature or ripe enough for resolution, and that consensus is appropriate. Earlier in this chapter, the idea of a life cycle for conflicts was discussed. Normally, only when a conflict reaches a mature stage can negotiations be effective. For example, the English–Wabigoon situation had reached a mature stage, which helped greatly in reaching a settlement. Furthermore, although consensus may be viewed as the preferred way to resolve a dispute, not every situation will be amenable to such an approach. If the conflict is not mature, the participants may believe there is not enough urgency to resolve the matter and may prefer to satisfy their interests in other ways. In the English–Wabigoon situation, there was a sense of urgency and a willingness to seek consensus.

Ground Rules

Before addressing what a conflict is about, it is important to establish how the negotiations will be conducted. Let us use the analogy of trying to play a game of football without first establishing the rules. A Canadian team would play by Canadian rules, and Americans would play by their rules. All the players would be participating in the same game, but there would be considerable confusion since they would be using different rules.

Establishing the ground rules should be done before the parties get caught up in the substance of the discussions. Decisions need to be made about the following matters: (1) scope of the deliberations (what is included and excluded); (2) definition of consensus, as well as when and how it is achieved; (3) responsibilities of participants regarding communication with their groups or organizations; (4) confidentiality of the discussions (who talks to the media in which ways); (5) sharing of information and data; and (6) deadlines (a target date to conclude the negotiations, as well as interim deadlines before the final deadline). Attention also needs to be given to the possibility that an agreement or consensus will not be reached. What would be a fall-back or contingency plan if an agreement cannot be reached?

Costs

The costs of participation in negotiations are usually not the same for all parties. Individuals employed by governments or corporations may participate as part of their work-related responsibilities. As a result, during the negotiations they continue to receive their salaries and have their expenses covered. In contrast, voluntary members of non-governmental agencies or private firms may only be able to attend by taking holiday time from their regular jobs. It is also possible that they will have to cover their own expenses for accommodation and travel.

It is important that the differential nature of the costs for participation be recognized before the process begins. If possible, efforts should be made to equalize the opportunities for participation. *Intervenor funding* has been created for this reason to assist individuals or groups who might otherwise be unable to participate.

The issues relating to ripeness, consensus, ground rules, and costs are best addressed before the parties in a dispute begin to discuss the other differences in a conflict. In some ways, attention to these matters is a bit like ensuring that the foundation for a house is solid before starting to erect the walls or roof. If the foundations are not well established and solid, the negotiations may break down due to matters that could have been addressed and resolved before they became major obstacles.

IMPLICATIONS

Several important implications deserve attention. First, resource and environmental management can often be characterized as management of conflict or disputes due to different values and interests. As a result, scientists, planners, and managers need to be aware that much of resource and environmental management is actually management of conflicts. Furthermore, conflict is not always undesirable or negative if it forces us to re-examine objectives, practices, and procedures over which there is disagreement.

Second, the emergence of *alternative dispute resolution* during the 1980s has provided a significant alternative to conventional judicial and quasi-judicial approaches to dealing with disputes. However, while the various methods of alternative dispute resolution—*negotiation*, *mediation*, and *arbitration*—will not always be preferable to judicial or quasi-judicial approaches, they do provide options that should be considered before automatically turning to the courts to address conflicts.

Third, alternative dispute resolution emphasizes the importance of *stakeholders* and *impacts*. In other words, alternative dispute resolution requires those affected by actions to work together to reach mutually acceptable decisions. ADR builds upon a belief that stakeholders are in a legitimate position to identify issues, and that many benefits can be achieved through a voluntary and joint search for mutually agreed-upon solutions. As a result, the ideas of alternative dispute resolution reinforce many of the concepts discussed in Chapter 11 regarding partnerships and stakeholders. In addition, alternative dispute resolution requires attention to the magnitude, extent, duration, and incidence of impacts of decisions, aspects of which are considered in Chapter 10.

Fourth, the initial costs in alternative dispute resolution can be significant, but there is a growing belief that this investment is more than returned because it eliminates the cost and time associated with delays, appeals, or deliberate obstructionism following decisions. Experience is still being developed with alternative dispute resolution approaches in Canada, and case-studies in Part D allow us to review how effective they have been.

SUMMARY

1. Conflict is a natural part of life and environmental management, and may emerge because of clashing or incompatible values, interests, needs, or actions.

2. Conflict has a life cycle of five stages: (1) absence, (2) emergence, (3) growth, (4) maturity, and (5) agreement and implementation.

3. Conflict is usually based on disagreements about rights, interests, and power.

4. Traditional means to resolve conflicts are political, administrative, and judicial.

5. The judicial approach uses the courts, emphasizes fact, precedent, and procedure; it is adversarial, and can be time consuming and expensive. However, it also uses well-established processes and standards of evidence, has accountability built in through provisions for appeals, and its decisions are enforceable.

6. Alternative dispute resolution (ADR) has been developed as an alternative (not a replacement for) to the judicial approach. It emphasizes joint conflict resolution, interests and needs, persuasion over argument, and decisions based on consensus.

7. The strengths of ADR include: (1) emphasis on issues and interests over procedures, (2) commitment to the agreed outcome, (3) attainment of a long-lasting settlement, (4) constructive communication and improved understanding, (5) effective use of information and experts, and (6) increased flexibility.

8. Preconditions for ADR include: (1) recognition of a dispute by all parties, (2) motivation to search for a joint solution, (3) definition of issues, (4) identification and representation of key interests, (5) scope for and willingness to compromise, and (6) capability to implement joint decisions.

9. Factors that help achieve effective negotiated agreements are: (1) equal opportunity for all participants, (2) participants are involved in designing the negotiation process, (3) agreement regarding what constitutes a consensus decision, and (4) use of a principled approach.

10. A principled approach focuses on the problem rather than the people, on interests rather than positions, identifies objective criteria against which to judge alternative solutions, and searches for solutions that benefit every party.

11. ADR can involve public consultation, negotiation, mediation, or arbitration.

12. The settlement between the Reed Paper Company of Dryden, Ontario, and the Ojibway people

of the Grassy Narrows and White Dog reserves regarding mercury contamination in the English-Wabigoon River system represents a benchmark settlement using ADR. Under the settlement, the Ojibway received just under $16.7 million.

13. Key factors leading to the settlement included the choice of an impartial and highly respected mediator, political pressure to reach a settlement, and mutual willingness to search for an agreement.

14. In preparing to use ADR, it is important to determine that the dispute is mature or ripe enough for resolution, a consensus approach is appropriate, ground rules are well established before substantive matters are addressed, and arrangements to cover costs are made.

REVIEW QUESTIONS

1. Why are conflicts almost always present in environmental management?

2. What are the positive and negative aspects of conflict or disputes?

3. What are the five stages of a life cycle for a conflict?

4. What are the characteristics of traditional dispute resolution mechanisms?

5. What are the advantages and disadvantages of an adversarial process for resolving disputes?

6. What are the characteristics of alternative dispute resolution methods?

7. What are the advantages of alternative dispute resolution?

8. What is the difference between positions and interests in disputes?

9. What are the seven preconditions for alternative dispute resolution?

10. What are the factors that have contributed to effective negotiated agreement?

11. Why is intervenor funding important for alternative dispute resolution?

12. What are the various ways to define consensus?

13. What are the characteristics of positional and principled negotiation?

14. What are the differences among public consultation, negotiation, mediation, and arbitration?

15. What aspects of the English-Wabigoon mercury pollution negotiations satisfy the preconditions for alternative dispute resolution?

REFERENCES AND SUGGESTED READING

Berger, T.R. 1977. *Northern Frontier, Northern Homeland: The Report of the Mackenzie Valley Pipeline Inquiry*, 2 vols. Ottawa: Minister of Supply and Services.

British Columbia Round Table on the Environment and the Economy (Dispute Resolution Core Group). 1991a. *Reaching Agreement, Vol. 1: Consensus Processes in British Columbia*. Victoria: Crown Publications.

_____. 1991b. *Reaching Agreement, Vol. 2: Implementing Consensus Processes in British Columbia*. Victoria: Crown Publications.

Dorcey, A.H.J. 1986. *Bargaining in the Governance of Pacific Coastal Resources: Research and Reform*. Vancouver: Westwater Research Centre, University of BC.

Downs, A. 1972. 'Up and Down with Ecology—the "Issue-Attention" Cycle'. *Public Interest* 28:38–50.

Emond, D.P. 1987. 'Accommodating Negotiation/Mediation within Existing Assessment and Approval Processes'. In *The Place of Negotiation in Environmental Assessment*, 45–52. Prepared for the Canadian Environmental Assessment Research Council. Ottawa: Minister of Supply and Services Canada.

Fisher, R., and W. Ury. 1981. *Getting to Yes: Negotiating Agreement without Giving In*. Harmondsworth: Penguin.

Jeffery, M.I. 1987. 'Commentary II'. In *The Place of Negotiation in Environmental Assessment*, 53–8. Prepared for the Canadian Environmental Research Council. Ottawa: Minister of Supply and Services Canada.

Johnson, P.J., and P.N. Duinker. 1993. *Beyond Dispute: Collaborative Approaches to Resolving Natural Resource and Environ-*

mental Conflicts. Thunder Bay, ON: School of Forestry, Lakehead University.

Ralston Baxter, R. 1992. *Conflict Resolution and Sustainable Development: An Alternative Dispute Resolution Policy for Ontario*. Toronto: A Study for the Ontario Round Table on Environment and Economy.

Rankin, M. 1989. 'The Wilderness Advisory Committee of British Columbia: New Directions in Environmental Dispute Resolution'. *Environment and Planning Law Journal* (March):5–17.

Sadler, B. 1993. 'Mediation Provisions and Options in Canadian Environmental Assessment'. *Environmental Impact Assessment and Review* 13, no. 6:375–90.

_____, and A. Armour. 1987. 'Common Ground: On the Relationship of Environmental Assessment and Negotiation'. In *The Place of Negotiation in Environmental*

Assessment, 1–6. Ottawa: Minister of Supply and Services Canada.

Shaftoe, D., ed. 1993. *Responding to Changing Times: Environmental Mediation in Canada*. Waterloo, ON: The Network: Interaction of Conflict Resolution, Conrad Grebel College.

Sherman, L., and J. Livey. 1992. 'The Positive Power of Conflict'. *Plan Canada* (March):12–16.

Shrybman, S. 1989. *Environmental Mediation*, 4 vols. Toronto: Canadian Institute for Environmental Law and Policy.

Szechenyi, J. 1992. 'Managing Environmental Issues and Conflict'. *Plan Canada* (March):17–19.

West, L. 1987. 'Mediated Settlements of Environmental Disputes: Grassy Narrows and White Dog Revisited'. *Environmental Law* 18:131–50.

Resource and Environmental Management in Canada

In parts B and C, attention was focused upon the science and management of resources and the environment. Part D illustrates how ideas and methods from science and management can be applied in problem solving.

Several comments about the structure of Part D should be helpful. It may appear contradictory to argue for an ecosystem approach in parts B and C, and then organize Part D on the basis of urban, agricultural, forestry, or wildlife issues. Indeed, it would be inappropriate to isolate components of resource or environmental systems and examine them on a sector by sector basis, as the chapter headings in Part D might suggest. However, we think it is appropriate to use one resource or environmental aspect as the starting-point for a discussion, as long as attention is given to other elements of the ecosystem and their linkages. This is the approach used in this section of the book.

For example, we thought it was important to have a chapter on issues related to urban environmental management since Canada is becoming an increasingly urbanized society, and insufficient attention is given to environmental issues within urban areas. Thus in Chapter 13 the focus is on three case-studies of different communities (Salmon Arm, Sudbury, and Saskatoon) in which people have tried to incorporate environmental understanding, concepts, and methods in planning, design, and remediation. In Sudbury, scientific research was needed to determine which nutrients had to be provided to the soil, and which types of trees, shrubs, and grasses could successfully grow on the reclaimed landscape in order to rehabilitate an area degraded by air pollution and surface disturbance related to mining. In addition to these scientific considerations, people involved in the Sudbury experience learned how to interrelate environmental, economic, and social considerations, and how to form partnerships among various interests in the community. As a result, the Sudbury case-study provides information and insights about many aspects of the ecosphere and planning and management in an ecosystem context.

◀ Saskatoon and two of the seven bridges that span the South Saskatchewan River (*Tourism Saskatchewan*).

Indeed, in many ways it exemplified a model for achieving sustainable development at a community level.

There is a similar example in Chapter 15, which focuses on forestry. In that chapter, Carnation Creek in British Columbia is a case-study. Scientific research was initiated there to determine the impacts of logging on the ecosystem, especially on soils and the terrestrial and aquatic wildlife. Particular attention was given to the impact of timber harvesting on the habitat of fish spawning areas. Thus while the Carnation Creek example appears in a chapter oriented to forestry, this experience considers issues related to soil erosion, habitat, and fish propagation, as well as the need for economic activity to support resource-based communities in British Columbia.

The use of case-studies also highlights the considerable complexity and uncertainty that is often associated with science and management. For example, in Chapter 18 the case-study of groundwater contamination in Elmira, Ontario, illustrates the difficulty of monitoring environmental contaminants and the differences in scientific opinion regarding safe thresholds of contaminants.

The James Bay case-study in Chapter 19 further reveals our inability to predict accurately the environmental and social impacts of resource development projects, often because of the complexity and uncertainty involved. Furthermore, the James Bay experience demonstrates that science can rarely be conducted in an objective manner. Indeed, one commentator suggested that one of the main lessons from the impact assessment work on energy development in James Bay was not that the predictions were incorrect, but that most of the impacts considered were simply irrelevant because they were not the impacts of greatest significance to the local people who would have to live with the consequences of the development.

The case-studies also demonstrate that management of an ecosystem can be (and often is) affected by decisions made by people living outside of that ecosystem. The Cree people living in northern Quebec found that their homeland was to be affected by hydroelectric development based on decisions made by people living in southern Quebec. In turn, the decision makers in southern Quebec found that their development plans changed from feasible to not feasible after other people living in New York and the New England states decided not to purchase the electricity from the proposed James Bay phase II development. In Chapter 20 the Canadian fisheries managers responsible for the Atlantic northern cod had to cope with decisions made by Spanish and Portuguese fishers to harvest fish outside Canada's territorial limit, which had ramifications for the integrity of the fish stocks on which Atlantic Canada's fishing industry depended. As a result, it becomes clear that to achieve sustainable development, it is necessary to think globally and act locally, and to think locally and act globally.

You should also keep in mind that the chapters in Part D do not always provide all the systematics of environmental science and management pertinent to the chapter. Thus, for example, Chapter 18, which includes two detailed case-studies on water quality and quantity problems, does not explicitly consider aspects related to the hydrological cycle or contaminant pathways, since these were systematically covered in chapters 6 and 7. Furthermore, the chapters in parts B and C also contain examples that provide insight about science and management, such as a discussion of research in the Experimental Lakes Area in northwestern Ontario in Chapter 7. This research has been recognized as of world-class standard and has contributed to the design of strategies to deal with acid precipitation and climate change.

The examples in Part D also raise fundamental questions about humans' relationship with the environment and resources. In these examples, we see a range of attitudes towards the environment and other living things, covering the gamut from humans dominating nature to humans striving to live in harmony with nature. The extinction of species, such as those reported in Chapter 16, reminds us that we have been (and can still be) incredibly arrogant in believing that it was acceptable for us to eliminate some species forever. On a more positive note, we see in Chapter 17 that conscious decisions are being made to protect valued areas, in some instances because we believe it is important to protect examples of different biomes. However, our current lifestyles are a major contributor to global or climate warming, as discussed in Chapter 20. The changes associated with such warming may confound attempts to identify and protect examples of various biospheres if those change in the future because of different climatic conditions.

Finally, the case-studies and examples illustrate how pervasive conflict can be in resource and environmental management. This reality reinforces our belief that resource and environmental management is not only a technical or scientific exercise. To be done effectively, resource and environmental management requires the ability to recognize, identify, and incorporate different values and interests. For this reason, managers often spend as much time attempting to mediate conflicts as they do on anything else. Given such a reality, scientists must also develop a greater appreciation that their work will usually be used in situations in which values and emotions can be as important or more so than theories, models, and quantitative evidence.

We hope that the mix of case-studies and examples in Part D and elsewhere in this book will help you appreciate the change, complexity, uncertainty, and conflict that are integral parts of resource and environmental management.

Chapter 13

Urban Environmental Management

More than three-quarters of Canadians live in urban places, and that proportion (already one of the highest in the world) can be expected to increase (Figure 13.1). (Statistics Canada defines an urban place as one with a population of least 1,000 people concentrated within a continuously built-up area, with a density of at least 400 people per square kilometre.) At the same time, the intensity of urbanization varies. For example, urbanization is close to 80 per cent in each of British Columbia,

Alberta, Ontario, and Quebec, and is about 50 per cent in the other provinces. The most urbanized area is the corridor between Windsor and Quebec City, which contains nine of the fifteen largest cities in the country, while the least urbanized is Prince Edward Island.

Given the urban nature of Canada, more attention should be given to managing urban environments and to considering the role of sustainable development in urban places. How do we encour-

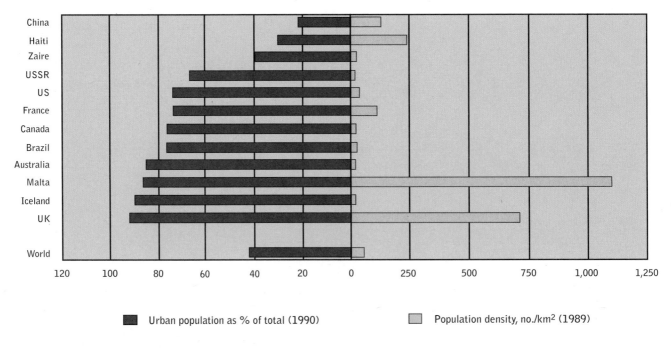

Urban population as % of total (1990)   Population density, no./km² (1989)

Figure 13.1 Canada in an urbanized world: population density and urban population for selected countries. NOTE: Out of 146 countries, Canada has the eighth lowest population density and is the twenty-eighth most urbanized. SOURCE: Environment Canada, *The State of Canada's Environment* (Ottawa: Minister of Supply and Services Canada, 1991).

age economic development for communities while protecting the integrity of their natural environment? Do present approaches to housing, transportation, energy and water use, green space, and waste management in local communities conform to the principles of sustainable development outlined in Chapter 2 (especially the idea of stewardship), or are they mostly unsustainable? To what extent are ideas from ecology being drawn upon in urban design? To what extent have the concepts of ecosystem and adaptive management, impact assessment, partnerships and stakeholders, and conflict resolution been incorporated at the urban level?

In this chapter, attention is first given to discussing the concept of sustainable development with regard to urbanization. Examples of the application of the idea are then reviewed regarding the design of urban developments, the rehabilitation and remediation of long-established industrialized areas, and the protection of environmentally sensitive areas.

SUSTAINABILITY AND CITIES

Richardson (1992) has argued that sustainable urban development must deal with the relationships among four critical aspects:

1. the natural environment and natural resources (air, water, land)
2. economic activity, especially the urban economic base
3. the 'built environment' of buildings, streets, and physical services
4. the 'human environment', or the totality of the above three aspects as a setting for human life

Regarding these four aspects, Richardson then suggested that:

. . . sustainable urban development in the context of human settlements in Canada means the continuing maintenance, adaptation, renewal, and development of a city's physical structure and systems and its economic base in such a way as to enable it to provide a satisfactory human environment with minimal demands on resources and minimal adverse effects on the natural environments (Richardson 1992:148).

If this definition is accepted, he believed that five components of urban sustainability require attention:

1. *Physical conditions*, including both the state of the urban fabric itself, as well as its vulnerability to external physical forces, such as climatic change
2. *Economic conditions*, particularly the stability of urban economic bases and their capacity to adjust to changing conditions
3. *Resource demands*, especially the extent to which cities draw upon renewable and non-renewable resources to maintain themselves
4. *Environmental effects* or the extent to which cities have adverse effects on their natural environments
5. *Urban management,* incorporating both political structures and processes and the planning of urban areas

These ideas are elaborated upon by Rees and Roseland (1991) in their discussion of the characteristics of sustainable communities in the twenty-first century. They argue that sustainable communities would have the following characteristics: efficient use of urban space, reduced consumption of material and energy resources, improved community livability, and sensitivity to ecological and socio-economic complexity.

However, they also noted that most cities in North America were designed and built on the assumption that abundant and cheap land would always be available. Consequently communities have grown quickly and inefficiently. The common urban form is one with lengthy distribution systems, spacious homes on individual lots, reliance on the automobile, and separation of workplaces from homes. The result has been urban sprawl, the use of abundant and inexpensive fossil fuel, and a perceived right to unrestricted use of the private car (Figure 13.2; Box 13.1).

They concluded that the *pattern* of growth has been more important than the *amount* of growth in determining the level and efficiency of resource use and congestion in an urban area. Thus achievement of sustainable communities will require much more attention to design and layout considerations, some significant changes in Canadians' lifestyles, as well as significant changes in local

Box 13.1 Urban Air Quality in Canada

Canadians are concerned about the quality of urban air. The indicator shown below is based on air quality objectives that have been set for the five common air pollutants. It is intended to allow Canadians to track trends in urban air quality.

Indicator: Number of Times Air Quality Objectives Exceeded

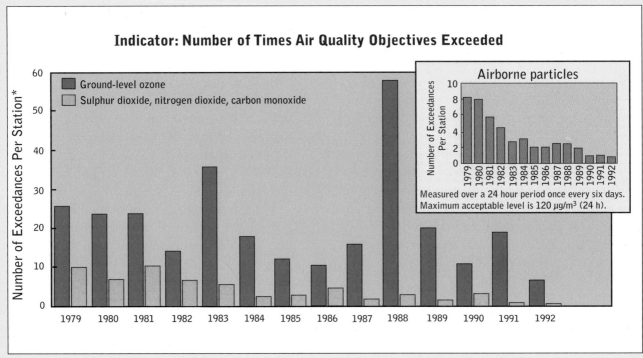

Refers to the average number of exceedances of the maximum acceptable level at stations across Canada. Measurements are generally taken hourly throughout the year. Maximum acceptable levels are: 82 ppb for ozone (1 h), 344 ppb SO_2 (1 h), 213 ppb for NO_2 (1 h), and 13 ppm for CO (8 h). 1992 data are preliminary. SOURCE: River Road Environmental Technology Centre, Environment Canada.

- Ground-level ozone (a major component of smog) and airborne particles are thought to be responsible for most of the serious air pollution in Canadian cities.
- There is no apparent trend for ground-level ozone since weather strongly influences its formation and movement. The hot, stagnant weather conditions experienced in central and eastern Canada in the summers of 1983 and 1988 favoured ozone formation.
- Ground-level ozone is primarily a problem in the Lower Fraser Valley of British Columbia, the Windsor to Quebec City corridor, and the southern Atlantic region (southern New Brunswick and western Nova Scotia).
- Levels of airborne particles have decreased significantly since 1979. However, they are still a major concern.
- The number of times that sulphur dioxide, nitrogen dioxide, and carbon monoxide exceeded their maximum acceptable levels fell from an average of 10 per station in 1979 to well below 1 per station in 1992. Most of

this decline has come from an improvement in the control of carbon monoxide emissions from vehicles.
- For most of the time pollutant levels are below the maximum acceptable levels. In 1992 airborne particles were below the objective for 98.6 per cent of the time and ozone for 99.9 per cent of the time.

HEALTH CONCERNS

Ground-level ozone: Measurable effects on breathing have been observed in individuals exposed to ozone at levels as low as the maximum acceptable level of 82 ppb.

Airborne particles: Many particles are small enough to be drawn deep into the respiratory tract, where they may decrease the ability of the lung to function properly.

The long-term impacts of air pollution and the combined effects of mixtures of pollutants are not well understood (Environment Canada 1994:1).

Characteristics of Unsustainable Urbanization in Canada

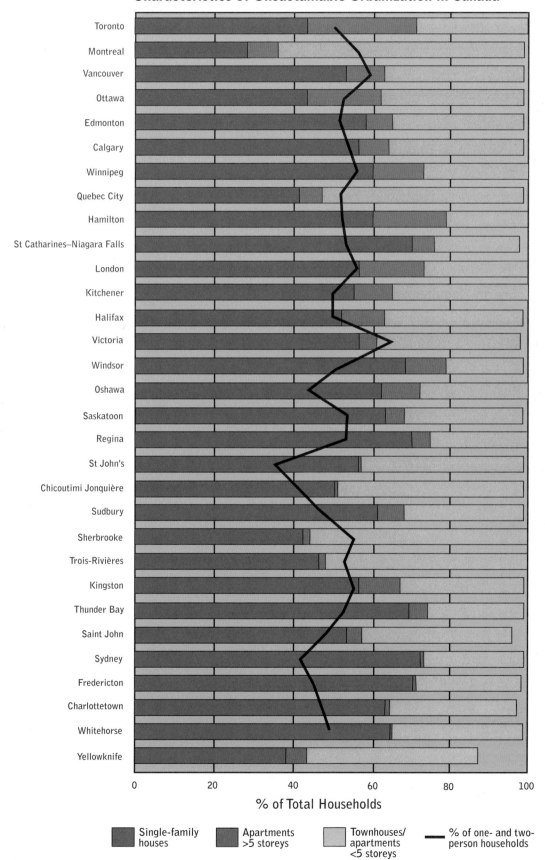

% of Total Households

Legend: Single-family houses | Apartments >5 storeys | Townhouses/ apartments <5 storeys | % of one- and two-person households

Figure 13.2 The dominance of the single-family house: percentage of dwelling type and one- and two-person households, 1986. SOURCE: Environment Canada, *The State of Canada's Environment* (Ottawa: Minister of Supply and Services Canada, 1991). Canadian Urban Transit Association; Statistics Canada; Ontario Ministry of Transportation and Communication.

Car Registrations and Transit Use

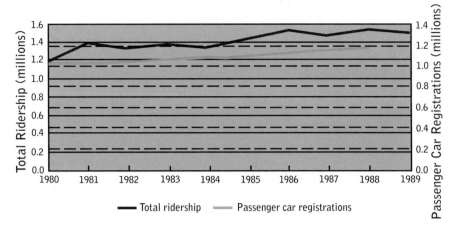

— Total ridership   — Passenger car registrations

Energy Requirements by Transit Mode

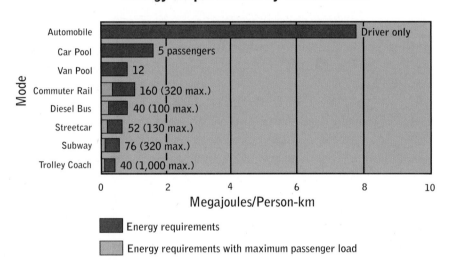

Energy requirements

Energy requirements with maximum passenger load

Figure 13.2 (continued)

decision-making processes. The following examples illustrate some of the ways in which urban decision makers are attempting to incorporate ideas related to sustainable development.

BALANCING ECONOMIC DEVELOPMENT AND ENVIRONMENTAL PROTECTION DURING NEW DEVELOPMENT

The community of Salmon Arm has addressed the issue of balancing economic and environmental concerns with regard to development along its shoreline. The Salmon Arm experience reflects many aspects that must be addressed in making the transition to sustainable development and is there-

fore a good example of both opportunities and problems (Callaway and Kipp 1991).

Salmon Arm is a community of about 12,000 people located on a bay of Shushwap Lake in the southern interior of British Columbia (Figure 13.3). As with many Canadian cities adjacent to lakes or rivers, the relationship of the people to the nearby water has been ambivalent. Initial development of lakeside land was for a railway, which effectively separated the lakeshore from the community. The waterfront area was also perceived as appropriate for industrial use, such as fuel-oil tanks and industrial buildings, and for the town's sewage treatment plant. There was steady silt deposition in the bay from the nearby Salmon River, fluctuating lake levels, and, in the 1980s, the appearance of Eurasian milfoil, an imported aquatic weed that has proven difficult to eradicate (Dearden 1983, 1984).

At the same time, however, the waterfront had other roles. The foreshore area, adjacent to what were called the mud-flats, provided a setting for recreational activities, such as picnicking and lawn bowling in summer or ice skating in winter. Before a network of roads existed in the area, the Salmon Arm wharf served as a distribution point, via boat, for goods and services to other communities. In addition, an active naturalists' group has existed for decades, reflecting community interest in the mud-flats or wetlands that provided a habitat for waterfowl and foreshore birds. Over time, other residents began to value the lakeshore habitat, especially since it was one of only two areas in the province in which the rare western grebe was known to nest.

In 1987 the District of Salmon Arm received a proposal for a residential-tourism-commercial development on 30 ha of lakeside land in the community. Of those 30 ha, 9.7 ha were sold by the developer to the Nature Trust of British Columbia, and another 6 ha were donated by the developer to protect foreshore areas. An adjacent 9.17 ha were purchased jointly by the District of Salmon Arm and the Nature Trust, allowing a significant area to be protected for a waterfowl reserve.

The reserve land is managed by a community society, to which the developer also donated $25,000. This initiative generated considerable enthusiasm and support in Salmon Arm, and many individuals and businesses purchased memberships. The intent was to build viewing platforms and walkways to facilitate nature observation. Provincial and lottery funding was given to the society to assist in the implementation of these plans.

The review of the proposed residential-tourism-commercial development revealed a strong community desire to protect the ecological sensitivity of the area. At the same time, there was cautious acceptance of the idea of developing a high-quality destination resort that emphasized natural themes and had the potential to contribute to community economic development. The review recommended some modifications to the initial proposal, including a connecting link across the railway tracks between the proposed development and the central part of the community, a pedestrian orientation in the development, several 'finger parks' to minimize the impact on the inland properties' views of the lake, and an improved buffer between the residential and tourism components of the development.

In the spring of 1988 the official community plan was amended to allow the development to occur, subject to its meeting conditions, all of which were acceptable to the developer. In the autumn of 1988 the developer hired a consultant to prepare detailed design guidelines for the area, called Harbourfront Village. Harbourfront Village was to include residential developments, a hotel, lodge, health spa, a marine sales and service facility, retail and office space, and spaces for events and festivals.

The consultant used a codesign workshop to help community people participate in the planning (King et al. 1989). In the codesign approach, community members describe their vision for a development, indicating what features they would like to see in it. Artists then quickly prepare sketches to

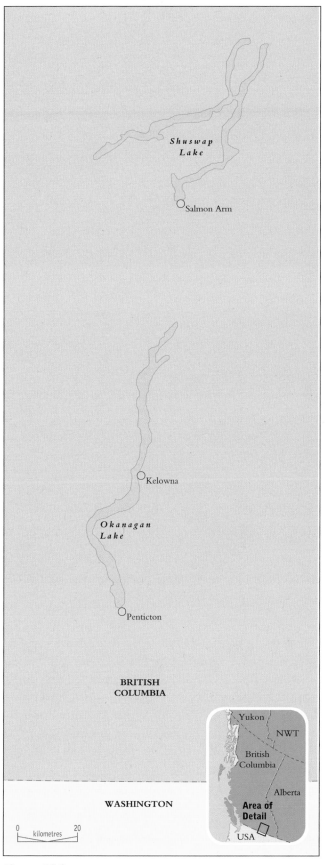

Figure 13.3 Location of Salmon Arm, BC source: Okanagan Study Committee, *Okanagan Basin Study Information Brochure* (Victoria: Canada–British Columbia Okanagan Basin Agreement, 1971).

Salmon Arm and Shuswap Lake (*S. Kipp*).

The Salmon Arm mudflats (*S. Kipp*).

problem. Each participant then generates ideas to overcome the problem. These ideas are recorded in a round-robin discussion in which group members read out their ideas until all ideas have been put forward.

In Salmon Arm, seven steps were used to structure the focus-group deliberations. They included consideration of: (1) existing image, (2) desired image, (3) attributes, (4) constraints, (5) concerns and issues, (6) opportunities, and (7) options. Once the ideas are listed for all seven of these topics, opportunity is provided for discussion and consensus building about each topic. This process can take some time. Kipp has explained that, for example, a group of about twelve people might require two sessions, each two to three hours long, to work through all of the questions and begin to reach a 'sensing' of main themes and subthemes.

By using these steps and by including participants who represented a range of interests (as outlined in Chapter 11), two central themes for the development emerged: (1) Harbourfront Village as a port-of-call destination, and (2) Harbourfront Village as a place for nature observation. Specific design ideas from the focus-group discussion were:

translate the verbal descriptions into visual images. Discussion then focuses on the sketches, which are continuously redrawn to incorporate the ideas from the community. As Callaway and Kipp (1991) state, codesign graphically illustrates the implications of alternative choices, and can rapidly help to clarify goals and priorities for elected officials, planners, and the public.

Following the codesign workshop, the consultants organized a day-long focus-group session to identify appropriate themes for the waterfront development. A focus-group approach is used to help a small group of people concentrate its discussion and reach consensus. In general, the procedure involves introducing the participants to a

1. primary emphasis on landscaping to highlight the penetration of nature into a built environment
2. use of natural landscaping with indigenous vegetation species
3. a marine architectural theme for all commercial buildings and, whenever possible, for street furniture
4. pedestrian features: walkways and benches, including an urban walkway to link the downtown to the wharf through the village's commercial core

5. water-theme areas: fountains as well as a children's play area with water features

The codesign artists were then hired to portray the two themes in sketches. The resulting illustrations and associated guidelines became significant aspects in transforming the Harbourfront Village idea from concept to reality. In addition, the use of codesign and focus-group approaches reflected the acceptance of partnerships and stakeholders by providing local community people with an opportunity to participate in the planning and design process.

Construction of Harbourfront Village began during 1990 with phase 1, which involved an office building and a set of residential units aimed at the upper-income market. Sales of the residential units were brisk, and the office building was occupied by a number of local government departments. Phase 2, scheduled to begin about 1992, was delayed, as explained later in this chapter.

Despite all of the community involvement and initial enthusiasm for Harbourfront Village, there was some resistance to the phase 2 development by 1990 (Callaway and Kipp 1991). These were concerns regarding the impacts on the view of the lake from existing properties inland from the development and about the impact on the wildlife area. Overall, there appeared to be a significantly increased appreciation of the value of the foreshore area in a natural state. The term mud-flats was being heard much less often. Instead, that area was increasingly referred to as a nature sanctuary.

In June 1990 the district council was presented with a petition containing 1,000 signatures, calling for a stop to phase 2. The district council tabled the petition while waiting for the completion of the revised official community plan. The waterfront issue, especially phase 2 of the Harbourfront Village project, dominated the entire public participation process associated with the revision of the official plan and almost overwhelmed the planning process to the extent that other legitimate issues likely did not receive adequate attention.

During the period (1991–2) when the official community plan was being prepared, many public meetings and an open house were held. Some focused specifically on the issue of waterfront development, particularly the proposed phase 2 of

Harbourfront Village. At the end of the planning process, the official community plan was completed without changing any policies regarding the downtown waterfront. One key reason was that the legal counsel for the district advised that any such change might make the district vulnerable to legal action. Thus while no changes were made, the district council stated that development could only proceed after the land was rezoned, and that the public would have the opportunity to comment upon any proposed zoning amendments.

During 1993 the district initiated a survey of residents in Salmon Arm to find out what their attitudes were regarding the proposed phase 2 development east of the wharf. The survey revealed that a majority of residents did not want to see residential development on those lands, but that they were willing to accept some commercial development, especially on the land closest to the wharf. Drawing upon the survey findings, the district staff developed a compromise proposal with the developer, who then submitted a zoning amendment to the council. However, there was strong opposition to the compromise proposal and the council rejected it, with the mayor casting the deciding vote.

Phase 1 was completed during 1992 and 1993, with some changes from the initially proposed uses. There was a greater emphasis on residential development. By early 1994, the non-residential component consisted entirely of government offices. From some perspectives, these changes illustrate the adaptive approach discussed in Chapter 9 as changes were made to reflect preferences and concerns. From another perspective, the changes represented modifications to accommodate the views of outspoken groups, who may or may not have reflected the spectrum of interests in the community. Opposition continued regarding some of the other commercial development concepts, particularly regarding a 'marine' pub. The outcome has been that the original tourism dimension of the development has been all but lost, at least in the phase 1 stage.

The phase 2 development of Harbourfront Village was again an issue during the November 1993 municipal election. Most of the elected councillors were committed to dealing with and resolving this issue. One idea was to hold a district-wide referendum focused on whether citizens supported or opposed development east of the wharf, or would like the district to purchase the lands. The developer had also offered to sell the land to the district.

The present zoning permitted uses such as a recreational-vehicle campground and a mini-golf course. The developer indicated that these options might be pursued if no other satisfactory solution was found. On a different front, between 1991–3 the Salmon Arm Bay Nature Enhancement Society constructed bird-watching blinds, trails, and boardwalks, and started transforming an island of dredged material to natural habitat.

Thus the Salmon Arm experience indicates that stakeholder and partnership involvement does not always lead to consensus and unanimity. In this case, there were different visions for the development and use of the waterfront. Furthermore, it also shows that a relatively small number of well-organized and articulate people can have a significant impact on defining the balance between economic development and environmental protection. In this situation, if the survey results are

valid, the majority of citizens were prepared to see some limited development on the waterfront as long as it did not include a residential component. However, it has been suggested that a few vocal residents who did not want any development were successful in delaying or stopping the project against what appeared to be the wishes of the majority in the community. We need to decide whether such adjustments represent the ideal of adaptive management or some distortion of it.

On the positive side, however, the Salmon Arm experience reflected a concern to address some of the components that Richardson identified for sustainable urban development, especially consideration of physical and economic conditions, as well as environmental effects. It also reflected interest in achieving the type of stewardship that Reed and Slaymaker (1993) outlined in Chapter 2: care, respect, and responsibility for the environment initiated at the local level.

REMEDIATION AND REHABILITATION

Located about 400 km north of Toronto, Sudbury was once 'notorious across the country for the air pollution and the barren, blackened landscape created by its smelters' (Richardson et al. 1989:4). However, during the 1970s and 1980s some remarkable initiatives were taken to revitalize and restructure the economy of Sudbury. The Sudbury story of remediation and rehabilitation of an industrialized landscape illustrates that challenging circumstances can lead to opportunities and positive changes.

A view of the Trans-Canada Highway in 1975 prior to reclamation (*Regional Municipality of Sudbury*).

A view of the same Trans-Canada Highway following reclamation (*Regional Municipality of Sudbury*).

Impacted terrain and water environments as a result of past mining activity (*Regional Municipality of Sudbury*).

Sudbury was established in the early 1880s as a construction camp for the Canadian Pacific Railway. During building of the railway, copper and nickel deposits were discovered only a few kilometres north of the construction camp. The result was that the construction camp evolved into a mining community, and Sudbury became the second largest world producer of nickel. In addition to Sudbury, other mining communities, such as Falconbridge, were established in the 30-km by 60-km Sudbury basin (Figure 13.4).

As with most mining-based communities, Sudbury experienced boom-and-bust cycles. Peak production occurred in 1970 when nickel output from the Sudbury basin reached an annual high of 200 000 tonnes at the same time as prices for nickel had climbed to their highest levels. At that time, Sudbury had one of the highest household incomes in Canada. In addition, the economy had begun to diversify as cultural and advanced educational facilities were built and good transportation links to southern Ontario were in place. The economic prospects of the Sudbury basin at the start of the 1970s appeared to be bright.

On the negative side, however, two aspects were of concern. First, the community had some serious social divisions as a result of bitter and prolonged labour-management and interunion conflicts within the mining industry. As a result, Sudbury was a deeply divided community. Second, the strong economic growth, based on the mining industry, was offset by a dramatically degraded physical environment in need of restoration due to previous mining activities. Ten thousand ha of the Sudbury area were devoid of significant vegetation due to air pollution and past mine practices (Figure 13.5). The tolerance limits of many species, as described in Chapter 4, had been exceeded. Many

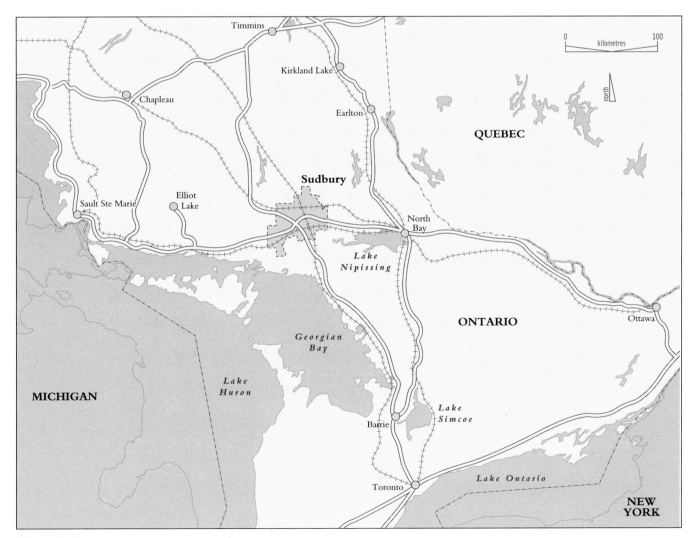

Figure 13.4 Location of Sudbury.

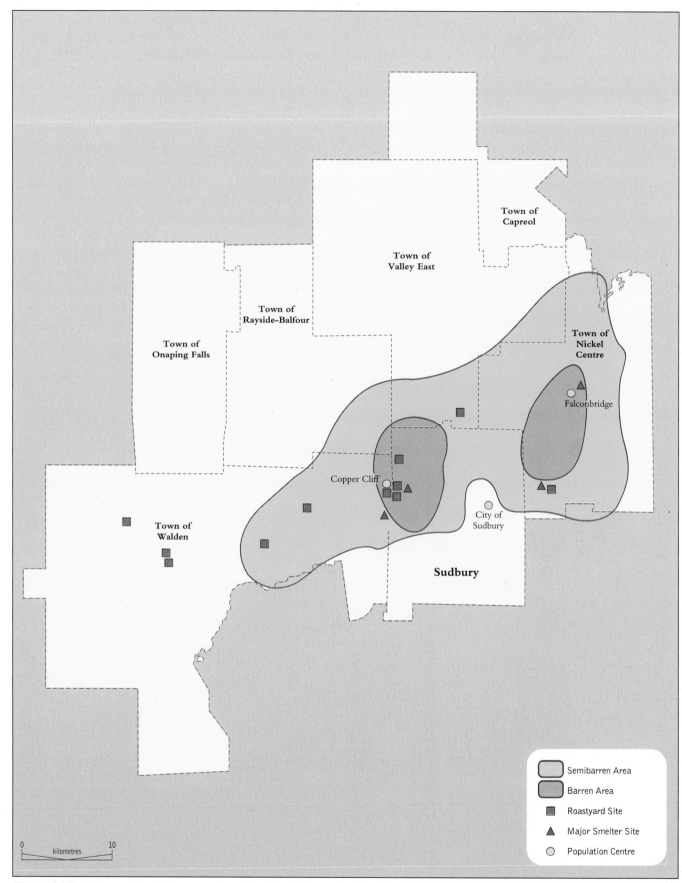

Figure 13.5 Sudbury: Extent of barren and semibarren landscape. SOURCE: W.E. Lautenbach, *Land Reclamation Program 1978–1984* (Sudbury, Ontario: Regional Municipality of Sudbury, Vegetation Enhancement Technical Advisory Committee, 1985):4.

lakes in the area had become highly acidic and degraded by metal contaminants, partially as a result of acid deposition associated with the smelter, as outlined in Chapter 7. The city periodically experienced episodes of choking air pollution from the mining smelters. Sudbury's notoriety was highlighted when the international media mistakenly reported that American astronauts went to Sudbury for training purposes since its stark landscape conditions most closely resembled the terrain conditions on the moon.

While Sudbury's economic growth had been strong and impressive, the reversal of its economic fortunes during the 1970s was equally dramatic. Increasing pressure from international and domestic nickel-producing competitors caused a steady drop in nickel prices, which in turn led to a significant drop in mining employment. For example, in 1970 the International Nickel Company (Inco) and the smaller Falconbridge Nickel Mines together employed nearly 30,000 people. By the late 1970s, that number dropped to less than 17,000. A short-term recovery slowed the lay-offs, but subsequent extensive shut-downs reduced employment at both companies to below 14,000 by 1983. Thus over a twelve-year period (1971–83), mining employment in the Sudbury basin fell from 47 per cent to 11 per cent of the labour force. This resulted in significant out-migration, but even with that, the unemployment rate in 1983 climbed to 17 per cent. By 1986 the mining employment fell further to 7,700, and between 1971 and 1986 the population of Sudbury dropped by 10 per cent, from 169,200 to 152,000. The prospects for the area appeared bleak.

Several factors contributed to the subsequent economic restructuring in the Sudbury basin. In 1973 the Regional Municipality of Sudbury was formed, creating for the first time a federation of the seven local communities, which allowed a regional perspective on planning and development. The daunting economic prospects led to much improved labour-management and inter-union relationships as people realized that it was imperative for them to work together. The provincial and federal governments relocated some of their offices to Sudbury, and regional hospital services were established. Machinery manufacturing also started to grow. In addition, significant growth occurred in service-oriented businesses.

Another important initiative was the development of a community-led strategy called Sudbury 2001 to help rejuvenate and diversify the local economy, with the ultimate goal of creating a sustainable community. A trilateral strategy was developed and focused on economic, social, and physical aspects (Katary 1982). The *economic goal* was to help Sudbury become a self-sustaining metropolis based on a diversified economic structure. The *social goal* was to allow individuals to discover their role and potential in the community, and to develop self-help and mutual-assistance programs. For the *physical dimension*, the goal was to enhance the quality of the physical environment, including the ecosphere. Particular attention was to be given to minimizing adverse environmental impacts and maximizing the aesthetic appeal in the community.

On the environmental side, some initiatives began before the negative consequences of the economic downturn became so evident. During the 1970s, Inco and Falconbridge had shut down part of their smelter capacity and significantly reduced emissions of sulphur dioxide. Inco also built its 381-m 'Superstack', which dispersed the emissions farther afield. The result was that the average level of sulphur dioxide in the air at Sudbury (in parts per million) dropped from 0.042 to 0.007 between 1970–88, a decrease of 83 per cent. The outcomes were tangible. The city was no longer periodically subjected to air-pollution fumigations. Vegetation could grow and began to return. At the same time, some of those positive aspects were offset by the fact that the Superstack simply shifted a portion of the air pollution problem from the Sudbury basin to areas downwind, sometimes as far away as parts of Quebec.

With local air quality significantly improved, there were opportunities for revegetation or regreening. Techniques were developed at Inco and in the Department of Biology at Laurentian University to grow grass, clover, and tree seedlings on formerly degraded land. There were setbacks with the initial plantings, but experience through trial and error led to procedures that resulted in healthy vegetation.

The regional municipality became involved in the reclamation program during 1974 with the establishment of the Vegetation Enhancement Technical Advisory Committee to the regional council. This committee consists of botanists, ecologists, landscape architects, horticulturalists, agriculturalists, planners, fisheries experts, as well as gardeners and interested citizens from the mining companies, university and college, provincial ministries, Ontario Hydro, municipalities, and the general public. This group has been charged with addressing Sudbury's vegetation restoration.

The results have been impressive. While bare and blackened rock as well as slag heaps are still present, 3200 ha of the most degraded land had

Experimental vegetation test plots demonstrated that extensive land reclamation was possible (*Regional Municipality of Sudbury*).

An experimental vegetation test plot successfully reclaimed (*Regional Municipality of Sudbury*).

been revegetated by 1994. In summer this formerly barren land is now green with grass and trees (boxes 13.2 and 13.3). Some 1.7 million trees comprising fifteen different evergreen or hardwood species were planted. All of this reclamation activity provided short-term employment for 3,300 students and unemployed individuals. The cost of this restorative work was $15 million ($11 million for labour and $4 million for capital costs), much of it covered by employment programs funded by the federal government. In addition, the mining companies (Inco and Falconbridge Limited) have restored an additional 2000 ha of degraded company-owned land.

The reclamation work in Sudbury is not complete. Between 1993 and 1998, the intent is to restore another 1500 ha of barren land. Attempts will also be made to reclaim several small watersheds and to encourage service clubs and school groups to adopt areas in need of restoration. Although still ongoing, Sudbury's rehabilitation work has already attracted international attention and recognition. In 1992 at the Earth Summit in Rio de Janeiro, the Regional Municipality of Sudbury received the United Nations Local Government Honours Award.

An intriguing question posed by Richardson et al. (1989) is to what extent the environmental enhancement contributed to the economic restructuring and recovery in Sudbury. They concluded that a definitive answer cannot be provided. However, they concluded that the changes in environmental quality and economic structure had a common cause: modernization of the production facilities at Inco and Falconbridge. The technological changes in the smelting process allowed the companies to become more efficient and capital intensive, significantly reducing the number

Students liming a barren site as part of the region's grassing program (*Regional Municipality of Sudbury*).

Students planting trees in a previously reclaimed area (*Regional Municipality of Sudbury*).

Box 13.2 Regional Land Reclamation through Grassing

In Lautenbach's (1987:230) view, to a considerable extent the regional land reclamation has been largely synonymous with grassing improvements. A key activity was site treatment of barren or semibarren landscapes to neutralize high soil acidity.

To counteract soil acidity, calcium carbonate or calcium and magnesium carbonate of an agricultural limestone grade was applied to the soil at an average rate of 5.08 tonnes per acre.

Bulk lime was delivered to staging areas where it was shovelled into bags for distribution. After bagging, the lime was hauled manually or transported by truck, railcar, or helicopter to distribution points. Then bags of lime were spread out at predetermined intervals for application in the required quantities.

Due to the large amounts of lime required at each site, about 80 per cent of the grassing and greening operations involved bagging, hauling, and spreading of lime.

Once lime was applied, the areas were fertilized at a rate of 159 kg per hectare, typically with fertilizer high in phosphorus, such as 6–24–24. Then in the early fall, a seed mixture of five grasses and two legumes was cyclone-seeded at a rate of 11–18 kg per hectare.

Box 13.3 Evaluation of the Grassing and Reforestation Activities

Evaluation of the initial reclamation activities provided the following results:

- Vegetation cover in limed areas remained in the range of 10–25 per cent.
- Herbaceous and woody species rapidly and spontaneously colonized.
- Over time, the percentage of grasses tended to decrease and that of woody species increased.
- The cover and vigour of legumes increased relative to grasses.

- Red pine and jack pine were the most vigorous of the planted trees.
- Surface soil pH increased from 3.5–4.5 before treatment to 4.0–5.5 following treatment and remained in that range.
- Vegetation in reclaimed areas contained elevated aluminium levels in grasses and trees relative to 'normal' sites, but levels of copper or nickel were low.
- The number of insects, birds, and some mammals increased in reclaimed areas (Lautenbach 1987:230).

of employees. Simultaneously, the modernization allowed the two companies to reduce substantially the emissions of sulphur dioxide into the atmosphere and thereby to meet increasingly more demanding environmental regulations. Once air quality started to improve, other improvements relating to revegetation, reduction of acidity in lakes, and re-establishment of fisheries were possible.

Richardson and his colleagues concluded that:

> . . . environmental improvement was probably not a direct cause of economic diversification, but rather a necessary condition without which the economy might have languished. In other words, while government policies and other factors certainly changed Sudbury's economic structure, we suggest that these policies might not have achieved their intended result without the improvements in environmental quality that took place. . . .
>
> In other words, while environmental improvement may not have generated economic benefits directly, it could have provided the conditions necessary for the positive economic changes in Sudbury to take root and flourish. In ecological terms, environmental quality could be a factor which does not directly *produce* economic benefits but which, when lacking, can *limit* economic development (Richardson et al. 1989:24–6).

To illustrate this argument, Richardson et al. (1989) considered the establishment of the regional Cancer Treatment Centre, one initiative to help diversify the economy. They considered what the role of environmental quality would have been from the viewpoint of health-care professionals whose training and livelihood provided them with considerable choice as to where they live and work. The medical doctor who was the director-designate of the cancer centre thought it would have been very difficult to establish such a centre had there been no improvement in the environment. From his perspective, he explained that he and his family would not have gone to Sudbury if the environment had been seriously polluted. While this is only one example, it does suggest that lack of environmental improvement could well have been a serious impediment to diversifying the community's economy.

Sudbury does provide a fine illustration of a community that was forced to address serious economic and environmental problems. In developing a strategy to revitalize its economic base, it was also presented with an opportunity to enhance the quality of the environment, which had been degraded through lack of foresight and management in return for high profit. In that manner, it addressed many of the ideas outlined by Richardson (1992) and Rees and Roseland (1991) at the beginning of this chapter.

Improvement of the environment alone was unlikely to be a key trigger to stimulate economic renewal, but, as has been pointed out, it could have become a serious constraint on renewal initiatives. As a result, the experience in Sudbury reinforces the argument from sustainable development that economic development and environmental management should be viewed as complementary, that each can support the other, and that new decision-making processes that consider a broader mix of factors and include the public are necessary.

ENVIRONMENTALLY SENSITIVE AREA PLANNING

Mathur (1989) has commented that conservation of river corridors in urban places is usually a challenge. In his view, such conservation requires long-term vision, a strategy for conservation, mobilization of public support, and joint participation from governments and the public. Other requirements include willingness to use regulatory authority to reduce land and water degradation, mediation to resolve conflicts between users and regulators, funds to initiate projects, and development of amenity features. The responsibility for such diverse tasks cannot normally be readily established since some are not assigned by law and others are divided among various jurisdictions. These challenges are significant, and Mathur concluded that Saskatoon was the only Canadian city in which an urban river corridor was being managed on a 'comprehensive basis'.

Saskatoon was started by the Temperance Colonization Society, and in 1883 its first plan was prepared by Land Commissioner John Lake. That plan focused upon land on the eastern side of the South Saskatchewan River. In a far-sighted decision, river-edge land was set aside as a public preserve and no lots were created along the edge of the river. This can be contrasted with the earlier example of Salmon Arm, which is more typical of most Canadian cities, in which land adjacent to a lake or river was usually set aside for railway lines or industrial use.

As the settlement expanded onto land on the western side of the river, Lake's initial plan was used as a model and river-edge land on that side of the river was also kept public. In 1930 when the first comprehensive plan was created for Saskatoon, the public preserves on both sides of the river were reaffirmed. By the end of the Second World War, the river-bank remained primarily in the public domain, with the exception of a few pockets of residential development and two industries (a brewery and a power-generating plant).

Pelicans in the South Saskatchewan River Valley (*Meewasin Valley Authority*).

Postwar development in Saskatoon did not impinge on the riverside land. However, during the 1960s and 1970s developments were proposed that threatened the concept of maintaining the river-banks as public land. Questions were also beginning to be raised about the river's capacity to support many uses and maintain its quality. As a result of these pressures, in 1974 the City of Saskatoon's Environmental Advisory Committee recommended to the council that a river-bank study should be completed. Subsequently, the City of Saskatoon, the provincial government, the Rural Municipality of Corman Park adjacent to the city, and the University of Saskatchewan (the largest land-owner along the river) collectively commissioned a series of studies, which culminated in a 100-year conceptual plan completed in 1978 by Raymond Moriyama Architects and Planners.

The Moriyama Conceptual Plan

The Moriyama plan was innovative and forward-looking, and incorporated many of the ideas later to be popularized by the Brundtland Report in 1987 with its emphasis on sustainable development. For that reason, some of the thinking that influenced the preparation of the conceptual plan is reviewed here.

In the introduction to the report, Raymond Moriyama Architects and Planners 1980:9) explained the approach. The objective was to achieve *balance*. The key ideas were *health* and *fit* (or compatibility). In other words, the continuing health of the river and all of its connecting parts had to be related directly and indirectly to individual and social health. To combine health and fit, the overriding theme was *linkage* or the 'physical linkage of the City and the rural area to each other and to the river and the natural system, the social linkage of people to people; the linkage of time—past, present and future'. In this manner, Moriyama anticipated many of the ideas of sustainable development, an ecosystem approach, and partnerships or stakeholders.

Raymond Moriyama Architects and Planners also anticipated many of the ideas of adaptive management outlined in Chapter 9. They explained that:

If we have learned anything in the past 30 years, it is this: the inflexible master plan, based on transient man-made 'facts' and technology, is an illusion that denies the facts of time and change. The word, *conceptual*, implies fundamental thinking about probabilities and possibilities. Such thinking provides a base or guideline while allowing for the dynamics of time—for later questioning, research, analysis, discussions, shifts in emphasis and modifications. In other words, the conceptual master plan offers a realistic, open base for creative future processes. It does not offer a fixed objective to be completed as conceived, without allowances for new awareness, new knowledge, inventions and social or technological change (Raymond Moriyama Architects and Planners 1980:9).

In the summary of the report, Moriyama and his colleagues emphasized that the thrust of the 100-year conceptual plan was to:

1. achieve *enrichment* of life
2. recognize that the base was the *natural system* and the spine was the *river*
3. facilitate *access* to and along the river by establishing *linkages*
4. implement through *nodes* and *links*
5. reflect the *dreams* and the *realities* from pioneers and their forebearers, the *perceived* and *stated needs* of the current people, and the *observed needs* by and *experiences* of the project team (Raymond Moriyama Architects and Planners 1980:75).

The Moriyama plan covered an 80-km portion of the South Saskatchewan River corridor, including the City of Saskatoon and the adjacent Rural Municipality of Corman Park. The plan identified a series of nodes connected by linkages (Figure 13.6), as well as relocating many recreational activities from environmentally fragile areas south of the city to more resilient areas to the north (Figure 13.7).

Moriyama concluded that the federal government, the provincial government, the city, the rural municipality, the university, and the private sector could not individually implement the concepts in the plan because of the need for developing a broad overview related to a large area, establishing a long-range time scale, transcending shared jurisdictions, establishing clear priorities within limited funding, and coordinating overall objectives. For these reasons, it was recommended that an 'authority', called the Meewasin Valley Authority, be established. The Moriyama report was approved in the summer of 1978.

Implementation: The Meewasin Valley Authority

During May 1979 the provincial government passed legislation to create the Meewasin Valley Authority (MVA), and the act was proclaimed in September of the same year. (*Meewasin* is a Cree word meaning lovely, beautiful, happy. Its general use has the more broad meaning of 'happy meeting place'.) Under the legislation, the authority was constituted as a partnership among the provincial government, the City of Saskatoon, the University of Saskatchewan, and the Rural Municipality of Corman Park.

The MVA was given jurisdiction over the South Saskatchewan River corridor for both public and private land in the city and the adjacent rural municipality. MVA's power was wide-ranging and included responsibility to plan the river corridor, regulate land and water use, acquire land through purchase, expropriate or have the right of first refusal, as well as to develop, maintain, and police its jurisdiction. The land and water under its responsibility was divided into two categories. The first was called a *conservation zone* in which all of its powers applied. The second, located within Saskatoon, was termed a *buffer zone*, where the MVA had limited power.

Reaction and Opposition

The extensive power given to the MVA caused a backlash. As Bolstad, Mathur, and MacKnight (1983) commented, as the existence and mandate of the MVA became known, apprehension developed among realtors, mortgage-lending institutions, developers, and land-owners regarding the MVA's intent and its possible impact upon their interests and activities. Particular concern was expressed about the regulatory power the MVA had been given for land use. The opposition became so vocal and intense that Mathur (1989:46) remarked 'at its height, the opposition to the Authority was so great that there were serious doubts as to whether the newly created agency would survive'.

The opposition was led by an organization called the River Edge Heritage Association (REHA) whose key members were individuals living in Saskatoon but with interests in land in adjacent Corman Park. This group led an effective campaign against the MVA and obtained the support of many councillors from Corman Park and Saska-

Hague Crossing Link
and Node

Cathedral Bluffs Node

Prairie Grass Terrace Node

Sutherland Beach –University Node

Saskatoon City
Core Link

Howe-
Diefenbaker
Node

Sand Dune and Islands Park Node

Meander Plain Sanctuary Node

Pike Lake
Provincial Park Node

north

0   kilometres   5

Figure 13.6 Nodes and linkages in the Moriyama plan. SOURCE: Raymond Moriyama Architects and Planners, 'The Meewasin Valley Project' (Toronto: Raymond Moriyama Architects and Planners, 1980).

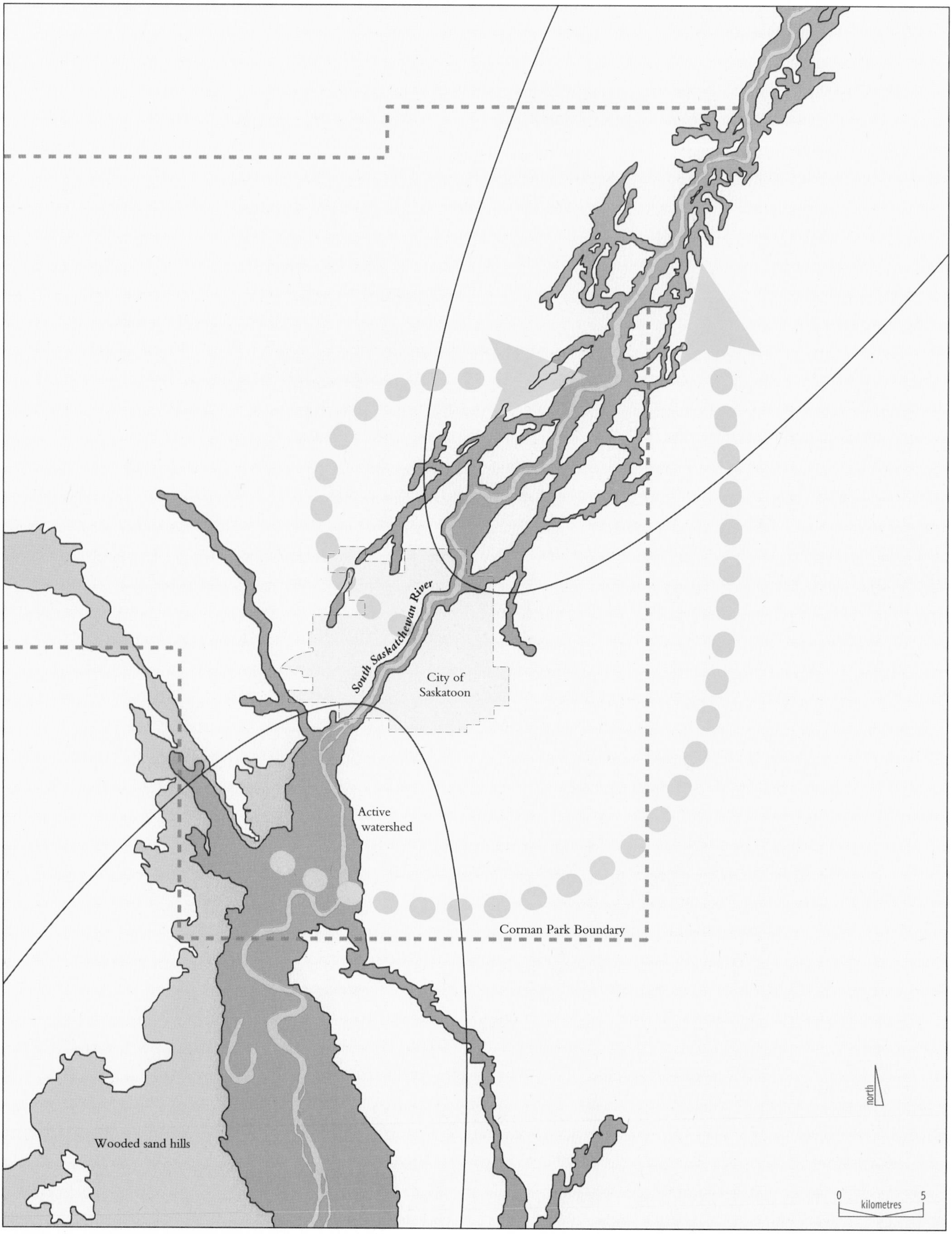

Figure 13.7 Proposed recreational areas in the Moriyama plan. SOURCE: Raymond Moriyama Architects and Planners, 'The Meewasin Valley Project' (Toronto: Raymond Moriyama Architects and Planners, 1980).

toon. Counterbalancing this opposition was consistent support from the provincial government as well as environmental and heritage groups.

The antagonism was fuelled by the rapidity with which the MVA had been established. When questions and worries were raised about how the mandate of the MVA would be interpreted, in many instances the newly appointed staff of the authority were unable to respond clearly because they had not had adequate time to examine either the 100-year conceptual plan or the legislation, nor to develop a set of policies, procedures, or plans to guide the activities of the MVA. In the absence of direct and tangible answers, critics attributed the worst possible motives to the MVA regarding heavy-handed regulations that would unduly restrict what land-owners could do with or on their property.

Moriyama had stressed that the conceptual plan identified a vision or direction for the future. It was not conceived as a fixed, inflexible master plan. Instead, the planning team viewed it as a general guideline to be modified in the context of further investigations and experience. However, as Bolstad, Mathur, and MacKnight (1983:263–4) explained, in practice such an approach created difficulties, raising some challenging questions about the concept of adaptive environmental management.

While the MVA staff treated the 100-year plan as a general guideline from the outset, critics viewed it as a fixed plan because it was the only document on which the authority could base its decisions. Rural land-owners cited drawings in the conceptual plan that indicated river-edge paths crossing their fields, and wondered when their land would be expropriated for such paths. The MVA staff's assurances that the drawings were only illustrative and that specific plans and actions would not occur without consultation with land-owners were not viewed as credible.

The 100-year plan also had at least two major limitations. First, it did not reflect any study of demand for recreation in the Saskatoon area. The ideas and proposals primarily reflected the ideas from the consultants regarding what could be done using the resources in the valley. As a result, the feasibility or timing of the projects contained in the conceptual plan were difficult to explain.

Second, the conceptual plan covered an extensive area that had been defined by the consultants as the 'natural system'. This 'system' included both an 'active' subsystem with a direct effect on the South Saskatchewan River, as well as an 'inactive' subsystem. The plan focused upon initiatives along a narrow band of land adjacent to the river edge, and indicated nothing for the considerable amount of other land over which the MVA had authority. An obvious question was why all the 'inactive' area had been included if there were no plans for it. In response, the MVA staff could only state that 'because it is part of the natural system'. As Bolstad, Mathur, and MacKnight (1983:264) remarked, such answers sounded empty and reinforced the apprehension of those who believed that the MVA represented an attempt by the city and the province for a rural land-grab.

Modification and Adjustment

As opposition grew and became more intense, it became obvious that some adjustments had to be made if the MVA were to survive. Decisions were subsequently taken to modify the legislation to exclude private lands in the Rural Municipality of Corman Park from the MVA's jurisdiction. In addition, the powers of expropriating land outside the Meewasin Valley and the right of first refusal of privately owned land were repealed. However, notwithstanding these changes, in June 1981 the rural municipality officially withdrew from the MVA, leaving the province, the city, and the University of Saskatchewan as the partners. The MVA was left with jurisdiction over the river land in Saskatoon and Corman Park, publicly owned lands adjacent to the river in both Corman Park and Saskatoon, and a small amount of private land in Saskatoon.

The idea to relocate recreational activities from the fragile areas south of the city to the north had to be set aside, since after the modifications, the MVA had no jurisdiction over most of the fragile areas and no means to acquire land to the north of the city.

Accomplishments

Following the controversy over its establishment, the Meewasin Valley Authority concentrated its activities on public lands within Saskatoon. These initiatives focused primarily upon restoring and revegetating land under public ownership that had become degraded, creating new river-bank parks, and developing a network of trails on both banks of the river. By 1993 the MVA had 17 km of primary trails and was responsible for numerous parks in Saskatoon. Furthermore, the MVA developed an environmental education program and had prepared plans for waterfront development in the central part of the city.

The MVA has also taken several significant initiatives on city-owned property in Corman Park, protected some areas, and created a centre for nature interpretation and environmental education.

North of Saskatoon, the MVA took the lead role in the establishment of Wanuskewin Heritage Park, a site of outstanding palaeontological, archaeological, historical, and natural significance. (*Wanuskewin*, an ancient Cree word, means seeking peace of mind or living in harmony.) This site was identified in the Moriyama 100-year plan of 1978 as a potential park and interpretive centre. In 1984 the area became a provincial heritage site, the first one designated at that time in the province to feature prehistoric resources. In 1992 a building, amphitheatre, and parking lot were created, costing $1.1 million.

One of the striking features of Wanuskewin has been the involvement of First Nations peoples in the project. There was considerable consultation with elders and other First Nations peoples. In

addition, twenty-six of the thirty staff at Wanuskewin when it opened in 1992 were Native peoples. All of the permanent and part-time interpretive positions are held by Native peoples.

Another major initiative of the MVA was its lead role in creating the Partners for the Saskatchewan River Basin. Both the North and South Saskatchewan rivers have their headwaters on the eastern slopes of the Rocky Mountains in Alberta, drain eastward to join near Prince Albert in Saskatchewan, and then flow into Manitoba (Figure 13.8). The basin covers an area of about 380 000 km^2, and a major problem is the lack of an effective mechanism to link the many jurisdictions and interests in such a large basin. The Partners initiative, officially launched in August 1993, is intended to develop awareness, knowledge, and commitment in sustaining the aquatic resources in the Saskatchewan River basin. Participants and partners are based in the three prairie provinces, as well as the federal government. The MVA acts as the managing partner for the project. It has a strong interest, since upstream activities on and adjacent to the South Saskatchewan River can have major consequences for the river and adjacent lands in the Saskatoon area. This initiative illustrates the notion that hierarchies of ecosystems exist, as was noted in Chapter 8. In other words, the Meewasin Valley Authority attempts to take an ecosystem approach for the river valley in the vicinity of Saskatoon, but also recognizes that the urban river valley is part of an ecosystem extending from the Rocky Mountains to Hudson Bay.

After the MVA's stormy beginning in 1979, it is interesting to note the way in which the residents of Saskatoon view the river and the MVA. An opinion survey in August 1993 showed that those polled stated the three aspects they liked the most about Saskatoon were its beauty (39 per cent of respondents), the people (21 per cent of respondents), and the river (11 per cent of respondents). With reference to the MVA, 83 per cent of the respondents believed it was a good investment of tax dollars, while 7 per cent disagreed and 10 per cent

Meewasin Valley Authority trail along the South Saskatchewan River (*Meewasin Valley Authority*).

were uncertain. Slightly over half of the respondents indicated that they would contribute to the Meewasin Foundation, a fundraising organization for the MVA, while 25 per cent said they would not and 19 per cent were unsure. Forty-four per cent stated that they would pay a 'green tax' to support the MVA, while 39 per cent would not.

After a difficult start, the MVA has the support of a large majority of people in Saskatoon. It has used an ecosystem approach by identifying interlinked natural systems adjacent to the South Saskatchewan River in Saskatoon. The 100-year plan was deliberately designed to be flexible and to be adapted as experience was gained. In that manner, even though critics of the MVA viewed the plan as a fixed document, it was conceived as an evolving guideline. However, it also provided a vision that was remarkably consistent with many of the core ideas related to the concept of sustainable development. In that regard, Mathur (1989:48) suggested that the Moriyama plan has been invaluable as it kept the activities of the MVA focused on the long-term

potential of the river valley despite the need also to address short-term implementation challenges.

The MVA has been a significant force in requiring developers to consider potential environmental and social impacts in their proposals. The authority has also emphasized the importance of establishing networks with other partners, ranging from interests within the City of Saskatoon to other governments and interests in both the Alberta and Manitoba portions of the Saskatchewan River basin. It has been particularly innovative in its leadership role related to the Wanuskewin Heritage Park, and has developed a sophisticated approach towards public involvement.

The MVA has had to deal with conflict, especially in its formative years. In several instances it demonstrated that the resolution of disputes requires willingness to re-examine and revise objectives and ways of operation to accommodate other societal interests. In that manner, the MVA has been a good example of an adaptive management agency, similar to the Kativik Environmental Quality Commission discussed in Chapter 9.

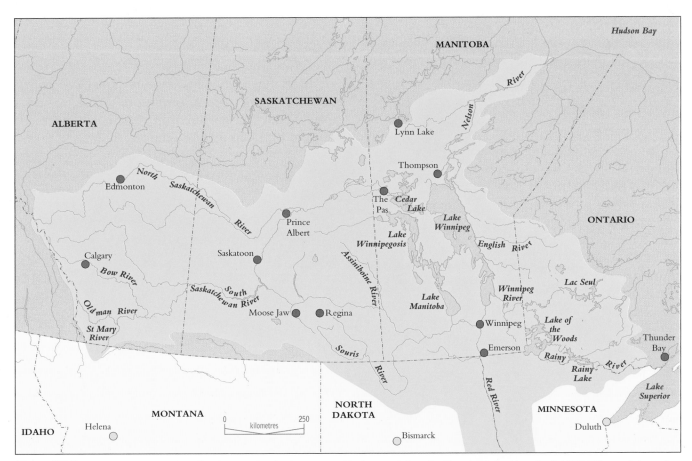

Figure 13.8 The North and South Saskatchewan River basins. SOURCE: Environment Canada, *Canada Water Year Book 1975* (Ottawa: Information Canada, 1975):45.

Mathur has summarized its accomplishments by suggesting that the MVA has been less effective as a regulator, but has succeeded as a steward of valley land. In his view, the agency has sensitized the public regarding the value and importance of the river and riverine lands, and 'in the long run, this is perhaps a more sustainable conservation strategy than is regulation alone' (Mathur 1989:48).

Meewasin's current approach has four cornerstones:

1. conserving and restoring the valley and its cultural and natural heritage
2. developing sites and facilities to meet ends
3. regulating development to ensure 'health and fit' in the valley
4. educating the community with the aim of developing river valley stewardship and a willingness to take action

This approach is being used to ensure the long-term sustainability and development of the South Saskatchewan River basin corridor in the Saskatoon area. In this manner, it is incorporating most of the ideas for urban sustainable development identified at the beginning of this chapter, as well as the principles for sustainable development discussed in Chapter 2.

IMPLICATIONS

The experiences of Salmon Arm, Sudbury, and Saskatoon highlight the argument that while it is important to think *globally*, it is essential to act *locally*. The economy of Sudbury was affected by global events (changing sources of nickel supplies and prices). However, the renewal of the economy and improvement of environmental quality were addressed through a partnership of the local community and the provincial government, with assistance from the federal government. The Sudbury experience confirms the importance of viewing economic development and environmental protection as complementary to each other.

An ecosystem perspective and adaptive management were both illustrated in Salmon Arm's Harbourfront Village, and in Saskatoon with the conservation of the South Saskatchewan River Valley. The importance of flexibility and a willingness to modify were demonstrated in Saskatoon as the

Meewasin Valley Authority was required to respond to criticisms and concerns from landowners in the Saskatoon area. Both of those experiences, as well as the one in Sudbury, also emphasized the value of developing partnerships and using a participatory approach in managing to achieve sustainability. In Sudbury and Saskatoon, scientific studies were also used as a basis for many of the initiatives introduced.

It is easy to overlook the need to improve our capability for environmental management in urban contexts. Although most Canadians live in urbanized places, there is a tendency to think about environmental management and sustainable development as applicable only to wilderness or less densely populated places. The examples in this chapter show that sustainability is equally applicable to densely populated habitats. The key is to appreciate that the specific guidelines for sustainable development may be different in urbanized and non-urbanized environments. Beyond that, however, the basic ideas about sustainability outlined in Chapter 2 appear to be relevant to a wide range of places and situations.

SUMMARY

1. More than 75 per cent of Canadians live in urban places (settlements of more than 1,000 people). Thus urban environmental management should be given more attention.

2. The five components of urban sustainability are: physical conditions, economic conditions, resource demands, environmental effects, and urban management.

3. Sustainable communities use urban space efficiently, achieve reduced consumption of material and energy resources, enhance community liveability, and are sensitive to ecological and socioeconomic complexity.

4. In Salmon Arm, BC, a codesign process was used to involve community people in the planning for a waterfront development.

5. A codesign process includes community members who describe their vision for a development,

including specific features. Then artists illustrate how the vision and features might appear in the development. Ongoing discussion leads to revisions of the sketches until consensus is reached. The codesign process thus graphically illustrates the implications of alternative choices.

6. Focus groups were also used in Salmon Arm. A focus group is a small group of people selected to represent various interests in the community, who meet to see if a consensus can be reached regarding a problem and possible solutions for it.

7. Despite considerable public involvement (codesign process, focus groups, surveys, and public meetings), strong opposition emerged against the proposed development in Salmon Arm, partially reflecting changing values and attitudes about the value of the foreshore area, and partially reflecting the influence of a small group of articulate and vocal citizens. A compromise was rejected by the council.

8. The urban landscape of Sudbury, Ontario, has been transformed through initiatives that started in the 1970s from a highly degraded one to a landscape that earned a Local Government Honour Award from the United Nations in 1992.

9. Sudbury developed as a major centre for copper and nickel mining, and has experienced boom-and-bust cycles. In the early 1970s it had one of the highest household incomes in Canada. In 1970 17,000 people were employed in copper and nickel mining and processing, but by 1986 only 7,700 people worked there, and over the same period the population of Sudbury declined by 10 per cent.

10. As a result of serious air pollution from the smelting operations, the Sudbury landscape was badly degraded, with over 10 000 ha devoid of significant vegetation.

11. As part of economic restructuring, a community-oriented strategy called Sudbury 2001 was developed, one component of which was to improve the quality of the physical environment.

12. The mining companies reduced emissions of sulphur dioxide by 83 per cent between 1970 and 1988, which allowed revegetation to be initiated.

13. By 1994 3200 ha of degraded land had been revegetated and some 1.7 million trees had been planted in programs that provided short-term employment for 3,300 people. The cost was $15 million. In addition, the mining companies restored about 2200 ha of their degraded property.

14. One conclusion has been that while the environmental improvements were not a direct cause of economic renewal in Sudbury, it was a necessary precondition to attract some of the professionals and service industries, which diversified the economy.

15. The City of Saskatoon has protected and rehabilitated the urban river corridor of the South Saskatchewan River through the leadership of the Meewasin Valley Authority.

16. Most of the river-edge property in Saskatoon is under public ownership and managed by the Meewasin Valley Authority, the city government, and major land-owners, such as the University of Saskatchewan.

17. A 100-year conceptual plan was created by a consultant in 1978. Its main purposes were balance, health, flexibility, and adaptability.

18. The MVA was created in 1979, and initially there was strong opposition from private land-owners who feared heavy-handed regulation and restrictions on what could be done on their properties.

19. The MVA viewed the 100-year conceptual plan only as a guideline, but its critics regarded it as a master plan and were not reassured when the MVA commented that the ideas in the plan were only illustrative rather than firm intentions.

20. Due to the strong opposition, the MVA legislation was modified so that private property in the adjacent rural municipality was excluded from the MVA's jurisdiction, and the MVA's powers of expropriation were removed.

21. The MVA has focused upon restoring and revegetating land, creating river-bank parks and a network of river-bank trails, developing an environmental education and interpretative program, and taking a lead role in creating a heritage park and facilitating a collaborative initiative for planning throughout the South Saskatchewan River basin (from Alberta to Manitoba).

REVIEW QUESTIONS

1. What are the five components of sustainable development for urban environments? How do those components relate to the principles of sustainable development discussed in Chapter 2?

2. What are the major problems or obstacles to achieving sustainability in urban environments? Are such problems or obstacles specific to urban environments, or are they encountered in almost all situations, urban or non-urban?

3. How were codesign workshops and focus groups used in Salmon Arm to improve the concept of a development so that it would achieve both economic development and environmental protection objectives? What are the general strengths and weaknesses of the codesign approach?

4. What design features can be used to incorporate natural features into built systems?

5. For what accomplishments did the Regional Municipality of Sudbury receive the United Nations Local Government Honours Award during the Earth Summit in Brazil? Can you think of other communities that have taken similar noteworthy initiatives regarding sustainable development? Can you see opportunities in your community for similar initiatives?

6. What were the key elements in Sudbury's regreening initiatives?

7. In what way does the experience in Sudbury confirm the argument that economic development and protection of environmental integrity are complementary aspects in sustainable development?

8. Why did the Meewasin Valley Authority initially experience strong opposition when it was created in Saskatoon? Do you think this reaction was unique to Saskatoon, or would you expect similar reactions in any other community? Why?

9. What are the lessons from the Meewasin Valley Authority's experience regarding the use of ecosystems, adaptive management, and stakeholder approaches at the community level?

10. If a sustainable development strategy were to be developed for your community, which key issues should be considered, and what should be the fundamental components of such a strategy?

REFERENCES AND SUGGESTED READING

Bolstad, W., B. Mathur, and H. MacKnight. 1983. 'A Case Study of Rural Opposition toward River Valley Planning'. In *River Basin Management: Canadian Experiences*, edited by B. Mitchell and J.S. Gardner, 253–65. Department of Geography Publication Series no. 20. Waterloo, ON: Department of Geography, University of Waterloo.

Buschak, L.A., and R.D. Brown. 1995. 'An Ecological Framework for the Planning, Design and Management of Urban River Greenways'. *Landscape and Urban Planning* 33:211–25.

Callaway, C., and S. Kipp. 1991. 'The Salmon Arm Waterfront: Balancing Economic and Environmental Objectives'. *Plan Canada* 31, no. 2:23–6.

Cook, E.A.I. 1991. 'Urban Landscape Networks: An Ecological Planning Framework'. *Landscape Research* 16:7–15.

Dearden, P. 1983. 'Anatomy of a Biological Hazard: *Myriophyllum spicatum* L. in the Okanagan Basin, British Columbia'. *Journal of Environmental Management* 17, no. 1:47–61.

_____. 1984. 'Public Perception of a Technological Hazard: A Case Study of the Use of 2,4-D to Control Eurasian Water Milfoil in the Okanagan Valley'. *Canadian Geographer* 28, no. 4:324–40.

Environment Canada. 1975. *Canada Water Year Book 1975*. Ottawa: Information Canada.

_____. 1991. *The State of Canada's Environment*. Ottawa: Minister of Supply and Services Canada.

_____. 1994. State of the Environment Bulletin no. 94–2. Ottawa: Minister of Supply and Services Canada.

Hough, M. 1994. 'Design with City Nature: An Overview of Some Issues'. In _The Ecological City: Preserving and Restoring Urban Biodiversity_, edited by R.H. Platt, R.A. Rowntree, and P.C. Muick, 40–8. Amherst: University of Massachusetts Press.

Katary, N. 1982. _Origins and Evolution of 2001: A Developmental Odyssey_. Sudbury, ON: Department of Planning and Development, Regional Municipality of Sudbury.

King, S., M. Conley, B. Latimer, and D. Ferrari. 1989. _Co-Design: A Process of Design Participation_. New York: Van Nostrand Reinhold.

Lautenbach, W.E. 1985. _Land Reclamation Program 1978–1984_. Sudbury, ON: Vegetation Enhancement Technical Advisory Committee, Regional Municipality of Sudbury.

_____. 1987. 'The Greening of Sudbury'. _Journal of Soil and Water Conservation_ 42, no. 4:228–31.

Mathur, B. 1989. 'Conserving the Urban River Corridor: Experience from Saskatoon'. _Plan Canada_ 29, no. 5:43–9.

Raymond Moriyama Architects and Planners. 1980. 'The Meewasin Valley Project'. Toronto: Raymond Moriyama Architects and Planners.

Reed, M.G., and O. Slaymaker. 1993. 'Ethics and Sustainability'. _Environment and Planning A_ 25, no. 4:723–39.

Rees, W.E., and M. Roseland. 1991. 'Sustainable Communities: Planning for the 21st Century'. _Plan Canada_ 31, no. 3:15–26.

Regional Municipality of Sudbury. 1990. _Five Year Land Reclamation Plan 1990–1994_. Sudbury, ON: Vegetation Enhancement Technical Advisory Committee, Planning and Development Department, Regional Municipality of Sudbury.

Richardson, N.H. 1991. 'Reshaping a Mining Town: Economic and Community Development in Sudbury, Ontario'. In _Urban Regeneration in a Changing Economy: An International Perspective_, edited by J. Fox-Przeworski, J. Goddard, and M. de Jong, 164–84. Oxford: Clarendon Press.

_____. 1992. 'Canada'. In _Sustainable Cities: Urbanization and the Environment in International Perspective_, edited by R. Stren, R. White, and J. Whitney, 145–67. Boulder: Westview Press.

_____, B.I. Savan, and L. Bodnar. 1989. _Economic Benefits of a Clean Environment: Sudbury Case Study_. Prepared for the Department of Environment, Canada. Toronto: N.H. Richardson Consulting.

Tomalty, R., R.B. Gibson, D.H.M. Alexander, and J. Fisher. 1994. _Ecosystem Planning for Canadian Urban Regions_. Toronto: ICURR Publications.

_____, and S. Hendler. 1991. 'Green Planning: Striving towards Sustainable Development in Ontario's Municipalities'. _Plan Canada_ 31, no. 3:27–32.

Chapter 14

Agriculture

Agriculture in Canada is responsible for about 12 per cent of gross domestic production, 10 per cent of employment, and about 10 per cent of export earnings. Yields have increased dramatically since the start of the century (figures 14.1a and 14.1b) as a result of improved agricultural technologies, such as mechanization, irrigation, genetic research, fertilizers, and pesticides. Canadian agriculture is also important on a global scale, producing enough food to feed twice the country's population and exporting the rest around the world. Canada donates about 12 per cent of the cereals given by all countries through aid programs.

Agriculture is fundamentally an ecological process, as solar radiation is converted through one or more transformations into human food supplies. Rapid growth in human population has increasingly disrupted natural systems in the effort to feed burgeoning populations. This chapter provides some context for this ecological process before considering some of the environmental challenges in Canadian agriculture.

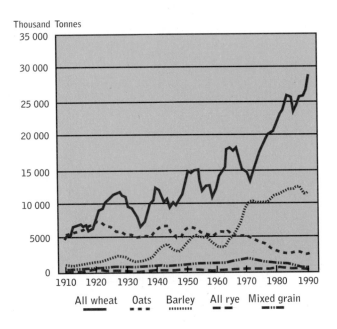

Figure 14.1a Selected grain crop production, 1910–91. NOTE: Five-year moving average. SOURCE: Statistics Canada, *Human Activity and the Environment*, cat. no. 11-509E (Ottawa: Statistics Canada, 1994):253.

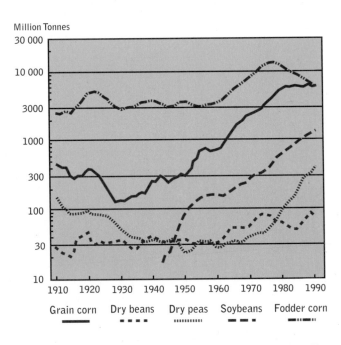

Figure 14.1b Selected field crop production, 1910–91. NOTES: Five-year moving average. This figure uses a log scale to compare trends of differing magnitude. SOURCE: Statistics Canada, *Human Activity and the Environment*, cat. no. 11-509E (Ottawa: Statistics Canada, 1994):253.

AGRICULTURE AS AN ECOLOGICAL PROCESS

Before the development of agriculture, the 'neolithic revolution', humans derived their food supplies by hunting and gathering produce from the surrounding ecosystems like other animals. Humans, along with other species such as bears, derive their energy supply from several different trophic levels, which supply a wide range of potential food sources, ranging from herbivorous to carnivorous. Within this food chain, humans also provide a potential food supply for other top carnivores—for example, we are potential prey for lions or sharks.

Humans have been hunter gatherers for approximately 75 per cent of the time they have existed on earth. Energy is derived directly from the food chain, from fire, and by human muscle power. This system of economy requires detailed knowledge of the ecosystem, which supplies all human needs, including food. Past human impacts on ecosystems were relatively minimal compared with the current situation, but there is evidence that humans caused greater impacts than other species, particularly because they used fire and cooperative hunting.

This was the mode of economy that dominated in Canada prior to the arrival of Europeans in the seventeenth century. Regional specializations developed, dependent upon ecosystem characteristics. West Coast Native peoples were heavily dependent upon salmon runs, Prairie people upon bison, the Inuit on sea mammals, and so on. Only in southern Ontario, in the deciduous woodlands of the Carolinian zone described in Chapter 3, did indigenous agricultural practices develop from about AD 500 onwards. These represent a diffusion northward of practices originating in Central America some 8,000 years earlier. Even here, the subsistence cultivation of crops such as maize was seen mainly as a supplement to, and not a replacement of, hunter-gatherer economies.

The development of agriculture does, however, have profound implications for ecosystems.

- Before agriculture, natural selection determined the plants and animals in a given area. Following the development of agriculture, humans became a major influence on species distributions. In modern agriculture this has led to large areas of monoculture cropping, often made up not only of identical crop species but individuals with the same genetic code. New species have been created to maximize the output of food for humans. Native species have been displaced. European cattle now dominate the plains of America and Australia where bison and kangaroos used to roam.
- Humans influenced not only distributions but also the numbers of each species. Plants and animals that were domesticated greatly increased in number and range. Other, less useful organisms often declined as their habitat was destroyed to make room for agricultural activities.
- Energy flows are increasingly directed into agricultural as opposed to natural systems. It is now estimated that humans take some 40 per cent of the net primary productivity of the planet.
- Natural food chains are truncated as humans destroy and replace natural consumers and predators at higher trophic levels.
- Auxiliary energy flows are used in agricultural systems to supplement the natural energy flow

Before the Europeans arrived, most of Canada's Native peoples were hunters and gatherers. On the prairie they depended so heavily on the bison that they almost decimated their numbers (*Nic Larter*).

Agriculture has had a profound impact on the distribution of species. Large areas of the agricultural landscape are dominated by monocultures—in this case sunflowers—in which each plant has the same genetic make-up (*Philip Dearden*).

changes in the structure and composition of the primary production system.

These changes reflect a transition from a world dominated by complex natural systems to one dominated by relatively simple control systems in which humans determine the species and numbers in a given area, as described in Chapter 1. This resulted in massive changes in ecosystems. However, unlike the later Industrial Revolution, these changes took place over an extended time period, allowing greater potential for adaptation to these changes. Furthermore, until the last 150 years or so, energy inputs were largely limited to photosynthetic energy from the sun, the energies of domesticated draft animals (such as oxen, camels, and horses), and human energy input. It was not until the Industrial Revolution when past deposits of photosynthetic energy in the form of coal and oil were used that industrial agriculture started and energy inputs and environmental impacts increased dramatically. Indeed, it was only during the latter half of the twentieth century that some of the most damaging impacts of agriculture (the use of synthetic biocides on a large scale) started.

AGRICULTURE AND ENERGY

Agriculture can be considered as a food chain, with humans as the ultimate consumers. Energy flows through this food chain as it does in natural food chains. As such, the second law of thermodynamics is also important to agricultural food chains: the longer the food chain, the greater the energy loss (Chapter 4). This constitutes one part of the argument for a vegetarian diet. By eating at the lowest level on the food chain, as a herbivore, humans will maximize the amount of usable energy in the food system. There are, however, other aspects of food production higher up the food chain that should also be taken into account.

from the sun. Such inputs in many advanced agricultural systems are in excess of those derived from natural sources.

- Natural successional processes are altered to keep the agricultural systems in an early seral stage. Auxiliary energy flows in the form of herbicides and mechanical weeding are often used to accomplish this.
- Biogeochemical cycles are interrupted as natural vegetation is replaced by domesticates, which are harvested on a regular basis. Auxiliary energy flows in the form of fertilizers are used to replenish some of the nutrients extracted through harvesting.
- The soil is altered not only chemically but also physically through ploughing. There is no natural process that mimics the disturbance created by ploughing.
- Agriculture in many areas supplements rainfall with irrigation to provide adequate water supplies. This has led to large-scale water diversions, changes in groundwater, soil characteristics, precipitation patterns, and water quality.
- Fire was a very important tool in early agriculture, and was used to clear extensive areas of land for growing crops and for grazing animals. Fire favours the growth of some species over others and also leads to the rapid mobilization of nutrients.
- The stocking densities of domesticated herbivores are often much higher than that of natural herbivores, leading to a reduction in the amount of biomass in grass growth and

- Food is more than energy requirements. There are also important protein and mineral demands. Animal products are, by and large, the main suppliers of these proteins

Box 14.1 Social Implications of the Development of Agriculture

Besides direct impacts on ecosystems, agriculture also had a profound influence on society, which in turn had further implications for ecosystems.

- More reliable food supplies permitted growth in populations.
- A sedentary life became more possible as a result of these food supplies and the ability to store food, which allowed larger, permanent settlements to become established.
- Permanent settlements allow greater accumulations of material goods than was possible under a nomadic lifestyle.
- Agriculture allowed food surpluses to be generated so that not all individuals or even families had to be involved in the food-generating process, and specialization of tasks became more clearly defined. The end result of this situation is that now only some 4 per cent of the population of Canada is directly involved in food production, allowing the others to direct their energies to other tasks, usually the process-

ing of raw materials into manufactured goods and thereby increasing the speed of flow of matter and energy in society. As indicated in Chapter 4, this high rate of throughput is at the core of many current environmental problems.

- The creation of food surpluses and increased material goods promoted increasing trade between the now sedentary settlements. This led to the development of road and later rail connections to facilitate the rapid transport of materials, thus involving the consumption of large amounts of energy.
- Land and water resources became more important, leading to increased conflict between societies regarding control over agricultural lands.
- Aggregation of large numbers of people in sedentary settlements also concentrated waste products in quantities over and above those that could be readily assimilated by the natural environment. Today we call this pollution.

and minerals and are also the major suppliers of elements such as calcium and phosphorus.

- Areas currently used as range land to support animal production often cannot be used as crop land to provide food for humans lower on the food chain. It is too dry or otherwise ecologically marginal, and severe problems (such as excessive soil erosion) have arisen in the past when humans have tried to cultivate such lands. Thus it is not always correct to presume that such lands could produce more food under tillage.
- Grazing animals in many parts of the world provide food supplies, are used as draft animals, and provide animal parts for clothing and a host of other uses.

However, there are valid arguments for reducing meat consumption in favour of a more vegetarian diet, especially in industrialized nations. The argu-

Horse power is one form of auxiliary energy used in agricultural production, but the environmental impacts of this system of cutting hay by hand and collecting it by horse in Newfoundland are much less than those resulting from the fossil-fuelled, mechanical processes favoured by most Canadian farmers (*Philip Dearden*).

ments for a vegetarian diet include not only ecological aspects but also health, moral, and religious considerations. A well-balanced vegetarian diet will provide protein and mineral requirements, so animal products are not necessary.

One major difference between energy flows in natural and agro-ecosystems is the large auxiliary energy flows in modern industrialized agriculture. These auxiliary energy flows are derived mainly from fossil fuels and constitute the fertilizers, bio-

---

Box 14.2 Food: A Global Perspective

The purpose of agricultural production, of course, is to produce food for humanity. Food supplies have increased remarkably over the last four decades, about two-and-a-half times. However, up to 20 million people still starve to death every year or die of diseases attributed to poor nutrition. The situation is particularly bleak in Africa, the only continent where food production has not outstripped population growth over the last few decades. If projections are correct, food supplies will have to double or triple over the next thirty years to keep pace with demands. However, supplies of grain are at an all-time low due to poor harvests and greater demand. The world price for wheat doubled between 1995 and 1996.

In the past, increases in food supplies have been brought about by various means. Low populations and a relative abundance of land meant that more land was brought under cultivation as populations grew. The potential for further increases in the area of cultivated land is now small. Most good land is already under cultivation and remaining lands may require high inputs of water or chemicals to make them productive. Furthermore, much of the remaining lands are in areas with unsuitable climates, such as the Arctic, or unsuitable soils, such as in large areas of the tropics. There is also a limit to the amount of the globe's surface we can turn into agricultural land without severely impairing the ecological processes that sustain life. Many experts feel that we have already reached these limits.

The other main way to increase food supplies is through greater food production or intensification per unit area. Starting in the 1940s, research stations in various parts of the world were able to increase dramatically the yields of crops such as wheat and rice through selective breeding and the application of auxiliary energy inputs, as described in the text. Yields of rice, for example, tripled in some locations. This was known as the *green revolution*. There were also some troubling aspects to these

increases, however. Most of these new breeds produced more poorly than the native breeds if fertilizers and biocides were not applied correctly. As these chemicals became more expensive, many poorer farmers were unable to produce high yields. The developments also encouraged a narrowing of the genetic base of the crop. With each farmer growing exactly the same strain of crop, any disease or pest that managed to adapt to the strain had an almost unlimited food supply. There are also concerns that the gains made through these techniques have already been maximized and will not show much future improvement. Macronutrients in the soil (for example, nitrogen, phosphorus, and potassium) can be replenished by fertilizers, but many of the micronutrients required in trace amounts cannot be replaced. As we take out more plant matter in crops from the soil, we may simply exhaust these nutrients, leading to greatly diminished returns in the future. There are already signs of such declines in some areas.

There are, however, hopes that bioengineering may offer further gains. For example, transgenic plants, which include an altered or borrowed gene, might be able to increase the yields of many species or their tolerance to adverse conditions. Many ecologists, however, fear that we will also inadvertently introduce these altered genes into wild species with unforeseen and potentially disastrous results.

There is no doubt that there will be severe challenges to the global food production system in the future if we are to avoid mass starvation on an unprecedented scale. There is no one solution to the problem, but approaches must embrace exploration of new food sources in both the marine and terrestrial ecosystems, continued research to increase yields of existing crops, population control, the breakdown of global agricultural tariffs, more efficient food distribution to those who need it, more moderate consumption in the already overfed countries, and the rationalization of land-ownership.

cides, gas for the tractor, investment in seed research, and a host of other energy subsidies to Western agriculture. Huge increases in crop yields have been achieved over the last century, particularly over the last two decades, mainly by using these energy subsidies. However, in the most energy-intensive food systems, such as in Canada and the United States, on average ten times as much energy needs to be invested through auxiliary energy flows for every unit of food energy produced. In contrast, subsistence farmers relying upon natural energy supplies may produce ten food units for every unit of energy invested.

Not all crops require the same level of chemical inputs. Tomatoes are particularly demanding compared with, for example, cereal crops (*Philip Dearden*).

Two main inputs to industrial farming are fertilizers and pesticides. New strains of crops, such as wheat, will only produce superior yields if fertilized adequately. Large increases in fertilizer inputs have occurred over the last few decades in Canada. Western Canada, for example, had a fivefold increase in fertilizer applications between 1970 and 1990, and the three prairie provinces now account for almost 60 per cent of Canada's total fertilizer sales, up from 28 per cent in 1970 (Figure 14.2). Both fertilizer sales and the amount of area fertilized increased up to 1985, followed by small declines. In some provinces fertilizer applications have declined over this period. In Ontario, for example, both the area fertilized and the amount of fertilizer applied per hectare declined between 1980 and 1990. In contrast, the Atlantic provinces, with the exception of Nova Scotia, have the highest rates of application per hectare (Table 14.1). Overall, Canada applies an average of 24 kg of nitrogen to each hectare of arable land. This is not high by international standards, with Japan and China both applying over 100 kg and the US and Mexico over 50 kg per hectare. Nonetheless, fertilizer application is of environmental concern as fertilizers help speed up the eutrophication process (which was discussed in more detail in Chapter 7) and are also a main contributor to groundwater pollution in some areas.

Agricultural biocides (pesticides) have also been used to boost crop yields by reducing the amount of energy flowing to the next trophic layer of the food chain through the respiration of heterotrophs (often insects), and by eliminating non-food plants that

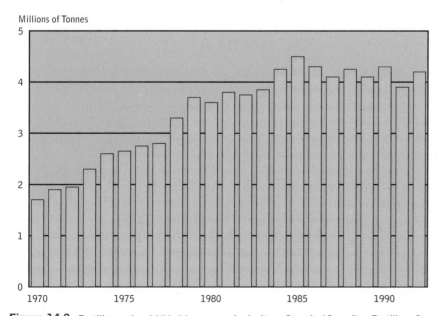

Millions of Tonnes

Figure 14.2 Fertilizer sales, 1970–92. SOURCE: Agriculture Canada, 'Canadian Fertilizer Consumption, Shipments and Trade 1991/1992', working paper (Ottawa: Farm Development Policy Directorate, 1993).

Table 14.1 Commercial Agriculture Fertilizer Application by Province, 1970, 1980, and 1990

Province	Commercial Fertilizer Applied				Area Fertilized				Fertilizer Per Hectare Fertilized			
	1970	1980	1990	Change 1980–90	1970	1980	1990	Change 1980–90	1970	1980	1990	Change 1980–90
	Thousand Tonnes			Per cent	Hectares			Per cent	Tonnes Per Hectare			Per cent
Nova Scotia	27.6	33.5	31.9	–4.8	38 131	88 537	82 267	–7.1	0.72	0.38	0.39	2.5
New Brunswick	38.3	52.5	43.5	–17.1	37 166	75 997	78 136	2.8	1.03	0.69	0.56	–19.4
Quebec	268.8	461.6	467.8	1.3	469 243	1 105 400	996 722	–9.8	0.57	0.42	0.47	12.4
Ontario	670.3	979.5	830.3	–15.2	1 252 259	2 533 823	2 273 262	–10.3	0.54	0.39	0.37	–5.5
Manitoba	116.4	542.9	577.3	6.3	1 186 101	3 196 455	3 688 332	15.4	0.10	0.17	0.16	–7.8
Saskatchewan	77.7	541.1	765.0	41.3	1 498 129	5 525 853	7 654 641	38.5	0.05	0.10	0.10	2.0
Alberta	242.4	734.2	908.9	23.8	2 259 356	5 505 173	6 349 884	15.3	0.11	0.13	0.14	7.3
British Columbia	50.2	96.8	121.9	25.9	129 519	362 104	330 937	–8.6	0.39	0.27	0.37	37.7
Canada	1539.1	3500.7	3810.5	8.9	6 928 292	18 505 200	21 561 732	16.5	0.22	0.19	0.18	–6.6

Notes: Changes in fertilizer tonnages may not precisely reflect the changing nutrient content of fertilizer. Quantity estimates for 1970 and 1990 were derived from fertilizer expense data. Figures may not add up due to rounding.

Sources: Statistics Canada, National Accounts and Environment Division and Agriculture.

compete with the crop plants for available growth resources. Between 1970 and 1990 the value of biocides applied to agricultural lands in Canada increased fourfold. The largest increase in value was in the Prairies where applications have increased tenfold. However, the *amount* of biocides applied per hectare in the Prairies is still considerably less than those applied in Ontario, for example, which had a 1990 application rate of $38 per hectare, almost double that of Manitoba, which was the prairie province with the highest application rate (Table 14.2). These large increases in the application of agricultural biocides over the last twenty years, with almost a threefold increase in expenditure of application per hectare in Canada, have profound environmental implications, as will be discussed in more detail later.

As these changes have taken place, farming systems have become increasingly dependent upon non-renewable resource use, often from sources far beyond the farm, and are dependent upon distant markets for selling their produce. With increasing specialization of product, reliance upon non-renewable energy sources, and dependence upon distant sources for inputs and a market for outputs, modern farms have begun to mimic the industrial manufacturing process rather than the ecological processes upon which they are ultimately based. Many problems have been created because of these changes, which will be discussed in greater depth in subsequent sections.

ENVIRONMENTAL CHALLENGES FOR CANADIAN AGRICULTURE

Urbanization of Agricultural Land

Although Canada is the second largest country in the world, the area of agricultural land is relatively small, only about 13 per cent of the land area of the ten provinces. The amount of arable land is even smaller, totalling less than 5 per cent of the land base. Saskatchewan has the largest amount of prime agricultural land, followed by Alberta, Ontario, and Manitoba (Figure 14.3). Unfortunately, most of the prime agricultural land is in southern Canada, as are most of the country's urban centres. This juxtaposition of prime agricultural land and the main urban centres has meant that expansion of the latter invariably leads to losses in the former.

Concern over the loss of agricultural land because of urbanization led to a major monitoring

Table 14.2 Agricultural Pesticide Expenditures and Application Rates by Province, 1970, 1980, and 1990

Provincial Sub-basin	Agricultural Pesticide Expenditures					Agricultural Pesticide Per Hectare of Cultivated Land				
	1970	1980	1990	Change 1980–90	Change 1970–90	1970	1980	1990	Change 1980–90	Change 1970–90
Newfoundland	185.2	154.7	262.7	68.3	40.3	26.73	16.70	23.80	42.50	-10.9
Prince Edward Island	3,718.5	6,654.1	8,962.3	34.7	141.0	19.35	33.69	51.40	52.60	165.6
Nova Scotia	2,691.9	3,097.0	4,448.2	43.6	65.2	18.65	18.90	32.20	70.60	72.6
New Brunswick	4,477.8	5,525.3	7,148.9	29.4	59.7	24.82	16.20	48.00	196.60	93.5
Quebec	20,295.1	29,231.6	43,069.3	47.4	112.2	8.18	12.98	22.40	72.50	173.7
Ontario	72,652.2	112,360.4	147,547.4	31.3	103.1	17.21	25.81	38.20	47.90	121.8
Manitoba	15,190.2	71,240.1	115,907.6	62.7	663.0	3.00	13.26	21.47	61.90	615.4
Saskatchewan	23,850.8	109,714.2	218,765.1	99.4	817.2	1.29	5.65	10.80	91.30	740.7
Alberta	20,466.3	85,179.1	158,442.9	200.0	674.2	1.19	6.97	12.40	77.60	943.1
British Columbia	9,477.1	12,029.0	16,333.1	35.8	72.3	14.08	13.40	19.10	42.70	35.6
Canada	173,005.0	435,185.5	720,887.6	65.7	316.7	4.04	9.60	15.80	64.40	291.2

Notes: Figures may not add up due to rounding. Farm input price indices were used to obtain 1990 constant dollar expenditures.

Sources: Statistics Canada, National Accounts and Environment Division and Agriculture.

program between 1966 and 1986 in seventy Canadian cities with populations over 25,000. During that period, 301 440 ha of rural land were converted to urban use (Figure 14.4). Furthermore, some 58 per cent of the converted land was of prime agricultural capability (Canada Land Inventory classes 1–3). To replace the productivity of this land would require cultivating twice as much land on the agricultural margins.

Whether these losses (conversions) should be of concern is open to debate. Some, particularly economists, might argue that these losses are not important, while others, particularly those concerned with the long-term food security of the nation, might argue that these losses *are* important. The Canada Land Inventory (CLI) indicates that only 18 per cent of the renewable resource base of the country has prime capability for crops, and less than 2 per cent have the highest capabilities (Class 1). This brings into question how we define lands that are important for agriculture (boxes 14.3 and 14.4).

To slow down the rates of conversion, several provinces have enacted legislation regarding the protection of agricultural lands. For example, in 1972 British Columbia enacted the Agricultural Land Reserve. At the time of its inception, some 6000 ha of prime agricultural land were being lost in the province each year. This has now fallen to less than one-seventh of this figure. Designation of lands to be included in the reserve was made largely on the basis of the CLI classification. Due to the imprecision of these assessments, provision was made for lands to be excluded if they had minimal agricultural potential. Unfortunately, this provision has sometimes been used for political reasons, but overall the legislation has been successful in slowing down the conversion of agricultural land to other uses. Similar programs also exist

Unfortunately, the policies and legislation applied [to slow losses of agricultural land] . . . have had limited success in the face of continued normal and speculative demands for urban development land and low prices for farmland per se.

M.J. Troughton, 'Agriculture and Rural Resources', 1995

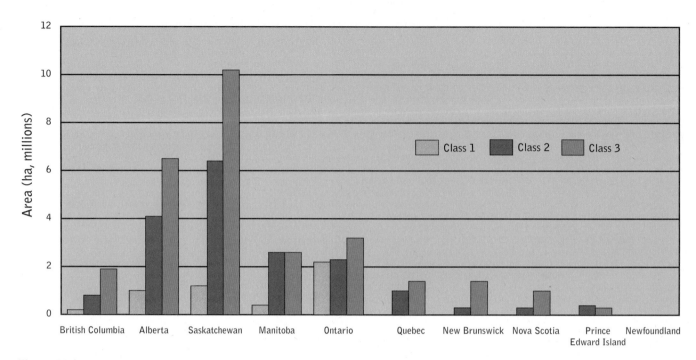

Figure 14.3 Distribution of prime agricultural land, by province. SOURCE: Environment Canada, *Land Capability for Agriculture: A Preliminary Report* (Ottawa: Environment Canada, 1976):9–6.

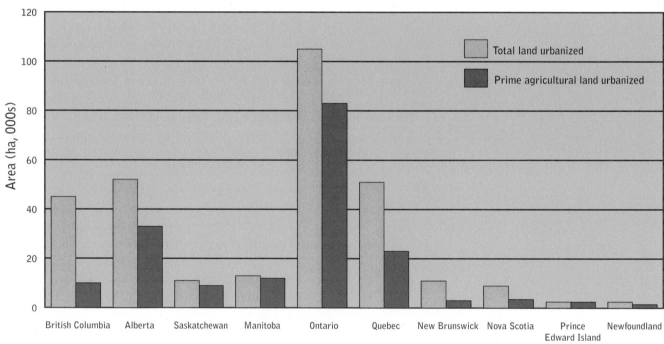

Figure 14.4 Urbanization of farm land, 1966–86. SOURCE: C.L. Warren, A. Kerr, and A.M. Turner, 'Urbanization of Rural Land in Canada, 1981–1986', SOE Fact Sheet no. 89-1 (Ottawa: Environment Canada, 1989).

now in other provinces, such as Quebec, New-foundland, and Prince Edward Island.

Land Degradation

Land degradation includes a number of processes that reduce the capability of agricultural lands to produce food. As agricultural activities have inten-

sified with increased cultivation and addition of agricultural chemicals to produce better yields, so has pressure on the soil resource. At a global scale, the Worldwatch Institute (1992) estimates that reductions in yield of some 11 million tonnes of grain output per year can be expected as a result of these processes. The Standing Committee on Agri-

Box 14.3 Agricultural Capabilities and the Canada Land Inventory

The Canada Land Inventory classifies the capability of land for certain uses. Definitions for classes 1 to 3 for agriculture are as follows:

Class 1: Soils in this class have no significant limitations for crops. These deep soils are level or have very gentle slopes, are well to imperfectly drained, and have a good water-holding capacity. They are easily maintained in good tilth and productivity. Damage from erosion is slight. They are moderately high to high in productivity for a wide range of field crops adapted to the region.

Class 2: Soils in this class have moderate limitations that restrict the range of crops or require moderate conservation practices. These deep soils have good water-holding capacity, can be managed with little difficulty, and are moderately high in productivity for a fairly wide range of field crops. The moderate limitations of these soils may be from any one of a number of factors, including mildly adverse regional climate, moderate effects of erosion, poor soil structure or low permeability, low fertility (correctable with lime), gentle to moderate slopes, or occasional overflow or wetness.

Class 3: Soils in this class have moderately severe limitations that restrict the range of crops or require special conservation practices. Under good management, these soils are fair to moderately fair in productivity for a wide range of field crops adapted to the region. Conservation practices are more difficult to apply and maintain. Limitations arise from a combination of two of the factors described under class 2, or from one of the following factors, including climate, erosion potential, low fertility, strong slopes, poor drainage, low water-holding capacity, or salinity.

culture, Fisheries and Forestry (1984) estimated that land degradation costs Canadian farmers over $1 billion per year.

Soil Erosion

Soil erosion is a serious land degradation problem in Canada and is estimated to cause up to $707 million damage per year in reduced yields and higher costs. In some parts of southwestern Ontario, erosion has caused a loss in corn yields of 30–40 per cent. Further costs are incurred off the farm when sedimentation blocks waterways, impairs fish habitat, lowers water quality, increases the costs of treatment, and contributes to flooding. One study in Ontario estimated these costs to be in excess of $100 million annually (Standing Committee on Agriculture, Fisheries and Forestry 1984).

Soil erosion is a natural process whereby soil is removed from its place of formation by gravitational, water, and wind processes. Under natural conditions in most ecozones, soil erosion is rela-

Most Canadian cities of any size are surrounded by good agricultural lands. As the cities, such as Montreal, increase in size, they invariably encroach on the surrounding lands (*Philip Dearden*).

tively minimal as the natural vegetation tends to bind the soil together and keep it in place. Agricultural activities may either totally remove this natural vegetation and replace it with intermittent crop plantings (thereby exposing the bare soil to erosive processes), or may keep the land under full vegetation for grazing purposes. The latter

Box 14.4 Agricultural Land Importance

Policies designed to protect agricultural lands need to define different qualities of land. Smit et al. (1987) suggest that three criteria should be involved: suitability of land, uniqueness, and demands on the land relative to productive capacity. They categorize existing approaches to define quality into eight types and assess them against the following criteria:

1. *Physical Quality*: These approaches assign land to different categories, based upon the physical qualities of the land. The Canada Land Inventory belongs to this group, as do rating scales devised in many other countries, such as the US and Britain. They do not, however, necessarily reflect the importance for agriculture, since they do not take uniqueness into account, nor the production capacities of lands relative to food needs.

2. *Productivity Indices*: Productivity indices attempt to provide a more precise measure of land productivity than physical quality schemes. They are based on actual yield measurements for various crops and are often expressed as a percentage of the highest possible yield. A high productivity index does not necessarily mean greater importance for agriculture because neither the uniqueness of the parcel nor the demand for agricultural products is considered.

3. *Integrated Suitability*: These schemes attempt to take not only physical characteristics into account but also relevant socio-economic variables that may influence importance, such as adjacent land use and parcel size.

4. *Economic Performance*: Economic performance approaches are similar to productivity indices, but the rating is based upon economic outputs. This may provide good indicators of importance over the short term, but due to long-term fluctuations in prices and costs, they are very time specific.

5. *Market Value*: In theory, the market (for example, what farmers would be willing to pay for a parcel of land) provides an assessment of importance for agriculture. However, in reality the agricultural importance is often difficult to separate from other factors potentially influencing the price of land.

6. *Land Sensitivity*: These approaches attempt to assess, often through mathematical modelling, the relative abilities of land to meet prescribed goals, such as maximizing profits or production, while taking into account constraining factors, such as the physical capabilities of the land or the demand for agricultural products. These approaches have seen little use as straight land-rating schemes due to the analytical complexity of the approach.

7. *Multiple Land Use*: The approaches outlined above focus on the relative qualities of parcels for agriculture. Multiple land-use schemes attempt to broaden the perspective by examining suitability for a number of different uses.

8. *Combination*: Some schemes do not fit neatly into any of the above categories, and may involve combinations of categories. Some approaches, for example, combine physical quality methods with more local information on specific economic and soil variables. These approaches may be very useful at the local scale, but are not as good in providing comparative assessments across provinces, for example.

All these schemes have strengths and weaknesses, but Smit et al. (1987) emphasize the need to continue developing such methods. Alternative potential users of agricultural land often argue for land conversion—to housing, for example—by pointing out the suitability of a particular parcel for housing, the limited supply of land for that purpose, and current and future demands for housing. Until there are well-accepted techniques to justify retaining land for agriculture, land conversion will continue.

approach provides much better protection for the soil, but may still result in erosion, particularly under conditions of high stock density.

The rate of soil formation varies as a function of different environmental factors. Due to Canada's latitude, soil formation is slow and an annual rate of 0.5–1.0 tonne per hectare may be considered average. Any soil erosion above this amount will result in some loss of productive capacity. Losses in excess of 5–10 tonnes per hectare per year may lead to serious long-term problems. These figures have often been exceeded and 30 tonnes per hectare have been recorded in BC's Fraser River Valley under row crops, while 20 tonnes per hectare are not uncommon in Prince Edward Island. Wind erosion is more difficult to measure, but is estimated as a significant problem in the prairie provinces, which have high wind speeds, dry soils, and cropping practices that often leave the soil unprotected. One study in Saskatchewan detected a net output of soil of 1.5 tonne per hectare for an almost level field due to wind erosion, whereas a field with a greater incline (3 degrees) was found to lose 6.6 tonnes per hectare due to a combination of water and wind erosion.

Large areas of tilled soil are particularly vulnerable to erosion (*Philip Dearden*).

To illustrate the complexities of dealing with soil erosion, the results from a study in the Avon River catchment in Perth County, Ontario, are revealing (McNairn and Mitchell 1991). Water erosion of agricultural land has been the main contributor to soil degradation and water quality impairment in that river basin. A study of seventy-five farms revealed that 17 per cent of them experienced high soil erosion (more than 10 tonnes of soil loss per hectare per year), 67 per cent had medium erosion problems (2–10 tonnes of soil loss per hectare per year), and 16 per cent had low erosion problems (less than 2 tonnes of soil loss per hectare per year).

Interviews with the farmers indicated that 79 per cent underestimated the soil erosion on their farms, 3 per cent overestimated the problem, and 18 per cent had an accurate perception. Furthermore, the findings showed that farmers with the most serious erosion problem had the poorest understanding of it. Finally, farmers had a better grasp of the overall soil erosion problem in their township than they did for their own property. Until farmers have a realistic appreciation of the nature of soil erosion on their land, it is unlikely that they will be inclined to adopt soil conservation methods.

Various soil erosion control practices are available. Figure 14.5 shows the percentage of farms in

Canada using various methods. Crop rotation, the most common, is a main means of recharging soil nitrogen through use of legumes, such as alfalfa and clover. Grassed waterways are used to control overland flow of run-off. They direct the flow and hence limit the formation of gullies on exposed soil surfaces. Contour cultivation involves cultivating the soil parallel to the contour of the slope. This reduces the speed of run-off by catching soil particles in the plough furrows. Over 16 per cent of PEI and Saskatchewan crop land is protected in this way. Strip cropping is a similar technique in which different crops may be planted in strips parallel to the slope. When one crop is harvested, leav-

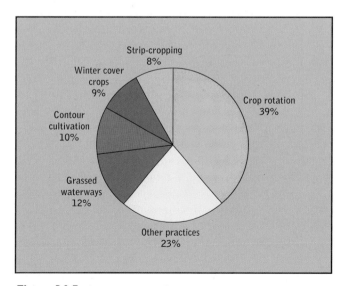

Figure 14.5 Crop rotation is the most common soil erosion control method, Canada, 1991. SOURCE: Statistics Canada, *Trends and Highlights of Canadian Agriculture and Its People*, cat. 96-303E (Ottawa: Statistics Canada, 1996).

ing bare soil, the other crop provides some protection. This technique is often used against wind erosion and is most prevalent in western Canada. The soil surface can also be protected in winter through growth of a winter cover crop. This is effective not only for wind erosion in winter but also to protect the soil from intense rainfall in the spring.

There is considerable regional variation in the use of these practices. Newfoundland, for example, has the smallest proportion of farmers using erosion control, largely because of the small amount of land devoted to seed crops in the province. In Saskatchewan, however, almost 72 per cent of farmers use some form of erosion control. Overall, 63 per cent of farms in Canada use one or more erosion control practices (Figure 14.6).

Farmers can also help reduce soil erosion through seed-bed preparation techniques. Conventional tillage turns over the soil and buries the crop residues. This exposes the soil to erosion, leads to the decline of soil organic matter, and encourages compaction. Conservation tillage helps maintain crop residues on the surface to help prevent erosion by minimizing the amount of tillage and using different implements. It is also possible to have no-tillage seed-bed preparation in which crops are planted using seed drills, and crop residues are spread over the field to help trap moisture. In Canada during 1991, 24 per cent of farms

reported using conservation tillage techniques, and a further 7 per cent used no-tillage. Again, there is considerable regional differentiation, with higher use of conservation and no-tillage in the Prairies.

When all these means of potentially controlling erosion are taken into account, only about 15 per cent of the total seeded area in Canada does not have any of the techniques in operation. These lands may be less prone to erosion, but they may also be the lands under greatest stress. Unfortunately, current information does not allow for the identification of these lands.

Compaction

Soil compaction results from frequent use of heavy machinery with wet soils or from too many cattle. Compaction breaks down the soil structure and inhibits the throughflow of water. Such conditions can reduce crop yields by up to 60 per cent. Soil compaction is a problem mainly in the lower Fraser River Valley in BC and in parts of central and eastern Canada. One estimate puts the annual cost of compaction at between $68 million and $200 million in Canada.

Acidification

Acidity in soils can occur naturally, but can also be augmented by fallout from acid precipitation (see Chapter 7) and the use of fertilizers. Nitrogen fertilizers undergo chemical changes in the soil, which result in production of positive hydrogen ions, causing greater acidity. In the Maritime provinces, where declines in soil pH have been measured, it is estimated that 60 per cent of the change can be attributed to fertilizer use and 40 per cent to acid precipitation. In the Prairies, concern over acidity is relatively recent due to the generally alkaline nature of the substrate. However, increased use of fertilizer has now led to acidification in some areas. Excess acidity reduces crop yields and also leads to nutrient deficiencies and the export of soluble elements, such as iron and aluminium, into waterways. The yields of crops, such as barley and alfalfa, fall sharply at soil pH of less than 6. Liming is a common agricultural practice to reduce acidity. It has also been used to rehabilitate land in the Sudbury area, as discussed in Chapter 13.

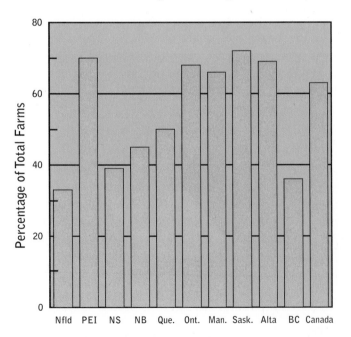

Figure 14.6 Farms using one or more methods of erosion control, 1991. SOURCE: Statistics Canada, *Environmental Perspectives 1993*, cat. 11-528E (Ottawa: Statistics Canada, 1993):49.

Salinization

Salinization is a major problem in many areas of the world where irrigation is common. As water evaporates, it leaves behind dissolved salts. Over time these can accumulate in sufficient quantities to render the land unusable. Ancient civilizations that designed complex irrigation systems were unable to counter the effects of salinization, which contributed to their eventual decline. Estimates suggest that 50–65 per cent of irrigated crop land worldwide will be less productive as a result of salinization by the turn of the century.

Alkaline soils occur naturally in areas of western Canada that have high sodium content and shallow water-tables. Salinization can also be exacerbated through cropping practices in which natural vegetation is removed and surface evaporation increases to concentrate salts at the surface. Summer-fallow has this effect. Summer-fallow is a practice common on the Prairies in which land is kept bare to minimize moisture losses through evapotranspiration. In the Prairies, crop yields have been reduced by 10–75 per cent. Despite the increased use of fertilizers, it is estimated that in some regions salinization is increasing in area by 10 per cent every year. Girt (1990) suggests that 2.2 million ha of dry land and 0.1 million of irrigated lands are affected on the Prairies, causing annual economic losses that are four to five times as great as the total losses due to erosion, acidification, and loss of nitrogen.

Organic Matter and Nutrient Losses

Cultivation involves a continuous process of removing plant matter from a field. In doing so, both the organic and nutrient content of the soil are reduced. Organic matter is critical for maintaining the structure of the soil, influencing water filtration, facilitating aeration, and providing the capacity to support machinery. It also helps to maintain water and nutrient levels.

On the Prairies, current organic matter levels are estimated to be 50–60 per cent of original levels, representing a probable annual loss of about 114 000 tonnes of nitrate nitrogen. This nitrogen is replaced by the addition of synthetic fertilizers (Figure 14.7), which, in turn, contribute to the problem of acidification. An alternative way of replacing the nitrogen is by growing leguminous crops to enhance biological nitrogen fixation, as discussed in Chapter 6. Before the increased use of fertilizers in the 1960s, it was estimated that nitrogen exports from prairie grain exceeded fertilizer applications by more than tenfold, and phosphate removals exceeded inputs by threefold. Current estimates still show a depletion of the soil, but with nitrogen now reduced to double the exports over inputs, and phosphorus inputs about 50-60 per cent of the export.

Biocides

Since the publication of Rachel Carson's classic book, *Silent Spring*, which outlined some of the environmental problems associated with the use of biocides, there has been considerable controversy regarding the relative benefits of their use. On the one hand, they have helped to boost yields throughout the world to meet the food demands of rising populations. Many more people would be starving without the use of biocides. They have also saved countless lives throughout the world by helping to control various diseases by attacking vectors, such as malaria-carrying mosquitos. On the other hand, scientific evidence indicates that many chemicals have profound negative impacts on ecosystems. Despite this evidence, biocide use has continued to grow at a global scale, but has shown some slight decline in Canada in recent years, as discussed earlier.

An ideal pesticide would be environmentally benign except to a certain target pest species, would break down harmlessly after doing its job, and would not cause genetic resistance in the target species. Unfortunately, such a chemical has yet to be synthesized. Existing pesticides can be thought of as biocides because they kill pests as well as other organisms that are not pests. Most chemicals now fall into four main groups (Table 14.3).

> *It was a spring without voices. On the mornings that had once throbbed with the dawn chorus of robins, catbirds, doves, jays, wrens and scores of other bird voices there was now no sound; only silence lay over the fields and woods and marsh.*
>
> *Rachel Carson,* Silent Spring, *1962*

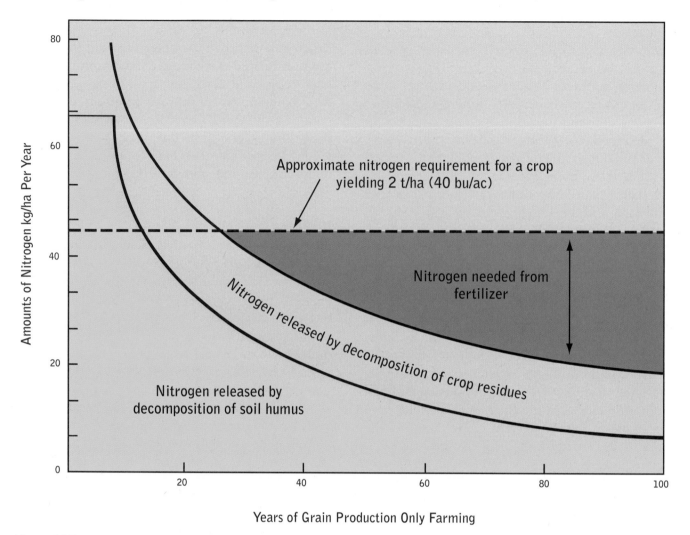

Figure 14.7 Diagrammatic illustration of approximate sources of nitrogen needed to maintain grain yields of about 2 tonnes per ha (40 bu/ac) of barley under a system of continuous grain production in the Prairie region. Note that this diagram illustrates plant requirements, not supply, i.e., the amount of fertilizer nitrogen applied would normally be greater than the plant requirements because of losses due to denitrification and/or leaching. SOURCE: C.F. Bentley and L.A. Leskiw, *Sustainability of Farmed Lands: Current Trends and Thinking* (Ottawa: Canadian Environmental Advisory Council, 1985):22.

Table 14.3 Major Types of Pesticides

Type	Examples
Insecticides	
Chlorinated hydrocarbons	DDT, aldrin, dieldrin, endrin, toxaphene, lindane, kepane, chlordane, methoxychlor, mirex, heptachlore
Organophosphates	malathion, parathion, diazinon, TEPP, DDVP
Carbamates	aldicarb, carbaryl (Sevin), carbofuran, propoxur, maneb, zineb
Botanicals	rotenone, nicotine, pyrethrum, camphor extracted from plants
Microbotanicals	bacteria (e.g., Bt), fungi, protozoans
Fungicides	
Various chemicals	captan, pentachorphenol, methyl bromide, carbon bisulphide
Fumigants	
Various chemicals	carbon tetrachloride, ethylene dibromide (EDB), methyl bromide (MIC)
Herbicides	
Contact chemicals	atrazine, paraquat, simazine
Systemic chemicals	2,4-D, 2,4,5-T, daminozide (Alar), alachlor (Lasso), glyphosate (Roundup)

Various environmental effects can be attributed to these biocides, but there is considerable variation among them in terms of these effects.

Resistance

When a population of insects is sprayed with a chemical, different individuals within the population will react in various ways. If the pesticide is effective, most of the population will be killed, but it is likely that a small number will have a higher natural resistance and will survive. This remnant resistant population may then grow rapidly in numbers due to the lack of competition from the dead insects. Seeing a resurgence of the pest insects, the farmer sprays again, and is again successful in killing some of the population, but not as high a proportion as before since the natural resistance has been passed on to a larger proportion of the population. As this process repeats itself, the use of chemicals creates a population that will ultimately be quite resistant to the chemical.

Over the last forty years, more than 500 insects have developed such resistant populations. Over half the major weed species are also resistant to herbicides, and more than ten species of rodents are resistant to poisons. Across Canada, farmers have experienced increasing difficulty in controlling pests through spraying. In New Brunswick, Colorado potato beetles that were killed with one spraying a season five years ago must now be sprayed five or six times, but they still cause substantial damage. Canola, the second most valuable crop on the Prairies (worth over $1 billion annually) is in danger because the pesticides used to protect it from flea beetles are having less effect. In British Columbia, the pear psylla, an aphid-like insect that feeds on pears, has become resistant to the main pesticide registered for use against it.

Non-selective

Many biocides are popular because they are broad-spectrum poisons. In other words, there is no need to identify the specific pest because a broad-spectrum poison will kill most insects. Unfortunately, they tend to eliminate not only the pest species but also other valuable species, including some that may control the population of the pest. This may result in a population explosion of the resistant members of the pest population after spraying, and also the development of new pests previously kept in check by natural predators. Biocides may also be extremely toxic to organisms other than insects, as described in Box 14.5.

Mobility

The purpose of biocides is to reduce the impact of a particular pest species or several species on a particular crop in a particular area. However, the effects of the chemical application are often felt over a much wider area, sometimes thousands of kilometres, due to the mobility of the chemicals in the earth's natural cycles, particularly the hydrological cycle, and the manner in which chemicals are applied. The US Department of Agriculture, for example, estimates that less than 2 per cent of the insecticide and less than 5 per cent of the herbicide in aerial spraying reach the target. The remainder finds its way into the ecosystem, where it may contaminate local water supplies or be transported by atmospheric processes to more distant sites (Box 14.6).

Between 1986 and 1990, studies undertaken on water quality at the mouths of the Grand, Saugeen, and Thames rivers in southern Ontario, some of the most important agricultural watersheds in that province, found that 72 per cent of the samples had atrazine present, with lesser frequencies of metolachlor, 2,4-D, cyanazine and a few other chemicals. Calculations suggest that between 342 and 2959 kg per annum of total atrazine passed through the river mouths in those years, entering lakes Erie or St Clair.

Persistence

Not only do biocides spread over vast areas, they also contaminate over time as many of them are very persistent. The time taken for some common organochlorine insecticides to decay is shown in Table 14.4. DDT is one of the best-known insecticides. First synthesized over 100 years ago, it was not until the 1940s that the chemical became widely used, first in health programs in the Second World War to control disease vectors and then later as an agricultural chemical. Production peaked by 1970 when 175 million kg were manufactured. By that time the environmental effects of DDT were becoming better understood, and its use (but not manufacture) was banned in the US in 1972. It was not until 1985 that registration of all DDT products was discontinued in Canada. More than 7 million

Box 14.5 Carbofuran and Birds

Carbofuran is a carbamate (Table 14.3) registered for agricultural use in Canada. It is available either as a liquid or in granular form on particles of grit. It is often applied in the latter form during seeding to protect recently germinated seedlings from insects. Considerable amounts of the chemical (up to 30 per cent of that applied) are often exposed on the soil surface following application. These granules are highly attractive to many birds, which ingest grit for use in their gizzards for grinding seeds. The chemical is highly toxic to birds, and the consumption of just one grain can be fatal to small seed-eaters. The carbofuran also contaminates invertebrates, such as earthworms, which in turn will poison organisms higher on the food chain. Flooded fields are also highly dangerous, as the chemical goes into solution. This is particularly serious in acidic fields, where the breakdown of carbofuran is very slow.

There have been many documented bird kills as a result of the use of carbofuran. Some Canadian examples include:

- *December 1973*: Fifty to sixty pintails and mallards killed in flooded turnip fields in BC
- *November 1974–January 1975*: Eighty ducks killed in flooded turnip fields in BC
- *October–December 1975*: 1,000 green-winged teal killed in flooded turnip fields in BC
- *May 1984*: Over 2,000 Lapland longspurs were killed in a rapeseed field in Saskatchewan
- *September 1986*: An estimated 500–1,200 seed-eating birds were killed in turnip and radish fields in BC
- June 1986: Forty-five gulls died after eating carbofuran-contaminated grasshoppers in Saskatchewan

Of particular concern is the impact of carbofuran on one of Canada's threatened species, the burrowing owl. The owl nests on the Prairies, feeding on small mammals and insects and nesting in abandoned mammal burrows. Two-thirds of the Canadian breeding population nest in Saskatchewan, where studies suggest that

breeding numbers declined by 50 per cent in the southcentral region between 1976 and 1987. Carbofuran spraying for grasshopper infestations is suspected as the main cause. There are significant declines in nesting success and brood size with increasing proximity of carbofuran spraying to the nests. A high percentage of adult owls also disappeared after spraying. There were particularly severe infestations of grasshoppers in the early 1980s, and it is estimated that in Saskatchewan alone over 3 million ha were sprayed in 1985 with 40 per cent of the area sprayed with carbofuran. The registration of carbofuran was finally cancelled in 1996. Meanwhile, a 1995 report on the status of the owl suggests that it be changed from threatened to endangered (see Chapter 16), and concludes that 'Unless the population trend is reversed, however, all indications are that the Burrowing Owl will be extirpated from Manitoba within a few years and from the entire country within a couple of decades' (Wellicome and Haug 1995:1).

The burrowing owl is an endangered species that has been particularly threatened by the use of agricultural biocides (*World Wildlife Fund Canada/T. Muir*).

Box 14.6 Arctic Pollution

We tend to think of the Arctic as a pristine wilderness. Recently, however, measurements have indicated that this is far from the truth. For example, at Ice Island, a floating ice-research station 1900 km above the Arctic Circle, concentrations of hexachlorocyclohexanes (HCHs), a family of pesticides, have been measured at twice the level of those in agricultural southern Ontario, yet there is not a single pesticide-dependent product grown in the North. The answer to this seeming anomaly is the chemical processes that transport chemicals from one place to another around the globe.

The so-called 'grasshopper effect' is one reason behind Arctic pollution. After the chemical first entered the environment, some of it is absorbed into lakes and the soil, or is deposited in plants. As the weather warms up yearly, some of these absorbed chemicals re-enter the environment as a gas (or 'de-gas'). The amount of de-gassing depends upon the volatility of the product, or the ease with which it undergoes de-gassing. The pollutants are then carried by weather patterns until increasing cold causes them to recondense and precipitate. The cycle then repeats itself. Chemicals emitted in southern Canada, for example, may go through the grasshopper cycle several times and take ten years to reach the Arctic. The more volatile the chemical, the more of it will travel further. Thus biocides such as lindane (an HCH) are quite volatile compared with DDT, for example, and usually reach the North in greater quantities.

The implications of this long-distance transport are serious. Arctic ecosystems are more vulnerable to toxic effects because toxins last longer in the North. Degradation processes are inhibited due to low temperatures and reduced ultraviolet radiation from the sun. The cold also condenses and keeps the chemicals locked up and slows evaporation rates. This has implications for the Inuit people in particular due to their high consumption levels of wildlife. Over 80 per cent of Inuit consume caribou, almost 60 per cent consume fish, and almost 40 per cent consume marine mammals, so they are at risk due to bioconcentration. It also has implications for the welfare of Arctic wildlife populations.

Concerns such as these merit the same kinds of approaches as those outlined in Chapter 20 on the global commons. They cannot be addressed unilaterally. One step in the right direction would be to get the other Arctic countries to upgrade the Arctic Environmental Protection Strategy into a legally binding treaty.

---

kg of DDT were sprayed on forests in New Brunswick and Quebec between the early 1950s and the late 1960s. DDT is extremely persistent, as can be seen in Table 14.4. Even now there are still considerable residues of DDT and its main breakdown product, DDE, in the environment. Because it is soluble in fat, it may also gradually accumulate over time in the tissues of organisms. This is known as *bioaccumulation*.

Organisms with long life spans are particularly susceptible to bioaccumulation. In British Columbia, for example, hexachlorocyclohexane, an insecticide banned in the late 1970s, was once used as a timber preservative and agricultural spray, and has now

Table 14.4 Persistence of Some Organochlorine Insecticides in Soil

Chemical	Typical Annual Dose (kg/ha)	Half-life (years)	Average Time for 95% Disappearance (years)
Aldrin	1.1–3.4	0.3	3
Isobenzan	0.3–1.1	0.4	4
Heptachlor	1.1–3.4	0.8	3.5
Chlordane	1.1–2.2	1.0	4
Lindane	1.1–2.8	1.2	6.5
Endrin	1.1–3.4	2.2	7
Dieldrin	1.1–3.4	2.5	8
DDT	1.1–2.8	2.8	10

Source: C.A. Edwards, *Persistent Pesticides in the Environment* (Cleveland: CRC Press, 1975):231.

been found in geoducks off the west coast of Vancouver Island and in Puget Sound. Geoducks are large clams that may live as long as 140 years. As filter feeders, they tend to be very susceptible to bio-accumulation, which, when combined with their lengthy life span, have resulted in concentrations of toxins high enough to render the clams unusable by processing plants. In another example of the persistence of biocides in aquatic environments, researchers have detected high levels of toxaphene in trout in Banff's Bow Lake from an application in 1959 to rid the lake of what were then seen as undesirable fish species. DDT and other chemicals were also detected in trout in many of the lakes in Waterton, Banff, Jasper, and Yoho national parks. You will see in Chapter 18 how the bioaccumulation characteristics of filter feeders have been used to monitor water pollution.

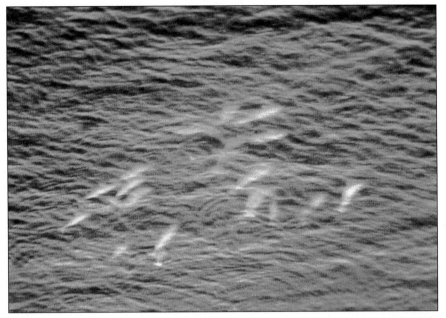

The beluga whale population has declined rapidly in the St Lawrence, and the toxic burden from biocides appears to be one cause (*Earthroots*).

Biomagnification

Another negative environmental effect of many biocides is also typified by DDT. Concentrations of fat-soluble chemicals increase as they are passed along the food chain. This process is illustrated in Figure 14.8, which shows the concentration of DDT and its derivatives along a food chain in the north Pacific Ocean. The relatively low concentrations at the lower end of the food chain are magnified many times by the time they reach top fish-eating predators.

Some species build up very large concentrations. The belugas of the St Lawrence estuary, for example, show concentrations of 70,000–100,000 ppb of DDT. The fact that their population has fallen to less than 10 per cent of the original population in the area, and that individual life spans are about half the normal life span for the species indicate that this level of toxic burden exceeds their levels of tolerance. Scientists examined seventy-three of 175 carcasses that washed ashore between 1983 and 1994, and found that 20 per cent of them had intestinal cancer. These whales account for more than half the cancer cases among all dead whales in the world. Of the 1,800 whales washed ashore and examined in the US, scientists found cancer in only one. Cancer has never been reported in Arctic belugas.

A major component of the toxic burden for the belugas is mirex and its by-products. Mirex is a biocide, now

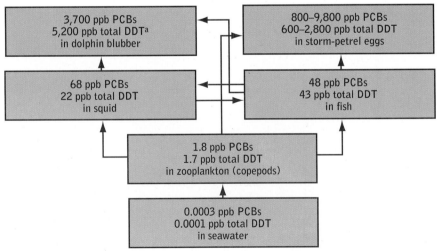

aTotal DDT=DDD+DDE+DDT.

Figure 14.8 Organochlorines in a north Pacific food chain. SOURCE: P.G. Noble, *Contaminants in Canadian Seabirds*, SOE Report no. 90-2 (Ottawa: Environment Canada, 1990).

banned, that was never produced along the St Lawrence. Biologists think that the source is Lake Ontario, where the chemical is accumulated by American eels, which, during their migration downstream, constitute a significant part of the whales' food supply. This example again emphasizes the need to take an ecosystem perspective, as outlined in Chapter 8, when considering the linkages between different parts of ecosystems and human activities and ecosystems.

During the 1960s and 1970s biomagnification was largely responsible for the drastic reductions in population of many birds, particularly birds of prey, such as ospreys, peregrine falcons, and bald and golden eagles, as well as fish eaters, such as double-crested cormorants, gannets, and grebes. Although some were killed directly through bioaccumulation and biomagnification, many were simply unsuccessful in breeding. DDT affects the calcium metabolism of these species, resulting in thinner eggshells, leading to breakage and chick mortality. The banning of DDT and similar chemicals has led to a recovery of many of these species in temperate countries (Figure 14.9). The continued use of the chemicals in some tropical countries affects populations in these areas and also the populations of migratory species, such as peregrines. Together, biomagnification and bioaccumulation are often known as *bioconcentration*.

Humans are also exposed to the harmful effects of biocides through bioconcentration. They may come into contact with chemicals through food products or the water supply. People more dependent upon fish or game-birds for food, as those in many First Nations communities are, may be particularly at risk because they consume large quantities of organisms that may already have concentrated significant amounts of toxic matter. Workers exposed to chemicals on an ongoing basis, such as farm workers or those in chemical manufacturing plants, may also be at risk due to bioaccumulation.

Synergism

When chemicals are tested for their harmful effects, they are tested individually in controlled situations. When applied in agriculture, however, they interact with each other and the environment in a myriad of ways. A single pesticide may contain up to 2,000 chemicals. As they break down, new chemicals are created that may again react with each other in unpredictable ways. Often the combined effects are greater than the sum of their individual effects. This is called synergism and can result in many unanticipated effects.

The concerns outlined earlier are significant. However, the value of biocides in agriculture in Canada is also very significant. Ending the use of biocides would not be feasible, but several steps, such as government regulations and integrated pest management, can be taken to minimize the damage.

New chemicals cannot be introduced into the market without undergoing scientific tests to determine if they would cause cancer, birth defects, and mutations. These tests have been undertaken by private laboratories on behalf of the manufacturers, and the results are submitted to Agriculture Canada for registration. This process has not been foolproof. For example, in 1986 Agriculture Canada cancelled the registration of two commonly used pesticides, barban and alachlor, after it was discovered that the laboratory tests for their registration had been falsified by one of the testing laboratories. Alachlor was subsequently found to be *carcinogenic* and *oncogenic* (tumour producing).

Under the Canadian Environmental Protection Act (1988), new chemicals must be approved by two federal departments (Environment Canada and Health Canada) before commercial release. There is also a Priority Substances List of the existing 35,000 chemicals in use in Canada. A chemical is placed on the Priority Substances List if it meets one or more of the following criteria: if it is released into the environment in significant quantities, if it is subject to bioconcentration, or if it causes or has the potential to cause adverse effects on human health or the environment. The review will determine the appropriate regulatory action to be taken for each chemical.

The application of chemicals is also a concern. Although they may be relatively benign when applied under prescribed conditions, the misuse of chemicals is rife. Many poisonings have been attributed to inappropriate application. This is one reason why farmers in Ontario have welcomed mandatory pesticide safety courses dealing with the use, mixing, handling, transportation, and laws governing pesticide use. The certificates from these courses must be renewed every five years, and after

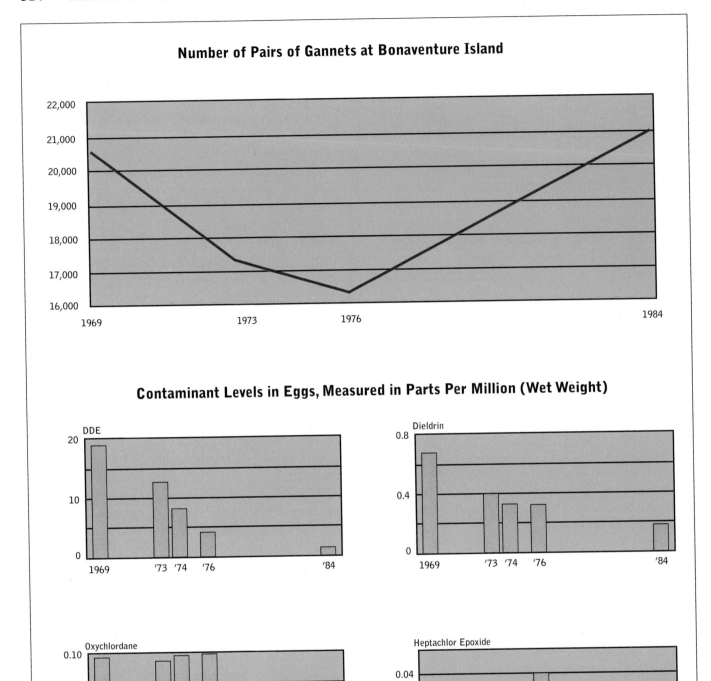

Number of Pairs of Gannets at Bonaventure Island

Contaminant Levels in Eggs, Measured in Parts Per Million (Wet Weight)

NOTE: Almost all contaminants appeared to decline after 1976. DDE, dieldrin, and PCBs declined significantly between 1969 and 1984. DDE, the main by-product of DDT, shows the most dramatic decline, probably reflecting the sudden elimination in 1969 of DDT from forest spray programs in Quebec and the Maritime provinces.

HCB and PCBs, which originate mainly from industrial processes, declined more slowly, as these chemicals were slowly phased out.

Oxychlordane and heptachlor did not change significantly over the period sampled. Chlordane, the main source of these two chemicals, was not restricted in Canada until 1978, which may explain the increasing levels between 1969 and 1976.

Figure 14.9 Numbers of pairs of gannets at Bonaventure Island, Quebec, 1969–84, and contaminant levels in gannet eggs at the colony, 1969–84.
SOURCE: P.G. Noble and S.P. Burns, *Contaminants in Canadian Seabirds*, SOE Fact Sheet 90-1 (Ottawa: Environment Canada, 1990).

Box 14.7 Are We at Risk?

Canadians buy about 225 kg of fresh fruits and vegetables per person per year, compared to 160 kg in 1970. About half the vegetables and almost 90 per cent of the fruit is imported, mostly from the US. Government sampling indicates that imported produce has higher pesticide residues than domestic produce. Between 1988 and 1991 Agriculture Canada found that 23 per cent of imported samples and 9.5 per cent of domestic samples had pesticide residues. However, only 3 per cent of imports and .5 per cent of domestic produce violated permissible levels. In 1994 sampling indicated that out of a total of seventy shipments of strawberries from Florida, forty had higher than allowable captan levels of 5 ppm. Some samples were as high 85 ppm.

Are we at risk? What do you think? What can you do? What should government regulatory agencies do?

1996 they must be presented in order to buy agricultural chemicals. Ontario also has a program to reduce the use of biocides by 50 per cent between 1987 and 2002. Average declines are in the order of 4 per cent every year, giving every indication that this target will be met.

Integrated pest management (IPM) is being considered as a strategy to reduce biocide use throughout the world. It means returning to a wider variety of techniques, such as crop rotation, biological control, pest monitoring, limited use of chemicals, and the development of pest-resistant species. IPM considers the crop and the pest as part of a larger agro-ecosystem, and seeks to keep pest damage below economically damaging thresholds. However, use of IPM requires more expertise than simply applying chemicals. For this reason, Ontario has established a formal system of IPM for twenty-one of the province's ninety-five agricultural commodities. Producers can obtain expert and current advice by phone on these products, their pests, and optimal courses of action. British Columbia also has an IPM program that has the rather modest goal of reducing pesticide application by 25 per cent by the year 2001. Internationally, countries such as Indonesia, which has developed an aggressive IPM program, has managed to cut pesticide use by as much as 90 per cent, whereas Sweden has reduced pesticide use by 50 per cent.

Towards Sustainable Agriculture

The foregoing has emphasized some of the environmental challenges confronting Canadian agriculture. However, socio-economic challenges—such as the decline of family farming, product price volatility, rising input costs, and macroeconomic policies—are also cause for concern. Allied with the environmental challenges outlined earlier, these have given rise to increasing interest in the concept of *sustainable food production systems*. Brklacich, Bryant, and Smit (1991) suggest that six key concepts underpin sustainability in agricultural systems:

1. *Environmental Accounting*: Environmental accounting concentrates on the biophysical limitations for agricultural production. It does not, however, develop explicit linkages between the quality of the environment and productive capabilities.

Genetic Improvement of Crops in Canada

Genetic diversity has allowed crop breeders around the world to improve many crops by adapting them to local conditions. For example, in Canada:

- *Agriculture Canada's Rust Research Laboratory has bred and released a series of wheat varieties that are genetically resistant to wheat stem rust, a fungus that wiped out spring wheat crops in 1916. As a result, there has been no stem rust epidemic in western Canada since 1954 and therefore no longer a need to use pesticides to control it.*
- *Since the 1950s, characteristics—such as high protein and energy, seed dormancy, and disease resistance—have been incorporated into new varieties of oats (Canadian Biodiversity Strategy 1995:33).*

Box 14.8 Biocides and You

Although most biocides in Canada are used by commercial producers, large amounts are also used domestically in homes and gardens to control unwanted organisms. How people use, store, and dispose of these chemicals is very important in terms of minimizing environmental damage. Here are a few tips:

- Use chemicals only as a last resort. Ask yourself why you need to kill the organism. If it is just for aesthetic reasons, such as removing dandelions on your lawn, then maybe you need to change your perceptions rather than automatically reach for a chemical solution. For each pest or weed, there are usually several other approaches you can take as part of your own integrated pest management strategy. These are too numerous to document here, but you can find out more specific strategies from government ministries, such as the Ministry of Agriculture in your area.
- Use the safest chemicals available in the minimum quantities. Many plant nurseries now sell products that are less toxic than traditional biocides. Often they need more skill in application, but are less environmentally damaging than regular chemicals. Examples of these include the Safer Soap line of products.

- Apply all chemicals in strict accordance with the manufacturer's instructions.
- Store unused chemicals so that they do not leak and cannot be accidentally spilled.
- Dispose of chemicals and containers in a safe manner. Contact your local Ministry of Environment to see what programs are in place in your province for safe disposal. Some provinces, for example, have specific disposal sites for biocides and other toxic chemicals. If your province does not have such a program or acceptable alternative, start lobbying for one!

You should also protect yourself against the risk of ingesting chemicals that have been applied to food:

- Grow your own food; don't use chemicals.
- Buy organically grown produce whenever possible.
- Fruit and vegetables that look perfect often do so because they have had heavier applications of fertilizers and pesticides. Choose products that show more natural blemishes; this is a sign that chemical use has not been as high.
- Carefully wash all produce in soapy water.
- Remove the outer leaves of vegetables, such as cabbage and lettuce, and peel all fruit.

2. *Sustained Yield*: For agriculture, the concept of sustained yield refers to output per unit area, and helps to describe the conditions that will maintain yields from year to year. There is thus a more explicit link between environmental degradation and productivity than in environmental accounting.

3. *Carrying Capacity*: Carrying capacity explores the relationship between the number of organisms that can be supported indefinitely in a given unit area. The concept was initially devised to determine the number of grazing animals that could be supported without impairing the quality of the range supporting them. It has since been extended into many different areas. It can, for example, be used to estimate the population carrying capacity of

different nations in terms of food production, although many different variables must be taken into account in addition to agricultural ones. Carrying capacity is also determined by considerations such as what constitutes adequate nutrition, and what kinds and amounts of environmental change are acceptable. In most cases no absolute number can be calculated to represent carrying capacity as many management variables must be taken into account.

4. *Production Unit Viability*: Sustainability requires farming to be biophysically and economically viable. Increasing costs of inputs and volatile markets for produce have created increasing stress for many farmers. The decline of rural communities has also meant a decline in the

social support for farming. Several authors have suggested that a vigorous rural community must underlie sustainable agriculture. Thought must also be given to the relationship between the short-term costs of achieving long-term environmental sustainability and how these can be met by society. Investigations into *resilience*—the ability of different kinds of farms to withstand stress (also discussed in Chapter 9)—is important here.

5. *Product Supply and Security*: A sustainable food production system must be able to provide food on a reliable and continuous basis to meet the demands of the population. Although food productivity has increased dramatically over the last few decades, increasing signs indicate that this trend is abating, and that population growth is outstripping food production in several areas of the world, especially sub-Saharan Africa. In 1970 Africa was virtually self-sufficient in food supply. Now about 30 per cent of the people south of the Sahara suffer from chronic hunger and malnutrition.

6. *Equity*: Although enough food is produced to feed all people on the planet, millions of people suffer from malnutrition due to the inequitable distribution of food. Not only must food be distributed more equitably, there are also concerns that we should not inhibit the abilities of future generations to produce food. Food production systems that cause environmental deterioration and pass the costs of this on to future generations are not sustainable.

On the basis of these concepts, Brklacich, Bryant, and Smit 1991) suggest that a sustainable food production system should have three qualities:

- maintain or enhance environmental quality
- produce adequate economic and social rewards to all individuals and firms in the production system
- produce a sufficient and accessible food supply

Achieving sustainable food production systems is complicated. Brklacich, Bryant, and Smit (1991) note that their definition is generic and does not

prescribe particular cultural or economic structures to achieve sustainability. There may be a number of different ways to proceed, depending upon the situation. One way to address a specific situation is to examine the factors that seem to be encouraging non-sustainable activities. Pierce (1993) suggests that these activities can be considered under market-induced factors (such as the commercialization of production, growing integration with agribusiness, heightened international competition, and division of labour) and policy-induced factors (such as price-support mechanisms for various crops). Many of these policies encourage practices that are rational in terms of the short-term interests of the farming community, but may deter long-term sustainability (Table 14.5).

Some progress is being made on reconciling these differences. Several initiatives have emerged from the Canadian Green Plan, such as the National Soil Conservation Program. However, in an analysis of the 1991 budget, Ralston-Baxter (1991) suggested that while $296 million went towards such new programs, $1.5 billion went towards the support of old, environmentally destructive programs. MacRae et al. (1990) suggest that the transition to sustainable practices will require three overlapping phases:

- *Efficiency*, in which conventional systems are altered to reduce their environmental impact and consumption of resources. These strategies are often favoured by decision makers as they are the least expensive and do not require new legislation, complicated technical analyses, or greater interdepartmental cooperation.

- *Substitution*, in which environmentally disruptive products are replaced by ones that are more environmentally benign. These strategies are relatively non-threatening to the status quo, although they may be more complex than efficiency strategies.

- *Redesign*, in which the farm is redesigned to emphasize integration, balance, and response to feedback.

The first two phases represent a continuation of dependence on externally derived curative solu-

Table 14.5 Effects of Canadian Agricultural Policies on the Environment

Type of Program	Program Example	Resource and Economic Effect	Environmental Impact
Price/Income support			
Safety net/ deficiency payment	Federal Agricultural Stabilization Act†	Reduces risk from crop failure	Producing on hazard-prone lands Land degradation
	Provincial price support/stabilization programs	Overprotection of certain crops Excessive allocations of land	Loss of wildlife habitat
Target prices	Western Grains Stabilization Act†	Maximize production on maximum area	Loss of habitat Crop bias Limits diversification Monoculture
	Special Canada Grains Program†	Intensification Increases land values	
Crop insurance	Crop Insurance Act	Influences crops growth Reduces risk from crop failure Encourages increased investment Greater competition against alternative uses of land	Cultivation of marginal/hazard-prone lands Drainage of wetlands
	Gross Revenue Insurance Plan‡	Based on yield and area Increases land under cultivation	Loss of organic matter Pollution of surface water
	Net Income Stabilization Account†		
Input			
Energy/chemical subsidies	Farm Fuel Tax Rebate	Energy inefficiency Overuse of fertilizer	Water pollution Loss of microbiological activity
	Federal Fertilizer Act	Reduces cost of basic farm inputs Increases capitalization	
Transport subsidies	Western Grains Transportation Act	Raises prices for eligible feed grains Discourages crop diversification and livestock breeding	Monocultures
	Feed Freight Assistance Program	Increases allocation of lands to feed grain use Encourages specialization and separation of livestock and crop production	Summer-fallowing
Credit	Farm Credit Cooperation	Permits low-cost loans for increased capitalization and expansion of land	Use of hazard-prone lands Loss of wildlife habitat
	Farm Improvement Loans Act	Industrial agriculture Increases debt Mechanization	

Table 14.5 Effects of Canadian Agricultural Policies on the Environment (continued)

Type of Program	Program Example	Resource and Economic Effect	Environmental Impact
Regional development	Federal/Provincial Agricultural Development Subsidiary Agreements	Expands income/ employment in agrifood sector Drainage and irrigation	Salinization Loss of wetlands Cultivation of hazard-prone land
Water subsidies	PFRA Water Conservation and Development Programs	Improves stable flow of water resources	On-farm salinization Off-farm salinization and sedimentation
Market			
Supply and quality control	Canadian Wheat Board	Delivery quota based upon preset yield and area cultivated for six major grains Discourages alternative crops and crop rotation	Encourages summer-fallow and saline seep
	Canadian Dairy Commission Canadian Egg Marketing Agency (numerous provincial boards and agencies)	Quota value capitalized into land and buildings Reduces competition and alternative production strategies	Excess concentration of animal wastes Discourages more environmentally positive practices, i.e., low input and organic agriculture
Research/extension	Agricultural Research on Natural Resources	Emphasizes technological solutions, maximizing production, and high input of agriculture production strategies	Ignores non-agricultural values and multiples uses Little research into alternatives

†To be replaced by the Farm Income Protection Act
‡New programs

Source: J.T. Pierce, 'Agriculture, Sustainability and the Imperatives of Policy Reform', *Geoforum* 24 (1993):388-9.

tions and inputs. The third phase involves making the farm more ecologically and economically diverse, self-reliant in resources, and self-regulating. It is a long-term strategy that requires a larger range of changes, but has the potential to generate more effective longer-term solutions. However, until the learning curve of the efficiency and substitution phases has been completed, people will probably not realize the need to redesign the system.

Box 14.9 presents the main goals and subgoals of the food production system, as outlined by MacRae et al. (1990), which might provide a focus for the redesign stage. These goals have far-reach-

ing implications related to the role of farmers and government in agriculture, the importance of self-reliance, the definition of an optimal diet and the means to achieve it, consumer choice, and the kinds of institutions and processes necessary to facilitate these changes. More detailed consideration of this last question (the role of political institutions in promoting sustainability) is discussed by Skogstad (1991). She suggested that there are two overriding needs for institutional reform. The first is to have a broader spectrum of interests involved in the formulation of agricultural policies (partners and stakeholders, as in Chapter 11). The second is for an internal reorganization of government to

---

Box 14.9 Goals for a Sustainable Food System and Their Relationship to Ecological Principles

PARAMOUNT GOALS OF THE FOOD SYSTEM

Nourishment
Human development and fulfilment
Environmental sustainability

SUBGOALS

Consumption
A. *Adequacy*: Should give every person access to sufficient food in quantity, quality, and degree of choice, to achieve optimal physical and mental health.
B. *Appropriateness*: Should be matched in production, consumption, recycling, thermodynamics, and technology to both limits and needs of its region and locality.

Security
C. *Dependability*: Should provide every person with a reliable food supply free from social, political, economic, and environmental disruption.
D. *Sustainability*: Should be culturally, environmentally, economically, and technologically sustainable with respect to production and all other aspects of the food system, including resource inputs, cultivation techniques, processing, and distribution.
E. *Efficiency*: Should practice resource efficiency by incurring minimal resource costs (energy, water, soil resources, genetic resources, forest, fisheries, and other wildlife).

Equity
G. *Wealth*: Should generate sufficient income to good producers to provide a quality of life (measured by a variety of indicators) equivalent to that of other sectors of the economy, to maintain vigorous rural communities and enable farmers to fulfil their land stewardship responsibilities.
H. *Flexibility*: Should be open to growth, evolution, creativity, and experimentation to deal with climatic, economic, and political stresses and variability.
I. *Participation*: Has its organization, decision-making process, and course towards the future determined by all sectors of the population that wish to be involved.
J. *Human development and fulfilment*: Must provide opportunities for creative and fulfilling paid and unpaid work, social interaction, psychosocial evolution, and social justice.
K. *Support*: Should interact with the food systems of other nations in such a way that they are able to achieve similar goals, including a sustainable food system.

Source: R.J. MacRae et al., 'Policies, Programs, and Regulations to Support the Transition to Sustainable Agriculture in Canada', *American Journal of Alternative Agriculture* 5 (1990):83.

---

achieve greater interdepartmental collaboration on issues of sustainability. However, Skogstad believed that the current system of government, which requires cooperation between the federal and provincial governments, does have some advantages over a more centrally directed system. In the Canadian system, there is more opportunity for learning between different levels of government and among the provinces.

Overshadowing the future of agriculture are the potential changes caused by climatic change. As pointed out in chapters 7 and 20, there is considerable uncertainty in predicting the effects of climatic change on various sectors of the economy in different parts of the country. What is true for agriculture in the Fraser River Valley may differ dramatically from growing grain on the Prairies, corn in Ontario, and potatoes in Prince Edward Island. In some areas, water supplies will increase and production may be stimulated by warmer temperatures. In other areas, there may be critical water shortages that preclude further agricultural production. In general, crop belts may move farther north, but often this would be into terrain with unsuitable soils. Water stress is likely to be a major factor on current agricultural lands in southern Canada. Factors such as wind erosion, decay of organic matter in the soil, and damage due to pests could all increase. However, the growing season would be longer and it should be possible to grow a wider variety of crops. All in all, the effects of climatic change for agricultural production in this country could be very complicated. The outcome will depend partially upon

Water deficiencies will probably increase on the Prairies as a result of global climatic change, which could have a serious impact on yields. On the other hand, this small garden in Inuvik in northern Yukon might do considerably better (*Philip Dearden*).

Box 14.10 What You Can Do

Although the challenges faced by agriculture at the global and national levels are vast, there are still some ways in which individuals can help:

1. Eat less. This entails finding out about good nutritional habits, so that we consume only the food we really need.
2. Eat lower on the food chain. Most North Americans eat far too much meat. Eating more vegetables will benefit not only the global food situation but also your own health.
3. Feed your pet foods that are lower on the food chain. Dogs and cats will also be healthier if they are fed balanced grain pet foods rather than meat. Be sure to consult your veterinarian.
4. Waste less food. Studies indicate that as much 25 per cent of food produced in

North America is wasted. Toronto has an organization called Second Harvest, which picks up unused perishable food from restaurants and stores to distribute to the hungry. The food would otherwise go to waste. Support such groups or try to start one in your community.
5. Grow at least some of your own food. If Canadians were to devote a fraction of the time and resources they spend on their lawns to growing food, this would leave more food for others elsewhere.
6. Support local food growers and food co-ops. This helps protect agricultural land in Canada from being transformed to other uses.
7. Join one of the NGO groups that specialize in rural development in lesser developed countries.

---

our capacities for adaptive management, as discussed in Chapter 9.

IMPLICATIONS

Agricultural modification is arguably the main impact that humans have had on natural ecosys-

tems. It is, however, also one of the oldest and one that is basically a modification of ecological systems to benefit humans. Over centuries, natural and human-modified agricultural landscapes have existed and transformed from one state to the other with little lasting damage to planetary life-support systems. However, as additional auxiliary energy flows were applied to boost the productivity of agriculture, the differences between these two ecological systems became more distinct, and agriculture's impacts on natural ecosystems increased. Agricultural production (certainly in Canada's commercial agriculture) is now more similar to industrial production than to the natural ecosystem from which agriculture was derived.

This industrialization has led to many environmental challenges for agriculture. Yields are declining in some areas as crops become less responsive to fertilizer input; biocides have helped eliminate many natural enemies of pests; and soils are eroded, salinized, and compacted. In response, researchers are suggesting that a fundamental restructuring is required in how agriculture is undertaken, with the emphasis changing from maximizing productivity to ensuring sustainability. This is a theme that will be repeated in the next chapter on forestry.

SUMMARY

1. Agriculture is an important industry in Canada, accounting for 12 per cent of gross domestic production, 10 per cent of employment, and about 10 per cent of export earnings.

2. Humans have been hunter gatherers for about three-quarters of the time they have existed as a species on this planet, deriving their energy from the food chain, fire, and human muscle power. Agriculture, entailing the domestication of various animal and vegetation species, fundamentally changed humans' interaction with the natural environment. Humans became the main determinant of the distribution and abundance of many species. Natural selection was replaced by human selection in one of the most significant examples of the replacement of a natural system by a control system. Auxiliary energy flows supplemented the natural flows described earlier. In agriculture today, these auxiliary flows may be in excess of those derived from natural sources. Water sources were harnessed to irrigate crops; fire became an important tool to clear land. These changes started to occur roughly 10,000 years ago in different parts of the world. However, it was only during the last 150 years or so that the use of large quantities of aux-

iliary energy flows in the form of fertilizers, biocides, and other products derived from fossil fuels started to increase some of the most damaging impacts of agriculture.

3. Agriculture is a food chain with humans as the ultimate consumers. The second law of thermodynamics dictates that the shorter the food chain, the more efficient it will be.

4. Fertilizer inputs have increased greatly in Canada over the last two decades. Western Canada, for example, had a fivefold increase in fertilizer applications between 1970 and 1990, and the three prairie provinces now account for almost 60 per cent of Canada's total fertilizer sales, up from 28 per cent in 1970.

5. Between 1970 and 1990, the expenditure on biocides applied to agricultural crops in Canada increased fourfold. The largest increases have been in the prairie provinces. The amount applied in Ontario is still more than double that of Manitoba, which has the highest application rate in the country.

6. Only 13 per cent of the land area of the provinces is suitable for agriculture, with the amount of arable land totalling less than 5 per cent of the land base. This land base is growing smaller as a result of conversion of agricultural lands to urban uses. Between 1966 and 1986, 301 440 ha of rural land were converted to urban use, with 58 per cent of that land having prime agricultural capability. Some provinces have enacted legislation to try to slow down this process.

7. Land degradation includes a number of processes that reduce the capability of agricultural lands to produce food. One study suggests that such processes cost Canadian farmers over $1 billion per year.

8. Soil erosion is estimated to cause up to $707 million damage per year in terms of reduced yields and higher costs. Soil formation in Canada is slow. A rate of 0.5–1.0 tonne per hectare may be considered average. Losses of 5–10 tonnes per hectare are common in Canada and figures of 30 tonnes per hectare have been recorded in the Fraser River

Valley. Farmers have various means to control soil erosion. In 1991 only 15 per cent of the total seeded area did not have any of these techniques in use.

9. Increasing acidity as a result of nitrogen fertilizers and acid deposition is also a problem that reduces crop yields. Salinization occurs where there are high sodium levels in the soils and shallow water-tables, such as in the Prairies. One estimate suggests that salinization causes economic losses four to five times as great as losses due to erosion, acidification, and loss of nitrogen.

10. Cultivation involves a continual process of removing plant matter from a field. In doing so, both the organic and nutrient content of the soil are reduced. On the Prairies, current organic matter levels are estimated to be 50–60 per cent of the original levels.

11. Biocides are applied to crops to kill unwanted plants and insects that may hinder the growth of the crop. They have boosted yields throughout the world and helped feed many hungry people, but there is also clear scientific evidence that they have serious negative impacts on ecosystem health.

12. Biocides have been shown to promote the development of resistance among target organisms. Over the last forty years, more than 500 insects have developed resistant populations. They are non-selective and tend to kill non-target as well as target organisms. They may also be highly mobile and move great distances from their place of application. They may also last a long time in the environment and accumulate along food chains. Such biomagnification has resulted in drastic reductions in the populations of some species at higher trophic levels, such as ospreys and bald eagles.

13. Chemicals, and their constituents as they break down, may interact synergistically.

14. Chemicals must be registered by the government before they can be used. Registration requires the chemical company to provide evidence that the chemical is not carcinogenic or oncogenic. Many chemicals have been registered, but subsequently found to fail these tests.

15. Integrated pest management is now more popular and several provinces have such programs. Ontario has a program aimed to cut the use of biocides by 50 per cent between 1987 and 2002.

16. Attention is now being devoted to sustainable food production systems that maintain or enhance environmental quality, produce adequate economic and social rewards to all individuals and firms in the production system, and produce a sufficient and accessible food supply. However, many current government policies encourage practices not consistent with these goals. Government reorganization for agriculture should involve a broader group of stakeholders and encourage more interdepartmental communication.

17. The possibility of future climatic change will present a major challenge to Canadian agriculture in the future. Adaptive management will be important.

REVIEW QUESTIONS

1. How do the laws of thermodynamics apply to agriculture?

2. Indicate how the law of tolerance and law of the minimum discussed in Chapter 4 relate to agriculture.

3. With regard to the principles for sustainable development presented in Chapter 2 and the discussion in this chapter, what do you think are some of the key building-blocks for a sustainable food system in Canada?

4. There is considerable interest in ecosystem management. What is the relevance, if any, of this concept for agriculture?

5. This chapter mentions the resilience of farm systems. In the discussion of adaptive management in Chapter 9, attention was also given to resilience. What value do the concepts of resilience and adaptive management have for agriculture?

6. How would you identify and assess the environmental and social impacts of agricultural policies

and practices in Canada? What ideas from Chapter 10 might be helpful in this?

7. In Chapter 11 attention focused on the ideas of partnerships and stakeholders. What relevance do such concepts have regarding agriculture if partnerships are to be formed among the federal and provincial governments, agribusiness, and family farms?

8. It has been suggested here that there can be conflict or tension between the needs of urban and adjacent rural areas. To what extent do you think alternative dispute resolution ideas might be applied to deal with such tensions?

9. In Chapter 18 there is a discussion of the Rafferty-Alameda Dam and reservoir development in southern Saskatchewan. One of the reasons for this project was to reduce farmers' vulnerability to drought. Do you think the Rafferty-Alameda project reflects the ideas necessary for a sustainable food production system in Canada?

10. If you were a commercial farmer in Canada, to what extent would it be important for you to think about the implications of climate change for your farming operations?

REFERENCES AND SUGGESTED READING

Agriculture Canada. 1993. 'Canadian Fertilizer Consumption, Shipments and Trade 1991/1992'. Working paper. Ottawa: Farm Development Policy Directorate.

Battison, L.A., M.H. Miller, and I.J. Shelton. 1987. 'Soil Erosion and Corn Yield in Ontario: I. Field Evaluation'. *Canadian Journal of Soil Science* 67:731–45.

Bentley, C.F., and L.A. Leskiw. 1985. *Sustainability of Farmed Lands: Current Trends and Thinking*. Ottawa: Canadian Environmental Advisory Council.

Brklacich, M., C.R. Bryant, and B. Smit. 1991. 'Review and Appraisal of the Concept of Sustainable Food Production Systems'. *Environmental Management* 15:1–14.

Bryant, C.R. 1986. 'Agricultural and Urban Development'. In *Progress in Agricultural Geography*, edited by M. Pacione, 167–94. Beckenham, UK: Croom Helm.

Canadian Biodiversity Strategy. 1995. *Canada's Response to the Convention on Biological Diversity*. Ottawa: Minister of Supply and Services Canada.

Carson, R. 1962. *Silent Spring*. Boston: Houghton Mifflin.

Dearden, P., and P. Bloodoff. 1981. 'The British Columbia Agricultural Land Reserve: The Case of the Langley Appeal'. In *Geographical Research in the 1980's: The Nanaimo Papers*, edited by N. Waters, 39–51. Vancouver: Tantalus Press.

Dumanski, J., D.R. Coote, G. Luciuk, and C. Lok. 1986. 'Soil Conservation in Canada'. *Journal of Soil and Water Conservation* 41:204–10.

Edwards, C.A. 1975. *Persistent Pesticides in the Environment*. Cleveland: CRC Press.

Environment Canada. 1976. *Land Capability for Agriculture: A Preliminary Report*. Ottawa: Environment Canada.

Frank, R., H.E. Braun, B.S. Clegg, B.D. Ripley, and R. Johnson. 1990. 'Survey of Farm Wells for Pesticides, Ontario, Canada, 1986 and 1987'. *Bulletin of Environmental Contamination and Toxicology* 44:410–19.

_____, S. Logan, and B.S. Clegg. 1991. 'Pesticide and PCB Residues in Waters at the Mouth of the Grand, Saugeen and Thames Rivers, Ontario, Canada, 1986–1990'. *Archives of Environmental Contamination and Toxicology* 21:585–95.

Furuseth, O.J., and J.T. Pierce. 1982. 'A Comparative Analysis of Farmland Preservation Programmes in North America'. *The Canadian Geographer* 26:191–208.

Girt, J. 1990. 'Land Degradation Costs in Canada: A Recent Assessment'. In *Dryland Management: Economic Case Studies*, edited by J.A. Dixon, D.E. James, and P.B. Sherman, 294–303. London: Earthscan Publications.

Glenn, J.M. 1985. 'Approaches to the Protection of Agricultural Land in Quebec and Ontario: Highways and Byways'. *Canadian Public Policy* 11:665–76.

Gregor, D.J., and W.D. Gummer. 1989. 'Evidence of Atmospheric Transport and Deposition of Organochlorine Pesticides and Polychlorinated Biphenyls in Canadian Arctic Snow'. *Environmental Science and Technology* 23:561–5.

Hoberg, G., Jr. 1990. 'Risk, Science and Politics: Alachlor Regulation in Canada and the United States'. *Canadian Journal of Political Science* 23:257–77.

McCuaig, J.D., and E.W. Manning. 1982. *Agricultural Land Use Change in Canada: Process and Consequences*. Land Use in Canada Series no. 21. Ottawa: Lands Directorate, Environment Canada.

McNairn, H.E., and B. Mitchell. 1991. 'Farmers' Perceptions of Soil Erosion and Economic Incentives for Conservation Tillage'. *Canadian Water Resources Journal* 16:307–16.

MacRae, R.J., S.B. Hill, J. Henning, and A.J. Bentley. 1990. 'Policies, Programs, and Regulations to Support the Transition to Sustainable Agriculture in Canada'. *American Journal of Alternative Agriculture* 5:76–92.

Noble, D.G. 1990. *Contaminants in Canadian Seabirds*. SOE Report no. 90-2. Ottawa: Environment Canada.

_____, and S.P. Burns. 1990. *Contaminants in Canadian Seabirds*. SOE Fact Sheet 90–1. Ottawa: Environment Canada.

Pierce, J.T. 1990. *The Food Resource*. London: Longman.

_____. 1993. 'Agriculture, Sustainability and the Imperatives of Policy Reform'. *Geoforum* 24:381–96.

Ralston-Baxter, R. 1991. *Environmental Assessment of the Federal Budget for the Department of Agriculture*. Ottawa: Resource Futures International.

Russwurm, L.H. 1980. 'Land in the Urban Fringe: Conflicts and Their Policy Implications'. In *Essays on Canadian Urban Process and Form II*, edited by R.E. Preston and L.H. Russwurm, 457–505. Waterloo, ON: Department Geography, University of Waterloo.

Singh, B., and R.B. Stewart. 1991. 'Potential Impacts of a CO_2-Induced Climate Change Using the GISS Scenario on Agriculture in Quebec, Canada'. *Agriculture, Ecosystems and Environment* 35:327–47.

Skogstad, G. 1991. *Political Institutions and a Sustainable Agriculture*. Ottawa: Science Council of Canada.

Smit, B., and M. Brklacich. 1992. 'Implications of Global Warming for Agriculture in Ontario'. *The Canadian Geographer* 36:75–8.

Smit, B., L. Ludlow, T. Johnston, and M. Flaherty. 1987. 'Identifying the Important Agricultural Lands: A Critique'. *The Canadian Geographer* 31:356–65.

Smith, E.G., and C.F. Shaykewich. 1990. 'The Economics of Soils Erosion and Conservation on Six Soil Groupings in Manitoba'. *Canadian Journal of Agricultural Economics* 38:215–31.

Standing Committee on Agriculture, Fisheries and Forestry. 1984. *Soil at Risk: Canada's Eroding Future*. Ottawa: Senate of Canada.

Statistics Canada. 1993. *Environmental Perspectives, 1993*, cat. 11-528E. Ottawa: Statistics Canada.

_____. 1994. *Human Activity and the Environment*, cat. no. 11-509E. Ottawa: Statistics Canada.

_____. 1996. *Trends and Highlights of Canadian Agriculture and Its People*, cat. 96-303E. Ottawa: Statistics Canada.

Sutherland, R.A., and E. de Jong. 1990. 'Estimation of Sediment Redistribution within Agricultural Fields Using Caesium-137, Crystal Springs, Saskatchewan, Canada'. *Applied Geography* 10:205–21.

Trant, D. 1993. 'The 1991 Census of Agriculture: Land Management for Soil Erosion Control'. In *Environmental Perspectives 1993: Studies and Statistics*, 47–51. Ottawa: Statistics Canada.

Troughton, M.J. 1995. 'Agriculture and Rural Resources'. In *Resource and Environmental Management in Canada*, edited by B. Mitchell, 151–82. Toronto: Oxford University Press.

Warren, C.L., A. Kerr, and A.M. Turner. 1989. 'Urbanization of Rural Land in Canada, 1981–1986'. SOE Fact Sheet no. 89–1. Ottawa: Environment Canada.

Wellicome, T.I., and E.A. Haug. 1995. 'Updated Report on the Status of the Burrowing Owl'. Report to COSEWIC. Ottawa: Canadian Wildlife Service.

Worldwatch Institute. 1992. *Vital Signs: The Trends That Are Shaping Our Future*. New York: W.W. Norton.

Chapter 15

Forestry

Canada is a forest nation. We are the only country in the world with a leaf on our national flag. Along with our northern latitude, the forests have provided the historical context for our national identity. Canada has one-tenth of the world's forests, which cover almost half of the nation's land area and a much higher proportion of southern Canada where most Canadians live. The forests not only influence our collective conscience, they also provide a livelihood for 800,000 Canadians in the forest and associated industries, with over 350 communities totally dependent upon forestry and another 600 partially dependent. We are the largest exporter of forest products in the world, the sec-

ond largest producer of wood pulp, the third largest producer of lumber, and the fourth largest producer of paper and paperboard, accounting for almost 20 per cent of the global trade in forest products. These activities were responsible for 3 per cent of GDP and 15 per cent of exports in 1992. The industry contributes over $20 billion per year to the Canadian economy ($28 billion in 1994, for example), and is the largest single contributor to Canada's balance of trade.

Although all the provinces are dominated by forest land, forest types differ significantly (as described in Chapter 3). The volume of wood produced per unit area also differs, rising to over 700 m^3 per hectare on the most productive sites in coastal British Columbia where mild temperatures, deep soils, and abundant rainfall create some of the most productive growing sites in the world. Volumes harvested vary by province. For example, in 1994 British Columbia harvested more than double the volume of any other province while cutting less area than either Quebec or Ontario (Table 15.1).

The industry is also enormously controversial. Names such as Carmanah, Temagami, and Clayoquot are known across the country after they appeared as headlines in newspapers and on national news broadcasts. All these conflicts

The use of wood products is an integral part of the livelihood of many Canadians, as illustrated by this local boat building in Newfoundland (*Philip Dearden*).

Table 15.1 Net Merchantable Volume of Roundwood Harvested by Province/Territory, 1994

Year	Nfld	PEI	NS	NB	Que.	Ont.	Man.	Sask.	Alta	BC	Yukon	NWT	Canada
1994	2,445	519e	5,106e	9,269	38,414E	25,952E	1,786	4,918	17,921p	75,093	421	181	182,025e

Notes: Net merchantable volume = Volume of the main stem, excluding stump and top but including defective and decayed wood; applies to trees or stands
e = Estimated by provincial or territorial forestry agency
E = Estimated by Natural Resources Canada, Canadian Forestry Service, or Statistics Canada
p = Preliminary figures

Source: Canadian Council of Forest Ministries, *Compendium of Canadian Forestry Statistics* (Ottawa: Canadian Forest Service, 1995):79.

revolve around questions of whether particular areas of the forest land base should be logged or preserved. Such conflicts are due to increasing appreciation of the forests' intrinsic value as well as the economic ones mentioned earlier. Few of these values are easy to calculate in monetary terms and compare against the financial returns of the forest industry. However, these values are gaining more recognition from the public and decision makers as the process of converting climax forests across Canada into managed forests continues.

The sheer scale of the industry has also attracted international attention. Concern over the destruction of the tropical forests and resulting impacts on earth processes and biodiversity has spread from countries such as Malaysia, Indonesia, and Brazil to temperate forest nations—Canada in particular—which has been dubbed, rightly or wrongly, 'the Brazil of the North'. Governments in Canada have launched large public relations campaigns around the world to combat this perception. They have also started to look more closely at the basis of these accusations, and substantial changes are underway in many aspects of the forest industry in Canada. This chapter provides some environmental context for the forest industry and outlines some of these emerging management challenges.

FORESTRY AS AN ECOLOGICAL PROCESS

Of the total of 417.6 (cf., 416.2) million ha of forest land in Canada, some 118.9 million ha are currently managed for timber production (Table 15.2). On these lands, forest ecosystems are being transformed from relatively natural systems to control systems (as described in Chapter 1) in which humans, not nature, influence which species will grow there and the age that they will grow to. This change in species and age distributions has a major

British Columbia has some of the largest trees in the world. This is Cathedral Grove, a stand of large Douglas fir and western red cedar on Highway 4 *en route* to Pacific Rim National Park Reserve (*Philip Dearden*).

impact upon ecological processes, such as energy flows, biogeochemical cycles, and the hydrological cycle, and the habitat that is provided for other species. We have only a rudimentary knowledge of how these complex forest ecosystems function. It is hence difficult to be precise about the possible impacts of wholesale conversion from complex natural to more simple human-controlled systems.

There are also important differences among forest ecosystems. Some, such as the boreal forest, have naturally evolved with periodic disturbances, such as fire or insect attack, which stimulate forest

Box 15.1 Forests: A Global Perspective

- Some 40 per cent of the land surface of the earth supports trees or shrubs. Approximately 27 per cent is covered in trees, with the rest in open scrub.
- The largest forest areas are in the former USSR (942 million ha), North America (749 million ha), Europe (195 million ha), and Australia, Japan, and New Zealand (178 million ha). The four countries of Canada, Russia, US, and Brazil contain over 50 per cent of the world's forests.
- Forests occupy tropical, temperate, and boreal zones, with approximately 1.67 billion ha in the last two zones and 1.79 billion in the tropics.
- There are about 100 million ha of forest plantations, of which 30 million are in tropical countries.
- The area of tropical forests declined by an average of 15.4 million ha annually during the 1980s.
- Estimates suggest that over 80 per cent of the world's terrestrial species are found in forests. The tropical forests are our richest terrestrial biome.
- The volume of global wood consumption more than doubled between 1961 and 1991.
- Developing countries consume more than

80 per cent of their wood as fuel; developed countries only 16 per cent, with the rest processed as wood products.
- Most wood products (85 per cent) are used domestically.
- Wood products, including fuel wood, are valued at an estimated US $400 billion, while exports account for about 3 per cent of world trade. Between 1961 and 1990 global trade in forest products increased 3.5 times. Developing countries are responsible for only about 13 per cent of world trade in wood products.
- Global forests provide wage employment and subsistence equivalent of 60 million work years worldwide, 80 per cent of which is in developing countries.
- People in developing countries consume much less wood products (30 m^3 per 1,000 people) and paper (12 tonnes per 1,000 people) than people in developed countries (300 m^3 per 1,000 people and 150 tonnes of paper).
- It is estimated that an additional 50–100 million ha of forests will be needed to meet the projected wood demands of developing countries by the year 2010 (Brown 1994; World Resources Institute 1992).

Table 15.2 Canada's Forests

		Million Hectares
Heritage forests[a] (protected from harvesting by legislation)		22.8
Commercial forests (capable of producing timber and non-timber products)		236.7
• managed forests (capable of producing timber and non-timber products)	118.9	
• unallocated forests (currently managed for timber products)	90.3	
• protection forests (unavailable for harvesting by policy)	27.5	
Open forests (small trees, shrubs, and muskegs)		156.2
Total forest land		**417.6**

[a]Preliminary estimates

Source: Natural Resources Canada, *The State of Canada's Forest, 1995–1996* (Ottawa: Canadian Forest Service, 1996):7.

renewal. Others, such as the rainforests of the West Coast, have little history of disturbance. The importance of disturbance by forestry must be considered against this background as forestry, with some important differences, mimics natural distur-

bance (see Box 15.2). One essential difference is that natural disturbances, such as fire or insect attack, do not result in the physical removal of the biomass from the site; it is merely converted from one form to another at that site. In contrast, log-

Box 15.2 Forest Disturbance: Natural Versus Clear-cut

As indicated in the text, some forests are more susceptible to disturbance than others, and different disturbances have different impacts. There are similarities between some disturbances and clear-cutting, but there are also important differences. Some of the differences between the effects of fire and clear-cutting are summarized below.

- Openings created by fire are generally irregular in shape with high perimeter-to-edge ratios, which facilitate natural reseeding. Boundaries tend to be gradual rather than the abrupt edge of a clear-cut.
- Fires will leave standing vegetation in wet areas that continue to provide habitat for wildlife and act as a natural seed source. Clear-cuts remove all trees.
- Fires tend to kill pathogens. Clear-cutting allows many pathogens to survive.

- Fire releases nutrients to the soil. Clear-cutting removes nutrients in the bodies of the trees.
- Fire helps break up rocks, which aids in soil formation. Clear-cutting physically disturbs the site, leading to compaction and erosion.
- Fire stimulates growth of nitrogen-fixing plants that help maintain soil fertility.
- Fire encourages the continued growth of coniferous species in many areas by stimulating cone opening. Clear-cutting often leads to dominance by shade-intolerant hardwoods. Ultimately, this changes species composition, as was found in studies in Ontario where regenerating poplar and birch increased by 216 per cent and spruce fell by 77 per cent in 1,000 sampled boreal sites (Hearnden, Millson, and Wilson 1992).

ging results in the physical translocation of nutrients from the site. Nonetheless, the closer forest harvesting approximates the conditions of natural perturbations, the less disturbing it will be to ecosystem processes.

Just as agriculture is fundamentally an ecological process (as described in Chapter 14), so is forestry. Chapter 4 described how energy flows through ecosystems. It is also stored in various compartments. Trees, for example, are energy stored in the autotrophic component of the ecosystem; deer represent energy storage in the herbivorous component. Thus a forester may wish to maximize the energy storage in trees and minimize energy losses to herbivores, whereas a wildlife or recreation manager may wish to move the energy storage further up the food chain to support more wildlife. Conflicts therefore arise about where energy should be stored in ecosystems to optimize societal values.

Forestry and Biodiversity

Most natural forest land is dominated by forests with old-growth characteristics, although the age of the trees and degree of structural and compositional attributes described for old growth in Box 15.3 vary greatly across Canada. These forests have ecological attributes that, by and large, are much

reduced or absent for a long time from forests that have been harvested. These changes include physical modifications to the ecosystem, as well as changes in biomass, plant species mixtures, and productivity. These changes affect biodiversity directly and indirectly by changing the nature of the habitat for other species. As there are more than 200,000 species (two-thirds of all the species found in Canada) dependent upon forest habitats, this is obviously of concern.

Direct changes include the effects on the genetic and species richness of the community. Most species have a wide range of genetic variability that helps them adapt to changes in the environment. As natural forests are replaced by plantations, this natural variability becomes reduced since most plantation-grown trees are specially selected from the same genetic base for their desirable characteristics. This makes them more susceptible to pest infestations and diseases, and less able to adapt to future environmental changes. For example, it has been suggested that the emergence of the spruce forest moth in central and eastern Canada since 1980 is at least partially due to the establishment of white spruce plantations (Neilson 1985). Over 90 per cent of the forest harvesting in Canada is done by clear-cutting in which large blocks of forest are removed at the same time. As

Box 15.3 Old-Growth Forests

There are various definitions of old-growth forests. This one was suggested by the Forest Land Use Liaison Committee of British Columbia (as quoted in Harding 1994:253):

Old growth forests include climax forests, but do not exclude sub-climax or even mid-seral forests. The age structure of old growth varies significantly by forest type and from one biogeoclimatic zone to another. The age at which old growth develops the specific structural attributes that characterise old growth will vary according to forest type, climate, site characteristics and disturbance regime. However, old growth is typically distinguished from younger stands by several of the following attributes:

1. large trees for species and site;
2. wide variation in tree sizes and spacing;
3. accumulations of large size dead, fallen and standing trees;
4. multiple canopy layers
5. canopy gaps and understorey patchiness, and
6. decadence in the form of broken tops or boles and root decay.

A clear-cut area (*Philip Dearden*).

may survive the harvest and, when freed from the competitive suppression of the harvested trees, may dominate the community for some time. Trees, such as red maple, yellow birch, and white pine, often fall into this category in central and eastern Canada. Most of these species will eventually be out-competed as the canopy closes. Some species that may survive the harvest or invade from surrounding areas may be found through all stages of succession. They have low light compensation thresholds, allowing survival under heavy shading. Balsam fir, hemlock, sugar maple, and beech fall into this category.

Once clear-cuts have had the opportunity to start regenerating, species diversity increases rapidly and usually results in a plant community that is more species rich and diverse than the harvested community. The exception is when replanting activities involve few species and steps are taken to reduce competition for these plantation trees. Herbicides, such as Roundup, are commonly applied to achieve this and may be applied consistently until the planted trees become established. Under these circumstances, an artificial lack of diversity is created, just as a farmer creates a similar system to maximize the amount of energy stored in the particular component that he or she wishes to harvest.

Naturally regenerating clear-cuts may also produce a higher biomass of herbivorous species, such as whitetail deer and mule deer, which require brushy habitats for at least part of the year. Both the quantity and quality of browse (twigs, young shoots) is greater in regenerating clear-cuts until the canopy starts to close, and is usually optimal in the eight-to-thirteen-year period. In many parts of their range, these species are more abundant than

many foresters have pointed out, this does not destroy the ecosystem *per se* as that ecosystem still exists on that unit of land, but it does dramatically alter the attributes of that ecosystem. For vegetation, this change includes removal of the previously dominant trees and their ecological influence, followed by the vigorous growth of other assemblages of plants as the successional process starts again.

Much depends upon a species's competitive abilities. Early successional species dominate in the immediate postharvesting phase, but are much reduced over time as the canopy closes. Tree species (such as alder, birch, cherry, poplar, aspen, and jack pine) often fit into this category along with semiwoody shrubs (such as elderberry and blackberry), annual and short-lived perennial herb species (such as members of the aster family and various grasses and sedges). Other species that existed in low abundance in the original forest they were before European colonization when the landscape was covered mostly in mature forest. However, even these species benefit most from a pattern of small clear-cuts as they prefer the edge habitats where the protective cover of the forest is not too far distant. Optimal clear-cut size for deer in southern Ontario was recommended by Euler (1978) as 2 ha.

Unharvested forests in most areas in Canada comprise a patchwork of forest stands of different ages and diversities regenerating from the effects of various disturbances, such as fire and insect attack. Clear-cutting at a certain rate and scale may be appropriate in some of these ecosystems. However, some forests, such as the coastal forests of BC or the mixed deciduous forests of southern Ontario

Box 15.4 Forestry and Grizzlies

The grizzly bear's range once extended across most of North America. Unlike its smaller cousin, the black bear, however, it is not able to tolerate human disturbance, and as human populations grew, grizzly populations shrank. Some of the densest remaining populations are in the lowlands of the coastal valleys in BC where the bears are attracted not only to the annual salmon runs but also to the abundant berry-producing shrubs and other nutritious vegetation found on the flood plains. These nutrient-rich sites also constitute some of the most productive forestry sites.

Studies over the last ten years have documented declines in grizzly populations following logging activities. One of the main problems was the changes in vegetation that occurred as a result of logging. The grizzly's favourite forage foods are those that compete with the re-establishment of trees when secondary succession takes place (Chapter 5). As a result, the forage foods are often suppressed using chemicals. In turn, this helps create dense stands of conifers, which have the long-term effect of shading out preferred forage for the grizzly. The result is a critical shortfall in grizzly food over an extended period of time.

As a result of these concerns, new approaches are being tried. For example, lower replanting densities for conifers and more open space between tree clusters are being tried in an effort to mimic the natural environment. The use of chemicals is restricted to the tree stands. A new adaptive management approach has been formulated for silviculturists dealing with these grizzly bear habitats. This involves not only the measures outlined earlier for new logging activities but also revisiting past sites to undertake remedial activities. The approach is adaptive and exemplifies an ecosystem approach (as discussed in Chapter 8), which takes a more holistic view towards resource management in considering the limits of tolerance (Chapter 4) of one of our most spectacular animal species in Canada. Although it is too early to assess the success of these efforts, in combination with the absolute protection of some of these sites (such as the grizzly reserve established in the Khutzeymateen Valley), they should help ensure that grizzly bear populations survive in the coastal valleys of BC.

The grizzly bear is not very tolerant of human disturbance and has suffered large declines in number and distribution since European colonization (*World Wildlife Fund Canada/Karl Sommerer*).

and Quebec are more influenced not by large-scale disturbances but by the pattern of death of individual canopy trees. More than half of the coastal rainforest is over 250 years old. Much smaller interventions would be required to mimic natural processes. Furthermore, some animal and bird species depend upon forests that have old-growth characteristics, such as ample lichen growth for woodland caribou or dead trees to provide nesting cavities for birds such as woodpeckers. If forest harvesting takes place at a rate and scale that eliminates stands with these characteristics from the landscape, these species will decline in number and may become extirpated (Table 15.3). Many authors have discussed the incompatibility between forest harvesting and maintenance of woodland caribou (Cummings and Beange 1993). Bunnell and Kremsater (1990) suggest that approximately 65 per cent of the bird species found on Vancouver Island, for example, require habitat with snags or logs. The case-study of the spotted owl (discussed later) is a good example of the difficulties associated with maintaining populations of species dependent upon old-growth habitat. Other species that have suffered in this regard include the fisher, pine marten, and wolverine in central and eastern Canada (Thompson 1988).

It is not only the elimination of high-profile species that should be of concern when considering the destruction of old-growth forests but also less well-known species, even unknown ones, and the kinds of ecological functions undertaken by these species. About 85 per cent of the species in some forests are arthropods, such as insects and spiders. Only recently is the richness of this fauna being appreciated. There may be over 1,000 species of invertebrates within a single forest stand (Franklin 1989). Many of these species are new to science and we have little idea of their ecological role in maintaining healthy ecosystems. However, research has indicated a clear pattern of different spider species associated with different stages of the successional process, with at least thirty years required in even the fastest-growing forests for the spider community to recover from clear-cutting (McIver et al. 1990). Studies of the upper branches of coastal forests in BC are revealing very complex predator-prey relationships among many different species. These relationships are not replicated in managed forests, suggesting that continued elimination of this habitat will lead to species extinctions, a decrease in genetic diversity in these communities, and removal of natural controls on forest pests.

From this brief review of the impacts of forest harvesting activities on biodiversity, we can conclude that substantial changes can occur. These changes are beneficial to some species and detrimental to others. It is therefore critical to understand the complex ecological relationships involv-

---

Table 15.3 Endangered and Vulnerable Species That Have Some Reliance on Old Growth in British Columbia

Endangered	Vulnerable
Keen's long-eared myotis bat (*Myotis keeni*)	Ancient murrelet (*Synthliboramphus antiquus*)
Marbled murrelet (*Brachyramphus marmoratus*)	Bald eagle (*Haliaeetus leucocephalus*)
Sharp-tailed snake (*Contia tenuis*)	Caribou (*Rangifer tarandus*)
Spotted owl (*Strix occidentalis*)	Cassin's auklet (*Ptychoramphus aleuticus*)
	Fisher (*Martes pennanti*)
	Flammulated owl (*Otus flammeolus*)
	Fork-tailed storm petrel (*Oceanodroma furcata*)
	Great blue heron (*Ardea herodias*)
	Rhinoceros auklet (*Cerorhinca monocerata*)
	White-headed wood-pecker (*Picoides albolarvatus*)
	Williamson's sapsucker (*Sphyrapicus thyroideus*)

Sources: L.E. Harding, 'Threats to Diversity of Forest Ecosystems in British Columbia', in *Biodiversity in British Columbia: Our Changing Environment*, edited by L.E. Harding and E. McCullum (Vancouver: Canadian Wildlife Service, 1994):255.

ing forests if these changes are to be fully evaluated and taken into account when planning forest harvesting activities. At a regional scale, it is essential to maintain some areas in their original state to maintain landscape biodiversity. The size, number, location, and shape of these remnants is a critical issue that will be discussed more thoroughly in the next chapter.

Forestry and Site Fertility

Forest harvesting removes nutrients from the harvested site (Figure 15.1). The amount of nutrients removed depends upon the kind and extent of harvesting. Selective whole-stem tree harvesting will remove relatively few nutrients compared with large clear-cuts of complete-tree (above and below ground biomass) harvesting. The latter will maximize the short-term yield of biomass from the forest, but may compromise the potential of that site over the long term to produce further harvests. More than half of the harvesting in Canada is done with full-tree systems.

To judge the potential effects of forest harvesting on site fertility, it is necessary to consider the amount of nutrients being removed, the size of the soil nutrient pool, the net accretions and depletions of nutrients in the forests, and the ways in which these variables interact. We have relatively little experience in assessing such long-term changes over one or more rotation cycles. The process of site impoverishment over time as a result of harvesting is shown in Figure 15.2. On some sites, the proportion of nutrient capital removed in the biomass will be relatively minor, while in others it may be substantial. This depends, to a large degree, on the existing nutrient capital of the site. Areas with high soil fertility will be less affected. Some sites will recover quickly from harvesting and can sustain relatively short rotations. Other sites will not recover adequately between rotations and site nutrient capital will fall, making tree regeneration difficult and in some cases impossible. Thus nutrient-deficient sites should have long rotations with just stem harvesting to maintain productivity.

The amount of nutrients removed by harvesting is also influenced by tree species, age, season of harvesting, and other factors. Older trees contain larger amounts of nutrients—such as nitrogen,

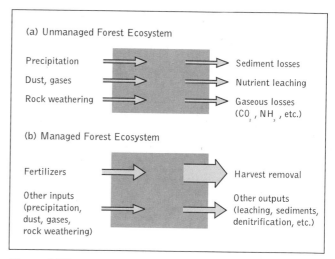

Figure 15.1 Nutrient inputs and outputs from managed and unmanaged forest ecosystems.

phosphorus, potassium, calcium, and manganese—than younger trees. There is also considerable variation among species in the amount of nutrients organically bound, and the nutrients preferentially held by different species. Alban et al. (1978), for example, found poplars to be the most nutrient-demanding trees. The differences between whole-tree and stem-only harvests can also be significant. In a study of black spruce in Nova Scotia, Freedman et al. (1981) found that there was almost a 35 per cent increase in biomass take for whole-tree harvesting. This resulted in an increased loss of 99 per cent nitrogen, 93 per cent phosphorus, 74 per cent potassium, 54 per cent calcium, and 81 per cent manganese derived mainly from the nutrient-rich foliage and small branches. When deciduous trees are harvested, the loss of nutrients can be significantly reduced by cutting in the dormant period when leaves are not present.

It is also important to consider nutrient inputs. Substantial amounts of nutrients will be added over time by precipitation. For a maple-birch stand in Nova Scotia, for example, Freedman et al. (1986) calculated that it would take ninety-six years of precipitation to replace the nitrogen lost through whole-tree removal, eighty-three years for potassium, 166 years for calcium, and forty-one years for manganese. Other nutrient inputs occur through dry deposition of gases and particulate matter, the weathering of minerals, and the fixation of atmospheric dinitrogen. Soil mycorrhizae are particularly important for the last-mentioned

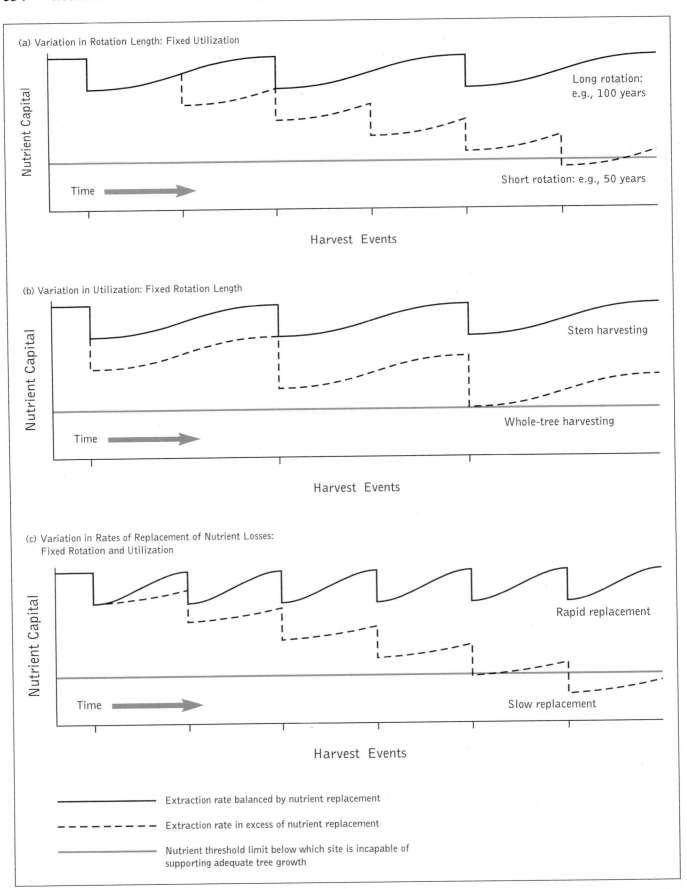

(a) Variation in Rotation Length: Fixed Utilization

Nutrient Capital

Time

Long rotation: e.g., 100 years

Short rotation: e.g., 50 years

Harvest Events

(b) Variation in Utilization: Fixed Rotation Length

Nutrient Capital

Time

Stem harvesting

Whole-tree harvesting

Harvest Events

(c) Variation in Rates of Replacement of Nutrient Losses: Fixed Rotation and Utilization

Nutrient Capital

Time

Rapid replacement

Slow replacement

Harvest Events

Extraction rate balanced by nutrient replacement

Extraction rate in excess of nutrient replacement

Nutrient threshold limit below which site is incapable of supporting adequate tree growth

Figure 15.2 Site impoverishment as a result of forest harvesting. SOURCE: Adapted from J.P. Kimmins, 'Evaluation of the Consequences for Future Tree Productivity of the Loss of Nutrients in Whole-Tree Harvesting', *Forest Ecology and Management* 1 (1977):169–83.

input (as discussed in Chapter 6), and may be adversely affected by clear-cut logging and subsequent land treatments, such as slash burning and pesticide use.

Nutrients are also lost to the system other than by removal in harvest. In the earlier example of the maple-birch forest in Nova Scotia, whole-tree harvest removal is equal to the loss through leaching that would occur in sixty-eight years for potassium, thirty-one years for calcium, and over four years for manganese, with losses of soluble nitrogen and phosphorus at much lower rates. The difference between these figures and those given for inputs in the previous paragraph gives some indication of the nutrient flux over time in this particular forest type. Both nitrogen and phosphorus are aggrading, so fifty-one years of nitrogen input and 128 of phosphorus would equal the losses of whole tree harvesting. Other nutrients such as potassium, calcium, and manganese are degrading, however, and in 305, sixty-four, and twenty-three years, respectively, would approximate the loss caused by whole-tree harvesting. Freedman (1995) suggested that this could lead to possible problems, particularly with calcium deficiency, if logging were to occur.

Forest harvesting also leads to dramatically increased rates of nutrient loss through leaching to the hydrological system in most ecosystems. The amount of loss varies according to the intensity and scale of the harvest and the particular ecosystem under consideration. Loss of nitrate is of most concern since it is it often a dominant limiting factor (see Chapter 4). It is also the nutrient that is lost most often in large quantities. One reason for this is the disturbance in the nitrogen cycle (Chapter 6) caused by logging, which results in an increase in the bacterial process of nitrification turning ammonium to nitrate. Nitrate is highly soluble, resulting in significant losses of nitrogen site capital in some ecosystems, particularly if the soil is not too acidic. Other factors—such as warmer soil temperatures, decreased uptake by vegetation, and abundant decaying organic matter on the forest floor—also contribute to increased losses of soluble nutrients following logging. Furthermore, old-growth trees on the West Coast of BC, for example, support a large biomass of lichens with nitrogen-fixing abilities, whereas in younger stands (less than 145 years old) they were virtually absent.

Forestry and Soils

Besides the direct influence on nutrients described earlier, forest harvesting can also have substantial impacts on soil through erosion, especially on steep slopes in areas of heavy precipitation. Such losses also contribute to losses of site fertility, remove substrate for further regrowth, and contribute to aquatic problems, such as flooding and the destruction of fish habitat. Poor design and maintenance of roads are often key factors behind accelerated soil erosion. Compacted road surfaces encourage overland flow with high erosive power. Cutting roads across steep terrain exposes large banks of unprotected topsoil. Much stricter regulations for road construction are now in place in many jurisdictions.

Forestry and Hydrological Change

Large-scale forest harvesting can have a significant impact upon hydrology. Under natural conditions, trees evapotranspirate large amounts of water into the atmosphere. Removal of the trees significantly reduces this evapotranspiration mechanism and other storage capacities, releasing large amounts of water into stream flow. Flooding in high discharge periods can often be the result. Furthermore, without the delaying mechanism of the trees, and with soils that have suffered compaction through harvesting, the speed of flow is often increased, again raising the potential for flooding as well as for erosion. In turn, sediment from erosion can damage spawning beds used by fish.

Logging has resulted in severe changes in the morphology of many coastal streams in BC, with large amounts of logging debris accumulating following flooding (*Philip Dearden*).

CASE-STUDIES ON THE IMPACTS OF LOGGING ON WEST COAST ECOSYSTEMS

Coastal BC has the highest timber volumes per unit area in Canada and one of the highest in the world. Harvesting methods have evolved over the last 100 years to maximize the number of big trees that can be taken from steep terrain with high economic returns. Until recently, little attention was devoted to the impacts of these harvesting methods on the environment. The following examples illustrate some of the impacts that have occurred, and further attention is devoted to the topic in Chapter 21, in which one of the best-known case-studies, Clayoquot Sound, is examined in more detail.

Carnation Creek

Coastal BC is home not only to a valuable forest industry but also to one of the most valuable fisheries in the world, those for the Pacific salmon. A study of one watershed was initiated on the west coast of Vancouver Island in 1970 to increase understanding of the coastal rainforest-salmonid interaction and the likely effects of timber production on streams and their abilities to support salmonids.

Carnation Creek is a 10 km² watershed on the southeastern side of Barkley Sound. Precipitation is heavy, ranging from 2100–5000 mm annually, producing a rainforest typical of the West Coast, which is dominated by western hemlock, western red cedar, and amabilis fir. Topography is also typical for the West Coast, with steep valley walls inter-

spersed by rock outcrops. The creek is spawning habitat for chum and coho salmon, with lesser numbers of steelhead and cutthroat trout. Sockeye and pink salmon also use the estuary to spawn.

The study design involved three five-year periods: a prelogging phase, a logging phase, and a postlogging phase. In addition, other streams outside the watershed were monitored to assess long-term changes in climate or fish populations. Thirteen areas covering 41 per cent of the watershed were cut using typical West Coast logging methods involving road construction, falling, yarding, hauling, prescribed burning, scarification, and herbicide use. The loggers were, however, aware that special attention was being directed towards their activities.

This latter point is important in determining how valid it is to generalize from Carnation Creek to other West Coast watersheds. For example, more stringent guidelines for streamside logging activities were introduced in BC in 1988. In 1992 a consultant examined how well these guidelines were being implemented on Vancouver Island. He reported that overall compliance with the guidelines and prescriptions written into logging plans for specific areas was poor (Tripp et al. 1992). On average, there was one major or moderate impact for *every* cut block inspected, and half of these were on the most highly rated salmon-producing streams. Debris load increase was the most frequently reported problem (Figure 15.3). Most of

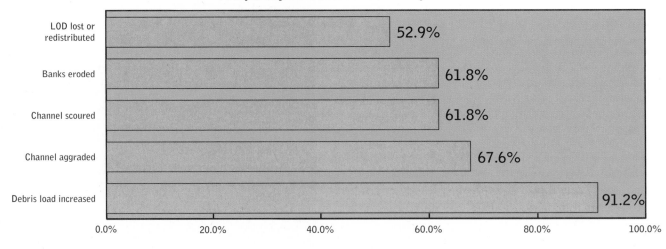

Frequency of Occurrence in Impacted Streams

LOD lost or redistributed — 52.9%
Banks eroded — 61.8%
Channel scoured — 61.8%
Channel aggraded — 67.6%
Debris load increased — 91.2%

Figure 15.3 Frequency of occurrence of different impacts in streams in recent cut blocks on Vancouver Island, March 1992. SOURCE: D. Tripp et al., *The Application and Effectiveness of the Coastal Fisheries Forestry Guidelines in Selected Cut Blocks on Vancouver Island* (Nanaimo: Ministry of Environment, Lands and Parks, 1992):8.

these problems could have been prevented had the guidelines and prescriptions been followed. The main problem has been a lack of compliance in situations in which forest companies are not constantly monitored.

Some of the results from the Carnation Creek study are summarized in boxes 15.5, 15.6, and 15.7. Many of the interactions were complex, again making clear the need to take an adaptive approach to management interventions, as described in Chapter 9. For example, as the mean particle size of the gravel declined, the size of the fry emerging from the gravel declined, and subsequently the number of eggs successfully developing into fry declined for coho and chum. Reduced chum fry size is correlated with reduced ocean survival. However, for coho, lower densities of fry resulted in increased growth rates and higher survival rates, enabling a greater proportion of coho to attain the size necessary to undergo the changes to enter the ocean during the postlogging (as compared to the prelogging) stage.

The scientists involved, who were mostly from federal agencies, suggested that logging created higher productivity and fish growth during the freshwater stages of the salmon's life history due to increased temperatures and nutrient release (Hartmann and Scrivener 1990). However, this was offset by changes in large organic debris, channel morphology, and stream bed stability and compo-

---

Box 15.5 Hydrological Changes in the Carnation Creek Watershed

- In the tributary where up to 90 per cent of the 12-ha drainage was logged, there were increases in water yield in six out of seven years, increases in summer flows by 78 per cent in two out of three years, and increased peak flows in early autumn. Peak rain to peak flow times were reduced as a result of road construction.

- In the watershed overall where only 41 per cent was clear-cut, such effects were less pronounced, although there was an increase in water yield and groundwater levels on the flood plain.

- During high flows, nutrient levels increased at least 40–80 per cent for two to four years following logging, and for one to two years following herbicide application. Nitrate flows increased twofold in the watershed that had been 41 per cent clear-cut and sevenfold in the watershed that was totally clear-cut and burned. These declined in postlogging years as they became absorbed by stream algae and the regenerating forest. There was little impact upon phosphorus export from the watershed.

Box 15.6 Changes in Stream Morphology and Debris in the Carnation Creek Watershed

- Logging activity reduced the stability of the debris and sediment in the channel prior to logging.
- Logging removed trees from the stream side, weakened root structure, and added logging debris to the channels.
- Logging exposed mineral soil to erosion.
- Logging had an important influence in reducing the amount, size, and stability of large organic debris (LOD) and increasing the amounts of smaller debris. LOD was reduced to approximately 30 per cent of prelogging volume within two years wherever logging had occurred up to the stream bank. These

changes were important in affecting channel morphology, fish cover, stream bed composition, and ultimately fish production. For example, increased stream velocity and erosion resulted in channel straightening that led to a loss of fish habitat; increased scouring and deposition during run-off that reduced egg-to-fry survival; downstream movement of sand that altered spawning gravel quality and reduced egg-to-fry survival (since cutting, sand and pea gravel content of the stream bed doubled); and bedload deposition that filled the channel and reduced summer rearing habitat.

Box 15.7 Changes in Temperatures and Trophic Response in the Carnation Creek Watershed

- Increases in diurnal and seasonal variability of stream temperatures.
- Increases in mean monthly stream temperatures. Mean temperatures were 3° C higher during summer and 0.5° C higher during winter for the first decade following logging. The proportion of a watershed that had been logged was a main predictor of stream temperature. This has important implications for fish, including encouraging their early emergence from the gravel, which reduced the survival rate of chum salmon during the period immediately after entering the sea. Coho salmon were also swept downstream due to this early emergence and many perished. However, the

salmon that managed to remain in the stream had a longer period to grow due to their early emergence, resulting in higher survival levels in the stream. The time at which the salmon head for the salt water (smolting time) was also advanced as a result of temperature increases, which reduced their marine survival.
- After logging and silvicultural treatments, litter input to the stream was reduced 25–50 per cent of prelogging levels.
- Macroinvertebrates' densities were reduced 40–50 per cent in stream sections clear-cut to the bank. They were also lower in leave strips (strips of forest left along a river-bank) after logging.

sition that depressed overall fish survival and production. For chum, the net effect of logging was calculated as a 26 per cent reduction in adult returns and increased interannual variability in fish production. For coho, a 6 per cent net reduction was suggested, with an increased interannual variability in fish abundance. The trend of future changes was anticipated to be what is shown in Figure 15.4. The major factor influencing return of most processes to a stable condition was the regrowth of the forest. Hartmann and Scrivener (1990) suggest that there will be reduced rates of change in these ecosystems about twenty years after logging.

The Northern Spotted Owl

In the latter part of the 1980s the northern spotted owl in the western US became the focal point of high-profile conflicts between conservationists and logging interests. Following several reviews and appeals, the owl was accorded threatened status throughout its entire range in the US under the US Endangered Species Act. This designation means that the owl is likely to become an endangered species within the foreseeable future throughout all or a significant part of its range. This designation required that its critical habitat be identified and a recovery plan be implemented. Due to the owl's dependence on old-growth forest, this decision has led to severe conflicts with forest harvesting activi-

The northern spotted owl (*John and Karen Hollingsworth/US Fish and Wildlife Service*).

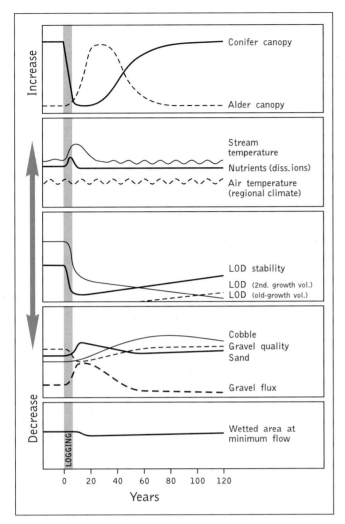

Figure 15.4 Anticipated patterns of change in physical conditions in a small coastal stream, like Carnation Creek, following logging (cross hatch). The patterns for stream temperature and stream insolation are presumed to be similar. SOURCE: G.F. Hartmann and J.C. Scrivener, *Impacts of Forestry Practices on a Coastal Stream Ecosystem, Carnation Creek, BC* (Ottawa: Department of Fisheries and Oceans, 1990).

ties. In 1994 the Clinton administration established reserves on over 4 million ha where harvesting would be severely restricted.

The spotted owl is also found in Canada in the old-growth forests of southwestern BC. This northern extension of its range is important since it is often individuals at the extremes of a species's range that are most valuable to protect as they may have the genetic diversity best suited for future adaptability. Furthermore, just as in the US, the old-growth forest upon which the owl depends was allocated for harvest. The spotted owl therefore provides a good case-study of the impacts of forest harvesting on biodiversity.

Biology and Range

The northern spotted owl's range is from northern California to southwestern BC (Figure 15.5). Between 1985 and 1993, thirty-nine active spotted owl sites and at least seventy-one adult owls were recorded in BC. There are probably less than 100 owls in British Columbia. It is difficult to assess population trends as there are no historic estimates of abundance, but it is thought that numbers have declined with reduced habitat. The historic range of the species in the province is probably not that much different than the current range, although the distribution of the owl within that range has changed significantly due primarily to destruction of its prime habitat, old-growth forests.

Superior habitat for the owls would have old-growth characteristics such as 'an uneven-aged, multi-layered multi-species canopy with numerous large trees with broken tops, deformed limbs and large cavities; numerous large snags, large accumulations of logs and downed woody debris; and canopies that are open enough to allow owls to fly within and beneath them' (Dunbar and Blackburn 1994:19).

Main prey items for the owls are small mammals, such as the northern flying squirrel and dusky-footed and bushytail woodrats. In BC the squirrel is the most important prey, and deer mice are the next most important. Both these species are abundant in old-growth forests.

Box 15.8 Why Do Spotted Owls Prefer Old-Growth Forests?

- greater abundance of prey
- better foraging conditions for the owl due to the relative openness of the canopy and abundance of foraging perches at many different levels
- better protection against predators, such as the great horned owl and goshawk
- more nesting structures available
- better for the thermoregulatory requirements of the owl
- evolutionary conditioning causing specializations that preclude use of other types of habitat

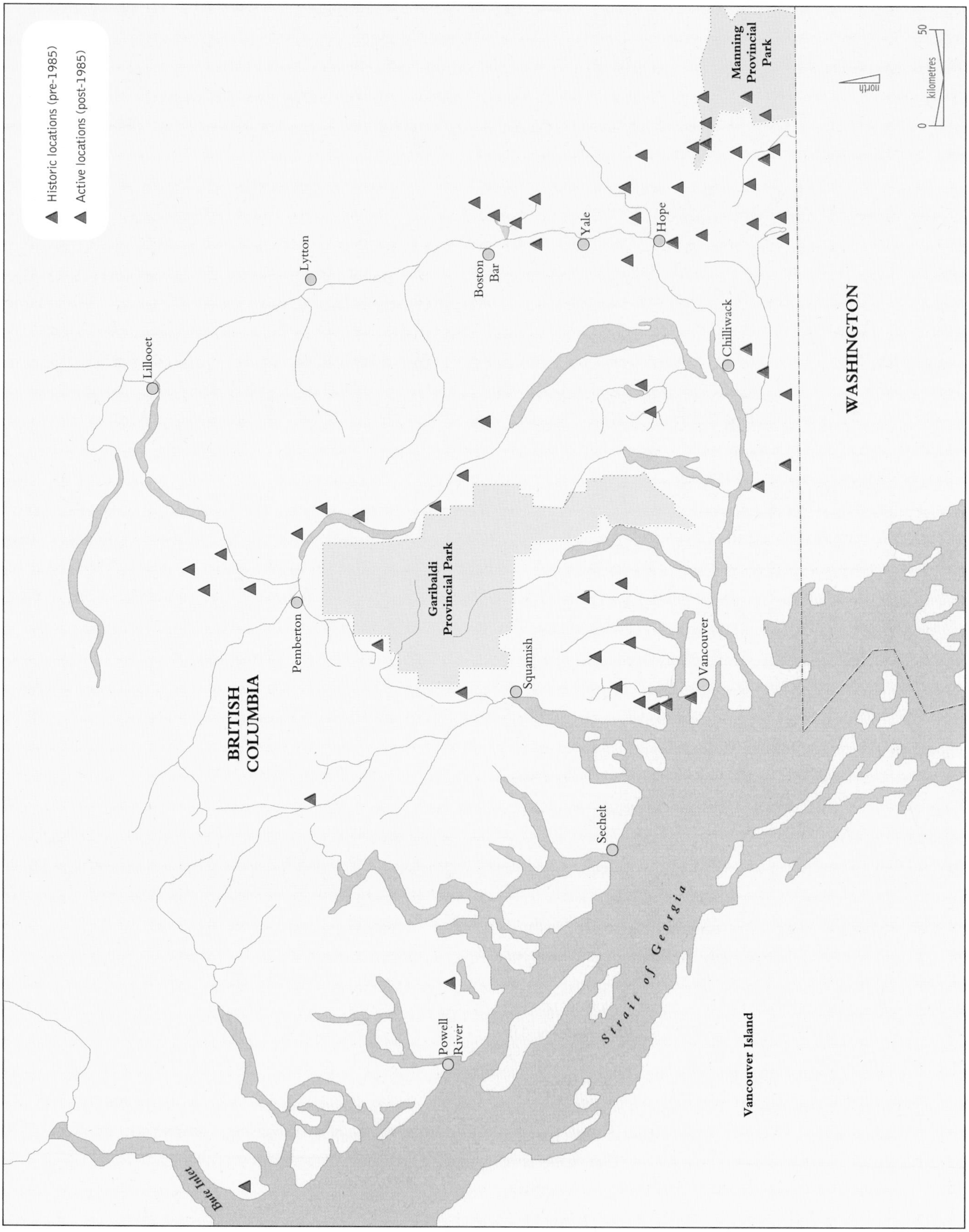

Figure 15.5 Distribution of spotted owls in British Columbia. SOURCE: D. Dunbar and I. Blackburn, *Management Options for the Northern Spotted Owl in British Columbia* (Victoria: Ministry of Environment, 1994).

Threats

The single greatest threat to the spotted owl is the continued logging of old-growth forests, leading to loss of suitable habitat. Estimates suggest that probably less than 30 per cent of the preharvesting habitat remains, much of which is highly fragmented. Some protection is provided by provincial and regional parks in two main blocks, covering some 110 000 ha in total. Unfortunately, these blocks are some 85 km apart with little interconnecting habitat. It is extremely unlikely that these two populations will be able to interbreed. Provincial forests managed for timber production contain much larger amounts of suitable habitat, but much of this land is scheduled to be harvested over the next 100 years.

Management Options

Concern over the future status of the spotted owl resulted in the formation of the Spotted Owl Recovery Team to develop a recovery plan for the species. The primary goal is to outline a course of action to stabilize the current population and improve the status of the species, allowing it to be upgraded from the endangered species category. Management options were derived from consideration of these three aspects:

- *The range of the owl that should be managed*: The main question here relates to whether the owl populations should be managed over their entire range in the province or just part of that range. Managing the entire documented range provides the greatest chance for maintaining viable populations.
- *The configuration of the management areas for the owl*: Designated management areas (DMAs) were outlined in provincial forests. Major questions relate to the size, shape, and location of these areas. Desirable characteristics would be to have the largest possible areas with a shape (such as a circle) that minimizes edge effects and are located close enough together so that juveniles can disperse to suitable habitat close by.
- *The timber management practices to be employed within the management areas*: Forest harvesting is the main threat to the owls. Options include a complete ban on harvesting in the DMAs, the

The rate of conversion from natural to managed forests has alarmed many environmentalists, who claim that forest companies have been allowed to extract too much wood too quickly. This will result not only in environmental problems but also in providing adequate fibre for industrial use in the future (*Philip Dearden*).

use of modified practices, or a continuation of current practices.

Consideration of these aspects led to sixteen different management options that correspond with the different categories of abundance status outlined by the Committee on the Status of Endangered Wildlife in Canada (COSEWIC) and discussed in detail in Chapter 16 (Box 15.9). The categories and criteria used in the case of the spotted owl to move it from one category to the other are shown in Box 15.10. The options range from banning all timber harvesting that might degrade suitable owl habitat within all forested habitats within the owl's entire range to no owl management beyond existing parks and protected areas. The status of forest land for five of the six options is shown in Figure 15.6. No data are available for the most protective option. However, the second option shows that 72 873 ha of operable forest

---

Box 15.9 Description of the Five Status Categories of COSEWIC

Delisted: A species not at risk (no designation required).

Vulnerable: A species particularly at risk because of low or declining numbers, small range, or for some other reason, but not a threatened species.

Threatened: A species likely to become endangered in Canada if factors threatening its vulnerability are not reversed.

Endangered: A species threatened with imminent extinction or extirpation throughout all or a significant portion of its range in Canada.

Extirpated: A species no longer existing in the wild in Canada, but occurring elsewhere.

---

would be included in spotted owl conservation areas where no harvesting would be allowed on owl habitat. The costs of withdrawing this amount of land from the operable forest have yet to be assessed, but they are likely to be significant. However, foregone timber revenues is the price that society must pay if an old-growth dependent species, such as the spotted owl, is going to survive.

FORESTRY AS A MANAGEMENT PROCESS

Over the last decade, increasing awareness and concern about the environmental impacts of forest harvesting described earlier have raised questions about the environmental sustainability of forest harvesting and about the different kinds of management approaches that might lead to sustainability. Key questions relate to the amount of forest reserved from logging, the amount of fibre harvested over a specific time period, the way in which it is logged, and what happens to the land after harvesting. The provincial and territorial governments are responsible for 71 per cent of the

---

Box 15.10 Down-listing Criteria for the Northern Spotted Owl in Canada

No change in status, species remains Endangered.

1. Quantity of and quality of habitat is very low, and habitat conditions are expected to decline at a significant rate.
2. Populations are extremely low and likely declining with the isolation of sub-populations or individuals.
3. Current conservation measures are judged to be insufficient to protect the owl from extirpation throughout all or a significant portion of its Canadian range.
4. Populations are extremely vulnerable to stochastic events.

Down-listing from Endangered to Threatened.

1. Quantity or quality of habitat is low, but habitat conditions are stable or increasing.
2. Populations are low, but are well distributed and most are stable or increasing.
3. Current conservation measures are judged to be sufficient to protect the owl from becoming Endangered throughout all or a significant portion of its range.
4. Populations are still vulnerable to stochastic events, and threats are still evident in most of their Canadian range.

Down-listing from Threatened to Vulnerable.

1. All or the majority of habitats potentially capable of supporting owls are occupied.
2. Populations are well distributed, and have been stable or increasing for at least 7 years.
3. Populations are recovered, and current conservation measures are judged to be sufficient to protect the owl from becoming Threatened throughout all or a significant portion of its range.
4. Populations are still vulnerable to major stochastic events.

Down-listing from Vulnerable to Not at Risk.

1. Population currently occupies all or most of its former geographic range.
2. Populations and habitat conditions are stable or increasing and well distributed throughout its Canadian range.
3. Conservation measures are in place which ensure continued healthy habitat conditions, population distribution, and demographic performance, such that the species is no longer at risk.
4. Population size is sufficient to withstand major stochastic events (Dunbar and Blackburn 1994).

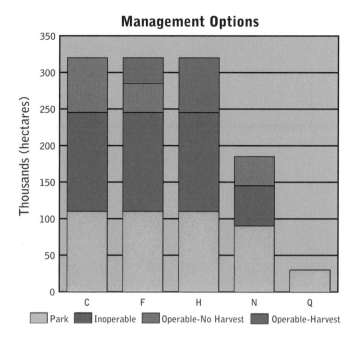

Management Options

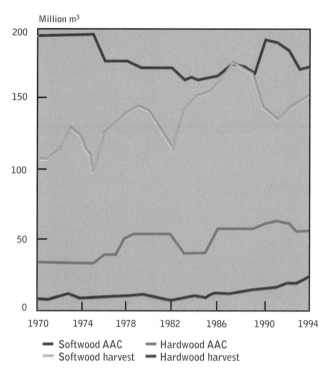

Million m³

Figure 15.6 Forested land use within DMAS for five management options. NOTES: C = no timber harvesting that degrades/destroys suitable owl habitat in all spotted owl conservation areas (SOCAS) throughout the bird's documented provincial range; F = no timber harvesting that degrades/destroys suitable owl habitat in all SOCAS within the Sasquatch Range and retain at least 67% of suitable owl habitat in each activity centre within all remaining SOCAS throughout the bird's documented provincial range (outside the Sasquatch Range); H = retain at least 67% suitable owl habitat in each activity centre within all SOCAS throughout the bird's documented provincial range; N = no timber harvesting that degrades/destroys suitable owl habitat in all SOCAS within the Sasquatch Range, no owl management beyond existing parks and protected areas outside the range; Q = no owl management beyond existing parks and protected areas. SOURCE: D. Dunbar and I. Blackburn, *Management Options for the Northern Spotted Owl in British Columbia* (Victoria: Ministry of Environment, 1994):69.

Figure 15.7 Softwood and hardwood AACs and harvesting. SOURCE: Natural Resources Canada, *The State of Canada's Forests, 1995–1996* (Ottawa: Canadian Forest Service, 1996):101.

nation's forests, and the federal and territorial governments for 23 per cent. The remaining 6 per cent is managed by 425,000 private land-owners. Forest management is hence primarily the responsibility of the provinces and territories, which manage the forest resource on behalf of the public through agreements with private companies. There are different forms of tenure, but generally all these require the company to submit plans that specify where they intend to cut, the details of the harvesting process (including the location of roads), and reclamation plans. The governments provide regulations and guidelines for these practices and have the authority to enforce them.

Rate of Conversion

The rate of conversion of natural to managed forests is one of the most controversial issues in Canadian forestry. It has increased dramatically over the years (Figure 15.7). The amount that can be cut is established by each province as the annual allowable cut (AAC). This is based on the theoretical annual increment of saleable timber, after taking into account factors like the quantity and quality of species, accessibility and growth rates, and amounts of land protected from harvesting because of other use values, such as parks and wildlife habitat. The AAC should reflect the long-run sustained yield of a given unit of land or what that land should yield in perpetuity. This is ultimately limited by the growth conditions, the biological potential of the site, and how it can be augmented by silvicultural practices. It is not sustainable to have an AAC that consistently exceeds this biological potential. Economists, however, who have dominated thinking on the issue in the past, might argue for the need to maximize the monetary value of the cut to invest in other wealth-producing programs and to provide social services. The dominance of this line of thought has led to rates of conversion significantly higher than those that can be supported biologically.

The total AAC for Canada is calculated by adding all the provincial and territorial AACs (where these figures are available). In 1994 the total AAC was 229.8 million m³. In most provinces the AAC is greater than the harvest. However, in some regions the AAC approaches or exceeds the harvest for the most important component, the softwoods. This is leading many provinces to review their rates of conversion. The potential softwood supply under intensive and current management regimes is shown in Figure 15.8. In the face of increasing world demand, which is estimated at 70 million m³ per year (Natural Resources Canada 1995), there are obviously serious questions about Canada's ability to meet this demand and, indeed, whether Canada should even try to.

Forestry Practices

Perhaps no aspect of resource or environmental management has created as much conflict in Canada as the dominant forest harvesting practice, clear-cutting (Figure 15.9). In 1991 more than 90 per cent of harvesting was done by clear-cutting in the boreal, Great Lakes–St Lawrence, and coast forest regions. Clear-cutting was also used 87 per cent of the time in the Montane forest region and 82 per cent in the Acadian forest region (Minister of Supply and Services 1992). The size of the clear-cuts varies widely from approximately 15 ha to over 250 ha, and in some cases clear-cuts extend for many thousands of hectares. Although openings of this size have few ecological advantages, it should not be assumed that more but smaller clear-cuts are superior to fewer but larger clear-cuts. More clear-cuts create more fragmentation and disturb more forest area to the detriment of interior forest species.

Clear-cuts are aesthetically unappealing to many Canadians, and their environmental impacts, especially cumulative as they spread across the landscape (as described in the previous section), can be substantial. For this reason, Nova Scotia has implemented guidelines for the establishment of wildlife corridors to connect habitat. If clear-cuts exceed 50 ha, the guidelines recommend that at least one corridor be created, with irregular bor-

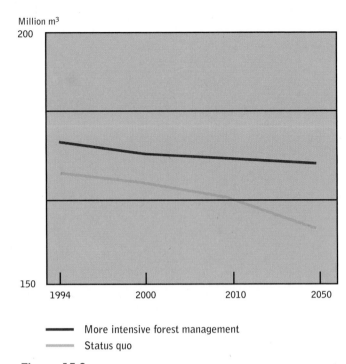

Figure 15.8 Potential softwood timber supply. SOURCE: Natural Resources Canada, *The State of Canada's Forests, 1994* (Ottawa: Canadian Forest Service, 1995):56.

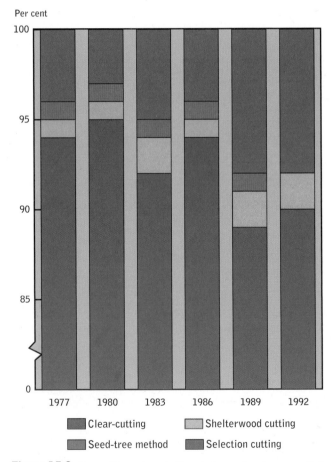

Figure 15.9 Harvesting systems in Canada (percentage). SOURCE: Natural Resources Canada, *The State of Canada's Forests, 1995–1996* (Ottawa: Canadian Forest Service, 1996):80.

Clear-cuts

There can be no generally applicable 'optimum' size of a clearcut, whether the main consideration is economics, aesthetics, habitat or biodiversity. Locality, ecosystem and topography will all affect the most desirable size of cut. Additional factors to be considered are large-scale spatial heterogeneity and the need to avoid uniformity in clearcut size (Boyle 1992:450).

ders and a minimum width of 50 m. While this is a step in the right direction, it is the implementation of such guidelines, not their specifications, that is important. Throughout the country there are numerous examples of situations in which guidelines were ignored in favour of profit maximization (Tripp et al. 1992).

Clear-cutting is the preferred means of harvesting on such vast areas because it is the most economical way for the fibre to be extracted and because in certain types of forests—such as those dominated by even-aged stands of fire origin like lodgepole and jack pine, black spruce, aspens, and poplars—clear-cutting may mimic natural processes more closely than selective or partial cutting systems. Clear-cutting also allows for easier replanting and tending of the regenerating forest. Nonetheless, other types of harvesting approaches, such as the 6.7 per cent of the land base where selective logging occurs, or the 4 per cent where seed tree and shelter-wood cutting dominate, are preferred by most ecologists for some of the reasons detailed in the previous section.

Reforestation

Until 1985, Canada's forests were considered so extensive that little effort was given to reforestation. Once logged, sites might be burned to facilitate rapid nutrient return to the soil, but were then abandoned in the hope that they would be satisfactorily recolonized by seeds from the surrounding area. Sometimes this was successful, often it was not. Thus with greater harvesting levels, the amount of land that no longer supported trees that could be harvested in the future gradually increased. This land, known as not satisfactorily restocked (or NSR) land grew from 585 000 ha in 1975 to 2.5 million ha by 1992. In 1992 933 177 ha of land were harvested, a further 868 388 were burned by fire, yet only 463 364 ha were planted and reseeded, and the success of these plantings cannot be guaranteed. Given this annual deficit between what is cut and what is replanted, many conservationists have difficulty understanding further allocation of old-growth timber to the forest industry. Nonetheless, the amount of land replanted over the last decade has increased considerably, largely due to the intervention of the federal government in subsidizing these efforts (Figure 15.10).

Site Preparation and Stand Tending

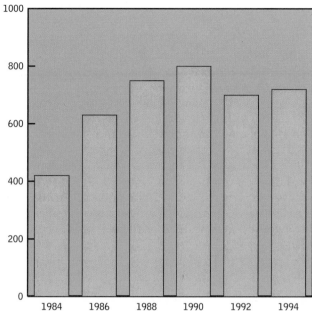

Figure 15.10 Site preparation and stand tending in Canada, 1994 (thousand hectares). SOURCE: Natural Resources Canada, *The State of Canada's Forests, 1995–1996* (Ottawa: Canadian Forest Service, 1996):102.

Biocide Use on Forest Land

Biocides have been used extensively on forest lands in Canada for the same reasons that they are used on agricultural lands: to reduce competition from unwanted plants and insects in replanted sites. Three herbicides (2,4-D, glyphosate, and hexazinone) are registered for forest management in Canada. Glyphosate (or Roundup), the most widely used, affects a broad spectrum of plants, but degrades quickly and is relatively non-toxic to terrestrial animals.

Chemical use is generally quicker, cheaper, easier, and more effective than using mechanical alternatives for weed suppression. Sites regenerating from forest harvesting have been returned to an earlier successional phase and, under natural conditions, a vigorous and diverse secondary succession will take place. This community will not, however, be dominated for many years by the commercial species desired by foresters. Chemicals are used to suppress these early species, compress the successional time span, and maximize the growth potential of the more commercially desirable species, usually conifers. In this process, species that might be ecologically advantageous, such as nitrogen fixers like red alder, are often eliminated. However,

Box 15.11 Forest Practices Act

Concern over how forest harvesting was practised in British Columbia led that province to introduce the Forest Practices Code in 1994, which established a consolidated package of legislation, regulations, standards, and field guides to govern forest practices. The code attempts to deal with three major problems: first, that existing requirements were scattered through hundreds of acts, regulations, policies, and guidelines; second, that substantial gaps occurred in these documents; and, third, that enforcement was often non-existent. The code sets minimum provincial standards for various practices. For example, clear-cutting will not be allowed on sites:

- with very unstable terrain
- within critical wildlife habitat areas for which the forest canopy is essential to maintain wildlife populations
- in streamside management zones
- in old-growth management areas
- where visual quality is a prime consideration
- on other sensitive sites

Also, new requirements maintain biodiversity and provide for higher-than-normal standards of forest practice in areas that have high non-timber values. Furthermore, all resource values must be considered in forest operations with full public involvement in all key plans and prescriptions. Although many environmentalists welcomed the code enthusiastically, this has waned considerably over the past two years as many loopholes have become apparent where less than rigorous forestry practices can still take place. There is, for example, considerable discre-

tion given to the regional manager for the Ministry of Forests about whether the Forest Practices Code has been applied and whether public involvement is necessary. In a few high-profile cases, the manager's judgement in making pro-industry decisions has been questioned, even by other government departments. As is so often the case with environmental regulations, it is not just the specification of the regulations that is important but also their enforcement.

Others, however, question whether it is realistic to have a province-wide set of practices in such a diverse province. The regulations form a stack of paper over 1 m high, but even this voluminous documentation cannot account for every diverse situation that might arise. There is also a very real cost to the forest industry and the province in implementing the code. Even now, BC logging costs are among the highest in the world, with BC pulp having the distinction of being *the* highest in the world, according to accountants at Price-Waterhouse. In their annual report on the pulp industry for 1995, the company determined that the code is costing companies $10 per cubic metre of wood sold on the coast and $8 in the interior of the province. In 1995 67 million m^3 were harvested from Crown land. It cost the companies $600 million to comply with the code, half as much as their 1995 earnings of $1.28 billion. In addition to these costs are others related to loss of revenue from trees not harvested as a result of the code, the costs of administration, retraining workers, and the research necessary to implement the code. UBC forest economist David Haley has estimated the total to be about $2.1 billion, or about 40 per cent of the province's education budget.

research on some sites has indicated the benefits to timber production of such control. A comparative study of several sites in Nova Scotia, some of which were sprayed and others not, found (twenty-eight years after control) that the volume of the balsam fir in the sprayed sites averaged three times that of the non-sprayed sites, and stem diameters were 21 per cent greater. The length of the harvest rotation can be shortened accordingly.

Insecticides are also commonly used to attack pests, such as the spruce budworm, jack pine budworm, hemlock looper, mountain pine beetle,

gypsy moth, and forest tent caterpillar. The amount sprayed varies considerably depending upon the population dynamics of these insects. Most spraying has occurred in eastern Canada, of which the spruce budworm spraying in the Maritime provinces is the most well known, and has caused considerable controversy (Box 15.12).

The same health and ecological concerns arise, however, as were described in Chapter 14 regarding agricultural use of chemicals, which has led to several high-profile confrontations. Attention is being increasingly directed towards the replace-

Box 15.12 The Spruce Budworm Controversy

Every year since 1952, aircraft have doused the forests of eastern Canada, particularly New Brunswick and Nova Scotia, with an array of chemicals to combat the eastern spruce budworm. The budworm feeds primarily on balsam fir, but will also eat spruce. Its range extends from Atlantic Canada to the Yukon wherever these species of trees can be found. Ordinarily the effects of the insect are unnoticeable in the forest, but it occasionally reaches epidemic levels and causes massive damage to the host trees. These build-ups last for six to ten years, and have been documented for the last two centuries as the product of a long-term budworm-fir ecological cyclic succession. They have, however, occurred with increasing frequency as a result of human intervention related to forest harvesting. These activities have reduced the natural diversity of the forest by removing preferred species, such as white pine, for example, and replaced them with a less diverse forest composed of large areas of mature balsam fir, the budworm's preferred food. Extensive mortality occurs in stands that have suffered defoliation for several years.

The long-term ecological stability of the budworm-fir system does not match the shorter-term demands of the economic system dependent upon the forests for products. As a result, the forests have been extensively sprayed to limit defoliation and mortality. Spray programs started in 1952. DDT was the chemical of choice, and some 5.75 million kg were sprayed in New Brunswick between 1952 and 1968 when use was suspended. Other chemicals took its place—such as phosphamidon, aminocarb, and fenitrothion—until questions were raised about their ecological and health impacts. Phosphamidon,

for example, is very toxic to birds. Until 1985 118.5 million ha (mostly in New Brunswick) were sprayed, some areas on an annual basis. Continual spraying of this nature appears necessary once natural controls are disrupted.

The organophosphate, fenitrothion, became the most popular chemical. It also became the source of heated controversy as to its health and ecological impacts. In particular, there was concern over the link between the chemical and Reye's Syndrome, a rare and fatal children's disease. As a result, and due to unfavourable reviews of the ecological impacts of the chemical, particularly on songbirds, aerial applications of fenitrothion will be cancelled as of 1998. Environmentalists have criticized this review process as they have been denied access to confidential information and the opportunity to participate directly in the process along with other stakeholders, despite an earlier multistakeholder review of the process, which recommended opening up the process to all interested parties (see Chapter 11).

The biological control *Bacillus thuringiensis* is now being used more extensively. Attention is also being devoted to other less toxic approaches, such as species and landscape diversification so that large areas of mature balsam fir do not dominate the landscape. This is a graphic example of the challenges presented by the conflict between longer-term ecological cycles and shorter-term economic dependencies.

An interesting paper by Miller and Rusnock (1993) uses the case of the spruce budworm to analyse the role of science in policy making. They conclude that the scientific uncertainties involved made scientists' advice of little use to policy makers.

ment of synthetic insecticides with biological control agents, such as *Bacillus thuringiensis* (Bt). Bt now accounts for some 60 per cent of the insecticides used in Canadian forests, and is non-toxic to humans and most wildlife, although it does affect moth and butterfly larvae of some non-target species. Plants also manufacture many chemicals as protection against insects. Several of these appear to be good prospects for the development of insecticides for forestry use in the future (Helson 1992).

Multiple Use

Although all legislation overseeing forest harvesting in Canada espouses principles regarding the forests as the source of many societal benefits, in reality uses other than timber extraction have often been given little recognition. In addition to timber products, forests are also a valuable source of commodities, such as meat and fish, wild rice, mushrooms and berries, maple syrup, edible nuts, furs and hides, medicines, Christmas trees, ornamental cuttings, and seeds. With careful management,

Hug an Ancient Tree

Do pause and hug an ancient tree.
Squeeze it with sensitivity.
A part of you may then transform.
And celebrate your liberty.

What ties you to the sheep-like swarms
Impairs your link with nature's forms.
A simple hug may break you free
From bonds of uniformity.

Jim Butler (1994:22)

these resource uses are renewable. Some are harvested commercially by licence. Others are freely available and contribute significantly to recreational values, including tourism. These commodities are also important in sustaining First Nations communities.

Besides these more tangible aspects of forest values, there are a host of less tangible ones related to cultural and spiritual fulfilment, and knowledge and understanding. Such values are even more difficult to assess, let alone manage. One such aspect that has achieved more attention over the last decade has been aesthetics. Our forest lands include some of our most scenic areas. Besides being a major attraction for tourists, they are also a main source of recreational and spiritual satisfaction for residents. Most provinces now have some procedures for trying to include assessments of aesthetic quality into harvesting plans. Unfortunately, many of these approaches still rely upon those in charge of timber extraction to make the judgements upon what viewpoints will be used for making scenic assessments and what harvesting prescriptions might follow. As a result, modifications to cutting plans often tend to be minimal.

The contributions of forest lands to the maintenance of ecological processes have already been discussed earlier. However, the sheer scale of Canada's forests means that they are significant contributors on a global scale. It is estimated that 20 per cent of the world's water originates in Canada's forests. The forests are also major carbon

Many people find forest lands a source of recreational and spiritual fulfilment (*Philip Dearden*).

sinks (sites of carbon storage), with an estimated 224 billion tonnes stored and a yearly accumulation of some 72 million tonnes. It is likely that this accumulation rate will grow in the future as a result of reforestation, improved fire suppression, reduced slash burning, and increased recycling of forest products.

Historically, these non-timber uses have received little attention. However, as timber harvest levels have increased and the public have become more aware and vocal about declines in these other forest values, forest companies are being increasingly required to take them into account in their cutting plans. In other words, they are being required to take a more ecosystem approach, as described in Chapter 8.

New Forestry

Concerns over the impacts and sustainability of forest practices have given rise to calls for what has been termed 'new forestry'. Critics have argued that band-aid solutions to each of the concerns discussed earlier will not solve the fundamental problems, and a totally new way of looking at our forests and forest management practices is required. Current approaches usually emphasize economic maximization over the short term through intensive forest management, subsidized (as in agriculture) by auxiliary energy flows and leading to a simplification of forest biology. This entails genetic simplification through the exclusion of non-commercial species from regrowth areas and genetic manipulations to homogenize the species grown. Intensive forestry emphasizes production of a young, closed-canopy, single-species forest, usually the least diverse of all successional stages. Furthermore, the strength and reliability of the wood produced from such forests has been questioned, and these plantations are susceptible to windthrow (the uprooting and throwing down of trees by the wind), insect infestations, and gradual nutrient depletion. Structural simplification also takes place as the range of tree sizes and growth forms is reduced, snags and fallen trees are removed, and trees are regularly spaced to optimize growth. At the landscape scale, simplification occurs as old growth is removed and the irregularity of wind and fire-created openings is replaced by the regularity of planned clear-cuts. Successional simplification also takes place. Intensive manage-

New forestry would recognize ecological roles by allowing old-growth trees to die and rot on the forest floor and return their nutrients to the ecosystem. This giant Sitka spruce is in Carmanah Provincial Park on Vancouver Island, the site of the tallest trees in Canada (*Philip Dearden*).

ment seeks to eliminate early and late successional stages from the landscape.

In contrast, new forestry embraces an approach that mimics natural processes more closely through emphasizing long-term site productivity by maintaining ecological diversity. This includes rotation periods that are sometimes longer than the minimum economic periods, reinvesting organic matter and nutrients in the site through snag retention, stem-only harvesting and minimizing chemical inputs, and a diversified range of trees and forest products that would take into account the fish in the streams and recreation as well as a broader range of wood products. Traditionally non-commercial species, such as alder and other early successional species (particularly nitrogen fixers), would be allowed to grow, and all stages of the successional process would be accommodated. Old-growth big-leaf maple, for example, provides

New forestry would devote much more attention to stand tending on commercial sites, such as this one, where stand limbing and thinning enhances growth (*Philip Dearden*).

large woody debris cannot be produced by a forest that no longer contains large trees.

New forestry also puts greater emphasis upon maintaining non-timber parts of the forest community. Special attention has to be directed towards the impacts on wildlife of the size, shape, and location of forest patches and the ways in which these may be connected to sustain wildlife populations. The ecological complexities of the forests are only just starting to be revealed. Recent research, for example, suggests that the younger the forests, the lesser the conifer seed production. Species that rely upon these seeds, such as crossbills, would also experience a decline. In western Canada, five species of crossbill have evolved, each of which depends on a particular species of conifer, and even different varieties of the same species. To protect this diversity of crossbills would require protecting old-growth stands and increasing rotation ages throughout the range of each conifer. Similar kinds of consideration must be given to the entire range of forest biodiversity if it is to be maintained in the future.

excellent growth sites for many epiphytes (plants that use other plants for physical support but not nourishment), which provide valuable nutrient accumulation and water retention. River-bank habitats would receive special attention; litter from streamside vegetation provides the primary energy base for the aquatic community, and management of coarse woody debris is particularly important for the structure of smaller streams. Needless to say,

---

Box 15.13 New Trees for New Forests

Although a considerable amount of attention has been devoted to developing trees with the most desirable genetic traits, new forestry also requires us to look at our existing trees in a new light. Two examples illustrate the concept.

Alder is a frequent natural recolonizer of logged sites across Canada. Its growth is often suppressed by the use of chemicals so that it will not interfere with the regrowth of conifers. Research now indicates that in some situations, conifers may actually grow better in association with alder than without alder (Hudson 1993). This is due to the alder's nitrogen-fixing abilities, as discussed in Chapter 6.

Alder also has many other uses, however, besides helping provide fertilizer for other trees. Ethnobotanist Nancy Turner (1992) points out that to some West Coast Native peoples, red alder was formerly a woman with red skin, who was transformed into a tree long ago as a gift to humans. Among the many examples of alder use that Turner cites are for smoking fish, carving bowls, rattles and masks, colouring fishnets to make them invisible to fish, and different medicinal products.

The western yew is another species that was routinely cut and left to rot in the woods because it had no commercial value. Investigations have revealed that the taxol produced from the bark is a potent cancer-treating drug. The yew is an understorey tree of western coniferous forests. The bark from six 100-year-old trees is required to treat one person. However, the tree grows very slowly, so supply is a problem. Unfortunately, only now have we realized the medicinal value of the species.

The kinds of changes suggested by new forestry make it unlikely that the dominant practices of today, such as extensive clear-cuts, can continue. Partial cutting is one preferred alternative where 85–90 per cent of the trees on a site may be harvested, after which the land may be left alone for a couple of decades to encourage a natural successional process. It may also not suit the needs for short-term economic return of the large corporations that now dominate the industry, and it is likely that more, smaller, and community-based companies will become involved (see Box 15.14). Monetary returns over the short term will probably also fall as the amount of wood fibre extracted from the forest declines. Proponents of new forestry argue, however, that these changes will have to occur anyway. Adoption of a new forestry approach would plan for these changes over time with a view towards the long-term value of the forest, while continuation of the old approaches will simply lead to an abrupt decline in the amount of timber available and provide little in the way of future prospects.

The new forestry would have to adopt an adaptive management approach, as described in Chapter 9. Indeed, this was the conclusion of the report of the Ontario Forest Policy Panel (1993). The first item in the priority agenda was the need

---

Box 15.14 New Forestry in Action

The ideas contained in new forestry must be capable of being implemented if change is to occur. There are several examples of alternatives to the dominant way of managing our forests on individual wood-lots and in more extensive areas managed by communities.

At the individual scale, one well-known example is that of Merv Wilkinson and his 55-ha wood-lot on southern Vancouver Island. Wilkinson has been practising sustained-yield forestry since 1936. Despite removing over 4000 m³ of timber, his wood-lot still contains as much wood as it did when it was first assessed in 1945. He removes forest products by cutting in five-year rotations. The straightest, most vigorous trees with good foliage and abundant cone production are left as seed trees, including some estimated to be as old as 1,800 years. There is no clear-cutting, slash burning, or use of chemicals. The canopy is left intact to shield seedlings, but thinned a little to promote good growth. Sheep are used for brush control. Wilkinson's model may not apply everywhere, but it has worked for him and is one demonstrable example of maintaining a forest while still retaining its essential ecological characteristics.

At a regional scale, attention has recently been focused on 'community forests'. Decisions on forest use are often made in boardrooms at the dictates of international capital flow. Such decisions may not be to the benefit of the local communities dependent upon the forests for their livelihoods. Concern over this situation has prompted interest in managing forests to maximize the benefits to local communities. In an overview of the situation in Canada, Duinker et al. defined a community forest as 'a tree-dominated ecosystem managed for multiple community values and benefits by the community' (1994:712).

There are many different manifestations of community forests in Canada with different types of land tenure and administrative arrangements. However, they all seek to achieve local benefits for the community and encourage local involvement in decision making. In Ontario, for example, the Sustainable Forestry Program, announced in 1991, includes one component in community forestry. Communities expressed interest in becoming pilot projects. From the twenty-two communities, four were selected. These are now receiving support from the government to develop as community forests. Harvey and Hillier (1994) provide a review of these projects and the lessons learned. Allan and Frank (1994) provide a similar review of the experience in BC, where some community forests have been in operation for over half a century.

It is clear that there are alternative approaches to forestry other than the ones that currently dominate operations. It is essential, however, that the goals of the forestry activities are specified before the most appropriate approaches can be chosen, as emphasized by the framework in Chapter 1. Current models have evolved to maximize economic returns over the short term; the alternatives described earlier have different goals that are more consistent with the demands of today.

for 'adaptive ecosystem management' to become the cornerstone of the new approach to forestry in the province. The knowledge base is incomplete and will continue to change, and factors other than harvesting will affect the amount of timber that may be cut as values and ecosystems change over time. An adaptive approach is particularly suitable under these conditions of uncertainty.

This process entails a change from the management of individual forest resources that dominated in the past to the ecosystem approach described in Chapter 8. A major challenge is lack of knowledge about the forest ecosystems' boundaries and characteristics of their major components, and the functional relationships among these components and their response to disturbance. Acquisition of this knowledge must be driven by the needs of managers implementing adaptive ecosystem management, and such research must be seen as a legitimate and necessary investment upon which to build sustainable forests for the future. Ontario has now passed the Crown Forest Sustainability Act (1994) to implement these suggestions.

The Canada Forest Accord

Some aspects of new forestry have been incorporated into the Canada Forest Accord, which was signed by over thirty forest-oriented associations, including all relevant government agencies and industrial and conservation interests. The accord established a goal to 'maintain and enhance the long term health of our forest ecosystems, for the benefit of all living things both nationally and globally, while providing environmental, economic, social and cultural opportunities for the

benefit of present and future generations.' In addition, nine strategic directions (Box 15.15) were outlined that encompassed ninety-six specific commitments to action (National Forest Strategy Coalition 1992). These included straightforward definable tasks, as well as more difficult initiatives, such as changing the ethical basis of sustainable forestry nation-wide.

A mid-term evaluation was conducted in 1994 to assess progress on these commitments (Blue Ribbon Panel for the National Forest Strategy Coalition 1994). Due to the complexity of the topic and the wide variation across the country, the panel found it difficult to make an overall characterization, but concluded that progress was being made on some aspects. Commitment 2.17, for example, involved establishing model forests in the major forest regions that would be:

- based on the management of forest ecosystems
- managed to achieve a full range of forest values
- open to public participation within a consultative framework
- capable of demonstrating and evaluating new and ecologically sound forest practices

Ten model forest agreements in six major forest regions across the country are now underway. A further development has these Canadian sites joined by sites in Mexico and Russia as part of the International Model Forest Program. In fact, the mid-term evaluation was particularly impressive, reflecting the fulfilment of commitments at the global level, and concluded that Canada has established itself as a global leader in fostering wise use of forest resources. The panel did, however, identify four issues that still need to be addressed if the strategy is to be a success:

- complete an ecological classification of forest lands
- complete a network of protected areas representative of Canada's forests
- establish forest inventories that include information about a wide range of forest values
- develop a system of national indicators of the sustainability of forest management

---

The Ecosystem Approach

To take an 'ecosystem approach' means that resource people shift their focus from parts to wholes, from the 'interest' to the 'capital'. From trees and other plants, animals, stream flow, esthetics and whatever else the earth's surface yields, to the three-dimensional landscape ecosystems and waterscape ecosystems that produce these valuable things. Acceptance of the ecosystem approach establishes common ground for those concerned with forestry, wildlife, water and recreation, thus encouraging partnerships in sustainability (Rowe 1992:222).

Box 15.15 Main Strategies of the Canada Forest Accord

- conserve the natural diversity of our forests, maintain and enhance their productive capacity, and provide for their continued renewal
- improve our ability to plan and practise sustainable forest management
- increase public participation in the allocation and management of forest lands and provide an increased level of public information and awareness
- diversify and encourage economic opportunities for the forest sector in domestic and international markets
- increase and focus research and technology

- efforts to benefit our environment and economy
- ensure that we have a highly adaptable workforce
- increase participation by, and benefits for, Aboriginal peoples in the management and use of forests
- assist private forest owners in continuing to improve their individual and collective abilities to manage and exercise stewardship of their land
- reinforce Canada's responsibilities as trustee of 10 per cent of the world's forests

Making progress on these aspects will be more difficult in the future, following the budget cuts announced in February 1995 when Natural Resources Canada, the federal department responsible for forestry, took a 57 per cent cut in budget, which resulted in the closure of nine forestry centres and district offices across the country and a reduction from 1,430 to 820 staff.

IMPLICATIONS

Forestry is at a watershed in Canada in terms of how forests, their values, and management are viewed. The next decade will be crucial in determining whether we as Canadians will still consider ourselves a forest nation in fifty years' time. Although society in general and government and industry in particular now have a much greater appreciation of the changes that need to be made in the industry to move towards more sustainable practices, actually making these changes will take some time. In British Columbia, for example—which has suffered from some of the worst practices and where strong measures have been implemented recently to try to reduce the environmental externalities created by logging, increase the industry returns to public purse, and maintain employment levels while reducing the rate of conversion—the industry's economic downturn as a result of depressed world prices for pulp has created great difficulties and calls to relax regulations so that the industry can carry on as before. It is likely that the government will, at least

partly, heed these calls to create more tax revenue. Whether there will be a satisfactory balance between the economic realities of the moment and the long-term view that sustainable forestry requires is open to question. History would suggest not.

SUMMARY

1. Canada is a forest nation. Not only do we have 10 per cent of the world's forests, we are also the largest exporter of forest products in the world. Some 800,000 Canadians and almost 1,000 communities rely upon forestry as a main source of income. The forests, along with the 'North', are dominant elements in the history and culture of the nation.

2. Forests are complex ecological systems. Forestry simplifies these systems by replacing uneven aged stands with high species and structural diversity involving young, closed-canopy, single-species plantations with little structural diversity.

3. As this transformation takes place, changes to the forest ecosystems occur. Species composition and abundance are changed. The proportion of forest with old-growth characteristics is reduced. Species such as the woodland caribou or marten, which depend upon these characteristics, decline in abundance. Other species, such as deer, may increase as

regenerating cut areas produce more forage for them.

4. Forest harvesting removes nutrients from the site. The significance of this for future growth varies, depending upon the nutrient capital of the site and the type of harvesting system used. Sites with abundant capital and/or selective harvesting systems that leave branches behind will suffer less chance of growth impairment of future generations than nutrient-poor sites or sites that are clear-cut with complete tree removal.

5. Forest harvesting may also contribute to increased soil erosion and water flows. Analysis of the impacts of logging on salmon production for a West Coast river indicated a 26 per cent reduction in adult return for chum salmon, and a 6 per cent reduction for coho salmon.

6. The spotted owl is perhaps the best-known example of the potential impacts of logging on biodiversity. The high-profile conflict in the US resulted in President Clinton placing severe harvesting restrictions on 4 million ha of forest land in the Pacific northwest. The owl is also present in southwestern BC where, as in the US, it requires old-growth forests to maintain populations. Management options to maintain populations are currently being considered.

7. The provincial and territorial governments are responsible for 71 per cent of the nation's forests and the federal government for 23 per cent on behalf of the owners, the people of Canada. The remaining 6 per cent is owned privately. The governments enter into contract arrangements with private companies and specify the forest management practices to be followed.

8. The rate of conversion from natural to managed forest is controlled by provincially established annual allowable cuts (AACs). In theory, the AAC should approximate what the land should yield in perpetuity. In some regions the harvesting level for softwoods is equal to the AAC, and many provinces are now reducing their AACs as future softwood supplies fall. In BC, for example, the mill capacity is still higher than the biological capacity to produce trees on a sustainable basis, resulting in that

province importing trees from as far away as Chile to satisfy the industrial demands of the mills.

9. Clear-cutting is the dominant harvesting system used in Canada. Not only is it the cheapest way to extract fibre, in some forest types it may also mimic natural processes more closely than selective or partial cutting systems. However, clear-cutting also has ecological and social costs, and may not be the most appropriate way to harvest timber in some areas.

10. Between 1975 and 1992, the area in Canada not satisfactorily restocked following harvest increased from 585 000 ha to 2.5 million ha.

11. Biocides are used in forestry to control populations of vegetation and insect species that compete with or eat commercial species. The amount of land treated has risen steadily through the years. Several high-profile conflicts have resulted over application of chemicals. Concern over the spraying of the spruce budworm in the Maritime provinces is one of the most significant. Increasing use is now being made of a biological control agent, *Bacillus thuringiensis*, against insect attacks in Canada.

12. Forests have much value for Canadians. In the past, attention has been focused almost exclusively on one type of value, the monetary returns from forest harvesting. However, as the amount of forest brought under management has increased, and as the public has become increasingly aware of the changes occurring in the forests, more attention is being devoted to the assessment and management of other values beside timber production. An ecosystem perspective is being adopted.

13. Concern over the impacts and sustainability of forest practices have given rise to calls for what has been termed 'new forestry'. Such an approach embraces an ecosystem and adaptive management perspective that mimics natural processes more closely and gives greater attention to the full range of values from the forests.

14. The Canada Forest Accord, signed by thirty forest-oriented associations, including government agencies, conservation interests, and private

companies, established a goal to 'maintain and enhance the long term health of our forest eco-systems, for the benefit of all living things, both nationally and globally, while providing environmental, economic, social and cultural opportunities for the benefit of present and future generations.' Nine strategic directions were outlined, encompassing ninety-six specific commitments to action.

REVIEW QUESTIONS

1. Outline some of the ways in which forests are important to Canada.

2. How is forestry an ecological process? Compare farming and forest activities in your province. In what ways are they the same, and in what ways do they differ?

3. Name some species that might increase in abundance as a result of forest harvesting, and others that might decline. What are the characteristics of these species that would encourage these responses?

4. What are the impacts of forest harvesting on site fertility and how do they differ between sites?

5. What are the potential impacts of forestry on fish?

6. What attributes of old-growth forests are necessary for spotted owl populations?

7. What are the five status categories of the Committee on the Status of Endangered Wildlife in Canada?

8. How is forest management administered in Canada? What are the main strengths and weaknesses of this approach? What alternatives might you suggest? Do examples of such alternatives exist in your region?

9. What is an AAC?

10. Outline some of the advantages and disadvantages of clear-cutting.

11. Is Canada reforesting all lands that are harvested?

12. Outline some of the pros and cons of using chemical sprays to control insect infestations in Canada's forests.

13. List all the different values that society gains from forests. What do you think the priorities should be between these different and sometimes conflicting uses?

14. What is new forestry?

15. Are there examples of community forestry in your province? If yes, what have been their strengths and weaknesses? If no, what changes would have to occur to allow community forestry to begin? Which other countries have had experience in community forestry from which we might learn?

16. What are the main strategies of the Canada Forest Accord? If you had to prioritize them, what would be the resulting order? How does this compare with the answer to question 13?

REFERENCES AND SUGGESTED READING

Alban, D.H., D.A. Perala, and B.E. Schlaegel. 1978. 'Biomass and Nutrient Distribution in Aspen, Pine, and Spruce Stands on the Same Soil Type in Minnesota'. *Canadian Journal of Forestry Research* 8:290–9.

Allan, K., and D. Frank. 1994. 'Community Forests in British Columbia: Models That Work'. *The Forestry Chronicle* 70:721–4.

Benkman, C.W. 1993. 'Logging, Conifers, and the Conservation of Crossbills'. *Conservation Biology* 7:473–9.

Blue Ribbon Panel for the National Forest Strategy Coalition. 1994. *Mid-Term Evaluation Report*. Ottawa: National Forest Strategy Coalition.

Boyle, T.J.B. 1992. 'Biodiversity of Canadian Forests: Current Status and Future Challenges'. *The Forestry Chronicle* 68:444–52.

Brown, L.R. 1994. *State of the World*. New York: W.W. Norton.

Bunnell, F., and L.L. Kremsater. 1990. 'Sustaining Wildlife in Managed Forests'. *Northwest Environment Journal* 6:243–9.

Burton, P.J., A.C. Balinsky, L.P. Coward, S.G. Cumming, and D.D. Kneeshaw. 1992. 'The Value of Managing for Biodiversity'. *Forestry Chronicle* 68:225–37.

Busby, D.G., L.M. White, and P.A. Pearce. 1990. 'Effects of Aerial Spraying of Fenitrothion on Breeding White-throated Sparrows'. *Journal of Applied Ecology* 27:743–55.

Butler, J. 1994. *Dialogue with a Frog on a Log*. Edmonton: Duval Publishing.

Canadian Council of Forest Ministers. 1995. *Compendium of Canadian Forestry Statistics*. Ottawa: Canadian Forest Service.

_____. 1996. *Forest Regeneration in Canada 1975–1992*. Ottawa: Canadian Forest Service.

Cummings, H.G., and D.B. Beange. 1993. 'Survival of Woodland Caribou in Commercial Forests of Northern Ontario'. *The Forestry Chronicle* 69:579–87.

Dearden, P. 1983. 'Forest Harvesting and Landscape Assessment Techniques in British Columbia, Canada'. *Landscape Planning* 10:239–53.

Drushka, K., B. Nixon, and R. Travers, eds. 1993. *Touch Wood: BC Forests at the Crossroads*. Madeira Park, BC: Harbour Publishing.

Duchesne, L.C. 1994. 'Defining Canada's Old-Growth Forests—Problems and Solutions'. *The Forestry Chronicle* 70:739–44.

Dufour, J. 1995. 'Towards Sustainable Development of Canada's Forests'. In *Resource Management and Development: Addressing Conflict and Uncertainty*, 2nd ed., edited by B. Mitchell, 183–206. Toronto: Oxford University Press.

Duinker, P.N., P.W. Matakala, F. Chege, and L. Bouthillier. 1994. 'Community Forests in Canada: An Overview'. *The Forestry Chronicle* 70:711–20.

Dunbar, D., and I. Blackburn. 1994. *Management Options for the Northern Spotted Owl in British Columbia*. Victoria: Ministry of Environment.

Euler, D. 1978. *Vegetation Management for Wildlife in Ontario*. Toronto: Ministry of Environment.

Franklin, J.F. 1989. 'Toward a New Forestry'. *American Forests* 95:37–44.

Freedman, B. 1981. 'Intensive Forest Harvest—a Review of Nutrient Budget Considerations'. Information Report, M-X-121. Fredericton: Maritimes Forest Research Centre.

_____, P.N. Duinker, and R. Morash. 1986. 'Biomass and Nutrients in Nova Scotia Forests, and Implications of Intensive Harvesting for Future Site Productivity'. *Forest Ecology and Management* 15:103–27.

_____, R. Morash, and A.J. Hanson. 1981. 'Biomass and Nutrient Removals by Conventional and Whole-Tree Clear-cutting of a Red–Balsam Fir Stand in Central Nova Scotia'. *Canadian Field Naturalist* 95:307–11.

Harding, L.E. 1994. 'Threats to Diversity of Forest Ecosystems in British Columbia'. In *Biodiversity in British Columbia: Our Changing Environment*, edited by L.E. Harding and E. McCullum, 245–78. Vancouver: Canadian Wildlife Service.

Hartmann, G.F., and J.C. Scrivener. 1990. *Impacts of Forestry Practices on a Coastal Stream Ecosystem, Carnation Creek, BC*. Ottawa: Department of Fisheries and Oceans.

Harvey, S., and B. Hillier. 1994. 'Community Forestry in Ontario'. *The Forestry Chronicle* 70:725–30.

Hearnden, K.W., S.V. Millson, and W.C. Wilson. 1992. *A Report on the Status of Forest Regeneration*. Toronto: The Ontario Independent Forest Audit Committee.

Helson, B. 1992. 'Naturally Derived Insecticides: Prospects for Forestry Use'. *The Forestry Chronicle* 68:349–54.

Hudson, A.J. 1993. 'The Importance of Mountain Alder on the Growth, Nutrition, and Survival of Black Spruce and Sitka Spruce in an Afforested Heathland Near Mobile, Newfoundland'. *Canadian Journal of Forestry Research* 23:743–8.

Kimmins, J.P. 1977. 'Evaluation of the Consequences for Future Tree Productivity of the Loss of Nutrients in Whole-Tree Harvesting'. *Forest Ecology and Management* 1:169–83.

_____. 1995. 'Sustainable Development in Canadian Forestry in the Face of Changing Paradigms'. *The Forestry Chronicle* 71:33–40.

McIver, J.D., A.R. Moldenke, and G.L. Parsons. 1990. 'Litter Spiders as Bio-indicators of Recovery after Clear-cutting in a Western Coniferous Forest'. *Northwest Environmental Journal* 6:410–12.

Miller, A., and P. Rusnock. 1993. 'The Ironical Role of Science in Policymaking: The Case of the Spruce Budworm'. *International Journal of Environmental Studies* 43:239–51.

Minister of Supply and Services. 1992. *The State of Canada's Forests, 1991*. Ottawa: Minister of Supply and Services Canada.

_____. 1994. *The State of Canada's Forests, 1993*. Ottawa: Minister of Supply and Services Canada.

National Forest Strategy Coalition. 1992. *Sustainable Forests: A Canadian Commitment*. Ottawa: Canadian Council of Forest Ministers.

Natural Resources Canada. 1995. *The State of Canada's Forests, 1994*. Ottawa: Canadian Forest Service.

_____. 1996. *The State of Canada's Forests, 1995–1996*. Ottawa: Canadian Forest Service.

Neilson, M.M. 1985. 'Spruce Budworm—a Case History: Issues and Constraints'. *The Forestry Chronicle* 61:252–5.

Ontario Forest Policy Panel. 1993. *Diversity: Forests, People, Communities—a Comprehensive Forest Policy Framework for Ontario*. Toronto: Queen's Printer for Ontario.

Restino, C. 1993. 'The Cape Breton Island Spruce Budworm Infestation'. *Alternatives* 19:29–36.

Reynolds, P.E., J.C. Scrivener, L.B. Holtby, and P.D. Kingsbury. 1993. 'Review and Synthesis of Carnation Creek Herbicide Research'. *The Forestry Chronicle* 69:323–30.

Rowe, J.S. 1992. 'The Ecosystem Approach to Forestland Management'. *The Forestry Chronicle* 68:222–4.

Sandberg, L.A., ed. 1992. *Trouble in the Woods: Forest Policy and Social Conflict in Nova Scotia and New Brunswick*. Fredericton: Acadiensis Press.

Swanson, F.J., and J.F. Franklin. 1992. 'New Forestry Principles from Ecosystem Analysis of Pacific Northwest Forests'. *Ecological Applications* 2:262–74.

Thompson, I.D. 1988. 'Habitat Needs of Furbearers in Relation to Logging in Boreal Ontario'. *The Forestry Chronicle* 64:251–61.

Tripp, D., A. Nixon, and R. Dunlop. 1992. *The Application and Effectiveness of the Coastal Fisheries Forestry Guidelines in Selected Cut Blocks on Vancouver Island*. Nanaimo: Ministry of Environment, Lands and Parks.

Turner, N.J. 1992. 'The Earth's Blanket: Traditional Aboriginal Attitudes towards Nature'. *Canadian Biodiversity* 2:5–7.

Wilderness League. (n.d.) *Forest Diversity—Community Survival Project*. Toronto: Canadian Parks and Wilderness Society.

World Resources Institute. 1992. *World Resources 1992–3*. New York: Oxford University Press.

Endangered Species

The previous chapter described some of the limitations that might be placed on logging activities in an effort to protect the preferred habitat of the spotted owl, the old-growth forests of southwest British Columbia. These measures will lead to less income from logging. There will be less money and fewer jobs. Many species have been made extinct because it was economically profitable to do so. They have been hunted, replaced by more profitable species, and had their habitat destroyed and polluted by industrial wastes just because their commercial value was deemed 'insufficient'.

Forces such as these have driven current global extinction levels to their highest rate in probably

65 million years, as described in Chapter 5. This chapter looks at some of the reasons why we should care about species extinctions. It then looks at some of the main pressures that cause extinctions and how species are protected in Canada. This last topic is expanded in greater detail in the next chapter, which deals with endangered spaces and protected areas.

SHOULD WE BE CONCERNED ABOUT SPECIES EXTINCTION?

Extinction is a natural process that has been taking place since life first evolved on this planet, as discussed in Chapter 5. Concern then is not really about the process itself but about what humans are doing to speed up the process. Before looking at some of the pressures that are causing this to happen, we need to have a clear idea of why these high extinction rates are undesirable.

Ecological Reasons

If we eliminate species, ecosystem functions are affected. Sometimes this can lead to unfortunate consequences, such as the impact on coastal marine ecosystems on the Pacific coast when the sea otter was extirpated, as discussed in Chapter 5. Species become extirpated when they have been eliminated from one part of their range but still exist somewhere else. Some other defin-

Extinction is a natural process that has been occurring since life first evolved on earth. However, it is the speed of current extinction rates across many different forms of life that concerns scientists (*Philip Dearden*).

itions relating to extinction have already been outlined in Chapter 15 (see Box 15.9). One aspect of extinction not mentioned earlier is the concept of *ecological extinction*. This occurs when species exist in such low numbers that, although they are not extinct, they can no longer fulfil their ecological role in the ecosystem. The eastern mountain lion, for example, may still exist in the Maritime provinces and has not been declared extinct. However, if it still does exist, it is in such low numbers that it no longer acts as a significant control for species in the preceding trophic level. It is ecologically extinct.

All species combine to maintain the vital ecosystem processes that make human life possible on this planet: oxygen to breathe, water to drink, and food to eat. We are part of this web. If we continue to eliminate components, the web may break and humans will be one of the main victims. Caus-

> *If the land mechanism as a whole is good, then every part is good, whether we understand it or not. If the biota, in the course of aeons, has built something we like but do not understand, then who but a fool would discard seemingly useless parts? To keep every cog and wheel is the first precaution of intelligent tinkering.*
>
> *Aldo Leopold,* Round River, *1953*

ing extinctions has been likened to removing rivets from an airplane. The systems may continue to function after losing a few components, but sooner or later the system will crash. Furthermore, one of the roles of science is to understand how these systems work. It is difficult to achieve this understanding if components are missing due to extinction.

We should also not lose sight of the importance of protecting species because of their intrinsic and evolutionary value. Species evolve over time, as discussed in Chapter 5, and as more species become extinct, this reduces the amount of variation in the ecosphere on which to base future adaptability. Fewer species mean a more impoverished biosphere on which to base evolutionary adaptability for future generations of all species.

Economic Reasons

The preservation of ecosystem functions and components also has economic benefits. Countless products used in agriculture and industry originate in the natural world. Naturally occurring plants in the tropics are the source of 90 per cent of the world's food today. Corn, or maize, is a crop now estimated to be worth at least $50 billion per year around the world and feeds millions of people. It was first domesticated by indigenous people in South America some 7,000 years ago. Many other important potential food sources may occur in the 99.8 per cent of plants that have never been tested for human food potential. We should also not forget that in many rural areas of Canada (and especially among First Nations communities) wild animals still provide an important source of food. Other products besides food are also of economic importance. Rubber, without which driving in a car might not be quite so comfortable, is just one example of the chemicals produced by tropical plants to prevent insect damage.

The great hornbill is found throughout the tropical forests of Southeast Asia where it has been extirpated from many areas by hunting. Even where it exists in small numbers, it is often considered ecologically extinct as there are no longer sufficient numbers to crack and distribute the seeds of many of the tree species. Ultimately this will also cause changes in the tree species composition of the forests (*Philip Dearden*).

The Catastrophozoic

Soon a millennium will end. With it will pass four billion years of evolutionary exuberance. Yes, some species will survive, particularly the smaller, tenacious ones living in places too dry and cold for us to farm or graze. Yet we must face the fact that the Cenozoic, the Age of Mammals, which has been in retreat since the catastrophic extinctions of the late Pleistocene, is over, and that the 'Anthropozoic' or 'Catastrophozoic' has begun. Our task now is to salvage some samples of the megafauna and protect enough habitat to give future human beings an opportunity to restore a semblance of evolutionary integrity in the 22nd century (Soule 1996:25).

One special category of products is medicinal. The role of wild species as active ingredients is not evident as you line up at the counter in your local pharmacy. However, one-quarter of the prescription drugs dispensed in the US between 1965 and 1990 contained ingredients from angiosperm plants. These contributed some $14 billion per year to the US economy and over $40 billion per year worldwide. Next time you take an aspirin tablet, say thank you to the white willow, the species wherein the active ingredient was first discovered. Another well-known example of a species that has made a significant contribution is the rosy peri-winkle, a small plant found only on the island of Madagascar and which is used in treating various cancers. A more recent discovery used for a similar purpose is the taxol found in the bark of the western yew, a small understorey tree found in the forests of the Pacific northwest. The species was of little commercial value until this discovery and was routinely cut down in clear-cuts and left to rot, as described in Box 15.13. Many animal species are also used to test these drugs, although this kind of testing is increasingly challenged by animal rights groups.

It is difficult to assess the economic value of many of the products we derive from nature. For example, natural gene pools provide a source of material to help create new genetic strains of crops to feed burgeoning populations. Wheat, the main-stay of the western Canadian agricultural economy, is a plant originating in Mediterranean countries, where most of its wild forebears have disappeared. Preservation of such wild strains is necessary to allow selective breeding based upon the widest range of genetic material to continue.

What is the value of natural pollinators for commercial crops? Mosquin and Whiting (1992) calculated an annual figure of $1.2 billion for Canada. When New Brunswick switched from spraying the forests with DDT to fenitrothion in 1970 to control spruce budworm (see Chapter 15), there was a devastating impact upon pollination of the blueberry crop due to the high toxicity of fenitrothion for bees. The commercial crop fell by 665 tonnes per year and the growers successfully sued the government. What is the role of natural predator-prey relationships in helping food production? It has been shown, for example, that woodpeckers provide an economically important

Whether a modern or traditional pharmacy, many of our medicinal products are based on products found in nature (*Philip Dearden*).

service in controlling pests such as codling moths in the orchards of Nova Scotia.

Other Extrinsic Reasons

The ecological and economic reasons mentioned earlier do not encompass all the values associated with protecting other species. How many of us will be permanently enriched and emotionally uplifted by a wildlife encounter sometime in our lives? Wildlife help contribute to the joy of life, which in turn may also contribute to economic values, as shown in Box 16.1. In fact, watching wildlife is a major reason for travel to some areas, both in Canada and elsewhere. This form of tourism—ecotourism—can become a significant force in encouraging conservation when managed to enhance biodiversity and educational values, and at the same time provide a sustainable livelihood for local residents.

All the reasons outlined earlier, however, are essentially *extrinsic*. That is, they refer to the values humans derive from other species. But are there not other reasons for protecting wildlife, reasons that depend more on *intrinsic* values?

Intrinsic Reasons

The extrinsic reasons for species protection discussed earlier, while important, should not be allowed to dominate our thinking on the matter. If we emphasize such utilitarian views, does this mean that species with less use values than others should not be protected? We should protect species for their own sake instead of seeing everything in life in such a utilitarian perspective. And what about ethical reasons? Do other species not have a right to exist just as humans do? We should be concerned about other species's existence, which is a more *biocentric* rather than an arrogant and *anthropocentric* view of life in which everything revolves around humans. A thought-provoking book written by a Canadian on this topic is *The Fallacy of Wildlife Conservation* by John Livingstone. This line of thought has also given rise to a different type of ecology, known as deep ecology, which is discussed in more detail in Box 16.2.

In the past, extinction has been viewed simply as a biological problem. The points raised earlier emphasize the necessity of making links among the biological process of extinction and the economic and ethical reasons why extinction is undesirable, why human-caused extinctions are increasing, and what measures have to be taken to prevent this from happening. In other words, extinction is not just the domain of biologists but involves consideration from a broad range of perspectives, including all the social sciences, geography, and law.

Box 16.1 Wildlife and Recreation

Wildlife is an important component of recreational activities for many Canadians. In 1991 over 90 per cent of the population took part in some wildlife-related activity, spending over 1.3 billion days and $5.6 billion. Total expenditures on wildlife-related activities increased by 32.9 per cent between 1981 and 1991. Almost 20 per cent of Canadians took a special trip to watch, photograph, feed, or study wildlife, spending almost $2.5 billion on travel expenses and equipment. About 7 per cent took part in some hunting activity, spending $1.2 billion, and just over a quarter of the population took part in recreational fishing, spending an estimated $2.8 billion. When this amount is added to the amount spent on wildlife, the $8.3 billion represents a considerable investment by Canadians. Canadian wildlife also attracts attention internationally, and a survey of Americans showed that 1.8 million visited Canada for fish and wildlife in 1991 and spent $842 million. These economic values are further underlined by the willingness of over 50 per cent of Canadians to pay more to protect wildlife from the effects of air pollution, acid rain, oil spills, and pesticides.

These expenditures should not be taken as the total value of wildlife. Most of them are based upon what people spend to take part in wildlife-related activities. The benefits of the activities must be greater than these costs or people would not continue to invest in such transactions. Furthermore, it is impossible to express in monetary terms the thrill of seeing a peregrine falcon stoop on its prey, a salmon jumping a waterfall, or feel the excitement of being in grizzly country, of knowing the 'wild' in wilderness.

Box 16.2 Deep Ecology

Deep ecology is a term that has come to mean different things to different people. It emerged as an environmental philosophy in the 1970s seeking to promote fundamental changes in how humanity sees and interacts with nature. Jim Butler, a professor at the University of Alberta and a well-known Canadian conservationist, has written a book of poems based upon this philosophy. In the book (Butler 1994), he describes deep ecology as having the following characteristics: the concepts of changing anthropocentric (human-centred) attitudes towards a new biocentred philosophy (biocentrism); a belief in the intrinsic worth and right to life of other species; an ecosystem perspective; an appreciation for the importance of biodiversity; the courage to take personal actions; the questioning of ourselves and the traditional ways we have interacted with nature; a return to our roots; a cultivation of meaningful interactions with nature through our intuitions and sensitivities; the search for meaning in an age of nihilism; the search for ecosophy or earth wisdom; the defence and restoration of ecosystems; a strong identification with our own bioregion (bioregionalism); affirming our identification and solidarity with nature; the encouragement of introspection, purification, harmony, and a celebration of life; an ecological, philosophical, and spiritual approach to environmental threats; a criticism of damaging resource practices and suggestions for positive alternatives; consistent personal and environmental ethics; a preference for non-exploitive practices in science and technology; an appreciation for the introspection of solitude and the rediscovery of observation and listening; an ecological consciousness that recognizes connectedness, interdependence, fragility, biodiversity, and a personal sense of wonder and place.

Whales were hunted to the point of extinction around the world because of their commercial value. Baleen, such as what is shown here, was used for such 'indispensable' things as umbrellas, buggy whips, and ladies' corsets (*Philip Dearden*).

VULNERABILITY TO EXTINCTION

All species are not equally vulnerable. Biologists have determined that certain characteristics tend to make some species more vulnerable to extinction than others:

- *Specialized habitats for feeding or breeding*: The northern spotted owl discussed in Chapter 15 is a good example.
- *Migratory species*: For example, many of the songbirds experience declines in population in Canada as a result of their long and hazardous migrations to South and Central America.
- *Species with insular and local distributions*: Dawson's caribou, endemic to the Queen Charlotte Islands, became extinct because it was easily hunted in a very restricted habitat, and there was no hope of an emigrating population to replace the herd.
- *Valuable species*: Many organisms are overharvested to the point of extinction because they have high economic values. The American ginseng was once abundant in forests of eastern North America, but is now much rarer due to demands for the dried roots for medicinal purposes in Asian countries. Similarly, North

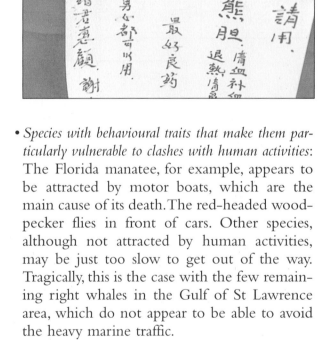

Although habitat destruction is a main cause of biodiversity declines in tropical countries, so is hunting. Species are shot to put on display, as in this hotel foyer in Vietnam, and are also killed for various superstitious and so-called 'medicinal' purposes, such as these bear parts offered for sale in a hill tribe village visited by Taiwanese tourists in northern Thailand (*Philip Dearden*).

American bears are now hunted for their gall bladders for the same markets; populations of Asian bears have been all but eliminated. A single bladder can be worth over $5,000; one illegal dealer in BC was found with 1,125 gall bladders in his possession. BC has now passed a law making possession of endangered animal contraband an offence.

• *Species at high trophic levels*: Such species, of necessity, depend upon the abundance of many different earlier links in the food chain. If the numbers of prey species are disrupted, then so too will be the predators. The proposed bison slaughter in Wood Buffalo National Park (discussed later in Chapter 17), for example, would have a catastrophic effect upon the wolf population of the park. Furthermore, as discussed in Chapter 14, these species are also vulnerable to the concentration of toxic materials through the food chain. Many birds of prey suffered precipitous declines in the 1960s and 1970s due to the concentrations of pesticides in fish and the small animals they caught for food.

• *Large species with low reproductive potential*: Extreme K-strategists are particularly vulnerable. When numbers of species, such as whooping cranes, drop to a low level, it takes a long time for them to recover, as will be discussed later.

• *Species with behavioural traits that make them particularly vulnerable to clashes with human activities*: The Florida manatee, for example, appears to be attracted by motor boats, which are the main cause of its death. The red-headed woodpecker flies in front of cars. Other species, although not attracted by human activities, may be just too slow to get out of the way. Tragically, this is the case with the few remaining right whales in the Gulf of St Lawrence area, which do not appear to be able to avoid the heavy marine traffic.

MAIN PRESSURES CAUSING EXTINCTION

Chapter 5 described the increasing rates of extinction largely as a result of human activities. This section outlines some of these activities in greater detail. Rarely are these the only factors; they are part of the overall stress that human demands are placing on the biosphere. Human use currently appropriates about 40 per cent of the net primary productivity of the planet. With global populations expected to at least double over the next century, this figure can only increase. The more productive capacity required to support one species, *homo sapiens*, the less that will be available to support others. The extinction vortex of Figure 16.1 is, therefore, ultimately driven by these demands.

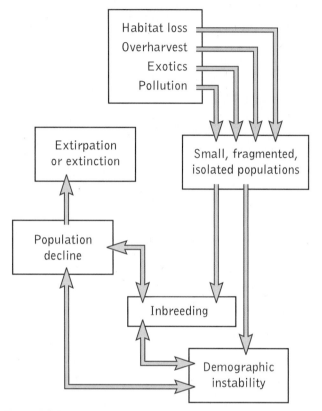

Figure 16.1 Extinction vortex.

Most attention regarding extinction has been focused on the tropical countries, not only because of the wealth of biodiversity to be found there but also because they face the brunt of many of these pressures. Of the 10 million terrestrial species estimated to be on earth, perhaps up to 90 per cent are thought to live in the tropics, particularly the tropical rainforests. In these environments, evolution over long-term periods, fuelled by high energy levels from the sun and adequate moisture, has created the most diverse ecosystems on earth.

Unfortunately, the rate of destruction in these regions is also very high. Estimates suggest that perhaps 50 per cent of tropical rainforests have already disappeared, and at current rates there will be only a few remaining fragments left after the next thirty years. Between 142 000 and 200 000 km^3 of tropical forest, an area about the size of Costa Rica, is destroyed every year. This destruction is a result of many different factors, including rapidly growing population levels, unequal wealth distribution, the abject poverty of many people who are forced to destroy their environment just to stay alive, and economic pressures from consumers elsewhere, such as you and me, for products like exotic pets, hardwood veneers, and tiger prawns, which lead to the impoverishment of tropical ecosystems.

Extinctions also occur in developed and temperate countries. Since the European colonization

Alive or dead, exotic species do not belong in your home. Do not be tempted to buy such souvenirs because by doing so, you only create a market for their death. These dead hawksbill turtles are hanging outside a souvenir shop in Hanoi. Other common tourist offerings in many locations are seashells. Although some may have washed up on beaches, many species are killed just to supply the tourist trade (*Philip Dearden*).

of North America began, more than 500 species and subspecies of native plants and animals have become extinct. The Committee on the Status of Endangered Wildlife in Canada has listed nine extinctions and eleven extirpations in Canada since the arrival of Europeans. Most of the following examples have been chosen to illustrate extinction pressures in Canada.

Overharvesting

There are many examples in Canada of species under pressure due to overharvesting. A couple of historical examples that took place at least partly in Canadian territory are well known.

The Great Auk

The great auk, a large flightless bird, inhabited the rocky islets of the north Atlantic. For many years fishers in these waters used the great auk as a source of meat, eggs, and oil. It was reasonably easy to catch and club to death, and once the feathers became an important commodity for stuffing mat-

tresses in the mid-1700s, extinction soon followed. On Funk Island off the east coast of Newfoundland, the species was extirpated by the early 1800s. The last two were clubbed to death off the shores of Iceland in 1844.

The Passenger Pigeon

The passenger pigeon had a continental distribution, whereas most extinct species were on islands. It was also a bird found in great abundance. There are claims that it was the most abundant land bird on earth, totalling up to 5 billion birds. These great flocks used to migrate annually from their breeding grounds in the oak, beech, and chestnut forests of southeastern Canada and the northeastern US to their wintering grounds in the southeastern US. So great were their numbers that tree limbs would break due to their weight, and trees would die due to the amount of guano deposited. There are many eyewitness accounts of flocks so huge that they blotted out the sun for hours, sometimes for days, providing an easy target for hunters. They were

Great auks, painted by John James Audubon (*Metropolitan Toronto Reference Library*).

A pair of passenger pigeons, painted by John James Audubon (*Metropolitan Toronto Reference Library*).

slaughtered in huge numbers for food to feed growing urban populations. It is estimated that over 1 billion birds were killed in Michigan alone in 1869. They were shot, netted, and clubbed into extinction. This hunting also occurred with a reduction in their breeding habitat as it was converted into agricultural lands. The last pigeon sighted in Canada was at Penetanguishene, Ontario, in 1902; the last one died in a zoo in Cincinnati in 1914. The world will never again experience the sound and sight of millions of passenger pigeons darkening the heavens.

This astonishing tale of a species going from such abundance to extinction is not that unusual. Three fish species of the Great Lakes—the blue walleye, the deepwater cisco, and the longjaw cisco—were once very abundant, and millions of kilograms of the fish were taken by the commercial fishery. They were fished into extinction only since 1950. Perhaps the extinction of the passenger pigeon and the great auk occurred so long ago that we can argue a lack of knowledge as an excuse for their demise. Such a case cannot, however, be made for the recent extinctions of the three fish species. They represent the failure of resource managers to understand the natural dynamics of the species they are supposed to be managing.

One aspect of overharvesting that is generally given little attention is the demand for captive species. In the past, zoo collectors' zeal to exhibit various species—especially rare species that people would come to see, such as pandas and orangutans—was of serious concern. International regulations, such as the Convention on International Trade in Endangered Species (CITES), of which Canada is a signatory, now make it difficult for this kind of trade to occur. It still does happen, however, because of demand for private collections. Exotic species, such as parrots and tropical fish, belong in their native homelands and not in people's homes. Some of the most sought-after Canadian species are our falcons, particularly the gyrfalcon and the peregrine falcon, which can be worth thousands of dollars each on the international market, although live capture of wild falcons in some provinces and territories is allowed (e.g., 212 Peale's peregrine falcons were taken from the Queen Charlotte Islands between 1952 and 1987). Given the market value of the birds, poaching is also a problem.

The Passenger Pigeon

The noise they made, even though still distant, reminded me of a hard gale at sea, passing through the rigging of a close-reefed vessel. As the birds arrived and passed over me, I felt a current of air that surprised me. Thousands of the Pigeons were soon knocked down by the pole-men, while more continued to pour in.... The Pigeons, arriving by the thousands, alighted everywhere, one above another, until solid masses were formed on the branches all around. Here and there the perches gave way with a crash under the weight, and fell to the ground, destroying hundreds of birds beneath, and forcing down the dense groups of them with which every stick was loaded. The scene was one of uproar and confusion. I found it quite useless to speak, or even to shout, to those persons nearest me. Even the gun reports were seldom heard, and I was made aware of the firing only by seeing the shooters reloading (Audubon 1913).

These examples illustrate the impacts of direct consumption of individuals from populations. There are, however, more subtle cases in which supposedly non-consumptive activities can have detrimental impacts upon species that may cause displacement from valuable habitat and even lead to death. For example, research is underway on both the Atlantic and Pacific coasts of Canada to assess the potential impacts of the boats used for whale-watching ecotourism on the well-being of

Sometimes the pressures on species that may cause them to decline are subtle and difficult to understand. For example, researchers have been studying the impacts of whale-watching on whales to see if it causes whales to alter their behaviour in ways that may be detrimental to their overall well-being. Results suggest that, by and large, this is not the case (*Dave Duffus*).

Box 16.3 Recovery of the Swift Fox

Not all species subject to heavy pressures are pushed to the brink of extinction. Some, such as the beaver (as described in Chapter 3) and the swift fox, may recover in numbers and start to repopulate their old range. The beaver did this with relatively little help. The swift fox, on the other hand, is the target of a twenty-year, $20 million reintroduction program that highlights the difficulties and costs associated with trying to reverse extinction trends.

The swift fox is so called because of its ability to run down rabbits and other prey in its home terrain, the dry, short-grass prairie. Speeds of over 60 km per hour have been recorded. It is a small fox, about half the size of a red fox, that at one time roamed all the way from Central America to the southern Prairies of Canada. Unlike most other members of the dog family, the swift fox uses dens throughout the year, which are located preferably on well-drained slopes close to a permanent water body. This may be for protection due to their small size. Natural enemies include coyotes and birds of prey, such as eagles and red-tailed and rough-legged hawks.

The last swift fox in the wild was seen in Alberta in 1938. A combination of factors led to its demise. It was heavily trapped in the middle and late 1800s for its soft, attractive pelt. The Hudson's Bay Company sold an average of 4,681 pelts per year between 1853 and 1877. By the 1920s the take was down to just 500 per year. The Prairies also came under heavy pressure from cultivation, which reduced the num-

bers of many species. However, it was probably the predator control programs against the coyote and wolf that finally removed it from the Canadian Prairies, coupled with the severe droughts during that period. As with many species, it is often not just one factor that drives a species to extinction but the combined effects of many factors, often interacting synergistically.

Since 1978 efforts have been made to return the little fox to the Prairies. Populations still existed in the US. Foxes have been bred in captivity and also relocated from these populations to Alberta. It is interesting that the major effort to initiate the captive breeding program was by two private citizens and not by government agencies. This again illustrates the kind of positive impacts that individuals can have on environmental problems. Since the first reintroductions in 1983, despite difficulties encountered *en route*, there are now over 200 swift foxes living and breeding in the wild on the Canadian Prairies once more.

The swift fox is an example of *ex situ* conservation, where the animal was reintroduced after populations were rebuilt *outside* the natural habitat in places such as zoos. *In situ* conservation is when conservation efforts focus on the organism *within* its natural habitat. The Thelon Game Sanctuary, for example, was established in 1927 to help protect the remaining population of musk oxen. Since that time, much of the animals' mainland habitat has been recolonized by out-migration from this sanctuary.

The swift fox feeds mostly on small mammals, but also on insects (© *Wayne Lynch*).

the whales (Table 16.1). Such research requires thorough knowledge of natural distributions and behaviours before it can be ascertained whether changes have occurred as a result of disturbance. These impacts are usually much more difficult to document than the more direct effects of consumptive use.

Predator Control

Several species have been totally eliminated by humans due to competition between the species and humans for consumption of the same resource. One North American example is the Carolina parakeet, the only member of the parrot family native to North America. It was exterminated in the early part of the century because of its fondness for fruit crops. Another example is the prairie dog, which was extensively poisoned because horses and cattle sometimes broke their legs after stepping into prairie dogs' burrows. Estimates suggest that these extermination programs were so successful that there has been a 99 per cent decline in prairie dog populations. Unfortunately, as discussed earlier, once such a decline occurs, there will be repercussions elsewhere in the food chain. In this case, the drastic decline in prairie dogs led to the collapse of their main predator, the black-footed ferret, 90 per cent of whose diet is made up of prairie dogs. Fortunately, the ferrets have been successfully bred in captivity and are now being reintroduced to the wild in places where prairie dogs are protected.

Predator control is *not* a thing of the past, as illustrated by the quote from Gro Harlem Brundtland, the chairperson of the World Commission on Environment and Development. For example, should deer be culled to reduce their pressure on vegetation (Box 16.5)? One of the most controversial environmental questions in Canada is whether wolves should be killed to increase the numbers of deer that humans can kill (Box 16.4). The seal hunt off the East Coast is also regaining momentum as a predator control program to protect cod to revive the fishing industry (Chapter

As a country dependent upon marine resources, Norway must adopt a scientific approach. Our scientists tell us that the minke whale stock in the North East Atlantic eat as much biomass, crustacea, krill and fish each year as all the Norwegian fishermen catch. Sustainable management of marine living resources requires that we take into account the relationship between the various species of the oceans, how they interact, how depletion or increase of one stock affects the health of other stocks, etc.

Gro Harlem Brundtland,
prime minister of Norway,
chair of the World
Commission on Environment
and Development, 1993

Table 16.1 Potential Influence and Consequences of Disturbance of Whales

| INFLUENCE | | CONSEQUENCES |
Direct (Mainly behavioural)	Indirect (Mainly ecological/population)	
Change in an individual's behavioural status, for example, the result of a collision	Individual may die, which may have consequences for the success of the breeding group	Immediate
Interference with important behaviour, such as feeding, courtship, or care of juveniles	May temporarily shift use of range, which may develop into permanent range reduction	Short term
Alteration of range	Potential reduction of fitness and reproduction, leading to a decline in population	Long term

Source: After D.A. Duffus and P. Dearden, 'Recreational Use, Valuation and Management of Killer Whales (*Orcinus orca*) on Canada's Pacific Coast', *Environmental Conservation* 20 (1993):149–56.

Carolina parakeets, painted by John James Audubon (*Metropolitan Toronto Reference Library*).

Box 16.4 The Wolf Kill

Few animals generate such emotions as the wolf. For centuries it was vilified as a rapacious killer and enemy of humans. It was shot, poisoned, and extirpated throughout large areas of its range. It is listed as an endangered species in the US. Canada has some of the healthiest wolf populations in the world, numbering around 58,000. Some of these are being used for reintroductions, such as the recent transfer of wolves from Alberta to Yellowstone National Park. However, wolves are still being shot and poisoned as part of predator control programs. Some 400,000 cattle graze in the area of the East Slopes of southern Alberta, for example, and in 1994 fourteen were killed by wolves. In response, at least twenty-nine wolves were killed by the government and ranchers. Biologists estimate that it could take decades for wolf numbers to recover in the area. Until 1993 the Alberta government had a program that would compensate ranchers by as much as $1,000 per cow killed by a predator. Due to budget restrictions, this program was discontinued until it was taken up on a voluntary basis by an NGO. It remains to be seen what kind of relationship will evolve between the wolves and ranchers in this area over the next few years.

There has also been considerable tension in the Yukon over wolf-kill programs. The conflict is over whether humans *or* wolves should harvest various ungulate species, such as moose and caribou. The conflict has been a long one, with the Yukon Game Branch conducting a systematic wolf poisoning program throughout the territory during the 1950s and 1960s in response to apparently declining ungulate numbers. Private aerial hunting was initiated in the 1980s in the Coast Mountains, although subsequent research indicated that grizzly bears were the primary predators on moose calves. Between 1983 and 1985, wolf numbers were reduced from 161 to forty-seven. This resulted in a decline of only 20–30 per cent in the amount of prey biomass killed during winter and biologists concluded that short-term removal of wolves would have no positive effects on prey populations.

Nonetheless, another wolf-kill program was initiated in 1993 in the Aishihik region east of Kluane National Park Reserve, a World Heritage Site. The three-year program was to remove some 140 wolves from the area to boost caribou and moose populations in response to requests from Native peoples. Although a 2–12-km-wide buffer zone was established between the killing area and the park, by 1994 no intact wolf packs remained in the park. In late 1994 the government decided to extend the wolf kill for another five years. By 1995 127 wolves had been killed through the program. After using aerial shooting in previous years, the government switched to snares in 1995. One moose, five wolverines (listed as a vulnerable species by COSEWIC), and twelve coyotes were caught in addition to the wolves.

Many conservationists feel that excessive human hunting, and not wolves, is the main cause of the ungulate declines. The controversy increased when the Yukon government leader, John Ostashek, a former big-game outfitter, revealed that he charged his clients by the number of animals killed rather than a set price for the hunt, thereby creating an incentive to kill more. He also feels that outfitters should be immune from hunting quotas. Conservationists feel that there should be a complete ban on hunting in the Aishihik area before wolf kills are implemented. What do you think?

Wolves have not fared well wherever humans dominate the landscape and have been extirpated from many areas of the world. Even in lightly populated regions in the Yukon there are calls for their control (*World Wildlife Fund Canada/Frank Parhizgar*).

Box 16.5 Culling Deer

During the summer of 1995, the Ontario Ministry of Natural Resources proposed that hundreds of deer at the Pinery Provincial Park, adjacent to Lake Huron and just north of Sarnia, should be killed as part of a parks management plan.

Pinery Provincial Park contained some 732 deer, while park officials estimated that the vegetation could only support about 175, an example of the kinds of carrying capacity calculations alluded to in Chapter 5. The ministry was concerned that overbrowsing by the deer posed a major threat to the integrity of the park ecosystem. In addition, nearby farmers had been complaining about crop damage from deer browsing. One reason for the large deer population was that its main predator, the wolf, had effectively been removed through human occupance and activity in the area. The culling was intended to 're-balance the ecosystem'.

In November 1993, at Rondeau Provincial Park on the north shoreline of Lake Erie, the Ministry of Natural Resources had authorized a cull in which 320 deer were shot and killed. During that cull, there were protests by both those opposed to the killing of the deer and by those who objected to not being allowed to participate in the cull. At Rondeau, the culling was done by First Nations peoples, and sports hunters wanted to participate. For the Pinery culling, park officials indicated First Nations and sports hunters could participate.

The ministry recognized that the proposal for the Pinery Provincial Park's deer cull would generate controversy. However, a spokesperson stated that 'What we're setting ourselves up for is protection of a priceless ecosystem' (Van Break 1995:B3).

The whitetail deer browses on twigs, shrubs, acorns, grass, and herbs, and can be found in groups of twenty-five or more in the winter (*Earthroots*).

This whitecoat seal pup will shed its white fur when it is approximately three weeks old. Canadian taxpayers subsidize the annual seal hunt, which has been the subject of much controversy and outrage (*International Fund for Animal Welfare/David White*).

20). Recently there have been calls in Prince Edward Island and Ontario to kill more double-crested cormorants, which are accused of taking too many fish. As the human population grows and our demands increase, these kinds of conflicts regarding whether people or another species will consume another organism will escalate.

Habitat Change

Habitat change is the most important factor causing the erosion of global biodiversity. Human demands are causing both physical and chemical changes to the environment. Physical changes, such as deforestation, are removing important habitat components, while chemical pollution of those habitats renders them no longer able to support animal populations, even if the physical structure of the habitat remains. Humans place further pressure on species by introducing alien species, as discussed in Chapter 5.

Physical Changes

Some impacts of physical changes in the natural environment have already been discussed within the context of agricultural and forestry practices. It is difficult historically to separate these influences from the more general impacts of colonization in North America. Large areas of forests in central and eastern Canada were cleared to make way for agriculture. Species dependent upon such forests, such as the eastern cougar and wolverine, suffered accordingly. Before the mid-1850s there were some 101 million ha of long-grass prairie in central North America; now less than 1 per cent remains. Other ecozones have not fared much better, with only 13 per cent short-grass prairie, 19 per cent of mixed-grass prairie, and 16 per cent of aspen parkland remaining. Millions of bison and antelope once grazed these lands, and the earth shuddered with their migrations. They have now been replaced by cattle. Not surprisingly, half of Canada's endangered and threatened mammals and birds are from the Prairies.

The Prairies have been virtually completely transformed into an agricultural landscape, with the result that many species native to this habitat are now endangered (*Philip Dearden*).

Accompanying the transformation of the natural prairie grassland to agricultural purposes has been the draining of thousands of hectares of wetlands to create more agricultural land. Until the early 1990s the Canadian Wheat Board Act made it financially attractive for farmers to expand crop land as much as possible rather than manage their land more effectively, so much marginal land was brought under the plough. It is now estimated that over 70 per cent of prairie wetlands have been drained. Of the remaining wetlands, 60–80 per cent of the habitat surrounding the basins are affected by farming every year. Such changes were a major factor in the declining numbers of waterfowl breeding on the Prairies (Figure 16.2), a trend that is only just being reversed by wildlife management practices (Box 16.6). Draining wetlands is not restricted to the Prairies. Eighty per cent of the wetlands of the Fraser River delta have been converted to other uses, as have 68 per cent of the wetlands in southern Ontario and 65 per cent of the Atlantic coastal marshes. Drainage not only has a negative impact upon marsh-dwelling species but also increases pollutant loads and sediment inputs in drainage water. High pesticide, fer-

Another form of habitat degradation caused the demise of this pilot whale off the east coast of Newfoundland. It drowned after getting caught in a fish trap (*Philip Dearden*).

Box 16.6 Rebuilding 'the Duck Factory'

The winter of 1995 provided prairie residents with the most spectacular migration of water-fowl seen since the mid-1950s. Biologists estimated that over 83 million ducks were involved, 24 million more than in 1993, which was a clear indication that intensive management practices were beginning to pay off following the catastrophic declines in populations of many ducks during the 1980s. The Prairies have long been known as the 'duck factory' of North America, with huge areas of breeding habitat spread over millions of hectares of agricultural land. As farm intensification increased, these areas were drained and ploughed, and duck numbers declined precipitously. However, the continent-wide North American Waterfowl Management Plan involving governments, conservation groups, hunters, farmers, and other land-owners is trying to reverse these trends through habitat conservation, such as maintaining nesting areas on land close to shallow water for land-breeders like mallards, pintails, teal, gadwalls, wigeons, and shovellors, and ensuring that water levels are

managed for diving ducks like redheads and canvasbacks. The goal is to reach a total breeding population of 62 million by the year 2000, with a fall migration of 100 million birds.

How close 1995's fall migration of 83 million for a smaller survey area (including just the Prairies) is to meeting this goal has not yet been established. Biologists also caution that the exceptionally good rains for a couple of years may have swelled populations to unsustainable figures, and that a couple of dry years may reduce populations once again. There are also political threats to these gains as proposals to the US Congress aim to cut funding for habitat restoration and exempt the smaller areas that, in total, constitute most of the prairie potholes. Nonetheless, the sight of the sky filled with so many waterfowl was a first-time experience for many younger prairie residents and reminded older ones of times when such sights were expected every year, without the need for a multibillion-dollar management project.

tilizer, sediment, and salt levels may negatively affect organisms further downstream.

Another ecozone that has been particularly hard hit by habitat destruction is the Carolinian forests of southwestern Ontario. These southern deciduous forests support a greater variety of wildlife than any other ecosystem in the country,

including 40 per cent of the breeding birds. More than 90 per cent of this habitat has now been transformed by forestry, agriculture, and urbanization; less than 5 per cent of the original woodland remains. It is estimated that 40 per cent of Canada's species at risk are in this zone. Most of the remaining forests are in privately owned wood-lots or tracts belonging to regional conservation authorities and are potentially open to logging. Even sites of special ecological significance may be threatened, as illustrated by the case of Moulton Tract in the Long Point Region Conservation Authority, which is home to seven COSEWIC-listed species, and yet has little protection from existing legislation.

Concern regarding the loss of habitat for Canadian species also involves assessing habitat loss in other countries. The harsh winters and productive summers that characterize much of Canada mean that many species are migratory, espe-

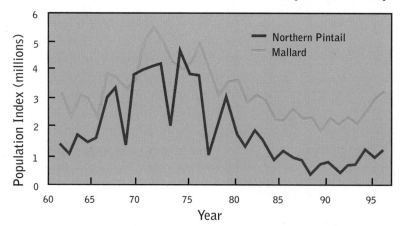

Figure 16.2 Population index for mallard and northern pintail. SOURCE: Environment Canada, *Status of Migratory Game Birds in Canada* (Ottawa: Environment Canada, 1996): figures 4 and 6.

cially birds. Over the last few decades, there have been significant declines in species that spend most of the year in tropical habitats but migrate to Canada to breed. In BC, for example, statistically significant declines have occurred between 1968 and 1988 in species such as northern flickers, Swainson's thrushes, chipping sparrows, yellow warblers, and dark-eyed juncos. The reasons for these declines are difficult to pin down, but probably involve several factors, including loss of winter range through tropical deforestation and increased fragmentation within their northern breeding habitat. More long-term data and detailed studies are required to sort out the complexities of these changes.

Chemical Changes

Chemical degradation of habitat is becoming more important as increasing numbers of chemicals enter the environment. Over 35,000 chemicals are in use in Canada. The effects of chemical pollution are often less obvious than those of physical destruction. Unless there is a catastrophic spill, often the signs of declining populations go unnoticed for several years. Even after declining numbers have been detected, it may take many years of careful analysis before they can be linked to chemical pollution. This was the case with the decline in the numbers of top trophic level birds, such as peregrines, ospreys, and eagles, when organochlorine chemicals changed the calcium metabolism of the birds, leading to thinner eggshells and ultimately lower breeding success. Although bald eagles had been persecuted for a long time around the Great Lakes, it was their total breeding failure due to high chemical levels that led to their extirpation from the Ontario side of Lake Erie in 1980. Figure 16.3 shows the concentrations of chemicals in the eggs of bald eagles on the north shore of Lake Erie. Research indicates that if DDE levels exceed 5.1 ppm, there is a marked impact on reproduction, and at levels over 15 ppm, there is complete failure. Eggs with more than 33 ppm of PCBs produce less than one young for every five pairs. Table 16.2 shows the associated rise in the number of successful young raised as chemical concentrations declined. A return to normal productivity is anticipated in the future. A similar result of using the agricultural chemical, carbofuran, on the endangered burrowing owl was dis-

cussed in Chapter 15, as was the impact of toxic chemicals on the population of beluga whales in the estuary of the St Lawrence River.

Alien Species

Introduced species can also have a significant impact by out-competing native species for necessary resources or by direct predation on native species. Examples of some successful aliens competing for resources were described in Chapter 5.

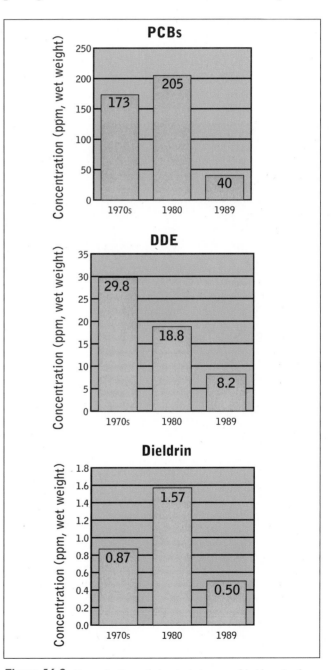

Figure 16.3 Concentrations of chemicals in eggs of bald eagles from the north shore of Lake Erie in the 1970s and 1980s. SOURCE: L. Mc-Keane and D.V. Weseloh, 'Bring the Bald Eagle Back to Lake Erie', SOE Fact Sheet 93-3 (Ottawa: Environment Canada, 1993).

Box 16.7 Return of the Raptors

If you have access to the Internet, try http://www.on.doe.ca/falcon. This site has been established to show a live video of a pair of peregrine falcons as they raise their young in downtown Toronto. The pair first nested here in 1995 and raised a brood, the first successful nesting in southern Ontario in over forty years. In 1996 several organizations got together to see if they could give more people a view into the peregrines' household and established the video-Internet link. Such a sight is symbolic of the decline and recovery of this majestic bird of prey.

The peregrine falcon once bred across the country. They are extremely powerful birds of prey, catching other birds in flight while attaining speeds as high as 300 km per hour. The prey is killed by a direct blow of the closed talons, delivered at great speed. Favourite prey include songbirds, waterfowl, pigeons, shorebirds, sea birds, and (especially in the Arctic) small mammals, such as lemmings. They nest on cliffs and sometimes trees where they can overlook water bodies. Tall buildings may serve as nesting sites where the main prey is urban pigeons. They are quite territorial during the breeding season, and their nests are seldom closer together than 1 km.

Populations of peregrines have been well studied around the world for centuries due to the interest of falconers. Until the 1940s these populations seemed to be remarkably stable. Then populations started to crash. Over 300 known nesting sites in the eastern US became unused. It became apparent that this precipitous decline was caused by the bioaccumulation of biocides, particularly organochlorines, such as DDT, at the end of long food chains (Chapter 14). DDT has now been banned in North America, although the migrating populations are still vulnerable because DDT is still used in Latin America.

By this time, peregrines had been extirpated from large areas of their previous range. However, a captive breeding program was started by the Canadian Wildlife Service at their facility in Wainwright, Alberta, and, together with university-based programs in Quebec and Saskatchewan, they have released over 700 birds to the wild at over twenty sites from the Bay of Fundy to southern Alberta. Over a dozen released birds are now breeding in the wild, and peregrine populations are on the road to recovery, although the southern subspecies (*Falco peregrinus anatum*) is still on the endangered species list.

The peregrine falcon suffered badly as a result of the biomagnification of agricultural chemicals, but is now recovering in numbers in many areas (*World Wildlife Fund Canada/Edgar T. Jones*).

Table 16.2 Number of Bald Eagle Nest Sites and Young Raised on the North Shore of Lake Erie, 1980–1992

Parameter	1980	1981	1982	1983	1984	1985	1986	1987	1988	1989	1990	1991	1992
Number of nests	3	4	5	5	6	7	7	7	8	8	9	10	9
Number of naturally raised young	0	1	5	5	7	6	6	6	10	7	11	11	14
Number of young hatched	0	0	0	6	4	6	8	8	0	0	0	0	0
Total number of young (Natural plus hatched)	0	1	5	11	11	12	14	14	10	7	11	11	14

Sources: Ontario Ministry of Natural Resources, Environment Canada, Canadian Wildlife Service.

Box 16.8 Purple Loosestrife Threatens Wetlands

Purple loosestrife was introduced to North America from Europe in ballast water from ships over a century ago. An aggressive invader of aquatic systems, this 'weed' has spread through thousands of hectares of wetlands in Ontario, and it is estimated that 190 000 ha of land in North America are lost to purple loosestrife each year. After its woody root systems are established, native plants and the animals that depend on these native plants for food are forced out.

At the University of Guelph, experiments on test sites with the *Galerucella pusilla* beetle have showed promising results in using this beetle to control this invader plant. The beetles have a voracious appetite for purple loosestrife. They ate the metre-high plant at such a rate that the plant's capacity to produce seed (about 2.5 million seeds per plant per year) has been reduced by up to 99 per cent.

The use of the beetles is promising, as previous control efforts—such as pulling out the plant, or burning, mowing, dusting, and spraying—have been unsuccessful. Furthermore, the beetles do not seem to have an appetite for native plant species. The use of beetles to combat this invader plant is a good example of how integrated pest management can allow managers to reduce reliance on chemical means of control.

The introduction of new species to insular habitats provides some graphic examples of the destruction that can be wrought. On the Queen Charlotte Islands, for example, the introduction of both raccoons and Norway rats is having a catastrophic impact on ground-nesting sea birds. For example, the breeding population of ancient murrelets on Langara Island, off the north coast of the Queen Charlottes, declined by approximately 40 per cent between 1988 and 1993 to a level less than 10 per cent of its original size. Farther south in the new national park reserve of Gwaii Hanaas, the main ancient murrelet colony on Kunghit Island decreased in size by approximately one-third between 1986 and 1993. In both cases predation by Norway rats appear to be mainly responsible for the declines. Gaston (1994) suggests that unless rats can be eliminated or the spread of raccoons halted, a significant proportion of the ancient murrelet population will be extirpated within a few decades.

Alien species can affect native species in other ways besides direct predation. In Newfoundland, for example, the Newfoundland crossbill has declined probably due to competition for resources, in this case pine cones, with the introduced red squirrel. The impact of control mechanisms for alien species on native species is often not considered. By the late 1980s, for example, about 11 million ha of land in western Canada was sprayed to control wild oats, a Eurasian weed species. Little is known about the impact of such spraying on native species. A further example of the complexity of natural systems is provided by the brown-headed cowbird, which was found originally on the grasslands of the Prairies. With the expansion of cleared agricultural lands to the east, the cowbird has expanded its range into the Maritimes. This range expansion is of note due to the parasitic breeding habit of the cowbird, which lays its eggs in the nests of a wide variety of songbird species, thereby significantly reducing the breeding success of those species. The cowbird chicks compete with the other nestlings for food and eject them from the nest. Parasitism rates of over 70 per cent have been recorded in the nests of the endangered Kirtland's warbler.

WILDLIFE MANAGEMENT

Wildlife in Canada is managed by numerous agencies, including those responsible for marine mammals, fish, terrestrial animals, birds, and forests. Each of these agencies at the provincial and federal levels has its own policies. In 1990 the Wildlife Policy for Canada was formulated to guide wildlife management:

- Expand the scope of wildlife policy, for example, to include any species of wild organism rather than just the so-called 'game' species that had dominated in the past, and to expand the concern for wildlife to include wildlife habitat.
- Recognize the needs of wildlife in economic and environmental policies.
- Involve Aboriginal peoples in wildlife management.
- Improve wildlife conservation by protecting habitat and populations, controlling introduced and genetically engineered species, and investing in wildlife research.
- Involve the public in wildlife policy determination and improving public access to wildlife.

A more specific focus on endangered species is provided by the Federal-Provincial Wildlife Conference's standing committee, the Committee on the Status of Endangered Wildlife in Canada, which, since 1976, has been responsible for deter-

---

Box 16.9 What You Can Do

Although the challenges of saving endangered species can seem quite daunting to the individual, there are several things that you can do:

1. If you own land, even your own backyard, try to encourage the growth of native species there and promote high diversity among these species. Provide the three staples: food, water, and shelter. Plant perennials, such as fruit and nut trees, nectar-producing flowers, and berry bushes. Don't use chemicals!
2. Write letters to politicians at all levels encouraging them to adopt specific measures. For example, write to the federal minister of environment or the prime minister to support the passage of endangered species legislation, or write to local politicians urging protection for a natural habitat in your area.
3. Join and support an environmental group that takes a special interest in endangered species. Some organizations and addresses can be found in the appendix.
4. Take part in an active biodiversity monitoring project, such as the Christmas Bird

Count, the Canadian Lakes Loon Survey, Frogwatch, Project FeederWatch, or one of the other host of organized activities that take place across the country. Details on these projects will be available from local NGOs and university and college departments.
5. Don't keep exotic pets.
6. If you have a pet, try to make sure that it does not injure or harass wildlife. Put a bell on your cat. Domestic cats kill large numbers of songbirds every year.
7. Don't buy products made of endangered animals or plants.
8. Vote for political candidates who share your views on conservation matters.
9. Stay informed about biodiversity issues by watching nature programs on TV, reading books, attending talks, and having discussions with local conservationists.
10. Actively try to learn more about wildlife, not just by reading and watching television but also by becoming more aware of the wildlife in your region through field observation. Encourage others, especially children, to do likewise.

mining the status of species. Species can be assigned to one of five categories: extinct, extirpated, endangered, threatened, and vulnerable (as defined in more detail in Box 15.9). By 1996 COSEWIC had assigned 275 species (Table 16.3). The status of species may change. For example, in 1995 the ferruginous hawk's status was changed from threatened to vulnerable. The burrowing owl, however, was upgraded from threatened to endangered, as efforts to halt population declines have been unsuccessful.

In 1988 another program was implemented to ensure some follow-up to COSEWIC's designations. This program, Recovery of Nationally Endangered Wildlife, requires that a recovery plan be devised for each new species designated as threatened,

endangered, or extirpated within two years of the designation. Recovery plans for eleven mammals, fifteen birds, and one fish had been initiated by 1995. Unfortunately, the implementation of the plans has no set timetable and the program does not yet include plants and invertebrates.

Besides the national perspective on endangered species, Canada is also party to several international treaties and agreements. For example, the North American Waterfowl Management Plan, with the US and Mexico, seeks to restore waterfowl populations to those of the 1970s by the turn of the century, largely through habitat enhancement, as described in Box 16.6. Another international convention, the Convention on the Conservation of Wetlands of International Importance (formulated

---

Box 16.10 Does Canada Need Endangered Species Legislation?

The 1992 Convention on Biological Diversity, which Canada signed, requires nations to develop legislation to protect threatened species and populations. Furthermore, public surveys in Canada have consistently shown very high levels of support for such legislation. Unlike the US, Canada does not yet have legislation to protect endangered species. If legislation is introduced, much will also depend upon the amount of discretion allowed in the legislation. In Canada, for example, much legislation is at the discretion of the minister involved who 'may' rather than 'shall' protect various aspects of the environment. This is the case in four provinces that have endangered species legislation (New Brunswick, Quebec, Ontario, and Manitoba). They also have discretionary legislation with wide variation on aspects such as recovery plans and habitat protection. Although endangered species have been seen mainly as a provincial responsibility in the past, many people feel that extinction is of national and international concern.

Effective legislation should:

- set national standards protecting all endangered species and populations in Canada
- provide a mechanism to ensure that species at risk are free from political influence
- prohibit actions that would harm endangered species or damage their habitat
- emphasize a preventive approach to keep

species from becoming endangered (the precautionary principle)
- require ecosystem-based species recovery plans
- impose stiff penalties
- work cooperatively with First Nations peoples and the provinces

In August 1995 a legislative proposal for endangered species legislation was introduced. However, conservationists were highly critical as the legislation would only apply to species in national parks and specified 'federal lands', and left a large amount of discretion in the preparation and implementation of species recovery plans. The proposal contained nothing on mandatory habitat protection nor on advance reviews of projects that might disrupt endangered species. In short, it failed to meet many of the criteria outlined earlier. When the bill was tabled in October 1996, many of these weaknesses were still evident. The bill will apply to about 40 per cent of the 276 species now at risk in Canada, including forty-four birds now protected under the Migratory Bird Convention Act and about sixty fish. Most of the rest will fall under some form of proposed provincial protection. The bill also gives the federal cabinet, not scientists, the final say on what species are considered endangered. The bill has now been withdrawn, although it is intended to reintroduce it in the future.

Table 16.3 Species at Risk in Canada

	Mammals	Birds	Fish	Reptiles	Molluscs	Plants	Lichens	TOTAL
Extinct	2	3	4	—	1	—	—	10
Extirpated	5	1	2	1	—	2	—	11
Endangered	11	16	4	4	—	27	1	60
Threatened	9	6	12	3	—	36	—	66
Vulnerable	25	20	38	8	—	31	3	125
TOTAL	52	46	60	16	1	96	4	275

Source: COSEWIC, *Canadian Endangered Species and Other Wildlife at Risk* (Ottawa: Canadian Wildlife Service, 1996).

The king eider, whose habitat is along rocky coasts, is found in the arctic regions of the northern hemisphere (© *Wayne Lynch*).

in Ramsar, Iran, and known as the Ramsar Convention) protects sites covering over 130 000 km² in Canada, more than any other country in the world.

Canada was also the first industrialized country to sign the Biodiversity Convention at the Earth Summit in Rio de Janeiro in 1992 (Chapter 20). A strategy has now been prepared for implementation in Canada based upon five goals:

- to conserve biodiversity and to sustainably use biological resources
- to improve Canada's understanding of ecosystems and increase Canada's resource management capacity

- to promote greater understanding of the need to conserve biodiversity and to use biological resources in a sustainable way
- to maintain or develop incentives and legislation that support the conservation of biodiversity and sustainable use of biological resources
- to work with other countries to achieve the objectives of the Biodiversity Convention

Each of these goals contains more specific measures for implementation. However, it is up to the discretion of individual governments as to how these might be implemented 'in accordance with their policies, plans, priorities and fiscal capabilities' (Government of Canada 1995:9). In other words, implementation is voluntary. Once the strategy is approved, the federal and provincial governments will have one year to announce the measures that they are taking, and must then publicly report on progress every two years thereafter. Given government cuts in spending, particularly in environment, it will be interesting to see whether any new or more effective programs will result in actual improvements in the conservation of biodiversity due to these measures.

IMPLICATIONS

At the World Conservation Congress in Montreal in 1996, the IUCN released the most thorough inventory and assessment ever undertaken on global biodiversity compiled by over 7,000 scientists (IUCN 1996). The results were chilling. For example, it was reported that over a quarter of the earth's known mammal species are threatened with extinction. Half of these may disappear in the next decade. Some groups are more threatened than others. Over one-third of the 275 primate species are at risk.

With figures like these, there is no doubt that we are entering a new period of relationships between humans and other life on earth, a period Michael Soule so aptly characterized as the 'Catastrophozoic' in the quote given earlier in the chapter. There are profound negative implications of this evolution that stretch from the ecological and utilitarian to the moral and ethical. The bottom line is not complex nor difficult to understand. It is this: If we do not take radical proactive steps to protect wildlife NOW, then it will be too late for many species. Extinction IS forever!

SUMMARY

1. Extinction levels have reached unprecedented levels, according to many biologists. There are several reasons why we should be concerned. Life-supporting ecosystem processes depend on ecosystem components. As we lose components through extinction, these processes become more impaired. We also derive many useful and valuable products from natural biota, including medicines. In addition to these utilitarian reasons, there are ethical and moral reasons why we should be concerned about species extinctions.

2. Not all species are equally vulnerable to extinction. Species that have specialized habitat requirements, migratory species, species with insular and local distributions, valuable species, species at higher trophic levels, and species with low reproductive potential tend to be the most vulnerable.

3. Many factors are behind current declines. The underlying factor is human demand as population and consumption levels grow. Much attention has concentrated on the tropics due to the high biodiversity levels and high rates of destruction there. However, Canada has also experienced nine extinctions and eleven extirpations since European colonization.

4. Some of the main pressures causing extinction include overharvesting, predator control, and habitat change. Habitat change includes not only physical changes but also those caused by chemicals and the introduction of alien species.

5. Wildlife in Canada is managed by numerous agencies at the federal and provincial levels. The Wildlife Policy for Canada was formulated in 1990 to encourage a more integrated approach.

6. The Committee on the Status of Endangered Wildlife in Canada is responsible for determining the status of rare species and categorizing them as extinct, extirpated, endangered, threatened, and vulnerable. By 1996 275 species had been classified. The Recovery of Nationally Endangered Wildlife program devises recovery plans for species designated as threatened, endangered, or extirpated.

7. Canada does not have endangered species legislation. Non-governmental organizations are currently pushing for such legislation to be passed.

8. Canada is also party to several international treaties for the protection of wildlife, and was the first industrialized nation to sign the Biodiversity Convention at the Earth Summit in 1992. A strategy to implement the Convention in Canada is now being developed.

REVIEW QUESTIONS

1. What are the main reasons why we should be concerned about species extinction?

2. Why are some species more vulnerable to extinction than others?

3. Are there any endangered species in your province? If so, what is being done to protect them?

4. Discuss what is being done in terms of management to assist the recovery of one endangered species in Canada.

5. What are the main pressures on wildlife in your province?

6. Does Canada need endangered species legislation? What are some arguments against it?

7. Do you agree with the wolf-kill program that is being implemented in the Yukon?

REFERENCES AND SUGGESTED READING

Bath, A. 1991. 'Public Attitudes about Wolf Restoration in Yellowstone National Park'. In *The Greater Yellowstone Ecosystem*, edited by R.B. Keiter and M.S. Boyce, 367–76. New Haven: Yale University Press.

Burnett, J.A., et al. 1989. *On the Brink: Endangered Species in Canada*. Saskatoon: Western Producer Prairie Books.

Butler, J. 1994. *Dialogue with a Frog on a Log*. Edmonton: Duval Publishing.

Canadian Wildlife Service. 1994. 'Endangered Species Legislation in Canada: A Discussion Paper'. Ottawa: Environment Canada.

COSEWIC (Committee on the Status of Endangered Wildlife in Canada). 1996. *Canadian Endangered Species and Other Wildlife at Risk*. Ottawa: Canadian Wildlife Service.

Duffus, D.A., and P. Dearden. 1993. 'Recreational Use, Valuation and Management of Killer Whales (*Orcinus orca*) on Canada's Pacific Coast'. *Environmental Conservation* 20:149–56.

Environment Canada. 1996. *Status of Migratory Game Birds in Canada*. Ottawa: Environment Canada.

Filion, F.L., et al. 1993. *The Importance of Wildlife to Canadians: Highlights of the 1991 Survey*. Ottawa: Minister of Supply and Services Canada.

Foster, J. 1978. *Working for Wildlife: The Beginning of Preservation in Canada*. Toronto: University of Toronto Press.

Gaston, A.J. 1994. 'Status of the Ancient Murrelet, *Synthliboramphus antiquus*, in Canada and the Effects of Introduced Predators'. *Canadian Field-Naturalist* 108:211–22.

Government of Canada. 1995. *Canadian Biodiversity Strategy: Canada's Response to the Convention on Biological Diversity*. Ottawa: Minister of Supply and Services Canada.

Groombridge, B. 1992. *Global Biodiversity: Status of the World's Living Resources*. London: Chapman and Hall.

Harding, L.E., and E. McCullum. 1994. *Biodiversity in British Columbia: Our Changing Environment*. Vancouver: Environment Canada, Pacific and Yukon Region.

Harfenist, A. 1994. *Effects of Introduced Rats on Nesting Seabirds of Haida Gwaii*. Technical Report Series no. 218. Vancouver: Canadian Wildlife Service, Pacific and Yukon Region.

Hayes, R.D., A.M. Baer, and D.G. Larsen. 1991. *Population Dynamics and Prey Relationships of an Exploited and Recovering Wolf Population in the Southern Yukon*. Whitehorse: Yukon Department of Renewable Resources.

Hummel, M., and S. Pettigrew. 1991. *Wild Hunters: Predators in Peril*. Toronto: Key Porter Books.

IUCN (World Conservation Union). 1996. *IUCN Red List of Threatened Animals*. Gland, Switzerland: IUCN.

Larsen, D.G., D.A. Gauthier, and R.H. Markel. 1989. 'Causes and Rates of Moose Mortality in the Southwest Yukon'. *Journal of Wildlife Management* 53:548–57.

Livingstone, J.A. 1981. *The Fallacy of Wildlife Conservation*. Toronto: McClelland and Stewart.

McKeane, L., and D.V. Weseloh. 1993. 'Bring the Bald Eagle Back to Lake Erie'. SOE Fact Sheet 93–3. Ottawa: Environment Canada.

Mosquin, T., and P.G. Whiting. 1992. *Canada Country Study of Biodiversity: Taxonomic and Ecological Census, Economic Benefits, Conservation Costs and Unmet Needs*. Ottawa: Canadian Centre for Biodiversity, Canadian Museum of Nature.

Soule, M. 1996. 'The End of Evolution?' *World Conservation* (April):8–9.

Terborgh, J. 1992. 'Why American Songbirds Are Disappearing'. *Scientific American* 266:98–104.

Theberge, J.B. 1991. 'Ecological Classification, Status and Management of the Gray Wolf (*Canis lupis*) in Canada'. *Canadian Field-Naturalist* 105:459–63.

_____, and D.A. Gauthier. 1985. 'Models of Wolf-Ungulate Relationships: When Is Wolf Control Justified?' *Wildlife Society Bulletin* 13:449–58.

Van Break, D. 1995. 'Pinery Provincial Park: Controlled Hunt Proposed to Reduce Growing Deer Herd'. *Kitchener-Waterloo Record* (1 August):B3.

van Kooten, G.C. 1993. 'Bioeconomic Evaluation of Government Agricultural Programs on Wetlands Conversion'. *Land Economics* 69:27–38.

Wildlife Ministers' Council of Canada. 1990. *A Wildlife Policy for Canada*. Ottawa: Canadian Wildlife Service.

Wilson, E.O. 1992. *The Diversity of Life*. Cambridge, Mass.: The Belknap Press of Harvard University.

Chapter 17

Endangered Spaces

The previous chapter explored some of the challenges associated with protecting biodiversity. One of the main ways to do this is through habitat protection. This is one of the main functions of protected area systems, such as national and provincial parks. Such areas also provide many other values to society, and just as it is important to understand why we should protect species from extinction, it is important to appreciate the many roles parks and other protected areas fulfil.

WHY HAVE PROTECTED AREAS?

The functions of protected areas can be likened to different buildings in a city. This analogy is useful because it emphasizes the many different roles played by protected areas and helps people understand that they are not 'single-use', a common criticism from industrial interests when protected designations are suggested.

- *Art gallery*: Many parks were designated for their scenic beauty, which is still a major reason why people visit parks.
- *Zoo*: As one component of the art gallery, parks are usually easy places to watch wildlife in rel-

The one process ongoing in the 1990's that will take millions of years to correct is the loss of genetic and species diversity by the destruction of natural habitats. This is the folly that our descendants are least likely to forgive us.

E. O. Wilson,
Harvard University

atively natural surroundings. Most park wildlife are protected from hunting and are not as shy of humans as wildlife outside parks.
- *Playground*: Parks provide excellent recreational settings for many outdoor pursuits.
- *Movie theatre*: Like a movie, parks are able to transport us into a different setting.
- *Cathedral*: Many people derive spiritual fulfilment from nature, just as others go to human-built structures, such as churches, temples, and mosques.
- *Factory*: The first national parks in Canada were designated with the idea of generating income through tourism. Since these early beginnings, the economic role of parks has been recognized, although it is a controversial one due to the potential conflict with most of the parks' other roles. A 1994 study of the contribution of BC provincial parks to the provincial economy, for example, calculated the annual revenue in excess of $400 million, and parks created more jobs than the mining and newsprint industries.
- *Museum*: Parks protect the landscape as it must have been when European colonists first arrived in North America. As such, parks act as museums to remind us of these conditions. These museums also serve a valuable ecological function as they provide important areas against which to measure ecological change in the rest of the landscape.
- *Bank*: Parks are places in which we store and protect our ecological capital, including threatened and endangered species. From these

accounts we can use the interest to repopulate areas with species that have disappeared, as with the musk oxen from the Thelon Game Sanctuary that recolonized terrain.

• *Hospital*: Ecosystems are not static and isolated phenomena but linked to support processes all over the planet. Protected areas constitute some of the few places where these processes still operate in a relatively natural manner. As such, they may be considered ecosystem 'hospitals' where air is purified, carbon is stored, oxygen is produced, and ecosystems are recreated.

• *Laboratory*: As relatively natural landscapes, parks provide outside laboratories for scientists to unravel the mysteries of nature. For example, Killarney Provincial Park in Ontario provided an important laboratory for early research on acid precipitation in Canada.

• *Schoolroom*: Parks can play a major role in education as outdoor classrooms.

The 'playground' role is one of the most controversial issues for park managers. This Parks Canada golf course sign in Pacific Rim National Park, for example, is advertising a facility that is not actually located on park lands. Nonetheless, by advertising it, a link is created in the public mind between national parks and certain recreational activities. What about traditional activities in some parks, such as horseback riding, which can have a significant impact on the environment, or are national parks places where we should rely on our own feet to explore the beauty of the landscape (*Philip Dearden*)?

These roles are often difficult to evaluate in economic terms. How do we calculate the profits derived from the cathedral role or the bank role? Yet virtually all land-use decisions are based upon market economics. Whether or not we log an area, plant crops of a certain type, or build a supermarket, all these decisions are fundamentally based upon the amount of profit that can be derived

---

Box 17.1 Canada Facts

First Park: Mount Royal, Montreal, 1872

First National Park: Banff, Alberta, 1885

First Wildlife Sanctuary: Last Mountain Lake, Saskatchewan, 1887

First Provincial Park: Algonquin, Ontario, 1893

from one land use relative to another. The values ascribed to parks are often termed market failures due to this problem, and are the main reason why governments must become involved if we want to maintain these values for future generations. Parks, then, are one type of protected area that the government has had to protect from market forces. In turn, these areas afford protection for the kinds of values described earlier. However, parks are just one type of protected area. The next section will discuss the different types in more detail.

DIFFERENT KINDS OF PROTECTED AREAS

There are different types of protected areas in many jurisdictions. National parks, for example, spread across the country and are governed by the same legislation and policies. Provincial and territorial parks, on the other hand, although they exist in all jurisdictions, may differ considerably in management emphasis. In some provincial park systems, for example, logging is still allowed. Can such parks be expected to fulfil many of the roles outlined earlier if they are still subject to market and political forces that allow resource extraction to occur? Algonquin Park in Ontario is a good example of where logging still takes place. In addition to the national and provincial park systems, there are a host of other protective designations ranging from national wildlife areas to regional and municipal parks that differ in their goals and associated management practices (Table 17.1).

The problems of making comparative assessments among different jurisdictions in Canada when, for example, some provincial parks allow logging and others do not is difficult enough, but the challenge is greatly magnified at the global scale. How, for example, can we compare the national parks of England where large areas are under private ownership and grow cash crops, with the national parks of North America, which are under national control and managed to maximize the values discussed earlier? In response to this problem, the World Conservation Union has established six different categories of protected areas:

  I Strict protection (i.e., I.a. Strict Nature Reserve/I.b. Wilderness Area)
 II Ecosystem conservation and recreation (i.e., National Park)
III Conservation of natural features (i.e., Natural Monument)
 IV Conservation through active management (i.e., Habitat/Species Management Area)
  V Landscape/seascape conservation and recreation (i.e., Protected Landscape/Seascape)
 VI Sustainable use of natural ecosystems (i.e., Managed Resource Protected Area)

Under this scheme, although the national parks of England might be called national parks in England, they would qualify as category V (protected areas by international standards). As the categories progress from category I (Strict Nature Reserve/ Wilderness Area) to category VI (Managed Resource Protected Area), they represent a gradual change in the balance between the different roles for which they are being managed (Table 17.2). Higher category areas are managed with priority attention given to ecological values. As we progress along the scale, the weight given to other values—particularly recreation, tourism, and even resource extraction—increases.

In addition to lands held by governments as protected areas, non-governmental organizations are also beginning to play a more prominent role in purchasing and managing conservation areas. In Canada, the largest of these by far is Ducks Unlimited Canada, a hunter-supported organization that owns almost 1 million ha of land devoted mainly to habitat enhancement for waterfowl. Altogether in Canada, the National Conservation Areas Data Base identifies 3,500 publicly owned protected areas covering about 800 000 km^2, with another 10 000 km^2, held by non-governmental groups.

In addition to these various types of protected areas, there may also be differences in management priorities *within* any given area. For example, in the Canadian national parks system, there are five different zones that planners may use:

Zone I	Special preservation	3.25%
Zone II	Wilderness	94.01%
Zone III	Natural environment	2.16%
Zone IV	Outdoor recreation	.48%
Zone V	Park services	.09%

Zone I (special preservation) occurs where the ecological values are so high that very little allowance is made for any other use of the zone

Box 17.2 Protected Areas: A Global Perspective

From the origins in the American west, the idea of having areas protected by the government for conservation, public benefit, education, and enjoyment has spread throughout the world. According to the 1992 summary prepared by the World Conservation Monitoring Centre, there were 8,641 areas covering almost 8 million km^2. The growth has been most rapid over the last decade in many areas, with over 50 per cent of the area protected since 1982 in regions such as North Africa/Middle East, East Asia, North Eurasia, Central America, and the Caribbean.

While the idea has spread, the implementation had differed to reflect local conditions. Nevertheless, managers from Canada to England to Southeast Asia are concerned about the relationship between protected areas and local populations. As the numbers of people grow, so do pressures for increased use of protected areas. In the UK this overuse might be mainly recreational, and significant biophysical impacts may result just from the sheer numbers of people enjoying the parks. In many tropical areas, such as Thailand, conflicts arise as local people, who are often driven by poverty and land-use pressures, encroach on the parks in large numbers, cutting down trees for sale and to make agricultural land, and hunting wild animals. Estimates suggest that there are over half a million people illegally occupying national parks and wildlife sanctuaries in Thailand, with almost half the area of some parks suffering environmental degradation.

These kinds of management problems are very challenging. There is little point in emphasizing the area of land officially designated as 'protected' if in actuality it is not protected from resource use. For example, in the past in Thailand, management activities have focused upon a preventive approach with the use of armed guards to patrol boundaries. The fact that most large remaining areas of forest and most wild animal populations are within the protected area system suggests that there have been some benefits to this approach. On the other hand, it is clear that large-scale poaching continues in many areas and that shoot-outs between poachers and park guards are not an ideal management tool. Attention has, therefore also spread to try to address some of the underlying motives behind poaching, such as poverty. Again, there are substantial challenges. In some villages where economic development programs have been initiated, this has caused land prices to rise, leading some villagers to sell their lands and encroach further into the park lands. Unscrupulous local leaders may also use the villagers in this way so that they can gain control over more land area.

As with many environmental management problems, the answers do not lie in one single solution. Each case is different and an adaptive management approach to the protected area ecosystem is essential. In the long term, education must play a lead role. Many people are unaware of the vital functions played by protected areas. It is better to achieve voluntary compliance with more flexible management regimes in the future than have armed stand-offs and mass non-compliance, as has often occurred in the past.

Park wardens in Thailand receive little pay and risk their lives to protect what remains of the wildlife. Every year lives are lost in battles with poachers (*Philip Dearden*).

Table 17.1 Update on Conservation Lands and Waters

March 1995, area (km²), Number of reserves

	National Parks	National Wildlife Areas/ Migratory Bird Sanctuaries	Provincial/Territorial Parks	Provincial/Territorial Wildlife Management Areas	Provincial/Territorial Wilderness Areas	Ecological and Nature Reserves/Zones	Provincial Park/ Ecological Reserves	Provincial Forest Reserves	Private Reserves	Natural Areas Protection Act Properties	Conservation Reserves	Louisbourg Game Preserve	Wildlife Refuges	Municipal Lands Conservation Authorities
FED. GOV.	*218 903 / 36	115 034 / 144												
YT	*36 304 / 3		501 / 2			181* / 2							5918 / 2	
NWT	*107 197 / 4.2	112 868 / 16	2775 / 51											
BC	*6299 / 6	54 / 12	*64 853 / 406	276 / 14	*2651 / 5	*1587 / 131			*121 / 89					
ALTA	*54 064 / 4.8	146 / 7	1427 / 65	943 / 12	*1008 / 3	*271 / 14			1 / 2	378 / 124				
SASK.	*4781 / 2	776 / 23	*10 993 / 34	*1154 / 24		*8 / 3		*2720 / 31	*77 / 109				*31 / 24	*173 / 122
MAN.	*2976 / 1	1 / 2	36 061 / 131	31 866 / 72		*574 / 13		21 985 / 15	*51 / 63		+*325 / 3		2061 / 50	10 / 11
ONT.	*2190 / 5	418 / 20	*52 531 / 176	271 / 23		*3810 / 352	746 / 17		*30 / 27		*5 / 4			1334 / 38
QUE.	*927 / 3	653 / 41	*4249 / 17	68 257 / 24		*642 / 50	*1152 / 23	391 / 14	*123 / 43				4 / 2	421 / 34
NB	*445 / 2	43 / 6	*213 / 10	*132 / 2	67 / 1	*11 / 13			*2 / 5				50 / 6	
NS	*1351 / 2	66 / 13	*43 / 7	1417 / 26		*12 / 7			*9 / 14			*52 / 1		
PEI	*26 / 1	1 / 1	*13 / 8	*25 / 8					*4 / 13	*33 / 36				
NFLD	*2342 / 2	10 / 33	*351 / 76	*619 / 2	*3965 / 2	*465 / 8	31 / 2							
CAN.	**218 902 / 36	**115 036 / 144	174 010 / 983	104 960 / 207	7655 / 11	7561 / 593	1929 / 42	25 096 / 60	253 / 220	411 / 160	330 / 7	52 / 1	8064 / 84	1938 / 205

*   Lands that contribute to 'total area protected'.
**   Totals may differ from the federal government (e.g., national parks) due to rounding.
***   Includes 1612 km² of marine waters in BC not factored into 'per cent of jurisdiction protected' calculation.
n/a   Not available.

Source: M. Hummel, ed., *Protecting Canada's Endangered Spaces: An Owner's Manual* (Toronto: Key Porter Books, 1995).

Table 17.1 Update on Conservation Lands and Waters (continued)

	Conservation Land Tax Reduction Program	Conservation Easements	Stewardship Agreements	Provincial/Territorial Park Reserves	Reserves for Campground and Recreation Sites	Provincial Recreation Areas	Critical Wildlife Habitat Program Land	Wildlife Land	Habitat Trust/Foundation Land	Special Lands	Other Reserves	Total Area Protected	Per cent of Jurisdiction Protected	Footnotes
FED. GOV.												218 903 36	2.2	
YT					14 535 54					+12 173 1	++80 1	36 485 5	7.5	+ Old Crow Flats special management area ++ Horseshoe Slough habitat protection area
NWT										+*52 925 1		160 122 5.2	4.7	+ Thelon Game Sanctuary
BC										+13 332 6		***75 475 637	7.8	+ Land use decisions/plan for Clayoquot Sound, Vancouver Isl., Cariboo-Chilcotin, Commonwealth Nature Legacy, East/West Kootenays, Kitlope
ALTA					1214 173	598 236				+420 1	++4597 1	55 343 21.8	8.4	+ Suffield National Wildlife Reserve ++ Willmore Wilderness
SASK.					75 123	*347 27	13 760 n/a		225 433	+*57 23	++554 n/a	20 341 399	3.1	+ Protected Areas under the *Parks Act.* ++ Wildlife Development Fund Lands
MAN.			80 168				27 32	*10 500 1	114 96	++*21 480 4	+++150 39	35 906 85	5.5	+ Special Conservation Areas ++ Four new provincial parks, 1995 +++ Voluntarily protected ecologically significant areas
ONT.	809 6,000	*13 10	248 1,887	42 1						+6024 1	++1191 56	58 579 574	5.5	+ Recreation zones in Algonquin Park ++ Managed forest agreements
QUE.				*57 372 18		81 6	42 236 496			685 109	*376 7	68 841 161	4.2	
NB		5 2	*7 2		3 24	13 34	63 13	3174 13		+5883 10	++*56 9	866 43	1.2	+ Land base restrictions on 10 Crown timber licenses ++ Includes international park
NS			*79 3							+*54 17	++118 220	1600 51	2.9	+ Unconstituted parks contributing to representation ++ Parks/park reserves not contributing to representation
PEI						6 24					+*26 1	127 66	2.2	+ Newly-protected lands adjacent to PEI National Park
NFLD				75 5							52 494 8	7742 90	1.9	
CAN.	809 6,000	13 10	333 2,057	57 575 29	15 827 374	1045 327	58 086 >541	13 674 14	339 529	99 701 167	72 974 >348	***517 542 2,175	5.2	

*   Lands that contribute to 'total area protected'.
**   Totals may differ from the federal government (e.g., national parks) due to rounding.
*** Includes 1612 km² of marine waters in BC not factored into 'per cent of jurisdiction protected' calculation.
n/a Not available.

Source: M. Hummel, ed., *Protecting Canada's Endangered Spaces: An Owner's Manual* (Toronto: Key Porter Books, 1995).

Table 17.2 Matrix of Management Objectives and IUCN-Protected Area Management Categories

Management Objective	I.a.	I.b.	II	III	IV	V	VI
Scientific research	1	3	2	2	2	2	3
Wilderness protection	2	1	2	3	3	—	2
Preservation of species and genetic diversity	1	2	1	1	1	2	1
Maintenance of environmental services	2	1	1	—	1	2	1
Protection of specific natural/cultural features	—	—	2	1	3	1	3
Tourism and recreation	—	2	1	1	3	1	3
Education	—	—	2	2	2	2	3
Sustainable use of resources from natural ecosystems	—	3	3	—	2	2	1
Maintenance of cultural/traditional attributes	—	—	—	—	—	1	2

Key: 1 Primary objective
    2 Secondary objective
    3 Potentially applicable objective
    — Not applicable

Source: IUCN, *Guidelines for Protected Area Management Categories* (Gland, Switzerland: IUCN, 1994):8.

other than to protect those values. However, in Zone V (the park services zone) there may be very little of ecological value, and most of the land base is given over to providing services for visitors, such as campgrounds and visitor centres.

HOW MUCH PROTECTED AREA DO WE NEED?

The question of how much land we need in protected areas is a difficult one with no one answer. There are various ways of addressing the question. The most popular one is to look at how well the system of protected areas in any jurisdiction represents the landscape variety to be found there. Parks Canada, for example, has divided the Canadian landscape into thirty-nine terrestrial natural regions. The goal of the national parks system is to have at least one national park in each of these regions (Figure 17.1). These regions are determined on the basis of differences in physiography, vegetation, and other environmental conditions. Within each region an assessment is undertaken to determine a natural landscape that best represents the different conditions of the region. Negotiations must then be undertaken between the various levels of government, First Nations peoples (if there is a land claim in the area), and other stakeholders to determine where the boundaries should be. When negotiations are complete, the area becomes protected under the National Parks Act and is under the jurisdiction of the federal government. This can be a very long and complicated process. Pacific Rim National Park Reserve, for example, has been the subject of a memorandum of understanding between the federal and provincial governments since 1970, and has been managed as a national park since that time. However, due to ongoing squabbles

Pacific Rim National Park (*Parks Canada*).

between the different levels of government, by 1996 it was still not officially protected under the National Parks Act! A national park reserve is where a park is designated, but that designation will not prejudice future land claims agreements.

An assessment undertaken during the centennial celebrations for the national parks in 1985 indicated that despite the long history of parks in Canada, we were less than halfway towards this goal. Furthermore, the longer the delay, the more difficult it is to achieve this goal because the competition for land resources increases. The Endangered Spaces campaign was started by the World Wildlife Fund and the Canadian Parks and Wilderness Society to mobilize political support for completion of the national parks system. Part of this campaign was the Canadian Wilderness Charter, shown in Box 17.3. By 1996 over 600,000 Canadians had signed the charter. The federal government finally bowed to this pressure, and formally committed to making sure that each region in the system plan would be represented by the turn of the century. In late 1992 an agreement between the Canadian Council of Ministers of the Environment, the Canadian Parks Ministers' Council, and the Wildlife Ministers' Council also committed the provincial and territorial governments to complete their parks systems by the turn of the century.

It remains to be seen whether this ambitious target can be met, given the amount of resources that are being invested, but progress has been made. Three new national parks have been created, and lands have been withdrawn to create several

---

Box 17.3 The Canadian Wilderness Charter

- Whereas humankind is but one of millions of species sharing planet Earth and whereas the future of the Earth is severely threatened by the activities of this single species,
- Whereas our planet has already lost much of its former wilderness character, thereby endangering many species and ecosystems,
- Whereas Canadians still have the opportunity to complete a network of protected areas representing the biological diversity of our country,
- Whereas Canada's remaining wild places, be they land or water, merit protection for their inherent value,
- Whereas the protection of wilderness also meets an intrinsic human need for spiritual rekindling and artistic inspiration,
- Whereas Canada's once vast wilderness has deeply shaped the national identity and continues to profoundly influence how we view ourselves as Canadians,
- Whereas Canada's aboriginal peoples hold deep and direct ties to wilderness areas throughout Canada and seek to maintain options for traditional wilderness use,
- Whereas protected areas can serve a variety of purposes including:
  — preserving a genetic reservoir of wild plants and animals for future use and appreciation by citizens of Canada and the world,
  — producing economic benefits from environmentally sensitive tourism,
  — offering opportunities for research and environmental education,
- Whereas the opportunity to complete a national network of protected areas must be grasped and acted upon during the next ten years, or be lost,

We agree and urge:
1. That governments, industries, environmental groups and individual Canadians commit themselves to a national effort to establish at least one representative protected area in each of the natural regions of Canada by the year 2000,
2. That the total area thereby protected comprise at least 12 per cent of the lands and waters of Canada as recommended in the World Commission on Environment and Development's report, *Our Common Future*,
3. That public and private agencies at international, national, provincial, territorial and local levels rigorously monitor progress toward meeting these goals in Canada and ensure that they are fully achieved, and
4. That federal, provincial and territorial government conservation agencies on behalf of all Canadians develop action plans by 1990 by achieving these goals by the year 2000.

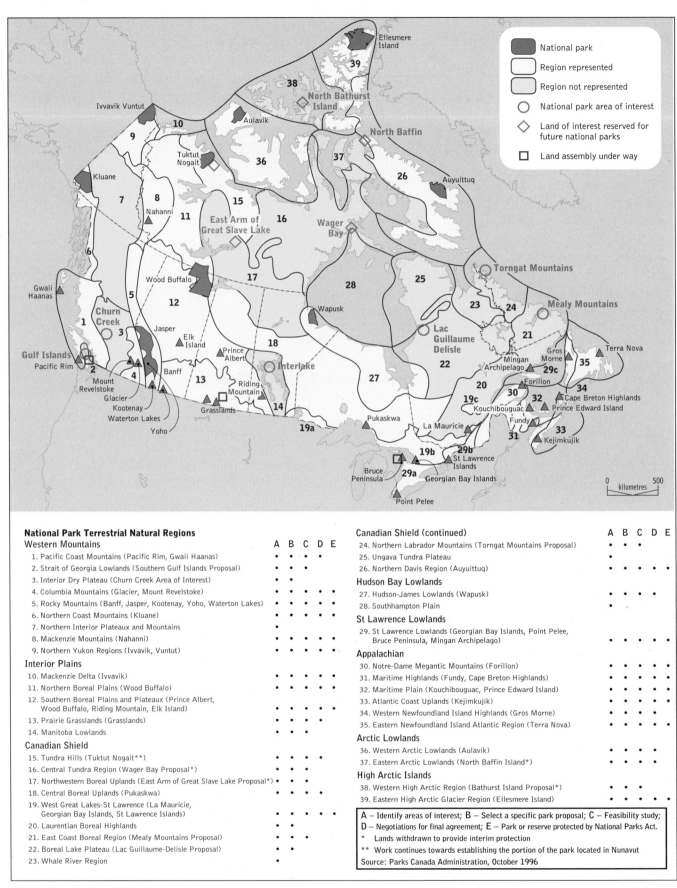

National Park Terrestrial Natural Regions

Western Mountains	A	B	C	D	E
1. Pacific Coast Mountains (Pacific Rim, Gwaii Haanas)	•	•	•	•	
2. Strait of Georgia Lowlands (Southern Gulf Islands Proposal)	•	•	•		
3. Interior Dry Plateau (Churn Creek Area of Interest)	•	•			
4. Columbia Mountains (Glacier, Mount Revelstoke)	•	•	•	•	•
5. Rocky Mountains (Banff, Jasper, Kootenay, Yoho, Waterton Lakes)	•	•	•	•	•
6. Northern Coast Mountains (Kluane)	•	•	•	•	•
7. Northern Interior Plateaux and Mountains	•				
8. Mackenzie Mountains (Nahanni)	•	•	•	•	•
9. Northern Yukon Regions (Ivvavik, Vuntut)	•	•	•	•	•

Interior Plains	A	B	C	D	E
10. Mackenzie Delta (Ivvavik)	•	•	•	•	•
11. Northern Boreal Plains (Wood Buffalo)	•	•	•	•	•
12. Southern Boreal Plains and Plateaux (Prince Albert, Wood Buffalo, Riding Mountain, Elk Island)	•	•	•	•	•
13. Prairie Grasslands (Grasslands)	•	•	•	•	
14. Manitoba Lowlands	•	•	•		

Canadian Shield	A	B	C	D	E
15. Tundra Hills (Tuktut Nogait**)	•	•	•	•	
16. Central Tundra Region (Wager Bay Proposal*)	•	•	•		
17. Northwestern Boreal Uplands (East Arm of Great Slave Lake Proposal*)	•	•	•		
18. Central Boreal Uplands (Pukaskwa)	•	•	•		
19. West Great Lakes-St Lawrence (La Mauricie, Georgian Bay Islands, St Lawrence Islands)	•	•	•	•	•
20. Laurentian Boreal Highlands	•	•			
21. East Coast Boreal Region (Mealy Mountains Proposal)	•	•	•		
22. Boreal Lake Plateau (Lac Guillaume-Delisle Proposal)	•	•			
23. Whale River Region	•				

Canadian Shield (continued)	A	B	C	D	E
24. Northern Labrador Mountains (Torngat Mountains Proposal)	•	•	•		
25. Ungava Tundra Plateau	•				
26. Northern Davis Region (Auyuittuq)	•	•	•	•	•

Hudson Bay Lowlands	A	B	C	D	E
27. Hudson-James Lowlands (Wapusk)	•	•	•	•	•
28. Southhampton Plain	•				

St Lawrence Lowlands	A	B	C	D	E
29. St Lawrence Lowlands (Georgian Bay Islands, Point Pelee, Bruce Peninsula, Mingan Archipelago)	•	•	•	•	•

Appalachian	A	B	C	D	E
30. Notre-Dame Megantic Mountains (Forillon)	•	•	•	•	•
31. Maritime Highlands (Fundy, Cape Breton Highlands)	•	•	•	•	•
32. Maritime Plain (Kouchibouguac, Prince Edward Island)	•	•	•	•	•
33. Atlantic Coast Uplands (Kejimkujik)	•	•	•	•	•
34. Western Newfoundland Island Highlands (Gros Morne)	•	•	•	•	•
35. Eastern Newfoundland Island Atlantic Region (Terra Nova)	•	•	•	•	•

Arctic Lowlands	A	B	C	D	E
36. Western Arctic Lowlands (Aulavik)	•	•	•	•	•
37. Eastern Arctic Lowlands (North Baffin Island*)	•	•	•		

High Arctic Islands	A	B	C	D	E
38. Western High Arctic Region (Bathurst Island Proposal*)	•	•	•		
39. Eastern High Arctic Glacier Region (Ellesmere Island)	•	•	•	•	•

A – Identify areas of interest; B – Select a specific park proposal; C – Feasibility study;
D – Negotiations for final agreement; E – Park or reserve protected by National Parks Act.
* Lands withdrawn to provide interim protection
** Work continues towards establishing the portion of the park located in Nunavut
Source: Parks Canada Administration, October 1996

Figure 17.1 Completing the national parks system. SOURCE: Auditor-General of Canada, *Report to the House of Commons, Chapter 31 'Canadian Heritage—Parks Canada: Preserving Canada's Natural Heritage'* (Ottawa: Minister of Public Works and Government Services Canada, 1996):31–16.

Moraine Creek in Banff National Park (*Parks Canada*).

others. There are now thirty-eight national parks in Canada, although these are not all in different regions. Currently, twenty-four of the thirty-nine regions are represented (Figure 17.1). When completed, the national parks system will cover some 3.4 per cent of the land area of Canada.

There is an even greater challenge with marine parks. Only two of the twenty-nine marine regions are currently represented by parks (Figure 17.2). The federal government aims to create six new marine parks by the year 2000. By 1995 none had been declared. Furthermore, these parks would not provide the same strict protection of biological resources as terrestrial parks because commercial fishing activities will still be allowed. To reflect this difference, such areas are no longer called marine parks but marine conservation areas.

Not only does the federal government have a system plan for park representation, so also does each province and territory. Of course, at the provincial scale there are many more regions per province than at the national scale. British Columbia, for example, has 110 natural regions defined by the provincial government, Nova Scotia has seventy-seven, and Alberta has twenty. It is also the goal of each province to represent each of their regions with provincial parks. They have met with various degrees of success. A good overview as of 1995, including maps of each province, is provided in a book from the World Wildlife Fund (Hummel 1995), which charts how well Canada is doing in meeting the goals of the Endangered Spaces campaign mentioned earlier. Every year the campaign releases a national report card indicating how well each jurisdiction is doing. The scores from recent report cards are shown in Table 17.3. More details on the criteria for each grade can be found in the 'Endangered Spaces Progress Report', available from the World Wildlife Fund (World Wildlife Fund 1996).

Canada has endorsed the international goal of protecting 12 per cent of its area within protected area systems. However, several comments need to be made. First, the 12 per cent was not a precise figure based upon a sound scientific base. It was a global estimate based upon a very poor data base, first suggested in the well-known report, *Our Common Future*, produced by the World Commission on Environment and Development (1987), which was discussed in more detail in Chapter 2. Many scientists feel that 12 per cent is a much too conservative figure for conserving biodiversity in the future, and have suggested figures as high as 35 per cent. Second, it is a global average, and some diverse regions with wide-ranging species may need much more than this. However, in highly humanized landscapes, such as Europe, there may no longer be opportunity to protect even this small

Canada has not done very well in protecting the marine environment. The national marine conservation areas system has only one legislated representative at the moment, Fathom Five, which is off the tip of the Bruce Peninsula in Lake Huron (*Philip Dearden*).

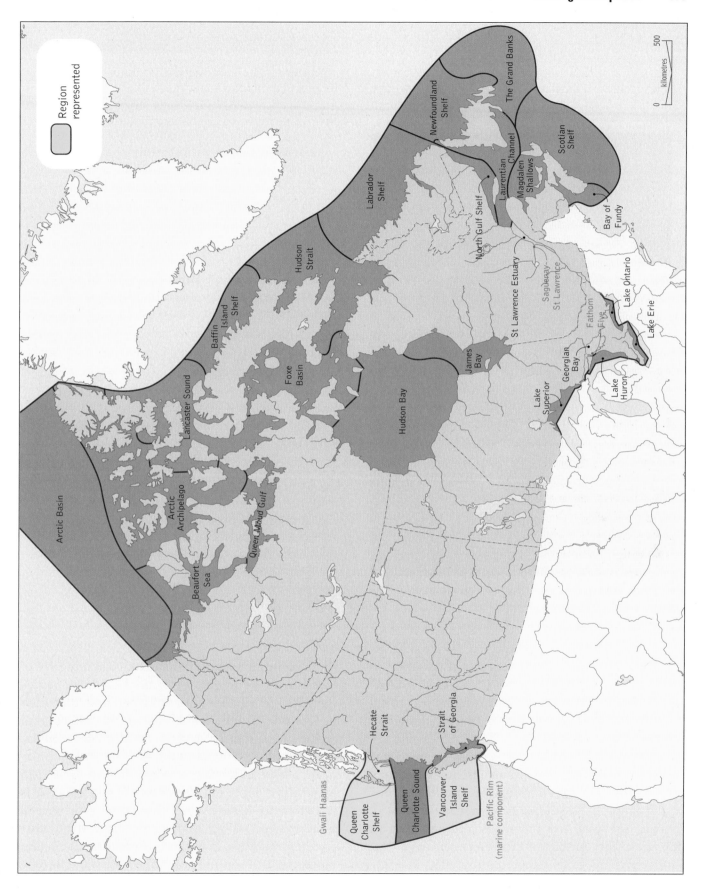

Figure 17.2 National marine conservation area, natural regions. SOURCE: Parks Canada.

Box 17.4 Does Canada Need an Oceans Act?

As was pointed out in Chapter 3, Canada has the longest coastline of any country in the world; we also have the second largest area of

Ocean vegetation off Port Hardy, BC (*World Wildlife Fund Canada/Frank Parhizgar*).

continental shelf. However, we have done little to protect this heritage, particularly in terms of marine protected areas. Such areas would perform a number of functions, including:

- conservation and protection of commercial and non-commercial species, including marine mammals and their habitats
- conservation and protection of endangered or threatened marine species and their habitats
- conservation and protection of unique habitats
- conservation and protection of areas of high biodiversity or biological productivity
- conservation and protection of any other marine resource or habitat as required

Unfortunately, the legislative means to establish such areas in Canada is limited. The National Parks Act, in practice, is limited to coastal areas, and the Wildlife Act is designed primarily to protect birds and bird habitat. Although in theory the Fisheries Act covers marine species and habitats, in reality it has been used solely for the management of commercial fish species. These deficiencies have caused many groups with an interest in marine conservation to press for an Oceans Act to facilitate the creation of marine protected areas and deal with threats such as pollution. In 1995 the Canada Oceans Act was brought to Parliament for first reading. It was proclaimed in January 1997.

proportion. Third, the degree of protection afforded to species within this 12 per cent is critical. Is it justified to include in this figure provincial parks that neither protect trees from logging nor animals from hunting?

This discussion has focused upon the establishment of protected areas in representative natural areas as the basis for choosing the location of protected areas and calculating how many we need. This is one way, and the most common way, to approach the problem. However, scientists also know that by simply representing the enduring features of a natural region in a protected area network does not necessarily guarantee that all native species will survive in that region. We also have to

think about maintaining the ecological integrity of these areas. We might, for example, achieve representation in all those regions, but find that each park is not of sufficient size to protect far-ranging species, or is too isolated to link different populations, leading to inbreeding and possible population collapse in the future. Many questions need to be resolved. How big should protected areas be? Is it better to have a single large area or several smaller areas? What is the best shape for a protected area? What is the best pattern for arranging protected areas on the landscape? How might they be linked? These are not questions that can be addressed here, but they do require a sound knowledge of the basic ecological principles detailed in

Table 17.3 Endangered Spaces Campaign for Each Jurisdiction Since 1992 and Total Area Currently Protected in Each Jurisdiction

Jurisdiction	1992	1993	1994–5	1995–6	% of Jurisdiction Protected
Fed.–Terrestrial	A–	B	C–	C	2.0
Fed.–Marine	C–	D	D–	C	n/a
Yukon	C	D+	D	D	7.6
Northwest Territories	B–	C	D	D	4.1
British Columbia	B–	B+	A–	A	9.2
Alberta	D	C	F	B	9.3
Saskatchewan	C	B–	D+	C	5.2
Manitoba	D	B	C–	D–	5.5
Ontario	C+	B	D+	F	6.5
Quebec	C+	B–	D+	C–	4.2
New Brunswick	D–	B–	D–	F	1.3
Nova Scotia	C	C	C+	A	8.0
Prince Edward Island	A–	B+	B–	C+	2.2
Newfoundland/Labrador	C	C+	C–	D	1.7

Source: World Wildlife Fund, 'Endangered Spaces Progress Report Number 6 1995–6' (Toronto: World Wildlife Fund, 1996).

chapters 4 to 7 and management approaches that are adaptive (Chapter 9), with a strong ecosystem focus (Chapter 8). The reading by Shafer (1990) provides a useful summary of the literature in this field.

NATIONAL PARKS

National parks are established under the National Parks Act to protect landscapes judged to be of national importance. Provincial parks, on the other hand, protect landscapes judged to be of provincial significance. In practice, however, the distinction is not always so clear-cut, and there are many provincial parks that are of national significance. One example is Dinosaur Provincial Park in Alberta, which, due to its rich palaeontological history, is deemed to be of even global significance as a World Heritage Site. There are seven other natural World Heritage Sites in the country, and twelve in total, including the sites recognized for their cultural significance, such as Head-Smashed-In Buffalo Jump in Alberta (Table 17.4). These sites are judged to be of global significance, and join over 400 other sites so designated by UNESCO's Conven-

tion on the Protection of the World Cultural and Natural Heritage.

National Park History

The national park idea evolved in the United States where Yellowstone, the first national park in the world, was designated in 1872. Shortly after, the Canadian government declared Canada's first national park in 1885, centred around the hot springs in Banff. The main purpose of the park

Gros Morne National Park in western Newfoundland is a World Heritage Site (*Philip Dearden*).

Table 17.4 Canadian World Heritage Sites

Site	Designated
Nahanni National Park Reserve	1978
L'Anse aux Meadows National Historic Site	1978
Dinosaur Provincial Park	1979
Kluane National Park Reserve/Tatshenshini Provincial Park	1979/1994
Anthony Island	1981
Head-Smashed-In Buffalo Jump	1981
Wood Buffalo National Park	1983
Rocky Mountains	1984/1990
Quebec City	1985
Gros Morne National Park	1987
Waterton National Park	1995
Lunenberg	1996

I do not suppose in any portion of the world there can be found a spot, taken all together, which combines so many attractions and which promises in as great a degree not only large pecuniary advantage to the Dominion, but much prestige to the whole country by attracting the population, not only on this continent, but of Europe to this place. It has all the qualifications necessary to make it a place of great resort. . . . There is beautiful scenery, there are curative properties of the water, there is a genial climate, there is prairie sport, and there is mountain sport; and I have no doubt that it will become a great watering-place.

Sir John A. Macdonald
on Banff, 1887

National Parks Act (1988)

5.1.2 Maintenance of ecological integrity through the protection of natural resources shall be the first priority when considering park zoning and visitor use in a management plan.

Guiding Principles and Operational Policies (1994)

Protecting ecological integrity and ensuring commemorative integrity take precedence in acquiring, managing, and administering heritage places and programs. In every application of policy, this guiding principle is paramount.

was to attract international tourists to the area to generate income. The president of Canadian Pacific Railway declared that since we couldn't export the scenery, we would have to import the tourists, and set about the construction of facilities, such as the Banff Springs Hotel, to promote the resort. There was little thought of wilderness preservation or ecosystem protection. The idea was to construct a spa resort that would compete with other spa resorts around the world for rich tourists. Over time, however, ideas changed. As more parks were created, there was more concern for environmental protection. By the time of the first National Parks Act in 1930, environmental protection had become part of the official mandate of the parks. Mineral exploration was prohibited, all game species were protected, and only limited use of green timber for essential parks management purposes was allowed.

The National Parks Act also created a dilemma that was to trouble parks management for the next half century. Not only were the parks to be 'protected unimpaired for . . . future generations', they were also dedicated to the people of Canada for their 'benefit, education and enjoyment'. Ever since that time, parks managers have wrestled with the task of providing opportunities for use and enjoyment while still leaving the parks unimpaired. Is the development of ski hills, gondolas, golf courses, tennis courts, hotels, shopping arcades, and restaurants leaving the parks unimpaired for future generations? At one time this kind of development was thought to be consistent with the National Parks Act. Amendments to the act in 1988 and the publication of a new policy to interpret the act and guide developments in the parks in 1994 make it clear that it is the conservation and not the use mandate of the parks that should be the top priority.

Current Management Challenges

Management concerns have evolved over time from the exclusion of resource extraction and protection of the parks' boundaries in the early years

Box 17.5 Riding Mountain National Park

Riding Mountain National Park is a good example of the kinds of external pressures that now threaten the ecological integrity of many of our national parks. The park is located on the Manitoba Escarpment and is an isolated boreal forest area totally surrounded by agricultural land. Nonetheless, the 3000-km^2 park provides habitat for 5,000 elk, 4,000 moose, over 1,000 black bears, and populations of cougars and wolves. These populations have become increasingly threatened by the intensification of agricultural activities surrounding the park. Between 1971 and 1986 the amount of land under agriculture within 10 km of the boundary increased from 77 to 93 per cent. Not only is the amount of farm land increasing, so is the intensity of use, with a 42 per cent increase in crop land area in the same zone over the same time period. Within the same time period, the area of woodland has declined an average of 63 per cent within a 70-km radius of the boundary, the volume of agricultural fertilizers used has quintupled, and pesticide expenditures indicate an increase of 744 per cent in pesticide applications.

Agricultural expansion is not the only challenge confronting park wildlife. Until recently, bear baiting was allowed directly on the park boundary. Farmers would condition the bear population to feeding from barrels full of meat. In the hunting season, the bears then make easy targets for so-called 'sportsmen'. Some seventy such stations exist around the park, causing unnatural bear distributions, very large bears, and irregularities in breeding behaviour. It also results in dead animals. Annual average mortality is 122 animals, a figure that scientists conclude cannot be maintained if the bear population of the park is to survive. Farmers are now required to move the bait barrels back from the park boundary itself. However, bears are highly mobile creatures and will have little difficulty in locating these barrels.

These problems illustrate the failure of one of the most highly touted means of trying to deal with external threats to protected areas through the Biosphere Reserve Program of the United Nations Educational, Scientific and Cultural Organization. The concept behind this program is sound. Biosphere reserves exist to represent global natural regions and should consist of a protected core area, such as a national park, surrounded by a zone of cooperation where socio-economic activities may take place, but which are modified to help protect the integrity of the core area. The reserves also have important educational and scientific roles. Unfortunately, there is no legislation to ensure cooperation on the privately owned lands in the zone of cooperation. Continued hunting around the park boundary at Riding Mountain National Park, designated as a Biosphere Reserve in 1900, is a graphic illustration of this dependency on local land-owners.

to concerns with accommodating and mitigating the tremendous pressures inside the parks created by the postwar recreation boom, and now to current concerns regarding the effects of external influences on the ecological integrity of the parks (figures 17.3 and 17.4). Current challenges include the following.

External Threats

Parks do not exist in isolation. They are intimately linked to surrounding and global ecosystems. In order to protect the ecological integrity of the parks, it is necessary to be

A Biosphere Reserve surrounding Waterton Lakes National Park has been set aside to allow for joint decision making between park authorities and local land-owners. The agreement has been particularly useful in reducing conflicts between grizzly bears and local ranchers (*Philip Dearden*).

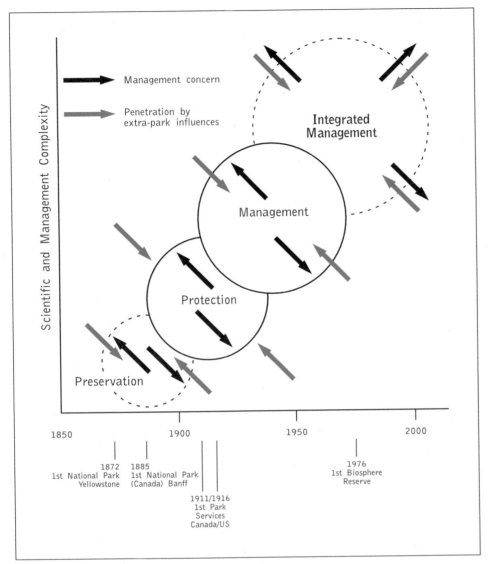

Figure 17.3 Protected areas: evolving relationships from isolation to integration. The circles represent the growing size of the protected area system over time. Boundaries (circle circumferences) were initially of little importance, but assumed greater significance in the protection and management phases. It is now realized that for park management (arrows) to be effective, it must also pay equal attention to environmental changes outside park boundaries. SOURCE: P. Dearden and R. Rollins, eds, *Parks and Protected Areas in Canada: Planning and Management* (Toronto: Oxford University Press, 1993):10.

ing their practices for the benefit of the park. The ideas discussed in Chapter 11 regarding stakeholders and comanagement are very relevant here.

Fragmentation

Parks are becoming islands of natural vegetation totally surrounded by human-modified landscapes, as illustrated by the Riding Mountain case discussed in Box 17.5. Studies of the area surrounding Fundy National Park in New Brunswick showed that only 20 per cent of the area remained in forest patches large enough to be 500 m from disturbed areas. This situation creates several problems since many animal and some bird species cannot cross modified landscapes. Thus they become an isolated breeding population, leading to genetic inbreeding and higher susceptibility to extinction. This raises questions regarding how large populations must be to survive, and how large an area of their habitat is required to sustain the population. The first question, related to the *minimum viable population* (mvp) of a species, can be estimated using genetic and demographic models. This is then multiplied by the area required to support each animal. In western Canada, for example, calculations suggest that 15 000 km² would be required to support a viable wolf population. The size of the four mountain parks that are contiguous (Banff, Yoho, Jasper, and Kootenay) is 20 000 km². Calculations by Newmark in 1986 for all parks in western North America suggested that this block of parks is the only area in the entire continent large enough to maintain populations of all native fauna. This finding is of special concern,

aware of influences from outside the park that may have a detrimental effect upon it. Sometimes these can be readily identified and even managed, such as timber cutting along a park boundary; at other times, the influences are too distant and diffuse for park managers to influence, such as the effects of global climate change. Increasingly then, the parks are turning to concepts of ecosystem management (as described in Chapter 8) and trying to manage along ecosystem rather than legal boundaries. This entails more cooperation with surrounding landowners to persuade them of the value of modify-

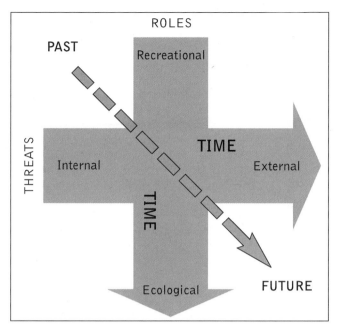

ROLES

PAST

Recreational

THREATS

Internal

TIME

External

TIME

Ecological

FUTURE

Figure 17.4 The changing emphasis in park roles over time. SOURCE: P. Dearden and R. Rollins, eds, *Parks and Protected Areas in Canada: Planning and Management* (Toronto: Oxford University Press, 1993):8.

At the top of the environmental agenda must be the establishment of many more large parks and reserves and the forging of broad and strongly-protective regional conservation strategies to protect them. With all the environmentally-destructive forces controlling resource and land management in Canada, there is little chance that we will ever protect too much.

John Theberge, 'Ecology, Conservation and Protected Areas in Canada', 1993

given the level of tourism development in the parks, particularly Banff, as will be discussed later.

These concerns have made it obvious that most of our parks are too small, too few, and too far apart to sustain populations of many species throughout the next century. More attention is now being directed towards ways of linking the parks together through corridors of natural habitat. One such scheme, for example, seeks to link American parks, such as Yellowstone, north through the Canadian Rockies up into Yukon and Alaska. In fact, one wolf that was marked for tracking in Montana was actually shot along this corridor on the Alaska Highway. These schemes explicitly acknowledge the limitations of our parks systems, and encourage a more integrated perspective of resource management on lands outside the parks, as discussed in Chapter 15. They also explicitly embrace the ecosystem approach described in Chapter 8.

Other Stresses

There are many other stresses that affect different parks to varying degrees. In 1992 a survey was sent to thirty-four national parks to assess the state of park ecosystems and identify the main sources

of stress. The answers pertain to the larger ecosystems of each park, and a national summary is shown in Figure 17.5. As can be seen, visitor and tourism facilities still constitute a source of significant stress for twenty-two of the parks. The case-study on Banff (discussed later) helps to illustrate this problem. Exotic vegetation was also of concern in a large number of parks. Purple loosestrife, for example, has invaded Point Pelee National Park, and knapweed is a problem in Elk Island National Park. Utility corridors cross nineteen of the parks, and can have very significant impacts where major transportation routes cross animal migration routes, as the Banff example again illustrates.

Only 11 per cent of the stresses originated within the parks, while 36 per cent came from outside the parks, with 56 per cent of stresses occurring in both. Furthermore, over 50 per cent of the stresses affected over 100 km², with less than 2 per cent localized to under 1 km². Forty-three per cent of stresses increased, and 12 per cent decreased. It is very difficult to identify and measure the stresses, let alone their impacts on park ecosystems. Figure 17.6 summarizes the most frequently reported impacts. Changes in community structure and population reductions are the most mentioned, but the rating of 'unknown impacts' as the third most frequent indicates how little we know about many ecosystem changes.

First Nations Involvement

Over time, various outside groups have been able to influence parks management decisions. Entrepreneurs, mainly tourist facility operators, were the

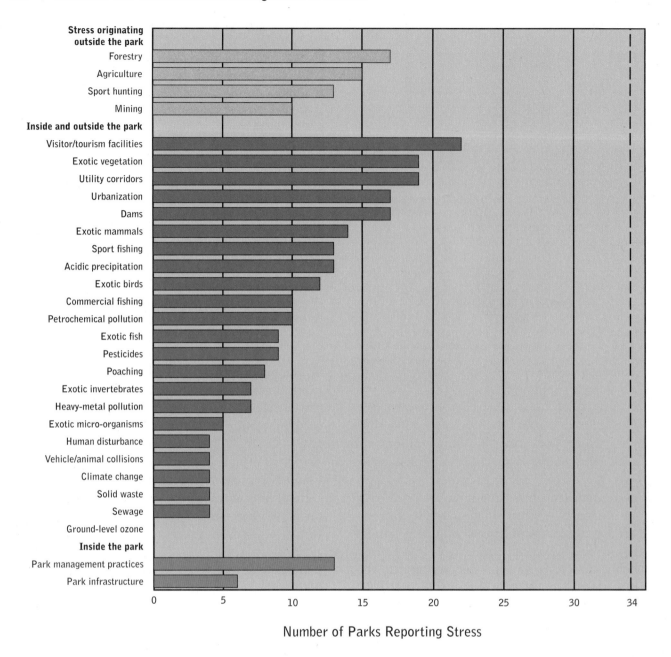

Figure 17.5 Number of national parks reporting significant ecological impacts from various human stresses (thirty-four national parks surveyed). SOURCE: Parks Canada, *State of the Parks, 1994 Report* (Ottawa: Minister of Supply and Services Canada, 1995):35.

first such group. Their influence was increasingly challenged by environmental interests seeking stronger protection for the parks. In the early 1970s a proposal by Imperial Oil, endorsed by Parks Canada, to expand the community of Lake Louise in Banff National Park and build a twelve-storey high-rise, bars, discotheques, movie theatres, a convention centre, restaurants, tennis courts, and over 2,800 parking spaces, was turned down following a public outcry. This event symbolizes the start of a period in which environmental interests

tended to be the most influential (Figure 17.7). It was also in this period that Parks Canada first explicitly stated that preservation would take precedence over use in its 1979 policy statement.

This same document also recognized the role that First Nations peoples should play in parks management. However, it was not until a decade later that this influence was demonstrated by the Haida's ability to delay significantly the designation of Gwaii Hanaas as a national park, over the objections of many environmentalists, who had fought

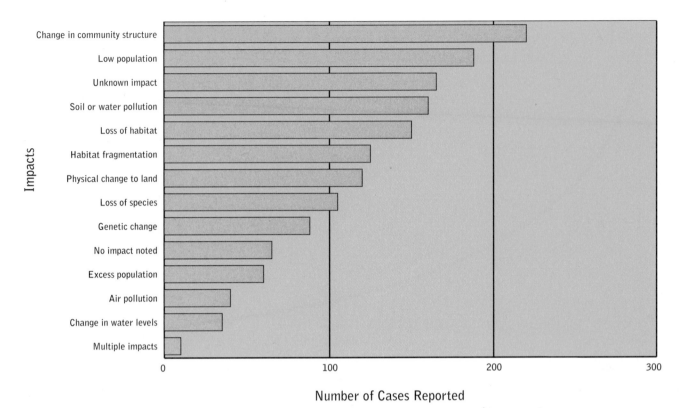

Number of Cases Reported

Figure 17.6 Number of reported cases for a variety of ecological impacts from a survey of twenty-nine ecological stressors in Canadian national parks (thirty-four national parks surveyed). SOURCE: Parks Canada, *State of the Parks, 1994 Report* (Ottawa: Minister of Supply and Services Canada, 1995):40.

alongside the Haida to prevent logging in the area. Previous to this disagreement, First Nations peoples had been influential in parks designation and management only in the North. An amendment to the National Parks Act in 1972 created a special designation of national park reserve, which could be established without prejudice to future land claims. This designation led to the creation of several parks in the North where the federal government has jurisdiction over land resources. It was not until after several important court cases that the legitimacy of Native claims over lands and resources in southern Canada began to be taken more seriously. The negotiations over Gwaii Hanaas followed these cases, and gave the Haida legal support for taking a strong stand on who should manage the new national park. A joint committee of Haida and park personnel has been

formed to make decisions. The Haida are also guaranteed continued access to Gwaii Hanaas for a host of traditional activities that might not have

Although the environmental lobby was successful in its fight against the construction of an upper and lower village at Lake Louise funded by Imperial Oil and supported by Parks Canada about twenty years ago, many would argue that the subsequent incremental developments have achieved almost the same result. This photograph shows the enlarged Château Lake Louise in front of what is advertised as the largest ski hill in Canada. Is this a national park landscape (*Philip Dearden*)?

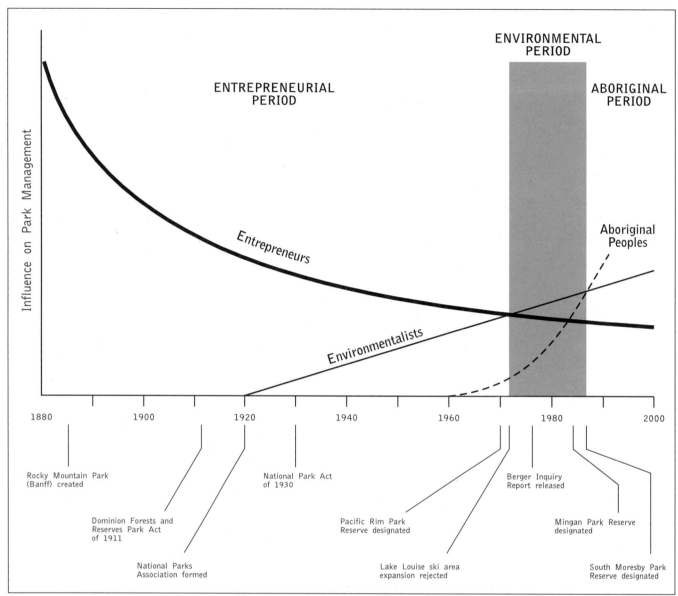

Figure 17.7   Administrative penetration model. SOURCE: P. Dearden and L. Berg, 'Canada's National Parks: A Model of Administration Penetration', *The Canadian Geographer* 37, no. 3 (1993):198.

been allowed under previous park policy. It is likely that similar agreements will be forthcoming in all parks where land claims exist, and that First Nations peoples will play a major role in the management of those parks.

CASE-STUDIES

Two case-studies exemplify some of the benefits and challenges of the national parks system. The first is Wood Buffalo National Park, which has provided the necessary habitat for the recovery of two of the continent's best-known endangered species, the wood bison and the whooping crane. The other is Banff, which, more than any other park,

epitomizes the difficulty of balancing the use and preservation mandate of the national parks. Both these parks are recognized as being of global significance and have been designated as World Heritage Sites by UNESCO.

Wood Buffalo National Park

Wood Buffalo is the second largest terrestrial national park in the world after Greenland. Covering 44 807 km², it represents mainly the Northern Boreal Plains in the systems plan of Parks Canada, and straddles the Alberta/Northwest Territories boundary (Figure 17.8). It was designated in 1922 and contains critical habitat for two endangered

The totem poles of Ninstints, an abandoned Haida village on Gwaii Hanaas (Moresby Island in the Queen Charlotte Islands), give some impression of the Haida's spiritual connection with the environment (*Philip Dearden*).

species, as well as the wetlands of the Peace–Athabasca delta, protected under the Ramsar Convention. Traditional harvesting of wildlife is allowed by Native Peoples.

Wood Buffalo was also distinguished as the last national park in Canada to permit commercial logging operations. The minister of environment renewed the lease as late as 1983, even though logging was against the National Parks Act and Canadian Parks Service's policy. Logging was finally terminated in 1992 as a result of a court action by the Canadian Parks and Wilderness Society on the grounds that such activities were illegal. Without such bold action, logging would still continue to this day in the park.

Attention will now turn to two of the endangered species that have

been saved from extinction by the habitat protected by the park.

Bison

Early French voyageurs called bison *les boeufs*, which became corrupted to 'buffle' and finally 'buffalo' by the English settlers, although the species are not related to the true buffaloes of Africa or Southeast Asia. The image of vast herds of bison, estimated at up to 60 million animals, ranging back and forth along the central plains of North America, is one that will never be seen again. These herds probably constituted the greatest large mammal congregations that ever existed on earth and were the main source of food for the Métis, Blackfoot, Cree, and Assiniboine in western Canada. One of the annual hunting parties from the Red River settlement in 1849 was recorded as including 700 Métis, 200 Indians, 603 carts, 600 horses, 200 oxen, and 400 dogs. Within half a century, the herds had been reduced to a few ragged remnants, the vast majority having been shot by

The vast herds of plains bison had been extirpated from Canada until efforts were made to reintroduce them from the US and eventually transport them to Wood Buffalo National Park, where they mixed with the wood bison population (*Philip Dearden*).

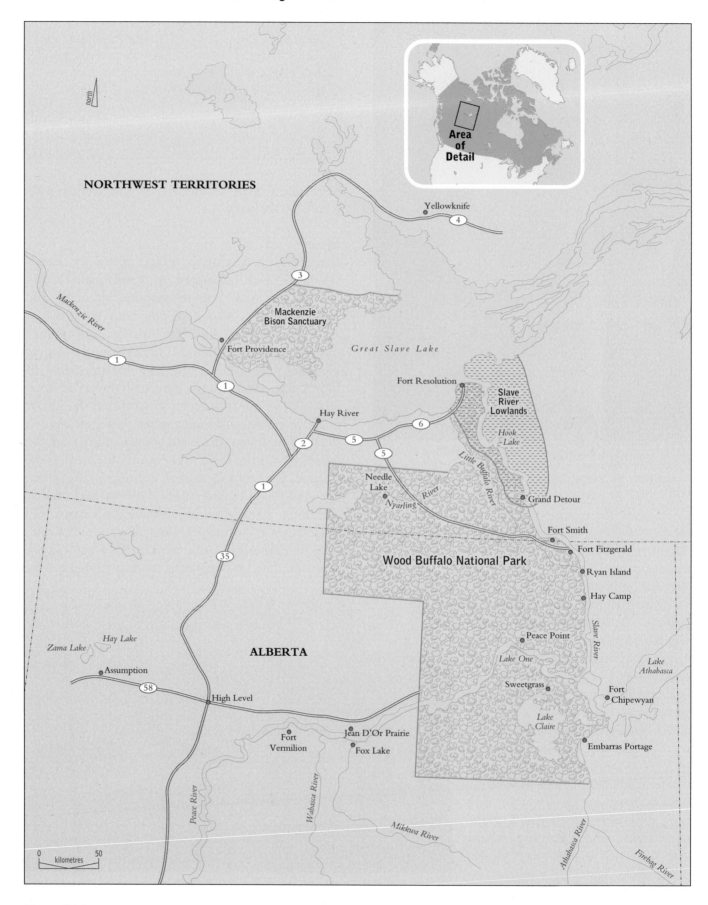

Figure 17.8 Map of Wood Buffalo National Park.

commercial hunters in the US. In the 1850s starving Assiniboine Indians were reported by officials at Rocky Mountain House due to a shortage of bison, and by the 1860s the bison had been extirpated from the plains of Manitoba. Trainloads of meat and hides were sent back east. As American Indians flooded into Canada to seek protection of the 'Great White Mother' (Queen Victoria), the pressure on the remaining herds increased dramatically. In terms of carrying capacity, there were simply too many people dependent upon the bison and not enough animals to support them. The wild bison herds were extirpated from the Canadian Prairies.

Several remnants remained, however. A small number had been protected by the earlier establishment of Yellowstone National Park. Indeed, Yellowstone is the only place where wild, free-ranging plains bison have survived since colonial times. Banff also had a growing population kept as a tourist attraction in an animal compound. There were also two remnants that had been brought together by an American rancher in Montana. He offered to sell this herd to the American government, but it refused to buy them and the herd was bought by the Canadian government, which transported the 703 animals to a national park created for that purpose adjacent to the railway near Wainwright, Alberta. In the mid-1920s the herd, then numbering 6,673, was relocated to Wood Buffalo National Park.

These were the plains bison (*Bison bison*). Less well known are their non-migratory, taller, and darker cousins, the wood bison (*Bison bison athabascae*), which are thought to be more closely related to the European (*Bison bonasus*) than the plains bison. The wood bison was once widely distributed throughout the aspen parklands of Saskatchewan and Alberta to the eastern slopes of the Rockies and British Columbia, and north to the coniferous forests of the Mackenzie Valley. It was never as abundant as the plains bison, and by 1891 it is estimated that fewer than 300 of them survived. Wood Buffalo National Park was established at least partially to protect this remnant, and by 1922 the herd had grown to 1,500–2,000 animals. Shortly thereafter, the herd of plains bison from Wainwright was imported, and interbreeding led to the disappearance of the distinctive wood bison characteristics. Wood bison were believed to have become extinct.

All through today's journey, piled up at the leading stations along the road, were vast heaps of bones of the earliest owners of the prairie—the buffalo. Giant heads and ribs and thigh bones, without one pick of meat on them, clean as a well washed plate, white as driven snow, there they lay, a giant sacrifice on the altar of trade and civilization.

Traveller on CP Rail, 1888

In 1957, however, an isolated group of wood bison was located in a remote area of this vast park. Relocations of this herd were made to two other locations to guard against further interbreeding within Wood Buffalo. Some animals were removed to the Mackenzie Bison Sanctuary in the North. They now number over 200 and have expanded their range outside the sanctuary. Others were removed to Elk Island National Park near Edmonton where, due to the small area available, their numbers have to be closely controlled. This herd has provided animals for satellite herds in the Yukon, the Northwest Territories, northwestern Alberta, and Manitoba. In 1988 the status of the wood bison was changed from endangered to threatened by COSEWIC.

The dangers for the bison are not yet over, however. When the plains bison were imported from Wainwright, they brought with them bovine diseases, such as brucellosis and tuberculosis. For this reason the move was opposed by conservation interests, but their concerns fell on deaf ears in Ottawa. The diseases are now taking their toll, and the animals have decreased from over 12,000 animals to a quarter of this number by the early 1990s, although several factors besides the diseases are implicated in this decline. The Peace-Athabasca delta, for example, supported the highest concentrations of bison. However, since the construction of the W.A.C. Bennett Dam upstream in British Columbia, water levels on the delta have fallen considerably, causing habitat changes that have negatively affected many animal species, including the bison. This impact from outside the park again emphasizes the need to take an ecosystem-based perspective on parks management (see Chapter 8).

There is strong concern from the agricultural industry. Bison represent the last focus for brucellosis and tuberculosis in Canada, and as agriculture has impinged upon the western boundary of the park, farmers are concerned that domestic stock will become infected. This has led the agricultural lobby to call for elimination of the herd, a move that was supported by the Federal Environmental Assessment Review in 1990. Environmentalists strenuously opposed this course of action on the grounds that the impact assessment review was scientifically flawed and morally corrupt.

In April 1995 the Canadian heritage minister announced that the herd would not be slaughtered and introduced a management plan calling for more research on the impact of habitat change on bison ecology, a buffer zone between the diseased animals and the Mackenzie Bison Sanctuary to the north, and more investigation of the impact of the diseases on the park's ecosystem.

Whooping Crane

Unlike the bison, whooping cranes (*Grus americana*) were never very numerous. Historical accounts suggest a population of 1,500. What they lacked in numbers they made up for in presence. Over 1.5 m high and pure white (except for black wing-tips, black legs, and a red crown) and with wing-spans in excess of 2 m, every year these majestic birds migrate from wintering grounds on the Gulf coast of Texas to the Northwest Territories. This is all that remains of a breeding summer range that at one time stretched from New Jersey

in the east to Salt Lake City in the west and as far north as the Mackenzie delta, and a winter range that included marshes from southern Louisiana into central Mexico (Figure 17.9). The cranes, which require undisturbed breeding habitat, soon declined under the expansion of agriculture. Unrestricted hunting along their long migration routes also contributed. By 1941 there were only twenty-two of them left on the planet.

The governments of the US and Canada agreed to a joint program to try to save the species from extinction, and have devoted considerable resources to the cause, as have NGOs, particularly the Audubon Society. The 1916 Migratory Bird Treaty between the US and Canada was used to stop legal hunting. In 1937, the US government bought the Aransas National Wildlife Refuge to protect the wintering habitat on the Gulf coast. In 1954 the only known nesting area was discovered in the northern part of another protected area, Wood Buffalo. Finding the breeding grounds also allowed for direct interventions, such as artificial incubation of eggs, to be attempted. Whooping cranes generally lay two eggs, but usually only one chick survives. A captive propagation program in the 1960s and 1970s moved one of the eggs for incubation, allowing the parents to devote more attention to the remaining chick. Removed eggs have also been incubated by the more abundant sandhill crane, although this program has now been discontinued due to high mortalities and failure to reproduce. By the late 1980s the world population of whooping cranes rose to over 200, including captive populations. In 1992 there were 136 individuals in the Wood Buffalo population and forty breeding pairs.

There is cautious optimism regarding the future of the cranes. However, dangers still exist. The wintering ground at Aransas is vulnerable to hurricane damage, which could conceivably wipe out the entire flock. The migration route is long and becoming increasingly modified by humans every year. Several cranes have died after colliding with overhead wires. Changes in the water regime in

The whooping crane, the tallest North American bird and one of the rarest, makes its habitat in muskeg, prairie pools, and marshes (*Parks Canada/R.D. Muir*).

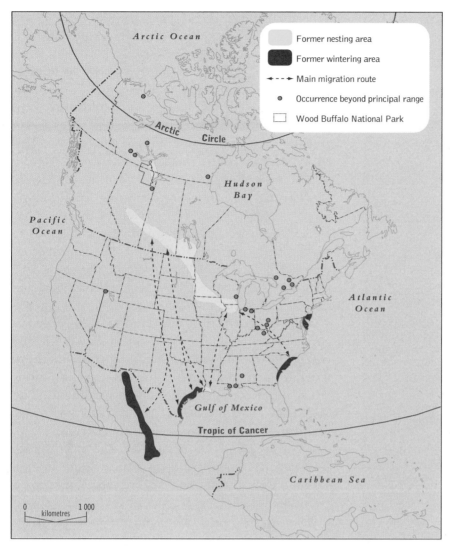

	Former nesting area
▉	Former wintering area
◄----►	Main migration route
●	Occurrence beyond principal range
⬚	Wood Buffalo National Park

Figure 17.9 The original range of the whooping crane in recent times.

pairs) by the year 2020. If these conditions are met for ten consecutive years, the species will be upgraded.

The cases of the bison and the whooping crane illustrate the critical role that protected areas play in saving endangered species from extinction. It is also not coincidental that the park in question, Wood Buffalo, is Canada's largest national park. Attention will now turn to Banff, another large park and Canada's most famous, to examine some of the different challenges facing the national parks system.

Banff National Park

'Banff Needs You', screamed the headlines of a news-sheet distributed by the Canadian Parks and Wilderness Society in 1993 as they launched an international campaign to prevent any further development in Banff National Park. Despite apparently more concern for environmental protection in legislation and policy, the level of development at Banff's townsite escalated rapidly throughout the 1980s. The total development in

Wood Buffalo with increasing drought conditions will make the cranes more vulnerable to terrestrial predators. It has been confirmed that one chick was eaten by a wolf.

The Canadian Recovery Plan of 1994 outlined a course of action that should result in upgrading the species from its present endangered status. This involves establishing a stable or increasing population in Wood Buffalo with a minimum of forty breeding pairs by the year 2000, and establishing and supporting two other wild whooping crane populations (each with a minimum of twenty-five breeding

The town site of Banff, although not large compared to the total area of the park, occupies one of the most valuable ecological sites in the park due to its low elevation, size, and location at a critical transportation point (*Parks Canada/W. Lynch*).

From 'Pilgrims in a National Park Townsite'

Into townsite gift-shops
 they pour,
pilgrims
 from our urban centres.
They linger here
in restaurants, shops, arcades.
As if by preference
 to mingle
among paintings of mountains,
 carvings and mounts of wildlife.
To thumb through books
 of words and photographs,
 when the reality
  is waiting near at hand.
Unworthy substitutes.
 they pilfer precious time
 in trade for banality.

Yet here in shops
they linger.

Wandering aimlessly before store windows;
 where the mountains
 humbly bid
 for their attention,
  as reflections in glass.
 An amorphous
 shimmering phantom
  beckoning.
. . . .

                    Jim Butler (1994:36)

Yellowstone, Grand Canyon, Yosemite, and Great Smoky Mountain national parks in the US is less than the development in Banff. It is the most developed national park in North America. It is also the core of the four mountain parks (Banff, Yoho, Jasper, and Kootenay), which were accorded World Heritage Site status in 1985 due to their importance for continental wildlife protection.

Although Banff started as a tourist town, at that time vast wilderness still remained in Canada. This is no longer the case, and in southern Canada the national parks provide one of the few protected refuges for wild animals. It is thus especially critical for animal habitat to be protected in parks. This is of special concern in Banff. Although the park itself is large, most of it is rock and ice. The Banff townsite is situated in one of the most productive

and rare habitats in the park and throughout the Rocky Mountains (the Montane zone), which covers between 2–5 per cent of the park. This zone has frequent chinook winds, low snow accumulation, warm winter temperatures, migration routes, and diverse habitats. These conditions are favourable for a dense concentration of wildlife and the zone has critical winter refuge areas. The Montane zone is the smallest ecoregion in Alberta and occupies less than 1 per cent of the land area.

The zone is also very attractive for development purposes. Over 70 per cent of the zone has already been occupied by highways, golf courses, towns, and resorts, and more development is planned. The Bow Valley corridor, extending east from its headwaters at Bow Glacier through Lake Louise, past Banff and Canmore to Morley (Figure 17.10), is crucial for maintaining the Montane zone. To the wildlife, it's all one valley. Unfortunately, this space is administered by many different federal, provincial, and municipal administrations with different ideas about development. The Wildlife Corridor Committee now acts as a forum for negotiations to protect wildlife corridors.

Within Banff National Park, the 4-km-wide Bow Valley contains the Trans-Canada highway, the 1A highway, a national railway, an airstrip, a twenty-seven-hole golf course, three ski resorts, the village of Lake Louise, and the town of Banff (population 7,500). Banff townsite takes up more than three-quarters of the largest block of montane habitat in the park. Outside the park, the population of Canmore has been growing almost as quickly as that of Banff (Figure 17.11). Between 1985 and 1992, building permits worth over $360 million were issued in Banff. Shopping space almost doubled between 1986 and 1994, a time when increases of between 10–15 per cent were normal in the rest of the country. There is now three times more the amount of shopping space per person in Banff than there is in Toronto!

East of Banff many developments are planned in the Bow Valley outside the park. These include four eighteen-hole golf courses; over 2,000 new hotel rooms and 6,000 housing units by one developer; an eighteen-hole golf course, 1,250 hotel rooms, and 1,600 housing units by another developer, both in the vicinity of Canmore, just outside the park boundary. Other developers have tabled plans for three more eighteen-hole golf courses,

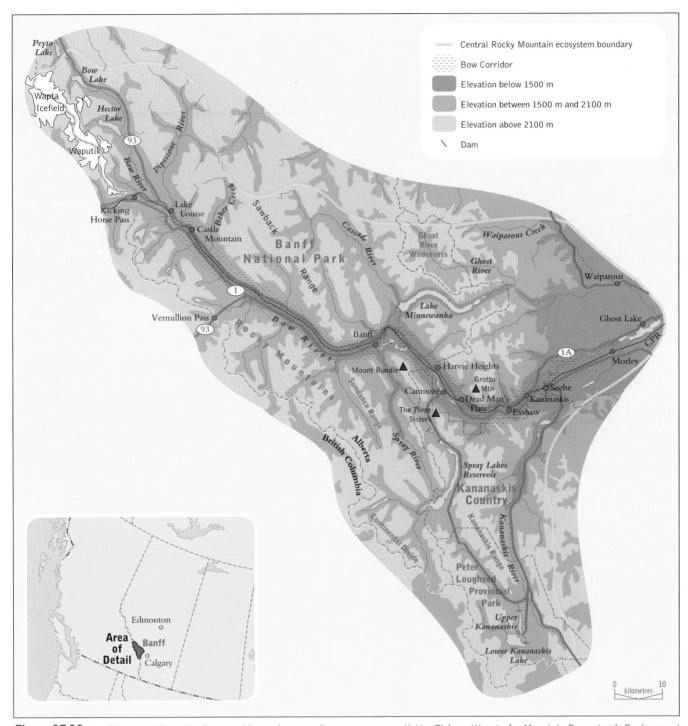

Figure 17.10 Banff National Park, the Bow corridor, and surrounding areas. SOURCE: K. Van Tighem, 'Heart of a Mountain Ecosystem', *Environment Views* 16 (Fall 1993):5.

140 recreational vehicle sites, and fifty chalets in the area. Developments such as these are strong reminders of the need for parks to adopt an ecosystem management approach that will consider all stresses within the ecosystem, not just within the park boundary.

Canadian Pacific, owner of the Banff Springs Hotel, has proposed adding nine more holes of golf to the existing twenty-seven holes at the hotel, which would destroy 26 ha of scarce montane habitat; 200 additional guest rooms and associated staff housing at the hotel, which is already the largest in any national park in North America; a convention centre and additional staff housing at the Chateau Lake Louise; relocating the swimming pool and health club and constructing tennis

The fact that the major cross-Canada transportation corridor bisects the park has created enormous problems, particularly in the face of additional demands for tourist infrastructure. The Trans-Canada Highway cuts a large swath down the Bow Valley, posing considerable challenges for wildlife. As a result, underpasses and overpasses have been created for wildlife. It remains to be seen as to how successful these will be in the long term in maintaining wildlife populations and their migratory routes (*Philip Dearden*).

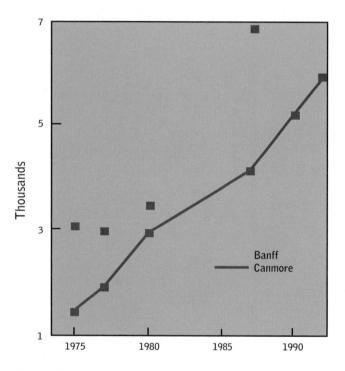

Figure 17.11 Population growth in the Bow corridor. SOURCE: R.W. Sandford, 'Blue Mountains, Black Coal', *Environment Views* 16 (Fall 1993):5.

courts at Chateau Lake Louise; and unspecified development of 4 ha of land in Banff.

There is no doubt that Banff is of major economic importance. Over 3.5 million people visit every year, and Banff tourism operators employ thousands of workers and generate hundreds of millions of dollars in annual revenue. Operators argue that they have to compete on an international basis, and if further developments are not

allowed, this source of income will dry up. A major goal is to cater to the international year-round convention market.

Reports have documented the cumulative impacts of all these developments. Fresh faeces have been found floating in the Bow River. Fish populations are badly depleted. Bear, wolf, and cougar populations are falling. Experts predict that the grizzly will be extirpated from the park over the next twenty-five years. The longnose dace has already become extinct (Box 17.6). Road development to accommodate more visitors has split some populations in half. Between 1989 and 1994, 176 animals were killed in the park on the railway tracks alone. Many species, such as elk, have been displaced from their natural habitat.

There are also many indirect influences of development on park ecosystems. Fire suppression, for example, has been regularly carried out since the earliest days of the park to protect tourist facilities. Fire is, however, a natural process within this habitat, and to suppress it is to change the ecosystem. Prescribed burning has now been reintroduced into the park's management to try to return to a more natural regime of vegetation renewal. This is a good example of the kind of adaptive management approach discussed in Chapter 9.

The question of the amount of development acceptable in Banff is obviously a controversial one. Historically, pro-development interests have been most influential and, given the amount of development that has taken place, many would

Box 17.6 The Banff Longnose Dace: Extinction in a National Park

The Banff longnose dace was a small minnow found in only one location on the planet, a marsh downstream from the famous Cave and Basin hot spring in Banff. Discovered in 1892, it was declared extinct by COSEWIC in 1987. As with many extinctions, it was likely a combination of factors that led to its demise. Tropical fish were introduced into the warm waters of the marsh, where they flourished and were able to out-compete the dace. The dace may also have interbred with other species of dace, and col-

lecting for scientific purposes probably also helped to reduce numbers. However, it seems that the most important factor in its extinction was the decision to allow a hotel to chlorinate hot spring water and discharge it into the marsh. The chlorine reacted to form chlorinated hydrocarbons, which are very toxic to fish, even at low concentrations. Shortly thereafter, the Banff longnose dace disappeared forever, hence it was pressure from tourism that led to Banff's first extinction.

argue that they still are, despite the vociferous opposition by environmental groups, such as the Alberta Wilderness Association and the Canadian Parks and Wilderness Society. This raises the question of how to determine the legitimate stakeholders in this kind of decision, as discussed previously in Chapter 11. It should not be forgotten, however, that the parks are dedicated to the people of Canada. A 1993 Angus Reid poll of Canadian attitudes to national parks found that when asked specifically about the development of townsites in national parks, a majority of Canadians believe that *no* further development should be allowed; this antidevelopment sentiment was highest among recent visitors to Banff.

In response to the controversy, the minister responsible called for a moratorium on development within the Bow Valley corridor, dependent upon the results of a two-year special enquiry, the Banff-Bow Valley Study. The objectives of the study were threefold:

- to develop a vision and goals for the Banff-Bow Valley that will integrate ecological, social, and economic values
- to complete a comprehensive analysis of existing information, and to provide direction for future collection and analysis of data to achieve ongoing goals
- to provide direction on the management of human use and development in a manner that will maintain ecological values and provide sustainable tourism (Banff-Bow Valley Study 1996:9)

However, major developments—such as the expansion of the Sunshine ski area, CP's proposal

for nine more holes of golf and a convention centre, Skiing Louise's new hotel project, and a new housing subdivision in Banff—were excluded from the enquiry.

The five-person task force attempted extensive consultations of stakeholders through a round table process involving fourteen different interest sectors. All the recommendations that achieved consensus from this group were adopted. Public consultations were also held in a few cities across the country and written submissions were solicited, of which 261 were received. People were kept informed of the progress of the study through newsletters, public presentations, news releases, community television, workshops, and the Internet. The study also commissioned a number of research projects to supplement existing information.

On the basis of this input, the task force came to the following conclusions from the Banff-Bow Valley Study:

1. While Parks Canada has clear and comprehensive legislation and policies, Banff National

Skiing in Banff

*The issue is basically the philosophy of use: do we actually reach out and touch the mountains, or do we just look at them? . . . Banff has been recognized worldwide for years as a major skiing area, but the Parks Service seems determined to kill that reputation. It inhibits sensible development and threatens to turn Banff into a second-class ski centre. It ought to allow Banff to enhance its status as an international ski destination (*Globe and Mail *1989:A4).*

Park suffers from inconsistent application of the National Parks Act and Parks Canada's Policy. Some of the explanation lies in the evolution of Banff National Park, some in *ad hoc* decision-making, and some in weak political will in the face of a range of interest-based lobbying.

2. Despite the fact that ecological integrity is the primary focus of the National Parks Act and Parks Canada's Policy, we have found that ecological integrity has been, and continues to be, increasingly compromised. Park management, human use, development, the highway and the railway have contributed to this situation, despite well-intended remedial actions.

3. While scientific evidence supports conclusion #2 above, a significant percentage of the population find it difficult, based on the beauty they see, to understand the ecological impacts that have occurred.

4. The current rates of growth in visitor numbers and development, if allowed to continue, will cause serious, and irreversible, harm to Banff National Park's ecological integrity. Stricter limits to growth than those already in place must be imposed if Banff is to continue as a national park. Growth also threatens the Park's cultural importance and its ability to inspire not only artists, but all Canadians. The built heritage that gives the Town of Banff its cottage atmosphere is disappearing fast under the pressure for new construction.

5. More effective methods of managing and limiting human use in Banff National Park are required. This will involve adjustments by visitors, residents, the tourism industry, park management and adjacent jurisdictions. While recognizing the need to manage growth in the number of visitors, restricting access should not replace creative visitor management programs that would allow more visitors to enjoy the Park, while maintaining ecological integrity.

6. To maintain natural landscapes and processes, disturbances such as fire and flooding must be restored to appropriate levels in Banff National Park.

7. There are existing anomalies in the Park, such as the Trans-Canada Highway, the Canadian Pacific Railway, and the Minnewanka dam. In their continued existence, they must update their designs in accordance with the most advanced science, and ecological and engineering practices.

8. We are proposing the refocussing and upgrading of the role of tourism. Tourism in Banff National Park will, to a greater extent, reflect the values of the Park and contribute to the achievement of ecological integrity. There will continue to be many attractive and profitable economic opportunities for sustainable tourism.

9. We acknowledge that mountain tourism in Alberta will continue to expand. Any new, related facilities will have to be located outside national park boundaries. In coming to this conclusion, we have been sensitive to Banff National Park's place in the regional ecosystem and understand that these developments will affect this ecosystem. The Task Force feels that regional coordination is essential and must start with discussions between senior officials of each of the jurisdictions.

10. Current growth in the number of residents, and in the infrastructure they require, is inconsistent with the principles of a national park. Revisions to the General Municipal Plan for the Town of Banff must address these inconsistencies and the need for limits to growth. Growth management must continue in the Hamlet of Lake Louise, and in other residential and commercial areas in the rest of the Park. In some areas, facilities must be downsized, relocated or removed.

11. Public scepticism and lack of trust in the decision-making process have led to a polarization of opinion. We are recommending new forms of broader-based public involvement and shared decision-making, with clear links to Parks Canada's decision-making and accountability. Such involvement will address national, regional, and local interests.

12. Visitors must be better informed about the importance of the Park's natural and cultural heritage, the role of protected areas and the challenges that the Park will face in the third millennium. It is also important for visitors to understand both the value and the cost of ecological integrity, so as to promote feelings of greater personal responsibility and steward-

ship. Improvements in education, awareness, and interpretation programs are required.

13. Improvement in Parks Canada's management is central to the success of Banff National Park. This should begin with a comprehensive revision of the Banff National Park Management Plan.

14. Current allocation of funding is inadequate to meet the requirements for maintaining ecological integrity and visitor management (Banff–Bow Valley Study 1996:14).

In response to the task force's report, the Honourable Sheila Copps, deputy prime minister and minister of Canadian heritage, announced that there would be no new land made available for commercial development within the park, the population of the townsite would be capped at 10,000 permanent residents, the airport would be closed, and the management plan for the park would be revised. In addition, another team of advisers was announced to study the implementation of the other recommendations and report back to the minister in January 1997. Although environmentalists were generally pleased about the report, the response has, to date, promised little that is new. Canadian Pacific Hotels voluntarily agreed to withdraw its application for expansion of the Banff Springs golf course, however, and to limit hotel capacity. Only time will tell if some of the major recommendations will actually be implemented in such a fashion as to make a significant contribution to retaining the ecological integrity of Banff in the future.

IMPLICATIONS

These case-studies of Wood Buffalo and Banff national parks illustrate some of the roles played by protected areas, particularly the bank role in providing a place to store and protect ecological capital such as endangered species, and also some of the conflicts that arise between the factory role of supporting tourism and this ecological role. The 1990s is crucial for determining the fate of many species and natural places in Canada. Most countries no longer have this option and, once lost, species and natural places cannot be regained. Many of the approaches to environmental management described earlier in this book—adaptive management, ecosystem approaches, working with

---

Box 17.7 What You Can Do

1. Visit parks and other protected areas often throughout the year. Enjoy yourself. Tell others that you have enjoyed yourself. Encourage them to visit.

2. Always follow park regulations regarding use. Feeding wildlife, for example, may seem kind or harmless, but it can lead to the death of the animal.

3. If you have questions regarding the park's management or features, don't be afraid to ask. A questioning public is a concerned public.

4. Many park agencies have public consultation strategies relating to topics ranging from park policy to the management of individual parks. Let them know your interests so you can be placed on the mailing list to receive more information.

5. Join a non-governmental organization, such as the Canadian Parks and Wilderness Society or the Canadian Nature Federation, which has a strong interest in parks issues.

6. Many parks now have cooperating associations in which volunteers can help with various tasks. Find out if a park near you has such an organization.

7. Write to politicians to let them know of your park-related concerns, such as completing the Endangered Spaces campaign.

---

various stakeholders, and being able to resolve conflict—are vital for the future of endangered species and spaces in this country.

SUMMARY

1. Protected areas fulfil many roles in society, including species and ecosystem protection, maintenance of ecological processes, areas for recreation and spiritual renewal, aesthetic appreciation, tourism, and science and education in natural outdoor settings.

2. There are many different kinds of protected areas in Canada, including national and provincial parks, wilderness areas, First Nations parks, wildlife

refuges, ecological reserves, and regional and municipal parks.

3. The amount of protection given to ecosystem components varies among these different types. National parks have one of the strongest mandates for protection.

4. Canada has endorsed the international goal of protecting 12 per cent of the land base by the year 2000. However, due to these differences in levels of protection, progress is not easy to measure. The World Wildlife Fund estimates that in 1995 Canada had 5.2 per cent of the land base protected in designations that did not allow activities such as logging. However, in terms of representing the 453 defined natural regions of the country, only 4 per cent were judged by World Wildlife Fund Canada in 1995 to be adequately represented by protected areas.

5. National parks are outstanding natural areas protected by the federal government because of their national significance. Banff, the first in Canada, was protected in 1885. Since that time the national parks have fulfilled a dual mandate that required protection of park resources in an unimpaired state, but also required their use. This conflicting mandate was first clarified in a policy statement in 1979, then enshrined in an amendment to the National Parks Act in 1988 and further clarified in the current policy document of 1994. All of these make it clear that the first and overriding priority of Canada's national parks is to protect the natural environment.

6. There are thirty-eight national parks in Canada. The goal is to have a least one national park in each of the thirty-nine regions of the national system plan. Since several of the current parks are in the same region, only about half of the system plan is completed. The government has committed to completing the plan by the turn of the century.

7. There is also a system plan for marine parks covering twenty-nine different natural marine regions. Only two of these are currently represented by parks. The government aims to create another six by the year 2000. No new marine parks have been created since this commitment.

8. Current management challenges to the national parks system include external threats and fragmentation, which require more of an ecosystem approach. The involvement of First Nations peoples also represents a broader view of the stakeholders for many national parks.

9. Case-studies of Wood Buffalo and Banff national parks indicate the valuable role of parks in protecting endangered species, and also some of the difficulties encountered in trying to control tourism development.

REVIEW QUESTIONS

1. What do you think should be the relative importance of the various roles played by protected areas?

2. What different classifications of protected areas exist in your province, and what kinds of protection are offered by these different systems?

3. What is your province doing to achieve the 12 per cent protected area that all jurisdictions have committed to establishing?

4. What NGOs are active in your area regarding protected areas? What are their main concerns?

5. What do you think is the appropriate balance between wildlife management and agricultural production in areas such as Riding Mountain and Wood Buffalo national parks?

6. What ecozone is Banff townsite situated in, and what is the importance of this zone?

7. What do you think is an acceptable level of tourism development for Banff?

REFERENCES AND SUGGESTED READING

Abbey, E. 1968. *Desert Solitaire: A Season in the Wilderness.* New York: Simon and Schuster.

Auditor-General of Canada. 1996. *Report to the House of Commons, Chapter 31 'Canadian Heritage—Parks Canada: Preserving Canada's Natural Heritage'.* Ottawa: Minister of Public Works and Government Services Canada.

Banff-Bow Valley Study. 1996. *Banff-Bow Valley: At the Cross-roads*. Summary report of the Banff-Bow Valley Task Force. Prepared for the Honourable Sheila Copps, minister of Canadian heritage. Ottawa: Auditor-General of Canada.

Bella, L. 1987. *Parks for Profit*. Montreal: Harvest House.

Binkley, C.S., and R.S. Miller. 1988. 'Recovery of the Whooping Crane *Grus americana*'. *Biological Conservation* 45:11–20.

Butler, J. 1994. *Dialogue with a Frog on a Log*. Edmonton: Duval Publishing.

Dearden, P. 1988. 'Protected Areas and the Boundary Model: Meares Island and Pacific Rim National Park'. *The Canadian Geographer* 32:256–65.

_____. 1989. 'Wilderness and Our Common Future'. *Natural Resources Journal* 29:149–57.

_____. 1995a. 'Parks and Protected Areas'. In *Resource and Environmental Management in Canada*, edited by B. Mitchell, 236–58. Toronto: Oxford University Press.

_____. 1995b. 'Park Literacy and Conservation'. *Conservation Biology* 9:1654–6.

_____, and L. Berg. 1993. 'Canada's National Parks: A Model of Administrative Penetration'. *The Canadian Geographer* 37, no. 3:194–211.

_____, and J. Gardner. 1987. 'Systems Planning for Protected Areas in Canada: A Review of Caucus Candidate Areas and Concepts, Issues and Prospects for Further Investigation'. In *Heritage for Tomorrow: Canadian Assembly on National Parks and Protected Areas*, vol. 2, edited by R.C. Scace and J.G. Nelson, 9–48. Ottawa: Environment Canada-Parks.

_____, and R. Rollins, eds. 1993. *Parks and Protected Areas in Canada: Planning and Management*. Toronto: Oxford University Press.

Dunlap, T.R. 1991. 'Organization and Wildlife Preservation: The Case of the Whooping Crane in North America'. *Social Studies of Science* 21:197–221.

Eagles, P.F.J. 1984. *The Planning and Management of Environmentally Sensitive Areas*. Harlow: Longman.

Edwards, R., et al. 1994. *National Recovery Plan for the Whooping Crane*. Ottawa: Recovery of Nationally Endangered Wildlife Committee.

Gauthier, D.A. 1994. 'The Buffalo Commons on Canada's Plains'. *Forum* 9:118–20.

Globe and Mail. 1989. 'Editorial' (3 April).

Hummel, M., ed. 1989. *Endangered Spaces*. Toronto: Key Porter Books.

_____, ed. 1995. *Protecting Canada's Endangered Spaces: An Owner's Manual*. Toronto: Key Porter Books.

IUCN (World Conservation Union). 1993. *Parks for Life: Report of the IVth World Congress on National Parks and Protected Areas*. Gland, Switzerland: IUCN.

_____. 1994. *Guidelines for Protected Area Management Categories*. Gland, Switzerland: IUCN.

Killan, G. 1993. *Protected Places: A History of Ontario's Provincial Park System*. Toronto: Dundurn Press.

Labatt, L., and B. Littlejohn. 1992. *Islands of Hope: Ontario's Parks and Wilderness*. Toronto: Firefly Books.

Leopold, A. 1949. *A Sand County Almanac*. New York: Oxford University Press.

Littlejohn, B., and J. Pearce, eds. 1973. *Marked by the Wild: An Anthology of Literature Shaped by the Canadian Wilderness*. Toronto: McClelland and Stewart.

McNamee, K. 1994. *The National Parks of Canada*. Toronto: Key Porter Books.

_____. 1995. 'The Federal Government: National Parks'. In *Protecting Canada's Endangered Spaces: An Owner's Manual*, edited by M. Hummel, 163-93. Toronto: Key Porter Books.

Marty, S. 1978. *Men for the Mountains*. Seattle: The Mountaineers.

_____. 1985. *A Grand and Fabulous Notion*. Toronto: McClelland and Stewart.

Nash, R. 1967. *Wilderness and the American Mind*. New Haven: Yale University Press.

Noss, R.F. 1992. 'The Wildlands Project Land Conservation Strategy'. *Wild Earth* 10–21.

_____, and A.Y. Cooperrider. 1995. *Saving Nature's Legacy: Protecting and Restoring Biodiversity*. Washington, DC: Island Press.

Parks Canada. 1994. *Guiding Principles and Operational Policies*. Ottawa: Minister of Supply and Services Canada.

_____. 1995. *State of the Parks, 1994 Report*. Ottawa: Minister of Supply and Services Canada.

_____. 1997. *Charting the Course: Towards a Marine Conservation Areas Act*. Ottawa: Canadian Heritage.

Sax, J.L. 1980. *Mountains Without Handrails: Reflections on the National Parks*. Ann Arbor: University of Michigan Press.

Scace, R., and J.G. Nelson, eds. 1986. *Heritage for Tomorrow: Canadian Assembly on National Parks and Protected Areas*. Ottawa: Minister of Supply and Services Canada.

Shafer, C.L. 1990. *Nature Reserves: Island Theory and Conservation Practice*. Washington, DC: Smithsonian Institution Press.

Theberge, J.B. 1993. 'Ecology, Conservation and Protected Areas in Canada'. In *Parks and Protected Areas in Canada: Planning and Management*, edited by P. Dearden and R. Rollins, 137–57. Toronto: Oxford University Press.

Trant, D. 1993. 'Land Use Change around Riding Mountain National Park'. In *Environmental Perspectives 1993: Studies and Statistics*, 33–46. Ottawa: Statistics Canada.

Woodley, S. 1993. 'Monitoring and Measuring Ecosystem Integrity in Canadian National Parks'. In *Ecological Integrity and the Management of Ecosystems*, edited by S. Woodley, J. Kay, and G. Francis, 155–76. Delray Beach, FL: St Lucie Press.

World Commission on Environment and Development. 1987. *Our Common Future*. Oxford: Oxford University Press.

World Wildlife Fund Canada. 1995. 'Endangered Spaces Progress Report Number 5, 1994–5'. Toronto: World Wildlife Fund Canada.

_____. 1996. 'Endangered Species Progress Report Number 6, 1995–6'. Toronto: World Wildlife Fund Canada.

Chapter 18

Water

Water is the life-blood of the environment. Without water no living thing, plant or animal, can survive. It is along the rivers and around the sea coast and lakes that life of all kinds is richest and most varied. Water plays a unique role in the traditional economy and culture of the Native peoples. It lies deep in the concept of Canada held by all Canadians.

Water is also a commodity: a renewable resource. The availability of an adequate and usable supply underpins our whole economy. Water is used for transportation and power generation, for waste disposal, recreation, agriculture, and fisheries, and is essential both in manufacturing and in the service sector. Happily, it is possible to use water without depleting its supply. But water is a fundamental component of a complex ecosystem. Its maximum sustainable yield is set by the effects of its exploitation in this total system.

Evidence of over-exploitation and, with it, evidence of the linked effect of environmental stress, is all too clear (Science Council of Canada 1988:8).

Nine per cent of the world's fresh water, as discussed in Chapter 6, is in Canada. That enormous amount of water, combined with Canada's relatively small population, has created a 'myth of superabundance'. However, the reality is that the water is often in the wrong place at the wrong time, hence the impetus to develop irrigation for agriculture in places such as the Okanagan Valley in British Columbia and in southern Alberta, for

flood damage reduction works to protect many communities, as well as initiatives to transfer water to areas with shortages. In addition, human activities have frequently caused serious degradation of water quality, as illustrated by concerns about pol-

The Myth of Superabundance

It was perhaps during the period of fur trading and exploration that Canada first developed the image of a land with unlimited water resources. The notion still prevails. Today the country is portrayed in both popular and official literature as a land of countless lakes and a myriad of untamed rivers. Attention is often drawn to the fact that some 7.6 per cent of Canada's vast territory of 9.97 million square kilometres is covered by lakes. It is also pointed out that several rivers drain areas of hundreds of thousands of square kilometres in extent, discharging huge volumes of water into the Pacific, Arctic and Atlantic oceans. . . . with less than 1 per cent of the world's population, Canadians are extremely well endowed with water resources. Only a tiny fraction of these assets has been exploited so far. Superficially, therefore, there not only appears to be a superabundance of water available now, but little prospect of there ever being a major Canadian shortage.

To an important extent, however, this view of superabundance is exaggerated. A more sober assessment suggests that much of the water is in the wrong place or is available at inappropriate times. It also reveals that part of the wealth that has been utilized so far has been developed relatively cheaply, and that future resources can only be harnessed at rapidly escalating costs (Foster and Sewell 1981:7).

Canada faces many water resource challenges with severe flooding in some areas and the need to have irrigation in many others (*Philip Dearden*).

lution in the Fraser River estuary, the Great Lakes, the St Lawrence River, and Halifax harbour. The perception of water abundance has contributed to Canadians' inefficient water use averaging 360 L per person per day, second only to that of Americans (Figure 18.1). In the remainder of this chap-

ter, two experiences with water management and development are examined with regard to the ideas presented in parts B and C.

MUNICIPAL WATER SUPPLY: ELMIRA, ONTARIO

On 13 November 1989 a headline on the front page of the *Kitchener-Waterloo Record* stated 'Wells in Elmira, St Jacobs closed after chemical tests'. Two of five municipal wells serving those communities had been closed after tests on 19 September revealed that they contained a chemical linked to cancer. N-nitroso dimethylamine (NDMA) had been detected in the two wells. The levels were as high as 40 ppb. N-nitroso dimethylamine is widely used as an industrial solvent, as an additive in gas and lubricants, and in plasticizing rubber. It can also be produced by combining herbicides with nitrogen fertilizers. No factory in Elmira used NDMA in its production process.

A spokesperson for the provincial Ministry of the Environment explained that no one had expected to find NDMA since it was not something for which testing was generally done. It had been found in a test (conducted only

The myth of Canada's superabundance of water comes from scenes like this of the Mackenzie Delta. However, most of the demand for water in Canada is thousands of kilometres away (*Philip Dearden*).

once a year due to the very high cost) that scanned for hundreds of chemicals. Officials had no idea where the contamination came from or how long it had been in the wells, and were unsure of how to remove it. With two of five wells shut down, residents and industries in Elmira were asked to reduce their water use. Uniroyal, a major manufacturer, cut its daily water consumption by about 5.9 million L and slashed production. Concern further increased when NDMA was discovered shortly afterwards in private wells in Elmira, and their owners were told not to drink water from those wells.

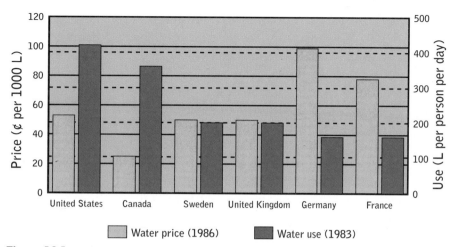

Figure 18.1 A comparison of municipal water prices and use in selected countries. SOURCE: Environment Canada, *The State of Canada's Environment* (Ottawa: Minister of Supply and Services Canada, 1991):3–8.

The medical officer of health for the Regional Municipality of Waterloo explained that there were no drinking water guidelines for this chemical in Canada (Box 18.2). American guidelines were not necessarily helpful because they were not intended strictly for drinking water. While Ontario had no legal guidelines for NDMA, it recommended a maximum concentration of .014 ppb as an interim drinking water standard. Elmira residents expressed considerable concern that results from tests completed in mid-September had not been made available until two months later.

By mid-December 1989 the Ontario Ministry of the Environment had detected NDMA at levels of 2,000 ppb entering the Elmira sewage treatment plant in waste water from the Uniroyal Chemical plant. Subsequent tests indicated that effluent leaving the treatment plant and entering the Canagagigue Creek and subsequently the Grand River south of Elmira had concentrations of 50 ppb. This

---

Box 18.1 Canada Facts: Water Use

- A reasonable quality of life needs about 80 L of water per person daily; Canadians' average use is 360 L.
- Municipal use in the mid-1980s accounted for 11.2 per cent of all withdrawals, agriculture 8.4 per cent (with the majority occurring in Alberta), manufacturing 14 per cent, and thermal (coal, oil, nuclear) power generation 60 per cent (mostly used for steam generation and cooling).
- One in four Canadians depends on groundwater for domestic supply of water. Over 85 per cent of water consumed by livestock is taken from the ground.
- Non-consumptive use is also very important for recreational activities, such as swimming, boating, and skating.
- Water is also essential in maintaining ecosystem integrity.

Box 18.2 Drinking Water Guidelines

Drinking water should not contain disease-causing micro-organisms, harmful chemical substances, or radioactive matter. Guidelines for drinking water have been established by Health Canada. Environment Canada has commented that:

> Regardless of the approach used to arrive at the guideline, the adopted limit normally balances health benefits against such socioeconomic factors as the cost and feasibility of attaining a minimum acceptable concentration. Further, it should be recognized that socioeconomic factors may be given more weight when it comes to deciding legally binding standards than in developing ideal guidelines (Environment Canada 1991:3.24).

generated concern in downstream communities (Figure 18.2). Wells adjacent to the Grand River in Kitchener were closed in mid-December while testing for NDMA was done. Nine wells in Kitchener were found to be contaminated.

At the same time, a Ministry of Environment spokesperson stated that NDMA was not a raw material in or a product from the Uniroyal plant. He speculated that the NDMA had been formed through reaction when different waste streams combined in the sewage system. Downstream communities, such as Brantford (with a population of 80,000) and Cayuga (with a population of 2,000), were both totally dependent on the Grand River for water and considered safe, since it was believed that if any NDMA contamination reached those communities, it would be so diluted as to be harmless.

However, before the end of December, readings of NDMA had been found in Kitchener wells (.069 ppb), in the drinking water of Brantford (.010 ppb) and Cayuga (.018 ppb), as well as at Oshweken (.078 ppb), a village on the Six Nations Indian reserve south of Brantford. The Six Nations residents were ordered not to drink water from their community water supply system, and the ban was later extended to use of water for bathing, cooking, and laundry. By this stage, various experts

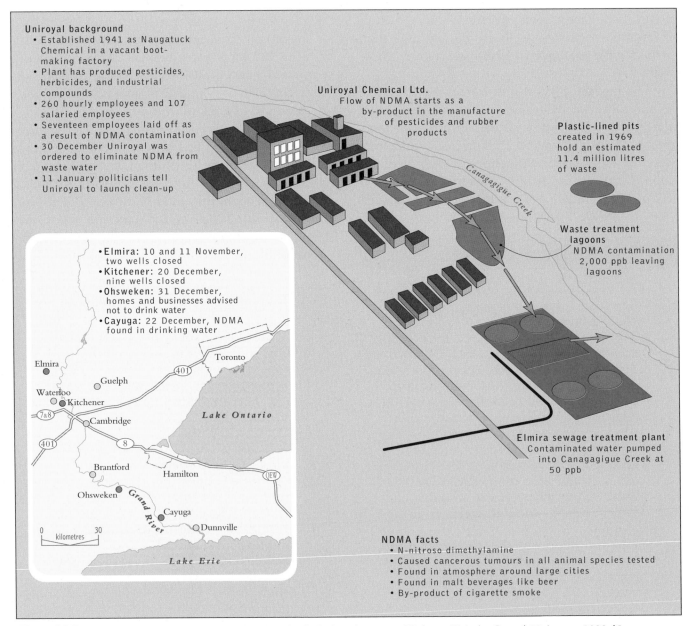

Uniroyal background
- Established 1941 as Naugatuck Chemical in a vacant boot-making factory
- Plant has produced pesticides, herbicides, and industrial compounds
- 260 hourly employees and 107 salaried employees
- Seventeen employees laid off as a result of NDMA contamination
- 30 December Uniroyal was ordered to eliminate NDMA from waste water
- 11 January politicians tell Uniroyal to launch clean-up

Uniroyal Chemical Ltd.
Flow of NDMA starts as a by-product in the manufacture of pesticides and rubber products

Canagagigue Creek

Plastic-lined pits
created in 1969 hold an estimated 11.4 million litres of waste

Waste treatment lagoons
NDMA contamination 2,000 ppb leaving lagoons

- **Elmira:** 10 and 11 November, two wells closed
- **Kitchener:** 20 December, nine wells closed
- **Ohsweken:** 31 December, homes and businesses advised not to drink water
- **Cayuga:** 22 December, NDMA found in drinking water

Elmira sewage treatment plant
Contaminated water pumped into Canagagigue Creek at 50 ppb

Elmira
Toronto
401
Guelph
Waterloo
Kitchener
7&8
Cambridge
401
8
Lake Ontario
Brantford
Hamilton
QEW
Ohsweken
Grand River
Cayuga
Dunnville
0 kilometres 30
Lake Erie

NDMA facts
- N-nitroso dimethylamine
- Caused cancerous tumours in all animal species tested
- Found in atmosphere around large cities
- Found in malt beverages like beer
- By-product of cigarette smoke

Figure 18.2 Location of Elmira and other communities and pollution's path. SOURCE: *Kitchener-Waterloo Record,* 12 January 1990:A1.

were providing conflicting views regarding what were safe levels for NDMA. A chemist at an American institute for cancer research stated that he disagreed with the Ontario safety guideline of .014 ppb on the basis that not enough was known about this potential carcinogen to determine an acceptable level (Box 18.3).

At the end of December 1989 the Ontario Ministry of the Environment ordered Uniroyal to ensure that NDMA did not enter the Elmira sewage treatment plant. Uniroyal was ordered never again to use dimethylamine (DMA) in its production process as it was believed to be the source of NDMA. Uniroyal suspended all manufacturing activity that involved use of DMA.

Chemical Manufacturing in Elmira

During 1941 Naugatuck Chemical, a division of Dominion Rubber Company, which in turn was a subsidiary of United States Rubber, bought an old shoe factory in Elmira that had been unused for a decade. The company chose Elmira because, among other reasons, the abandoned footwear plant was on the eastern edge of the community beside a creek that could carry away wastes. In addition, there was abundant land for dumping chemicals, and the prevailing westerly winds would carry odours or other air pollution away from the town. The company was welcomed by the community. The local newspaper stated that it was the first good break in ten years for Elmira, following the difficult Depression years in the 1930s.

Naugatuck produced a material used as a stabilizer for explosives. The associated tar residue

---

Box 18.3 Guidelines for Carcinogenic Substances

Environment Canada (1991:3–23) states that for carcinogenic substances and ionizing radiation, there is no level of exposure at which a hazard does not exist.

When developing drinking water guidelines for such substances, researchers and regulators assume that any level of exposure carries some risk. Guidelines are usually set at levels at which the risk is thought to be acceptably low or negligible, such as one person in a million developing cancer.

---

was buried in the town dump. By January 1942 Naugatuck was making aniline oil for explosives used in the Second World War. Sludge from that production was used as fill on the company property. Before the end of the war, the company started to make a rubber chemical and a synthetic rubber. By the end of the war, it was also manufacturing pesticides, such as the weed killer 2,4-D and the fungicide, Vitavax. During 1945 Naugatuck became the first Canadian factory (and one of the first in North America) to produce the insecticide DDT.

With the blessing of local politicians, Naugatuck promoted a herbicide spraying campaign to eradicate weeds in the town. During June 1947 Elmira proudly proclaimed itself as the first 'weedless town' in Canada, thanks to the miracle of Naugatuck's herbicides. This announcement received attention across North America, and residents believed that Naugatuck and its herbicides had put Elmira on the map. Notwithstanding the enthusiasm with which Naugatuck was welcomed, there were some concerns. As early as 1946, some res-

The old Naugatuck stack at Uniroyal Chemical (*Bruce Mitchell*).

idents complained that locally grown tomatoes tasted of chemicals.

Later in the 1950s the company began to store its solid wastes in steel drums, which were then buried. Liquid wastes continued to be released into Canagagigue Creek. This two-pronged procedure was deemed satisfactory enough that in 1953 the local council expressed concern about spending $150,000 on a new sewage treatment plant because the councillors believed the creek was 'almost safe'.

By the mid-1950s, Naugatuck employed 200, and in 1957 it announced a half-million-dollar expansion. That news was welcomed. However, in the same year a neighbourhood on the downwind side of Elmira was blackened by greasy soot from the company's smokestack. Mothers complained that they were changing their children's clothes two or three times a day because of the dirt. During the early 1960s, some local farmers complained that cattle had died after drinking from Canagagigue Creek, or that their crops were killed by run-off from the Naugatuck site.

In 1966 the US Rubber Company changed its name to Uniroyal and, except for the lettering on the smokestack, the name Naugatuck disappeared from Elmira. At about the same time, Uniroyal began to make the defoliant Agent Orange for the US Defense Department's use in Vietnam, as well as for Ontario Hydro and Canadian National Railway. By 1969 the company was examining safer ways to dispose of its wastes. Under provincial supervision, Uniroyal prepared a 11.36 million L chemical dump site on its own property. Many of the old drums were dug up and placed in two plastic-lined disposal pits, then covered with clay. For most people, this approach was the solution to any contamination problem.

In 1978 the Ontario Ministry of the Environment warned Uniroyal that it had to improve the quality of the effluent being sent to the municipal sewage treatment plant in Elmira. The next year, the company announced that it would spend more than $4 million on new equipment for waste treatment. Provincial testing indicated NDMA concentrations of 1,163 ppm in a process lagoon on the Uniroyal property, but no traces were found in the outflow from the community sewage treatment plant. By 1980 the company and the Ministry of the Environment announced a joint monitoring program to determine if chemicals were entering the groundwater. Monitoring subsequently indicated high pollution levels in Canagigague Creek.

During 1981 pesticide traces were discovered in the monitoring wells around the Uniroyal property and in the air around the plant. The following year, Uniroyal was fined $2,500 for polluting the creek and not reporting the spill. In 1984 the Elmira sewage treatment plant was expanded at a cost of $3.8 million, of which Uniroyal paid $1 million. The Ministry of the Environment established new standards to control wastes from the chemical plant.

Between 1985 and 1987 Uniroyal started to clean up its property by digging up 1,691 drums of buried toxic waste. The drums held traces of dioxin, DDT herbicide, arsenic, and cyanide. In addition, the open chemical lagoons were excavated. In early 1989 the company announced plans to build a toxic waste incinerator in Elmira. Then in November 1989 the Regional Municipality of Waterloo shut the two municipal wells after test results revealed high levels of NDMA in the water.

The NDMA Issue

The NDMA issue involved technical, social, and legal/administrative aspects. Some selected examples illustrate the challenges in the Elmira situation. One important technical dimension involved disagreements about what constituted appropriate environmental quality standards. This issue illustrates some of the practical problems related to monitoring and thresholds, discussed earlier in Chapter 10.

There were different views regarding what was an appropriate standard for NDMA in drinking water. When the chemical was detected in the groundwater in September 1989, neither Ontario nor the federal government had a legal regulation for NDMA in drinking water. An interim guideline of 0.014 ppb was adopted by the Regional Municipality of Waterloo for waste water *entering* the sewage treatment plant at Elmira. In contrast, a provincial Ministry of Environment order stipulated that the NDMA limit was 0.5 ppb for waste water *leaving* the sewage treatment plant. A toxicologist with a consulting firm hired by the regional government stated that 0.011 should be the limit for NDMA, while another expert noted that a standard of 8.12 ppb was used in the state of Georgia.

Conflicting Evidence and Arguments

After listening to conflicting evidence, interpretations and arguments from experts representing Uniroyal, and the regional and provincial governments, the chairman of the Environmental Appeal Board stated that he was concerned about 'the tremendous range of opinion and data', and also that 'it just doesn't make sense anymore'. (Kitchener-Waterloo Record, 18 July 1990:B3).

Given this range of views, it is not surprising that members of the provincial Environmental Appeal Board investigating the NDMA situation in Elmira were reported to be 'baffled by the staggering array of scientific data which has been thrust at them' (*Kitchener-Waterloo Record*, 18 July 1990:B3). After eighteen days of hearings in which expert testimony was given by toxicologists and other experts on behalf of Uniroyal, Waterloo Region, and the federal and provincial governments, the chairperson of the Environmental Appeal Board was alleged to have stated that he did not know if the experts were on the same continent, let alone the same province. He directed the experts from all sides to meet to determine whether they could agree on a common method of data collection. At an earlier meeting, the experts had been unable to agree upon such a common strategy.

The lawyers representing Uniroyal and the governments before the hearing indicated that they understood and sympathized with the board members' dilemma. However, they argued that the board had to reach a decision based on burden of proof, as is done in criminal courts. The lawyer for the Ministry of the Environment stated that the interests of the company and the government were diametrically opposed, and the government was not going to back down. The lawyer for Uniroyal was equally firm. He explained that the experts on behalf of Uniroyal had shared their information with everyone as much as possible. In his view, after reviewing the conflicting evidence and interpretations, the Environmental Appeal Board members would simply have to reach a decision. This type of situation emphasizes the uncertainty about scientific understanding that can occur in attempting to develop solutions to environmental problems.

Socially, the NDMA situation created some divisions within Elmira. One of the splits was between those who were concerned about Uniroyal's pollution track record, and those who were concerned about the possible loss of jobs if the company decided that pollution regulations were too burdensome. As noted in Chapter 2, for many people this is the central challenge regarding sustainable development: what it is and how it should be achieved.

Three citizens' groups formed in Elmira. The first was the Citizens' Environmental Advisory Committee (CEAC). It was established in the early 1980s to work with Uniroyal and the provincial regulatory agencies. CEAC had played a central role in getting the community involved in pollution battles in the 1980s. It had contributed to reductions in air pollution emissions from Uniroyal and to removal of drums of toxic waste and old processing lagoons from Uniroyal property. CEAC stressed a scientific and technical approach, and indicated that its position was to address the problem rather than to take sides. CEAC believed that emotions and simplistic solutions would not resolve pollution problems. Working with the key participants in a behind-the-scenes manner was considered the approach most likely to result in improvements.

Some citizens in Elmira viewed CEAC as too complacent and cautious and too close to Uniroyal and the provincial agencies. Consequently, another group was established after May 1989 when Uniroyal announced plans to construct a toxic waste incinerator in Elmira. This group, called Assuring Protection for Tomorrow's Environment (APT), had 200 members by the spring of 1990. Its approach was more aggressive and visible than that of CEAC, and was viewed as either a saviour or a loose cannon. Strong feelings were generated against APT to the extent that vandals spray-painted the home of a high-profile APT member, as well as the car and garage of another who had criticized the way the water problems were being addressed. When APT wanted to participate in Elmira's major annual event, its Maple Syrup Festival each spring, the festival committee told its members that APT was not welcome.

A third group was created to counter the charges being presented by APT. Called FACT, or Friends Addressing Concerns Together, this group was formed primarily by Uniroyal employees and their families. FACT's position was that Uniroyal undoubtedly caused some pollution, but that it was not the only contributor, and therefore it was inappropriate to focus only upon Uniroyal and ignore other manufacturers handling chemicals in Elmira. Furthermore, FACT argued that it was not conclusive that Uniroyal was the source of NDMA found downstream in the Grand River. FACT also expressed unhappiness with the media's coverage of the NDMA problem, and suggested that the media had sensationalized matters far out of proportion.

The existence of these three citizens' groups illustrates the challenges in trying to incorporate stakeholders into a partnership approach to environmental problem solving. There was not one public interest but a variety of concerns in Elmira regarding NDMA. When those various interests were combined with the interests of the company and the different government regulatory agencies, and these were all placed in the context of the previous decade of pollution events in the community, the obstacles to creating an open, participative, and consensus-based approach became formidable.

An example of the legal or administrative aspect of the NDMA situation was disagreement about who was ultimately responsible. The provincial Ministry of the Environment issued a control order requiring Uniroyal to pay for the estimated multimillion dollar clean-up of the aquifer and for the removal of the tonnes of buried waste on company property. Uniroyal officials argued that it had not been conclusively established that the company was solely responsible for the contamination of Elmira's groundwater, and that therefore it should not be expected to cover all the costs. In addition, Uniroyal spokespersons argued that the Ministry of the Environment had to share responsibility for the tonnes of toxic waste buried on company property. In the words of a company spokesperson, 'we think the ministry has some responsibility for the buried waste because it was buried under the direction and approval of government agencies'. Uniroyal's position was that it had complied with directions and orders from the provincial government in disposing of its wastes, so if there was a problem, then the approving agency had to share responsibility for it.

Given the costs of remediation, the arguments regarding responsibility were significant. By the autumn of 1990 the Regional Municipality of Waterloo had spent some $12–13 million on Elmira's water problems, and estimated that the total cost would be $20 million. Those costs did not include the expense of cleaning up the aquifer or the wastes on the Uniroyal property. In turn, Uniroyal had spent $11 million on environmental clean-up in the six years leading up to 1990, and anticipated considerable further expenditures.

Uniroyal and its employees were also concerned about the potential implications for production and employment. In January and February 1990 the company was operating at about 60–70 per cent of its production capacity because of the need to reduce the waste water going to the Elmira sewage treatment plant. Uniroyal, employing 360 employees and thereby the largest employer in a community of about 5,000 people, had laid off seventy-five people as a result of the production cuts in the winter of 1990. Uniroyal and its employees were concerned that if all of the

'The site won't be restored during my career, maybe not for 100 years,' said David Ash, Uniroyal Chemical's manager of operations. 'The legacy of the site and past practices will be with us for the foreseeable future', Ash said, adding that one of the important challenges is to make sure that what's done now doesn't produce a negative legacy for future generations. Ash said that cleanup operations at the site since 1984, together with those expected over the next several years, are expected to cost the company about $37 million.

Kitchener-Waterloo Record,
7 June 1996

If the ongoing controversy about the cleanup has tainted the plant's image, it hasn't deterred society's demand for agricultural and industrial chemicals.

The Elmira plant is just coming off a record year for production in 1995, and worldwide, the parent company Uniroyal Chemical Inc. reported sales of $1.1 billion last year.

Kitchener-Waterloo Record,
7 June 1996

clean-up expenses were borne by Uniroyal, the company might have to alter its production process and reduce the number of employees.

Solutions

By the end of February 1993, after fourteen months of negotiations among Uniroyal and the regional and provincial governments and after various appeals and court challenges by Uniroyal, agreements were reached out of court to develop a reliable water supply for Elmira and to begin the clean-up of the contaminated aquifer. The estimated cost to Uniroyal was at least $10 million in cash payments and loss of its share of the capacity of the Elmira sewage treatment plant for its waste water. Through two agreements signed in late March 1993, the following actions were agreed to:

- The last two operating wells in the north end of Elmira were to be shut down, and the 8,500 residents of Elmira and St Jacobs would be supplied with piped water from Waterloo.
- Uniroyal would pay for the cost of a water pipeline from St Jacobs to Elmira, building a booster pumping station in St Jacobs and modifying a reservoir in Elmira. The costs were unknown, but it was estimated the total could reach $4 million.
- Uniroyal would transfer some of its capacity in the Elmira sewage treatment plant, jointly owned by the municipality and Uniroyal, to the region. This transfer of capacity by Uniroyal would allow the region to defer expansion of the Elmira sewage treatment plant. The transfer of capacity would be in the form of a one-time payment of $700,000, as well as a transfer by Uniroyal of 636 425 L a day of sewage treatment capacity to the region. The 636 425 L a day was estimated to be worth between $2.5–5 million. This arrangement was to compensate the region for costs it incurred in helping to offset the problems after NDMA was discovered in 1989.

- Uniroyal would pay the Regional Municipality of Waterloo $3 million in compensation.
- The Ministry of the Environment would pay the region to offset its costs incurred since 1989, in addition to the $8.4 million the ministry already paid to the region in 1991. The region had spent $21 million since 1989 on projects associated with the Elmira water problem and received a total of $15 million from the province.
- A plan was included to treat the contaminated groundwater. The goal was to have the groundwater suitable for drinking in thirty years. The plan included building an above-ground storage facility on the Uniroyal property, and then the tonnes of buried toxic waste on the company's property would be exhumed and stored in that building.

Progress

Several additional municipal wells were closed in late 1989, and the water supply to Elmira was supplemented with Waterloo water in early 1991. In April 1994 an upgraded pumping station in St Jacobs was commissioned, and the remaining wells in Elmira were closed. From that time, 100 per cent of the water to Elmira was supplied from Waterloo.

Uniroyal's Envirodome, constructed to store contaminated wastes from the plant (*Bruce Mitchell*).

Buried waste was recovered and sprayed with material to keep it from leaking while being transported to the Envirodome for storage (*Murray Haight*).

In terms of cleaning up the contaminated soil at the Uniroyal site, construction of the storage facility was completed during the summer of 1993, the recovery of the buried wastes began in September 1993, and the last of the waste was removed in early December 1993. Some 27 000 m³ of buried waste were excavated and placed in the special storage building. The building, about the size of a football field and called the Envirodome by local people, was erected on a 5-m-deep concrete foundation and designed to contain the waste for at least thirty years. The waste is stored on a floor of sand and two plastic liners. Vapours from it are filtered through carbon scrubbers and then discharged from the building. The estimated cost to Uniroyal for this clean-up was $4.5 million, making it one of the biggest and most costly environmental clean-ups in the history of Ontario.

In late December 1994 Uniroyal requested provincial approval to continue storing the contaminated soil in its facility for another five years. The company had been given until the end of December 1994 to submit a plan to treat or dispose of the toxic waste in storage. However, the company indicated that it needed more time to prepare a good solution. Uniroyal explained that it was not easy to find a way to treat or dispose of over 20 000 m³ of contaminated earth. The company had considered trucking the waste to another

disposal site, using bioremediation (in which bacteria would consume the chemical contaminants) and incineration. There were advantages and problems for each of these approaches, and more time was needed to make a choice. The contaminated soil was secure in the storage facility, and therefore time was needed to make the best choice.

By the end of 1994 Uniroyal had spent $12.5 million over the previous four years in the clean-up, while the province had spent $12 million. It was apparent that significant further expenditures would have to be made.

Regarding rehabilitation of the groundwater, Uniroyal considered seven alternative approaches. The preferred option was to drill eleven wells on the southwestern edge of its property, pump the polluted groundwater, treat it, and then pump the treated water to Canagagigue Creek. This approach would cost Uniroyal $725,000 to construct and $240,000 annually to operate, and was estimated to stop about 95 per cent of the contamination from getting into the creek. Uniroyal suggested that this approach was the appropriate way to begin, and the Ministry of the Environment later indicated that this level of contaminant control would meet provincial requirements.

If the wells did not do the job satisfactorily, however, then it would be necessary to drill additional wells and/or to construct a protective in-ground wall to protect the creek from contamination. The cost of including the extra wells and the wall would be $4.2 million for construction and $308,000 for annual operation. A third solution included the wells and in-ground wall, as well as building a new dam to raise the level of the creek to keep polluted water from flowing into it. That option was estimated to cost $6.2 million.

At a public meeting in Elmira during mid-January 1995 to discuss this proposal, residents from Cambridge, Brantford, Dunnville, and the Six Nations Reserve, as well as officials from the Regional Municipality of Waterloo, expressed concern about the recommended option. The main

Box 18.4 Molluscs as Indicators of Aquatic Contamination

In mid-September 1994 a dozen cages of ordinary mussels were placed in Canagagigue Creek. Dioxins can be extremely toxic, even at very low levels, but they are also difficult to detect. However, such dioxins do build up or bioaccumulate in the tissue of living creatures, whether people, fish, or shellfish.

Cages of mussels were placed at thirteen spots in the creek above, beside, and below the Uniroyal site. Mussels filter significant amounts of water. After three weeks, the mussels were examined to determine what chemicals had built up in them.

The highest levels of dioxins and breakdown products of DDT were found in mussels that had been placed in a side canal of the creek, adjacent to the southwest corner of the Uniroyal prop-

erty. The findings showed a concentration of 1.7 ppt of 2,3,7,8-TCDD, the most toxic kind of dioxin, in the mussels there. Low levels of less toxic dioxins were found in mussels placed on the east side of the creek near the northern part of the Uniroyal property. The tests did not detect PCBs (polychlorinated biphenyls) or organochloride pesticides at any of the sampling sites. No dioxins were found in the cages of mussels placed in the creek upstream of the Uniroyal property.

The use of the mussels provided important information regarding some of the most critical questions associated with the multimillion dollar clean-up at the Uniroyal plant. What dioxins were entering Canagagigue Creek? From what locations were they entering the creek?

arguments were that Uniroyal should be dealing with contaminated groundwater on the northwest and east sides of the creek by drilling the extra wells there at the beginning of the remediation program, and that the containment wall should also be built. Respondents were insistent that it was wrong to allow any contamination to enter the creek, but a Ministry of the Environment spokesperson stated that there never could be 100 per cent containment of pollution. In his words, 'we can't achieve that'. This exchange of views illustrates the difficulties and costs of groundwater pollution, as pointed out in Chapter 6.

The Uniroyal plant, the Envirodome (top left), and the city of Elmira (*Ontario Ministry of Environment and Energy/Ron E. Johnson*).

Two other aspects related to the agreements to resolve the groundwater contamination deserve mention. First, since early 1992 the Uniroyal Public Advisory Committee participated in the discussions regarding clean-up plans. This committee included representatives from health, business, conservation, and community groups. The Ministry of the Environment and Uniroyal had non-voting representatives on the advisory committee.

A Uniroyal spokesperson commented in July 1993 that the advisory committee had been a learning experience for everyone. He hoped that the company was learning to be a good listener and to understand the concerns of the committee members. In turn, he hoped that committee members were learning more about the company, its interests, and motivations. Various other members of the committee agreed that the entire clean-up initiative was predicated on *trust*. However, they believed that trust would take some time to build, especially since the company had resisted clean-up orders through the courts during 1991 and 1992, and in 1992 had been convicted of violating the Fisheries Act due to polluting Canagagigue Creek and had paid a $16,000 fine.

This trust was further tested in March 1994 when Uniroyal told the Public Advisory Committee that an underground pool of 45 500 000–182 000 000 L of toluene-based liquid was contaminating the groundwater and moving into the Canagagigue Creek. The *Kitchener-Waterloo Record* reported that Uniroyal had started investigating the problem in July 1993, and by December 1993 had considerable information about the pool of toluene-based liquid. Uniroyal's manager of manufacturing explained in March 1994 that the company's lawyers had advised the company to study the problem further before making the information public. This action led the vice president of APT Environment, one of the public interest groups represented on the Public Advisory Committee, to write to the *Kitchener-Waterloo Record* to comment that:

> . . . at the March 28 meeting we were appalled to discover that Uniroyal withheld information about a seriously contaminated area from us and from the ministry for several months, as the company sought legal advice. It looks as if Uniroyal was seeking to protect itself rather than the environment.
>
> To us, the issue is betrayal of trust. We wonder what other information the company is withholding and why the ministry was unaware of the problem. . . . As members of the advisory committee, we have to trust in the integrity of information provided by the company and its consultants.
>
> This will no longer be possible after the shock we received on March 28. It was like being at a green light and trusting the cars on the red light to stop. Uniroyal ran the red light and lost our trust. It will be difficult to keep on driving forward (*Kitchener-Waterloo Record*, 5 April 1994:A5).

Second, arrangements were made for an 'independent' person to monitor the clean-up operation for the buried toxic wastes on the Uniroyal property. A professor at the University of Waterloo, a specialist in waste management and also a resident of Elmira, was seconded from the university to be a 'watchdog'. In this role he worked for the provincial Ministry of the Environment, but his salary was paid by Uniroyal. Under this arrangement, the professor was bought out of his duties at the university for the autumn of 1993 so that he could be a full-time observer of the waste recovery operations at the Uniroyal site in Elmira. The idea behind this arrangement was to ensure that the clean-up process was perceived as credible and sensitive to the residents of Elmira.

IMPLICATIONS

The experience with contaminated groundwater at Elmira highlights many of the issues addressed in this book. It certainly emphasizes the complexity, uncertainty, and conflict that often characterize much environmental management. The lack of agreed standards for drinking water in general and

---

Box 18.5 Water Pollution

- Minimum levels of nutrients, particularly nitrogen and phosphorus, are essential for plant production in aquatic systems (see Chapter 6).
- Too much nutrient leads to algae blooms and eventually to eutrophication (Chapter 7).
- Water pollution originates from both point (e.g., municipal sewage treatment plants, factories) and non-point (e.g., farm fields, your lawn) sources; the latter are more difficult to regulate.
- Since the early 1970s, programs have been implemented to reduce phosphorus content from detergents and sewage plants. After ten years of such controls, levels of phosphorus in the Bay of Quinte, close to Kingston, Ontario, dropped by nearly 50 per cent.
- Toxic contaminants are increasingly of concern. Some toxics of greatest concern are dioxins, furans, PCBs, and pesticides. Heavy metals, such as mercury, often produced as a by-product of pulp and paper production, are also of concern (Environment Canada 1991:3-15 to 3–22).

NDMA in particular, and the conflicting views of toxicological experts, underlined the inadequate understanding characteristic of environmental problems. The concerns from downstream communities illustrated that ecosystems exist at different scales, varying from the aquifer underlying Elmira, the creek draining from the Uniroyal property into the Grand River, and the Grand River itself. As a result, it is not always clear which is the most appropriate ecosystem to use in environmental management. Conflict resolution was central in the Elmira groundwater situation. When issues become polarized and trust has been dissipated, it becomes a challenge to settle disagreements. In this situation, while the courts were used, an out-of-court settlement was ultimately reached.

The situation also illustrates the practical implications of the law of conservation of matter discussed in Chapter 6, where it was pointed out that matter will always exist somewhere, and that it cannot be consumed but merely changed in form. In this case, chemical transformations have taken place to create the toxic pollutant NDMA. The pollutant dissolves, reflecting one of the major properties of water discussed in Chapter 6, and is carried downstream as part of the hydrological cycle to affect other users. A major challenge for industrialized society is obtaining the benefits of the goods being produced while minimizing the environmental damage created by the waste products. Whenever you buy a product, think of all the environmental goods and services (e.g., raw materials and the waste disposal provided by water or the atmosphere) required for its production. The actions of countless numbers of people, making consumer decisions that increase impacts of these waste products, challenge the environment's ability to absorb pollutants.

Reducing consumption is one of the most effective ways to deal with the problem.

MULTIPLE-PURPOSE DEVELOPMENT

The Rafferty-Alameda project in southern Saskatchewan and the Oldman Dam in southern Alberta provided important testing grounds for environmental impact assessment processes and procedures in the late 1980s and early 1990s. The experience with the Rafferty-Alameda project, announced by the government of Saskatchewan in

Rafferty Dam under construction (*Ecologistics Ltd*).

Rafferty Dam's completed spillway (*SaskPower*).

the summer of 1986, is examined here because this regional project became a landmark test of environmental assessment procedures, particularly at the federal level. The Rafferty-Alameda project was characterized by the Canadian Press as 'one of the longest and most convoluted battles in Canadian environmental law'.

Background

The Souris River begins in southeastern Saskatchewan, flows south into North Dakota and through the city of Minot, before turning north and re-entering Canada in southwestern Manitoba (Figure 18.3). The Rafferty-Alameda project, conceived to control the flow of the Souris River, was designed to have multiple purposes: flood control, particularly in North Dakota; water for irrigation of 6500 ha, and cooling for a new $600 million 300-megawatt coal-fired power station; and wetland systems to enhance recreation and wildlife habitat in southeastern Saskatchewan.

The project included two dams in Saskatchewan. The Rafferty Dam was to be built across the Souris River near Estevan, while the Alameda Dam was to be constructed on Moose Mountain Creek, a tributary of the Souris, close to the village of Alameda. The Rafferty reservoir would extend 57 km and flood 4900 ha, while the Alameda reservoir would be 23 km long and cover 1240 ha. The two reservoirs would provide a fivefold increase in the water-storage capacity in the basin. Such storage was considered important since flows of the Souris River and Moose Mountain Creek are highly variable and unpredictable (figures 18.4a and 18.4b). The highest annual recorded flow for the Souris River is about 600 times more than the lowest flow, and over 80 per cent of the flow occurs in the spring (from March to the end of May) as a result of snow-melt. The erratic flow of the river in the Souris basin had led to a proposal as early as 1907 for a dam on Moose Mountain Creek. The initial estimated cost of the Rafferty-Alameda project was $120 million, with the United States paying one-third of the costs in return for the flood control benefits it would receive.

Constructing water-control structures on the Souris River in southeastern Saskatchewan was not a new idea. Virtually every flood in the Souris River Valley had been followed by demands for flood-control structures, and each drought inevitably led to calls for measures to store and redistribute water, which was perceived as 'wasted' each time the river flooded. Prolonged drought during the 1930s stimulated renewed requests for water-control structures, and it was during the Depression that projects similar in magnitude to the Rafferty-Alameda project were proposed. In contrast, the 1970s were relatively wet on the Prairies, with major floods occurring in 1969, 1974, 1975, 1976, and 1982. Those floods, followed by the return of serious drought conditions for the rest of the 1980s, triggered further pressure for a water-control system in the Souris River basin.

A new provincial government was elected in Saskatchewan in 1982. Grant Devine, the premier, was elected in the riding of Estevan, in which the Rafferty Dam would be built. Eric Berntson, the deputy premier, was the representative from the adjacent riding of Souris-Cannington, in which the Alameda Dam was planned. Substantial support for both men came from groups that had long advocated the concept of water control in the Souris basin (McConnell 1991). During the election campaign, Devine had promised that if his party formed the government, it would develop the region.

By 1984 planning for the Rafferty-Alameda project was underway, and during 1985 it was registered with Environment Canada's Regional Screening and Coordinating Committee. At the same time, planning was initiated for a new coal-fired power station in the Estevan area. In 1986 the provincial government gave final approval for the Shand Thermal Generating Station. Water from the two reservoirs would provide cooling water for the generating plant.

Initial Review and Approval

In March 1986 the provincial government created the Souris Basin Development Authority (SBDA) as a Crown corporation to plan and build the Rafferty-Alameda project. The SBDA prepared an 1,800-page environmental impact statement (EIS) and submitted it in August 1987 to the Saskatchewan minister of the environment and public works. In September an independent board, appointed by the government, held public meetings to review the EIS. The board recommended that the project should proceed, but specified con-

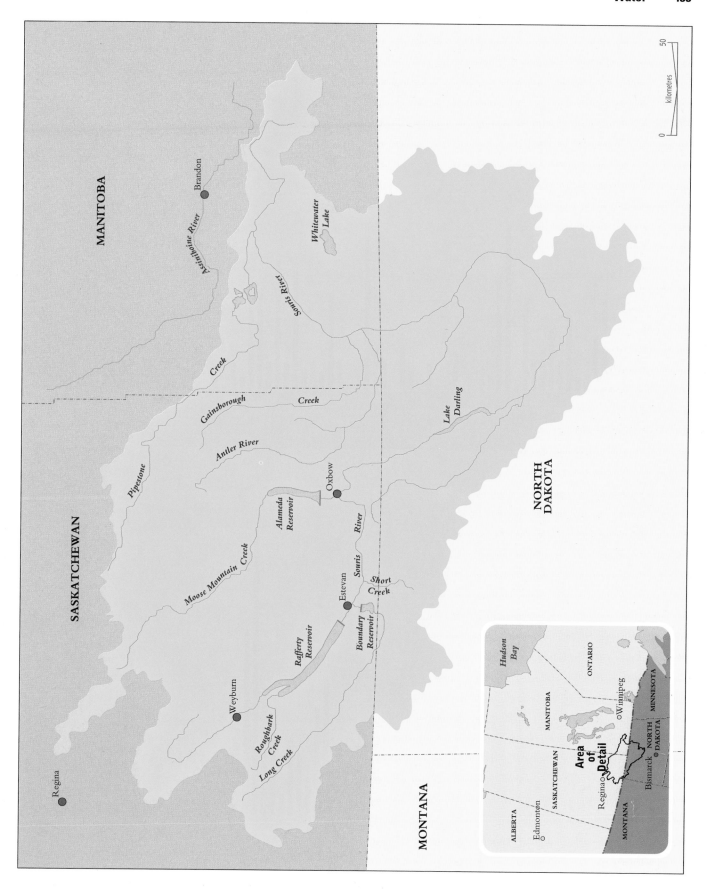

Figure 18.3 Souris River basin and the Rafferty-Alameda project. SOURCE: 'Water Management Plan for the Souris River Basin, Saskatchewan', April 1990.

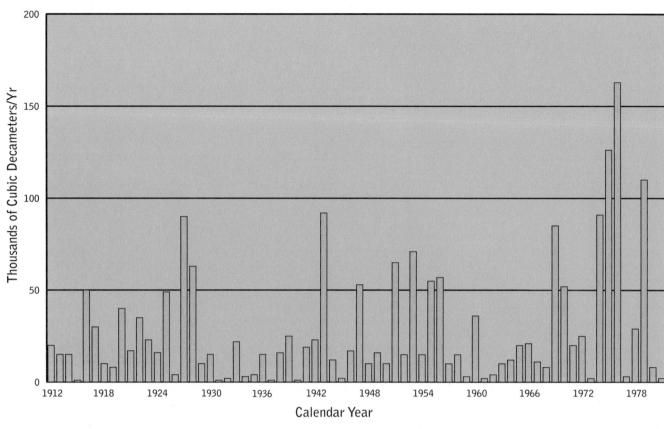

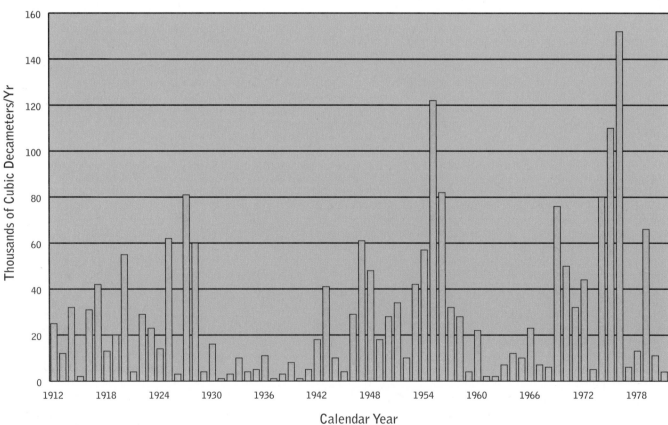

Figure 18.4 (a) Souris River near Rafferty, naturalized flows (*top*). (b) Moose Mountain Creek near Alameda, naturalized flows (*bottom*). SOURCE: Federal Environmental Assessment Review Office, *Rafferty-Alameda Project: Report of the Environmental Assessment Panel* (Ottawa: Federal Environmental Assessment Review Office, 1991):19.

ditions. The minister of the environment subsequently authorized construction to begin with twenty-three conditions to mitigate environmental impacts (Box 18.6).

The Rafferty-Alameda project also needed approval from the federal government because the Souris is an international and interprovincial river. The project was considered under the 1955 International River Improvements Act, which gives the federal minister of the environment authority to issue licences for water-management activity that might affect water quality and quantity beyond international boundaries. Environment Canada coordinated the federal review of the project's environmental implications. Following that review and consultations with representatives from agencies in Saskatchewan, Manitoba, and the United States, the federal minister issued a licence in June 1988.

However, in January 1989 the Canadian Wildlife Federation and two area farmers filed an application with the Federal Court to have the federal licence revoked. Their main argument was that the federal government's environmental assessment and review process (EARP) had not been properly applied. In addition, they were concerned that the project would harm water quality downstream and flood valuable wildlife habitat. In response, the federal government argued that during 1987 and 1988 the provincial government had conducted an environmental assessment to satisfy Saskatchewan's environmental assessment process,

and that the federal minister of the environment had concluded that the provincial process had met the federal EARP requirements. The federal government contended that it also had completed its own environmental review before issuing the licence.

The Federal Court decided that the provincial review had not satisfactorily met the responsibilities of the federal government, particularly regarding potential environmental impacts *outside*

Box 18.6 Significant Conditions from the Saskatchewan Minister of the Environment in Approving the Rafferty-Alameda Project

- no net loss of wildlife habitat as a result of the development
- establishment of an ecological reserve to protect unique natural areas
- extraction and stockpiling of all economically extractable gravel
- replacement of community pasture lands lost due to the development
- completion of detailed implementation plans regarding water management, water-quality monitoring, parks, and fisheries

---

Shared Responsibilities for Water Resources in Canada

The [Canadian] constitution does not mention water. It does deal with some water uses, such as navigation, fisheries and, more recently, electrical energy generation. But most questions of jurisdiction must be inferred from the constitution's treatment of other issues, like property rights, criminal law, foreign relations, and so on. Moreover, few jurisdictional disputes have been litigated in Canada, no doubt partly because most of the population has relatively abundant water supplies. The combination of indirect reference to water in the constitution and limited guidance from the courts makes it impossible to define precisely the respective roles of the federal and provincial governments in water management.

. . . The provinces' law-making power over water derives from specific clauses in the constitution that assign to them jurisdiction over 'property and civil rights' and 'the management and sale of public lands'. Water is traditionally regarded as a form of property, and the term 'public lands is taken to include water: so these provisions convey a general provincial power over water'.

. . . The constitution provides for exclusive federal jurisdiction over some matters that bear on water management. Among these are the federal authority over fisheries, navigation, relations with foreign governments, federal lands, Indians, works for 'the general advantage of Canada' and 'peace order and good government'. The authority to legislate on these matters limits the provinces' authority over water.

[Another matter] . . . is the authority of the federal government to intervene where waters flow from one province or territory into another, or along their common boundaries. This is an important issue because many lakes and rivers abut or cross boundaries within Canada, and because activities in one jurisdiction can impair the flow or quality of water available to other jurisdictions (Pearse, Bertrand, and MacLaren 1985:63–5).

Saskatchewan (see Box 2.11). The court indicated that such potential impacts had to be considered under the federal EARP procedure. As a result, the Federal Court revoked the licence and ordered the federal minister of the environment to conduct a review, following the procedures stipulated under EARP.

Subsequent Review and Conflict

During the summer of 1989 the federal government initiated a review of the Rafferty-Alameda project, as directed by the court. Following the EARP guidelines, Environment Canada completed an initial environmental evaluation and organized public meetings in Saskatchewan, Manitoba, and North Dakota to discuss the results. Based on the findings from this process, the federal minister of the environment issued a new licence, with conditions, in August 1989. Overlapping this work were negotiations between Canada and the US regarding management of the Souris River. Those negotiations led to a signed agreement in October 1989 between the Canadian and American governments in which terms for both water allocation and quality at the international border were set out. The United States also agreed to pay $41.1 million (1985 US dollars) for the flood-control storage, which would be provided by the Rafferty and Alameda dams.

With the licence issued and the agreement signed with the United States, the federal government decided that the second phase under EARP, a public review by an independent environmental assessment panel, was unnecessary. However, the Canadian Wildlife Federation and the two residents appealed this decision in a second court action during October 1989. The outcome was a Federal Court decision stating that an independent panel review had to be completed.

A five-person environmental assessment panel was appointed by the federal minister in January 1990. An agreement was also signed between the governments of Saskatchewan and Canada to suspend work on the project while the independent review was conducted. The federal government agreed to pay $1 million per month to Saskatchewan to compensate for the delay, up to a maximum of $10 million. However, it was agreed that construction could continue until the safety of the structure was secure.

As the review continued, it became obvious that the federal government and Saskatchewan had different views regarding the nature of construction activity to ensure the safety of the Rafferty Dam. The review panel repeatedly expressed its concerns to the federal minister about the ongoing construction, and then suspended its work in mid-October 1990. The following day, when the Saskatchewan government announced that construction on *all* aspects of the project would continue without delay, the panel resigned. The federal government announced that it would obtain a court injunction to stop work on the project. However, in mid-November 1990 the Saskatchewan Court of Queen's Bench denied the injunction.

A new, three-person review panel was established by the federal minister of the environment in early February 1991 to review the environmental and social impacts of the project. Construction continued on both projects. The second review panel submitted its report in September 1991. The panel concluded that the flood-control objectives in North Dakota would be satisfied, legal obligations regarding required flows for the North Dakota apportionment would be met, and cooling water for the Shand power station would be supplied much of the time. The panel noted that there was considerable uncertainty regarding whether the other objectives would be met, including fulfilment of future irrigation demands, achievement of water-quality objectives at the international border, and provision of water for wildlife habitat and recreation. The panel also commented that another drawback was that the project had proceeded to the advanced stage before either the first or second review panels became involved, which was viewed as inconsistent with the idea that impact assessment should begin early in project planning.

Construction on the Rafferty Dam was completed in the spring of 1991, while work had started in October 1990 on the Alameda Dam. The Alameda Dam was 90 per cent completed by September 1993. By that time, the water-reservoir level behind the Rafferty Dam had reached a depth of 0.66 m in a dam designed to create a reservoir 15 m deep. There were substantial differences of opinion regarding how long it would take the reservoir to fill to its design depth, with some estimates extending up to thirty years.

Implications of the Rafferty-Alameda Project

As discussed in Chapter 10, the government of Canada deliberately chose to use a non-legislative approach regarding environmental impact assessment when it introduced its procedures in 1973 and then revised them in 1984. This choice was a conscious attempt to avoid what were perceived as undue delays in the United States after it introduced legislation in 1969 requiring environmental assessments. In that way, the federal government in Canada explicitly intended to build considerable discretion and flexibility into the federal assessment procedure. From a positive perspective, this choice incorporated basic ideas of an *adaptive management approach*. On the other hand, it also created a process that could lead to inconsistencies and inequity, since various projects could be treated in quite different ways.

According to the chairperson of the federal environmental review process, the relatively small Rafferty-Alameda project became symbolic. In his words, 'It's a symbol of: Does the government follow the law strictly, or does it not' (*Vancouver Sun*, 16 October 1990:B4)? The interpretation of the Federal Court significantly altered the original intent of the federal review process. What had been conceived and used as a set of *discretionary* guidelines was interpreted by the courts as regulations that had to be implemented. As a result, guidelines were transformed into firm and compelling orders legally binding on the federal government and any project over which it had jurisdiction.

By December 1990 the chairperson of the Federal Environmental Assessment Review Office stated that the transformation of guidelines into legally binding orders had created an impossible situation. In his view, the change implied that the federal government literally could become involved in every culvert in the land. He commented that 'We are about to appoint a review panel for a $150,000 dump in Nova Scotia and the process we'll have to follow will be the same as for Grande Baleine in Quebec, an $8-billion project. It's insane' (*Kitchener-Waterloo Record*, 15 December 1990:A8).

A second major dilemma of the initial design of the assessment process was a perception of unfairness because of inconsistency in handling different projects. Premier Devine was particularly

---

Box 18.7 Impact Assessment: Mandatory or Discretionary?

In the first court case, the Federal Court ruled that the federal environmental assessment guidelines were mandatory rather than discretionary and had to be followed. This meant that many more projects would have to be reviewed under the EARP procedure. The second judgement upheld a petition for a full public review, even if construction had to be halted on what had become a $140 million dollar project, which by that time was 60 per cent completed.

---

angry when he heard in late October 1990 that the federal government had decided it would not complete an environmental review of the $350 million Oldman Dam in Alberta until after that project was completed. Devine argued that 'This latest development demonstrates the complete [*sic*] inconsistent and unfair manner by which the federal bureaucracy applies the ill-conceived Environmental Assessment Review Process guidelines' (*Saskatoon Star-Phoenix*, 25 October 1990:A14). He went on to note that other major projects, such as the $800 million Alcan Kemano hydroelectric project in British Columbia and the Port Arconi project in Nova Scotia, were exempt from the federal assessment process.

This inconsistent application of the assessment process was also noted at the same time regarding the federal minister of the environment's decision not to object if Hydro-Québec began work on roads and airports for phase 2 of the James Bay hydro complex before an environmental review of the entire project was completed (see Chapter 19). The minister had stated that he had no problem with splitting the environmental assessment into two parts. Critics of this decision argued that such an approach would preclude a credible environmental review of the entire project. One critic suggested that once the infrastructure to service the main dams was constructed, it would be virtually impossible to stop the entire project, regardless of the findings of an environmental assessment. Bill Namagoose, a spokeperson for the Grand Council of Crees, said 'Can you imagine that after spending $600 million on infrastructure they [the Quebec government] will accept a ruling that the project not proceed' (*Saskatoon Star-Phoenix*, 26 October 1990:A17)?

Other uncertainties arose because of the discretionary and flexible nature of the provincial and federal assessment processes. For example, McConnell (1991) remarked that early critics of the Rafferty-Alameda project had difficulty in challenging the provincial assessment findings because there were no published regulations or guidelines for environmental impact statements or boards of inquiry in Saskatchewan. Consequently, an impact statement's adequacy was difficult to challenge, and it was equally problematic to raise requests for the board of inquiry to seek or confirm any information for an environmental assessment.

As an example of a data problem and different interpretations of data, McConnell cited the issue of whether or when the two reservoirs would reach their storage capacity. The historical data used by the Souris Basin Development Authority (SBDA) in its environmental assessment ended with the last significant run-off year (1982) in the 1980s and excluded all of the subsequent dry years in that decade. Using such data, the SBDA simulation model indicated a 60 per cent probability that the Rafferty reservoir would be more than half full within ten years, and the Alameda would be more than half full within five years. In contrast, Environment Canada's analysis in 1987, using a different data set, concluded that under the SBDA simulation model, the Rafferty reservoir might never reach its full supply level during its design life. Furthermore, anticipated climate change and decreased precipitation levels on the Prairies (as discussed in Chapter 7) reduced the likelihood of the reservoirs filling as predicted by the SBDA model.

The problems arising from the discretionary nature of the assessment process, and the questionable data and assumptions used in analysis, led to sharp criticism of the assessment process. A newspaper columnist commented that the Federal Court's interpretations of the federal assessment process had resulted in 'All of a sudden, federal-provincial relations [were being] sent topsy-turvy' (*Saskatoon Star-Phoenix*, 25 October 1990:A4). As a result, the Rafferty-Alameda project became a significant test for environmental assessment in

---

Impact Assessment: A Flawed Process?

Canada's environmental review process is costly, inefficient and frustrating. . . .

Neither side is blameless. Saskatchewan should not have flouted Ottawa's prohibition order. And Ottawa should have figured out long before now that the nearly complete dam 'is causing and will continue to cause irreparable harm' to the environment (in the words of today's court challenge).

But the whole process is flawed. Ottawa cannot interrupt a multimillion-dollar project repeatedly without creating tension and disrespect for federal authority (Saskatoon Star-Phoenix, *26 October 1990:A4).*

Canada. In Part E we will consider the implications of this test, particularly against the evaluative criteria for assessment presented in Chapter 10 and the ideas about adaptive environmental management considered in Chapter 9.

IMPLICATIONS

Two case-studies from a country as diverse as Canada provide a fragile basis on which to generalize. Nevertheless, regarding the key themes of change and challenge in this book, there are several important considerations. The experiences in Elmira and southeastern Saskatchewan highlight the considerable uncertainty that often prevails regarding scientific understanding of natural systems. In Elmira, discovery of NDMA was a surprise. Subsequently, scientists had difficulty establishing its source, agreeing upon a safe standard, and developing an appropriate solution. At Rafferty-Alameda, there were different views regarding the likelihood that the reservoirs behind the two proposed dams would be filled, given the erratic and highly variable flow in the Souris River. Both these examples relate to the system diagram in Chapter 1 (Figure 1.6), indicating the importance of understanding the natural systems involved before trying to control those systems. The simplified models of the systems proved inadequate as a basis upon which to predict the impact of various management interventions.

The two examples also illustrate the difficulty in selecting an appropriate ecosystem for planning in management. In Elmira, candidate ecosystems were the aquifer beneath Elmira, the Canagigue

Box 18.8 What You Can Do

Many areas in Canada suffer from water shortages. Traditionally we have tried to address water deficits through *supply management*; that is, increasing the supply of water available for use to humans by increasing storage capacities and diverting water from one area to another. These management strategies lead to ecological disruptions and encourage wasteful use of this precious resource. Increasing attention is now being given to *demand management*, in which the emphasis is on reducing the demand for water. There are many things you can do as an individual to help:

- Don't keep the water running while you are brushing your teeth or shaving.
- Reduce the amount of water used by installing water-saving toilets that use as little as 6 L per flush (compared with the 20 L or so used by conventional toilets). Put a plastic container weighted with stones in the tank of existing toilets to reduce tank capacity.
- Install water-saving shower heads.
- Wash only full loads in your washing machine. When choosing a new machine, make water consumption a main factor in your decision.
- If using a dishwasher, use only for full loads and with the air-dry cycle.
- Create a garden landscape that needs little watering rather than extensive areas of lawn.
- Check for leaks frequently. A tap leaking only one drop per second wastes more than 25 L of water a day or 9000 L per year. The cause of a leak is often a worn-out washer, which costs less than 10 cents to replace.

Creek, or the entire Grand River basin. For the Rafferty-Alameda project, the government of Saskatchewan limited its impact assessment to that portion of the Souris River within its provincial boundaries, whereas the Federal Court concluded that federal responsibility extended to downstream impacts in both North Dakota and Manitoba. In contrast, the Canadian Wildlife Federation was concerned about the protection of wetland systems, which provided habitat for migratory birds whose movements extended from the Canadian Arctic to the southern United States.

The appropriateness of adaptive environmental management as an approach has some strong challenges based on these experiences. In Elmira, the regulating agency, the Ontario Ministry of the Environment and Energy, had changed its standards and regulations over the years, an approach consistent with adaptive environmental management. However, when the NDMA contamination was detected, Uniroyal argued that it had always met the prevailing standards set by the ministry, so that the provincial regulatory agency should have to bear some of the responsibility for clean-up costs. In this manner, while it is appropriate to change practices as experience is gained, a problem arises when a firm alleged to be causing pollution can counterargue that contemporary standards

cannot be used to judge past practices. For the Rafferty-Alameda project, the assessment guidelines had been generalized to provide discretion and flexibility to meet specific circumstances. The weakness of that approach was the lack of clear and explicit statements about process and expectations, making it difficult to know when the guidelines had been satisfied.

Both experiences indicate the opportunities and pitfalls for a stakeholder approach. In Elmira, there were several organized stakeholder groups, some of which were formed because of dissatisfaction with what existing groups had been doing or not doing. This situation emphasizes that there is not a single public interest but many interests and numerous stakeholders. Rafferty-Alameda confirmed this situation, and also showed the potential complications when some groups have local concerns while others have concerns that are regional, national, and even international. In such situations, devising an appropriate way to achieve conflict resolution becomes a major challenge.

Finally, the experiences at Elmira and Rafferty-Alameda provide an opportunity to consider whether they represent sustainable development, as discussed in Chapter 2. More specifically, you are invited to consider the extent to which these two examples, and others in the country, reflect the sustainability principles for water management in Canada that have been developed by the Canadian Water Resources Association (Box 18.9). Which of the principles were satisfied in these experiences? What changes would have to occur for all of the principles to be satisfied? Do you have any ideas for modifications or additions to these principles? What can you do as an individual in your daily life to try and reflect these principles?

SUMMARY

1. Canada has 9 per cent of the world's fresh water, but less than 1 per cent of its population.

2. On average, Canadians use 360 L of water per capita each day, making us the second highest (after the United States) per capita water users in the world.

3. About one in four Canadians depends on groundwater for domestic water supplies.

4. Contamination of groundwater can threaten municipal or individual water supplies, as illustrated by the experience in Elmira, Ontario.

5. When N-nitroso dimethylamine (NDMA) was discovered in the municipal water supply of Elmira, officials found that there were no drinking water guidelines for this chemical (a known carcinogen), and that there was wide-ranging advice from experts as to what was an acceptable amount of NDMA to have in water. There was also disagreement among technical experts regarding an appropriate method for testing the waste water in Elmira for NDMA.

6. The Elmira experience emphasizes that water-quality issues are usually not just technical matters but also involve legal, economic, social, and administrative considerations.

7. The main source of the NDMA contamination was the Uniroyal chemical plant in Elmira, one of

---

Box 18.9 Sustainability Principles for Water Management in Canada

SUSTAINABILITY ETHIC

Wise management of water resources must be achieved by a genuine commitment to:

- ecological integrity and biological diversity to ensure a healthy environment;
- a dynamic economy; and
- social equity for present and future generations.

WATER MANAGEMENT PRINCIPLES

Accepting this sustainability ethic, we will:

1. Practice integrated water resource management by:

- linking water quality, quantity and the management of other resources;
- recognizing hydrological, ecological, social and institutional systems; and
- recognizing the importance of watershed and aquifer boundaries.

2. Encourage water conservation and the protection of water quality by:

- recognizing the value and limits of water resources and the cost of providing it in adequate quantity and quality;
- acknowledging its consumptive and non-consumptive values to both humans and other species; and
- balancing education, market forces and regulatory systems to promote choice and recognition of the responsibility of beneficiaries to pay for the use of the resource.

3. Resolve water management issues by:

- employing planning, monitoring and research;
- providing multi-disciplinary information for decision making;
- encouraging active consultation and participation among all affected parties and the public;
- using negotiation and mediation to seek consensus; and
- ensuring accountability through open communication, education and public access to information.

Source: Canadian Water Resources Association (1994).

the main employers in the community, leading to debates about the appropriate balance between protecting jobs versus protecting the environment.

8. Social divisions also arose in the community regarding groundwater contamination between supporters of the chemical company and environmental groups. There was also disagreement between two environmental groups regarding the best approach for the contamination problem. Many people in the community were critical of the media's role, believing that the media had been irresponsible in oversensationalizing aspects of the incident.

9. The Elmira incident also emphasized that decisions or actions in one part of a watershed (or ecosystem) can have implications for other places in the watershed. Downstream communities were concerned that NDMA from the Elmira waste water was appearing in their stretches of the river on which they were dependent for their water supply.

10. A partnership approach to resource and environmental management requires trust among partners. Members of the citizens' Public Advisory Committee explained that it was difficult for them to trust officials from Uniroyal because of the legalistic approach they took in dealing with the groundwater contamination problem.

11. The Rafferty-Alameda project in southern Saskatchewan was a benchmark test of federal environmental impact assessment procedures.

12. The Rafferty-Alameda project was designed with multiple purposes: flood control, irrigation, industrial cooling water, and wetland enhancement.

13. The two reservoirs, estimated to cost $120 million, would increase storage capacity five times in the basin. The United States would pay one-third of the cost in exchange for flood-control benefits for North Dakota.

14. The two reservoirs were constructed in the electoral ridings of the provincial premier and the deputy premier.

15. Planning was underway by 1984. Provincial approval was received in 1987, and federal approval was given in 1988. In January 1989 the Canadian Wildlife Federation challenged the federal approval in the courts, claiming that the federal environmental impact process had not been followed properly. This challenge was upheld, and the court directed the federal government to complete an environmental assessment.

16. While the federal environmental assessment was being conducted, construction continued on the Rafferty Dam.

17. Major issues related to the Rafferty-Alameda project were: (1) whether environmental assessment procedures should be viewed as guidelines or as legally binding requirements, and (2) how concerns about lack of fairness can be avoided if discretion is allowed in the application of guidelines when some projects appear to have more stringent conditions placed on them.

18. The Rafferty-Alameda project also demonstrated how selectivity in choice of data can lead to startlingly different conclusions. Two separate reviews used markedly different time periods regarding water flows for their calculations, leading to strongly divergent conclusions about whether or if the reservoirs would be filled by natural inflow.

REVIEW QUESTIONS

1. What is the myth of superabundance regarding water in Canada? Do you think such a myth exists in your area?

2. What challenges does the groundwater experience in Elmira raise about determining the most appropriate ecosystem on which to base management decisions? What are the implications of the Rafferty-Alameda project in this regard? What ecosystem do you think is the most appropriate in your community or region?

3. Why is there considerable uncertainty regarding appropriate standards for drinking water quality or the variability of flow in a river system? How vari-

able or reliable is the water flow in the rivers in your area?

4. What are some of the difficulties in determining who should be responsible for the costs of cleaning up a site that was contaminated many years earlier? Do you think there are contaminated sites in your community? If you are not sure, how would you go about determining if there are any?

5. What are some of the opportunities and problems when a number of citizens' groups are established and all of them want to participate in an environmental management situation? Do you have a number of citizens' or non-governmental organizations in your community or area? If so, do they work together or compete with one another?

6. Why has the Rafferty-Alameda project been viewed as a landmark regarding environmental assessment in Canada?

7. What were the aspects of due process of concern in the debate over the assessment for the Rafferty-Alameda project? What aspects regarding due process do you think should be included in an environmental impact assessment process?

8. What is the best time for an environmental assessment to be initiated relative to development of a project, program, or policy? Have impact assessments been used in your area? Were they initiated at an appropriate time?

9. Is impact assessment best implemented on the basis of a statute or on discretionary guidelines? What are the arrangements in this regard for impact assessment in your province or territory?

10. What are the implications of the experiences with the Elmira groundwater contamination and the Rafferty-Alameda project for the use of adaptive environmental management? Are there examples of adaptive environmental management in your community or area? Do you think they have been successful?

REFERENCES AND SUGGESTED READING

Canadian Water Resources Association. 1994. 'Sustainability Principles for Water Management in Canada'. *Water News* 13, no. 2.

Dominion Ecological Consulting Ltd. 1989. *Rafferty-Alameda Project: Summary Report of the Initial Environmental Evaluation*. Ottawa: Environment Canada, Conservation and Protection.

Environment Canada. 1975. *Canada Water Year Book, 1975*. Ottawa: Information Canada.

_____. 1986. *Canada Water Year Book, 1985: Water Use Edition*. Ottawa: Minister of Supply and Services Canada.

_____. 1991. *The State of Canada's Environment*. Ottawa: Minister of Supply and Services Canada.

Foster, H.D., and W.R.D. Sewell. 1981. *Water: The Emerging Crisis in Canada*. Ottawa: Canadian Institute for Economic Policy.

Healy, M.C., and R.R. Wallace, eds. 1987. *Canadian Aquatic Resources*. Ottawa: Minister of Supply and Services Canada.

Hood, G. 1992a. 'Developing Communication Strategies for Major-Project Development: The Rafferty-Alameda Project'. In *Changing Political Agendas*, edited by M.L. McAllister, 126–41. Centre for Resources Studies Proceedings no. 25. Kingston, ON: Centre for Resource Studies, Queen's University.

_____. 1992b. 'Upstream All the Way: The Rafferty-Alameda Project'. *CRS Perspectives* no. 41:2–13. Kingston, ON: Centre for Resource Studies, Queen's University.

_____. 1994. *Against the Flow: Rafferty-Alameda and the Politics of the Environment*. Saskatoon: Fifth House.

McConnell, J.G. 1991. 'The Rafferty-Alameda Project in Relation to Environmental Impact Assessment'. *Operational Geographer* 9, no. 4:14–18.

Mitchell, B., and D. Shrubsole. 1994. *Canadian Water Management: Visions for Sustainability*. Cambridge, ON: Canadian Water Resources Association.

Pearse, P.H., F. Bertrand, and J.W. MacLaren. 1985. *Currents of Change: Final Report of the Inquiry on Federal Water Policy*. Ottawa: Environment Canada.

Science Council of Canada. 1988. *Water 2020: Sustainable*

Use for Water in the 21st Century. Ottawa: Science Council of Canada, Publications Office.

Shpyth, A.A. 1991. 'An Ex-post Evaluation of Environmental Impact Assessment in Alberta: A Case Study of the Oldman River Dam'. *Canadian Water Resources Journal* 16, no. 4:367–79.

Shrubsole, D., and B. Mitchell, eds. 1997. *Practising Sustainable Water Management: Canadian and International Experiences*. Cambridge, ON: Canadian Water Resources Association.

Stolte, W.J. 1993. 'The Hydrology and Impacts of the Rafferty-Alameda Project'. *Canadian Water Resources Journal* 18, no. 3:229–45.

_____, and M.H. Sadar. 1992. 'The Application of the Canadian Environmental Assessment Process to Dams: A Case Study of the Rafferty-Alameda Project'. In *Resolving Conflicts and Uncertainty in Water Management*, edited by D. Shrubsole, 8.1–8.17. Cambridge, ON: Canadian Water Resources Association.

_____, and M.H. Sadar. 1993. 'The Rafferty-Alameda Project and Its Environmental Review: Structures, Objectives and History'. *Canadian Water Resources Journal* 18, no. 1:1–13.

Sundstrom, M. 1994. 'Oldman River Dam'. In *Public Issues: A Geographical Perspective*, edited by J. Andrey and J.G. Nelson, 221–37. Department of Geography Publication Series no. 41. Waterloo, ON: Department of Geography, University of Waterloo.

Chapter 19

Energy

Energy is as essential as water and food for the survival and well-being of humankind. However, ever-increasing rates of fossil fuel use have resulted in many environmental- and health-related problems, which appear to be threatening the very life-support systems of local and global environments (see chapters 7 and 20 regarding climate change).

Canadians have become particularly dependent upon fossil fuels for transportation in this huge country, for industrial production, and for heating in the generally cold winters (Figure 19.1). Canadians are also among the highest per capita energy consumers in the world. Does our pattern of energy use have to be what it has been? Not necessarily. While total energy consumed in Canada tripled between 1958 and 1992, most of this increase had occurred by 1975 when energy prices were relatively low and economic growth was high (Environment Canada 1994).

It is difficult for Canadians to know what to believe about our energy needs and supplies. As

Although Canadians use energy in the same ways as people in other industrialized countries, on a per-person basis Canada consumes more energy than any other country in the world, except Luxembourg. This can be attributed to a number of factors: vast distances that encourage car use, a cold climate, an energy-intensive industrial base, and relatively low energy prices.

Environment Canada
'Energy Consumption:
Environmental Indicator Bulletin', 1994

Bott, Brooks, and Robinson (1983) explained, we have had confusing signals. In the 1950s and 1960s energy prices, after being adjusted for inflation, actually fell in Canada. The country appeared to have an abundance of inexpensive coal, oil, natural gas, and hydroelectricity, as reflected in the agreement between the province of British Columbia and the Aluminum Company of Canada to develop power at Nechako for an aluminium plant at Kitimat (Chapter 11). Even at the start of the 1970s, it was possible for the federal minister responsible for energy to state that Canada had 392 years of natural gas supplies and 923 years of oil.

However, within three years of the minister's optimistic statement, Canadians heard that we were running out of oil and natural gas. For the remainder of the 1970s, energy prices skyrocketed, various energy policies were introduced and withdrawn, the phrase 'energy crisis' became part of our vocabulary, and Canadians became resigned to ongoing 'energy shocks', such as when world oil prices tripled during 1979. Major projects were proposed, such as pipelines from the Arctic, to ensure that Canada did not become vulnerable to apparently unreliable sources of supply from other countries.

Almost as suddenly, a recession and energy glut occurred in the early 1980s. Energy megaprojects, often viewed as major sources of jobs, were put on hold or cancelled. Utilities found themselves with an unanticipated surplus of generating capacity and growing debt. By the end of 1993, for example, Ontario Hydro had a $35 billion debt. From 1992 until the end of 1993, Ontario Hydro's

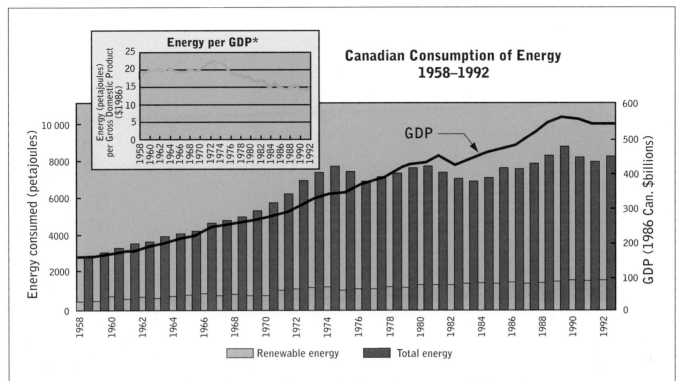

NOTE: Total energy consumption (production plus imports minus exports) refers to primary energy rather than end use.
* Energy used per unit of economic activity, as measured by gross domestic product (GDP), is a measure of the energy intensity of the economy.

- Non-renewable energy (mostly fossil fuels) made up 82 per cent of the total energy consumed in 1992. Nuclear energy, introduced in the 1970s, had risen to 11 per cent of the total by 1992.
- Renewable energy sources comprised 18 per cent of the total in 1992. Two-thirds of this renewable energy was hydroelectric and one-third was wood. Much of the wood was waste wood chips from pulp mills.

- Alternative energy technologies (e.g., solar and wind power) make up less than 1/10,000th of total energy consumption.
- The decline in energy per \$GDP suggests that the economy is becoming less energy-intensive. The principal reasons for this are more energy-efficient technology and more energy conservation.

Changes in Types of Energy Consumption 1958–1992

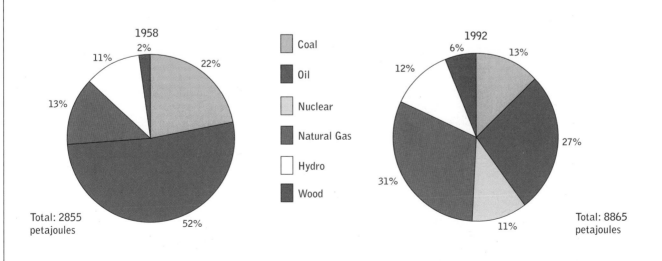

Figure 19.1 Canadian consumption of energy. SOURCE: Environment Canada, 'Energy Consumption Environmental Indicator Bulletin', SOE Bulletin no. 94-3 (Ottawa: Environment Canada, 1994):2, 3.

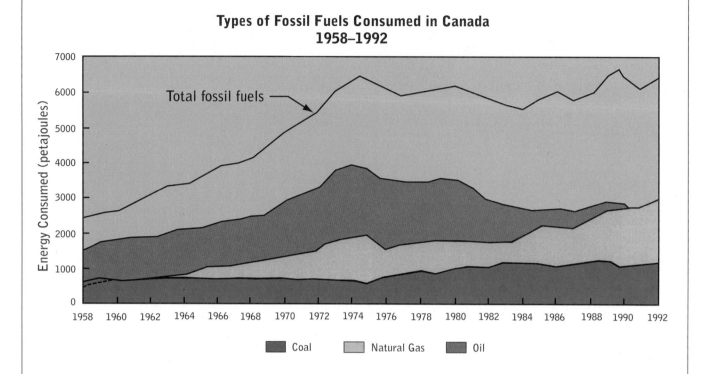

**Types of Fossil Fuels Consumed in Canada
1958–1992**

- The amount of fossil fuel consumed in Canada in 1992 was more than two-and-a-half times that consumed in 1958.
- In 1992, the fossil fuel contribution to total Canadian energy consumption was 71 per cent.
- The fossil fuel mix is changing. The natural gas contribution to fossil fuel consumption increased from 15 per cent to 44 per cent between 1958 and 1992. This increase is largely due to the increased availability of natural gas sup-

plies, encouraged recently by a move towards more environmentally benign fuels. Compared with other fuels, natural gas emits less carbon dioxide, sulphur dioxide, and nitrogen oxides, gases that contribute to smog, acid rain, and global warming.
- The relative contribution from oil has decreased since 1958. This reflects the higher oil prices of the early 1980s and government energy policies at that time that encouraged a shift away from oil for security reasons.

How Canada Uses Its Energy

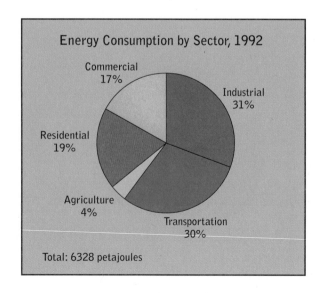

- This graph shows the relative proportion of energy that is actually used by the different types of users (*known as secondary energy*).
- The energy share of each sector has remained consistent between 1958 and 1992.
- The industrial and transportation sectors consume nearly two-thirds of delivered energy.

Figure 19.1 (continued)

workforce was cut by 6,300 through massive downsizing, and it was considering various privatization options to reduce its financial difficulties. In the summer of 1995 the chairperson of Ontario Hydro recommended to the provincial government that the utility should be privatized. Notwithstanding these measures, Ontario Hydro experienced a loss during 1993 estimated to be between $2.9–3.5 billion, one of the biggest financial losses in Canadian corporate history.

Change, challenge, uncertainty—all of these aspects have been part of the energy scene in Canada over the past several decades, and managers will undoubtedly have to continue to deal with them. In the remainder of this chapter, we will consider some of the issues and responses. The following section focuses upon issues that need to be considered when determining future demand for and supply of energy. In that context, a distinction is drawn between *forecasting* and *backcasting*. In the next section, the experience with hydroelectric development at James Bay is examined, with particular regard to issues related to the role of *stakeholders* and *impact assessment*.

Ontario Hydro's hydroelectric plant at Cameron Falls (*Ontario Hydro*).

> *The period ahead must be regarded as transitional from an era in which energy has been used in an unsustainable manner. A generally acceptable pathway to a safe and sustainable energy future has not yet been found.*
>
> World Commission on
> Environment and Development,
> Our Common Future, *1987*

ANTICIPATING ENERGY SUPPLY AND DEMAND

As already indicated, a major challenge is anticipating energy supply and demand. Traditionally, analysts have used forecasts or predictions of energy supply and demand as part of the process of developing energy policies and approaches. Policy makers rely on forecasts since they want to know the implications of alternative choices regarding various patterns of supply and demand. However, forecasts are based on many assumptions, and may therefore constrain the range of options considered.

Three Perspectives on the Future

Kuhn (1992) reviewed three energy forecasts that were national in their focus: (1) Economic Council of Canada (1985), (2) Energy, Mines and Resources Canada (1990), and (3) Brooks, Robin-

son, and Torrie (1983), updated by Torrie (1988). Characteristics of each of these forecasts indicate the opportunities and challenges such analyses present.

The Economic Council of Canada's forecast was completed in the mid-1980s after a ten-year period of high energy prices in Canada and following the withdrawal of the National Energy Program (1979–83) (Box 19.1). The purpose of the forecast was to help develop an energy policy until the year 2000. It gave particular attention to security of energy supply, economic efficiency, sustained economic growth, and increased Canadian ownership and control. It did *not* systematically consider the nuclear industry, environmental issues, or land claims by First Nations peoples. The Department of Energy, Mines and Resources prepared its forecast to the year 2020 to help anticipate development prospects for the country. Energy, Mines and Resources recognized the

Box 19.1 National Energy Program, 1979–1983

Objectives

- Reduce the consumption of oil and discontinue imports by 1990.
- Increase the supply of domestic oil and the overall efficiency of use.
- Enlarge the existing oil pipeline from the west to Montreal, and extend the gas line to Quebec City and the Maritime provinces.
- Establish a 'made in Canada' price structure that is below the world price but rising towards it, and increase the gas price but keep it below the oil price.
- Create fairer distribution of revenues among the federal government, governments of the producing provinces, and corporations.
- Assert 'national interest', while recognizing provincial jurisdiction over resources.
- Increase Canadian ownership of the hydrocarbon industry to 50 per cent by 1990 (after Chapman 1989).

Establishing a Price for Energy

The gradual decline in real oil prices (i.e., after inflation) during most of the 1950s and 1960s, and the failure of the market to reflect fully the environmental costs of energy production and use in consumer prices, led to higher levels of energy consumption in all industrial countries than would have occurred otherwise.

It is very difficult to quantify the environmental costs of energy use because many of these costs, such as the loss of wildlife habitat, are difficult to estimate in economic terms. In other cases, the impacts themselves, such as climatic change, are not well understood. Because they are not being borne by today's energy consumers, these costs are being passed on to future generations in the form of a degraded environment (Environment Canada 1994:12–29).

importance of environmental issues and the need to develop energy-saving approaches, but these received relatively limited attention.

Kuhn concluded that both the Economic Council of Canada and the Department of Energy, Mines and Resources' approaches accepted the existing energy supply, demand, and delivery systems. Both assumed that demand for energy would increase substantially, and that this demand would be satisfied by increased oil production and importation, expanded hydro and nuclear capacity, and increased reliance upon natural gas.

In contrast, the studies by Brooks, Robinson, and Torrie (1983) and Torrie (1988) covered a longer period (1978–2025) and took a different approach. They focused upon the feasibility and consequences of more efficiency in energy use and increased use of renewable energy supplies instead of upon increased growth in energy demand and dependence upon non-renewable sources (Table 19.1). They emphasized that their studies were *not* intended to be predictions. Instead, they were interested in four questions: (1) How technically feasible was a renewable energy future? (2) How

economically viable was it? (3) What would a renewable energy future look like? (4) What decisions must be taken to achieve a renewable energy future if Canadians decided they should try to achieve one?

The primary focus in the forecasts of the Economic Council of Canada and the Department of Energy, Mines and Resources was on energy supply to satisfy anticipated demand. The Economic Council of Canada developed three projections based upon different assumptions about world oil prices. The Department of Energy, Mines and Resources provided two projections. The first assumed high worldwide economic growth and demand for oil, restricted production from OPEC countries, low production from non-OPEC nations, no severe restrictions on oil products use, and modest changes in oil product taxes. Both forecasts assumed that current institutional and policy arrangements regarding energy development, supply, and delivery would not change. As a result, their projections did not reflect any significantly different policies regarding energy but rather provided a baseline against which new policy initiatives could be compared.

Brooks, Robinson, and Torrie (1983) took a sharply different approach. They did not try to predict patterns of supply and demand in the future. Instead, their interest was in developing two scenarios of what the Canadian energy situation might be like by 2025. More specifically, they were

Table 19.1 Sources of Alternative Renewable Energy

Source	Description
Small low-head hydro	Run-of-the-river devices, small hydraulic turbines (under 15 mW), tidal power: half of the 400 hydroelectric stations in Canada are small hydro stations and generate only 10 mW of the more than 60 000 mW of hydroelectric generating capacity in Canada.
Active solar	Direct conversion of solar radiation into thermal energy; their greatest potential is remote locations.
Passive solar	Collection, control, storage, and distribution of solar energy through building design.
Photovoltaics	Direct conversion of light into energy.
Bioenergy	The collection of biomass residues and the harvesting of woody biomass for conversion into an energy form.
Wind energy	The conversion of kinetic energy into mechanical and/or electrical energy; shows great potential for remote communities.
Geothermal energy	The use of the heat in the earth's crust; very little active development in Canada.

Source: Environment Canada, *The State of Canada's Environment* (Ottawa: Minister of Supply and Services Canada, 1991):12–31.

interested in defining the characteristics of an energy future based on renewable sources and then exploring its feasibility. In other words, they were more interested in what was *feasible* and *desirable* than what was *probable*.

Indeed, Brooks (1985) was very critical of the Economic Council of Canada's forecast because of its emphasis upon supply questions and relative neglect of conservation and demand issues. Brooks argued that energy demand should be considered before supply to establish what level of demand needed to be satisfied. In this manner, conservation opportunities would shape both patterns of supply and demand, which he believed was more logical than addressing conservation opportunities after decisions had been made regarding supplies.

After reviewing these three approaches, Kuhn (1992) concluded that the forecasts by the Economic Council of Canada and the Department of Energy, Mines and Resources reflected the dominant social world view reviewed in Chapter 2. This was particularly evident in the focus upon economic growth and the relative lack of consideration of the forecasts' environmental implications. Market forces were used as the main basis for determining what might happen. No adjustments were made regarding future levels of consumption or sources of energy relative to the possible consequences for the environment. These two approaches also accepted continuing reliance upon present arrangements, which emphasized large-scale, technologically driven supply and delivery systems.

On the other hand, Brooks, Robinson, and Torrie's approach reflected the new environmental world view reviewed in Chapter 2. Their concern was with long-term environmental integrity. Energy supply sources were accepted or rejected on the basis of whether or not they were environmentally benign. They also believed that decentralized and small-scale development was feasible and preferable, and that a goal of demand reduction was both feasible and desirable.

The important point that emerges from comparing these three studies is that perspectives on the future can be based on significantly different world views. Users of such studies should take the time and care to determine which world view underlies an analysis to be sure that an important alternative has not been overlooked or set aside *only* because it is incompatible with the world view incorporated in the study. Which world view do you think is the most appropriate? What are the major problems with your preferred world view?

Feasibility, Desirability, and Probability

Bott, Brooks, and Robinson (1983) explained that one major problem is the time horizon for energy planning, which is usually longer than the life span of most people. As a result, it is not possible for us to know what *will* happen. Consequently, they believed it is more useful to focus on three critical aspects and related questions to help cope with an uncertain future (Box 19.2).

They concluded that too many energy studies concentrate on the third aspect (probability) and ignore the other two (feasibility and desirability). Or, even worse, all three aspects are combined as if they are one. They argued that the best analysis should consider each aspect separately and sequentially, otherwise if only probability is addressed, alternatives that might be very feasible are not considered because they have not been identified. Furthermore, emphasis on probability leads to an assumption that the only desirable futures are those from among the small set with a high likelihood.

To illustrate, in the mid-1970s officials in energy utilities and governments believed that demand for electricity would grow at about 7 per cent annually since that had been the pattern for the previous two decades. As a result, since 7 per cent was the most probable rate, it was used for future planning. No thought was given as to whether 7 per cent was desirable, or whether a 2 per cent rate would be more desirable and, if so,

whether it were feasible and probable. Experience in the 1990s subsequently showed that a rate of 2 per cent was indeed possible, so a fixation on the higher 7 per cent rate was not appropriate. Furthermore, if greater attention had been given to lower rates, many of the major utilities might have been able to avoid massive overbuilding, which later caused financial headaches.

For example, in September 1994 BC Hydro released an electrical energy conservation report. Prepared over two years by a committee that included more than thirty interest groups, the report concluded that a significant portion of the unconstrained potential for electricity conservation was realistically achievable. Indeed, the report suggested that conservation measures phased in between 1994–2010 could reduce provincial energy consumption in 2010 by 14–23 per cent of what it might be otherwise. Such a saving would be equivalent to the annual consumption of electricity by 1 million homes. The study also noted that if consumers adopted a 'sustainable lifestyle' (such as washing clothes in warm or cold rather than hot water), power consumption in British Columbia could be less in the year 2030 than it was in 1990 (Box 19.3).

Bott, Brooks, and Robinson (1983) concluded that one of the major problems of focusing on probability or likelihood is that what you believe is most likely may not be the most desirable, or even desirable at all. If we assume that 7 per cent will be the annual rate of growth of electricity use, then pricing and other policies will be based on that expectation and will most likely transform it into reality without systematic consideration about its desirability.

Some people have argued that instead of extrapolating from the past, we should decide what kind of energy use should characterize our society in the future, and then make decisions now to reach that desired goal (Robinson 1988a). Such an approach is referred to as *backcasting* (Robinson 1983a, 1988b, 1990; also see Chapter 1).

Many provincial electrical utilities drastically expanded their production capacities in the 1970s and 1980s as a result of poor planning. This is the W.A.C. Bennett Dam on the Peace River in northeastern British Columbia. Most of the power from this dam is consumed over 1000 km away in Vancouver. However, the impacts of the dam are felt mainly by the Native peoples who live on the Peace-Athabasca Delta in Wood Buffalo National Park, where their subsistence, wildlife-based trapping economy was largely destroyed by the dam hundreds of kilometres away (*Philip Dearden*).

Box 19.3 Sustainable Energy Use: What You Can Do

In BC, energy consumption rates are among the highest in the world—at 18 000 kilowatt-hours (kWh) per person—about three times the per capita consumption in Great Britain, and 30 per cent more than the energy used by the average American. BC Hydro has installed more energy-efficient operating facilities and saved 900 million kWh (enough to meet the annual needs of 90,000 homes) as a result. They also initiated a program to encourage energy efficiency in BC homes. If 500 TVs are turned off for an extra hour per day, for example, this would meet the total electricity needs of 100 homes. Already the domestic Power Smart program has saved 1100 million kWh. These savings represent the economies resulting from thousands of changes made by small consumers.

Have you ever considered what changes you could make to reduce your energy consumption? Could you conduct an inventory in your home to determine the energy efficiency of appliances and fixtures? When the time comes to replace them, are you aware of the energy-efficient alternatives that might be considered? If it is cool in your room, do you ever think of putting on a sweater instead of turning up the thermostat? If you drive to university or work, have you considered some combination of car pooling, public transportation, or using a bicycle? Has an energy audit been conducted at your university or place of work? If not, what might you propose as a first step?

For example, in Chapter 2 the characteristics of a sustainable society were outlined. If we would like to achieve a sustainable society, then continuing many present energy practices is unlikely to satisfy criteria for sustainability. Instead, it will be necessary to make conscious decisions to modify contemporary energy use patterns. Preoccupation with questions of probability will not lead to the appropriate questions related to achieving sustainability.

Consideration of some of the issues associated with forecasting, based on past or present energy-use patterns, illustrates why backcasting should be used together with forecasting. The former is *prescriptive* (what should be), and the latter is *predictive* (what is expected to be). We need both approaches when anticipating future energy supplies and demands, but have relied on or overused forecasting until now.

THE HARD PATH

The 'hard path' in energy describes approaches that emphasize developing energy sources to meet demands. In contrast, a 'soft path' gives priority to modifying patterns of demand, as discussed in the preceding section. Canada has used many energy megaprojects to meet anticipated energy demands, and virtually every region in the country has had energy megaprojects (Figure 19.2). One of the most high-profile megaprojects has been James

Bay in Quebec, while others have been Churchill Falls in Labrador, the Nelson-Churchill in Manitoba, the Columbia and Nechako rivers in British Columbia, nuclear powerplants in Ontario, and Hibernia in Newfoundland. Selected aspects of the James Bay project are considered next.

James Bay: Phase I

Background

In 1971 Premier Robert Bourassa proposed hydroelectric development using the rivers on the eastern side of James Bay. The purpose was to satisfy future electricity needs in Quebec. The estimated cost was $2 billion. Alternative projects were examined, and the decision was taken to develop La Grande River basin to double the flow in that river by diverting water from adjacent catchments (Figure 19.3). Other river systems north and south of La Grande were to be developed in later phases.

Two major diversions channelled water into La Grande basin. From the south, 845 m^3 per second (m^3/sec) or 87 per cent of the flow measured at the mouth of the Eastmain River was redirected into the LG2 reservoir on La Grande. Twenty-seven per cent (790 m^3/sec) of the Caniapiscau River was diverted into the LG4 reservoir, so that this water would flow through all four power stations to be built on La Grande.

As Day and Quinn (1992:134) explained, these diversions together add an average of 1635 m^3 per

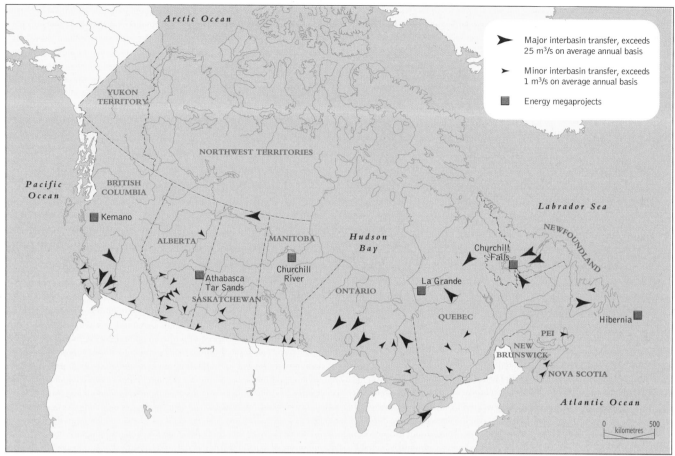

Figure 19.2 Energy megaprojects in Canada. SOURCE: J.C. Day and F. Quinn, *Water Diversion and Export: Learning from Canadian Experience* (Waterloo, ON: Department of Geography, University of Waterloo, 1992):16.

The James Bay hydroelectric project (*Hydro-Québec*).

second to La Grande, almost doubling the natural flow in that river. The diversions into La Grande account for 36 per cent of all the volume of interbasin water transfers in Canada (Table 19.2). Over a fifteen-year period, the cost estimates increased to $14.6 billion, compared to the $2 billion announced by Premier Bourassa in 1971.

In stage 1 of the development, three hydroelectric plants (LG2, LG3, and LG4) with a combined 10 283 megawatt (mW) capacity were built. LG2, with 5328 mW capacity, became the most powerful underground generating station in the world. The first electricity was generated from LG2 in 1979, and LG4 was completed in 1986. LG1 and other dam construction were deferred to stage 2 due to decreasing demand for energy. The overall scope of phase I and phase II of the James Bay hydroelectric development is shown in Box 19.4.

The scope and magnitude of the James Bay development has been described as 'breathtaking'. It would produce electricity from rivers flowing in a 350 000 km² area of Quebec, more than one-fifth of the province, or an area equivalent to France. The provincial government and Hydro-Québec have justified the James Bay development for the jobs it will create, the industrial growth it will attract to the province, and for the stability it will create. Premier Bourassa, in his book *Power from the North*, stated that 'Quebec is a vast hydroelectric plant-in-the-bud, and every day millions of potential kilowatt-hours flow downhill and out to sea. What a waste!' Does this viewpoint reflect the dominant social world view or the new environmental world view discussed in Chapter 2?

In the enthusiasm for the perceived benefits from hydroelectricity in James Bay, an area remote from the settled part of the province, little regard was given to the fact that this area was the homeland for about 10,000 Cree and Inuit, who had lived and hunted in this region for centuries. In the following section, attention will be given to the way in which the concepts of *stakeholders* and *partners* were addressed in the James Bay development. The subsequent section will consider selected aspects of *impact assessment*.

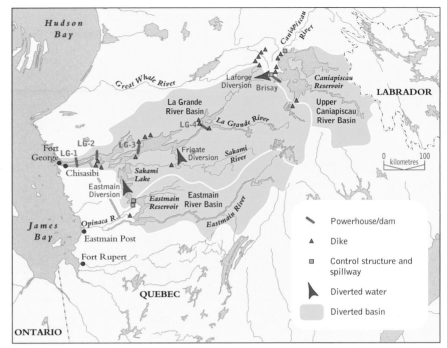

Figure 19.3 La Grande River hydroelectric development project, phase I. SOURCE: J.C. Day and F. Quinn, *Water Diversion and Export: Learning from Canadian Experience* (Waterloo, ON: Department of Geography, University of Waterloo, 1992):134.

Stakeholders

The James Bay and Northern Quebec Agreement is considered by many to be the first 'modern' Native land claims agreement in Canada. However, in April 1971 when Premier Bourassa announced the construction of the hydroelectric megaproject on the La Grande River system, no systematic ecological or social impact assessments had been completed, although such a large-scale development could be expected to have major impacts. The Cree in northern Quebec, who had not been consulted, soon organized themselves to fight the project.

The Cree, joined by the Inuit and other First Nations groups, initiated legal proceedings against Hydro-Québec and the government of Quebec to stop the development on La Grande. By May 1972 the matter was before Justice Malouf of the Superior Court of Quebec. Justice Malouf issued an injunction, ordering a stop to all operations related to La Grande due to unsettled First Nations land claims in the affected area. However, in the following week, the Quebec Court of Appeal suspended the injunction. The Supreme Court of Canada

We have had fifteen years of constant struggle to try to force Quebec and Canada to respect their commitments under the overall James Bay Agreement. If I had known in 1975 what I know now about the way solemn commitments become twisted and interpreted, I would have refused to sign the Agreement.

Billy Diamond,
chief of the Waskaganish
Band of James Bay Crees,
'Villages of the Dammed', 1990

Table 19.2 Major Interbasin Water Diversions in Canada

No.	Jurisdiction	Project	Contributing Basin(s)	Receiving Basin	Average Annual Diversion (m³/s)	Uses	Operational Date	Owner
1.	British Columbia	Kemano	Nechako (Fraser)	Kemano	115	Hydro	1952	Alcan Ltd
2.	British Columbia		Bridge	Seton Lake	92	Hydro	(1934) 1959	BC Hydro
3.	British Columbia		Cheakamus	Squamish	37	Hydro	1957	BC Hydro
4.	British Columbia		Coquitlam Lake	Buntzen Lake	28	Hydro	(1902) 1912	BC Hydro
5.	Saskatchewan		Tazin Lake	Charlot (Lake Athabasca)	25	Hydro	1958	Eldor Nuclear
6.	Manitoba	Churchill Diversion	Churchill (Southern Indian Lake)	Rat-Burntwood (Nelson)	775	Hydro	1976	Manitoba Hydro
7.	Ontario		Lake St Joseph (Albany)	Root (Winnipeg)	86	Hydro	1957	Ontario Hydro
8.	Ontario		Ogoki (Albany)	Lake Nipigon (Superior)	113	Hydro	1943	Ontario Hydro
9.	Ontario		Long Lake (Albany)	Lake Superior	42	Hydro/ Log.	1939	Ontario Hydro
10.	Ontario		Little Abitibi (Moose)	Abitibi (Moose)	40	Hydro	1963	Ontario Hydro
11.	Ontario	Welland Canal	Lake Erie	Lake Ontario	250	Hydro/ Navig.	(1829) 1951	Govt of Canada
12.	Quebec	James Bay	Eastmain-Opinaca	La Grande	845	Hydro	1980	J.B. Energy Corp.
13.	Quebec	James Bay	Fregate	La Grande	31	Hydro	1982	J.B. Energy Corp.
14.	Quebec	James Bay	Caniapiscau	La Grande	790	Hydro	1983	J.B. Energy Corp.
15.	Newfoundland	Churchill Falls	Julian-Unknown	Churchill	196	Hydro	1971	Nfld & Lab. Hydro
16.	Newfoundland	Churchill Falls	Naskaupi	Churchill	200	Hydro	1971	Nfld & Lab. Hydro
17.	Newfoundland	Churchill Falls	Kanairktok	Churchill	130	Hydro	1971	Nfld & Lab. Hydro
18.	Newfoundland	Bay d'Espoir	Victoria, White Bear Grey and Salmon	Northwest Brook (Bay d'Espoir)	185	Hydro	1969	Nfld & Lab. Hydro

Source: J.C. Day and F. Quinn, *Water Diversion and Export: Learning from Canadian Experience* (Waterloo, ON: Department of Geography, University of Waterloo, 1992).

subsequently decided that it did not have authority to consider the case.

Following the decision of the Quebec Court of Appeal, the government of Quebec offered compensation to the Cree and Inuit. While the Cree and Inuit were opposed to the hydroelectricity development, they concluded that they had little option but to negotiate. The outcome was the

Box 19.4 James Bay Hydroelectric Development: Completed and Proposed

PHASE I

La Grande: Stage 1

Completed in 1985 after twelve years of construction and at a cost of $16 billion, La Grande stage 1 has three reservoirs and powerhouses: LG2, LG3, and LG4. The combined production is 10 282 mW. Five smaller rivers were diverted into La Grande. The Eastmain River was reduced to a fraction of its former flow. The average flow of La Grande has been doubled, and the flow in the winter has been increased by a factor of four. LG2 is being expanded with a new powerhouse that will produce 1998 mW.

La Grande: Stage 2

LG1, near the mouth of La Grande, will be the main powerhouse. Five other powerhouses (Brisay, Eastmain 1 and 2, and Laforge 1 and 2) will be built on rivers diverted during phase 2.

PHASE II

Great Whale (Grande rivière de la baleine)

The Great Whale basin is immediately to the north of La Grande and drains into Hudson Bay. Development will include three power stations with a total capacity of 2890 mW, and diversion of two other rivers. Three or four reservoirs would be built on the Great Whale River.

The Nottaway, Broadback, and Rupert Rivers

The Nottaway, Broadback, and Rupert are all large rivers that are immediately to the south of La Grande and flow into the southern part of James Bay. The Nottaway and Rupert are to be diverted into the Broadback, where as many as eight powerhouses would produce 8700 mW.

Shelving

On 18 November 1994 Premier Jacques Parizeau announced that the $13 billion Great Whale hydroelectric project had been shelved. The premier stated that the project was being set aside solely for economic reasons. In his words, 'This government has not made it a priority because we don't need the power.' This announcement occurred at the same time as the release of a report by federal and provincial committees, which stated that an eleven-year, $256 million environmental impact study of the project was inadequate, and that Hydro–Québec should correct 'major inadequacies' in the assessment (After Gorrie 1990:23-4; *Kitchener-Waterloo Record*, 19 November 1994:A3).

The Great Whale River (*Earthroots*).

James Bay and Northern Quebec Agreement, which was signed on 11 November 1975 and subsequently approved by the government of Canada and the Assemblée Nationale du Québec.

Mainville (1992) has argued that the negotiations were conducted too quickly to result in thoughtful solutions. He believed that Hydro-Québec and the provincial government's main interest was obtaining a rapid agreement to ensure that James Bay phase I would be protected from further court actions. In contrast, the Cree were concerned about obtaining commitments to preserve their traditional ways of life; clarify their rights to land, hunting, fishing, and trapping; provide control over the activities of public institutions on their lands; protect the environment; provide compensation; and ensure their participation in development in the territory.

The agreement is complex and often ambiguous. However, it does provide for land rights and guarantees a process to deal with future hydroelectric developments. The agreement included provisions for environmental and social impact assessment for future developments, monetary compensation, economic and social development, and income security for Cree hunters and trappers.

Regarding the provisions for environmental and social impact assessment, Mainville (1992) concluded that they have been disappointing. This view was shared by Billy Diamond, who stated that:

> Protection of the environment in Northern Quebec has been a farce. The regime set out in the Agreement does not work. It has not been well implemented because the provincial government has put almost no resources into it. . . . No independent expertise is brought to bear on environmental questions posed by development. There are no public hearings. There is no funding for third parties to study the questions (Diamond 1990:28).

Mainville argued that neither the provincial nor federal government appeared to respect the provisions of the agreement. No adequate secretariat or other personnel were allocated for the impact assessment process outlined in the agreement, and neither government has shown interest in creating an adequate ecological data base for the area. The federal government took a very narrow perspective regarding its role in impact assessment proceedings, and even then was criticized by the government of Quebec for infringing on provincial jurisdiction. The provincial government was supposed to provide human and financial resources for the committees responsible for the impact assessment of proposed developments on the Great Whale River and the Nottaway, Broadback, and Rupert rivers, but did not do so.

James Bay: Phase II

When Premier Bourassa announced phase II in 1985, he explained the development would have two benefits: (1) generate revenue for Quebec through exports of electricity to the United States under long-term contracts, and (2) attract energy-intensive industries (such as aluminium and magnesium smelters) as a result of competitively priced electricity. James Bay phase II involved completing the development in La Grande basin, particularly the building of LG1, as well as new hydroelectricity development in the Great Whale and Nottaway, Broadback, and Rupert river systems. During 1986 the Cree agreed to the completion of the development in La Grande basin, but opposed the projects to be started in the basins to the north and south of La Grande.

The projects in the Great Whale basin would provide just under 3000 mW of new power by diverting several adjacent rivers. One outcome would be an 85 per cent reduction in the flow of the Great Whale River at the community of Whapmagoostui (Figure 19.4). The Nottaway, Broadback, and Rupert development would produce 8000 mW of additional power and, as with development on La Grande, would involve flooding land due to dam construction.

The Cree have opposed these projects, primarily because they are not being built to meet energy needs in Quebec but to provide electricity for export to the United States, or to attract multinational smelting companies to Quebec by providing subsidized electricity.

Hydro-Québec first signed a long-term contract with a utility in Maine, but that contract was later cancelled because of opposition in the state to the estimated environmental damage that would be caused by the James Bay megaprojects. Other contracts with Vermont and New York state also became challenged in the United States because of the environmental implications and the lack of consideration for impacts on the First Nations peoples in the project area.

Regarding impact assessment, Hydro-Québec had maintained for over fifteen years that it would treat the Great Whale River megaproject as a single entity for assessment purposes. However, in the late 1980s it attempted to avoid a complete environmental impact assessment by proposing to divide the project into segments. Thus it argued that basic infrastructure (airports, roads) should be assessed separately from the hydroelectric dams and reservoirs. This position was taken to accelerate the construction and therefore meet export commitments made in contracts with the American states.

The Cree used the courts to challenge the Great Whale River project. They demanded that the courts require a full environmental impact assessment of the entire project, as provided for under the James Bay and Northern Quebec Agreement. They also argued that to be credible, any impact assessment should include an independent and expert panel, provision of information and findings to the public, and public hearings.

As a result of this pressure, in late January 1992 the federal and provincial governments signed a memorandum of understanding with the Cree to have an impact assessment completed for the entire Great Whale River project rather than a series of assessments of various segments of the project. The understanding also stipulated that public meetings and hearings would be part of the review, and that $5 million would be made available to fund participants. In

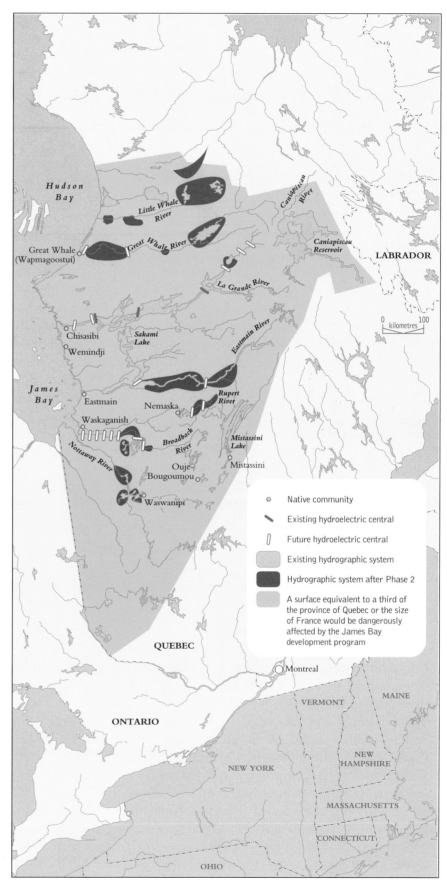

Figure 19.4 The Great Whale project. SOURCE: B. Diamond, 'Villages of the Dammed', *Arctic Circle* (November/December 1990):32.

addition, the review would address the overall justification for the development.

Several months later (on 27 March 1992), the governor of New York announced that the state was cancelling its 1000 mW firm export contract (worth $17 billion) with Hydro-Québec because of concerns about environmental impacts and implications for the First Nations peoples living in the development area. At the time of that announcement, Grand Chief Matthew Coon-Come of the Grand Council of the Crees of Quebec stated that he welcomed such a review by New York, but remarked that 'Mayor Dinkins is not doing this out of the goodness of his heart or because he cares about the Crees. He realizes that buying hydroelectricity from Quebec is a bad economic decision. . . .' (*Globe and Mail*, 7 August 1991:B2). Coon-Come was referring to the combined results of a recession and energy conservation efforts, which had reduced the demands for electricity forecast earlier.

Hydro-Québec had not remained passive during the review period. In October 1991 it hired one of the largest public relations firms in the world to help it promote James Bay phase II. The task of New York-based Burson-Marsteller was to publicize the benefits of the then $12.6 million Great Whale project and to counter the Cree's claim that James Bay phase II would destroy their land, way of life, and culture.

The Cree from Quebec were being supported by more than 100 environmental and Native rights groups in New York, which were lobbying the New York Power Authority to cancel its $17 billion contract with Hydro-Québec. These groups had a significant victory in August 1991 when the New York governor announced the state would take a year to consider the contract, leading Premier Bourassa to state that the Great Whale project would be delayed by one year.

The public campaign to oppose the Great Whale project had started in 1989 after Hydro-Québec signed the contract to supply electricity to New York state for twenty-one years, starting in 1995. When the Cree found that they could not create a public debate in Quebec, they began to concentrate on Hydro-Québec's customers in the United States. The Cree hired a major public relations firm at a cost of $129,000, and were successful in gaining support from a broad mix of influ-

ential lobby groups, such as the National Audubon Society and the Sierra Club. The Cree took out a full-page advertisement at a cost of $28,000, which was paid for mostly by Greenpeace, in the *New York Times* and called the Great Whale project the 'most destructive project in North America'.

During the widespread protests, the New York Power Authority announced it was postponing final approval of the $17 billion export contract with Hydro-Québec for one year, and would not make its decision until November 1992. The chairperson of the New York Power Authority indicated that they decided to delay their decision because they were displeased that Hydro-Québec was moving ahead with the project without having conducted a complete impact assessment. Subsequently, and following some victories by the Cree in the courts, the federal government and Quebec agreed to conduct a joint impact assessment in which the Cree and Inuit would be involved. The impact process, started in January 1992, was expected to last for two years, further delaying Hydro-Québec's construction plans.

In April 1992 Quebec's Minister of Energy Lise Bacon stated that the Great Whale project would be modified because the export contract with New York state was cancelled. While the project was originally scheduled to be finished by the year 2000, Bacon explained that the first of three stages would now be completed by 2000, with the second and third stages completed by 2005 and 2008, respectively. She blamed the Cree for the cancellation of the contract with New York, and claimed they were threatening the economic well-being of Quebec. She was quoted by Canadian Press as asking if the Cree 'are Quebecers or not' and of saying that 'It's unacceptable for Quebecers to see their economic situation reduced or put in danger by natives. I blame them for discrediting Quebec all over the world. Do you think Quebecers can accept that? I don't think so' (*Kitchener-Waterloo Record*, 1 April 1992:B7).

In September 1993 Hydro-Québec released a 5,000-page report on the Great Whale hydroelectric project. Public hearings were anticipated to begin during the spring of 1994 and to continue for at least one year. When releasing its report, Hydro-Québec announced it was prepared to offer compensation of more than $130 million to the Native peoples in northwestern Quebec for flooding 1667 km² of their land. Bill Namagoose, executive director of the Grand Council of the Cree, stated that the Cree would not accept a compensation package because they were totally opposed to the Great Whale project and their goal was to stop it.

As noted in Box 19.4, in late November 1994 the $13.3 billion Great Whale hydro project was shelved by the government of Quebec. At the time of the announcement, Premier Parizeau explained that 'We're not saying never but that project is on ice for quite a while' (*Kitchener-Waterloo Record*, 19 November 1994:A3). It should also be noted that in April 1994 Matthew Coon-Come, the grand chief of Quebec's northern Cree, had been awarded the Goldman Environmental Prize, worth US $60,000, in New York City's Central Park for his tenacious defence of Cree land from Hydro-Québec's dam projects.

Implications

The Cree and Inuit peoples' experience with James Bay phases I and II offer a number of insights into resource and environmental management in Canada.

First, establishing who is responsible for impact assessment is problematic. Provincial governments, as we already saw in Chapter 18 regarding the Rafferty-Alameda project in Saskatchewan, are quick to charge the federal government with intruding into provincial jurisdiction. In Quebec, Lise Bacon stated in 1991 that Quebec would 'never submit' to Ottawa's proposal that it would unilaterally conduct an impact assessment of the Great Whale project.

Second, the importance of *scoping* an impact assessment was highlighted at James Bay phase II. Since September 1990, the government of Quebec had insisted that it would assess the impacts of the roads and airports separately from the dams and reservoirs. This division of the Great Whale project was criticized sharply by the Cree and environmental groups as contrary to principles of sound environmental management. However, in October 1991 Quebec an-

nounced that it would submit the $12.6 billion project to one complete impact assessment. Lise Bacon explained that the one-year delay of the Great Whale project allowed time for a complete assessment. However, as the cartoon in Figure 19.5 illustrates, many people felt that the pressure from New York and the Cree were significant factors in the decision.

Impacts

An earlier section of this chapter covered issues related to forecasting or predicting energy supply and demand. Many of the same questions arise when trying to estimate impacts from development. Berkes (1988) shared many of the concerns associated with predictions. He noted that many projections of environmental impacts assume that with enough facts, the future can be predicted. His view is that it is impossible to foresee even *most* of the impacts. He examined the experience at James Bay since 1971 to illustrate the challenges in predicting impacts related to aspects important to the local people of the area, the Cree Indians of Chisasibi, the community closest to the dams built on La Grande River.

Berkes first noted that the concerns regarding impacts changed over time (Table 19.3). During the initial years of the development, at a regional scale the Cree were concerned mainly about the loss of riverine habitat and resources, especially beaver, in the areas to be flooded, and about the

Table 19.3 Changing Importance of Impact Assessment Issues

Time Period	Primary Concerns of the Local People
Court case 1972 to 1973	Reduction of wetland resources Loss of hunting areas Fishery at the First Rapids of La Grande
Construction Phase I 1974 to 1984	Relocation of Fort George village Drinking water quality Difficulty in harvesting areas flooded by reservoirs Travel safety in the lower La Grande Access to James Bay's north coast
Construction Phase II 1984 onwards	Access problems regarding various areas High mercury of fish in reservoir system

Source: Modified after F. Berkes, 'The Intrinsic Difficulty of Predicting Impacts: Lessons from the James Bay Hydro Project', in *Environmental Impact Assessment Review* 8, no. 3 (1988):201–20.

Figure 19.5 Rationale for having an impact assessment. SOURCE: *Kitchener-Waterloo Record* (11 October 1991):A6.

loss of hunting and trapping land, with the inevitable implications for traditional harvesting areas and ways of life. At a community scale, there was also concern about the flooding of burial sites and potential damage from construction activity to equipment and supplies left in seasonal hunting camps.

During the construction period from 1974–84, a different set of concerns became important. These included the relocation of Fort George to the new site at Chisasibi, the quality of the drinking water in the new community, the problems in maintaining traditional hunting activity in areas that became accessible from the new roads built for the construction of dams, and, due to the altered patterns of ice breakup on the lower river and estuary because of the release of relatively warmer

water from the reservoirs in winter and early spring, the difficulty for hunters in travelling to the north coastal area across the river from Chisasibi.

There were other community concerns, for example, the increased erosion along the banks of La Grande because of fluctuating water levels in the river caused by releases from the upstream reservoirs, which at one point threatened the site of the new community. The newly built road exposed the community to other people and values, contributing to problems such as alcohol abuse for some individuals.

Following completion of the first three dams on La Grande, another set of problems emerged. The major problem was the very high levels of mercury in fish caught in the reservoirs or connecting rivers. As a result, by the end of 1985 the

Cree completely stopped fishing in the LG2 area. Another problem was that hunters from Chisasibi purchased vans to travel to distant inland hunting grounds. However, after construction of LG4 was completed, maintenance of the road network east of LG4 was stopped, and the vans could no longer be used.

The challenges in estimating impacts on some specific matters against this changing mix of issues and concerns are considered below. These changing issues and concerns also reinforce the arguments for an adaptive management approach, as discussed in Chapter 9.

Mercury in Reservoirs

None of the environmental impact assessment studies predicted the appearance of mercury in reservoir fish. This is a puzzling aspect of the impact assessment work since evidence about elevated mercury levels in fish was available from earlier hydroelectric projects at the Smallwood Reservoir in Labrador and from Southern Indian Lake in Manitoba. Such impacts were apparently dismissed by investigators as being only short term and not significant for the La Grande.

As Gorrie (1990:27–8) explained, mercury is common in rocks throughout the North. It exists in an insoluble form that does not affect air or water. However, when such rocks are inundated by a reservoir, bacteria associated with the decomposition of organic material in the reservoir water transform the insoluble mercury into methyl mercury, which vaporizes, is released to the atmosphere, and returns to the water. Once in the water, the mercury enters the food chain, reaching the highest trophic levels in fish species that prey on other fish (biomagnification, as reviewed in Chapter 14). Such predator fish—pickerel, pike, and lake trout—had been an important source of high-quality protein food for the local people. Berkes indicated that in most years about one-quarter of the community's total wild food harvest came from fishing, averaging about 60 kg per year for every man, woman, and child.

A burst of decomposition in new reservoirs often accelerates the release of mercury. In the La Grande river system, few trees were removed prior to the flooding of the reservoir area, so there was a lot of organic matter to decompose. Downstream from the dams on La Grande, levels of mercury in

fish climbed to six times their normal levels within months of completion of the dams. By the sixth year following the impoundment, concentrations of mercury were four to five times higher in all species sampled. A survey in 1984 of the Cree at Chisasibi showed that 64 per cent of the villagers had unsafe levels of mercury in their bodies. The highest levels of mercury were in walleye and northern pike, which also eliminated the possibility of a recreational fishery in the reservoirs behind the three dams.

As time passes and the drowned vegetation completely decomposes, the release of mercury should return to normal (and safe levels). However, as Gorrie (1990:27) observed, the length of time for that to occur is not known. In studies completed at the La Grande until 1981, the mercury problem was not even acknowledged. Once mercury was recognized as an issue, Hydro-Québec estimated that the high levels would last for about six years. However, in March 1988 a Hydro-Québec study of one of the reservoirs stated that mercury levels could remain high for ten to twenty years. The uncertainty about the magnitude and duration of the mercury problem makes development of any compensation a challenge since it could be a generation or more before the fish are safe to eat again.

La Grande Estuary Fish

The preconstruction impact assessments indicated that the estuarine fishery in the La Grande was unlikely to survive the development of the dams and reservoirs. The impact study predicted that salt water would move farther up into the river as a result of interrupted water flow in the river during construction and when the reservoirs began to fill behind the dams, and also because of the resulting absence of ice cover to dampen the impact of ebb and flow of the tidal water from James Bay. The consequence would be the elimination of the freshwater overwintering fish habitat for species important to the local fishery (Box 19.5). On the other hand, if the river water flow were reduced *after* the formation of ice cover, then salt water intrusion would be impeded and a critically important pocket of freshwater could be maintained in the key habitat area.

Partially as a result of pressure exerted by the local fishers, the river flow was not cut off until

---

Box 19.5 Limiting Factor Principle

In Chapter 4 we discussed the limiting factor principle, which indicates that all factors necessary for growth must be available in certain quantities if an organism is to survive. We also noted that the weakest link is known as the dominant limiting factor. Are these ideas helpful in understanding the impact from interrupting river flow and changing patterns of ice cover on the overwintering fish habitat in La Grande estuary?

---

after an ice cover had formed on the river. Monitoring revealed that this action did create the necessary freshwater pocket, which remained in place throughout the winter. The freshwater pocket was created even though no minimum flow was provided, other than what entered the river from very minor tributaries. The outcome was that the predicted fish kill did not occur, and subsequent fish populations were about the same as in the preconstruction period. In this situation, the impact prediction was incorrect, but the outcome was positive for the fishery.

Caribou Drowning

In late September 1984 some 10,000 caribou were drowned while crossing the Caniapiscau River. No impact study had considered the hydro development's possible disruption of the caribou's migration routes, and from an impact perspective the incident would likely have been considered a unique event. Nevertheless, the drowning event became controversial and illustrated the difficulties in anticipating impacts.

Limestone Falls, known to be a regular crossing point in the river for the caribou, had been a traditional hunting site for the Cree from Chisasibi, Great Whale River, and Mistassini, as well as the Naskapi from the east and the Inuit from the north. Even at low water levels, the crossing was a difficult one for the animals. In late September 1984 the river had abnormally high water levels due to heavy rainfall. When caribou were crossing the river at Limestone Falls during their traditional migration, extra water was released from the newly filled Caniaspiscau reservoir. After the event, Hydro-Québec and the Société d'énergie de la

Baie James maintained that it was an 'act of God' since the area had been receiving heavier than normal rainfall, and the actual flow in the river had been lower than what it would have been if the reservoir had not been in place. Native hunters rejected that position, and argued that the release of water from the reservoir was the primary cause of the drowning of the caribou.

Impact of Roads

The social impact of roads built to service the hydroelectric development was not addressed in any of the official EIA studies. The road construction had both positive and negative impacts. On the positive side, the roads made for easier access to distant hunting grounds. The downside was that the easier access created difficulties in maintaining the integrity of the traditional hunting territories because they became accessible to anyone with a vehicle. In the Cree resource allocation system, maintenance of sustainable harvests is dependent upon the territorial use of land and the control of access to hunting and trapping areas. The introduction of roads contributed to a partial breakdown in the traditional system of access and control, leading to overharvesting of some wildlife.

Improving Impact Assessment

One of the main lessons from the James Bay experience relative to impact assessment is that care must be taken when deciding which impacts to consider. Berkes suggested that much of the impact work in James Bay could not be characterized as either successful or unsuccessful but mainly as irrelevant because it did not focus on those impacts of greatest concern to the Cree. For example, there was awareness of the likelihood of increased mercury levels in reservoir fish, yet this issue was not addressed until well after the reservoirs had filled and unsafe mercury levels had been documented in the Cree, who relied on fish as part of their food supply. As another example, lack of consideration about the consequences of the roads built to service the construction activity indicated a poor understanding of the importance of traditional hunting areas and the traditional system of control over the hunting in such areas. This experience with James Bay is a reminder of the importance of including local stakeholders as partners in the design, performance, and monitoring of impact work.

The James Bay experience also reinforces the viewpoint that surprises should be expected in impact assessment and other aspects of environmental management. Berkes (1988) concluded that some sources of uncertainty are hardly unexpected. Impacts do not occur simultaneously but over an extended time period and are the results of the decisions made. Critical decisions are made regarding matters such as filling schedules for reservoirs, release of excess water through spillways, and including more or fewer turbines and powerhouses. Furthermore, impacts do not happen in isolation but often combine to form *cumulative impacts*. Such a situation is particularly important in situations such as James Bay for which development is contemplated over decades for adjacent catchments on the eastern side of James Bay. Impact assessment is still inadequate when dealing with such cumulative impacts and the subsequent mitigation and compensation strategies.

IMPLICATIONS

Energy provides an excellent example of some of the challenges to be encountered in dealing with change and uncertainty. The discussion on estimating future conditions and energy needs highlights the importance of different *world views* (reviewed in Chapter 2) in influencing our judgements as to what is the most appropriate course of action. Also important is the distinction between forecasting and backcasting, and the need to differentiate among that which is probable, desirable, and feasible.

The experience with James Bay, illustrative of many megaproject developments in hinterland regions, emphasizes the difficulties in conducting complete and accurate impact assessments, and in developing appropriate mitigation and compensation packages. Furthermore, such developments raise difficult questions about *equity*. How should decisions be made about massive energy projects that are justified on the basis that they will generate significant benefits to a province or country, but which also create major negative impacts on a relatively small number of people, who are almost invariably indigenous peoples? What moral, as opposed to legal, obligations do scientists and policy makers have in such situations? These same questions apply to the South Moresby experience, noted in Chapter 11. Such issues cannot be answered by more scientific research, but require a society to clarify the values and principles on which it makes decisions to allocate or develop resources.

SUMMARY

1. Canadians are very dependent on fossil fuel for energy supplies, and we are among the highest per capita energy consumers in the world.

2. Forecasting energy supply and demand is a challenge, as many assumptions must be made about population growth, the nature of the economy, technology, relative prices, changing values, national security, environmental assessments, and access to offshore supplies.

3. Forecasts focus upon the most probable or likely futures, whereas we also need to consider the most desirable and most feasible future.

4. A hard energy policy or path, reflecting the dominant social world view, focuses upon increasing supplies of non-renewable energy sources to meet growing demand. A soft energy policy or path, reflecting the new environmental world view, attempts to reduce the demand for or use of energy by focusing upon use of renewable sources of energy supply, reducing demand or use, and achieving more efficiency in patterns of use.

5. Backcasting has been suggested as one way of shifting attention to what are desirable futures rather that what are only most probable futures.

6. The development of hydroelectricity on the eastern shores of James Bay represents a hard path approach to energy, in which the emphasis is on increasing energy supplies. It involved doubling the flow of La Grande River by diverting water from two adjacent catchments (Eastmain and Caniapiscau). These two diversions account for 36 per cent of the volume of all interbasin water diversions in Canada.

7. The initial cost estimate was $2 billion. Over fifteen years, the cost estimate increased to just under $15 billion.

8. The James Bay project has been justified for the jobs it created, the province's industrial growth that it stimulated, and the stability it created.

9. The James Bay project is located in the homeland of some 10,000 indigenous peoples. The Cree, in cooperation with the Inuit and other First Nations peoples, initiated legal proceedings against Hydro-Québec and the government of Quebec to stop development on La Grande.

10. The outcome was the James Bay and Northern Quebec Agreement, signed in November 1975 and often identified as one of the first modern Native land claims agreements. The agreement provided for environmental and social impact assessment for future development, monetary compensation, economic and social development, and income security for Cree hunters and trappers.

11. The indigenous peoples have argued that the province has not allocated adequate resources to implement the terms of the agreement.

12. James Bay phase II was announced in 1985 and involved new hydroelectric development in the Great Whale and Nottaway, Broadbeck, and Rupert river catchments. The benefits were identified as earning new revenue for Quebec through exports of electricity to the US and attracting energy-intensive industry (aluminium and magnesium smelters) to the province.

13. A long-term contract was negotiated with Maine, but this was later cancelled by the state. Other long-term contracts with Vermont and New York were challenged in the US on the basis that insufficient attention had been given to the environmental and social impacts of the hydroelectricity developments.

14. The Cree insisted on a complete environmental impact assessment for the Great Whale River project, while the provincial government wanted to break the impact assessment into smaller components of the overall project.

15. In January 1992 the federal and provincial government signed a memorandum of understanding to have an environmental impact assessment con-

ducted for the complete Great Whale River project, and agreed that the process would include public meetings and hearings, and that $5 million would be provided to fund participation by local people.

16. Both Hydro-Québec and the Cree hired public relations firms to present their arguments to the general public, and both used advertisements to present their positions.

17. There was confusion regarding the roles of the federal and provincial governments in the impact assessment process, which also happened in the Rafferty-Alameda project discussed in Chapter 18.

18. The impact assessment process revealed that the analysts either did not include predictions or were incorrect in predicting the release of mercury into the reservoirs from bedrock, the impact on the habitat of freshwater fish in the estuary, the drowning of caribou, and the social consequences of road construction.

19. It has been suggested that one of the key lessons of the impact assessment experience with James Bay is the importance of considering the impacts of greatest concern to the local people, something that was not usually done at James Bay.

REVIEW QUESTIONS

1. Canadians are among the highest per capita energy consumers in the world. Why do Canadians use so much energy? Is it necessary for Canadians to be such intensive energy users?

2. What are some of the usual key assumptions made when forecasts of energy supply and demand are calculated? Why are they necessary? What problems do such assumptions create?

3. To what extent does acceptance of either the dominant world view or the new environmental world view likely lead to quite different perceptions about what the future will be like and what energy strategies will be most appropriate?

4. Why is the most likely future not necessarily the most desirable? Explain the implications with regard to energy.

5. What is the difference between forecasting and backcasting? What are their relative advantages and disadvantages for energy planning?

6. What would be the characteristics of energy use in a sustainable society in the developed world?

7. Why have the Cree and Inuit in Quebec opposed the hydroelectric projects in La Grande and Great Whale River basins? Why were they not consulted by the government of Quebec during the planning of the hydroelectric projects?

8. How should meeting the energy needs of southern Canadians be balanced against the needs of First Nations peoples and the Inuit living in northern areas where energy megaprojects are usually built?

9. When people oppose resource development projects, such as those in James Bay phases I and II or South Moresby (Chapter 11), what means can they use to express their opposition? In your view, which means are the most effective? Why?

10. When the hydroelectric projects were being planned for La Grande River basin in northwestern Quebec, no systematic impact assessments were completed. Why?

11. A number of the predictions made in subsequent impact assessments for La Grande hydroelectric projects were incorrect. Why were they wrong? What are the implications for how impact assessments should be conducted?

12. How can some of the principles of ecology discussed in chapters 4, 5, 6, and 7 be used to improve predictions of impacts?

13. How would you propose compensation be determined for negative impacts, such as the effects of elevated mercury levels in fish, which have been an important source of food for First Nations peoples in the James Bay area?

14. If it is not possible to identify all key impacts prior to development, what role might the concept of adaptive environmental management have in providing flexibility to modify developments in the light of experience?

15. Given the criticisms of hydroelectric megaprojects, do you think nuclear energy projects offer a better or worse alternative source of energy supplies?

REFERENCES AND SUGGESTED READING

Berkes, F. 1981. 'Some Environmental and Social Impacts of the James Bay Hydroelectric Project, Canada'. *Journal of Environmental Management* 12, no. 2:157–72.

_____. 1988. 'The Intrinsic Difficulty of Predicting Impacts: Lessons from the James Bay Hydro Project'. *Environmental Impact Assessment Review* 8, no. 3:201–20.

Bott, R., D. Brooks, and J. Robinson. 1983. *Life After Oil: A Renewable Energy Policy for Canada*. Edmonton: Hurtig Publishers.

Bourassa, R. 1985. *Power from the North*. Toronto: Prentice-Hall Canada.

Brooks, D.B. 1985. 'Energy Conservation and Energy Strategy: A Connection Not Made'. *Canadian Public Policy* 11 (Supplement):438–42.

_____, J.B. Robinson, and R.D. Torrie. 1983. *2025: Soft Energy Futures for Canada*, vol. I National Report; vols II–XII Provincial and Territorial Reports. Report of the Friends of the Earth to the federal Department of Energy, Mines and Resources and the Department of Environment. Ottawa: Energy, Mines and Resources Canada.

Chapman, J.D. 1989. *Geography and Energy: Commercial Energy Systems and National Policies*. Harlow: Longman Scientific and Technical.

Day, J.C., and F. Quinn. 1992. *Water Diversion and Export: Learning from Canadian Experience*. Waterloo, ON: Department of Geography, University of Waterloo.

Department of Energy, Mines and Resources. 1990. '2020 Vision: Canada's Long-Term Energy Outlook'. Working paper. Ottawa: Department of Energy, Mines and Resources, Energy and Fiscal Analysis Division.

Diamond, B. 1985. 'Aboriginal Rights: The James Bay Experience'. In *The Quest for Justice: Aboriginal People and Aboriginal Rights*, edited by M. Boldt and J.A. Long, 265–85. Toronto: University of Toronto Press.

_____. 1990. 'Villages of the Dammed'. *Arctic Circle* (November/December):24–34.

Economic Council of Canada. 1985. *Connections: An Energy Strategy for the Future*. Ottawa: Minister of Supply and Services Canada.

Energy, Mines and Resources Canada. 1990. *2020 Vision: Canada's Long-Term Energy Outlook*. Ottawa: Energy and Fiscal Analysis Division of Energy, Mines and Resources Canada.

Environment Canada. 1991. *The State of Canada's Environment*. Ottawa: Minister of Supply and Services Canada.

_____. 1994. 'Energy Consumption: Environmental Indicator Bulletin'. SOE Bulletin no. 94-3, State of the Environment Reporting. Ottawa: Government Services Canada.

Feit, H.A. 1980. 'Negotiating Recognition of Aboriginal Rights: History, Strategies and Reactions to the James Bay and Northern Quebec Agreement'. *Canadian Journal of Anthropology* 1, no. 2:159–72.

Gorrie, P. 1990. 'The James Bay Power Project: The Environmental Cost of Reshaping the Geography of Northern Quebec'. *Canadian Geographic* 110, no. 1:21–31.

Kuhn, R.C. 1992. 'Canadian Energy Futures: Policy Scenarios and Public Preferences'. *The Canadian Geographer* 36, no. 4:350–65.

McCutcheon, S. 1991. *Electric Rivers: The Story of the James Bay Project*. Montreal: Black Rose Books.

Mainville, R. 1992. 'The James Bay and Northern Quebec Agreement'. In *Growing Demands on a Shrinking Heritage: Managing Resource Conflicts*, edited by M. Ross and J.O. Saunders, 176–86. Calgary: Canadian Environmental Law Association.

Mitchell, J.G. 1993. 'James Bay: Where Two Worlds Collide'. *National Geographic* Special Edition (November):66–75.

Raphals, P. 1992. 'The Hidden Cost of Canada's Cheap Power'. *New Scientist* no. 1808 (15 February):50–4.

Robinson, J.B. 1983a. 'Energy Backcasting: A Proposed Method of Policy Analysis'. *Energy Policy* 10, no. 4:337–44.

_____. 1983b. 'Pendulum Policy: Natural Gas Forecasts and Canadian Energy Policy, 1969–1981'. *Canadian Journal of Political Science* 16, no. 2:299–319.

_____. 1988a. 'Loaded Questions: New Approaches to Utility Forecasting'. *Energy Policy* 16, no. 1:58–68.

_____. 1988b. 'Unlearning and Backcasting: Rethinking Some of the Questions We Ask About the Future'. *Technological Forecasting and Social Change* 33, no. 4:325–38.

_____. 1990. 'Futures under Glass: A Recipe for People Who Hate to Predict'. *Futures* 22, no. 8:820–42.

Torrie, R. 1988. *2025: Soft Energy Futures for Canada—1988 Update*. Report prepared for the Energy Policy Options Review, Ottawa.

World Commission on Environment and Development. 1987. *Our Common Future*. Oxford and New York: Oxford University Press.

Chapter 20

The Global Commons

Chapters 13 to 19 in Part D focused primarily upon environmental science and management within Canada. However, as highlighted in one of the case-studies in Chapter 1, Canada shares a moral responsibility for global environmental problems. These are referred to as global commons issues in this chapter because often no single country can resolve them through unilateral action.

To illustrate some of the global issues, problems, and opportunities, this chapter first reviews the events leading up to and associated with the Earth Summit in Brazil during 1992. Two specific environmental management issues with commons attributes are then considered. One is climate change, which is truly global in nature. The other is the collapse of the groundfishery, which has been one of the economic foundations of Atlantic Canada for centuries. While the Atlantic ground-fishery is regional rather than global, it still illustrates how, even at a regional scale, many environmental or resource issues require multilateral attention.

THE EARTH SUMMIT

In Chapter 2 the concepts associated with sustainable development were introduced and reviewed. The World Commission on Environment and Development, also known as the Brundtland Commission, had advocated sustainable development in its 1987 report entitled *Our Common Future*. Five years later, the United Nations Conference on Environment and Development (UNCED), more popularly known as the Earth Summit, was held in Rio de Janeiro to assess progress since the publication of *Our Common Future* and to chart a strategy for the twenty-first century. The context for the Earth Summit is reviewed and its main initiatives are identified in this section.

An Initial Step

The Earth Summit of 1992 occurred twenty years after the UN Conference on the Human Environment was held in Stockholm in June 1972. Over 2,000 delegates from 114 countries participated in the Stockholm conference whose objectives were to 'stimulate international awareness and understanding of global, international and common national environmental problems, and, based on this understanding, to evolve agreements in substance or in principle to deal with these problems'. The conference, the first such global meeting on the environment, was not intended to solve specific problems but to be a first step towards addressing them. Furthermore, while the conference was originally designed to focus on the environment, at the urging of Third World countries, development issues were also incorporated.

The Stockholm conference produced several results. It developed the Declaration on the Human Environment, a statement with twenty-six principles concerning the environment and development, some of which are presented in Box 20.1. These principles reflect many of the basic ideas of sustainable development: addressing basic human needs, keeping options open for future generations, reducing injustice and achieving equity, increasing self-determination, maintaining ecolog-

Box 20.1 Selected Principles from the UN Declaration on the Human Environment, 1972

- Man [sic] has the fundamental right to freedom, equality and adequate conditions of life, in an environment of a quality that permits a life of dignity and wellbeing, and he [sic] bears a solemn responsibility to protect and improve the environment for present and future generations. In this respect, policies promoting apartheid, racial segregation, discrimination, colonial and other forms of oppression and foreign domination stand condemned and must be eliminated.

- The natural resources of the earth including the air, water, land, flora and fauna and especially representative samples of natural ecosystems must be safeguarded for the benefit of present and future generations through careful planning or management, as appropriate.

- The capacity of the earth to produce vital renewable resources must be maintained and wherever practicable, restored or improved.

- Man [sic] has a special responsibility to safeguard and wisely manage the heritage of wildlife and its habitat which are now gravely imperilled by a combination of adverse factors. Nature conservation including wildlife must therefore receive importance in planning for economic development.

- The non-renewable resources of the earth must be employed in such a way as to guard against the danger of their future exhaustion and to ensure that benefits from such employment are shared by all mankind [sic].

- The discharge of toxic substances or of other substances and the release of heat, in such quantities or concentrations as to exceed the capacity of the environment to render them harmless, must be halted in order to ensure that serious or irreversible damage is not inflicted upon ecosystems. The just struggle of the peoples of all countries against pollution should be supported.

- States shall take all possible steps to prevent pollution of the seas by substances that are liable to create hazards to human health, to harm living resources and marine life, to damage amenities or to interfere with other legitimate uses of the sea.

- Economic and social development is essential for ensuring a favourable living and working environment for man [sic] and for creating conditions on earth that are necessary for the improvement of the quality of life.

- Environmental deficiencies generated by the conditions of underdevelopment and natural disasters pose grave problems and can best be remedied by accelerated development through the transfer of substantial quantities of financial and technological assistance as a supplement to the domestic effort of the developing countries and such timely assistance as may be required.

- For the developing countries, stability of prices and adequate earnings for primary commodities and raw material are essential to environmental management since economic factors as well as ecological processes must be taken into account.

- In order to achieve a more rational management of resources and thus to improve the environment, States should adopt an integrated and coordinated approach to their development planning so as to ensure that development is compatible with the need to protect and improve the human environment for the benefit of their population.

- Planning must be applied to human settlements and urbanization with a view to avoiding adverse effects on the environment and obtaining maximum social, economic and environmental benefits for all. In this respect projects which are designed for colonialist and racist domination must be abandoned.

- Demographic policies, which are without prejudice to basic human rights and which are deemed appropriate by Governments concerned, should be applied in those regions where the rate of population growth or excessive population concentrations are likely to have adverse effects on the environment or development, or where low population density may prevent improvement of the human environment and impede development.

- Appropriate national institutions must be entrusted with [the] task of planning, managing or controlling the environmental resources of States with the view to enhancing environmental quality.

- Science and technology, as part of their contribution to economic and social development, must be applied to the identification, avoidance and control of environmental risks and the solution of environmental problems and for the common good of mankind [*sic*].

- Education in environmental matters, for the younger generation as well as adults, giving due consideration to the underprivileged, is essential in order to broaden the basis for an enlightened opinion and responsible conduct by individuals, enterprises and communities in protecting and improving the environment in its full human dimension.
- States have, in accordance with the Charter of the United Nations and the principles of international law, the sovereign right to exploit their own resources pursuant to their own environmental policies, and the respon-

sibility to ensure that activities within their jurisdiction or control do not cause damage to the environment of other States or areas beyond the limits of national jurisdiction.

- International matters concerning the protection and improvement of the environment should be handled in a co-operative spirit by all countries, big or small, on an equal footing. Co-operation through multilateral or bilateral arrangements or other appropriate means is essential to effectively control, prevent, reduce and eliminate adverse environmental effects resulting from activities in all spheres, in such a way that due account is taken of the sovereignty and interests of all States (Environment Canada 1972:19–21).

ical integrity and diversity, and integrating conservation and development. They incorporate many of the concepts introduced in chapters 8 to 12 (ecosystem approach, adaptive management, impact assessment, partnerships, and conflict resolution).

In addition, the UN Conference on the Human Environment resulted in an action plan containing recommendations for six general topics: human settlements, natural resource management, pollution of international significance, educational and social aspects of the environment, development and environment, and international organizations. The conference also led to the establishment of the UN Environment Programme and prompted the creation of environmental ministries or departments in many countries, a call for cooperative action to reduce marine pollution, and the initiation of a global monitoring network. In contrast, while development issues had been included in the conference agenda and were addressed in the principles, postconference activity gave them much lower priority than the environmentally oriented concerns.

Twenty Years Later: The Earth Summit, June 1992

There were actually two conferences associated with the twelve-day Earth Summit in early June 1992. The UN Conference on Environment and Development involved almost 170 countries and finished with a two-day summit that included

more than 100 heads of state, leading to the label of the Earth Summit. A second meeting of many non-governmental organizations, called the Global Forum, was a combination of meetings, NGO networking, trade show, street fair, political demonstrations, and various general events. It involved more than 1,400 NGOs accredited by UNCED and attracted 18,000 participants, as well as over 200,000 local Brazilians who visited the Global Forum site. These two events (UNCED and the Global Forum) attracted more than 8,000 journalists from 111 countries.

The UNCED process produced five major documents that were signed by almost all the heads of state: (1) the Rio Declaration, (2) the Convention on Climate Change, (3) the Convention on Biodiversity, (4) Forest Principles, and (5) Agenda 21. The Rio Declaration contains twenty-seven principles to guide countries' activities regarding environmental protection and economic development. Agenda 21, a forty-chapter document, provides goals and priorities for a broad range of major environmental, resource, social, legal, financial, and institutional issues. Agenda 21 is not a binding legal document but an agenda for action or a 'greenprint' for the future, with a political commitment from heads of state to pursue a related set of goals and activities.

To facilitate a comparison with the principles from the Conference on the Human Environment in 1972, Box 20.2 provides selected principles in

Box 20.2 Selected Principles from the Rio Declaration, 1992

• Human beings are at the centre of concerns for sustainable development. They are entitled to a healthy and productive life in harmony with nature.

• States have, in accordance with the Charter of the United Nations and the principles of international law, the sovereign right to exploit their own resources pursuant to their own environmental and development policies, and the responsibility to ensure that activities within their jurisdiction or control do not cause damage to the environment of other States or areas beyond the limits of national jurisdiction.

• The right to development must be fulfilled so as to equitably meet development and environmental needs of present and future generations.

• In order to achieve sustainable development, environmental protection shall constitute an integral part of the development process and cannot be considered in isolation from it.

• All states and all people shall cooperate in the essential task of eradicating poverty as an indispensable requirement for sustainable development, in order to decrease the disparities in standards of living and better meet the needs of the majority of people of the world.

• The special situation and needs of developing countries, particularly the least developed and those most environmentally vulnerable, shall be given special priority.

• States shall cooperate in a spirit of global partnership to conserve, protect and restore the health and integrity of the Earth's ecosystem. In view of the different contributions to global environmental degradation, States have common but differentiated responsibilities. The developed countries acknowledge the responsibility that they bear in the international pursuit of sustainable development in view of the pressures their societies place on the global environment and of the technologies and financial resources they command.

• To achieve sustainable development and a higher quality of life for all people, States should reduce and eliminate unsustainable patterns of production and consumption and promote appropriate demographic policies.

• Environmental issues are best handled with the participation of all concerned citizens, at the relevant level. At the national level, each individual shall have appropriate access to information concerning the environment that is held by public authorities, including information on hazardous materials and activities in their communities, and the opportunity to participate in decision-making processes. States shall facilitate and encourage public awareness and participation by making information widely available. Effective access to judicial and administrative proceedings, including redress and remedy, shall be provided.

• States shall enact effective environmental legislation. Environmental standards, management objectives and priorities should reflect the environmental and developmental context to which they apply. Standards applied by some countries may be inappropriate and of unwarranted economic and social cost to other countries, in particular developing countries.

• States shall develop national laws regarding liability and compensation for the victims of pollution and other environmental damage.

• States should effectively cooperate to discourage or prevent the relocation and transfer to other States of any activities or substances that cause severe environmental degradation or are found to be harmful to human health.

• In order to protect the environment, the precautionary approach shall be widely applied by States according to their capabilities. Where there are threats of serious or irreversible damage, lack of full scientific certainty shall not be used as a reason for postponing cost-effective measures to prevent environmental degradation.

• National authorities should endeavour to promote the internalization of environmental costs and the use of economic instruments, taking into account the approach that the polluter should, in principle, bear the cost of pollution.

• Environmental impact assessment, as a national instrument, shall be undertaken for proposed activities that are likely to have a significant adverse impact on the environ-

Box 20.2 Selected Principles from the Rio Declaration, 1992 (continued)

ment and are subject to a decision of a competent national authority.

- Women have a vital role in environmental management and development. Their full participation is therefore essential to achieve sustainable development.
- Indigenous people and their communities, and other local communities, have a vital role in environmental management and development because of their knowledge and traditional practices. States should recognize and duly support their identity, culture and interests and enable their effective participation in the achievement of sustainable development.
- Warfare is inherently destructive of sustainable development. States shall therefore respect international law providing protection for the environment in times of armed conflict and cooperate in its further development, as necessary.
- Peace, development and environmental protection are interdependent and indivisible (United Nations Conference on Environment and Development 1992).

---

the Rio Declaration from the Earth Summit in 1992. It should also be noted that the preamble to the Rio Declaration reaffirmed the Declaration of Principles from the Stockholm Conference.

Regarding climate change, systematic international consideration had started during 1988 with the creation of the Intergovernmental Panel on Climate Change, an advisory group of officials and scientists from many countries, which examined causes of, impacts from, and potential responses to climate change.

However, subsequent international negotiations regarding appropriate actions to ameliorate or reduce the effects of climate change were stalled due to a fundamental difference of opinion between the United States and other industrialized countries, especially those in western Europe. The Europeans favoured an international convention that included specific commitments to limit emissions of carbon dioxide (judged to be the major contributor to human-induced climate change) to 1990 levels by the year 2000. In contrast, the United States' position was that such limits were premature, that there was inadequate scientific understanding about climate change, and that controls should extend to *all* gases contributing to climate change.

The Convention on Climate Change at the Earth Summit represented a compromise between those two positions. It stipulated that industrialized countries would develop national emission standards and limits, and would report periodically on progress. However, no targets or dates were specified. Rather than specific commitments, countries accepted a more ambiguous goal of reducing their greenhouse gas emissions to 'earlier levels' that would prevent dangerous interference to the climate system. In addition, the convention states that this goal 'should be achieved within a time frame sufficient to allow ecosystems to adapt naturally to climate change. . . .' Representatives from 166 countries signed this convention in Rio de Janeiro.

Discussions about biodiversity had started during 1988 through the United Nations Environment Programme. Biodiversity and biotechnology were initially considered as separate topics, but were later merged in 1991 when addressed by a single working group appointed by the UN. The Convention on Biodiversity, signed by 153 nations (the first of which was Canada) at the Rio conference, has several goals, including conservation and sustainable use of biological diversity, and fair sharing of products from gene stocks. The countries signing the convention agreed to: (1) develop plans for protecting habitat and species (see chapters 16 and 17); (2) provide funds and technology to assist developing countries in protecting species and habitat; (3) ensure commercial access to biological resources for development and share revenues fairly among source countries and developers; and (4) establish safety regulations and accept liability for risks associated with biotechnology development.

The major stumbling-blocks in the biodiversity negotiations were the funding mechanism, the sharing of benefits, and biotechnology regulation.

Tropical rainforest vegetation in Costa Rica (© Mark Hobson).

Although some countries advocated the development of a formal forest treaty, the end result was a set of principles that committed the signatory governments only 'to keep regard to' the principles.

Haas, Levy, and Parson (1992:7) concluded that the Earth Summit's success should not be judged on how many treaties were signed or which specific actions were agreed upon. Instead, they believed that its effectiveness should be judged against the extent to which it increased attention, sophistication, and effectiveness in the management of environment and development issues. They concluded that the potential impact of the Earth Summit against such criteria was considerable. In addition to the Rio Declaration, Agenda 21, the two conventions (on climate and biodiversity), and the Forest Principles, the Earth Summit created a new United Nations organization called the Sustainable Development Commission, which would carry out the institutional objectives in Agenda 21. UNCED also created a procedure for national reporting, with participating countries encouraged to develop national reports on their environment, development, and activities designed to reduce emissions of greenhouse gases.

Financial arrangements for the implementation of Agenda 21 were addressed by the creation of the Global Environmental Facility, but most observers concluded that the final arrangements were weak. UNCED subsequently stated that developed countries reconfirmed a commitment to reach a UN target of 0.7 per cent of gross national product (GNP) for official development assistance. The cost of implementing Agenda 21 in the developing countries was estimated to be about $600 billion per year, but the commitment from the developed countries would provide only $125 billion annually. Furthermore, in 1992 the actual level of funding from the developing countries totalled only $60 billion, and it was not clear where the difference between $60 billion and $125 billion would come from.

At the time of signing the convention, only the United States refused to sign because of disagreement related to these three matters.

The United Nations Food and Agricultural Organization and the Environment Programme provided a focus for discussions about the protection of global forests. The initial discussions regarding a forest treaty were hindered by significant differences in viewpoints between industrialized nations, which favoured a treaty focused upon tropical rainforests and developing countries, which argued that temperate and boreal forests should also be included. The seventeen non-binding Forest Principles accepted at the Earth Summit explicitly incorporate all three types of forest.

Various developed countries, such as the United States and Switzerland, have made no commitment to increase their official development aid to 0.7 per cent of GNP. Other countries, while making that commitment, have not shown signs of fulfilling it. Canada falls into this second category, with different federal governments having steadily reduced foreign aid since the late 1980s, notwithstanding the commitment made by Canadian Prime Minister Mulroney (1993:73) at the Earth Summit to increase Canadian aid to developing countries. For example, in its budget in February 1994, the Liberal government, which had been elected after the Earth Summit, announced that it would reduce international aid by 2 per cent for 1994–5 and freeze international aid expenditures in 1996–7. Thus Canada and other developed nations still seem to have a long way to go in accepting and implementing leadership roles regarding global commons issues.

CLIMATE CHANGE

There is a consensus among the scientific community that increases in greenhouse gas emissions will affect climate. Considerable uncertainty, however, exists with regard to the magnitude of the effect, its timing, and its regional pattern. In addition, there is great uncertainty about changes in climate variability and regional impacts (Wong et al. 1989:1).

The major difficulties in dealing with the current generation of international and global environmental problems are, first, scientific uncertainty about their effects and second, the need for agreements among many governments to deal with them. Economists refer to this latter difficulty as the 'free-rider' problem, since costly controls are necessary to achieve improvements, and countries that fail to institute controls will receive some of the benefits resulting from the efforts of others.

Scientific uncertainty is a difficult issue because uncertainty means that it is always possible to argue that better policies can be developed by waiting until a broader scientific consensus emerges. Unfortunately, there are costs as well as benefits associated with waiting, and sometimes these costs and benefits cannot be

measured with enough accuracy to be useful (Smith 1990:112).

. . . I argue that the level of uncertainty is not enough to prohibit action. Things can be done now that will not bankrupt the world economy, and that may even be beneficial on other grounds—conceivably profitable. This attitude is not really optimistic; I prefer to call it realistic (Hare 1995:23).

The review of the Earth Summit in the previous section referred to the Convention on Climate Change. Here attention is given to the evolution of interest in climate change, and to the manner in which science and policy have interacted.

As explained in Chapter 7, greenhouse gases—mainly carbon dioxide, methane, nitrous oxide, chloro-fluorocarbons (CFCs), and tropospheric ozone—are produced as by-products of human activity. Once released and combined with the gases occurring naturally in the atmosphere, they remain in the atmosphere for decades or centuries. They absorb much of the infrared radiation released from the earth and emit some of that energy back to the earth, thereby warming its surface. A certain amount of heat trapped by greenhouse gases in the atmosphere keeps the planet warm enough to allow humans and other species to exist. However, as the proportion of greenhouse gases increases in the atmosphere, the global climate will likely change significantly, which in turn may have a variety of impacts on human activity and other life forms.

Awareness of greenhouse gases is not new, at least not to the scientific community. In the second half of the nineteenth century, scientists warned that increases in concentrations of carbon dioxide could contribute to climate change. Later, in a 1938 presentation to the Royal Meteorological Society in Britain, a scientist reported on the increase in carbon dioxide that had occurred since the beginning of the century. In 1979, as a consequence of the growing scientific interest in and concern about the possibility of human-induced climate modification, the World Climate Programme was established by the World Meteorological Organisation, the United Nations Environment Programme, and the International Council of Scientific Unions. The purpose was to coordinate national climate research programs around the world.

The results from this research led to a number of global initiatives. One of the first actions resulted from the Vienna Convention in 1985, which initiated coordinated international monitoring and assessment activities regarding greenhouse gases. The first major policy initiative to control the human release of greenhouse gases was the 1987 Montreal Protocol on Substances That Deplete the Ozone Layer. Signed by thirty-two nations, including the members of the European Community (which accounted for more than 85 per cent of the world consumption of CFCs and halons), the protocol outlined a schedule for reducing the consumption of CFCs and halons:

- reduce consumption of CFCs in 1989–90 to 1986 consumption levels
- reduce consumption of CFCs in 1993–4 to 80 per cent of 1986 levels
- reduce consumption of CFCs in 1998–9 to 50 per cent of 1986 levels
- freeze consumption of halons in 1992 at 1986 levels

The protocol directly addressed the free-rider issue by including provisions for trade sanctions against countries that chose not to participate in implementing the protocol agreement. It also incorporated provisions to recognize the special needs in developing countries and establish a slower reduction schedule for such countries, as well as the possibility for financial assistance from the developed countries.

Provision was made for revising the 1987 agreement based on new scientific understanding of the role and impact of CFCs and halons. Revisions were made at a summer 1990 meeting in London of the countries that had signed the Montreal protocol. The revisions specified an accelerated reduction schedule. For example, controlled CFCs were to be reduced to 50 per cent of 1986 levels by 1995, to 15 per cent by 1997, and to be phased out entirely by 2000. At the same time, halons were to be cut to 50 per cent by 1995 and to be phased out by 2000. A 1992 meeting in Copenhagen further accelerated these schedules and added new ozone depleters, such as methyl bromide.

The next major international action regarding climate change occurred when representatives from forty-six nations met in Toronto during June 1988 at the World Conference on the Changing Atmosphere. At the conclusion of that meeting, the participants released a statement declaring that 'humanity is conducting an unintended, uncontrolled, globally pervasive experiment whose ultimate consequences could be second only to a global nuclear war'. In terms of action, the conference delegates recommended an international policy of reducing emissions by 20 per cent from their 1988 levels by the year 2005. Furthermore, given the important role of energy consumption patterns on climate change, it was recommended that greater efficiency in energy use and reliance on less polluting fuels was necessary. Two years later, in 1990, at the Second World Climate Conference, governments from around the world recognized that a convention should be negotiated to protect the global climate system from human activity. The outcome was the signing of the Convention on Climate Change at the Earth Summit in Rio de Janeiro in June 1992.

Alternative Responses to Climate Change

Scientific investigations have indicated that the composition of the earth's atmosphere has been changing. Emissions from human activities have significantly increased atmospheric concentrations of greenhouse gases. Scientists have also concluded that the changes are rapid and unprecedented in magnitude and extent. However, there is much less certainty regarding the effects or impacts of climate change on regions or countries, and the uncertainty is even greater regarding any specific location. Thus who will be affected, how they will be affected, and to what extent they will be affected are still questions that have not been answered with certainty. The outcome has been debate over the most appropriate type of response.

One type of response has been characterized as a *remedial approach*. The idea is to wait until there is conclusive evidence of environmental degradation associated with climate change before allocating human and financial resources to correct such damage. Another consideration underlying this approach is that it may be unnecessary to take expensive action now as future technological innovations will provide solutions to the problems.

However, criticisms of the remedial approach are that it is inappropriate because of humankind's inability to restore atmospheric greenhouse gases to preindustrial levels in less than a century, as well as its costliness because related damage will often be irreversible and foreclose opportunities for future generations.

Those opposing remediation have argued for *preventive* and *anticipatory* approaches. As Hewson (1993:13) explained, 'these aim at *preventing* the impacts that cannot be accommodated and taking measures based on *anticipating* the impacts that cannot be prevented. These two classes of responses are called *limitation* and *adaptation*'. The preventive strategy would involve restricting or reducing activities that contribute to concentrations of carbon dioxide in the atmosphere. Regulation is the key tool for prevention, which requires well-defined enforcement procedures. The anticipatory strategy focuses upon either neutralizing carbon dioxide concentrations through a combination of regulations and technology, or adapting human activities to changing conditions. A combination of preventive and anticipatory strategies would reflect the adaptive approach to environmental management discussed in Chapter 9.

Implications for Canada

The federal government has commissioned a set of investigations through its Climate Change Program. The intent is to identify the impacts of possible climate change and to develop alternative responses. For example, studies have examined the consequences of a 1-m rise in sea level as a result of climate change in both Charlottetown, Prince Edward Island (P. Lane and Associates Ltd 1988) and St John, New Brunswick (Martec Limited 1987). Other studies, listed at the end of this chapter, have addressed the implications of climate change for agriculture in the prairie provinces, natural resources in Quebec, the boreal forest and forest economics in western Canada, transportation, water management, and sea ice in the Arctic relative to the petroleum industry. These studies indicate that climate change could have significant implications for regional economies, and that a *remedial approach* is unlikely to ameliorate the consequences. As a result, attention is increasingly turning to preventive and anticipatory strategies.

THE ATLANTIC FISHERIES

. . . northern cod off Newfoundland recently declined slightly, although TACs [total allowable catches] were reduced substantially when it was realized that previous exploitation rates had been higher than the target levels. The stock is now increasing as a result of improved recruitment (Environment Canada 1991:8–7).

The state of our ignorance is appalling. We know almost nothing of value with respect to behaviour of fish. We don't even know if there's one northern cod stock, or many, or how they might be distinguished. We don't know anything about migration patterns or their causes, or feeding habits, or relationships in the food chain. I could go on listing what we don't know. . . . Our technology has outstripped our science. We have under-estimated our own capacity to find, to pursue, and to kill (Leslie Harris, chairman of the Northern Cod Review Panel, 1990, quoted in Cameron 1990:29, 35).

We are facing an unprecedented crisis in the groundfish stocks. Immediate action is needed to permit this fragile resource to rebuild (Federal Fisheries Minister Ross Reid, quoted in *Kitchener-Waterloo Record*, 31 August 1993:A3).

The marine fishery has been an essential component of the economy and culture of Atlantic Canada for centuries. After 1977 when Canada declared a 200-nautical mile (370-km) fishing limit off its coasts, codfish were the mainstay for more than 50,000 fishers and 60,000 fish-plant workers in Atlantic Canada. In Newfoundland and Labrador alone, about 700 communities depended entirely upon the fishery. In 1991 the value of the harvested cod to fishers in the Atlantic provinces was $226 million.

However, between July 1992 and December 1993, decisions to cut back on the rate of harvesting resulted in lost jobs for 40,000–50,000 fishers or shore workers in Newfoundland, the Maritime provinces, and Quebec. How could this dramatic collapse of a renewable resource occur in such a relatively short time? Why was it not anticipated by fisheries scientists? Why was action not taken earlier to avoid such degradation of the fishery and

It is difficult for urban dwellers to imagine the close relationship that built up over the centuries between the people in the outports of Newfoundland and the sea. Virtually every family would be involved in some way with fishing. When the fish were exposed to such fishing pressure that they could no longer be caught in any numbers, it was not just the economy that suffered but a whole way of life (*Tourism Newfoundland and Labrador/W. Sturge*).

species: cod, haddock, flounder, pollock, hake, herring, redfish, crab, scallop, and lobster.

In contrast, the Labrador coast fishery is dominated by the northern cod stock and extends east of the Labrador coast and north and east of Newfoundland. The northern cod traditionally yielded about half of Atlantic Canada's cod catch and one-quarter of all groundfish landings in the region. In many ways it has been the backbone of the Atlantic fishery. That is why so much attention has focused upon it and why the collapse of the northern cod stocks has been such a blow to the regional economies in which fishing provides an important percentage of the jobs (up to 25 per cent in parts of Newfoundland). The northern cod are caught by larger inshore vessels, and especially by offshore draggers, multimillion dollar vessels that drag a huge net across the bottom of the ocean. However, the northern cod migrate to the shores of Newfoundland in the summer, and thus also support an inshore fishery that relies on much smaller boats using traps, hooks, and nets. As Cameron (1990:30) explained, the inshore fishery has been an important one. Until the late 1950s, the inshore catch was usually never less than 150 000 tonnes. By 1974, as a result of overfishing by boats from more than twenty nations, the inshore catch had fallen to 35 000 tonnes. After Canada declared its exclusive fishing zone in 1977 and banned fishing by foreign draggers in that area, the inshore catch increased. It peaked at 115 000 tonnes in 1982, but fell to 68 000 tonnes by 1986, and inshore fishers were complaining that the fish being caught were very small. The first signs of serious problems were identified by local fishers. Unfortunately, the models used by the fishery scientists indicated that the stocks were still abundant, so these early warnings were not heeded.

disruption to the economy of a region? Is it possible for the fishery to rebound and once again become a key mainstay in the regional economy? If recovery is not possible, what are the people in communities dependent upon the fishery supposed to do? What new jobs could they be retrained and educated for? Is it realistic for residents of Joe Batt's Arm or Twillingate in Newfoundland to participate in the new economy and remain in their communities?

The Nature of the Collapse

Box 20.3 presents a chronology of events and decisions that led to and followed the virtual closing of the Atlantic groundfishery by the end of 1993. In viewing these events, however, it is important to recognize that there is not a single groundfishery in Atlantic Canada. Indeed, the fishery has been focused upon four different areas: the Scotian shelf, the Gulf of St Lawrence, the Grand Banks, and the Labrador coast. Two areas most affected in the harvesting cutbacks have been the Scotian shelf, which extends from the mouth of the Bay of Fundy to the northern tip of Cape Breton Island, and the Labrador coast (Figure 20.1). These areas supported two different kinds of fishery. The fishery on the Scotian shelf is readily accessible to the inshore fishers along the coast of Nova Scotia and New Brunswick, and includes a wide mix of

Some Reasons for the Collapse

A single cause is unlikely to account for the depletion of the groundfishery stocks. It is more probable that a combination of reasons triggered the problem. The main factors are likely to have been

Box 20.3 The Path to the Collapse of the Atlantic Groundfishery

1958–78: Northern cod stocks reach a historic low. Following a peak high of 810 000 tonnes caught in 1968, the catch drops to 137 000 tonnes in 1978.

1977: Canada declares a 200-*nautical* mile (370-km) exclusive fishing zone, and requires foreign fleets to fish beyond that limit. Stock rebuilding starts.

1982: Michael Kirby is appointed as head of a task force to study the difficulties of the East Coast fishery. The task force concludes that northern cod is a bright spot in an otherwise discouraging situation, and its total catch is expected to reach 380 000 tonnes by 1987.

1986: Catches of northern cod continue to increase, reaching 252 000 tonnes, almost twice what they were in 1978. Scientists are mainly optimistic about the stock, but inshore fishers in Newfoundland express concern about what they consider depletion of the stock.

1987: A task group appointed by the federal fisheries minister concludes that the northern cod stock is still increasing but at a lower rate than had been anticipated several years previously.

12 February 1989: Based on new scientific advice, the federal fisheries minister reduces the total allowable catch (TAC) for northern cod to 235 000 tonnes, down from 266 000 tonnes in 1988. He also appoints Leslie Harris, president of Memorial University in St John's, to head another independent review.

1 January 1990: The TAC for northern cod is reduced to 197 000 tonnes.

7 May 1990: The federal government announces a $584 million, five-year program to help communities affected by the falling fish quotas.

October 1990: Harris's final report recommends that the TAC for northern cod be reduced to between 100 000–150 000 tonnes for 1991. The federal government decides the 1991 quota will be 188 000 tonnes.

1991–2: Northern cod catches drop sharply. Scientists agree that the stock is in serious trouble, but they do not know why.

24 February 1992: The federal fisheries minister reduces the northern cod TAC to 120 000 tonnes, and curtails offshore fishing for six months.

Table 20.1 Atlantic Groundfish Canadian Allocations and Catches[1]
Thousand Tonnes 1978–1993

Year	All Groundfish Allocations	Catches	All Cod Allocations	Catches	Northern Cod Allocations	Catches
1978	472	535	204	271	100	102
1979	562	634	270	359	130	131
1980	705	615	353	400	155	147
1981	790	741	400	422	185	133
1982	924	775	490	508	215	211
1983	997	728	561	505	240	214
1984	1005	700	553	466	246	208
1985	1003	738	576	477	250	193
1986	973	748	530	475	250	207
1987	969	723	512	458	247	209
1988	985	688	523	461	266	245
1989	942	652	478	422	235	215
1990	812	604	408	384	197	188
1991	812	572	399	311	188	133
1992	808	418	333	182	120	21
1993	512	N/A	121	N/A	Moratorium	

Note: [1]All groundfish species and stocks regulated and managed under the annual Atlantic Groundfish Management Plan (i.e., excluding non-regulated and commercially marginal species, such as cusk and catfish).

Source: Task Force on Incomes and Adjustments in the Atlantic Fishery, *Charting a New Course: Towards the Fishery of the Future* (Ottawa: Minister of Supply and Services Canada, 1993):124.

Box 20.3 **The Path to the Collapse of the Atlantic Groundfishery** (continued)

2 July 1992: The federal minister of fisheries announces a moratorium on northern cod fishing for two years (until May 1994), and explains that Ottawa will provide $500 million (later to become $772 million) to compensate the 20,000 fishers and plant workers expected to lose their jobs because of the moratorium. In March 1994 the federal government announces another $140 million to assist Atlantic fishers until 15 May 1994, at which time a new cod compensation program would take effect. Thus the total cost in assistance from 1992 to May 1994 is $912 million.

July 1993: The Fisheries Resource Conservation Council (an advisory group for the federal minister) report blames the crisis on overfishing, changes in migration, harsh climatic conditions, poor feeding, and seals and other predators. The relative importance of these causes is not known.

August 1993: Another Fisheries Resource Conservation Council report recommends sweeping cuts in the Atlantic cod fishery, which would affect another 9,000–12,000 workers. This recommendation would all but close the cod fishery.

The federal government bans cod fishing in five more areas and sharply reduces quotas for other valuable species. Fishers have seven days to remove their nets from the newly affected areas.

The result is a total loss of about 35,000–40,000 fisher and fish plant jobs in Atlantic Canada since the closures began in July 1992. Analogies are made to closing the automotive industry in Ontario and losing 800,000 jobs there, plus crippling cities such as Windsor and St Catharines.

30 November 1993: The Fisheries Resource Conservation Council urges that the two-year fishing ban on northern cod off Newfoundland be extended and broadened to include the Gulf of St Lawrence and areas off Nova Scotia.

December 1993: The Task Force on Incomes and Adjustment in the Atlantic Fishery report calls for halving the number of fishery workers by buying licences from fishers and fish-processing plants, offering early retirement packages, and retraining people.

The federal fisheries minister closes all but one Atlantic cod fishery and curtails catches of other species. Approximately another 5,000 people have lost their jobs.

31 January 1994: The federal government announces a ban in most of Newfoundland on catching cod for personal use because the fish stocks are so depleted that they are threatened even by this small hook-and-line fishery. The federal government suggests that this ban was necessary because too many Newfoundlanders were not fishing for personal use but were catching fish for private sale, often to restaurants.

18 February 1994: The Northwest Atlantic Fisheries Organization (NAFO) votes by a majority to stop fishing for one year the declining stocks that congregate on the southeast tip or tail of the Grand Banks, beyond the 200-nautical mile limit. Of the eleven NAFO members at the meeting, eight approved the ban and three (Denmark, Norway, and the European Union) abstained. The three members that abstained had argued that there were enough young cod to allow fishing to continue. The NAFO members that fish the most in the area are the EU countries, especially Spain, Portugal, Russia, and the Baltic states.

23 February 1994: The Liberal government's federal budget indicates that Atlantic cod fishers will receive $1.7 billion over five years.

1 April 1994: Canadian fisheries officials arrest the freezer trawler, *Kristina Logos* (registered in Panama and carrying an all-Portuguese crew, but owned in Nova Scotia) for fishing beyond the 200-nautical mile limit on the southern Grand Banks in restricted waters. Technically, Canada has no authority beyond the 200-mile limit, but the fisheries minister declares that Canada is prepared to seize ships operating just outside the limit in order to protect endangered species. The *Kristina Logos* had 107 tonnes of fish on board, more than half of which was the protected cod and flounder.

19 April 1994: The federal government announces a new five-year, $1.9 billion aid package for unemployed Atlantic fishers and plant workers after the closure of the Atlantic groundfishery. The new aid package is designed to help some 30,000 people in Newfoundland, the Maritimes, and Quebec. The fisheries minister also explains that some people will have to leave the fisheries.

Box 20.3 The Path to the Collapse of the Atlantic Groundfishery (continued)

12 May 1994: A new bill was quickly passed into law by the federal government, giving fisheries officers legal authority to arrest ships fishing beyond the 200-mile limit in areas covered by agreements to curtail fishing. Ships fishing in contravention of the agreement can be arrested and towed to St John's, where the captain and owners could face prosecution and penalties, including fines up to $500,000 and seizure of the vessel.

22 July 1994: Fisheries Minister Brian Tobin announces an easing of the fishery ban for personal use. He also acknowledged that it had been a mistake to allow tourists special rights to jig for cod while preventing Newfoundlanders from doing the same thing. Between 26 August and 30 September, Newfoundlanders were to be allowed to catch ten groundfish a day, twice a week.

27 July 1994: Two American scallop draggers suspected of fishing illegally about 15 km outside the 200-mile limit were intercepted by HMCS Fraser and a fisheries patrol boat, and escorted to St John's. The captains of the two vessels were to be charged for fishing without a Canadian licence, which could result in a maximum $750,000 fine. The US government issued a strong protest and demanded that the vessels be released. Trials were set for October and December.

11 August 1994: Two other American fishing vessels were seized in separate actions for allegedly fishing in Canadian waters. The fisheries minister indicated the following week that Canada would continue to act unilaterally to protect depleted fish stocks until a strong international agreement was developed. Canada advocated a binding convention that would allow regional fisheries organizations to set total catch limits, create mechanisms to resolve dis-

putes, and enforce rules. Japan and the European Union countries were resisting a binding agreement and preferred a less stringent UN resolution.

14 October 1994: The fisheries minister announced that $300 million was being allocated to allow the federal Department of Fisheries and Oceans to buy licences from fishers who caught the endangered cod, flatfish, and haddock stocks. The Department of Fisheries and Oceans would also pay a pension for retirement prior to age sixty-five. The intent was to reduce the more than 20,000 Atlantic fishers who had traditionally depended on the threatened fish for their living. It was believed that even if the stocks recovered, less than half of the present number of fishers could expect to make a living from the sea. It was hoped that this program would reduce the fishing power of the Atlantic fleet by half.

1 January 1995: Some 3,500 former fishers and plant workers in Atlantic Canada are cut off from the federal aid package as eligibility rules are tightened. Over 3,000 of the affected people are in Newfoundland.

19 May 1997: Some 5,000 fishers, mostly from Newfoundland but some from Quebec, became involved in a limited test fishery in the waters off southern Newfoundland and the northern Gulf of St Lawrence. This was the first commercial fishery allowed since the closure of the cod fishery four years previously.

The catch limit was set at a modest 16 000 tonnes, and the purpose of the fishery was to allow officials to obtain more information regarding the state of the cod stocks. The decision to start the test fishery was made despite advice from some scientists who argued that the stocks were still too fragile to support a commercial fishery.

foreign and domestic overfishing, faulty science and management, inappropriate incentives for processing plants, changing environmental conditions, and predators. As the regional director of science for the federal Department of Fisheries and Oceans in St John's is alleged to have said: 'No one in this fishing industry, either federally, provincially, science, management or whatever, should be pointing fingers unless they are facing the mirror' (Hall 1994:D4).

Foreign Overfishing

Once Canada established the 200-nautical mile fishing limit in 1977, foreign fleets were required to fish outside that boundary or fish in Canadian waters only for that portion of domestic quotas not taken by Canadian vessels. During the 1970s and early 1980s foreign fleets mostly followed the rules for harvesting and were monitored by the Northwest Atlantic Fisheries Organization. How-

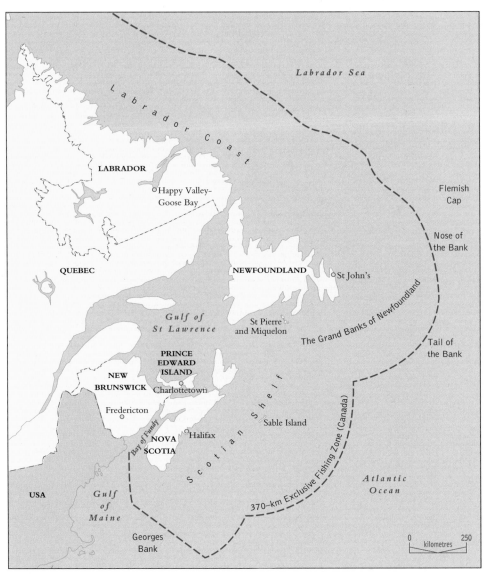

Figure 20.1 Major fishing areas in Atlantic Canada. SOURCE: S.D. Cameron, 'Net Losses: The Sorry State of Our Atlantic Fishery', *Canadian Geographic* 110, no. 2 (199):30.

Thus there is strong evidence that foreign vessels, especially those from Spain and Portugal, were overfishing at least during the mid- and late 1980s. Since cod migrate towards the coast in summer and then move offshore in winter to spawn in deeper waters, this behaviour made the fish vulnerable to foreign (over)fishing, which operated year round.

Domestic Overfishing

Despite the foreign fishing vessels' pressure on the stocks, the majority of the key fishing grounds have been under Canadian control since the 200-mile fishing limit was created in 1977. The challenge is managing at least two fisheries (inshore and offshore).

The inshore fishery has many fishers, especially from Newfoundland, who rely on small wooden boats, lines, traps, and nets during the spring and summer when the cod move close to shore. Until the mid-1950s, that fishery, combined with limited offshore fishing by Canadian boats, resulted in annual landings of 200 000 tonnes or more. Foreign fishers were harvesting another 30 000–50 000 tonnes each year. Such harvesting did not appear to affect adversely what was estimated to be a breeding stock of 1.6 million tonnes in the North Atlantic.

In the mid-1950s the introduction of large offshore trawlers, which operated year round in the North Atlantic, was a significant change to this pattern. Initial catches were very high, but the spawning stocks were placed under great pressure. For example, in the 1970s yields reached a high of 800 000 tonnes per year before they started to drop. Until 1977, most of the offshore fishing was done by foreign trawlers. Following that year and

ever, in 1986 Spain and Portugal entered the European Community (EC), and that year the EC unilaterally established quotas that were considerably higher than those set by NAFO (Figure 20.2). Furthermore, the EC boats harvested fish well beyond the EC limits. The EC then raised the quota the next year, which was again exceeded by the actual catch. In 1988 just half of the EC target was achieved, even though it was 4.5 times higher than the recommended target set by NAFO. The EC, now known as the European Union, later rejected NAFO's northern cod moratorium. In 1993, however, the EU finally accepted all NAFO quotas, after having set its own quotas at a much higher level since the mid-1980s.

the establishment of the fishing limit, the Canadian offshore fleet expanded and it was Canadian off-shore trawlers that became the main harvesters of northern cod. By the time the moratorium was placed on the fishery in the summer of 1992, Newfoundland was the base for fifty-five large and thirty medium-size offshore trawlers. Thus Canadian offshore draggers, operating on a year-round basis, placed considerable pressure on the groundfishery stock.

Critical in this regard are the ecology and behaviour of the northern cod. Initially, the harvesters caught a mix of ages and sizes of fish. However, market demand and net mesh sizes led to a focus on larger fish, which are also the older ones. It is important to understand that cod swim in groups or schools of similar ages, primarily because the larger cod will eat smaller and younger cod. The emphasis on larger fish had two consequences. First, the northern cod normally do not begin to spawn until they reach seven years. Second, older fish produce more eggs. As the larger fish became scarce, the fishery then concentrated on fish in the five- to seven-year age range. The outcome was that by the early 1990s most of the older fish had been overharvested, and attention shifted to fish that would be categorized as preadolescent. They were being caught before they had spawned. The consequence was a dramatic decline in the fish stock.

A further complication was that some domestic fishers overharvested. As Cameron (1990:35) was told, fishers readily admitted that cheating was easy and profitable. Fishers could bring their catch to a local processor without reporting it, or truck it directly to the Boston market. Some estimates were that up to 50 per cent more fish were being landed than were being reported, and, as one fisher commented, 'There is no shame to getting caught and paying a $400 fine' (Cameron 1990:35). Often this cheating was driven by a need to catch enough fish to pay for the costs of increasingly sophisticated vessels. As Cameron (1990:35) noted, greed and desperation are powerful motivators.

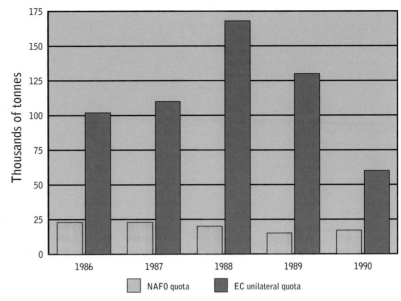

Figure 20.2 NAFO and EC quotas, 1986–90. SOURCE: P. Hall, 'Crisis in the Atlantic Fishery', *Canadian Business Review* 17, no. 2 (1990):47.

Imperfect Science and Management

Fishery scientists did not anticipate the collapse of the Atlantic fishery, especially the northern cod stocks. Why? There are a variety of reasons.

As Cameron (1990) commented, the sampling procedures do not provide sufficient ecological information about the fish stocks. An American scientist outlined the difficulties by drawing an analogy with estimating the number of cattle on a ranch. If you prepared such an estimate by dragging a large bag hung from a helicopter across the ranch at night, and then counted how many cows were in the bag by morning, you would be making your estimate in the same way that fishery scientists make their estimates about fish stocks. Few people would put much faith in an estimate of cattle derived from such a sampling method, yet that is basically the type of information fishery scientists work from. As Robert Fournier, an ocean scientist at Dalhousie University is reported to have said, 'We're dealing with animals you can't see, in an environment which is basically opaque. So you're trying to understand the biology of these things from a distance' (Spears 1993:A9). As a result, in his view, the problem is that we simply do not know enough about why fish populations rise or fall.

A dramatic illustration of this problem of accurately estimating numbers of fish was presented by Gwyn (1993). He noted that the scientists had to accept a special blame in the collapse of the fish stocks since they 'lost 500 000 tonnes of cod'.

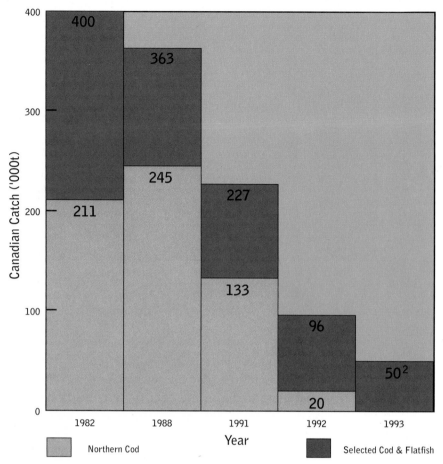

Figure 20.3 Collapse of Newfoundland's groundfish base: decline in catch from major stocks[1]. NOTE: [1]Canadian catch of six major groundfish stocks: 2J3KL cod, 4RS, 3Pn cod, 3Ps cod, 3NO cod, 3LNO American plaice, 3LNO yellowtail founder. [2]Based on 1993 quotas and catch to date (31 527t as of 15 September 1993) SOURCE: Task Force on Incomes and Adjustments in the Atlantic Fishery, *Charting a New Course: Towards the Fishery of the Future* (Ottawa: Minister of Supply and Services Canada, 1993):30.

viewed as credible evidence by the scientists, which made this a classic situation in which local or indigenous knowledge was not considered equal to science-based knowledge. The scientists had other sources of information about the fishery beyond their own samples, but initially they chose to ignore them.

Inappropriate Incentives for Processing Plants and Fish Workers

The fish-processing plants on shore are licensed by the provincial governments. By the early 1990s Newfoundland had about 100 large and small fish-processing plants, two-thirds of which processed northern cod. In Atlantic Canada as a whole, the number of plants increased from about 500 in 1977 to nearly 900 in 1988. Employment grew from some 25,000 full-time jobs to about 33,000. It has been suggested that the provinces provided incentives for creating new processing plants as a way to create new jobs for small communities, which put political pressure on the federal Department of Fisheries and Oceans to keep increasing the total allowable catch. Another incentive for people to enter or stay in the fishing industry was the federal unemployment insurance program. In 1956 the federal government introduced the '10/42 scheme'. For people working in fish-processing plants, it was possible to collect unemployment insurance for forty-two weeks of the year after working for ten weeks. This arrangement resulted in several individuals in a community sharing one job, but all qualified for separate benefits. For fishers, the unemployment benefits were based on the sale value of fish caught during the May to November season (twenty-six weeks), which created the potential to collect unemployment benefits for the other twenty-six weeks of the year.

Until 1991, the total allowable catch was based on the assumption of a biomass of 1.1 million tonnes of cod. However, in the autumn of 1991, the sampling from the federal Department of Fisheries and Oceans' research ships indicated only 600 000 tonnes. Two years later, the chief scientist of the federal Department of Fisheries and Oceans in St John's could not explain what had happened. Sampling in distant areas did not reveal that the cod had migrated to other areas. Significant numbers of diseased or dead fish had not been found. The scientists simply did not know what had happened. This led to the development of different theories to account for the missing fish.

On the other hand, the scientists had been receiving warnings from the inshore fishers during the mid-1980s that the fish being caught were fewer in number, smaller in size, and lower in weight. However, these observations were not

This program, intended to provide a social safety net, had several undesirable consequences. It encouraged more people to become involved in the Atlantic fishery than could be justified economically. There was little incentive to consider other types of work. It also helped to reinforce an outlook in which little value was placed on education, often because little financial support was available for educational facilities in the outports. A result was that by time the fishery was closed, 50 per cent of Newfoundland's nineteen year olds were already on unemployment insurance, and 80 per cent of the fishery workers did not have a high school diploma. Thus the fishery had too many people than could be realistically supported over the long term, yet unemployment insurance programs provided little incentive for individuals to consider alternatives. This made the trauma of the fishery closure in the summer of 1992 even greater than it might have been otherwise.

Changing Environmental Conditions

One theory for the groundfishery depletion is based on the idea of environmental change. Records show that in 1991 the ocean temperatures off Newfoundland were the coldest ever measured, and it has been suggested that there had been several years of very cold water temperatures for several years before that. The water warmed slightly in 1992, then cooled again in 1993. It is possible that colder waters, combined with overfishing from the previous two decades, have been preventing or inhibiting the shrunken stocks from regenerating. However, because relatively little is known about the migratory patterns of the northern cod, it is difficult to determine what the specific implications of changing water temperatures may have been.

Predators

Another possible cause is seals, identified because of their so-called 'voracious appetites' and their growing numbers due to the closure of the seal hunt in the early 1980s as a result of pressure from animal rights activists, especially

> *This government will work with the sealers and fishermen to rebuild a viable and expanded commercial seal hunt next year.*
>
> Brian Tobin,
> minister of fisheries
> and oceans, 28 June 1995

those in Europe. However, as Hall (1990:47–8) indicated, the evidence is not conclusive.

Harp seals have been the greatest concern in this regard for several reasons: they eat commercially important fish; they are a predator of capelin, the main food for cod; and they are believed to carry parasites, which are passed to commercially important fish species. Available data do not confirm the first two concerns, but have suggested that a growing increase in parasites in fish may be attributable to the larger number of harp seals. Grey seals have also been viewed as a threat to the fishery. However, David Lavigne (1986), a zoologist at the University of Guelph, has concluded from his studies of the grey seal that there is no simple relationship between the numbers of seals and the numbers of parasites in fish, and that scientific evidence, at least in the mid-1980s, did not support killing seals to relieve the problem.

The northwest Atlantic harp seals give birth off the east coast of Canada from late February until mid-March (*International Fund for Animal Welfare*).

Box 20.4 Seals: Deadly Predators or Political Scapegoats?

The annual hunt for seals off the East Coast was probably Canada's biggest foray on the international stage regarding environmental matters. Famous celebrities, such as Brigitte Bardot, were shown nightly on televisions across the world as they tried to protect helpless white-coated seal pups from being clubbed to death. Eventually, following bans by the US and the European Union on the import of seal pelts, the Canadian government banned the hunt in 1987. However, it now appears that a new hunt will be promoted, ostensibly to help in the recovery of the endangered cod stocks.

Seals eat fish. Everyone agrees on that. However, how many seals eat how many fish of various species is highly contentious. Federal scientists estimate that each seal eats about 18 kg of Atlantic cod per year. If their estimate of the seal population is correct (about 4.8 million), that amounts to a take of over 81 000 tonnes per year, enough to put a considerable dent in the recovery of the fish stock. However, there is widespread disagreement about these figures. Some estimates of seal numbers are as low as 2 million. Federal researchers have found that virtually all the cod eaten by harp seals are the smaller and not commercially fished arctic cod. Nonetheless, in 1995 sealers killed 60,000 seals out of a government quota of 186,000. In 1996 the government increased the quota and also paid seal hunters a bounty of 20 cents a pound for seal meat. The cost to taxpayers was esti-mated at $1.5 million. This is, in effect, a make-work program.

International environmental groups have suggested that the real reason behind the desire to increase the seal hunt is not related to seal predation on cod but more related to the need for a political scapegoat in economically depressed areas and the demand for seal penises in the Asian market.

What You Can Do

If you want more information on the seal hunt, the International Fund for Animal Welfare (IFAW), which has been working hard to raise public awareness of the hunt, suggests doing the following:

1. Contact your local MP and the prime minister to ask them to state their position on the hunt and to voice your opinion. Make it known that, based on their position, you will vote accordingly in the next election.
2. Contact the local media and push for more attention to the issue. If you feel that media coverage is biased, bring it to the media's attention and insist on fair coverage.
3. Contact IFAW for more information on the seal hunt and make a donation: International Fund for Animal Welfare, 400–410 Bank Street, Ottawa, Ontario K2P 1Y8; (1-888-500-4329 or 1-613-233-8458, FAX 1-613-233-9602).

Lessons

The collapse of the Atlantic groundfishery has highlighted how some contemporary resource management practices may not be sustainable. It has also emphasized that management of the fishery requires more than scientific understanding of the biophysical resource system. The Atlantic fishery is an example of how this resource can be managed only with a parallel understanding of the history, culture, economy, and politics of the region, and of federal and provincial fisheries and regional development policies. The Atlantic fishery also shows how inexact, complex, and uncertain science often is. It also demonstrates conflict among different values and interests, and how conditions can change dramatically over a relatively short time. As with forestry, the situation is readily comparable with the framework introduced in Chapter 1. Not only were our simplified models of the complex biophysical system and our resulting attempts to assess the status of the system inadequate, societal expectations and management directions were not clear. Some of the approaches suggested in Part C relating to identification of stakeholders, resolving conflict, and taking more ecosystem-based and adaptive approaches are applicable to these aspects.

IMPLICATIONS

Change, challenge, uncertainty, complexity, and conflict. Issues linked to global commons issues reflect all of these attributes. The basic purpose of the 1992 Earth Summit was to prepare the global community for changing conditions in the twenty-first century. However, different interests and priorities in developed and developing countries created significant conflict and disagreement, which often hindered decisions. Climate change is a function of changing greenhouse gas concentrations in the atmosphere. There is considerable uncertainty and complexity about the timing, magnitude, and extent of the impacts. The Atlantic fishery seemed to 'collapse' in a couple of years, despite long-term monitoring of fish populations, illustrating both the complexity and uncertainty that must be addressed by environmental managers.

The global commons issues highlight the challenges in implementing the environmental management concepts reviewed in Part C. For example, which *ecosystems* are the appropriate ones for dealing with climate change? How can Canada manage the ecosystem pertinent to the Atlantic northern cod when those fish migrate from Canadian to international waters? What approach should be taken when no single authority has clear-cut jurisdiction over the cod? In developing agreements at meetings such as the Earth Summit, what is the best way to recognize significant regional differences regarding matters such as forests (boreal forests and rainforests) while maintaining a global perspective?

In terms of *adaptive management*, which emphasizes learning by trial and error, how many errors like the collapse of the Atlantic groundfishery can we afford in the process of learning what to do? What are the costs to humans and to other species associated with such lessons? How should we handle a trial-and-error approach to climate change when it appears that it could take more than a century to reverse concentrations of greenhouse gases in the atmosphere? Regarding *impact assessment*, do we have the understanding and capability to complete impact assessments at global scales relative to policy and other initiatives associated with climate change?

At a global scale, is it practical to think in terms of *partnerships* and *stakeholders*? Is it possible to include all relevant interests when dealing with global phenomena or processes? Who decides which groups and interests get to participate? Given the inevitability of *conflict* due to different conditions, perceptions, and needs within and among countries, what mechanisms are likely to be most appropriate to address and resolve differences?

Finally, what global changes are needed to facilitate *sustainability* or *sustainable development*? Are current practices and strategies sustainable? The experiences with the Atlantic groundfishery and climate change strongly suggest that much of contemporary human behaviour is *not* sustainable. If that is indeed the case, then what fundamental changes are required to achieve sustainable development as outlined in Chapter 2?

SUMMARY

1. The United Nations Conference on Environment and Development, or the Earth Summit, occurred in Rio de Janeiro in the summer of 1992. Its purpose was to assess progress since publication of the Brundtland Report (*Our Common Future*) in 1987 and to create a strategy for the twenty-first century. It was held twenty years after the UN Conference on the Human Environment in Stockholm, the first global meeting on the environment.

2. The Stockholm Conference led to the establishment of the UN Environment Programme, as well as the creation of environmental departments or ministries in many countries, a call for cooperative action to reduce marine pollution, and the initiation of a global monitoring network.

3. In 1992 representatives of about 170 countries met at the Earth Summit. Five major agreements were signed by most countries: (1) Rio Declaration (twenty-seven principles regarding environmental protection and economic development), (2) Convention on Climate Change, (3) Convention on Biodiversity, (4) Forest Principles, and (5) Agenda 21 (a forty-chapter report with goals and priorities for a mix of environmental, economic, and social issues requiring attention in the twenty-first century).

4. There was a major disagreement at the Earth Summit between the United States and other industrialized countries, especially those in western Europe, regarding the best approach to deal with the problems from climate change. The Conven-

tion on Climate Change was signed by 166 countries, but not by the United States.

5. The Convention on Biodiversity was signed by 153 countries and has goals for achieving conservation and sustainable use of biodiversity, and the fair sharing of products from gene stocks.

6. The development of Forest Principles generated strong disagreements between representatives of developing and developed countries, with the latter arguing for special attention to tropical rainforests, and the former arguing that all forest systems needed attention.

7. A new United Nations Sustainable Development Commission was created to implement the institutional objectives of Agenda 21.

8. The cost of implementing the recommendations of Agenda 21 was estimated to be $600 billion annually, but developed countries committed themselves to providing only $125 billion annually, and in 1992 the developing countries had actually provided $60 billion.

9. Greenhouse gases (carbon dioxide, methane, nitrous oxide, and chloro-fluorocarbons) remain in the atmosphere for a long time and absorb much of the infrared radiation released from the earth. In the second half of the nineteenth century, scientists warned that increased concentrations of carbon dioxide could result in climate change.

10. The Vienna Convention of 1985 established coordinated international monitoring of greenhouse gases.

11. The 1987 Montreal Protocol on Substances That Deplete the Ozone Layer, signed by thirty-two countries, was the first major international initiative to control release of greenhouse gases. Revisions were made in 1990 at London and in 1992 at Copenhagen to accelerate the schedule identified in the protocol.

12. There is a mix of possible responses to climate change: remedial, preventive, and anticipatory.

13. The Climate Change Program in Canada was established to identify the impacts of possible climate change and to develop alternative response strategies.

14. The marine fishery has been one essential foundation of Atlantic Canada's economy for centuries, and the northern cod stock was the backbone of the industry, but between July 1992 and December 1993, decisions to reduce the rate of fishing resulted in 40,000–50,000 fishers and shore workers losing their jobs.

15. By 1986 inshore fishers were complaining that fewer fish were being caught, and those being caught were smaller in size and lower in weight. However, because the fishery scientists' models indicated that the fishery stocks were abundant, this early warning was disregarded.

17. It is likely that several causes led to the collapse of the fish stock, with the most likely being foreign and domestic overfishing, too many processing plants, poor science, changing environmental conditions, and predators.

REVIEW QUESTIONS

1. Compare and assess the principles from the 1972 UN Conference on the Human Environment with those from the 1992 UN Conference on Environment and Development.

2. What are the strengths and weaknesses of the two conventions on climate change and biodiversity developed at the Earth Summit?

3. What progress has been made since 1992 in Canada and in your province or territory relative to the principles in the Rio Declaration?

4. What does science have to say about the change in greenhouse gases in the atmosphere during this century?

5. What were the main ideas contained in the Montreal protocol of 1987 and the World Conference on the Changing Atmosphere in Toronto in 1988?

6. What are the differences among remedial, preventive, and anticipatory approaches to climate change? What is the Canadian government's strat-

egy? What is your provincial or territorial government's strategy? Which approach, or mix of approaches, do you think is best?

7. To what extent are present patterns of energy use contributing to climate change? What could be done to modify such patterns of energy use? What could you do as an individual to change either your or society's patterns of energy use?

8. Why is so little known about the ecology of the northern cod in the north Atlantic Ocean?

9. What were the main factors causing the 'collapse' of the groundfishery in the north Atlantic Ocean? Why are there disagreements regarding the relative importance of different possible factors for the collapse?

10. How can we explain the suddenness of the 'collapse'? What lessons should be learned from this experience?

11. If you were the federal fisheries minister, what would be your approach to the groundfishery in Atlantic Canada? Would your perspective be different if you were premier of Newfoundland or Nova Scotia? If you were a fisheries scientist? If you were a fisher or plant worker?

12. What would be the key components of a strategy to ensure a sustainable future for the groundfishery in Atlantic Canada?

REFERENCES AND SUGGESTED READING

Antler, E., and J. Faris. 1979. 'Adaptation to Changes in Technology and Government Policy: A Newfoundland Example'. In *North Atlantic Maritime Cultures*, edited by R. Andersen, 129–54. The Hague: Mouton.

Arthur, L.M. 1988. *The Implications of Climate Change for Agriculture in the Prairie Provinces*. Climate Change Digest 88–01. Ottawa: Minister of Supply and Services Canada.

Brown, R.D. 1993. *Implications of Global Climate Warming for Canadian East Coast Sea-Ice and Iceberg Regimes over the Next 50–100 Years*. Climate Change Digest 93-03. Ottawa: Minister of Supply and Services Canada.

Caldwell, L.K., ed. 1990. *International Environmental Policy: Emergence and Dimensions*, 2nd ed. Durham, NC: Duke University Press.

Cameron, S.D. 1990. 'Net Losses: The Sorry State of Our Atlantic Fishery'. *Canadian Geographic* 110, no. 2:28–37.

Clarke, R., and L. Timberlake. 1982. *Stockholm Plus Ten*. London: Earthscan.

Department of Fisheries and Oceans. 1989a. *Underwater World: Atlantic Cod*. Ottawa: Department of Fisheries and Oceans.

_____. 1989b. *Today's Atlantic Fisheries*. Ottawa: Minister of Supply and Services Canada.

_____. 1990. *Economic Overview of the Fishing Industry in Atlantic Canada*. Ottawa: Minister of Supply and Services Canada.

DPA Group Inc. 1988. *CO_2 Induced Climate Change in Ontario: Interdependencies and Resource Strategies*. Climate Change Digest 88-09. Ottawa: Minister of Supply and Services Canada.

Environment Canada. 1972. *Conference on the Human Environment*. A report on Canada's preparations for and participation in the United National Conference on the Human Environment, Stockholm, Sweden, June 1972. Ottawa: Information Canada.

_____. 1991. *The State of Canada's Environment*. Ottawa: Minister of Supply and Services Canada.

Finlayson, A.C. 1994. *Fishing for Truth: A Sociological Analysis of Northern Cod Stock Assessment from 1977 to 1990*. St John's: Institute of Social and Economic Research.

Finlayson, C. 1991. 'The Social Construction of a Scientific Crisis: A Sociological Analysis of Northern Cod Stock Assessments from 1977 to 1990'. MA thesis. St John's: Memorial University.

Goos, T.O. 1989. *The Effects of Climate and Climate Change on the Economy of Alberta*. Climate Change Digest 89-05. Ottawa: Minister of Supply and Services Canada.

Grubb, M., M. Koch, A. Munson, F. Sullivan, and K. Thomson. 1993. *The Earth Summit Agreements*. London: Earthscan.

Gwyn, R. 1993. 'A Way of Life in Peril'. *Toronto Star* (4 September):B1, B5.

Haas, P.M., M.A. Levy, and E.A. Parson. 1992. 'Appraising the Earth Summit: How Should We Judge UNCED's Success?' *Environment* 34, no. 8:7–11, 26–33.

Hall, J. 1994. '"Greed" Alone Led to Cod Disaster'. *Toronto Star* (13 February):D4.

Hall, P. 1990. 'Crisis in the Atlantic Fishery'. *Canadian Business Review* 17, no. 2:44–8.

Hare, F.K. 1995. 'Contemporary Climatic Change: The Problem of Uncertainty'. In *Resource and Environmental Management in Canada: Addressing Conflict and Uncertainty*, edited by B. Mitchell, 10–28. Toronto: Oxford University Press.

Harris, L. 1995. 'The East Coast Fisheries'. In *Resource and Environmental Management in Canada: Addressing Conflict and Uncertainty*, edited by B. Mitchell, 130–50. Toronto: Oxford University Press.

Hewson, M.D. 1993. 'Climate Change—the International Response'. *Delta*, newsletter of the Canadian Global Change Program, 4, no. 4:12–14.

House of Commons, Subcommittee on Acid Rain of the Standing Committee on Fisheries and Forestry. 1981. *Still Waters*. Ottawa: Minister of Supply and Services Canada.

Hurst, L. 1994. 'And No Fish Swam'. *Toronto Star* (13 February):D1, D6–8.

Hutchings, J.A., and R.A. Myers. 1994. 'What Can Be Learned About the Collapse of a Renewable Resource?: Atlantic Cod, *Gadus morhua* of Newfoundland and Labrador'. *Canadian Journal of Fisheries and Aquatic Sciences* 51, no. 9:2126–46.

_____, C. Walters, and R.L. Haedrich. 1997. 'Is Scientific Inquiry Incompatible with Government Information Control?' *Canadian Journal of Fisheries and Aquatic Sciences* 54, no. 5:1198–210.

Innis, H.A. 1978. *The Cod Fisheries: The History of an International Economy*, rev. ed. Toronto: University of Toronto Press.

Kerr, S.R., and R.A. Ryder, 1997. 'The Laurentian Great Lakes Experience: A Prognosis for the Fisheries of Atlantic Canada'. *Canadian Journal of Fisheries and Aquatic Sciences* 54, no. 5:1190–7.

Kirby, M.J.L. 1982. *Navigating Troubled Waters: A New Policy for the Atlantic Fisheries*. Ottawa: Minister of Supply and Services Canada.

Klemes, V. 1990. 'Sensitivity of Water Resource Systems to Climatic Variability'. In *Innovations in River Basin Management*, edited by R.Y. McNeil and J.E. Windsor, 233–42. Cambridge, ON: Canadian Water Resources Association.

Lavigne, D.M. 1986. 'Killing 40,000 Grey Seals Won't Improve the Fishing'. *Canadian Geographic* 106, no. 5:73.

McCormick, J. 1989. *Reclaiming Paradise*. Bloomington: Indiana University Press.

McGillivary, D.G., T.A. Agnew, G.A. McKay, G.R. Pilkington, and M.C. Hill. 1993. *Impacts of Climatic Change on the Beaufort Sea-Ice Regime: Implications for the Arctic Petroleum Industry*. Climate Change Digest 93-01. Ottawa: Minister of Supply and Services Canada.

Martec Limited. 1987. *Effects of a One Metre Rise in Mean Sea Level at Saint John, New Brunswick and the Lower Reaches of the Saint John River*. Climate Change Digest 87-04. Ottawa: Minister of Supply and Services Canada.

Mathews, R., and P.A. Phyne. 1988. 'Regulating the Newfoundland Inshore Fishery: Traditional Values Versus State Control in the Regulation of a Common Property Resource'. *Journal of Canadian Studies* 23, no. 1 and 2:158–76.

Matthews, D.R. 1993. *Controlling Common Property: Regulating Canada's East Coast Fishery*. Toronto: University of Toronto Press.

May, A. 1967. 'Fecundity of Atlantic Cod'. *Journal of the Fishery Research Board of Canada* 24, no. 7:1531–51.

Mulroney, B. 1993. 'Statement by the H.E. Mr Brian Mulroney, Prime Minister of Canada'. In *Report of the United Nations Conference on Environment and Development, Rio de Janeiro, 3–14 June 1992*, vol. III, Statements Made by Heads of State or Government at the Summit Segment of the Conference, A/CONF.151/26/Rev.1, 72-4. New York: United Nations.

Myers, O. 1992. 'Collapse: The Northern Cod Have Disappeared'. *Earthkeeper* 2, no. 5:20–5.

Neis, B. 1992. 'Fishers' Ecological Knowledge and Stock Assessment in Newfoundland'. *Newfoundland Studies* 8, no. 2:155–78.

Northern Cod Review Panel. 1990. *Independent Review of the State of the Northern Cod Stock: Final Report* (the Harris Report). Ottawa: Department of Fisheries and Oceans.

Ontario Ministry of the Environment. 1982. *The Case Against the Rain: A Report on Acidic Precipitation and Ontario Programs for Remedial Action*. Toronto: Information Services Branch, Ministry of the Environment.

P. Lane and Associates. 1988. *Preliminary Study of the Possible Impacts of a One Metre Rise in Sea Level at Charlottetown,*

Prince Edward Island. Climate Change Digest 88-02. Ottawa: Minister of Supply and Services Canada.

Parson, E.A., P.M. Haas, and M.A. Levey. 1992. 'A Summary of the Major Documents Signed at the Earth Summit and the Global Forum'. *Environment* 34, no. 8:12–15, 34–6.

Phyne, J. 1990. 'Disputed Settlement in the Newfoundland Inshore Fishery: A Study of Fishery Officer Responses to Gear Conflicts in Inshore Fishing Communities'. MAST 3, no. 2:88–102.

Sanderson, M., and J. Smith. 1990. 'Climate Change and Water in the Grand River Basin, Ontario'. In *Innovations in River Basin Management*, edited by R.Y. McNeil and J.E. Windsor, 243–61. Cambridge, ON: Canadian Water Resources Association.

Singh, B. 1988. *The Implications of Climate Change for Natural Resources in Quebec*. Climate Change Digest 88-08. Ottawa: Minister of Supply and Services Canada.

Smith, D.A. 1990. 'The Implementation of Canadian Policies to Protect the Ozone Layer'. In *Getting It Green: Case Studies in Canadian Environmental Regulation*, edited by G.B. Doern, 111–28. Toronto and Calgary: C.D. Howe Institute.

Spears, J. 1993. 'Fishery's Collapse Has Scientists Baffled'. *Toronto Star* (28 December):A9.

Stephenson, R.L. 1990. 'Multiuse Conflicts: Aquaculture Collides with Traditional Fisheries in Canada's Bay of Fundy'. *World Aquaculture* 21, no. 2:34–45.

Task Force on Incomes and Adjustments in the Atlantic Fishery. 1993. *Charting a New Course: Towards the Fishery of the Future*. Ottawa: Minister of Supply and Services Canada.

Thiessen, V., and A. Davis. 1988. 'Recruitment to Small Boat Fishing and Public Policy in the Atlantic Canadian Fisheries'. *Canadian Review of Sociology and Anthropology* 25, no. 4:603–27.

United Nations Conference on Environment and Development. 1992. 'The Rio Declaration on Environment and Development'. Geneva: UNCED Secretariat.

Van West, J.J. 1989. 'Ecological and Economic Dependence in Great Lakes Community-Based Fishery: Fishermen in the Smelt Fishery of Port Dover, Ontario'. *Journal of Canadian Studies* 24, no. 2:95–115.

Weeks, E., and L. Mazany. 1983. *The Future of the Atlantic Fisheries*. Ottawa: The Institute for Research on Public Policy.

Wong, R.K.W., M. English, F.D. Barlow, L. Cheng, and K.R. Tremaine. 1989. *Towards a Strategy for Adapting to Climate Change in Alberta*. Edmonton: Resource Technologies Department, Alberta Research Council.

Environmental Change and Challenge Revisited

'The only person who likes change is a wet baby', observes educator Roy Blitzer. 'Two basic rules of life are: 1) change is inevitable; and 2) everybody resists change. Much of the world has its defences up to keep out new ideas.'

—R. von Oech,
*A Whack on the Side
of the Head*

This book has provided an overview of environmental change and challenge in Canada. It has emphasized the need to understand the ecological aspects of environmental change and the various management approaches that may be useful.

This final part consists of two chapters. The first chapter uses the example of Clayoquot Sound in British Columbia to illustrate the interrelationships among many of the different factors described in earlier chapters. By now you will be able to understand the biophysical context of the sound, why biological productivity is so high, what we mean by dominant limiting factors,

the relative lack of perturbations in these ecosystems, the importance of retaining nutrient capital, and many other factors. You will also understand something of the resource and environmental management implications of forestry as well as the changes to hydrological characteristics and biodiversity that might be anticipated, the importance of protecting endangered species, and the role of protected areas.

In the past, areas such as Clayoquot were routinely logged. There are records of trees over 30 m higher than the tallest trees now remaining that were rendered into wood products with little thought given to protection. But times have changed. An increasingly well-informed and vocal public now wants a say in the allocation of public resources. Conflict is generated, stakeholders identified, and processes implemented to represent the various perspectives. New perspectives on management that start from an ecosystem perspective and employ adaptive management approaches are recommended. The previous

◀ Carmanah Valley, BC (*World Wildlife Fund Canada/Adrian Dorst*).

chapters in the book have laid the groundwork for you to acquire a greater appreciation of all these factors, which are a part of the case-study described here, and which are in different forms in many of the resource management challenges we face in society.

This book has also tried to present a frank account of the environmental challenges we face in Canada. We have not shied away from discussing situations that do not look particularly optimistic, but we have also tried to incorporate examples representing innovation and progress. The first chapter in Part E illustrates the impact that ordinary people can have in changing government policies that do not seem to be in the best long-term interests of society. The final chapter in the book continues this theme by pointing out some of the things that individuals can do to have a positive impact upon reversing environmental degradation. We hope that you will not only read it but also take action to improve your balance sheet with the environment! Your efforts, combined with those of thousands of others acting individually, can make substantial changes.

Chapter 21

Imperatives for Change: Clayoquot Sound, the Challenge

On 13 January 1993 a full-page advertisement appeared in the *New York Times* with the question 'Will Canada do nothing to save Clayoquot Sound, one of the last great temperate rainforests in the world'. It was paid for by eight major international conservation groups. Six months later, Greenpeace International in London produced a seventeen-page colour booklet, 'British Columbia's Catalogue of Shame', outlining the background to the Clayoquot decision and demanding 'an end to all clearcut logging in Clayoquot Sound, full inventory of all plants and animals to be carried out and outstanding native land claim issues to be settled.' Robert Kennedy, Jr flew in as a lawyer representing the Natural Resources Defence Council of the US and promised the support of his organization. A resource and environmental management issue that had made nightly headline news in Canada also resulted in the arrest of over 800 protesters and captured worldwide attention.

At stake was the future of one of the greatest remaining temperate rainforests of the world and the largest remaining tract of old-growth forest on Vancouver Island, which was slated to undergo forest harvesting. The area is also under First Nations land claim negotiation, is coveted by the mining industry, is an arena for conflict between the rapidly expanding aquaculture industry and fishers and recreationists, and includes part of a national park.

THE BIOPHYSICAL CONTEXT

Clayoquot Sound is on the west coast of Vancouver Island (Figure 21.1), part of the Pacific Mar-

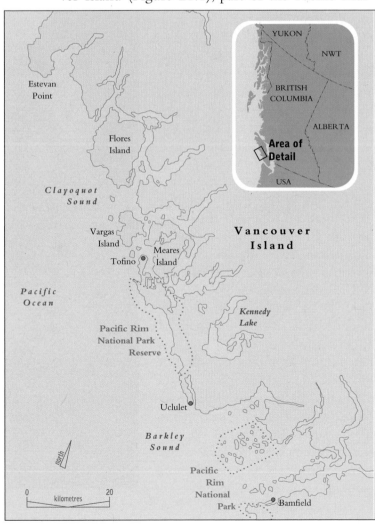

Figure 21.1 Location map of the Clayoquot Sound study area, west coast of Vancouver Island, BC.

itime ecozone described in Chapter 3. On the extreme western edge of this ecozone, the climate is strongly influenced by proximity to the Pacific Ocean. There is prolonged and heavy precipitation, particularly in the winter as a result of storms moving in from the ocean and winds rising over the mountains of Vancouver Island. Daily rainfalls of 10–15 mm are not unusual and extreme rainfall may be over 200 mm. The ocean also moderates temperatures, producing relatively cool summers and mild winters. Frost and snow are unusual at lower elevations. The heavy rains have helped to produce podzolic soils over most of the area, which are relatively acidic and nutrient poor. Nitrogen and phosphorus availability are often dominant limiting factors (Scientific Panel for Sustainable Forest Practices in Clayoquot Sound 1995a). Soils are generally thin, usually with only 10–30 cm of organic layer produced by forest litter over the bedrock. This thin layer absorbs and retains water, protects against erosion, supplies nutrients, and supports a diverse soil community. It is the product of successional processes over the last 14,000 years or so since the deglaciation of the area.

Clayoquot Sound forms part of the Coastal Western Hemlock biogeoclimatic zone within the broader Pacific Maritime ecozone. Dominant tree species include western hemlock, western red cedar, Sitka spruce, amabilis fir, Douglas fir, yellow cedar, and red alder. Much of the international attention focused upon Clayoquot has resulted from the classification of these forests as part of the Coastal Temperate Rainforest biome, which includes areas in Tasmania, Japan, Norway, and even the Black Sea coasts of Turkey and Georgia, but which is mainly concentrated on the west coasts of North and South America. Estimates suggest that 18–25 per cent of this biome is in British Colum-

bia, and about 60 per cent of the unlogged Coastal Temperate Rainforest biome is in BC and Alaska (Kellogg 1992; Weigand 1990).

The forests of Clayoquot are particularly important because of their productivity, lack of disturbance, and size. The sound has a total land area of 262 000 ha, of which 244 000 (93 per cent) are forested. About 160 000 ha are considered commercially productive, of which 30 500 ha have already been logged. Of the remainder, 39 100 ha are in protected areas, leaving 90 400 ha of primary old-growth forest (Figure 21.2). Trees in this forest reach exceptional proportions. The two tallest red cedar trees in Canada (59.2 and 56.4 m) are found

Clayoquot Sound is home to some of the largest trees in Canada and in the world, as illustrated by this giant western red cedar. Empirical research has demonstrated that it is possible for fourteen undergraduates to cram into a hole at the base of the trunk of one of these giants (*Philip Dearden*)!

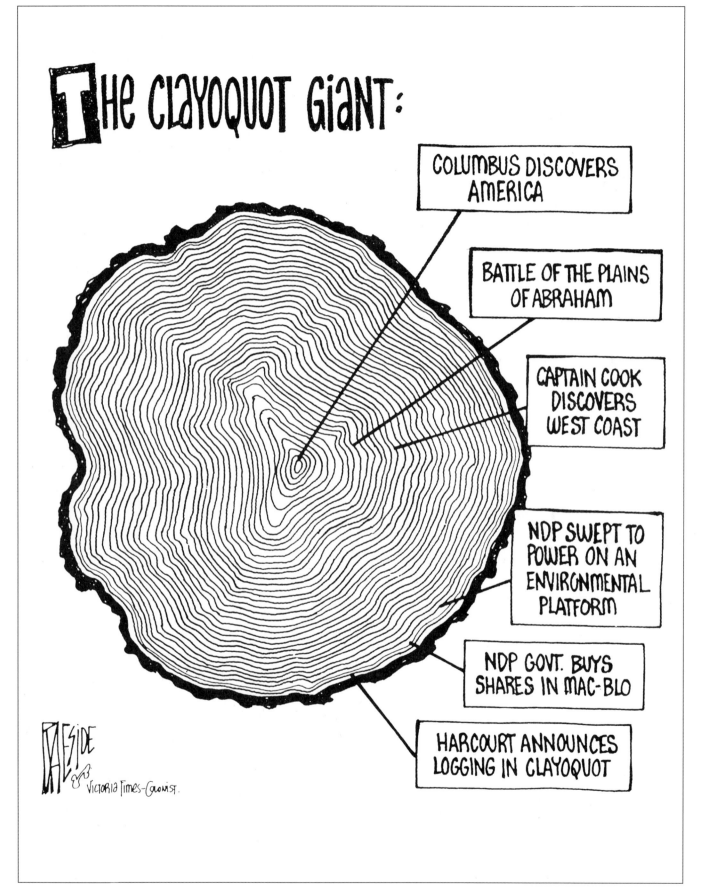

Figure 21.2 The Clayoquot Giant. SOURCE: *Victoria Times-Colonist* (18 April 1993):A4.

within adjacent lands in Pacific Rim National Park Reserve, the tallest Douglas fir (82.9 m) in adjacent Strathcona Provincial Park, the tallest tree in Canada, a Sitka spruce (95.7 m) is now protected in Carmanah Pacific Provincial Park, and the tallest western hemlock and yellow cedar are also close by (Stoltmann 1993). The volume of standing live trees at lower elevations is as high as 900 m^3 per hectare, with downed wood adding another 400 m^3 per hectare, and further contributions from standing dead and decaying trees.

The area has experienced relatively little disturbance due to natural or human causes. Extensive areas of blown-down trees or wildfires are very infrequent, occurring at 400–1,000-year intervals, if at all. Most areas are dominated by old trees that may be more than 1,000 years old. Forest replacement occurs as older trees die and fall to be replaced by young saplings. Some animal and plant species are limited to, or obtain their best growth in, these structurally complex forests. However, no comprehensive ecological survey of these forests has yet been undertaken. It appears that at least one lichen species is unique to the shores of Clayoquot (Goward 1994), and researchers in the same kind of forest just to the south in Carmanah Pacific

Provincial Park have discovered over sixty species of insects and spiders in the forest canopy that have never been scientifically described. In other words, if we return to the resource management model presented in Chapter 1 (Figure 1.6), the scientific models we build to try to understand natural systems, in this case, are quite deficient. We are still discovering new species, new components to the model, and do not yet understand how these species interact with other components of the ecosystem.

Clayoquot is also rich in vertebrate species (Table 21.1), with 297 of the 368 species found in the coastal temperate rainforest between coastal Alaska and Oregon recorded in the area. About 62 per cent of the vertebrate species are forest dwellers. Much of this richness can be ascribed to the old-growth component (Figure 21.3). Aquatic systems are also rich and varied, containing all species of Pacific salmon, seagoing trout, sculpins, char, minnows, sticklebacks, and lampreys, and with more than twenty marine mammal species in nearshore waters.

Most of these forests in Clayoquot Sound are virtually untouched by human disturbance. A survey of Vancouver Island indicated that there were

Table 21.1   Number of Native Land-Dwelling Vertebrates in Clayoquot Sound Region and Related Forest Types

Zone	Amphibians	Reptiles	Birds	Mammals	TOTAL
Forests of Coastal Western Hemlock zone[1]	11	6	138	64	219
Forests of Mountain Hemlock zone[1]	7	4	69	58	138
Coastal temperate rainforest (Alaska to Oregon)[2]	24	6	259	79	368
Clayoquot Sound:					
all species[3]	7	3	258	29	297
blue-listed species[4]	—	—	31	3	34
red-listed species[5]	—	—	8	3	11

[1]Breeding species only. Includes mainland British Columbia as well. All but eight species in the MH zone are also found in the CWH zone.
[2]Includes non-breeding species.
[3]Includes non-breeding species; many birds use the area primarily during migration.
[4]Species considered to be vulnerable or sensitive.
[5]Species that are candidates for designation as endangered or threatened.

Source: Scientific Panel for Sustainable Forest Practices in Clayoquot Sound, *Sustainable Ecosystem Management in Clayoquot Sound: Planning Practices*, Report 5 (Victoria: Ministry of Forests, 1994):26.

only five watersheds larger than 5000 ha that were undeveloped for resource extraction (Wilkinson 1990). Three of these are in Clayoquot Sound.

THE CONFLICT

Like the rest of BC, the forests of Clayoquot Sound (publically owned forests) had been leased to forest companies with the express purpose of cutting the old-growth timber and replacing it with even-aged, managed stands, as discussed in Chapter 15. Forestry provides about 6 per cent of all jobs in the province, and about 16 per cent of all employment when indirect jobs are included (Price Waterhouse 1994). The province contributes 34 per cent of the world's softwood exports, and forestry generally accounts for about 8 per cent of the provincial gross domestic product. To maintain this level of economic return, the industry cut ever-increasing amounts of timber through the 1960s, 1970s, and 1980s. Many critics suggested that these levels of cut were not consistent with the principle of sustained yield in which the rate of volumetric cut is limited to the long-term growth rate that the forest can yield in perpetuity. Thus there were ongoing and province-wide concerns about the sustainability of the industry. These have since proven well-founded, with annual allowable cuts reduced in many areas, especially on the coast (Box 21.1).

Other factors were also important, however. As the area of old-growth forests continued to decline, the public became more aware of the forests' many different functions in addition to the traditional concentration on forest harvesting. The popularity of outdoor recreation pursuits brought more people into contact with the forests. Tourism started to play an ever-increasing role in the provincial economy, with many tourists indicating that scenic

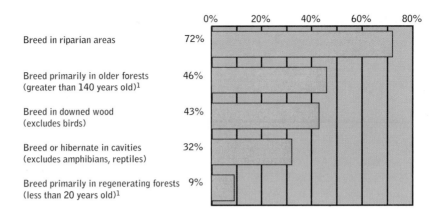

Breed in riparian areas	72%
Breed primarily in older forests (greater than 140 years old)[1]	46%
Breed in downed wood (excludes birds)	43%
Breed or hibernate in cavities (excludes amphibians, reptiles)	32%
Breed primarily in regenerating forests (less than 20 years old)[1]	9%

Figure 21.3 Percentage of forest-dwelling vertebrate species in Clayoquot Sound using different forest components for breeding. NOTE: [1]Few species breed only in one age class of forest, but many breed primarily in older or in younger forests. SOURCE: Scientific Panel for Clayoquot Sound, *Sustainable Ecosystem Management in Clayoquot Sound: Planning and Practices* (Victoria: Ministry of Forests, 1995a):27.

Logging in Clayoquot Sound (*Philip Dearden*).

Whale-watching has become an important part of the local tourist economy (*Philip Dearden*).

Box 21.1 Annual Allowable Cut in BC

In BC the timber harvest has grown exponentially between 1912 and 1989 when 73.5 million m³ were harvested from public lands that the ministry calculated to have a long-range sustained yield of about 60 million m³ (Figure 21.4), a figure that is thought to be about 10 million m³ too high by most outside observers. It is obviously impossible to maintain a level of cut that is in excess of the growth rate for an extended period. As a result, since 1989 the AAC in BC has been reduced each year. Of the areas so far reviewed for AACs, the average softwood reduction has been 11 per cent. These reductions confirm the environmentalists' allegations for many years, and which were hotly denied by the government and industry, that current cutting rates were not sustainable. The mill capacity is still in excess of the biological capacity, however, and BC imports trees from other provinces, such as Alberta and Saskatchewan, and from as far away as Alaska and Chile in an effort to keep its mills operating.

and wilderness values were a main motivation for visiting the province, and forest harvesting activities were a major detractor from these values (Berris and Bekker 1989; Dearden 1988).

The perspectives of First Nations peoples also began to be taken more seriously. A series of court decisions had considerably strengthened the rights of First Nations peoples to have a say in resource and environmental management decisions, as discussed in Chapter 11. The Nuu-Chah-Nulth people were the traditional land-owners and resource managers in Clayoquot. Even today, their population constitutes about half of the resident population, yet they were seldom consulted about any resource and environmental management decisions.

On the other hand, the forest workers also felt their concerns were not being met. Most of the wood in the Clayoquot area is processed in the nearby mill-town of Port Alberni. MacMillan Bloedel employs workers in logging, sawmills, and pulp and paper operations. In the decade 1980–90, 2,200 out of 5,300 permanent forestry-related jobs were lost in the town. Another 987 workers lost their jobs between 1991 and 1993. Industry people tend to blame reductions in annual allowable cuts (AACs) on park establishment, whereas outside critics point their fingers at the increasing mechanization of the logging industry. Whatever the balance between these causes, the people of Port Alberni and their political representatives were not happy about the threat of any further reductions in the AAC as a result of increased protection in Clayoquot Sound.

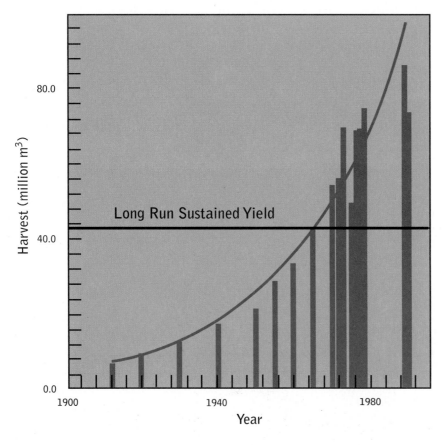

Figure 21.4 Total annual timber harvest. SOURCE: L.E. Harding, 'Threats to Diversity of Forest Ecosystems in British Columbia', in *Biodiversity in British Columbia: Our Changing Environment,* edited by L.E. Harding and E. McCullum (Vancouver: Canadian Wildlife Service, 1994):257.

Many factors thus came together in Clayoquot Sound to challenge the way the forest had traditionally been viewed by resource managers and how it might be managed to provide values to society. Thus there was conflict over the old way of doing things, dissatisfaction over the results, and a lack of consensus as to how things might improve. There were already conflicts over similar disputes between forest harvesting and protection in the area. Meares Island, in the centre of the sound (Figure 21.1), had been the focus of international protests in 1985 as First Nations peoples and environmentalists protested MacMillan Bloedel's right to log there. A court injunction subsequently halted logging operations pending settlement of the outstanding First Nations land claims. Further conflicts, in the north of the sound at Sulphur Passage, also attracted national headlines in 1988 as protesters once again suffered jail terms to draw attention to what they saw as unsustainable and destructive forest harvesting. After reviewing the situation, the provincial Ministry of Forests agreed and made the company alter its plans. Other incidents occurred throughout Clayoquot: equipment was vandalized, logging bridges were torched, and people were arrested as precursors to the mass demonstrations in the summer of 1993.

CONFLICT RESOLUTION

Efforts to resolve the land-use conflicts in Clayoquot Sound go back over a decade. The government had appointed a committee to examine the logging of Meares Island in the early 1980s, but then overruled its recommendations by choosing a more pro-logging alternative than any of those offered by the committee. Further efforts were initiated by the community of Tofino, which challenged the government to establish a more comprehensive process that would follow the prin-

> ### Our Home
>
> *It is truly unfortunate that there is not enough timber to go around but it is not morally correct to alter and change forever this magnificent area. We seem to be encouraging and feeding an international corporate giant that is rampaging out of control in our home. The watersheds of Clayoquot Sound that are in their natural state should be left that way (Comment on draft strategy document for Clayoquot Sound Sustainable Development Strategy, quoted in Prescott-Allen 1993:26).*

Meares Island in the Centre of Clayoquot Sound was one of the first conflicts to arise as MacMillan Bloedel proposed to log areas on this slope in the centre of the photograph. This is the view from Radar Hill, the major viewpoint in adjacent Pacific Rim National Park (*Philip Dearden*).

Local Native peoples protested and managed to delay the logging on Meares Island (*Philip Dearden*).

The Clayoquot Peace Camp became the centre for the Clayoquot protests (*Philip Dearden*).

> *Clayoquot Sound has seen a multitude of planning by 'public' processes since 1988. This is expensive time-consuming, inconclusive and inevitably results in clashes of fundamental values about land use which are exaggerated by special interest groups.*
>
> J. Ross Munro, chair,
> Share Our Resources

ciples of sustainability as espoused by the Brundtland Commission (see Chapter 2). The government established a task force in 1989 to devise a sustainable economic strategy for the area, but failed to reach consensus and disbanded.

In 1990 another committee, the Clayoquot Sound Sustainable Development Steering Committee, was formed to try again. The government agreed to implement any consensus agreement of the committee, subject to budgetary constraints. In terms of Arnstein's classification of the ladder of citizen participation discussed in Chapter 11, this initiative would be near the top, just one rung down from total citizen control.

The stakeholders (Chapter 11) on the committee represented First Nations peoples, municipal governments, provincial and federal government agencies, and nine interest groups (aquaculture, environment, fishing, labour, mining, small business, big timber companies, small timber companies, and tourism). In 1991 the environmen-

tal representatives withdrew from the committee because logging was allowed to continue in areas under discussion for protection. Furthermore, there were disagreements over the range of topics open for discussion. The representatives from Tofino, for example, wanted to discuss the long-term sustainability of forestry in the area, but were vetoed from doing so by the industry representatives. Since a topic needed agreement by all parties to be addressed, their concerns went unheeded.

In late 1992 the committee disbanded, unable to reach consensus about how much land area should be protected, despite having reached agreement on numerous broad principles related to sustainability (Box 21.2). This illustrates one of the difficulties of the alternative dispute resolution approach discussed in Chapter 12. Although diverse participants may agree on broad, global principles ('We all believe that there should be protected areas' or 'There should be no net job loss'), when such principles actually have to be applied to concrete situations regarding *which* areas should be protected, or how to maintain employment in the face of reduced AACs, intractable disagreements often occur.

Failure to reach agreement resulted in the land-use decision's reversion to the government. However, this outcome did not prevent industry from touting what they called the 'majority option' of the committee. This option called for one-third of the area to be protected and the rest to be open to logging. It was a position not endorsed by many of the stakeholders, including the community of Tofino, First Nations, tourism, and environmental interests. It did, however, appear to form the basis of the government's April 1993 decision, which sparked the main protest outlined earlier. The 1993 decision increased protected areas from 14.9 per cent of the area to 33.4 per cent. Forty-four per cent was open to logging, with a further 17.6 per cent open but with limitations due to wildlife, scenic, and recreational values (Figure 21.5). Four per cent was reserved for Native reserves and protection of Meares Island. The AAC

Box 21.2 Principles of Sustainable Development Agreed to by the Clayoquot Sound Sustainable Development Strategy Steering Committee

1. Keep human impact within earth's carrying capacity.
2. Maintain the stock of natural wealth by conserving life support systems and natural diversity and ensuring all uses of renewable resources are sustainable.
3. Minimize the depletion of non-renewable resources.
4. Promote long-term economic development that increases the benefits from a given stock of resources and maintains natural wealth.
5. Aim to share fairly the benefits and costs of economic development and environmental conservation.
6. Allow communities and interest groups to participate effectively in making the decisions that affect them the most.
7. Promote values that help achieve sustainable societies (Clayoquot Sound Sustainable Development Strategy Steering Committee 1992:4–1).

Figure 21.5 New forestry. SOURCE: *Victoria Times-Colonist.*

for the area was reduced from 900 000 m³ to 600 000m³, with the government estimating that as many as 400 jobs might be lost as a result of these changes. However, environmentalists pointed out that the 'protected' areas contained less than 50 per cent of old-growth forest, but included large areas of commercially useless bog and scrub forests, and even some areas that had already been logged.

Unfortunately, our country's demand for timber is fueling the destruction of the ancient forests of British Columbia and is undeniably a factor in the decision to log Clayoquot Sound. It is unconscionable to allow US demand for timber to destroy the remaining temperate rain forests of North America simply because they are outside our borders.

Congressmen John Kerry,
John Porter, and
Henry Waxman in letter
urging Vice-President Al Gore
to take action to
protect Clayoquot Sound

Rather than resolve the conflict as the government had hoped, this decision led to the extensive demonstrations already described and propelled the debate into the international arena. As a representative of the Sierra Club of western Canada commented, 'Clayoquot Sound does not just belong to the Alberni-Clayoquot district anymore. It belongs to the whole world' (*Victoria Times-Colonist*, 3 September 1994:A1). The BC ombudsman was asked to review the fairness of the government's decision-making process and concluded that the government had not followed due process, citing in particular the complete lack of consultation with one of the major stakeholders in the area, the First Nations peoples (Box 21.3). As stiff jail sentences (some up to six months) began to be handed out to protesters, five provincial government MLAs, including four cabinet members, wrote an open letter of protest about the sentences. A federal MP from the same political party as the provincial government wrote an open letter to the premier criticizing the decision. Environmental groups organized international boycotts of forest products. Counterdemonstrations were organized by pro-industry supporters, attracting some 4,000 people to one such gathering in Ucluelet, a logging town just south of the sound.

In efforts to control the escalating conflict, the government announced several new measures, including an agreement with First Nations peoples. The agreement provided for a $1 million training and employment program for First Nations peoples to work as forestry inspectors and park wardens, and for creation of a joint management board to make recommendations to the cab-

inet regarding the sound. In return, the First Nations peoples would agree to allow MacMillan Bloedel to extract timber from the Clayoquot River and Flores Island areas of the sound. While this move may have helped allay the concerns of First Nations peoples, the environmental community still had many concerns. The government finally moved to address these by appointing a scientific committee to examine the issue of where and how to log in Clayoquot Sound.

THE CLAYOQUOT SOUND SCIENTIFIC PANEL

In October 1993 the Scientific Panel for Sustainable Forest Practices in Clayoquot Sound was announced by the premier of BC, who defined its task as making 'forest practices in Clayoquot not only the best in the province, but the best in the world' (Scientific Panel for Sustainable Forest Practices in Clayoquot Sound 1994:2). The panel provides a good example of the 'new' forestry discussed in Chapter 15. The recommended practices were suggested by people specifically chosen for their expertise in all aspects of forest harvesting and ecosystem management, ranging from biodiversity, ethnobotany, and First Nations' concerns to worker safety and road design. The panel did not explicitly discuss economic and social sustainability, partly because they believed that good forest practices would automatically lead to long-term economic and community stability in the area.

The panel reviewed all harvesting standards in the area and discovered many problems. Different agencies had conflicting regulations. There were implicit assumptions regarding the primacy of fibre production over all other values and no inclusion

The Scientific Panel . . . has produced a document, parts of which are not in the best social economic interest of the province. For instance, a recommendation . . . is that . . . native hereditary chief ownership of the land be recognized. First, what does this have to do with science? And second, is this in the best interest of the province? Why is there no mention of non-native rights?

J. Ross Munro, chair,
Share Our Resources,
Port Alberni

Box 21.3 First Nations Peoples' Perspective on the BC Government's Decision

The Nuu-chah-nulth Tribes of
Hesquiaht, Ahousaht, Tla-o-qui-aht First Nations
Ucluelet and Toquaht
Are Opposed to the Clayoquot Sound
Land Use Decision
Made by the BC Government

We were not involved before or during the Clayoquot Sound decision-making process, even though we represent almost half the population of the region.

Our land base on government-created reservations represent less than half of one per cent of the Clayoquot Sound area. But we will one day begin treaty negotiations with the BC and Federal Governments for our territory, including Clayoquot Sound, which has never been given up, signed away or conquered.

The Harcourt Cabinet has promised a new Government-to-Government relationship with First Nations in BC. Yet, the same Harcourt Government did not even *involve* us before making this crucial decision. Was this Government-to-Government? No!

The BC Government has stated the Clayoquot Sound decision will not prejudice the future treaty process. How can this be when the Clayoquot decision affects some of the very same land and resources to be negotiated?

Until negotiations begin, the BC Government has committed to interim protection measures by accepting a report by the BC Claims Task Force. The report states, 'The parties negotiate interim measures agreements before or during the treaty negotiations when an interest is being affected which could undermine the process.' Clayoquot Sound is the perfect example of where an interim protection measure is needed.

We are opposed to decisions which jeopardize our future treaty negotiations with this same government which ignored our concerns and ignored us when making this decision.

We are opposed to the decision to log in some specific areas like the Hesquiaht territory, *Clayoquot Valley*, and Flores Island. We are opposed to methods of logging that threaten our unique way of life and destroy sacred sites.

But we are not opposed to logging some other areas, if proper logging methods are used to ensure the *protection of our environmentally sensitive salmon streams, cultural and spiritual sites.*

And we do not want all the resources. We want to share. Sharing is a major philosophy of our peoples. When the Europeans first arrived here, their survival was due in large part to the sharing hands of our peoples.

We are fishermen, loggers and environmentalists. But we suffer unemployment rates of at least 70 per cent. We want to work. Instead of a lifetime of hopelessness and despair, we seek to provide a lifetime opportunity to generate wealth for our future generations through successful Treaty making. Perhaps you now begin to understand why we are not opposed to logging jobs. Our concern is how and where the logging is done.

Please support us. Join us in opposing the decision on Clayoquot Sound by writing the Harcourt Government. Ask that the decision be reconsidered, and our peoples be fully involved in the development and decision-making process.

THIS DECISION AFFECTS OUR WAY OF LIFE.
WE WANT TO PROTECT OUR CHILDREN'S FUTURE.

---

of First Nations' perspectives. A new approach was required that would emphasize an ecosystem-based perspective rather than fibre production, and the panel set about articulating the principles and standards that might achieve this result. The principles for developing the new standards are shown in Box 21.4.

On the basis of these principles and the review of existing practices, fourteen general recommen-

dations were advanced, focused on the need to transform the perception of the forests from just potential products to the biological systems upon which these products depend. Four generic recommendations were outlined as prerequisites for this change:

- Planning must be long term and inclusive. There must be provisions for determining lev-

---

Box 21.4 Principles to Guide Resource Management in Clayoquot Sound

1. Responsible land stewardship, including forest management, must respect the land and all living things.
2. Ecosystems must be recognized as the functional base from which all goods and services are derived, and provisions must exist for determining and setting levels of resource extraction within the limits and capabilities of ecosystems.
3. Planning must be long term and inclusive, linking provincial, regional, and local levels. At each of these levels, sustaining ecosystem productivity must take precedence over specific product outputs.
4. Social, environmental, and economic dimensions of resource management must be incorporated into the planning process.
5. Inventories must be expanded to include the status, abundance, and distribution of resources and values in Clayoquot Sound and the critical factors (e.g., slope stability) that affect timber harvesting or other resource-extracting operations. Some unde-

veloped areas must remain as baseline reference areas against which managed areas can be compared.
6. An effective monitoring program must be implemented and adaptive management practised to improve forest practices and procedures as experience and knowledge are gained.
7. As part of adaptive management, research must be undertaken to ensure that the standards set are adequate to maintain long-term ecosystem integrity.
8. Information and education are essential for successful implementation of new forest practices standards.
9. Resource management policies reflect human values, understanding, and knowledge at a particular point and time. They must be reviewed and revised to keep pace with changes in these states (Scientific Panel for Sustainable Forest Practices in Clayoquot Sound 1994:6).

---

els of resource extraction within the limits prescribed by ecosystems. Planning must involve First Nations peoples.

- There must be active and rapid learning of the best way to manage the forest through adaptive management, as discussed in more detail in Chapter 9. This requires looking at forest management as a process involving three procedures: (1) the management practice is viewed as a replicated experiment; (2) the outcomes of the practice are monitored and compared with anticipated results; and (3) feedback is incorporated to modify future practices on the basis of the 'experiment'.

- Transition to sustainable ecosystem management will require considerable investment in education and training for those currently involved in forest harvesting, as their backgrounds and experience have not equipped them to understand and apply this approach.

- There must be wide-based support among all stakeholders for the transition to take place.

At a more detailed level, recommendations were made on four main themes:

Maintaining Watershed Integrity: Watershed integrity was recognized as necessary for the security of the forest and stream ecosystem. Four specific subsections were recognized and goals, objectives, findings regarding the current system, and recommendations for a new system were outlined. The goals and objectives of these subsections are shown in Box 21.5. The findings and recommendations for each of these sections are too detailed to be repeated here, but provide specific guidelines about what practices are required for ecosystem sustainability.

Maintaining Biodiversity: The panel acknowledged the paucity of information about biodiversity in the area, and the challenge that this presented in terms of protecting species and genetic variants from serious declines or extinctions caused by human activities. The panel members recognized that maintaining a full variety of habitats is critical in this process, and that old-growth forests were

Box 21.5 Goals and Objectives for Maintaining Watershed Integrity

Soils, Slope Stability, and Erosion:

Goal: To maintain the integrity of the soil component of terrestrial ecosystems.

Objectives: To retain the soil and surficial materials in place; that is, to manage the land and ecosystems so that modes and rates of landsliding and erosion are not significantly changed and remain within the range of natural variability.

To maintain the physical, chemical (nutritional), and biological characteristics of the soil and maintain the soil's capability to sustain a wide range of ecosystem states.

Water Flow:

Goals: To preserve the fundamental timing of ecological processes associated with water flows.

To minimize soil erosion.

Objectives: To maintain water flows within the range of natural variability on both a seasonal and event basis.

To maintain the natural drainage system on hill slopes.

Water Quality:

Goals: To manage land-use practices so that critical elements of water quality remain within natural ranges and follow natural patterns within the ecosystem.

To ensure that overall ecosystem productivity is maintained over the long term (hundreds of years).

To maintain the productivity of fish populations and other biota.

Objectives: To maintain the stream thermal regime within the natural range for the system, including maintaining the timing of seasonal thermal changes, the range of daily fluctua-

tions, and the levels of maximum temperatures.

To minimize both concentration and duration of suspended sediment in aquatic ecosystems.

To minimize the export of nutrients from the ecosystem in the period between logging and forest re-establishment.

To prevent the entry of toxic chemicals into the hydrological system.

Channel Integrity:

Goal: To manage watershed systems to prevent alterations in hydrological regimes, increases in sediment input, and loss of riparian vegetation, which result in loss of channel integrity and dependent biological productivity.

To maintain full-length stream channel integrity.

Objectives: To maintain the character of the riparian area and the integrity of the channel in the flood plain of the stream.

To minimize the deposition of the fine sediment and sand in the channel system, and maintain the quantity and quality of spawning gravel used by fish.

To maintain the structural diversity of the channel by maintaining the volume, stability, and distribution of large woody debris.

To manage the flood plain and the riparian area to assure a continuing supply of large woody debris to the channel.

To manage the slopes and gully systems to maintain at natural frequency the episodic input of large volumes of broken woody debris.

crucial to the area's ecological integrity. Box 21.6 outlines the goals and objectives for this section.

Recognizing First Nations' Values and Perspectives: The rights of indigenous peoples and the contributions they can make to resource and environmental management are being increasingly recognized on a global scale. The Nuu-Chah-Nulth have been in the area the longest, yet have benefited the least from the economic activity. Furthermore, their cultural sites have been damaged in the past by logging activities, and no attention has been given to their cultural and spiritual values within the context of their forest environment. In the future, First Nations peoples must be one of the most important participants in planning for the area. Goals include:

- Recognizing and supporting the long-standing aspirations and needs of the Nuu-Chah-Nulth, which are based on traditional occupation and use of the land and waters.
- Recognizing, supporting, and incorporating Nuu-Chah-Nulth traditional ecological knowledge and values into land-use planning and decision making.
- Recognizing and supporting the Interim Measures Agreement to engage Nuu-Chah-Nulth participation in Clayoquot Sound land and resource use, including aquatic and marine systems.

Maintaining Scenic Resources, Recreation, and Tourism: The scenic resources of Clayoquot Sound are spectacular. Unfortunately, they have been severely compromised in many areas by vast clear-cutting.

Box 21.6 Goals and Objectives for Maintaining Biodiversity

Goals: To maintain all naturally occurring species and genetic variants so they can survive over the long term and adapt to changes in their environment within the normal range of variation.

To maintain the functional integrity of ecosystems, recognizing the connections between terrestrial, freshwater, and marine ecosystems.

Objectives: To conserve water quality, hydrological processes, and soils.

To maintain ecosystem function by protecting the integrity of riparian areas from the terminus to the headwaters of watersheds.

To protect habitats of known importance to particular species.

To maintain old-growth and forest-interior habitats.

To use forest management techniques that produce stand structures, species composition, and landscape patterns similar to those generated by the natural disturbances of forests in Clayoquot Sound.

Box 21.7 Traditional Ecological Knowledge in Clayoquot Sound

Chapters 4 and 11 pointed out the increasing recognition of traditional ecological knowledge (TEK). The Nuu-Chah-Nulth of Clayoquot have developed complex systems for classifying and naming natural phenomena. Over 270 species have been recognized, including twenty trees, thirty shrubs, eighty herbs, twenty-five mosses and lichens, twenty land and sea mammals, twenty-five bird species, thirty-five fish species, thirty-six shellfish and other invertebrates, and other species. They had detailed knowledge of many ecological processes and had management systems in place to ensure that species were not excessively harvested. None of these factors were taken into account in resource and environmental management decisions prior to the recommendations of the Clayoquot Sound Scientific Panel.

All tourist surveys undertaken in the area have highlighted the importance of scenery to their enjoyment (Dearden 1988). Current methods of assessing scenic values have not been satisfactory. New goals include:

- Managing scenic resources to maximize people's enjoyment by ensuring that opportunities for tourism and recreation reflect the inherent quality of the area.
- Ensuring that First Nations and other residents are satisfied that the essential elements of the scenery around them are maintained.

Ucluelet, a predominantly logging town south of Tofino, had made little attempt to protect its viewscape, with the result that tourists inevitably prefer the setting of Tofino and spend most of their money there (*Philip Dearden*).

In terms of recreation and tourism, the panel recognized the outstanding resources of the sound and its potential to attract recreationists from around the world. Tourism is of major economic importance to the province. To be sustainable, this must involve protection of outstanding recreational resources. Goals for recreation and tourism in this area are to manage resources to protect features important to tourism and recreation and to provide recreation and tourism opportunities that reflect the inherent quality of the resources in Clayoquot and the recreational desires of residents and visitors to the area. Specific objectives to meet these goals are given in Box 21.8.

On 6 July 1995 the government announced that it would implement all of the panel's 120-plus recommendations and not allow logging in undisturbed watersheds until comprehensive ecological assessments were completed and all the recommendations were implemented. Key changes include:

- an end to conventional clear-cut logging in Clayoquot Sound
- new cutting permits will have to meet the panel's recommendations on cutblock size
- roads will be limited to 5 per cent of any watershed's harvestable area, and will be planned to minimize impacts on water flows and soil stability
- harvesting levels will be based on watershed planning rather than on a predetermined annual allowable cut

The First Nations' perspective has been consistently overlooked in many of the past dealings in the sound, even though their land claim, now under active discussion with the provincial and federal governments, covers the entire area. Recognizing this deficiency, the provincial government has appointed a twelve-person board, the Central

There are costs involved in adopting these new techniques—in planning, operating and in reduced harvest rates. We will make sure that they do not fall unfairly on workers and communities

Andrew Petter,
BC minister of forests

We think the environmentalists deserve credit for pointing out the poor logging practices that were going on here. But in 1996 we think it's time for blockades to end. They have already served their purpose. Now the challenge for environmentalists is to help us go forward.

Francis Frank,
chief of the Tla-o-qui-aht

Box 21.8 Goals and Objectives for Maintaining Scenic and Recreational Resources

Scenic Resources:

Goals: To manage scenic resources to maximize the enjoyment of those present by ensuring that opportunities for tourism and recreation reflect the inherent quality of resources in an area.

To ensure that First Nations and other residents are satisfied that the essential elements of the scenery around them are maintained.

Objectives: To provide for a range of visual landscape experiences ranging from completely unaltered or undeveloped settings to areas with various uses. To plan the experiences in relation to existing and potential recreation and tourism routes.

To conduct sustainable forest practices and related educational/interpretive programs because this will affect how people perceive forest practices and the landscape.

To apply landscape design principles in all areas so that the visual impacts of commercial forestry are minimized through proper location, size, and shape of alterations.

To maintain examples of different types of natural landscapes in a relatively unaltered state so that people can experience their landscape heritage (e.g., fjord, lake, archipelago).

To use procedures for landscape analysis and planning that are as thorough and objective as possible

and that involve recreation, tourist, and resident groups.

To consider all existing and potential uses of the landscape in inventory, analysis, and planning for scenic resources.

Recreation Resources:

Goals: To manage resources to protect features that are important to tourism and recreation.

To provide recreation and tourism opportunities that reflect the inherent quality of the resources in Clayoquot Sound and the recreational desires of residents and visitors to the area.

Objectives: To provide for a range of recreation and tourism opportunities from wilderness-based expeditions to high-end excursions that are sensitive to and based on the area's natural resources.

To protect valuable resources for recreation and tourism.

To use procedures for recreation and tourism analysis and planning that are as thorough and objective as possible.

To integrate into recreation planning the use patterns and needs of tourist and resident groups, including First Nations.

To involve recreation, tourist, resident, and First Nations groups in planning and managing recreation resources.

Region Board (with 50 per cent First Nations representation), as an interim measure to consider all logging plans in the sound until land claims are resolved. One of the group's main goals is to see if they can reduce the 70 per cent unemployment level in local First Nations communities.

Unfortunately, these seemingly comprehensive recommendations have not totally resolved problems in the sound. Heavy rains in January 1996

triggered 260 landslides in the area, many of which were attributed to logging practices. Interfor, one of the two logging companies in the sound, was subsequently assessed a $10,000 fine for inadequate road maintenance. An international advertising campaign was started by the Rainforest Action Network, based in the US, claiming that forestry changes were merely cosmetic and that substantial old-style clear-cutting was still underway. Members of Greenpeace and the Friends of Clayoquot Sound chained themselves to logging equipment during the summer of 1996 to draw attention once again to logging practices. They withdrew only when they were asked to do so by the First Nations peoples. The 'Fight for the Woods' does not appear to be over yet.

IMPLICATIONS

The Clayoquot Sound experience is instructive for many reasons. Not only was it a high-profile conflict that generated national and international attention, but it also heralded radical changes regarding how forest lands were seen and managed. There has been a change in public values that was not reflected in the way forests were being harvested. Old ways of operation emphasized fibre extraction at the lowest unit cost at the expense of other forest values. Much of the conflict was a result of this. To return to the framework presented in Chapter 1 (Figure 1.6), this conflict related to the change in the management objectives (shown in Box 5 of Figure 1.6) that would establish what kind of a control system would be implemented in the area. Crucial aspects of change included the proportion of the area and selection of areas to be logged, and the actual logging practices to be implemented on these sites.

The Scientific Panel also concluded that we had only a very rudimentary knowledge of the natural systems that we were seeking to manage (boxes 1.1 and 1.2, Figure 1.6). Much of the pressure to log in the sound arose because of the shortage of timber and unsustainable practices that had been followed in adjacent areas on Vancouver Island. Only recently, however, has the Ministry of Forests achieved a reasonably accurate estimate of the status of the timber supply and started to lower the AAC accordingly.

The case-study touches on virtually every aspect of environmental science and management discussed in the text. Major concerns were raised about sustainability (as discussed in Chapter 2) and these concerns were not merely local but

> ### The Recommendations
>
> *The Scientific Panel's recommendations are among the first to shift forestry from its historical focus on sustaining output levels for specific forest products to a focus on sustaining forest ecosystems. The scientifically based recommendations incorporate traditional ecological knowledge of First Nations' peoples in whose territories Clayoquot Sound is located. In scope, these recommendations represent the most complete attempt, globally, to synthesize principles of international agreements and express these as specific actions (Scientific Panel for Sustainable Forest Practices in Clayoquot Sound 1995c:vii).*

extended across the world as people became more concerned about the global commons discussed in Chapter 20. The experience in Clayoquot Sound particularly emphasizes the tensions and challenges for sustainable development when integrating protection of environmental integrity and support for economic development. The loggers from communities such as Tofino and mill workers in Port Alberni had difficulty in seeing the complementarity of environmental protection and economic development. For them protection of old-growth forests translated into loss of jobs and negative economic impacts on individuals, families, and communities. And, for the environmentalists, if

The Scientific Panel made many recommendations regarding the size and location of clear-cuts. Many people feel that these regulations should be extended to cover all of coastal BC (*Philip Dearden*).

Box 21.9 Clayoquot: The BC Context

There have been many forest land-use disputes all over BC and the rest of the country. The BC government realized that a more comprehensive, province-wide perspective was required. In 1992 the government established the Commission on Resources and Environment (CORE) to complete a land-use strategy for the province. Four regions were initially designated as high priority. Vancouver Island was to be the first. However, in a very controversial decision, the government *excluded* Clayoquot Sound from the mandate of CORE. This enraged conservationists, who felt that any decisions regarding land use in the sound should reflect the larger regional context of the island. For example, the chief forester's review of the long-term sustainability of forest harvesting in the timber supply area for Vancouver Island, including the northern part of the island, concluded that AACs would have to be reduced by 35 per cent to allow for the overcutting that had already occurred on those lands if a long-range sustained yield was to be obtained. On the other hand, pro-forestry forces wanted the sound to be considered in isolation without taking previous cutting in the surrounding area into account.

What are the arguments for and against taking the wider perspective?

The conflict in Clayoquot became so severe that some environmentalists resorted to spiking trees in an effort to protect them. By inserting metal spikes into the trees, they hoped to deter loggers from cutting them down since it would be extremely dangerous for fallers to risk having their saws come in contact with the spikes. The logging companies subsequently dispatched crews with metal detectors to find and remove the spikes, identified here with the SP sign on the cedar tree (*Philip Dearden*).

harvesting were to continue to protect jobs in the forestry-dependent industries and communities, then the integrity of the old-growth forests would be lost. For many of the environmentalists, compromise was not an option. Furthermore, if sustainable development requires us to consider issues related to justice, social equity, and empowerment, First Nations peoples had difficulties in seeing how management practices reflected such concepts. A dilemma often arises regarding who social justice is for. To accommodate the needs, interests, and expectations of First Nations peoples, it appeared to many non-Native people that they would have to carry a disproportionate share of the costs associated with the new forestry practices.

The concerns expressed by the loggers and mill workers dependent on the fibre from Clayoquot Sound, and the loss of permanent jobs in Port Alberni, were similar to the experience of Sudbury (documented in Chapter 13) as mining operations were reduced there. In the Sudbury case, however, imaginative ideas were pursued and introduced to diversify the region's economy to make it less dependent on mineral resources. In Clayoquot Sound, some have argued that tourism could be the starting-point to diversify the regional economy. In your opinion, what options might be explored to diversify the economy in the Clayoquot Sound and Port Alberni areas to provide greater resilience?

The case-study also helps to highlight our lack of knowledge about the ecological systems described in Part B. Although the main ecosystem components are known, we still don't know *all* of the components, as evidenced by the new insects and lichens discovered in the area. We have even less of an idea about the functional relationships between the components and are only recently beginning to understand the complexities of the biogeochemical cycles and links between biotic and abiotic ecosystem components. Given these inadequacies, it is difficult to predict the impacts of human activity, such as logging. Recent studies of the regeneration of second-growth forests on the West Coast have found many growing at levels far below what was anticipated. Scientists indicate that the main reason is nutrient depletion on these heavily leached sites. There are questions regarding the interactions between activities such as logging and fish survival. However, the studies undertaken at Carnation Creek (discussed in Chapter 15) provide some insight into these latter aspects.

Many different aspects of the management strategies discussed in Part C were involved. For example, while the debate in Clayoquot Sound has often been labelled a 'forestry problem', the review of experience indicates that planning and management decisions need to take an ecosystem perspective. Thus while the starting-point has often been the forests and the forestry industry, it quickly became apparent that scientists and managers had to consider other linked components of the ecosystem, such as watersheds, biodiversity, and amenity values of landscapes. As a result, this complex case-study is similar to those introduced in Part D, in that while the starting-point was often a specific resource sector or environmental component, it becomes clear that understanding requires attention to an ecosystem, its component parts, and the linkages among those parts.

The Scientific Panel for Sustainable Forest Practices in Clayoquot Sound explicitly identified adaptive management as an appropriate model or approach to guide activities in this area. There is still is not much experience in Canada in actually applying the adaptive approach in the manner outlined by Holling and others reviewed in Chapter 9. Thus in the future, Clayoquot Sound offers the potential to become a 'management laboratory' to determine both the benefits and the problems associated with an adaptive approach. In this way, it will be desirable in the future to monitor experience in Clayoquot Sound and compare it with

ongoing activities in places such as the Columbia River basin in the United States and the Mirimachi River valley in New Brunswick.

The challenges in conducting impact assessment were clear in this case-study as they were in chapters 18 and 19. There were differences of opinion regarding which impacts should be focused upon, what the impacts actually might be (given the rudimentary understanding of the biophysical and social processes), what the significance of those impacts would be, and what could be done to mitigate them. All of those issues were addressed in Chapter 10 and were further examined in some of the other case-studies in Part D.

The roles of participatory and stakeholder approaches (addressed earlier in Chapter 11) became a key part of the Clayoquot Sound story. Indeed, instances arose in which some people believed the effort to include a broader mix of stakeholders did nothing more than extend and prolong the planning and decision-making process (at great expense and with little tangible benefit) since not all interests could be met or satisfied. It becomes clear that a participatory approach is not a panacea for resource and environmental problems as there are many issues and problems that have to be dealt with.

Certainly the Clayoquot Sound experience does confirm the observation made in Chapter 12 that resource and environmental management is often about the management of conflict and disputes. There was a long history related to this experience, leading some people to have difficulty in trusting or respecting people or positions that were contradictory to their own. The problem was certainly ripe (as discussed in Chapter 12), but some of the parties in the dispute were unwilling to collaborate in a search for a solution that would provide mutual benefit. Thus the dispute had likely passed the stage in which alternative dispute resolution methods could have been successfully used. Nevertheless, that does not mean that such methods could not be used in the future to try and ensure that conflicts do not reach such an impasse.

The Clayoquot Sound experience illustrates how forest harvesting should adopt the tenets of the new forestry described in Chapter 15 and give more attention to the values of conservation practices discussed in chapters 16 and 17. The kinds of changes required in forest harvesting in Clayoquot

Box 21.10 A Biosphere Reserve for Clayoquot?

A decade ago, Dearden (1988) suggested that the Biosphere Reserve designation would be appropriate for the Clayoquot Sound area. At that time, the chances seemed remote. Biosphere Reserves are a United Nations' designation of a protected area. However, they differ from our traditional ideas of protected areas, the ones described in more detail in Chapter 17, because they involve local people much more and allow some sustainable resource extraction. Ideally, three components are included:

- *Core Area*: Each reserve must have one or more core areas, such as a national or provincial park that is fully protected and large enough to sustain biodiversity in the area.
- *Buffer Zone*: Surrounding the core area is a buffer zone where low-impact uses that do not negatively affect the core are allowed.
- *Zone of Cooperation*: This zone permits more intensive extractive uses as long as they are sustainable and do not have a negative impact on the other zones.

The main purpose of Biosphere Reserves is to allow for comparative research and education among these three zones. Local people are intimately involved with management decisions to achieve a balance between sustainable use and conservation of biodiversity in the reserve. All stakeholders must approve nomination as a reserve before the application will be considered by UNESCO in Paris. Canada currently has six such reserves, including Waterton Lakes (Alberta) and Riding Mountain (Manitoba) national parks, and the Niagara Escarpment (Ontario).

The changes in forestry practices and the inclusion of greater input from First Nations peoples initiated by the BC government may now make such a nomination a possibility, and the government itself has promised to pursue vigorously such a nomination. This would accord greater overall protection in the sound also for one section of Pacific Rim National Park Reserve, thereby also improving the national park's conservation of diversity.

also reflect the more generic concerns in rapid change and uncertainty and other aspects of environmental change and resource management, as illustrated in the chapters in Part D. This requires thorough knowledge of the environmental system being managed, whether it is being managed for agriculture, urban demands, or water and energy use. Adaptive management approaches that take into account a wider ecosystem perspective are gaining increasing recognition in many different types of environmental management. More people and more demands lead to more conflict over all resources, and make clear the need to adopt a broader range of dispute resolution approaches that embrace a wider clientele. Many of the points made earlier about the Clayoquot example and the generic significance also apply to the example that opened the book—the struggles of one family to come to terms with environmental change and uncertainty and the many health, legal, and economic challenges this posed for them. The issue—groundwater contamination—and scale of the problems differ, but there are many similarities.

SUMMARY

1. The dispute over land use in Clayoquot Sound on the west coast of Vancouver Island attracted international attention and led to the arrest of over 800 citizens protesting the government allocation of areas for forestry. The example illustrates many of the principles and concepts discussed throughout the book.

2. The sound is one of the largest remaining old-growth temperate rainforests in the world. It contains many trees of exceptional size and age, and a rich representation of animal species. Of the five watersheds larger than 5000 ha on Vancouver Island that have not been logged or otherwise disturbed, three are in Clayoquot Sound.

3. The sound is also an important source of raw materials for the logging industry. Port Alberni, where much of the labour is based, has had dramatic declines in employment in the forest industry over the last decade.

4. The First Nations peoples of the West Coast hold Clayoquot Sound in special reverence and it forms the core of their as yet unresolved land claim.

5. There have been several land-use conflicts within the sound in the last decade. Previous attempts at resolving these conflicts have generally been unsatisfactory as local committees have failed to reach consensus on important issues.

6. In 1993 the government announced a decision that would have protected about a third of the sound from logging. This enraged conservationists and large-scale protests resulted. In response, the government appointed a scientific committee to recommend the kinds of logging that could be allowed in the sound.

7. The Scientific Committee established new principles to guide logging practices that are ecosystem- and adaptive-management oriented. The government has announced that it will follow all the 120-plus recommendations from the committee. If implemented, this announcement will mean an end to clear-cut logging in the sound as it has been practised in the past. However, environmental groups remain sceptical about the changes and have held sporadic demonstrations in the sound to draw attention to current practices that they believe are not sustainable.

8. First Nations peoples have been given much more input into decision making regarding the location and amount of logging activity, pending resolution of their land claim.

REVIEW QUESTIONS

1. Why did the Clayoquot Sound issue attract so much attention?

2. Who were the main stakeholders in the dispute?

3. What approaches to achieve dispute resolution were attempted?

4. Describe the characteristics of the main natural systems, based on other chapters in the book, that you would find in the sound.

5. Which of the scientific concepts and methods reviewed in Part B do you think are most relevant to understanding the ecosystem in Clayoquot Sound?

6. Which of the management concepts reviewed in Part C do you think should be used in managing the ecosystem in Clayoquot Sound?

7. From the other case-studies in Part D, what lessons could be applied to Clayoquot Sound as scientists and managers attempt to develop a strategy that will achieve sustainable development?

8. Did the Scientific Committee take a partial or total view of the situation? What are the strengths and weaknesses of their approach?

REFERENCES AND SUGGESTED READING

Berris, C., and P. Bekker. 1989. *Logging in Kootenay Landscapes: The Public Response.* Victoria: BC Ministry of Forests.

Brush, S.B., and D. Stabinsky. 1995. *Valuing Local Knowledge: Indigenous People and Intellectual Property Rights.* Washington, DC: Island Press.

Clayoquot Sound Sustainable Development Strategy Steering Committee. 1992. *Clayoquot Sound Sustainable Development Strategy.* Victoria: Government of BC.

Dearden, P. 1983. 'Tourism and the Resource Base'. In *Tourism in Canada: Issues and Answers for the Eighties*, edited by P. Murphy, 75–90. Western Geographic Series, vol. 21. Victoria: University of Victoria.

_____. 1988. 'Protected Areas and the Boundary Model: Meares Island and Pacific Rim National Park'. *The Canadian Geographer* 32:256–65.

Drushka, K., B. Nixon, and R. Travers, eds. 1993. *Touch Wood: BC Forests at the Crossroads.* Madeira Park, BC: Harbour Publishing.

Goward, T. 1994. 'Status Report on the Seaside Centipede Lichen *Heteroderma sitchensis*'. Unpublished report. Ottawa: Committee on Status of Endangered Wildlife in Canada.

Harding, L.E. 1994. 'Threats to Diversity of Forest Ecosystems in British Columbia'. In *Biodiversity in British Columbia: Our Changing Environment*, edited by L.E. Harding and E. McCullum. Vancouver: Canadian Wildlife Service.

Hodgins, B.W., and J. Benidickson. 1989. *The Temagami Experience: Recreation, Resources, Aboriginal Rights in the Northern Ontario Wilderness.* Toronto: University of Toronto Press.

Kellogg, E., ed. 1992. 'Coastal Temperate Rainforests: Ecological Characteristics, Status and Distribution Worldwide'. Occasional Paper no. 1. Portland: Ecotrust Conservation International.

Prescott-Allen, R. 1991. 'Clayoquot Sound Sustainable Development Strategy'. *Forest Planning Canada* 8, no. 1:24–32.

_____. 1993. 'Where Loggers and Tree Huggers Play'. *People and the Planet* 2:25–8.

Price Waterhouse. 1994. *The Forest Industry in British Columbia in 1993.* Vancouver: Price Waterhouse.

Reed, M.G. 1995. 'Implementing Sustainable Development in Hinterland Regions'. In *Resource and Environmental Management in Canada: Addressing Conflict and Uncertainty,* edited by B. Mitchell, 335–59. Toronto: Oxford University Press.

Sandberg, L.A., ed. 1992. *Trouble in the Woods: Forest Policy and Social Conflict in Nova Scotia and New Brunswick.* Fredericton: Acadiensis Press.

Scientific Panel for Sustainable Forest Practices in Clayoquot Sound. 1994. 'Report of the Scientific Panel for Sustainable Forest Practices in Clayoquot Sound'. Victoria: BC Ministry of Forests.

_____. 1995a. 'Sustainable Ecosystem Management in Clayoquot Sound: Planning and Practices'. Report 5. Victoria: BC Ministry of Forests.

_____. 1995b. 'First Nations Perspectives Relating to Forest Practices Standards in Clayoquot Sound'. Report 3. Victoria: BC Ministry of Forests.

_____. 1995c. 'A Vision and Its Context: Global Context for Forest Practices in Clayoquot Sound'. Report 4. Victoria: BC Ministry of Forests.

Stoltmann, R. 1993. *Guide to the Record Trees of British Columbia.* Vancouver: Western Canada Wilderness Committee.

Weigand, J. 1990. 'Coast Temperate Rainforests: Definition and Global Distribution with Particular Emphasis on North America'. Unpublished report. Portland: Ecotrust Conservation International.

Wilkinson, J.F. 1990. *Undeveloped Watersheds on Vancouver Island Larger Than 1000 ha.* Victoria: BC Ministry of Forests.

Chapter 22

Making It Happen

When I call to mind my earliest impressions, I wonder whether the process ordinarily referred to as growing up is not actually a process of growing down; whether experience, so much touted among adults as the thing children lack, is not actually a progressive dilution of the essentials by the trivialities of life.
—Aldo Leopold,
A Sand County Almanac

This final chapter rests on the firm conviction that individuals like you and us can make a significant difference in how many of the environmental challenges presented in this book will develop over the next decade if we are aware of these problems and are willing to do something about it. From the last chapter you will have gathered some idea of what can happen when citizens band together to create change. It is very doubtful whether such stringent logging controls would have been implemented by the government had not over 800 citizens from all walks of life been prepared to go to jail and be branded with a criminal record to make that change happen. Not all changes come about by such dramatic means, however, and there are many other ways that you can help create positive environmental changes in your everyday life. This final chapter provides some ideas about how you can become involved in creating change.

PICK UP THAT DEGREE

'The future is increasingly a race between education and catastrophe'
—H.G. Wells

This century has witnessed many changes. It may be characterized as an age of diminishing imperial powers, ongoing wars, atomic bombs, the harnessing of the entire globe into an interconnected economic system, rising consumer demands, and an exploding human population. The next century will witness the continuation of some of these trends, but environmental scientists seem convinced that global climatic change, water shortages, biological impoverishment, declining food yields per capita, desertification, pollution, and overpopulation will constitute the backdrop for the events of the next century.

The challenge with most of these factors is the enormity of their scale. They are so widespread that most Canadians do not realize they are happening, especially because we are probably more sheltered from their effects than other populations. Most of them also have long lag times (the period between when the processes are set in motion and when the effects are felt), especially when the effects may have different impacts in different parts of the globe. Global warming, for example, may lead to cooling and increased precipitation in some areas, while the effects may be just the opposite in other areas. Changes of this complexity and magnitude require long-term study before they can be understood. The scale of change also suggests that, within the human life span, many of these trends may be irreversible. Extinction certainly is.

Is our educational system preparing you and the students who will follow you to understand and deal with these changes? Urgent actions are required if we are to help diffuse these trends, but actions do not occur in a vacuum. They require some understanding of the road we are on, where it leads, and how we can get on other, more desirable roads. Educational systems should help to bring about this understanding. Currently, many schools, colleges, and universities graduate students who have little or no idea about how the ecosphere functions and how human activities impair those functions. They shop, travel, eat, drink, work, and play in blissful ignorance of the impacts they may be having on life-support systems.

Every morning as one of the authors cycles to work, he is reminded of this. His route takes him through the grounds of a local high school. Outside, at the back, numerous boys (yes, always boys!) tinker with their cars. This is not recess time; they are taking a class on cars. A class on cars should, first and foremost, deal with the tremendous impacts that cars have on the environment. How much matter and energy does it take to build and maintain one? What are the impacts upon the carbon cycle and global warming? What are the effects of high temperature combustion on the ozone layer? Astonishingly, he found that none of the students had even considered such questions, let alone knew the answers or the significance of the answers. Our education system is backwards. Before being given time to play with cars in school, students should be required to take and pass a course on the environmental implications of cars, but they have no time for such things. Cars are part of their 'real' world.

Colleges and universities are frequently accused of not being part of the 'real' world. By 'real' world, people usually mean the economic realities of today's society. However, this is not the real 'real' world; it is a game that humans introduced to facilitate barter and exchange. Important? Yes! Is this the real world? Only partially! The real world is composed of the air we breathe, the water we drink, the organisms that keep the life-support systems going, and the ground we stand on. Without these things, there can be no other invented 'real' world. We concentrate on balancing these play budgets when, in reality, it is the

> *. . . without significant precautions, education can equip people merely to be more effective vandals of the earth. If one listens closely, it may even be possible to hear the Creation groan every year in May when another batch of smart, degree-holding, but ecologically-illiterate, Homo sapiens who are eager to succeed are launched into the biosphere.*
>
> David Orr,
> Earth in Mind, 1994

more significant budgets of energy throughflow and material balance that determine the future of society. The deficit? Yes, Canadians have a huge deficit, but it relates to the 60 million bison that no longer roam the Prairies, the skies cleared of all but a fraction of the birds that used to flock in such large numbers that day turned to night, and the seas that were once home to the largest animals ever to evolve on the planet but where now not even the smallest fish are safe from sonar detection, vacuum trawling, and human consumption. Yes, we have a deficit, not all of which is incorporated or even recognized in financial and economic accounting procedures.

A primary function of our educational system should be to give students a general level of understanding about the nature of the environment and resources. We have requirements for general levels of language and mathematical competence, but

Real students in the real world (*Philip Dearden*).

require nothing from our students in terms of this most fundamental challenge of the future. Indeed, there are now pressures, especially on universities, to put greater and greater emphasis on meeting the short-term economic demands of society. Business schools and faculties of commerce flourish, yet few additional resources are allocated for programs dealing with the environment.

Even in such programs, colleges and universities have seldom done a good job of instilling in students an appreciation for and love of the planet. Increasingly, science programs have become a process of learning more and more about less and less. They have produced technically competent scientists, but have often missed the mark considerably in terms of maintaining students' wonder about the natural world and combining the rigour of scientific enquiry with deep moral questioning. They have often mistaken the laboratory for the 'real' world, and cut students off from a more comprehensive understanding of and passion for their environment.

You can create change on your campus. Are there sufficient courses on the environment? Do these courses cover a wide spectrum from the technical to the philosophical, and, more important, are students encouraged or even required to select from courses all along this spectrum? You should also not forget that campuses are large consumers and processers of matter and energy. How efficient are they? Has anyone undertaken an environmental audit of your campus? How are wastes disposed of? How much recycling occurs? Are chemicals used for landscaping? Does the faculty pension fund invest in businesses with unsound environmental practices? There are many questions that can be investigated either through course work, environmental clubs, or individually. If you are interested in pursuing these ideas more, you may wish to draw on such experiences elsewhere (e.g., see Smith 1993; Thompson and van Bakel 1995) and start some activities on your own campus. For an example of how one interested student managed to get the ball rolling at a Canadian university, see Bardati (1995).

So, pick up that degree and encourage others to increase their understanding. Do not be intimidated by people who think that interest in the environment and higher learning is not the 'real' world. Challenge your teachers to inspire you. Be

This is the 'real' world (*Parks Canada/Brian Morin*).

Box 22.1 Greening a Campus

At the University of Waterloo in Ontario, the WATGREEN program has been underway since 1990. The goal is to create a sustainable campus through students studying the university campus as an ecosystem. The vision is for the university to become a 'true ecosystem in harmony with its environment'. WATGREEN has provided an opportunity for students, staff, and faculty to improve the quality of their environment while also decreasing the overall operating costs of the university.

The benefits have been identified as: (1) an opportunity for students to learn more about environmental issues while receiving a course credit; (2) a forum for all university members to make a positive contribution to the resolution of environmental issues facing UW; (3) mobilization of campus resources to find cost-effective solutions for dealing with environmental concerns and new environmental regulations; (4) an opportunity for members of the university community to make changes on campus and empower them to make changes in society at large; (5) awareness of the university infrastructure necessary to support activities, and a chance for students to work within an institutional infrastructure; and (6) graduates who will take with them into the broader society the knowledge and skills required to work towards an environmentally sustainable future. Examples of WATGREEN projects include:

- *Solid Waste Management*: The goal is to reduce waste by 50 per cent by the year 2000. Waste audits have been completed for the residences and various departments. Composting of food waste from cafeterias has been initiated. The use of disposable products in food outlets has been reduced. Paper is recycled.
- *Landscape Practices*: A plan for a self-sustaining landscape has been prepared, which suggests annual plants be replaced with perennials. A timeline was prepared for eliminating pesticides and herbicides.
- *Water*: Since 1990 water costs have increased by 40 per cent. Water-efficient devices (to minimize flushes on toilets and water pressure on showers) have been installed. Water audits are being conducted.
- *Energy*: Energy consumption per square metre has been reduced by 42 per cent since 1973, even though the campus grew substantially during that period. Much of these savings were achieved before WATGREEN started.
- *Environmental Awareness, Attitudes, and Habits*: In 1991 a package on how to 'green' an office was prepared and distributed. An attitude survey was conducted of second-year students. A Green Week has been held to profile WATGREEN initiatives.
- *Changing What Is Consumed*: 'Green' cleaning products have been identified, leading to use of biodegradable cleaning products and reuse of containers—and eliminating health complaints from cleaning staff about the previous cleaning materials. A handbook that provides a catalogue of recycled products available through the university purchasing department was developed. A strategy for retrofitting buildings for water use, lighting, and interior design has been prepared.

interested. Apply what you learn to your life. Don't be misled, however, into thinking that the formal education system is the only source of learning. Keep on reading. Many of the most inspiring works on the environment do not make their way onto college or university reading lists, and most of the rewarding environmental experiences are certainly not part of the curriculum. Challenge society to change and seek a new kind of relationship between humanity and our home, planet earth.

LIGHT LIVING

Not only is it possible to reduce your impact upon the environment, you can also usually save money by doing so and take satisfaction in your contribution towards sustainability. Many books and guides have been written on the topic, such as *The Canadian Green Consumer Guide*. What follows is just a brief selection of ideas that you may wish to try. Given the statistics presented elsewhere in the text regarding per capita energy consumption, waste production, and water consumption levels in

Canada, you can be assured that Canadians contribute greatly to overconsumption. Light living is often characterized by the four R's.

Refuse

We live in a society that is geared towards making consumption easy. Our newspapers are full of advertisements regarding the best buys. Turn on the radio and you hear from the sponsor, or you are bombarded with commercials on the TV. Estimates suggest that the average American will see 35,000 television ads every year. Most of us are surrounded by shopping opportunities on a daily basis. We can drive to one of several megamalls in most Canadian cities where parking is easy and cheap, consume from a wide variety of stores, pay by credit card, ATM, cheque, or even cash. We frequently shop not to fulfil basic needs but to indulge frivolous and petty whims. Clothes are discarded when no longer fashionable rather than when they lose their durability. Gadgets are discarded in favour of newer and shinier models.

Resist and refuse to buy anything that you don't really, really need. If the purchase is necessary, shop carefully. Buy items that are less harmful to

Resist the temptation to give in to the consumer binge (*Philip Dearden*).

the environment during all stages of the product life cycle from manufacture to use and disposal. There are products certified by the government for their low impacts, as discussed later. Buy one good item rather than a succession of several shoddy ones to fulfil your needs. Buy organic produce wherever available or, better still and if possible, refuse to buy any and grow your own.

It's difficult. Consuming is easy. All the messages that we receive from society extol the virtues of buying things (Figure 22.1).

Box 22.2 Dreaming of a 'Green' Christmas?

Christmas heralds the biggest consumer bash of the year, although merchants are also trying to persuade us to be equally excessive at other times. Take control of your consumer lifestyle at Christmas and the battle is half won. Consider the following gifts:

1. Arrange an event, an outing, or a personal service rather than giving a material item.
2. Increase 'green' education by giving a book or subscription to a magazine. You could also buy someone a membership in a 'green' organization, such as Pollution Probe, the Canadian Parks and Wilderness Society, or Greenpeace.
3. Give a houseplant, a backyard composting kit, or an unbreakable coffee mug to replace

the use of disposable ones. The World Wildlife Fund also enables you to protect an acre of rainforest by making a donation to fund a project that was started by unemployed students.

4. Give something that conserves energy, such as a bus pass or an energy-saving shower head.
5. Give second-hand items.
6. Make your own gifts, such as a sweater, dried flowers, or jam.
7. Give items that display the EcoLogo of three doves.
8. Choose gifts that require little wrapping. Reuse old wrapping paper or use reusable fabric gift bags instead.

Calvin and Hobbes

by Bill Watterson

Figure 22.1 Excess consumption. SOURCE: CALVIN AND HOBBES © 1995 Watterson. Dist. by UNIVERSAL PRESS SYNDICATE. Reprinted with permission. All rights reserved.

Reduce

Can you reduce your consumption of certain items? Energy is a good place to start. A lot of energy is taken up in space heating in Canada, but must you have that thermostat set so high? Canadians tend to keep their houses much warmer inside than northern Europeans, for example. These cultures are accustomed to setting a low thermostat and wearing more clothing in the house in winter. They would not expect to be comfortable wearing just a t-shirt. Turn down the thermostat, wear more clothes, and turn the thermostat down further when you go out or go to sleep.

Reduce lighting costs by replacing burnt-out bulbs with long-life bulbs. They may be more expensive, but last longer and use 75 per cent less electricity. When replacing electrical appliances, a major factor in your choice should be energy efficiency.

Do Less

We should at least entertain the idea of doing less, concentrating on just being in the world. How foolish to laud a work-ethic that keeps people nose-to-the-grindstone, busy making and doing far beyond their needs. Here is a monumental mistake. We ought to do less, simply because much of so-called 'productive labour' is destructive: it consumes resources, encourages over-population, creates garbage and weakens the earth-source. By working less, we free up time for the more worthy task of harmonizing with the surrounding reality from which we came and will soon return (Rowe 1993:8).

Transport is also a big energy consumer, accounting for one-quarter of all energy used in Canada. Road vehicles are responsible for 83 per cent of that share. Canadians' average per capita gasoline consumption is 1100 L per year, compared with 350–500 L in European countries. It is also estimated that each kilometre of road or highway takes up about 6.5 ha of land. In Ontario, which has 155 000 km of highways, roads, and streets, this would add up to 1 million ha for motorized vehicles. Try walking or riding a bicycle whenever feasible. If you have to use motorized transport, use public transport, such as buses and

Box 22.3 The Environment-Friendly Driver

How you drive is as important as what you drive in terms of minimizing negative environmental impacts. Here are some tips:

1. Slow down: Cutting speed from 112 km per hour to 80 km per hour reduces fuel consumption by 30 per cent.
2. Keep your tires inflated: By cutting tire drag, radials give you 6–8 per cent fuel saving.
3. Keep your car tuned: This will ensure maximum efficiency and minimum pollution.
4. Minimize idling times: Even −20° C requires only a couple of minutes of warm-up.

trains. If you must have a car, get a small economical one with standard transmission, use it sparingly, and try to carpool.

You can also think about reducing the waste associated with the things you buy. Many products are overpackaged. They look good on the store shelf, but will only add to the amount of waste sent to the landfill. Buy groceries in bulk to help cut down on packaging. Reduce your waste by starting a compost heap for kitchen wastes. Reduce and, if possible, eliminate your use of toxic materials. Products that may seem quite innocuous (paint, solvents, and cleaning agents, for example) become hazardous wastes when disposed of. Try to find alternatives to these products. Don't buy more product than you need, and dispose of them in full accordance with the instructions from your local municipality.

Reuse

Buy products, such as rechargeable batteries, that can be reused. Try to find another use for something no longer useful in its original state. Use plastic food containers to store things in your fridge or workshop. Return with the same plastic bags that you bought your groceries in for your next load of groceries. When you're finished with something, it may still be useful to someone else. Organize a garage sale or donate the items to charity rather than throwing them out.

Recycle

Recycling facilities have sprung up all across the country over the last decade. Recyclable materials include newspaper, cardboard, mixed paper, glass, and aluminium. In many areas you can now also recycle plastics, car batteries, tires, and oil. These materials can be reprocessed into new goods. It takes 30–55 per cent less energy, for example, to make new paper from old paper than it does to start fresh from a new tree. Estimates suggest that if we recycled all the paper we used in Canada, we would save 80 million trees. The amount of newsprint is over 50 kg per person per year— enough to account for one whole mature tree. Similar efficiencies can be obtained by recycling other materials where less energy is required for remanufacture. Oil, for example, fuels many industrial processes. Estimates suggest that if 1 per cent of the Canadian population recycled instead of

---

Box 22.4 Recycling Costs

In 1994 Don Young, the head of Blue Box recycling in Toronto, said that 'if we're recycling to conserve landfill space and natural resources, the Blue Box is working well. But if it was intended to be a profit maker, we've been greatly misled' (*Financial Post*, 11 June 1994:16). Politicians across the country echoed the sentiments as recycling programs needed to be subsidized. However, by 1995 the situation had changed dramatically. Newspaper pirates circled the streets trying to empty Blue Boxes before the recycling vehicles arrived. The reason? Diminishing forests and world demand had pushed the price of new newsprint to record levels of $1,040 a tonne. The price of old newspapers has quadrupled to $125 a tonne. Recycling has now become a major profit source for some municipalities.

---

trashing an aluminium can a day, the oil saved would make 21 million L of gasoline. However, the bottom line remains that while it is better to recycle than not recycle, reducing consumption levels in the first place is still the preferred option.

Managing Planet Earth

A second myth is that with enough knowledge and technology, we can, in the words of Scientific American (1989), 'manage planet earth.' Higher education has largely been shaped by the drive to extend human domination to its fullest. In this mission, human intelligence may have taken the wrong road. Nonetheless, managing the planet has a nice ring to it. It appeals to our fascination with digital readouts, computers, buttons and dials. But the complexity of earth and its life systems can never be safely managed. The ecology of the top inch of topsoil is still largely unknown as is its relationship to the larger systems of the biosphere. What might be managed, however, is us: human desires, economies, politics, and communities. But our attention is caught by those things that avoid the hard choices implied by politics, morality, ethics, and common sense. It makes far better sense to reshape ourselves to fit a finite planet than to attempt to reshape the planet to fit our infinite wants (Orr 1994:9).

There are also many other ways in which you can reduce the pressure on the environment. These are just a few suggestions to get you started. The key point to remember is that the accumulated actions of many concerned individuals acting together will make a difference. This is the concept of cumulative effects discussed in Chapter 10 relative to impact assessment.

A good example is provided by the experience of Port Elgin in Ontario regarding water conservation. Development pressures had created demand for a new water treatment plant. The municipality's calculations indicated that this would cost $5.5 million, which was too much for the municipality. Instead, they initiated a vigorous water-conservation campaign that involved the installation of water metres, so that people would be charged according to their consumption, and provided inexpensive water-saving devices, such as faucet aerators and toilet dams. Water consumption in summer (the critical period) was reduced by 50 per cent as a result of these measures. The expansion was not required and the development ban was lifted.

INFLUENCE

One of the best ways to influence business is through the purchasing power of consumers. If consumers band together and refuse to buy certain products because of their impacts upon the environment, then the manufacturer will either have to respond to these concerns or go out of business. There are many successful examples of these kinds of actions. During the 1970s and 1980s, for example, conservationists were able to exert increasing pressure on hamburger chains to change their source of beef supply to ensure that tropical rainforests were not being cut down to be replaced by grass to graze cattle for hamburgers. Following boycotts, all major chains were persuaded to ensure that their sources did not contribute to the destruction of the rainforests. Similar campaigns have been used on tuna canners to ensure that their tuna was not being caught by methods that killed dolphins, which often swim with the schools of tuna. Consumer boycotts can be a very effective way of influencing business practices.

Governments, especially those at provincial and municipal levels, can support initiatives to create and implement sustainable development strate-

gies. Box 22.6 provides the vision for sustainable development, which was endorsed unanimously by the Alberta legislature in June 1992, and subsequently by more than 100 municipalities in that province. Now the challenge is following through on the vision and principles to ensure action.

What can you do to encourage implementation of such a set of principles? Election times offer a major opportunity. Make a point of asking your local candidates at all levels of government for their views on certain environmental or sustainable development issues. Help publicize these views. It is much more difficult for politicians to change their positions if they are well known to the public. Make sure that the politicians follow through on their pre-election promises on environmental matters. Make sure you express your views on the environment to politicians between elections. Write letters, call them, go and see them, and organize demonstrations.

Government policies on the environment seem to be moving in the right direction, albeit slowly. Some of the actions being taken by governments at all levels have been described throughout the text. It is important to let politicians know that you appreciate their efforts when they have taken what might be unpopular stands for the sake of environmental interests. Sometimes government actions may also seem ambiguous, as, for example, described in Chapter 17 when national parks policies and legislation appear to emphasize increasingly the maintenance of ecological integrity, yet activities in the parks seem to be much more attuned to revenue generation.

Another example is provided by the much-heralded Canadian Environmental Choice product labelling program. This program aims to influence business practices by shifting consumer and institutional purchasing to products and services that conform to environmental guidelines. The guidelines are established by a board that develops the guidelines each category of product must achieve to obtain the recognition signified by the three doves logo. These guidelines consider factors such as recyclability, design of packaging material, use of recyclate in manufacturing, effluent toxicity, and energy consumption. However, in addition to this list, the government has now required the board to take into account the economic impacts of the guidelines to ensure that they do not impinge

Box 22.5 Corporate Contributions

Large corporations and retailers are highly visible targets for environmental action and attracted a lot of attention, particularly in the 1980s. However, many corporations do much better than we do personally, or our governments do, in taking a systematic look at some of their environmental impacts and moving to address them. As individuals, we often just let things slide. McDonald's, for example, has the Earth Effort program, which embraces the reduce, reuse, and recycle philosophy described in the text. Every year McDonald's is committed to buying at least $100 million worth of recycled products for building, operating, and equipping their facilities. Carry-out bags are made from recycled corrugated boxes and newsprint; take-out drink trays are made from recycled newspapers. New restaurants have been constructed with concrete blocks made from recycled photographic film and roofs made from computer casings. It has also cut down on the amount of waste created by, for example, switching from foam packaging to paper wraps for sandwiches, thus reducing sandwich packaging by over 90 per cent.

McDonald's also has a strict policy that it will not buy the beef of cattle raised on land that was converted from rainforests for that purpose. It also has programs to reduce energy consumption and takes part in a wide range of local initiatives ranging from tree planting to local litter drives. For its efforts McDonald's has won White House awards and the National Recycling Coalition's Award for Outstanding Corporate Leadership.

McDonald's is not, of course, alone in its efforts. Many well-known corporations have similar programs. Eddie Bauer, for example, has joined forces with an NGO, American Forests, in a tree-planting effort on damaged forest ecosystems in eight reforestation sites in the US and Canada. Consumers can opt to add a dollar to the price of their purchase at the store, and in turn the store will plant a tree for each dollar and donate another tree.

Also consider Valhalla Pure Outfitters of Vernon, BC, makers of high-quality outdoor clothing. It now makes all its polyester fleece jackets from recycled materials that look, feel, wear, and last the same as new fabric, even though they are made from recycled pop bottles. Each pop bottle diverted from a landfill site for recycling is sent to a factory in South Carolina where they are separated by colour and then reprocessed before being sent to finishing plants in the US and Quebec. The final cost is similar to making new material, but has the added environmental advantages. Each jacket is equivalent to about twenty-five pop bottles.

These are not the only ways in which the corporate and business world is contributing positively towards addressing environmental problems. Some of them also have substantial foundations to which environmental groups can apply for funding for specific problems. As government coffers are drained, more pressure is put on these sources. The Laidlaw Foundation, for example, used to give $50,000 a year for environmental projects, and requests rarely exceeded this amount. It now disburses $350,000 but has requests exceeding $1 million dollars. The Richard Ivey Foundation in London, Ontario, provides up to 80 per cent of its $2 million budget for environmental projects each year, but still manages to meet only 5 per cent of the amount requested. With such pressure on resources, it is essential to ensure that the money is used effectively to address the most pressing problems. If there are environmental problems in your area that you think need to be addressed, don't be afraid to approach local businesses for support. They may well agree with you and be happy to make a contribution to a carefully crafted solution.

upon the economic competitiveness of Canadian products. As David Cohen, a member of the board, concludes:

> The insistence on an economic analysis requirement for a non-regulatory program, which attempts to do little more than provide reasonably accurate consumer information,

reveals the ambiguity of official commitments to environmental recovery and the depth of government concern that environmental actions might bring economic damage. If governments believe environmental information for consumers is a serious economic threat, we have good reason to be concerned about the design and likely results of governments' more

Box 22.6 Alberta's Vision of Sustainable Development

Alberta, a member of the global community, is a leader in sustainable development, ensuring a healthy environment, a healthy economy, and a high quality of life in the present and in the future. Our vision encompasses all of the following elements:

- The quality of air, water, and land is assured.
- Alberta's biological diversity is preserved.
- We live within Alberta's natural carrying capacity.

- The economy is healthy.
- Market forces and regulatory systems work for sustainable development.
- Urban and rural communities offer a healthy environment for living.
- Albertans are educated and informed about the economy and the environment.
- Albertans are responsible global citizens.
- Albertans are stewards of the environment and the economy (Future Environmental Directions for Alberta Task Force 1995:7).

Box 22.7 National Action Plan to Encourage Municipal Water Use Efficiency

In 1994 the Canadian Council of Ministers of the Environment established a national action plan to achieve more efficient use of water in Canadian municipalities in order to save money and energy, reduce expansion of existing water and waste water systems, and conserve water.

Some of the key principles underlying the action plan are:

- *Harmonization:* There shall be consistent regulatory requirements related to water-use efficiency across Canada.

- *User Pays on Basis of Volume:* Consumers shall pay for water and waste water services on the basis of measured actual use.
- *Full-Cost Pricing:* Municipalities shall move towards water and waste water rate structures that reflect the full costs of delivery and treatment.
- *An Informed Public:* The public shall be informed of the real costs of water use and the savings that can be achieved through water efficiency, and of actions that they can take to reduce usage.

The three doves logo of the Canadian Environmental Choice Program (*Canadian Environmental Choice Program*).

direct regulatory initiatives for the environment (Cohen 1994:27).

His concerns echo those of many environmentally concerned people who feel that the government is more than willing to adopt environmentally friendly, voter-popular measures, as long as they do not involve any significant economic costs. The problem is that policies and the implementation of those policies are often too complicated for one individual to grasp and act on. For this reason, concerned people may band together in non-governmental organizations, as described in Chapter 11. Such organizations represent the collective concern and resources of many people, and are in a much better position to attack a problem than an individual. One of the best ways to spend your conservation dollar is to support such a group. They have had significant impacts upon govern-

Box 22.8 EcoLogo Facts

- Established in 1988, EcoLogo has given out 109 licences to use its logo on 1,400 products and services.
- Annual fees to use the logo range from $300 annually for companies with Canadian sales under $300,000 to $5,000 for firms with sales over $1 million.
- Products endorsed include recycled plastic

bags, batteries, water and solvent-based paints, cotton diapers, photocopier toner cartridges, and composting systems for household waste. Rerefined motor oil has now also been endorsed if it contains at least 50 per cent recycled oil. Canadians discard 300 million of 450 million L of motor oil used every year.

ment policies in Canada. A list of some organizations is provided in the appendix.

IMPLICATIONS

The overall implications of this book should be brutally clear by now. Humanity is facing some very major challenges over this next decade regarding our relationship with the earth and its life-support systems. So great are our capabilities to affect these systems that human-controlled influences now dominate many natural processes, resulting in the many critical environmental problems described throughout this book. But we are not powerless to change the direction of society. We, as individuals and as concerned individuals banding together, can effect many of the changes that need to be made. So, despite the severity of the challenges, it is important to remain optimistic.

Above all, stay cheerful, stay active, look after yourself, look after others, love this planet, and don't give up! We are on the most beautiful planet we know about. Canada has some of the most breathtaking wonders in the universe. We have a responsibility. We can think of no better advice than that offered by Edward Abbey:

Despite all the environmental challenges that confront society, it is important not to despair, to do whatever you can to help, to get out and enjoy yourself, and to wonder at the beauty of it all and be humbled (*Parks Canada*).

One final paragraph of advice: Do not burn yourselves out. Be as I am—a reluctant enthusiast . . . a part-time crusader, a half-hearted fanatic. Save the other half of yourselves and your lives for pleasure and adventure. It is not enough to fight for the land; it is even more important to enjoy it. While you can. While it's still here. So get out there and . . . ramble out yonder and explore the forests, encounter the grizz, climb the mountains, bag the parks, run the rivers, breathe deep of that yet sweet and lucid air, sit quietly for a while and contemplate the precious stillness, that lovely, mysterious and awesome space. Enjoy yourselves, keep your brain in your head and your head firmly attached to the body, the body active and alive, and I promise you this much: I promise you this one sweet victory over our enemies, over those desk-bound people with their ears in a safe deposit box and their eyes hypnotized by desk calculators. I promise you this: you will outlive the bastards.

SUMMARY

1. Ordinary citizens can have a positive impact in many ways on the environmental challenges facing society. A first step is building awareness of these problems. Universities and colleges should be intimately involved with this process by ensuring that all graduates have a measure of environmental literacy before they graduate. Universities and colleges are also large consumers of matter and energy and should lead by example in reducing their impacts on the environment.

2. Individuals can help by living in accord with the four R's: refuse to be goaded into overconsumption, reduce consumption of matter and energy where possible, reuse materials where possible, and recycle those that you cannot reuse.

3. Individuals can also wield influence by banding together and taking collective action, for example, through consumer boycotts of products and companies that engage in environmentally destructive practices. One of the most successful ways of doing this is by joining a non-governmental organization composed and supported by like-minded individuals. As an individual you will probably not be able to afford to support an environmental lobbyist in Ottawa. However, if thousands of people contribute, then this becomes a reality.

4. In spite of the serious nature of many of the environmental problems described in the book, it is important to stay optimistic about their solution and take time to get out and enjoy the beauty and challenge of one of the most splendid parts of the planet, Canada, our home.

REVIEW QUESTIONS

1. Discuss your institution's present and potential role in raising environmental awareness.

2. What are three concrete steps towards 'light living' that you are willing to make over the next month?

3. What are some initiatives that you could take as an individual or as part of your community to reduce negative environmental consequences from our activities?

4. Find out the names and mandates of the environmental NGOs in your area. Is there anything you can do to help them?

REFERENCES AND SUGGESTED READING

Bardati, D.R. 1995. 'An Environmental Action Plan for Bishop's University'. *Journal of Eastern Townships Studies* 6:19–38.

Cohen, D.S. 1994. 'Subtle Effects: Requiring Economic Assessments in the Environmental Choice Program'. *Alternatives* 20:22–7.

Future Environmental Directions for Alberta Task Force. 1995. *Ensuring Prosperity: Implementing Sustainable Development.* Edmonton: Alberta Environmental Protection.

Homer-Dixon, T. 1995. 'The Ingenuity Gap: Can Poor Countries Adapt to Resource Scarcity?' *Population and Development Review* 21, no. 3:587–612.

Lerner, S., ed. 1993. *Environmental Stewardship: Studies in Active Earthkeeping.* Department of Geography Publication Series no. 39. Waterloo, ON: Department of Geography, University of Waterloo.

Newman, P., S. Neville, and L. Duxbury, eds. 1988. *Case Studies in Environmental Hope*. Perth: Environmental Protection Authority for the Western Australian State Conservation Strategy.

Orr, D.W. 1994. *Earth in Mind: On Education, Environment and the Human Prospect*. Covelo, CA: Island Press.

Rowe, S. 1993. 'In Search of the Holy Grass: How to Bond with the Wilderness in Nature and Ourselves'. *Environment Views* (Winter): 7–11.

Shrubsole, D., and D. Tate, eds. 1994. *Every Drop Counts*. Based on Canada's First National Conference and Trade Show on Water Conservation. Cambridge, ON: Canadian Water Resources Association.

Smith, A. 1993. *Campus Ecology: A Guide to Assessing Environmental Quality and Creating Strategies for Change*. Los Angeles: Living Planet Press.

Starke, L. 1990. *Signs of Hope: Working Towards Our Common Future*. Oxford: Oxford University Press.

Thompson, D., and S. van Bakel. 1995. *A Practical Introduction to Environmental Management on Canadian Campuses*. Ottawa: National Round Table on the Environment and Economy.

Appendix: Conservation Organizations

In addition to the references provided at the end of each chapter, you should also consult two other sources of growing importance: CD-ROMs and the Internet.

Information is increasingly being made available on CD-ROMs. For example, Environment Canada has provided the 1996 State of the Environment Report on a CD-ROM. This source is a valuable reference and contains a remarkable breadth and depth of information about resources and the environment in Canada.

Home pages on the World Wide Web are a rich (if somewhat uneven in quality) source of information. It is difficult to provide a 'catalogue' of such home pages since new ones are created every day while others disappear. However, to provide a sample of home pages available in early 1997, the following are listed to encourage you to 'surf the net'.

Throughout this book, reference has been made to State of the Environment reporting documents in Canada. The State of Canada's Environment Infobase can be examined on http://www.doe.ca or on http://199.212.18.12/~soer/. At a global level, the World Resources Institute publishes an annual state of the world report, which can be checked at http://www/wri.org/wri/index.html.

The Commission for Environmental Cooperation was created by the North American Agreement on Environmental Cooperation to enhance regional cooperation, prevent potential environmental and trade disputes, and promote effective enforcement of environmental law. The agreement was signed by Canada, Mexico, and the United States, and complements the environmental provisions established in the North American Free Trade Agreement (see http://www.cec.org/index.html).

Canada's National Round Table on Environment and Economy was referred to in several chapters, especially in Chapter 11 (see http://www.nrtee-trnee.ca). Reference has also been made to the Canadian Council of Ministers of the Environment (see http://www.mbnet.mb.ca/ccme). Many provincial resource and environmental agencies also have home pages. For example, to see what the BC Ministry of Environment, Lands and Parks offers, see http://www.env.gov.bc.ca. Another good source to check is Green Product Policy and Life Cycle Assessment in Canada: http://rmit.edu.au?Pubs/ERSNews/ERD4/green Canada.html.

Other good sources to look into are the following:

Commissioner of the Environment and Sustainable Development:
  http://www/oag-bvg.gc.ca
Department of Fisheries and Oceans:
  http://www.ncr.dfo.ca
Earth Summit +5 Special Session of the General Assembly to Review and Appraise the Implementation of Agenda 21:
  http://www.un.org/dpcsd.earthsummit
Mining Association of Canada:
  http://www.mining.ca
Natural Resources Canada:
  http://www.nrcan.gc.ca
UN Division for Sustainable Development:
  http://www.un.org/dpcsd/dsd
World Bank Environment Department:
  http://www-esd.worldbank.org/html/esd/env/ envmain.htm
World Business Council for Sustainable Development:
  http://www.wbcsd.ch

These are only a start, but will give you an idea of the remarkably rich sources of information available by surfing the Internet.

INTERNATIONAL ORGANIZATIONS

Alliance for the Wild Rockies
PO Box 8731
Missoula, MT 59807
Phone: 406–721–5420
Fax: 406–721–9917

Antarctic & Southern Ocean Coalition
PO Box 76920
Washington, DC 20013
Phone: 202–544–0236
Fax: 202–544–8783

Center for Marine Conservation
1725 DeSales Street NW
Washington, DC 20036
Phone: 202–429–5609
Fax: 202–872–0619
E-mail: dccmc@ix.netcom.com

Center for Plant Conservation
PO Box 299
St Louis, MO 63166–0299
Phone: 314–577–9450
Fax: 314–664–0465

Clean Water Action Project
4455 Connecticut Avenue NW, Suite A300
Washington, DC 20008
Phone: 202–895–0420
Fax: 202–895–0438
E-mail: cleanwater@essential.org

Conservation International
2501 M Street, Suite 200
Washington, DC 20037
Phone: 202–429–5660
Fax: 202–887–5188

Cousteau Society
870 Greenbrier Circle, Suite 402
Chesapeake, VA 23320
Phone: 757–523–9335
Fax: 757–523–2747
E-mail: tcsva@igc.apc.org

Earth Council
Communications Coordinator
PO Box 2323–1002
San Jose, Costa Rica
Phone: 506–256–1611
Fax: 506–255–2197
E-mail: kcook@terra.ecouncil.ac.cr
Web site: http://www.ecouncil.ac.cr

Earth Island Institute
300 Broadway, Suite 28
San Francisco, CA 94133–3312
Phone: 415–788–3666
Fax: 415–7324
E-mail: earthisland@earthisland.org

EarthKind
2100 L Street NW
Washington, DC 20037
Phone: 202–778–6149
Fax: 202–778–6134
E-mail: earthknd@ix.netcom.com

Earthwatch
680 Mount Auburn Street
PO Box 9104
Watertown, MA 02272–9104
Phone: 617–926–8200
Toll free: 1–800–776–0188
Fax: 617–926–8532
E-mail: info@earthwatch.org

Ecological Agricultural Projects
Macdonald Campus
McGill University
Ste-Anne-de-Bellevue, Quebec
H9X 3V9
Phone: 514–398–7771
Fax: 514–398–7621
E-mail: info@eap.mcgill.ca

Environmental Action Foundation
6930 Carroll Avenue, Suite 600
Takoma Park, MD 20912
Phone: 301–891–1100
Fax: 301–891–2218

Friends of Animals
777 Post Road, Suite 205
Darien, CT 06820
Phone: 203–656–1522
Fax: 203–656–0267

Friends of the Earth International
International Secretariat
PO Box 19199
1000 GD Amsterdam
The Netherlands
Phone: 3120–622–1369
Fax: 3120–639–2181

Greenpeace Council
Keizersgracht 176
1016 DW
Amsterdam
The Netherlands
Phone: 31–20–523–6222
Fax: 31–20–523–6200
E-mail: greenpeace.international@green2.greenpeac

International Water Resources Association
University of New Mexico
1915 Roma NE
Albuquerque, New Mexico 87131–1436
Phone: 505–277–9400
Fax: 505–277–9405

International Wildlife Coalition
70 East Falmouth Highway
East Falmouth, MA 02536
Phone: 508–548–8328
Fax: 508–548–8542

IUCN (World Conservation Union)
rue Mauverney 28
CH 1196 Gland
Switzerland
Phone: 41–22–999–0155
Fax: 41–22–999–0015
Web site: http://iucn.org.

Jane Goodall Institute for Wildlife Research, Education & Conservation
PO Box 599
Ridgefield, CT 06877
Phone: 520–325–1211
Fax: 203–431–4387

Jersey Wildlife Preservation Trust
Les Augres Manor
Trinity, Jersey
Channel Islands
JE3 5BP
Phone: 1534–864–666
Fax: 1534–865–161
E-mail: ldexec@itl.net

National Audubon Society
700 Broadway
New York, NY 10003
Phone: 212–979–3000
Web site: http://www.audubon.org/
E-mail for memberships: join@audubon.org
E-mail for general information:
webmaster@list.audubon.org

Ocean Voice International
PO Box 37026
3332 McCarthy Road
Ottawa, Ontario
K1V 0W0
Phone: 613–264–8986
Fax: 613–521–4205
E-mail: ah194@freenet.carleton.ca

Rainforest Action Network
450 Sansome Street, Suite 700
San Francisco, CA 94111
Phone: 415–398–4404
Fax: 415–398–2732
E-mail: rainforest@ran.org

Sierra Club
85 Second Street, 2nd Floor
San Francisco, CA 94105
Phone: 415–977–5500
Fax: 415–977–5799
E-mail: information@sierraclub.org
Web site: www.sierraclub.org

Soil and Water Conservation Society
7515 NE Ankeny Road
Ankeny, IA 50021–9764
Phone: 515–289–2331
Fax: 515–289–1227
Toll free: 1–800-THE-SOIL
E-mail: swcs@swcs.org

United Nations Environment Programme
Information Officer
Regional Office for North America
2 United Nations Plaza, Room DC 2–803
New York, NY 10017
Phone: 212–963–8210
Fax: 212–963–7341
E-mail: uneprona@un.org
Web site: http://www.unep.ch/unep.html
Convention on Biodiversity:
http://www.unep.ch/biodiv.html/

World Commission on Forests and Sustainable Development
CP 51
CH–1219 Châtelaine
Geneva, Switzerland
Phone: 22–979–9165
Fax: 22–979–9060
E-mail: dameena@1prolink.ch
Web site: http://iisd1.iisd.ca/wcfsd
World Resources Institute
1709 New York Avenue NW, Suite 700

Washington, DC 20006
Phone: 202–638–6300
Fax: 202–638–0036
Web site: www.wri.org

World Society for the Protection of Animals
29 Perkins Street
PO Box 190
Boston, Massachusetts 02130
Phone: 617–522–7000
Fax: 617–522–7077
E-mail: wspa@world.std.com

Worldwatch Institute
1776 Massachusetts Avenue NW
Washington, DC 20036
Phone: 202–452–1999
Fax: 202–296–7365
E-mail: worldwatch@worldwatch.org
Web site: http://www.worldwatch.org

CANADIAN NATIONAL ORGANIZATIONS

Animal Alliance of Canada
221 Broadview Avenue, Suite 101
Toronto, Ontario
M4M 2G3
Phone: 416–462–9541
Fax: 416–462–9647
E-mail: aac@inforamp.net

Assembly of First Nations
1 Nicholas Street, Suite 1002
Ottawa, Ontario
K1N 7B7
Phone: 613–241–6789
Fax: 613–241–5808

Association for the Protection of Fur-Bearing Animals
2235 Commercial Drive
Vancouver, BC
V5N 4B6
Phone: 604–255–0411
Fax: 604–255–1491

Canadian Arctic Resources Committee
1 Nicholas Street, Suite 1100
Ottawa, Ontario
K1N 7B7
Phone: 613–241–7379
Fax: 613–241–2244
Web site: http://www.carc.org/pubs/

Canadian Council on Ecological Areas
Secretariat
c/o Lee Warren
Place Vincent Massey
351 St Joseph Boulevard
Hull, Quebec
K1A 0H3
Phone: 819–953–1444
Fax: 819–994–4445

Canadian Earth Energy Association
130 Slater Street, Suite 605
Ottawa, Ontario
K1P 6E2
Phone: 613–230–2332
Fax: 613–237–1480

Canadian Environmental Law Association
517 College Street, Suite 401
Toronto, Ontario
M6G 4A2
Phone: 416–960–2284
Fax: 416–960–9392

Canadian Global Change Program
225 Metcalfe Street, Suite 308
Ottawa, Ontario
K2P 1P9
Phone: 613–991–5639
Fax: 613–991–6996
E-mail: cgcp@rsc.ca
Web site: http://datalib.library.ualberta.ca:80~cgcp

Canadian Nature Federation
1 Nicholas Street, Suite 520
Ottawa, Ontario
K1N 7B7
Phone: 613–562–3447
Fax: 613–562–3371

Canadian Parks and Wilderness Society (CPAWS)
401 Richmond Street West, Suite 380
Toronto, Ontario
M5V 3A8
Phone: 416–979–2720
Fax: 416–979–3155

Canadian Water Resources Association
PO Box 1329
Cambridge, Ontario
N1R 7G6
Phone: 519–622–4764
Fax: 519–621–4844
E-mail: cwranat@worldchat.com
Web site: http://www.cwra

City Farmer Canada's Office of Urban Agriculture
318 Homer Street
Suite 801
Vancouver, BC
V6B 2V3
Phone: 604–685–5832
Fax: 604–685–0431
E-mail: cityfarm@unixg.ubc.ca
Web site: www.cityfarmer.org

Earth Day Canada
144 Front Street W., Suite 250
Toronto, Ontario
M5J 2L7
Phone: 416–599–1991
Toll free: 1–900–561–3300
Fax: 416–599–3100

Elsa Wild Animal Appeal of Canada
2482 Yonge Street
PO Box 45051
Toronto, Ontario
M4P 3E3
Phone: 416–489–8862
Fax: 416–489–4769

Energy Probe
225 Brunswick Avenue
Toronto, Ontario
M5S 2M6
Phone: 416–964–9223
Fax: 416–964–8239
E-mail: EnergyProbe@nextcity.com

Environment Bureau
Agriculture and Agri-Food Canada
Sir John Carling Building
930 Carling Avenue, Room 367
Ottawa, Ontario
K1A 0C5
Fax: 613–759–7238
Web site: http://www.aceis1.ncr.agr.ca

Environment Canada Inquiry Center
351 St Joseph Boulevard
Hull, Quebec
K1A 0H3
Phone: 819–997–2800
Toll free: 1–800–668–6767
Fax: 819–953–0966
E-mail: enviroinfo@cpgsv1.am.doe.ca
Web site: http://www.ec.gc.ca/
A Guide to Green Government:
http://www.doe.ca/grngvt_e.html
State of the Environment Infobase:
http://www1.ec.gc.ca/~soer/
Sustainable Development Strategy (draft):
http://www.ec.gc.ca/sd-dd_consult/sds/sdgtcc_e.html

Fisheries and Oceans Canada
200 Kent Street
Ottawa, Ontario
K1A 0E6
Phone: 613–993–0999
E-mail: info@www.ncr.dfo.ca
Web site: http://www.ncr.dfo.ca/home_e.htm

Friends of the Earth
251 Laurier Avenue West, Suite 701
Ottawa, Ontario
K1P 5J6
Phone: 613–230–3352
Fax: 613–232–4354

Greenpeace
185 Spadina Avenue, 6th Floor
Toronto, Ontario
M5T 2C6
Phone: 416–597–8408
Fax: 416–597–8422

International Fund for Animal Welfare (Canada)
400–410 Bank Street
Ottawa, Ontario
K2P 1Y8
Phone: 613–233–8458
Toll free: 1–888–500–IFAW
Fax: 613–233–9602

International Institute for Sustainable Development
161 Portage Avenue East, 6th Floor
Winnipeg, Manitoba
R3B 0Y4
Phone: 204–958–7700
Fax: 204–958–7710
E-mail: reception@iisdpost.iisd.ca
Web site: http://iisd1.iisd.ca

Marine Protected Areas Network
c/o Martin Willison
Biology Department
Dalhousie University
Halifax, Nova Scotia
B3G 3J4
Phone: 902–494–3514

National Parks Directorate Parks Canada
25 Eddy Street, 4th Floor
Hull, Quebec
K1A 0M5
E-mail: National-Parks_Webmaster@pch.gc.ca
Web site: http://parkscanada.pch.gc.ca/

Nature Conservancy of Canada
110 Eglinton Avenue W., 4th Floor
Toronto, Ontario
M4R 2G5
Phone: 416–932–3202
Fax: 416–932–3208

Probe International
225 Brunswick Avenue
Toronto, Ontario
M5S 2M6
Phone: 416–964–9223
Fax: 416–964–8239

Sea Shepherd Conservation Society
PO Box 48446
Vancouver, BC
V7X 1A2
Phone: 604–688–7325

Sierra Club of Canada
1 Nicholas Street, Suite 412
Ottawa, Ontario
K1N 7B7
Phone: 613–241–4611
Fax: 613–241–2292

Wildlife Habitat Canada
7 Hinton Avenue North, Suite 200
Ottawa, Ontario
K1Y 4P1
Phone: 613–722–2090
Fax: 613–722–3318

Wildlife Preservation Trust Canada
120 King Street
Guelph, Ontario
Phone: 519–836–9314
Fax: 519–824–6776
E-mail: wptc@inforamp.net

World Wildlife Fund Canada
90 Eglinton Avenue E., Suite 504
Toronto, Ontario
M4P 2Z7
Phone: 416–489–8800
Fax: 416–489–3611

ZOOCHECK Canada Inc.
3266 Yonge Street, Suite 1729
Toronto, Ontario
M4N 3P6
Phone: 416–696–0241
Fax: 416–696–0370

BRITISH COLUMBIA

BC Spaces for Nature
Box 673
Gibsons, BC
V0N 1V0
Phone: 604–886–8605
Fax: 604–886–3768

BC Wild
Box 2241, Main Post Office
Vancouver, BC
V6B 3W2
Phone: 604–669–4802
Fax: 604–669–6833

CPAWS, BC Chapter
611–207 West Hastings Street
Vancouver, BC
V6B 1H7
Phone: 604–685–7445
Fax: 604–685–6449
E-mail: cpawsbc@direct.ca

Friends of Clayoquot Sound
PO Box 489
Tofino, BC
V0R 2Z0
Phone: 604–725–4218
Fax: 604–725–2527

Ministry of Environment, Lands and Parks
Public Affairs and Communications
810 Blanshard Street, 1st Floor
Victoria, BC
V8V 1X4
Phone: 250–387–9422

Nature Trust of British Columbia
100 Park Royal South, Suite 808
West Vancouver, BC
V7T 1A2
Phone: 604–925–1128
Fax: 604–926–3482
E-mail: naturetrust@mindlink.bc.ca

Ocean Resource Conservation Alliance
PO Box 1189
Sechelt, BC
V0N 3A0
Phone: 604–885–7518
Fax: 604–885–2518

Sierra Club of Western Canada
1525 Amelia Street
Victoria, BC
Phone: 604–386–5255
Fax: 604–386–4453

Western Canada Wilderness Committee
20 Water Street
Vancouver, BC
V6B 1A4
Phone: 604–683–8220
Fax: 604–683–8229

Wildlife Rescue Association of British Columbia
5216 Glencairn Drive
Burnaby, BC
V5B 3C1
Phone: 604–526–7275
Fax: 604–524–2890

ALBERTA

Alberta Environmental Protection
Communications
9915–108 Street, 9th Floor
Edmonton, Alberta
T5K 2G8
Phone: 403–427–2739 or 403–944–0313

Alberta Recreation, Parks & Wildlife Foundation
Harley Court Building
10045–111 Street
Edmonton, Alberta
T5K 1K4
Phone: 403–482–6467
Fax: 403–488–9755

Alberta Wilderness Association
Box 6398, Station D
Calgary, Alberta
T2P 2E1
Phone: 403–283–2025
Fax: 403–270–2743

Bow Valley Naturalists
Box 1693
Banff, Alberta
T0L 0C0
Phone/Fax: 403–762–4160

CPAWS, Calgary-Banff Chapter
319 Tenth Avenue SW, Suite 306
Calgary, Alberta
T2R 0A5
Phone: 403–232–6686
Fax: 403–232–6988
E-mail: cpawscal@cadivision.com

CPAWS, Edmonton Chapter
Box 52031, 8210–109 Street
Edmonton, Alberta
T5X 4L6
E-mail: edmcpaws@edmonton.ab.ca

Federation of Alberta Naturalists
Box 1472
Edmonton, Alberta
T5J 2K5
Phone: 403–453–8629
Fax: 403–453–8553

SASKATCHEWAN

CPAWS, Saskatchewan Chapter
Please direct letters/inquiries to:
CPAWS National
401 Richmond Street W., Suite 380
Toronto, Ontario
M5V 2A8
Phone: 416–979–2720
Toll free: 1–800–333–WILD
Fax: 416–979–3155
E-mail: cpaws@web.net

Environment and Resource Management
Education and Communications
3211 Albert Street
Regina, Saskatchewan
S4S 5W6
Toll free: 1–800–667–2757

Nature Saskatchewan
1860 Lorne Street, Suite 206
Regina, Saskatchewan
S4P 3W6
Phone: 306–780–9273
Toll free: 1–800–667–4668
Fax: 306–780–9263

Saskatchewan Environmental Society
PO Box 1372
Saskatoon, Saskatchewan
S7K 3N9
Phone: 306–665–1915
Fax: 306–665–2128

MANITOBA

CPAWS, Manitoba Chapter
Box 344
Winnipeg, Manitoba
R3C 2H5
Phone/Fax: 204–237–5947
E-mail: pturenne@mbnet.mb.ca

Department of Environment
123 Main Street, Suite 160
Winnipeg, Manitoba
R3C 1A5
Phone: 204–945–7100

Manitoba Naturalists Society
63 Albert Street, Suite 401
Winnipeg, Manitoba
R3B 1G4
Phone/Fax: 204–943–9029

ONTARIO

Citizens Network on Waste Management
17 Major Street
Kitchener, Ontario
N2H 4R1
Phone: 519–744–7503
Fax: 519–744–1546
E-mail: jjackson@web.net

CPAWS, Wildlands League Chapter
401 Richmond Street W., Suite 380
Toronto, Ontario
M5V 3A8
Phone: 416–979–2720
Toll free: 1–800–WILD
Fax: 416–979–3155
E-mail: cpaws@web.net

Earthroots Coalition
401 Richmond Street W., Suite 410
Toronto, Ontario
M5V 3A8
Phone: 416–599–0152
Fax: 416–340–2429

Elora Center for Environmental Excellence
160 St David Street South
Fergus, Ontario
N1M 2L3
Phone: 519–843–7283
Fax: 519–843–1910

Energy Action Council of Toronto
16 Howland Road
Toronto, Ontario
M4K 2Z6
Phone: 416–461–9654
Fax: 416–461–9540

Environment North
704 Holly Crescent
Thunder Bay, Ontario
P7E 2T2
Phone: 807–475–5267
Fax: 807–577–6433

Federation of Ontario Naturalists
355 Lesmill Road
Don Mills, Ontario
M3B 2W8
Phone: 416–444–8419
Fax: 416–444–9866

Ministry of Environment and Energy Communications
135 St Clair Avenue West, 2nd Floor
Toronto, Ontario
M4V 1P5

Phone: 416–323–4321
Toll free: 1–800–565–4923

Northwatch
PO Box 282
North Bay, Ontario
P1B 8H2
Phone: 705–497–0373
Fax: 705–476–7060

Water Environment Association of Ontario
63 Hollyberry Trail
North York, Ontario
M2H 2N9
Phone: 416–502–1440
Fax: 416–502–1786

QUEBEC

La fondation pour la sauvegarde des espèces menacées
8191 Avenue du Zoo
Charlesbourg, Quebec
G1G 4G4
Phone: 418–622–0313

La fondation québécoise en environnement
800 boul. de Maisonneuve Est, 2e étage
Montreal, Quebec
H2L 4L8
Phone/Fax: 514–849–0028

L'Union québécoise pour la conservation de la nature
690 Grande-Allée Est, Bureau 420
Quebec City, Quebec
G1R 2K5
Phone: 418–648–2104
Fax: 418–648–0991

Ministry of Environment and Wildlife Institutional Affairs
675 boulevard René-Lévesque est, 8th Floor
Quebec, Quebec
G1R 5V7
Phone: 418–643–1853
Toll free: 1–800–561–1616

NEW BRUNSWICK

Conservation Council of New Brunswick
180 St John Street
Fredericton, New Brunswick
E3B 4A9
Phone: 506–458–8747
Fax: 506–458–1047

Department of the Environment Communications and Environmental Education
PO Box 6000
Fredericton, New Brunswick
E3B 5H1
Phone: 506–453–3700

Nature Trust of New Brunswick
c/o Hal Hinds
Biology Department
University of New Brunswick
Fredericton, New Brunswick
E3B 5A3
Phone/Fax: 506–457–2398

New Brunswick Federation of Naturalists
277 Douglas Avenue
Saint John, New Brunswick
E2K 1E5

New Brunswick Protected Natural Areas Coalition
180 St John Street
Fredericton, New Brunswick
E3B 4A9
Phone: 506–451–9902
Fax: 506–458–1047

PRINCE EDWARD ISLAND

Department of Environmental Resources Planning and Administration
PO Box 2000
Charlottetown, PEI
C1A 7N8
Phone: 902–368–5320

Island Nature Trust
PO Box 365
Charlottetown, PEI
C1A 7K4
Phone: 902–892–7513
Fax: 902–628–6331

NOVA SCOTIA

CPAWS, Nova Scotia Chapter
73 Chadwick Street
Dartmouth, Nova Scotia
B2Y 2M2
Phone: 902–466–7168
E-mail: ab538@chebucto.ns.ca

Department of the Environment
Media and Public Relations
PO Box 2107
Halifax, Nova Scotia
B3J 3B7
Phone: 902–424–5300

Ecology Action Center
1553 Granville Street
Halifax, Nova Scotia
B3J 1W7
Phone: 902–429–2202

Federation of Nova Scotia Naturalists
73 Chadwick Street
Dartmouth, Nova Scotia
B2Y 2M2
Phone: 902–466–7168

NEWFOUNDLAND AND LABRADOR

Department of Environment and Labour
Communications
PO Box 8700
St John's, Newfoundland
A1B 4J6
Phone: 709–729–2575

Natural History Society of Newfoundland and Labrador
c/o Len Zedel
PO Box 1013
St John's, Newfoundland
A1C 5M3
Phone: 709–737–3106
Fax: 709–737–8739

Newfoundland and Labrador Environmental Association
c/o Stan Tobin
140 Water Street, Suite 603
St John's, Newfoundland
A1C 6H6
Phone: 709–722–1740
Fax: 709–726–1813

Protected Areas Association of Newfoundland and Labrador
Box 1027, Station C
St John's, Newfoundland
A1C 5M5
Phone/Fax: 709–726–2603

Tuckamore Wilderness Club
11 Carty Place
Corner Brook, Newfoundland
A2H 6B5
Phone: 709–639–1770
Fax: 709–639–8125

Wilderness and Ecological Reserves Advisory Council
c/o Parks and Natural Areas Division
Department of Tourism and Culture
PO Box 8700
St John's, Newfoundland
A1B 4J6
Phone: 709–729–2421
Fax: 709–729–1100

YUKON

CPAWS, Yukon Chapter
30 Dawson Road
Whitehorse, Yukon Territory
Y1A 5T6
Phone/Fax: 403–668–6321
E-mail: peepre@web.net

Renewable Resources
Communications
PO Box 2703
Whitehorse, Yukon
Y1A 2C6
Phone: 403–667–5237

Yukon Conservation Society
PO Box 4163
Whitehorse, Yukon
Y1S 3T3
Phone: 403–668–5678
Fax: 403–668–6637

NORTHWEST TERRITORIES

Department of Resources, Wildlife and Economic Development
Government Information Office
PO Box 1320
Yellowknife, NWT
X1A 2L9
Phone: 403–669–2302

Ecology North
4807–49th Street, Suite 8
Yellowknife, NWT
X1A 3T5
Phone: 403–873–6019
Fax: 403–873–3654

Glossary of Selected Terms

abiotic factors: Non-living components of the ecosystem, including chemical and physical factors.

absorption: The incorporation of a substance into a solid or liquid body.

abyssal zone: The bottom waters of the ocean beyond the continental shelf, usually below a depth of 1000 m.

acid deposition: Rain or snow that has a lower pH than precipitation from unpolluted skies; also includes dry forms of deposition, such as nitrate and sulphate particles.

adaptive environmental management: An approach that develops policies and practices to deal with the uncertain, the unexpected, and the unknown. Approaches management as an experiment from which we learn by trial and error.

adaptive planning: An approach that includes collaboration of interest groups, identification of shared values, and continuous learning, monitoring, and evaluation.

aerobic: Requiring oxygen.

aerobic decomposition: The degradation of organic material by living organisms in the presence of oxygen.

aerosols: Minute mineral particles suspended in the atmosphere to which water droplets or crystals and other chemical compounds may adhere. Both natural processes and human activities emit aerosols.

Agenda 21: A forty-chapter report from the Earth Summit in 1992 that outlines goals and priorities for economic development and environmental protection in the twenty-first century.

air pollution: A condition of the air produced by the presence of one or more air contaminants, which endangers people's health, safety, or welfare, interferes with normal enjoyment of life or property, endangers the health of animal life, or causes damage to plant life or property.

alien (also known as exotic, introduced, or non-native): Any organism that enters an ecosystem beyond its normal range through deliberate or inadvertent introduction by humans.

alkaline: Having the properties of or containing a soluble base.

annual allowable cut: The amount of timber that is permitted to be cut annually from a specified area.

alternative dispute resolution: A non-judicial approach to resolving disputes that uses negotiation, mediation, or arbitration.

ammonia: A colourless gas composed of nitrogen and hydrogen (NH_3); the main form in which nitrogen is available to living cells.

anadromous: Characterizes migrating fish that grow in the sea and ascend freshwater streams to spawn.

aquatic: Pertains to marine and freshwater environments.

aquifer: Underground, water-saturated zone that may extend from a few square kilometres to several thousand square kilometres.

arbitration: A procedure for dispute resolution in which a third party is selected to listen to the views and interests of the parties in dispute and develop a solution to be accepted by the participants.

atmosphere: Layer of air surrounding the earth.

autotroph: Organisms, such as plants, that produce their own food, generally via photosynthesis.

auxins: Plant hormones responsible for stimulating growth.

backcasting: A futures method in which a desirable future endpoint is identified so that it is possible to discern the necessary decisions and actions to achieve the desired endpoint.

benthic: Of or living on or at the bottom of a water body.

benthos: The plant and animal life whose habitat is the bottom of a sea, lake, or river.

bioaccumulation: The storage of chemicals in an organism in higher concentrations than are normally found in the environment.

biocide: A chemical that kills many different kinds of living things.

biodegradable: Capable of being broken down by natural systems.

biodiversity: The variety of life forms that inhabit the earth. Biodiversity includes the genetic diversity among members of a population or species as well as the diversity of species and ecosystems.

biogeochemical cycle: A series of biological, chemical, and geological processes by which materials cycle through ecosystems.

biological control: Use of naturally occurring predators, parasites, bacteria, and viruses to control pests.

biological magnification: Build-up of chemical elements or substances in organisms in successively higher trophic levels.

biological oxygen demand (BOD): The amount of dissolved oxygen required for the bacterial decomposition of organic waste in water.

biomass: The sum of all living material in a given environment.

biome: A major ecological community of organisms, both plant and animal, that is usually characterized by the dominant vegetation type, for example, a tundra biome and a tropical rainforest biome.

bioregion: A place or area that its inhabitants relate to with regard to its natural and cultural characteristics.

bioregional perspective: Awareness that a place consists of interconnected parts, and that the role of those parts and their relationships should be understood and respected.

biosphere: Total of all areas on earth where organisms are found; includes deep ocean and part of the atmosphere.

biotic potential: The ability of species to reproduce regardless of the level that an environment can support. See Carrying capacity.

calorie: A unit of heat energy, the amount of heat required to raise 1 g of water by 1° C.

carcinogen: Material that has been shown to cause cancer.

carnivore: An organism that consumes only animals.

carrying capacity: Maximum population size that a given ecosystem can support for an indefinite period or on a sustainable basis.

CFCs (chloro-fluorocarbons): Gaseous synthetic substances composed of chlorine, fluorine, and carbon. They have been used as refrigerants, aerosol propellants, cleaning solvents, and in the manufacture of plastic foam. CFCs cause ozone depletion in the stratosphere.

change: The process of becoming different or altered. Change implies that the difference is fundamental or complete, whereas altered suggests a less drastic difference that is limited to some particular aspect.

chemoautotroph: A producer organism that converts inorganic chemical compounds into energy.

chemosynthesis: The process of producing energy from inorganic materials through simple chemical reactions.

chlorophyll: Pigment of plant cells that absorbs sunlight, thus enabling plants to capture solar energy.

clear-cutting: A forest management technique in which an entire stand of trees is felled and removed.

climate: The long-term weather pattern of a particular region.

climax community: Last stage of succession; a relatively stable, long-lasting, complex, and interelated community of organisms.

closed system: A system that can exchange energy, but not matter, with the surrounding environment.

codesign: A process in which community members describe their vision for development, and then artists quickly prepare sketches to translate the verbal descriptions into visual images.

coevolution: Process whereby two species evolve adaptations as a result of extensive interactions with each other.

coliform: A group of bacteria used as an indicator of sanitary quality in water. The total coliform group is an indicator of sanitary significance because the organisms are usually present in large tracts of humans and other warm-blooded animals, and exposure to them in drinking water causes diseases such as cholera.

collaboration: The art of working together.

comanagement: An arrangement in which a government agency shares some of its legal authority regarding a resource or environmental management issue with local inhabitants of an area.

commensalism: An interaction between two species that benefits one species and neither harms nor benefits the other.

competition: Vying for resources between members of the same or different species.

competitive exclusion principle: The principle that competition between two species with similar requirements will result in the exclusion of one of the species.

complexity: Something consisting of a number of parts, something that is complicated.

conflict: A disagreement, dispute, or quarrel, especially one that is prolonged.

conserver society: A society that puts less emphasis on material possessions and more on living in harmony with nature.

consumer: An organism that cannot produce its own food and must get it by eating or decomposing other organisms. In economics, one who uses goods and services.

consumerism: Wasteful consumption of resources to satisfy wants rather than needs.

contour farming: Soil erosion control technique in which row crops are planted along the contour lines in sloping or hilly fields rather than up and down the hills.

convergent evolution: The independent evolution of similar traits among unrelated organisms resulting from similar selective pressures.

critical population size: Population level below which a species cannot successfully reproduce.

crop rotation: Alternating crops in fields to help restore soil fertility and also control pests.

cumulative effects: The consequences from the accumulation of numerous individual activities or actions.

DDT (dichlorodiphenyltrichloroethane): An organochlorine insecticide used first to control malaria-carrying mosquitoes and lice and later to control a variety of insect pests, but now banned in Canada because of its persistence in the environment and ability to bioaccumulate.

decomposer food chain: A specific nutrient and energy pathway in an ecosystem in which decomposer organisms (bacteria and fungi) consume dead plants and animals as well as animal wastes. Essential for the return of nutrients to soil and carbon dioxide to the atmosphere. Also called detritus food chain.

demand management: An approach that focuses upon modifying human behaviour regarding resource use, such as energy or water, and often involves the use of pricing, information, or education to change patterns of use. The opposite of supply management.

denitrification: The conversion of nitrate to molecular nitrogen by bacteria in the nitrogen cycle.

desert: An area of land in which evaporation exceeds precipitation, usually where annual precipitation is less than 250 mm per year.

detritus feeders: Organisms in the decomposer food chain that feed primarily on organic waste (detritus), such as fallen leaves.

development: Achievement of potential, both quantitatively and qualitatively.

dynamic equilibrium: An ecosystem's ability to react to constant changes thereby maintaining relative stability.

Earth Summit: The International Conference on Environment and Development held in Rio de Janeiro during June 1992 when most of the nations of the world met to discuss issues related to sustainable development.

ecological footprint: The land area a community requires to provide its basic needs for food, water, and other essential elements.

ecosphere: Refers to the entire global ecosystem, which comprises atmosphere, lithosphere, hydrosphere, and biosphere as inseparable components.

ecosystem: Short for ecological system. A community of organisms occupying a given region within a biome. Also the physical and chemical environment of that community and all the interactions among and between organisms and their environment. The term may be applied to a unit as large as the entire ecosphere. More often it is applied to some smaller division.

ecotone: The transitional zone of intense competition for resources and space between two communities.

endangered: An official designation assigned by the Committee on the Status of Endangered Wildlife in Canada to any indigenous species or subspecies or geographically separate population of fauna or flora that is threatened with imminent extinction or extirpation throughout all or a significant portion of its Canadian range.

endemic species: A plant or animal species confined to or exclusive to a specific area.

energy: The capacity to do work. Found in many forms, including heat, light, sound, electricity, coal, oil, and gasoline.

entropy: A measure of disorder. The second law of thermodynamics applied to matter says that all systems proceed to maximum disorder (maximum entropy).

environmental impact assessment: Part of impact assessment that identifies and predicts the impacts from development proposals on both the biophysical environment and on human health and well-being.

environmental resistance: Abiotic and biotic factors that can potentially reduce population size.

epilimnion: Upper, warm waters of a lake.

estuary: Coastal regions, such as inlets or mouths of rivers, where salt and fresh water mix.

eutrophic: Pertaining to a body of fresh water rich in nutrients and hence in living organisms.

eutrophication (also known as nutrient enrichment): The overfertilization of a body of water by nutrients that produce more organic matter than the water body's self-purification processes can overcome.

evapotranspiration: Evaporation of water from soil and transpiration of water from plants.

evolution: A long-term process of change in organisms caused by random genetic changes that favour the survival and reproduction of those organisms possessing the genetic change. Organisms become better adapted to their environment through evolution.

exponential growth: Increase in any measurable thing by a fixed percentage. When plotted on graph paper, it forms a J-shaped curve.

externality: A spillover effect that benefits or harms others. The source of the effect (for example, pollution) does not pay for the effect.

extinction: The elimination of all the individuals of a species.

extirpated: An official designation assigned by the Committee on the Status of Endangered Wildlife in Canada to any indigenous species or subspecies or geographically separate population of fauna or flora no longer known to exist in the wild in Canada but occurring elsewhere.

First law of thermodynamics: Also called the law of conservation of energy. States that energy is neither created nor destroyed; it can only be transformed from one form to another.

focus group: A small group of people who use a structured discussion to reach consensus regarding a problem or issue.

food chain: A specific nutrient and energy pathway in ecosystems proceeding from producer to consumer. Along the pathway, organisms in higher trophic levels gain energy and nutrients by consuming organisms at lower trophic levels.

food web: Complex intermeshing of individual food chains in an ecosystem.

forecasting: Extrapolating from the present into the future, with emphasis on identifying the most likely or probable conditions in the future.

fossil fuel: Any one of the organic fuels (coal, natural gas, oil, tar sands, and oil shale) derived from once-living plants or animals.

greenhouse effect: A warming of the earth's atmosphere caused by the presence of certain gases (e.g., water vapour, carbon dioxide) that absorb radiation emitted by the earth, thereby retarding the loss of energy from the system to space.

greenhouse gas: A gas that contributes to the greenhouse effect, such as carbon dioxide.

Green Revolution: Development in plant genetics in the late 1950s and early 1960s resulting in high-yield varieties producing three to five times more grain than previous plants but requiring intensive irrigation and fertilizer use.

gross primary productivity (GPP): The total amount of energy produced by autotrophs over a given period of time.

groundwater: Water below the earth's surface in the saturated zone.

habitat: The environment in which a population or individual lives.

hard energy path: The reliance on technology to provide additional sources of energy to meet increasing demand. Contrasts with a soft energy path, which seeks to modify human demand for energy.

hazardous waste: Waste that poses a risk to human health or the environment and requires special disposal techniques to make it harmless or less dangerous.

herbivores: Animals that eat plants, that is, primary consumers.

heterotroph: An organism that feeds on other organisms.

hydrological cycle: The circulation of water through bodies of water, the atmosphere, and land.

hypolimnion: Deep, cold waters of a lake. Contrast with epilimnion.

indigenous knowledge: Also referred to as local knowledge or traditional ecological knowledge, this is understanding based on experiential knowledge from living or working in a particular area for a long period of time.

inertia: The tendency of a natural system to resist change.

interspecific competition: Competition between members of different species for limited resources, such as food, water, or space.

intraspecific competition: Competition between members of the same species for limited resources, such as food, water, or space.

keystone species: Critical species in an ecosystem whose loss profoundly affects several or many others.

kilocalorie: One thousand calories.

kinetic energy: The energy of objects in motion.

limiting factor: A chemical or physical factor that determines whether an organism can survive in a given ecosystem. In most ecosystems, rainfall is the limiting factor.

lithosphere: The earth's crust.

littoral zone: Shallow waters along a lakeshore where rooted vegetation often grows.

LULU: Locally unwanted land use, often the source of a NIMBY reaction.

macronutrient: A chemical substance needed by living organisms in large quantities (for example, carbon, oxygen, hydrogen, and nitrogen). Contrast with micronutrient.

mediation: A negotiation process guided by a facilitator.

micronutrient: An element needed by organisms, but only in small quantities, such as copper, iron, and zinc.

molecule: Particle consisting of two or more atoms bonded together. The atoms in a molecule can be of the same element, but are usually of different elements.

monoculture: Cultivation of one plant species (such as corn) over a large area. Highly susceptible to disease and insects.

monsoon: A seasonal wind of the Indian Ocean and southern Asia that blows from the southwest from April to October and brings large amounts of precipitation to many areas.

Montreal Protocol: Signed in 1987 by thirty-two nations, this agreement established a schedule for reducing use of chloro-fluorocarbons and halons. This is one of the first agreements to reduce the use of greenhouse gases.

mutualism: Relationship between two organisms that is beneficial to both.

negative feedback: Control mechanism present in the ecosystem and in all organisms. Information in the form of chemical, physical, and biological agents influences processes, causing them to shut down or reduce their activity.

net primary productivity (NPP): Gross primary productivity (the total amount of energy that plants produce) minus the energy plants use during cellular respiration.

niche: An organism's place in the ecosystem: where it lives, what it consumes, and how it interacts with all biotic and abiotic factors.

NIMBY: 'Not in my backyard' is a phrase used to describe local people's reactions when a noxious facility—such as a landfill site, sand and gravel pit, or expressway—is proposed to be built adjacent to or near their property.

nitrogen fixation: Conversion of gaseous (atmospheric) nitrogen (N_2).

non-point source: Source of pollution in which pollutants are discharged over a widespread area or from a number of small inputs rather than from distinct, identifiable sources.

nutrient: Any element or compound that an organism must take in from its environment because it cannot produce it or cannot produce it as fast as it needs it.

old growth: A descriptive term attributed to forests that generally have a significant number of huge, long-lived trees; many large, standing dead trees; numerous logs lying about the forest floor; and multiple layers of canopy created by the crowns of trees of various ages and species.

oligotrophic: Nutrient poor.

omnivores: Organisms that eat both plants and animals.

ozone: An atmospheric gas (O_3) that, when present in the stratosphere, helps protect the earth from ultraviolet rays. However, when it is present near the earth's surface, it is a primary component of urban smog and has detrimental effects on both vegetation and human respiratory systems. Used in some advanced sewage treatment plants.

ozone layer: Thin layer of ozone molecules in the stratosphere. Absorbs ultraviolet light and converts it to infrared radiation. Effectively screens out 99 per cent of the ultraviolet light.

parasitism: Relationship in which one species lives in or on another, which acts as its host.

particulates: Solid particles (dust, pollen, soot) or water droplets in the atmosphere.

partnerships: A sharing of responsibility and power between two or more groups, especially a government agency and a second party, regarding a resource or environmental issue. Comanagement is an example of a partnership.

permafrost: Permanently frozen ground.

pesticide: A general term referring to a chemical, physical, or biological agent that kills organisms we classify as pests, such as insects and rodents. Also called biocide.

photosynthesis: A two-part process in plants and algae involving: (1) the capture of sunlight and its conversion into cellular energy, and (2) the production of organic molecules, such as glucose and amino acids from carbon dioxide, water, and energy from the sun.

phototroph: An organism that produces complex chemicals through photosynthesis.

phytoplankton: Single-celled algae and other free-floating photosynthetic organisms.

point source (of pollution): Easily discernible source of pollution, such as a factory.

pollutant: A substance that adversely alters the physical, chemical, or biological quality of the earth's living systems or that accumulates in the cells or tissues of living organisms in amounts that threaten their health or survival.

population: A group of organisms of the same species living within a specified region.

positive feedback: A situation in which a change in a system in one direction provides information that causes the system to change further in the same direction.

precautionary principle: A guideline stating that when there are threats of serious or irreversible damage, lack of scientific certainty is not an acceptable reason for postponing a cost-effective measure to prevent environmental degradation.

predator: An organism that actively hunts its prey.

prey: Organism (e.g., deer) that is attacked and killed by a predator.

primary consumer: The first consuming organism in a given food chain, such as a grazer in grazer food chains or a decomposer organism or insect in decomposer food chains. Belongs to the second trophic level.

primary succession: The development of a biotic community in an area previously devoid of organisms.

principled approach: A dispute resolution approach that focuses on interests and needs rather than on positions, and which seeks to separate people from the problem.

producer: An autotroph capable of synthesizing organic material, thus forming the basis of the food web.

range of tolerance: Range of abiotic factors within which an organism can survive from the minimum amount of a limiting factor that the organism requires to the maximum amount that it can withstand.

reclamation: The process of bringing back an area to a useful, good condition—similar to rehabilitation.

reliability: The ability of an ecosystem to resist or avoid breakdown or collapse.

remediation: The process of removing or relieving any bad or undesirable condition, a remedy.

resilience: Ability of an ecosystem to return to normal after a disturbance.

respiration: The process by which organisms produce energy by capturing the chemical energy stored in food.

Rio Declaration: The twenty-seven principles related to sustainable development, developed at the Earth Summit in Rio de Janeiro during June 1992.

risk: A calculable chance of harm or loss. A situation in which the odds or probability of an event are known. In contrast to uncertainty, when the odds or probability are not known.

round table: A forum of people with competing and converging interests regarding economic development and environmental protection, brought together to generate ideas regarding how to put sustainable development into practice.

salinization: Deposition of salts in irrigated soils, making soil unfit for most crops. It is caused by a rising water table due to inadequate drainage of irrigated soils.

secondary consumer: Second consuming organism in a food chain; belongs to the third trophic level.

secondary succession: The sequential development of biotic communities after the complete or partial destruction of an existing community by natural or anthropogenic forces.

second growth: A second forest that develops after harvest of the original forest.

second law of thermodynamics: When energy is converted from one form to another, it is degraded; that is, it is converted from a concentrated to a less concentrated form. The amount of useful energy decreases during such conversions.

sere: A stage in succession.

smelter: A factory where ores are melted to separate impurities from the valuable minerals.

social impact assessment: The procedure used to identify and predict the impacts of development on people and communities and on their well-being.

soft energy path: Use of information, education, and pricing to reduce people's use of energy.

soil compaction: The compression of soil as a result of vehicle traffic, especially that of heavy equipment.

soil erosion: The detachment and movement of soil by the action of wind (wind erosion) and moving water (water erosion).

soil horizons: Layers found in most soils.

soil productivity: The soil's ability to sustain life, especially vegetation.

specialist: Organism that has a narrow niche, usually feeding on one or a few food materials and adapted to a particular habitat.

speciation: Formation of new species.

species: A group of individuals that share certain identical physical characteristics and are capable of producing fertile offspring.

stakeholders: Any person or group with a legal responsibility relative to a problem or issue, or likely to be affected by decisions or actions regarding the problem or issue, or able to provide an obstacle to a solution of the problem or issue.

stratosphere: The layer of the atmosphere (about 10–50 km above the earth's surface) in which temperatures rise with increasing altitude.

subsistence farming: The production of food and other necessities to satisfy the needs of the farm household.

supply management: Modification or manipulation of the natural system to provide additional supplies of a resource during a time of anticipated or actual shortage.

sustainable development: Economic development that meets current needs without compromising the ability of future generations to meet their needs.

sustained yield: The amount of harvestable material that can be removed from an ecosystem over a long period of time with no apparent deleterious effects on the system.

symbiosis: Any intimate association of two dissimilar species regardless of the benefits or harm derived from it.

sympatric speciation: Formation of new species without geographical isolation. Common in plants.

synergism: An interaction between two substances that produces a greater effect than the effect of either one alone; an interaction between two relatively harmless components in the environment.

taiga: The biome south of the tundra across North America, Europe, and Asia, that is characterized by coniferous forests, soil that thaws during the summer months, abundant precipitation, and high species diversity.

teratogen: A chemical or physical agent capable of creating birth defects.

terrestrial: Of or pertaining to land environments.

tertiary consumer: In a food chain, an organism at the top that consumes other organisms.

threshold: A point or limit beyond which something is unsatisfactory relative to a consideration, such as health, welfare, or ecological integrity.

transpiration: The loss of water vapour through the pores of a plant.

trophic: Relating to processes of energy and nutrient transfer from one or more organisms to others in an ecosystem.

trophic level: Functional classification of organisms in a community according to feeding relationships: the first trophic level includes green plants, the second level includes herbivores, and so on.

troposphere: Layer of the atmosphere that contains about 95 per cent of the earth's air and extends about 6–17 km up from the earth, depending upon latitude and season.

uncertainty: A situation in which the probability or odds of a future event are not known and therefore indicates the presence of doubt.

vision: An identified desirable direction or future condition or arrangement to which to strive or aspire.

vulnerable: An official designation assigned by the Committee on the Status of Endangered Wildlife in Canada to any indigenous species or subspecies or geographically separate population of fauna or flora that is particularly at risk because of low or declining numbers, because it occurs at the fringe of its range or in restricted areas, or for some other reason, but which is not a threatened species.

water table: The top of the zone of saturation.

zone of intolerance: Range of environmental conditions in which an organism cannot survive.

zone of physiological stress: Upper and lower limits of the range of tolerance in which organisms have difficulty surviving.

zooplankton: Non-photosynthetic, single-celled aquatic organisms.

Publisher's Acknowledgements

The publisher would like to thank the following for donating photographs: Cheryl Belt and A.J. Cadie, International Fund for Animal Welfare; John W. Chardine, Canadian Wildlife Service; Nick Dawe and Heather Cumming, Newfoundland Tourism, Culture and Recreation; Amber Ellis, Earthroots; Ursula Gattoc, Ontario Hydro; Jennifer Good and Nora McCarthy, Greenpeace; Murray Haight, University of Waterloo; Laura Hubenig, SaskPower; Joan Lipscomb, Parks Canada; Florence Miller, Grasslands National Park; Karen Rosborough, World Wildlife Fund Canada; and Pierre Vachon, Canadian International Development Agency.

Every effort has been made to determine and contact copyright owners. In case of any omissions, the publisher will be pleased to make suitable acknowledgement in future editions. The following publishers have granted permission to reproduce previously published materials:

ALBERTA ENVIRONMENTAL PROTECTION: 'Alberta's Vision of Sustainable Development' from *Ensuring Prosperity: Implementing Sustainable Development*, Future Environmental Directions for Alberta Task Force, Edmonton: Alberta Environmental Protection, 1995, p. 7.

S. ARNSTEIN: Table 'Rungs on the Ladder of Citizen Participation' from 'A Ladder of Citizen Participation' by S. Arnstein. Reprinted with permission from *Journal of the American Planning Association*, copyright July 1969 by the American Planning Association, Suite 1600, 122 South Michigan Avenue, Chicago, IL 60603–6107.

F. BERKES: Excerpt 'Changing Importance of Impact Assessment Issues' reprinted with permission from *Environmental Impact Assessment Review*, vol. 8, no. 3, F. Berkes, 'The Intrinsic Difficulty of Predicting Impacts: Lessons from the James Bay Hydro Project', pp. 201–20, 1988. Elsevier Science Inc.

JIM BUTLER: 'Hug an Ancient Tree' and extract from 'Pilgrims in a National Townsite' from *Dialogue with a Frog on a Log* by Jim Butler. Edmonton: Duval House Publishing, 1994. Reprinted by permission of Les Éditions Duval Inc.

S.D. CAMERON: Figure 'Major Fishing Areas in Atlantic Canada' from *Canadian Geographic*, April/May 1990. Reprinted by permission of Canadian Geographic.

CANADIAN COUNCIL OF FOREST MINISTERS: 'Areas Burned and Harvested' from Canadian Council of Forest Ministers, National Forestry Database Program. Reprinted by permission.

CANADIAN FOREST SERVICE: Table 'Net Merchantable Volume of Roundwood Harvested by Ownership, Species Group, and Province/Territory, 1993–1994' from *Compendium of Canadian Forestry Statistics*, Canadian Council of Forest Ministers. Ottawa: Canadian Forest Service, 1995, p. 79. Reprinted by permission of Canadian Forest Service. Table 'Canada's Forests' from *The State of Canada's Forest*, Natural Resources Canada. Ottawa: Canadian Forest Service, 1996, p. 7. Reprinted by permission of Canadian Forest Service. Figure 'Softwood & Hardwood AACs & Harvesting' from *The State of Canada's Forests, 1995–1996*, Natural Resources Canada. Ottawa: Canadian Forest Service, 1996, p. 101. Reprinted by permission of Canadian Forest Service. Figures 'Potential Softwood Timber Supply' and 'Harvesting Systems in Canada' from *The State of Canada's Forests*, Natural Resources Canada. Ottawa: Canadian Forest Service, 1995, pp. 56, 80. Reprinted by permission of Canadian Forest Service. Figure 'Site Preparation & Stand Tending' from *The State of Canada's Forests*, Natural Resources Canada. Ottawa: Canadian Forest Service, 1995, p. 102. Reprinted by permission of Canadian Forest Service.

CANADIAN GEOGRAPHIC: Figure 'Major Fishing Areas in Atlantic Canada' from 'Net Losses: The Sorry State of Our Atlantic Fishery' from *Canadian Geographic*, April/May 1990. Reprinted by permission of Canadian Geographic.

CANADIAN GLOBAL CHANGE PROGRAM: Figure 'Energy Consumption Per Capita'. Canadian Global Change Program, 1993, *Global Change and Canadians*, ISBN 0-920064-46-9. Ottawa: The Royal Society of Canada. Reprinted by permission of Canadian Global Change Program.

CANADIAN HERITAGE: Figure 'National Marine Conservation Area Natural Regions' from *Charting the Course: Towards a Marine Conservation Area Act*, Parks Canada, 1997, p. 1. Reproduced with the permission of the Minister of Canadian Heritage and the Minister of Public Works and Government Services.

CANADIAN WATER RESOURCES ASSOCIATION: 'Indicators and Thresholds' from 'Nitrate Pollution in England' from *Managing the Water Environment, Proceedings of the 48th Annual Conference* (Cambridge, ON: Canadian Water Resources Association, 1995), vol. 1. 'Sustainability Ethic' and 'Water

Management Principles' from *A Policy of the Canadian Water Resources Association*, June 1994. Reprinted by permission of the Canadian Water Resources Association.

J.D. CHAPMAN: 'National Energy Program, 1979–1983' from *Geography and Energy: Commercial Energy Systems and National Policies* by J.D. Chapman. Harlow, UK: Longman Scientific and Technical, 1989. Reprinted by permission of Addison Wesley Longman Ltd.

D.F. CHARLES AND S. CHRISTIE: Figure 'Acidification of Surface Waters in Eastern Canada' from 'Southwestern Canada: An Overview of the Effects of Acid Deposition on Aquatic Resources' by D.S. Jeffries, from *Acid Deposition and Aquatic Ecosystems: Regional Case Studies*, edited by D.F. Charles and S. Christie. New York: Springer-Verlag, 1991. Reprinted by permission of Springer-Verlag New York, Inc.

TERENCE CORCORAN: 'Too Much Green, Too Little Growth' from *The Globe and Mail*, 3 July 1993. Reprinted with permission from *The Globe and Mail*.

J.C. DAY AND F. QUINN: Figures 'The Kemano Diversion', 'Energy Megaprojects in Canada', 'La Grande River Project: Phases I and II', and table 'Major Interbasin Water Diversions in Canada' from *Water Diversion and Export: Learning from Canadian Experience* by J.C. Day and F. Quinn, Department of Geography Publication Series, University of Waterloo, Waterloo, Ontario.

P. DEARDEN AND L. BERG: Figure 'Administrative Penetration Model' from 'Canada's National Parks: A Model of Administrative Penetration' by P. Dearden and L. Berg from *The Canadian Geographer*, vol. 37, no. 3, 1993, p. 198. Reprinted by permission.

B. DIAMOND: Figure 'A Surface Equivalent to a Third of the Province of Quebec or the Size of France Would Be Dangerously Affected by the James Bay Development Program' reproduced from *Arctic Circle Magazine*, November/December 1990 issue.

D.A. DUFFUS AND P. DEARDEN: Table 'Potential Influence and Consequences of Disturbance of Whales' from 'Recreational Use, Valuation and Management of Killer Whales (*Orcinus orca*) on Canada's Pacific Coast' by D.A. Duffus and P. Dearden, *Environmental Conservation*, vol. 20, 1993. Reprinted by permission of Cambridge University Press.

D. DUNBAR AND I. BLACKBURN: Figures 'Distribution of Spotted Owls in British Columbia' and 'Forested Land Use within DMAs for Five of the Six Management Options', and table 'Down-listing Criteria for the Northern Spotted Owl in Canada' from *Management Options for the Northern Spotted Owl in British Columbia* by D. Dunbar and I. Blackburn, BC Ministry of Environment, 1994.

C.A. EDWARDS: Table 'Persistence of Some Organochlorine Insecticides in Soil' reprinted with permission from *Persistent Pesticides in the Environment* by C.A. Edwards. Copyright CRC Press, Boca Raton, Florida. © 1973.

ENVIRONMENT CANADA: Figure 'Drainage Regions of Canada' from *Currents of Change: Final Report, Inquiry on Federal Water Policy*, Environment Canada, Ottawa, Canada, 1985. Reproduced with the permission of the Minister of Public Works and Government Services Canada, 1997. Figures 'Projections for Climate Warming in Summer' and 'Projections for Climate Warming in Winter' from *Climate Change and Canadian Impacts: The Scientific Perspective*, Environment Canada, 1991. Reproduced with the permission of the Minister of Public Works and Government Services Canada, 1997. Figure 'Sulphur Dioxide Emissions in Eastern Canada' from *Acid Rain*, SOE Bulletin 96-2, 1996, State of the Environment Reporting Program, Environment Canada. Reprinted by permission.

D.W. GULLETT AND W.B. SKINNER: Figure 'Average Annual Temperature Departures from 1951–80, Average for 1980 to 1989' from *The State of Canada's Climate: Temperature Change in Canada 1895–1991* (State of the Environment Report 92–2) by D.W. Gullett and W.B. Skinner, State of the Environment Reporting Program, Environment Canada, 1992. Reproduced with the permission of the Minister of Public Works and Government Services Canada, 1997.

P. HALL: Figure 'European Overfishing, 1986–90' from 'Crisis in the Atlantic Fishery' by P. Hall from *Canadian Business Review*, The Conference Board of Canada, vol. 17, no. 2, 1990, p. 47.

G.F. HARTMAN AND J.C. SCRIVENER: Figure 'Anticipated Patterns of Change in Physical Conditions in a Small Coastal Stream, Like Carnation Creek, Following Logging (crosshatch). The Patterns for Stream Temperature and Stream Insolation Are Presumed to Be Similar' from *Impacts of Forestry Practices on a Coastal Stream Ecosystem, Carnation Creek, BC* by G.F. Hartman and J.C. Scrivener, Canadian Bulletin of Fisheries and Aquatic Sciences 223 (Ottawa: Department of Fisheries and Oceans, 1990).

M. HUMMEL: Table 'Update on Conservation Lands and Waters' from *Protecting Canada's Endangered Spaces: An Owner's Manual*, edited by M. Hummel (Toronto: Key Porter Books, 1995). Reprinted by permission of Key Porter Books.

INTERNATIONAL UNION FOR THE CONSERVATION OF NATURE AND NATURAL RESOURCES: Table 'Matrix of Management Objectives and IUCN Protected Area Management Categories' from *Guidelines for Protected Area Management Categories*. Gland, Switzerland: IUCN, 1994, p. 8. Reproduced by permission of the International Union for the Conservation of Nature and Natural Resources.

J.P. KIMMINS: Figure 'Site Impoverishment as a Result of Forest Harvesting' reprinted from *Forest Ecology and Management*, 1977, vol. 1, 'Evaluation of the Consequences for Future Tree Productivity of the Loss of Nutrients in Whole-Tree Harvesting' by J.P. Kimmins, pp. 169–83, with kind permission of Elsevier Science - NL, Sara Burgerhartstraat 25, 1055 KV Amsterdam, The Netherlands.

R.C. KUHN: Table adapted from 'Canadian Energy Futures: Policy Scenarios and Public Preferences' by R.C. Kuhn from *Canadian Geographer*, vol. 36, no. 4, 1992, p. 352. Reprinted by permission.

C.E. KUPCHELLA AND M.C. HYLAND: Table 'Relative Amounts of Chemical Elements That Make Up Living Things' from

546 Publisher's Acknowledgements

Environmental Science: Living within the System of Nature by C.E. Kupchella and M.C. Hyland. Copyright © 1989 by Allyn and Bacon. Reprinted by permission.

MACMILLAN MAGAZINES LIMITED: Figure 'Variation of Temperature Over Antarctica and of Global Atmospheric Carbon Dioxide and Methane Concentrations During the Last 160,000 Years, as Inferred from the Vostok Ice Core from Antarctica' from 'Ice Core Record of Atmospheric Methane Over the Past 160,000 Years' reprinted with permission from *Nature*, vol. 345, no. 6271, 1990. Copyright 1990 Macmillan Magazines Limited.

R.J. MACRAE et al.: Table 'Paramount Goals of the Food System' from 'Policies, Programs, and Regulations to Support the Transition to Sustainable Agriculture in Canada' by R.J. MacRae et al., *American Journal of Alternative Agriculture*, vol. 5, 1990, p. 83. Reprinted by permission of Henry A. Wallace Institute for Alternative Agriculture, Inc.

RAYMOND MORIYAMA: Figures 'Nodes and Linkages in the Moriyama Plan' and 'Proposed Recreational Areas in the Moriyama Plan' from 'The Meewasin Valley Project', Raymond Moriyama Architects and Planners, p. 46. Reprinted by permission of Raymond Moriyama.

T. MOSQUIN, P.G. WHITING, AND D.E. MCALLISTER: Figure 'Groups with Most Species (Excluding Viruses)' from *Canada's Biodiversity: The Variety of Life, Its Status, Economic Benefits, Conservation Costs and Unmet Needs* by T. Mosquin, P.G. Whiting, and D.E. McAllister, 1995. Canadian Museum of Nature, p. 58.

P.R. MULVIHILL AND R.F. KEITH: 'Design Features and Capabilities of Adaptive Organizations and Processes' and table 'Principles and Criteria for Designing Adaptive Environmental Organizations and Processes' reprinted with permission from *Environmental Impact Assessment Review*, vol. 9, no. 4, P.R. Mulvihill and R.F. Keith, 'Institutional Requirements for Adaptive EIA: The Kativik Environmental Quality Commission', pp. 402 and 409, 1989. Elsevier Science Inc.

NATURAL RESOURCES CANADA: 'Areas Burned and Harvested' from Canadian Council of Forest Ministers, National Forestry Database Program. Reprinted by permission.

E. ODUM: Adapted table 'Characteristics of Mature and Immature Ecosystems' from 'The Strategy of Ecosystem Development' reprinted with permission from 'The Strategy of Ecosystem Development' by E. Odum, *Science*, vol. 164, 1969, pp. 262–70. Copyright 1969 American Association for the Advancement of Science. Table 'Production and Respiration as (kcal/m²/yr) in Growing and Climax Ecosystems' from *Fundamentals of Ecology*, Third Edition, by Eugene P. Odum, copyright © 1971 by Saunders College Publishing, reproduced by permission of the publisher.

OFFICE OF THE AUDITOR GENERAL: Figures 31.3 and 31.4: 'Completing the National Parks System' and 'Progress in Completing the National Parks System' from Chapter 31, 'Canadian Heritage—Parks Canada: Preserving Canada's Natural Heritage', Report of the Auditor General of Canada, November 1996. Reproduced with the permission of the Minister of Public Works and Government Services Canada, 1997.

ONTARIO MINISTRY OF NATURAL RESOURCES: Table 'Example of an Integrated Approach: Subwatershed Planning in Ontario' from *Subwatershed Planning*, Ontario Ministry of Natural Resources. Copyright: 1997 Queen's Printer Ontario.

PARKS CANADA: Figures 'Number of National Parks Reporting Significant Ecological Impacts from Various Human Stresses' and 'Number of Reported Cases for a Variety of Ecological Impacts from a Survey of 29 Ecological Stressors in Canadian National Parks' from *State of the Parks 1994*, Parks Canada. Ottawa: Minister of Supply and Services Canada, 1995, pp. 35, 40. Reproduced with the permission of the Minister of Public Works and Government Services Canada, 1997.

J.T. PIERCE: Table 'Effects of Canadian Agricultural Policies on the Environment' reprinted from *Geoforum*, vol. 24, J.T. Pierce, 'Agriculture, Sustainability and the Imperatives of Policy Reform', pp. 388–9, copyright © 1993, with kind permission from Elsevier Science Ltd, The Boulevard, Langford Lane, Kidlington, OX5 1GB, UK.

PUBLIC WORKS AND GOVERNMENT SERVICES CANADA: Figure 'Ecosystems' from *Royal Commission on the Future of the Toronto Waterfront*. Ottawa: Privy Council Office, 1992. Figures 'Completing the National Parks System' and 'Progress in Completing the National Parks System' from Chapter 31, 'Canadian Heritage—Parks Canada: Preserving Canada's Natural Heritage', *Report of the Auditor General of Canada*, November 1996. Map 'Average Annual Rain and Snow for Canada' from *The Climate of Canada* by D. Phillips, Environment Canada, 1990. Figure 'The Potential of Soils and Bedrock to Reduce the Acidity of Atmospheric Deposition in Canada' from Environment Canada, 'Acid Rain: A National Sensitivity Assessment', Environmental Fact Sheet 88-1 (Ottawa: Environment Canada, 1988). Figure 'Nelson-Saskatchewan River Basin' from *Canada Water Year Book 1975* (Ottawa: Information Canada, 1975):45. Figure 'Years of Grain Production Only Farming' from *Sustainability of Farmed Lands: Current Trends and Thinking* by C.F. Bentley and L.A. Leskiw, Canadian Environmental Advisory Council, 1985, Environment Canada. Figure 'Number of National Parks Reporting Significant Ecological Impacts from Various Human Stresses' from *State of the Parks 1994*, Parks Canada (Ottawa: Minister of Supply and Services Canada, 1995):35. Figure 'Souris River Near Rafferty, Naturalized Flows' and 'Moose Mountain Creek Near Alameda, Naturalized Flows' from *Rafferty Alameda Project: Report of the Environmental Assessment Panel* (Ottawa: Federal Environmental Assessment Office, 1991). Table 'Atlantic Groundfish, Canadian Allocations and Catches, Thousand Tonnes, 1978-1993' from *Charting a New Course: Towards the Fishery of the Future*, Fisheries and Oceans Canada, 1993. Figure 'Trends in Lake Acidity by Region (1981–1994) from *Acid Rain*, State of the Environment Bulletin 96-2, State of the Environment Reporting Program, Environment Canada, 1996. Figure 'Indicator: Global and Canadian Average Temperatures (1895–1992)' from *Climate Change*, cat. EN1-19/94-4E, State of the Environment Reporting Program, Environment Canada,

1994. Figure 'Indicator: Global Atmospheric Concentration of Carbon Dioxide' from *Climate Change*, cat. EN1-19/94-4E, State of the Environment Reporting Program, Environment Canada, 1994. Figure 'Carbon Dioxide Emissions from Fossil Fuel Use in Canada (1958–1992)' from *Climate Change Environmental Bulletin*, State of the Environment Reporting, Environment Canada, 1994. Figure 'Who Is Producing the Carbon Dioxide in Canada? (1992)' from *Climate Change Environmental Bulletin*, State of the Envirnoment Reporting Program, Environment Canada, 1994. Figure 'Indicator: Stratospheric Ozone Levels over Canada' from *Stratospheric Ozone Depletion*, State of the Environment Reporting Program, Environment Canada, 1993. Figure 'Changes in Forest and Grassland Boundaries Resulting from a Typical Doubled-CO_2 Climate' from *Understanding Atmospheric Change* (State of the Environment Report 91–2), State of the Environment Reporting Program, Environment Canada, 1991. Figure 'Urbanization of Farm Land, 1966–1986' from *Urbanization of Rural Land in Canada, 1981–1986* by C.L. Warren, A. Kerr, and A.M. Turner, State of the Environment Fact Sheet 89–1, State of the Environment Reporting Program, Environment Canada, 1989. Figure 'Organochlorines in a North Pacific Food Chain' from *Contaminants in Canadian Seabirds* by P.G. Noble, State of the Environment Report 90–2, State of the Environment Reporting Program, Environment Canada, 1990. Figure 'Concentrations of Chemicals in Eggs of Bald Eagles from the North Shore of Lake Erie in the 1970s and 1980s' from 'Bring the Bald Eagle Back to Lake Erie', State of the Environment Fact Sheet 93–3, State of the Environment Reporting Program, Environment Canada, 1993. Figure 'Indicator: Canadian Consumption of Energy' from *Energy Consumption Environmental Indicator Bulletin*, State of the Environment Bulletin 94–3, State of the Environment Reporting Program, Environment Canada, 1994. Reproduced with the permission of the Minister of Public Works and Government Services Canada, 1997. Figures 'Projections for Climate Warming in Summer, Canada' and 'Projections for Climate Warming in Winter, Canada' from *Climate Change and Canadian Impacts: The Scientific Perspective*, Canadian Climate Program Board, Environment Canada, Ottawa, 1991. Reproduced with the permission of the Minister of Public Works and Government Services Canada, 1991. Figure 'National Marine Conservation Area Natural Regions' from *Charting the Course: Towards a Marine Conservation Area Act*, Parks Canada, 1997, p. 1. Reproduced with the permission of the Minister of Canadian Heritage and the Minister of Public Works and Government Services.

ADRIAN RAESIDE: Cartoons 'The Clayoquot Giant' and 'Logging in the Clayoquot Sound will now have to meet stricter environmental standards . . . ' from *Victoria Times-Colonist*. Copyright Adrian Raeside. Reprinted by permission.

REGIONAL MUNICIPALITY OF SUDBURY: Figure 'Extent of Barren and Semi-Barren Landscape within the Regional Municipality of Sudbury' from *Land Reclamation Program 1978–1984*, W.E. Lautenbach. Sudbury, ON: Regional Municipality of Sudbury, Vegetation Enhancement Technical Advisory Committee, 1985, p. 4. Reprinted by permission.

F.W. SCHUELER AND D.E. MCALLISTER: Map from 'Maps of the Number of Tree Species in Canada: A Pilot GIS Study of Tree Biodiversity', *Canadian Biodiversity* (now published under the title *Global Biodiversity*), Canadian Museum of Nature, vol. 1, no. 1, 1991, p. 23.

STATISTICS CANADA: Figure 'Population Growth Rate, 1971–1991' from Statistics Canada, *Human Activity and the Environment, 1994*, cat. no. 11–509, p. 44. Figure 'Selected Grain Crop Production, 1910–1991' from Statistics Canada, *Human Activity and the Environment, 1994*, cat. no. 11–509, p. 253. Figure 'Fertilizer Sales, 1970–1992' from Statistics Canada, *Human Activity and the Environment, 1994*, cat. no. 11–509, p. 137. Figure 'Crop Rotation Most Common Soil Erosion Control Method, Canada, 1991' from Statistics Canada, *Trends and Highlights of Canadian Agriculture and Its People*, cat. no. 96–303, p. 5. Figure 'Farms Using One or More Erosion Control, 1991' from Statistics Canada, *Environmental Perspectives, 1993*, cat. no. 11–528, p. 49. Statistics Canada information is used with the permission of the Minister of Industry, as Minister responsible for Statistics Canada.

D. TRIPP et al.: Figure 'Frequency of Occurrence of Different Impacts in Streams in Recent Cut Blocks on Vancouver Island, March, 1992' from *The Application and Effectiveness of the Coastal Fisheries Forestry Guidelines in Selected Cut Blocks on Vancouver Island* by D. Tripp et al., BC Ministry of Environment, Lands & Parks, 1992, p. 8.

M. WACKERNAGEL et al.: Table 'The Ecological Footprint of the Average Canadian in Hectares Per Capita' from *How Big Is Our Ecological Footprint? A Handbook for Estimating a Community's Appropriated Carrying Capacity* by M. Wackernagel et al., the Task Force on Planning Healthy and Sustainable Communities, 1993.

E.O. WILSON: Table 'Changes in the Numbers of Breeding Birds in Areas of Comparable Size with Latitude' from *The Diversity of Life* by E.O. Wilson. Copyright © 1992 by E.O. Wilson. Reprinted by permission of Harvard University Press.

WORLD CONSERVATION MONITORING CENTRE: Figures 'Number of Living Species Known and Estimated: World' and 'Number of Living Species Known and Estimated: Canada' from *Global Biodiversity: Status of the World's Living Resources*, World Conservation Monitoring Centre.

M. WYNDHAM AND K.M. DICKSON: Figure 'Population Index for Mallard and Northern Pintail' based on figures 4 and 6 from *Status of Migratory Game Birds in Canada*, 1 November 1996, Canadian Wildlife Service, Environment Canada, unpublished report.

Index

Note: Page numbers in italics denote an illustration.
An 'f' following a page number denotes a figure.
A 't' following a page number denotes a table.